NUTRITIONAL LIMITS TO ANIMAL PRODUCTION FROM PASTURES

EDITOR J.B.HACKER

Proceedings of an International Symposium held at St. Lucia, Queensland, Australia, August 24th-28th, 1981

Published on behalf of CSIRO Division of Tropical Crops and Pastures by the Commonwealth Agricultural Bureaux, Farnham Royal, UK

Published by

Commonwealth Agricultural Bureaux
Farnham House
Farnham Royal
Slough SL2 3BN
UK

ISBN 0 85198 492 4

CONTENTS

ORGANIZING COMMITTEE

E.F. Henzell

Chief

CSIRO Division of Tropical Crops and Pastures

D.J. Minson

Chairman

M.W. Silvey

Secretary

T.R. Evans

A.G. Eyles

J.B. Hacker

M.P. Hegarty

J.P. Hogan

R.J. Jones

R.M. Jones

L.J. Lambourne

D.A. Little

L. 't Mannetje

J.R. Wilson

The Editor is grateful to Mrs K. Pickering and Miss A. Ting for assistance in preparation of the final manuscript.

FOREWORD

An effective understanding of the nutritional limits to animal production is basic to research on pasture improvement. The low feeding value of pastures is a major problem in both tropical and temperate climates. It is particularly serious in those tropical regions where livestock must exist on mature forage during a long dry season and in those temperate environments that are unsuitable for growth of ryegrasses and clovers. When grown efficiently, as in New Zealand, these provide the standard of perfection against which other pasture feeds are assessed.

Although much early pasture research was done in the absence of grazing, it has long been obvious that animals provide the only reliable measure of pasture production. This fact imposes a major limitation on pasture plant improvement, in that the early stages of plant selection, whether from introduced wild material or from breeders lines, must be done in the absence of grazing, for logistical reasons. Only a few selected genotypes can be taken through to the point where the grazing animal can be used to measure their merit, or lack thereof. The scientists engaged in this research are well aware of the deficiencies in their practices but are in the "Catch 22" situation of having no practicable alternative.

The hope of overcoming this frustrating bottleneck in pasture research would alone be sufficient justification for the publication of this book. It develops the subject in a logical sequence, beginning with the worldwide role of pastures and their current deficiencies for animal production. Groups of chapters then deal with the present knowledge of factors limiting intake, digestion and use of digested nutrients. These are followed by a section that account for more than a third of the book, and which deals with methods for overcoming the limitations to animal production. The last two chapters deal respectively with the role of computer simulation and subjects identified as warranting further research.

Readers who are working on problems of pasture improvement will be particularly interested in the "state of the art" in relation to overcoming limitations to animal production. While much can be done with existing knowledge, it is quite clear that further progress is impeded by major gaps in understanding, particularly of the sward characteristics, feed characteristics and animal factors that control intake by grazing livestock, and of factors that influence the use of digested nutrients. Until this understanding can be obtained and translated into techniques that can be used by agronomists and breeders to test large numbers of plants, progress will be slow.

E.F. Henzell

OPENING SPEECH TO THE SYMPOSIUM

M.V. TRACEY

Director, CSIRO Institute of Biological Resources, P.O. Box 225, Dickson, ACT. 2602, Australia.

I feel honoured by having been invited to open this CSIRO International Symposium and naturally pleased to be given the traditional opportunity to do so by offering superficial and I am sure ill-based comments free of any editorial control with no opportunity for subsequent public rebuttal.

I have had the opportunity of reading not only the program but most of the synopses of the papers to be presented. On reflection two things are worthy of comment. First it is one of those unfortunately rare occasions on which animal and plant scientists have joined together to discuss this vital topic. Second it appears a remarkable fact that in a symposium devoting a week to the nutritional limits to animal production from pasture such an overwhelming emphasis is placed on feed rather than on the animal's means of using it. An easy answer to this perhaps naive surprise is that over a long period of time the mix of individual species of bacteria, protozoa, mycoplasmas and phages has become adapted to a very closely controlled environment in which random invaders are disadvantaged and rapidly eliminated, thus maintaining an overall dynamic stability through periods of repletion and want. Thus it is the wide variation in the composition of the feed that is of practical importance when viewed against the much less variable properties of the rumen liquor. It is the remarkable consistancy in the rumen liquor that makes possible its widespread use as an analytical tool in determining the digestibility of feeds. Nevertheless we shall be hearing later in the symposium of conditions as a result of which departures from the norm may occur or be induced.

It is true that there are a number of papers on the means by which digestion occurs in the rumen, but naturally (though remarkably) the complex ecological system of the rumen is commonly taken for granted as a stable system substantially invariant with time in the adult animal and from animal to animal, almost irrespective of the animal's environment. I hope that you will forgive me therefore for ruminating a little on the feeder rather than the feed.

I wonder whether the adaptations by the ruminant _of which we are aware_ were all that were necessary for such a successful innovation? Apart from the obvious physiological and anatomical adaptations to the process of antecedent microbial digestion we see in the ruminants, do they also engage in active intervention in response to swings in the population structure of the rumen contents by the secretion of inhibitors or promotors specific to important components of the population?

The argument that, given sufficient time all or nearly all is possible in the evolution of biological systems, gives rise to another question concerning ruminant digestion - why is lignin not more completely broken down? If it were solubilized in a significant amount there would be a benefit to both sides in that the ligninolytic organisms would make a living and while the host might not benefit from the breakdown of lignin _per se_ it would certainly do so by the increased availability of polysaccharides. The origin of lignin breakdown by micro-organisms must have been almost coeval with the origin of vascular plants, for were it a late innovation, plants would have been early divorced from soil by an accumulated layer of lignin too deep for their roots to penetrate. Is it possible that lignin decomposition products absorbed from the rumen, or for that matter from the bowel of monogastrics, would have deleterious effects on the metabolism of the animal and that the rumen environment has evolved as one hostile to lignin-olysis? Perhaps there is a simpler answer - such as that a peculiar difficulty of access by hydrolytic enzymes to labile bonds in this particular insoluble polymer so slows down the achievable rate of hydrolysis that a greatly increased throughput time would immobilize the animal by the sheer bulk of its rumen contents.

Having given my fancy rein there is another apparent convention for introductory speeches that I must observe - that of calling on famous names. I propose to use only two, Darwin and Linnaeus. Darwin in his Voyage of the Beagle writes of an excursion he made into the interior from Sydney. One of his comments may strike a chord with our visitors. He say, "The season it must be owned, had been one of great drought, and the country did not wear a favourable aspect, although I understand it was incomparably worse two or three months before" (this was written in the second half of January 1836) "The secret of the rapidly growing prosperity of Bathurst is, that the brown pasture, which appears to the stranger's eye so wretched, is excellent for sheep-grazing." The work of Linnaeus is more directly relevant. On a journey into a remote province of Sweden in the mid-eighteenth century he noticed that his horses would leave untouched many plants in the sward on which they grazed. Soon after his return he and his pupils conducted more than 2,300 trials, many of which were repeated ten times over on some twenty cattle, sheep, horses, goats and pigs, to see which plants the animals would accept and which reject. I will give only his findings on sheep and cattle. Cattle ate 276 species and rejected 141. He published his results later in Latin in thesis form in 1749 and in Amoenitates Academicae in 1751 listing 483 species and there are some blanks for some species of animal. His work was translated into English and published in the Gentleman's Magazine in 1758 and appeared in English twice more (as far as I know) first in 1781 and then oddly enough in the report by the Commissioner of Patents to the U.S. Senate in 1847. It is humbling to find at a time when we are beginning to realise the importance of animal preferences in determining their food intake to realise that a man known nowadays almost exclusively for founding plant systematics made such a major contribution to knowledge in quite another field more than two centuries ago.

I have indulged myself enough with idle speculation and it is time to turn from fancy to fact. Let me then congratulate all those who took part in the conception, pregnancy and birth of this symposium; thank and welcome those who have travelled far to ensure that it is a lively and healthy infant and express my hope that it will so satisfy those who play a part in it that this will be but the start of a long though necessarily intermittent life for the enterprise. It is with great pleasure that I declare the symposium open.

REFERENCES

"Pan Suecus" Amoenitatus Academicae volume 2, No. 25, pp.225-262 1751. (published as thesis 1749).

Translated in abbreviated from by Richard Pulteney. Gentleman's Magazine 28, 360-4, 407-9, 463-5, 515-7, and 567-8 (1758) and in longer form in "Grand View of the Writings of Linaeus" R. Pulteney, London 1781, pp. 374-409 (tables 384-409).

U.S. Patent Office: Annual Report of the Commissioner of Patents 1847. 512-523. (note: there are some disparities between the tabular material in this version and that of Pulteney, 1781).

PART 1

KEYNOTE ADDRESS

PASTURES AND ANIMAL PRODUCTION

P. MAHADEVAN

Animal Production and Health Division, FAO, Rome, Italy

ABSTRACT

Constraints to an effective use of the livestock resources of the developing countries are considered in relation to possible solutions. In the arid and semi-arid areas of Africa where nomadism and transhumance have been imposed by ecological conditions, the problems to be resolved include overgrazing of pastoral resources, crop cultivation on marginal lands, indiscriminate use of woody plants for building and fuel, uncontrolled burning of vegetation and haphazard drilling for water. The solutions would need to be strongly based on a combination of educational, technical, economic and social considerations. The FAO/EMASAR (Ecological Management of the Arid and Semi-Arid Rangelands) program is a step in this direction.

In the sub-humid and humid zones of both Africa and Latin America there are great expanses of land which could be highly productive livestock areas; the major obstacles are animal disease (especially trypanosomiasis) in Africa and the gap between technology development (in particular, the development of improved grass-legume pastures) and its on-farm application in Latin America. The FAO Programme for the Control of African Animal Trypanosomiasis and Related Development (in which ILCA (International Centre for Africa, Addis Ababa, Ethiopia), ILRAD (International Laboratory for Research on Animal Diseases, Nairobi, Kenya) and many bilateral agencies participate) has been designed to combat the former. In regard to the latter, the FAO field programme in Latin America and CIAT's (Centro Internacional de Agricultura Tropical, Cali, Colombia, S.A.) outreach activities constitute the principal efforts being made at the international level.

In Asia there is a marked imbalance between animal resources and available grazing lands - a situation which is aggravated by the pressure of human population growth and pronounced land scarcity. Greater reliance needs to be placed in this context on a more effective utilization of non-conventional feed resources and the selection and breeding of better forages for inclusion in small-holder crop rotations. In the Near East, cooperative grazing societies combined with the use of treated crop residues and the growing of leguminous forages as rotation crops on both irrigated and rainfed land seem to offer the best prospects for success.

INTRODUCTION

"Half a century of animal disease prevention and eradication measures, several decades of grassland and pasture investigations and of research into most aspects of animal nutrition and genetics have scarcely affected animal production." This experience of stagnating production and productivity in Africa, which was highlighted in a working paper (Blair-Rains & Kassam 1979) for the FAO/UNFPA (United Nations Fund for Population Activities) project on Land Resources for Populations in the Future, is shared by all developing regions of the world. It is associated with poor performance in agriculture generally, as well as in the overall economy, but it is particularly pronounced in the livestock sector. With few exceptions, any production increase has been mainly due to livestock numbers rather than productivity.

What are the major constraints that have contributed to poor performance and what measures may be taken to overcome them? A regional analysis of these questions is presented below.

AFRICA

Tropical Africa has one of the lowest per capita levels of meat and milk production in the world. However, there are significant differences between the various agroclimatic zones (see Table 1).

TABLE 1

Preliminary estimates of meat and milk production by agroclimatic zone in tropical Africa

Zone	Meat production			Milk production		
	% of total	000 t	per capita kg	% of total	000 t	per capita kg
Arid	28.3	913	50.7	45.3	3 169	176.1
Semi-arid	31.5	1 016	11.1	27.5	1 924	20.9
Sub-humid	17.3	558	8.0	2.6	182	2.6
Humid	3.9	126	1.3	1.0	70	1.1
Highlands	19.0	613	9.3	23.6	1 651	25.0
Total	100.0	3 226	9.5	100.0	6 996	20.5

Source: Hans E. Jahnke, 1980 - personal communication

Arid zone

The relatively high figures for milk production in the arid zone (where the growing period for pasture is less than 90 days) reflect the importance of milk as a subsistence commodity among nomadic and transhumant peoples. Pastoralism is the predominant production system, involving communal ownership of the range and its water resources and private ownership of the livestock, with the attendant consequences of overstocking, land degradation and pauperization. Different development approaches have been attempted, but the results suggest that the prospects for increased productivity are limited. Thus, although the establishment of private ranches by European settlers in Kenya, Zambia and Zimbabwe during the turn of the century was reasonably successful, it proved to be highly localized in its applicability. Likewise, although the setting up of group ranching schemes in the 1960s and 1970s by the governments of Kenya and Tanzania provided the Masai with a way of acquiring title to land, it turned out to be less effective as a means of commercializing beef production. The fundamental problem of overstocking has yet to be resolved by the group ranch approach. The principal reason for the lack of success appears to be that the animal production problems of the arid zone have as much to do with socio-economic considerations as with technical ones. Indeed, if the arid zone is to continue to be a valuable resource for the African economies, a concerted educational effort must be made to enable the nomad to discipline himself, keep animal numbers down, develop grazing reserves and, wherever possible, fit him into occupations other than pastoralism.

Semi-arid zone

In the semi-arid zone with a 90 to 180 day growing period, the year-round carrying capacity of pasture has been generally put at 10-12 hectares per head of cattle. But in parts of this zone, there is a progressive expansion of cropping based on the use of draught animals and their manure. This, in turn, is resulting in a gradual erosion of the pasture, forage and feed base. Crop residues in the semi-arid zone can at best supply only about 20 percent of the animals' feed requirements and this leads to the dilemma that it becomes impossible to maintain the very animals that make the expansion of cropping possible. Thus, mixed farming, involving the full exploitation of the cropping potential of the land and the integration of livestock into the farming system, has not proved to be feasible in these areas because of feed constraints. Likewise, the grazing of livestock away from the cropping areas has been found to be difficult because of the distances that have to be traversed and the associated labour constraints. In the face of this dilemma, a major thrust in the FAO Grassland, Range and Fodder Crops development programme is the screening, evaluation and multiplication of superior forages for the semi-arid and arid zones. Some of the most promising species screened so far are *Atriplex numularia*, *Maerua trifoliata*, *Stylosanthes fruticosa*, *S. hamata* and *S. scabra*. Propagation studies of some 25 of the most promising indigenous browse species are also underway in Kenya.

The success of these endeavours is critical to future development, because there are few options left for more effective use and management of the range resource in the semi-arid and arid zones of Africa. Nomadism and transhumance were imposed on these areas by ecological conditions; the human populations developed strategies for utilizing the available resources in a manner that enabled them to sidestep the constraints of climate. But, when basic thresholds of resource tolerance to human and animal pressures are exceeded, the result is generally more catastrophic in these areas than in the sub-humid and humid zones. A well known example is the Sahelian drought of 1970-73, which proved to be a disaster for livestock populations (Temple & Thomas 1973). The starting point of any corrective action to prevent calamities such as this is encouragement of social acceptance of the solutions proposed. These must be simple and also capable of coping with the multitude of problems encountered viz. over-grazing of pastoral resources, crop cultivation on marginal lands, indiscriminate use of woody plants for building and fuel, uncontrolled burning of vegetation and haphazard drilling for water. Clearly then, the solutions must be strongly based on a combination of educational, technical, economic and social considerations. An action programme conceived in this manner and initiated by FAO goes under the acronym EMASAR - the Ecological Management of Arid and Semi-Arid

Rangelands. Some 60 projects have been formulated under this programme and are being implemented in 30 countries; they have already resulted in a doubling of rangeland and livestock production in some of the project areas.

Sub-humid and humid zones

In the sub-humid zone where the growing period for pasture extends from 180 to 270 days, the problems are, as may be expected, quite different; the nutritional strategy is principally one of bridging the 90-180 day dry period in terms of feed quantity and quality (ILCA 1979). It is theoretically feasible to achieve this by one or more of the following measures: pasture improvement through the use of drought-resistant grasses and legumes (e.g. Cynodon, Cenchrus, Stylosanthes); appropriate adaptation of the harvesting and fertilizing regimes; discriminate use of burning; planting different grasses on a catena; silage making; hay making; planting fodder trees and shrubs; and feeding agricultural by-products. Although yield increases have been demonstrated through these various practices, it is their economic viability and social feasibility that will determine their large scale adoption at the national level.

It should be noted in this context that the sub-humid zone in Africa has a relatively higher crop potential than the other zones and, in common with the humid zone, is characterized by the prevalence of tsetse flies and trypanosomiasis. The result is that the average stocking density in terms of ruminant animals in the sub-humid and humid zones is less than half that of the semi-arid and arid zones (see Table 2). Thus, the availability of livestock products per head of human population is lower, although productivity per head of stock is somewhat higher.

TABLE 2

Estimated distribution of the ruminant livestock population in tropical Africa. (H.E. Jahnke, personal communication)

Zones	Cattle(000)	Sheep(000)	Goats(000)	Camels(000)	Livestock units (LU) (000)	LU/km^2
Arid	30 900	26 600	35 100	11 600	45 000	6.4
Semi-arid	47 200	28 000	36 800	-	40 565	9.8
Sub-humid	29 250	13 150	21 900	-	24 564	5.0
Humid	6 350	5 850	11 300	-	6 450	1.5
Highlands	30 000	23 500	17 500	-	25 341	25.4
Total	143 700	97 100	122 600	11 600	141 920	6.7

The FAO Programme for the Control of African Animal Trypanosomiasis and Related Development has been designed to develop the potential of the vast land resources in the sub-humid and humid zones that are infested by tsetse fly. It is a comprehensive rural development programme that aims at bringing order into the spontaneous, unplanned settlement which is occurring at an accelerating pace in areas where tsetse density is being reduced. The initial emphasis of the programme is on areas where population pressures are greatest, with livestock development strategies varying according to needs. Thus, in the humid zone, where large scale tsetse clearance may not be feasible, trypanotolerant breeds continue to provide the framework for development. Elsewhere, tsetse control and the application of trypanocidal drugs are the preferred strategies. A combination of these approaches in the context of the Sudanian and Guinean zones offers considerable possibilities because of the opportunities that these areas provide both for the development of improved pastures and for feed grain production. However, their economic viability would depend to a large extent on whether the prevailing low meat prices can be raised to make the investments attractive.

The highlands

The principal characteristic of the highlands is that the grassland, forage and feed resource base in that zone provides a high potential for the intensification of dairy, meat and wool production. Perennial fodder production based on temperate species is practised, which, when combined with the low incidence of the major animal epizootics, makes it possible to raise the high yielding breeds of cattle and sheep from the temperate regions of the world. The success of both the smallholder dairy development schemes and the large commercial holdings for meat and fine wool production in Kenya demonstrate the considerable potential of the highlands for increased production. However, in some parts of the highlands, increasing pressure of human populations has led to a drastic reduction in the grazing areas due to a fragmentation of holdings and reduced farm size. But, even here, it has been found that in terms of total feed availability, at least as much feed can be obtained from crop residues, by-products and stubble grazing as from unimproved pastures. It is therefore not surprising that although the highland zone accounts for less than 5 percent of the land area of tropical Africa, it contains almost 20 percent of the ruminant livestock population.

ASIA AND THE FAR EAST

The Asian and Far East region (including China) has about one-half of the world's human population, but contains less than one-fifth of the agricultural area. Two-thirds of the world's poorest people are to be found in this region. The region consists predominantly of small farmers and the core of rural poverty is to be found on farms in the rainfed areas where the majority of the rural poor live.

Livestock populations

Table 3 provides a summary of the livestock populations in Asia and the Far East in relation to total world populations (FAO 1980). The region, as defined in this table, has about three-quarters of the world's buffaloes and two-fifths of the ducks, but only one-quarter of the goats and one-fifth of the cattle. Chickens, sheep and pigs represent much smaller proportions.

Generally speaking, ruminants are far more important to the national economies than non-ruminants. The majority of the ruminants are maintained mainly for products or services of a non-food nature. The monetary value of these goods and services often exceeds that from meat or milk production. In these circumstances, milk is a secondary or incidental product and meat a terminal or salvageable commodity.

TABLE 3

Livestock populations in Asia and the Far East (excluding China). (FAO 1980)

Species	Asia and the Far East 000	World 000	Percent of world population
Cattle	260 687	1 212 017	21.5
Buffaloes	92 746	130 554	71.0
Sheep	72 639	1 083 954	6.7
Goats	123 500	445 919	27.7
Pigs	34 473	737 872	4.7
Chickens	662 358	6 705 934	9.9
Ducks	57 570	144 659	39.8

Draught animals constitute the most important single grouping among the cattle and buffalo populations of Asia and the Far East. For most of the region there is one draught animal per hectare of land, but in much of south-east Asia there is an acute shortage of work animals. Rollinson & Nell (1973) estimated that Indonesia needed 850 000 more work animals to achieve its food goals for 1985.

Grassland, forage and feed resources

Table 4 shows the context of permanent pasture land available in Asia and the Far East in relation to the rest of the world. When these figures are compared with the corresponding ruminant livestock populations presented in the same table, there is a marked imbalance between animal resources and available grazing lands in Asia and the Far East. In India, the National Commission on Agriculture (1976) estimated the shortfall in meeting the nutritional requirements of the country's livestock as being 38 percent for green fodders, 44 percent for dry fodders and 44 percent for concentrates. In Pakistan, the deficit, expressed in terms of energy and protein, is of the order of 49 percent for energy and 42 percent for digestible crude protein.

Livestock development strategy

One of the most successful development strategies in this setting of high population pressure and pronounced land scarcity is that adopted with smallholder dairy development in India. Its beginnings can be traced to the founding of the Kaira District Cooperative Milk Producers' Union Ltd. in Anand, Gujarat, in 1946 (Kurien 1975). It was recognized from that time on, that since all arable land must be used primarily for food crop production, the availability of cultivated forage for animal feeding would be limited. Thus, the total area devoted annually to forage crop production in India is about seven million hectares and this figure has shown no significant change over the past 35 years. On an average, therefore, 30 animals share the forage produced on a hectare of land, each animal getting some 2-3 kg per day. As a result, the main feed resource base of ruminants consists of crop residues (principally rice straw and maize stover), by-products of agricultural processing and the herbage that can be harvested from communal lands around areas of settlement, national grazing areas, forest reserves, wayside grazing, edible hedges, the stubble left over from the cultivation of crops and volunteer species growing on fallow land.

TABLE 4

Extent and distribution of permanent pastures and ruminant livestock (compiled from FAO Production Yearbook 1980)

Region	Permanent pastures (million km^2)	Ruminant livestock units (millions)
Developing market economies		
Africa	6.9	157.2
Asia and the Far East	0.4	329.2
Latin America	5.3	264.4
Near East	2.7	80.2
Total	15.3	831.0
Asian centrally planned economies	3.5	128.6
Total developing countries	18.8	959.6
Developed market economies		
North America	2.7	110.3
Western Europe	0.7	95.3
Oceania	4.6	48.8
Others	0.8	18.3
Total	8.8	272.7
European centrally planned economies	3.9	147.3
Total developed countries	12.7	420.0
World total	31.5	1 379.6

The initial emphasis in the strategy for livestock development consisted of transmitting the high urban demand for milk in the four major cities of India to the rural producer. This was done through producer cooperatives, by offering the dairy producer an assured and attractive year round price for his milk and by organizing the processing and marketing of that milk in an efficient manner. These measures alone enabled the producer to raise his income by 50 percent or more. The farm level inputs in these early years included little more than an effective harnessing of the low opportunity cost inputs of labour, grazing on non-arable land and feeding crop residues and waste by-products. A significant value-added effect was achieved through the feeding of concentrates that were supplied at cost through the producer cooperatives.

In more recent years, the role of these dairy producer cooperatives has been expanded to include the provision of veterinary care for the milch animals, artificial insemination services in each village and assistance to producers through a "minikit fodder demonstration programme" to promote the incorporation of suitable forages in crop rotations. The resulting high level of economic efficiency has benefitted over one million milk producers. It has been achieved without any significant increase in average size of farms, production per farm or production per animal. It underlines the fact that although moderate to low levels of animal performance may be inefficient from a biological standpoint, they could be economically more viable than high levels of performance under the conditions of land use prevailing in many Asian and Far East countries. This point is further illustrated in Table 5 where a comparison is made of the net annual food calorie balance between buffaloes subjected to a feeding regime based on cultivated forage and concentrates and those fed alkali-treated straw, agro-industrial by-products, minerals and non-protein nitrogen.

Smallholder dairy development in India has currently entered a second phase which involves an outlay of over US$600 million and aims at assisting some 10 million rural milk producers to build a viable, self-sustaining dairy industry by mid-1985 and to facilitate the rearing of a national milch herd of some 15 million crossbred cows and improved buffaloes. This project is an ambitious one. Critical to its success is the development of a forage and feed resource base appropriate to the maintenance of the 15 million crossbred cows and improved buffaloes. It would call for the selection and breeding of more productive forages (in particular, leguminous forages such as berseem and lucerne) that the smallholder could be persuaded to incorporate in his crop rotations. In addition, since a greater proportion of the available land area will be increasingly used to produce foods for direct human consumption, greater reliance would need to be placed on the development of new techniques for higher forage production from the poorer soils (for example the planting of leucaena patches in communal grazing areas, which will help to integrate animal production with nitrogen fixation and soil building). Equally important is the need to harness and utilize more effectively for animal feeding the vast amount of non-conventional feed resources which include the waste materials generated from crop and animal production and the residues from the processing of food for human consumption. Devendra (1981) has estimated the total availability of crop by-products from field, plantation and tree crops in the Asian and Far East region to be around 471 million tons, of which about 216 million tons are reported to be non-conventional feedstuffs. These do not, however, take into account tree leaves, fruit and animal processing wastes and animal excreta, which, if included, would increase substantially the total availability of non-conventional feed resources.

Viewed against the available feed resource base, the recent proliferation of crossbreeding programmes, for milk and meat production in the region gives cause for concern, because they are

TABLE 5

Net annual food calorie balance resulting from the adoption of two different feeding regimes to milch buffaloes (Mahadevan 1977).

Feeding regime	Annual milk production (kg)	Input (Mcal)	Output (Mcal)	Balance (Mcal)
Cultivated forage from 1/6 ha + 600 kg concentrate feed/year	2 000	2 800 + 2 400 = 5 200	Milk 2 200	-3 000
Alkali-treated straw (8-10 kg/day) + 700 kg agro-industrial by-products + NPN + minerals	1 400	3 000	Milk 1 500 + 1/6 ha crop 2 800 = 4 300	+1 300

so often accompanied by implicit acceptance of high levels of concentrate feeding. The role of the ruminant animal in the Asian setting is primarily to convert fibrous agricultural wastes into animal products. The concentrate that is fed should be in catalytic amounts so as to: (i) increase the intake of the basal forage; (ii) increase the efficiency of ruminal digestion (by increasing volatile fatty acid availability and microbes from the forage); (iii) increase the efficiency of utilization of the digestible energy available from the rumen; and (iv) provide extra amino acids post-ruminally in order to avoid the losses incurred in rumen fermentation (Leng *et al*. 1979). This does not imply that there is no role for crossbreeding under these conditions; rather it means that any breeding program should have as its objective the production of animals which utilize existing feed resources in the most efficient way. This applies not only to smallholder dairy development but also to the raising of other ruminant animals in a setting of pronounced land scarcity and high population pressure.

LATIN AMERICA

The cattle population of Latin America and the Caribbean is about 267 million (FAO 1980), of which some 160 million are estimated to be in the lowland tropics. Animal productivity in the lowland tropics is generally low. Calving is usually seasonal, most cows dropping their calves once every two years, and this results in a calving rate of 40-50 percent. Slaughter animals are marketed at 3 to 5 years of age at liveweights of 350-450 kg. Extraction rates are approximately 13 percent, with an annual production of carcass beef from the total beef herd of only 25 kg per head (FAO 1980). These figures have shown no significant change over the years.

Productivity per unit area of land is also low, with an average stocking rate of 4-5 hectares per animal and carcass beef production of 5.1 kg per hectare per year (Raun 1976).

Many beef cattle herds in proximity to the urban centres, where there is a demand for milk, serve as dual purpose beef/milk herds. But levels of milk production average only about 2-2.5 litres per day in the dry season and 3 litres in the rainy season. Continuous rather than seasonal breeding is used in these milk/beef systems.

In restricted areas where climatic and nutritional stresses are not severe, European breeds (now mainly North American Friesians) are the mainstay of the dairy industry. They are kept either on grassland in the mountains or in drylots around the cities and yield from 2,500 to 5,000 kg milk per lactation (Vaccaro 1974). Elsewhere in the lowlands, various Criollo and zebu breeds and miscellaneous crosses between them constitute the region's population of milked cattle; the best of these herds average 1 500-2 000 kg per lactation, principally from grazing.

Constraints to increased productivity

The constraints to increased productivity vary with soil type, amount and distribution of rainfall and management practices. In the infertile savanna areas which cover some 300 million hectares, allic soils predominate. These are primarily the red-yellow latosols, or oxisols and utisols. They are acidic, high in aluminium and deficient in nitrogen, phosphorus, calcium, magnesium and potassium. These lands cannot support sustained crop production without sizeable fertilizer inputs. Although grazing by ruminants appears to be the only feasible method of utilization, its success would depend on whether the management techniques developed by research are adopted more widely at the farm level.

Rainfall distribution limits nutrient intake in the dry season and reduces animal productivity. This tends to be accentuated in the low fertility areas. The effects of disease, parasitism and genetic potential are usually secondary to inadequate nutrition.

Management constraints are particularly important in the fertile soil areas which cover approximately 370 million hectares. Part of this is now grassland and part is forested land that could be cleared and developed as grassland. More intensive production systems and practices need to be adopted in these fertile areas, with particular attention to the selection of nutritious and high-yielding pasture species that can withstand heavy grazing and be competitive with weeds. The higher incidence of disease and parasitism associated with the increased concentrations of cattle also call for improved managerial and preventive medicine programmes.

Technology development and on-farm application

In developing suitable technology for raising animal production in the Latin American tropics, it needs to be recognized that beef is one of the staple foods of the region's urban and rural poor. Families in the lower 25 percent income strata currently spend 8-16 percent of their total budgets to buy beef (CIAT 1980). Unfortunately, during the past 15 years the annual growth in the demand for beef (5.6 percent) has exceeded production increases (3.6 percent) and this is contributing to real price increases which will cause a decrease in beef consumption.

Of equal concern is the situation concerning milk production trends in the region. Imports of milk and milk products have tripled over the past 10 years.

Against this background, technology development should be specifically directed at achieving production increases at lower unit cost. From a producer point of view, this would mean less dependence on purchased inputs. This minimum philosophy identifies both producers and consumers with limited resources as the major beneficiaries of technology development.

Application of this philosophy has led to a concentration of effort in two directions: (i) developing more productive and nutritive pastures as animal feed; and (ii) utilizing rapidly growing crops that are rich in fermentable carbohydrates, for example sugarcane, sweet potato, cassava and bananas, in combination with various by-pass proteins for animal feeding (Preston & Leng 1978). Some striking results have been obtained. One example is in the Peruvian tropics where traditionally the rain forest is burned and *Hyparrhenia rufa* is sown on the utisols. The resulting pastures, together with the less productive species that invade it, stabilize at a carrying capacity of less than 0.5 cows per hectare, with the animals exhibiting low growth rates and long calving intervals. In the UNDP/FAO project on pasture development, it was found (Santhirasegaram 1976) that pasture productivity could be improved by the inclusion of legumes (e.g. *Stylosanthes guianensis*) that are adapted to the high acidity and high aluminium content of the soils. Dry matter yields and protein content increased. Phosphorus deficiency was corrected by the application of 100 kg/ha of single superphosphate every six months. The resultant grass-legume pasture doubled both stocking rate and growth rate, and gave more than a fourfold increase in livestock gain per hectare (over 600 kg per year) compared with the traditional pastures (see Fig. 1).

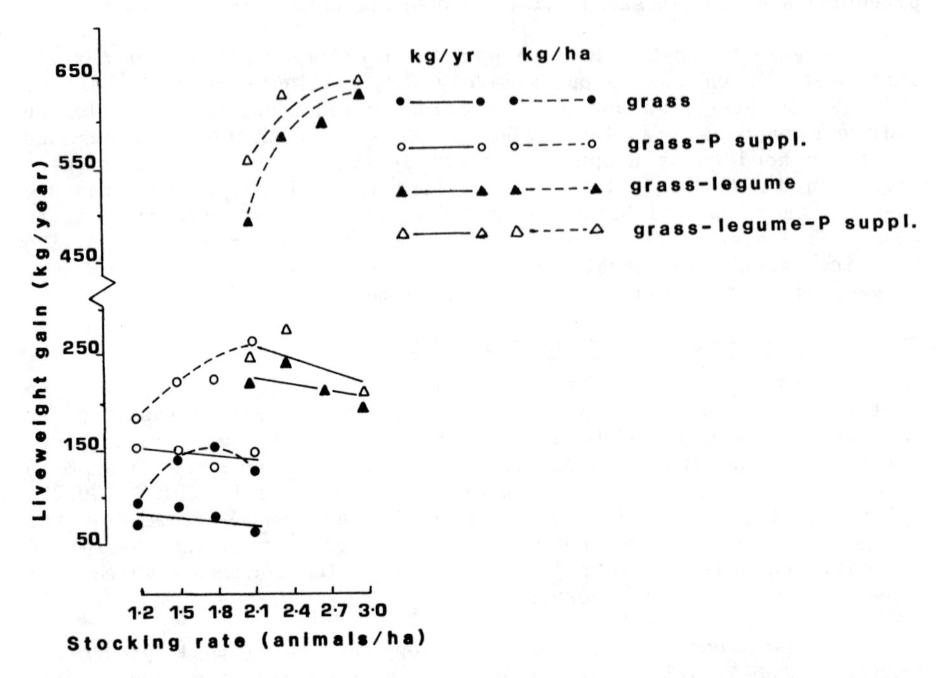

Fig. 1. Annual liveweight gains of Nellore type bull calves grazing different pastures at varying stocking rates in Pucallpa, Peru.

Source: Santhirasegaram (1976)

Preliminary economic estimates from the Peruvian study, based on 80 percent efficiency in farm management, suggested an 18 percent return on investment from the grass-legume pastures compared with negative returns from the traditional pastures. It was concluded that further improvements in productivity and profitability are possible through the selection of better grasses and legumes, greater attention to the efficiency of nitrogen fixation by the legumes, and the development of improved grazing and stock management systems within the whole farm economy.

In regard to the utilization of rapidly growing crops such as sugarcane, sweet potato, cassava and bananas for animal feeding, it is still too early to generalize about the economics of such use. Factors such as the local prices of products and inputs, as well as the management system employed, will exert a considerable influence. But the results obtained so far suggest that these crops do have potential relevance to animal production in many parts of Central America and the Caribbean.

Early results from the CIAT (1980) study on the strategic use of improved pastures and the length of the mating season on breeding herd productivity in the Colombian savannas have been encouraging. They have shown that cows with access to improved pastures for 3-5 months in the late dry and/or early rainy seasons had higher calving rates than cows grazing only native savanna. Furthermore, the calving rates of cows with access to improved pastures were unaffected by a shortening of the mating period from 4 to 3 months. By contrast, the calving rates of cows grazing only savanna decreased when mating periods were shortened. Preliminary results from the CIAT/EMBRAPA collaborative germplasm evaluations on two different soils in Brazil - the red-yellow latosol and dark-red latosol - indicate that Stylosanthes is the most promising genus for pasture improvement. A distinctive group of S. guianensis ecotypes originating in Brazil and Venezuela has been found to be outstanding. These ecotypes are late-flowering and fine-stemmed and have a very viscous pubescence, narrow leaflets and a characteristic flower-head structure. They are resistant to anthracnose. All of these results would suggest that the tropical savannas and jungles of Latin America could support a thriving beef industry, if work on the evaluation of forage species is extended to cover all ecological areas and if the development of associated pasture and cattle management systems at the farm level is pursued with vigour.

NEAR EAST

The Near East region has a strongly marked pattern of summer drought and winter rain, a high degree of aridity (with rainfall often below 100 mm per annum) and high summer temperatures. Because of the dominance of arid and semi-arid lands and the seasonal low rainfall, agriculture in the region is characterized by irrigation, dryland farming and pastoralism.

Land use and stock numbers

Land use in the Near East and the numbers and kinds of ruminant animals in the region are set out in Tables 6 and 7. It will be noted that a total of some 407 million hectares, consisting of permanent pasture land and forest and woodland, is used for range animal production. Altogether 280 million head of ruminant animals use these range lands. Of these, sheep and goats constitute 79 percent.

TABLE 6

Land use in the Near East

Kind of land use	Area (million hectares)	% of total
Permanent pasture land	267.2	22.4
Forest and woodland	140.1	11.8
Arable and permanent crop land	81.1	6.8
Unused or waste land	703.9	59.0
Total land area	1 192.3	100.0

Source: FAO (1980)

TABLE 7

Ruminant livestock populations in the Near East

Kinds of animals	Numbers (millions)	% of total
Sheep	152.9	54.6
Goats	68.3	24.4
Cattle	51.3	18.3
Buffaloes	3.8	1.4
Camels	3.7	1.3
Total	280.0	100.0

Source: FAO (1980)

Meat, milk and wool production

The total production of meat, milk and wool from the ruminant livestock industries of the Near East countries is given in Table 8. Of the total red meat production of 2,260,000 tons, over one-half is supplied by sheep and goats, but the contribution of the latter to total milk supplies is only about one-third. This shows that even under Near East conditions, the dominance of the cow and the buffalo as milk producers remains, despite the overwhelming superiority in numbers of sheep and goats in the region.

TABLE 8

Meat, milk and wool production in the Near East

Product	Source of product and amount (tons)		Total
	Sheep and goats	Cattle and buffaloes	
Meat	1,196,000	1,064,000	2,260,000
Milk	4,599,000	9,947,000	14,546,000
Wool (greasy)	180,760	-	180,760

Source: FAO (1980)

Grassland, forage and feed resources

The Near East is severely deficient in feed for ruminants. The productivity of most of the ranges in the vicinity of the cultivated areas has declined rapidly over the past few decades because of overgrazing. Livestock populations have increased substantially, triggered by the introduction of better disease control programmes, the development of water resources and an ever increasing human population. Nevertheless, there are opportunities for increasing forage production from the range, for a better utilization of available feed supplies and for the introduction of more cultivated pastures into the cereal-fallow rotation.

In Syria, an integrated programme of steppe, range and sheep development has paved the way for increased livestock productivity (Draz 1980). The programme is based on the re-introduction of the ancient "hema" system in a modified form to ensure that the users of the range assume responsibility for its continued productivity. Major components of the programme include reserving specific areas for the exclusive use of members of cooperative grazing societies and developing a number of basic facilities such as government demonstration ranges, a Bedouin Training Centre, a network of feedstores for emergency supplementary feeding of range ewes during drought years and credit to cooperative range and fattening societies. Encouragement is given to forage production on fallow land in rainfed areas and on newly developed irrigated land, as well as to intensive dairy development. The success of the programme has been demonstrated by Syria's ability to feed and finish enough sheep to meet her domestic mutton requirements (Demirüren 1974). It is relevant that all of this has been accomplished without any effort to settle the nomads as sedentary agriculturists. Indeed, if the pastoralists had been settled as crop producers, Syria would have lost one of her major assets, namely animal production from her extensive range areas, and would not have been able to achieve the reported increases in mutton production.

Improvements in the use of cereal straws for feeding ruminants also offer opportunities for a more effective utilization of available feed supplies. Grazing on the range in the Near East region normally occurs during the winter and spring when vegetation grows rapidly in response to the winter rains. In May and June the range vegetation dries up and the sheep flocks are moved to the cultivated areas to graze on fallow land before returning to the range once again. In this sequence of dependence on range, cereal stubble and fallow land, only part of the cereal straws are collected for hand-feeding; the rest is left to be grazed, burnt or ploughed under. Preliminary estimates (Jackson 1980, Arnason 1981) suggest that about one-third to one-half of the straw produced is currently not fed to livestock. A systematic study at the national level of straw utilization practices and the straw trade is necessary to determine what the specific constraints are. Equally important is the need to raise the level of protein in the diet, which would be aggravated by increasing the amount of straw fed. Various possibilities are in process of being explored in this context. These include both on-farm and factory-scale treatment of straw with ammonia/urea; the use of urea in concentrate rations (for fattening sheep) as a replacement for cottonseed cake (Markotic *et al.* 1976), thus sparing the latter for feeding to range and farm livestock; and the growing of leguminous forage crops in place of fallow in the present cereal-fallow rotation in areas with at least 300 mm rainfall per year. More than 60 000 hectares are now under a wheat/medic rotation in Algeria, Tunisia and Libya (Bakhtri 1979); average yields of 2 tons of wheat per hectare in years when the crop can be grown and stocking rates of 2-4 sheep per hectare on medic during favourable years have been recorded in the area around Tripoli. In Syria, the cultivation of vetches for hay on rainfed cereal lands in the steppe with 300-400 mm annual rainfall has been gaining increasing acceptance because of the profitable income derived from it. In the denuded steppe areas of Syria, Jordan and Iraq where marginal cereal cultivation is still practised, and where annual self-seeding legumes are unlikely to provide the answer because of wide fluctuations in rainfall from year to year, there appears to be a gradual change over to planting these areas with perennial shrubs and grasses such as Artemisia, Atriplex, Salsola and Stipa (M.D. Kernick, 1981, personal communication). In Cyprus, the adoption of cultivated pastures in place of fallow has facilitated the introduction of Friesian and Shorthorn cattle yielding an average of 5000 kg milk per lactation (Photiades 1979).

These developments serve to demonstrate that improved management of forage and feed resources would need to form an integral part of any strategy for achieving increased agricultural productivity instead of increased desertification in the Near East region. This includes grazing management to prevent further overgrazing, rehabilitation of overgrazed range, an increase in forage supplies from both rainfed and irrigated land (primarily in the form of forage legumes) and the controlled distribution of concentrate feeds, particularly for the fattening and finishing operations.

CONCLUSIONS

The problems of animal production in the developing countries are extremely diverse. Appropriate strategies for improvements in productivity differ between and within regions. Where livestock production is based on pastoralist systems, it is necessary to ensure that thresholds of resource tolerance are not exceeded by human and animal pressures, especially during drought years. On arable lands the most successful approaches are likely to be those based on integrated crop/livestock farming systems, including a more efficient utilization of crop residues and agro-industrial by-products for animal feeding. Where land is not a limiting factor to livestock development, the greatest need is for systems of grazing management that fit into the traditional patterns of farming. Viable solutions to these are not simply a matter of technology transfer, because both technical and social problems have to be overcome.

The strategies appropriate to each situation must be developed and pursued with growing urgency if the targets set for livestock production in the FAO (1979) study "Agriculture: Toward 2000" are to be achieved. The national efforts of the developing countries will need massive support from the world community if hunger is to be abolished within the time span covered by that study.

REFERENCES

Arnason, J. (1981) Consultant report PDY/77/003. FAO, Rome. 23 pp.
Bakhtri, N. (1979) Introduction of medic/wheat rotation in the North African and Near East countries. Regional Seminar on Rainfed Agriculture, Amman. Paper No. 15. Rome, FAO, 12 pp.
Blair-Rains, A.; Kassam, A.H. (1979) Land resources and animal production. Working Paper No. 8. FAO/UNFPA project INT/75/P13. Rome, AGLS, FAO, 28 pp.
CIAT (1980) CIAT report. 101 pp.
Demiruren, A. (1974). The improvement of nomadic and transhumance animal production systems. AGA/MISC/74/3. Rome, FAO, 46 pp.
Devendra, C. (1981) Non-conventional feed resources in Asia and the Far East. Consultant report. FAO Regional Office, Bangkok, 99 pp.
Draz, O. (1980) Range and fodder crop development - Syrian Arab Republic. Consultant report, SYR/68/011. Rome, FAO, 95 pp.
FAO (1979) Agriculture: Toward 2000. FAO, Rome. 257 pp.
FAO (1980) FAO Production Yearbook 1979. Vol. 33. 309 pp.
ILCA (1979) Livestock production in the sub-humid zone of West Africa. ILCA Systems Study No. 2. Addis Ababa, 184 pp.
Jackson, M.G. (1980) Treatment of straw for animal feeding: Cyprus and Syria. Consultant report. Near East Regional Office, Cairo. 27 pp. + annexes.
Kurien, V. (1975) The impact of animal production on developing countries: sociological impact, including nutrition. In: Proceedings of the Third World Conference on Animal Production. Editor R.L. Reid. Sydney, Sydney University Press, pp.304-308.

Leng, R.A.; Kempton, T.J.; Nolan, J.V. (1979) Recent advances in the feeding of NPN and by-pass protein meals to ruminants. Paper presented at First Research Coordination Meeting on the use of nuclear techniques to improve domestic buffalo production in Asia, Sri Lanka, 25-29 June 1979. Vienna, FAO/IAEA, 23 pp.

Mahadevan, P. (1977) Intercountry cooperative research program on the water buffalo: proposals for international support. Report to the Technical Advisory Committee for the Consultative Group on International Agricultural Research. Rome, FAO, 48 pp.

Markotic, B.; Karam, H.A.; Mitchell, I.W.; Kernick, M.D.; Yousif, M.Y.; Al-Ashairy, L.; Yousufani, F.G.; Yaqoob, N.I.; Potrus, P.T. (1976) Sheep fattening in northern Iraq. 3. Supplementing barley rations. Rome, FAO, 13 pp.

National Commission on Agriculture (1976) Part VII Animal Husbandry. Ministry of Agriculture and Irrigation, Government of India, New Delhi, India, 531 pp.

Photiades, T. (1979) Integration of livestock with rainfed agriculture in Cyprus. Regional Seminar on Rainfed Agriculture, Amman. Paper No. 14. Rome, FAO, 8 pp.

Preston, T.R.; Leng, R.A. (1978) Sugarcane as cattle feed. World Animal Review (FAO) 27, 7-12 & 28, 44-48.

Rau, N.S. (1976) Beef cattle production practices in the lowland American tropics. World Animal Review (FAO) 19, 18-23.

Rollinson, D.H.L.; Nell, A.J. (1973) The present and future situation of working cattle and buffalo in Indonesia. UNDP/FAO project INS/72/009. Rome, FAO, 40 pp.

Sathirasegaram, K. (1976) Recent advances in pasture development in the Peruvian tropics. World Animal Review (FAO) 17, 34-39.

Temple, R.S.; Thomas, M.E.R. (1973) The Sahelian drought - a disaster for livestock populations. World Animal Review (FAO) 8, 1-7.

Vaccaro, Lucia Pearson de (1974) Dairy cattle breeding in tropical South America. World Animal Review (FAO) 12, 8-13.

PART 2

PROBLEMS WITH PASTURES

PROBLEMS OF ANIMAL PRODUCTION FROM TEMPERATE PASTURES

R.L. REID

West Virginia University, Morgantown, West Virginia 26506, USA.

G.A. JUNG

USDA Pasture Research Laboratories, Pennsylvania 16802, USA.

ABSTRACT

Temperate grasslands presently constitute some 30 percent of the total world area in permanent range and pasture and maintain approximately 35 percent of the world's ruminant livestock units. However, they produce between 65 and 70 percent of the worlds supply of beef, 50 to 55 percent of mutton and lamb, and 75 percent of whole milk. Highly diverse systems of pasture and forage utilization have evolved in different temperate areas in response to climatic, soil, economic and social conditions. In recent years a distinct trend, particularly in North American and western European countries, has been the substitution of high energy concentrate feeds for forage in the diets of domesticated ruminant livestock. In light of diminishing energy resources, availability of land and the rapidly expanding world population, it seems probable that there will be a reversal of this trend in the future.

The principal nutritional constraint on animal productivity on a world-wide basis is the intake of digestible nutrients, particularly available energy. Pasture in temperate climates, as compared with tropical or sub-tropical regions, has the inherent advantage of a higher nutritional quality in terms of digestibility and intake potential; this enables high levels of individual animal production to be attained. The primary problem, relating to the seasonal nature of pasture growth and composition, is to maintain an acceptable compromise between individual animal performance and animal output per hectare by adjustment of grazing pressure, fertilization, conservation and supplementary feeding. A comparison is made of levels of performance reported in grazing trials in the United States with those obtained in other temperate and tropical areas. Regional differences are apparent, relating mainly to the pasture species employed and the marked effects of climate on forage composition and quality.

Intensification of pasture production in many temperate regions by use of improved species and cultivars of grasses and legumes and higher fertilization levels has resulted in an apparently increased incidence of metabolic disorders and nutritional imbalances in grazing animals. Such disorders require further definition if the maximum potential of the ruminant on pasture is to be obtained.

INTRODUCTION

In considering the topic of limitations to livestock productivity on the temperate grasslands of the world, it is first necessary to define briefly the distribution and characteristics of such grassland areas, the environmental factors controlling their occurrence, and modifications which have evolved through the development of agricultural systems by man.

While differences of opinion exist between geographers on the classification of temperate climates, it might generally be agreed that they lie between the subtropical and the boreal or subarctic climates, and approximately within the latitudes of 30 to 60°. Two principal types of temperate climate have been defined, the relatively mild oceanic or marine, and the more severe continental (Trewartha & Horn 1980). Annual rainfall in the humid temperate areas tends to vary between 500 and 2000 mm (Bula et al. 1977). The world's natural grasslands developed under conditions of restricted precipitation effectiveness (ratio of precipitation to evaporation) with, as described by McCloud & Bula (1973), the tall grass prairies found in the subhumid mesothermal and microthermal temperature provinces, and the drier grassland, or steppes, in the semi-arid humidity zones. Within the temperate regions the International Biological Program (Coupland 1979) has further distinguished natural grasslands from semi-natural or successional types, in recognition of modification of existing ecosystems by man. Distinct from these are arable grasslands, in which vegetative cover is maintained by introduced grasses and legumes through periodic reseeding, and cropland, on which annual seeding is practised.

Climatic and vegetation zones obviously do not follow political boundaries. However, Bula et al. (1977) have provided a classification of political entities falling largely within the humid, temperate climates of the world. This has been modified to include Canada and the United States on the North American continent, western and eastern Europe, the Union of Soviet Socialist Republics (USSR), mainland China and Japan, and New Zealand. It is recognized that Australia supports a considerable proportion of its cattle and sheep population in a temperate zone south of latitude 30°S with a mean average rainfall of more than 380 mm. This area includes New South Wales, Victoria and Tasmania.

FAO Production Yearbooks (1958-1979) indicate that, of an estimated world population of 1512 million forage consuming livestock units (cattle, buffalo, camels, sheep, goats; 1 cattle unit = 10 sheep or goats) in 1979, 534 million, or 35 percent of the total, were located in temperate countries. This represents an increase of 51.5 percent in animal numbers since 1950. World meat production increased steadily over the same period, with major increases in beef and veal production and little change in mutton and lamb. As would be expected, the relative contribution of temperate countries to the world supply of meat was markedly higher than their contribution to livestock numbers. McCloud (1974) has commented that, since the total area in grassland has remained fairly stable during the last 20 years, the increase in meat output represents an intensification of livestock production from pasture - recognizing that countries such as the United States have in recent years utilized high levels of feed grains in the finishing phase of beef cattle production. Estimates of meat production by region, and of the regional contribution to world supply, are summarized in Fig. 1. Given the vagaries of recording systems, it is still apparent that a high proportion of the world's meat production, 65 to 70 percent for beef and 50 to 55 percent for mutton and lamb, originates in the temperate countries of the world.

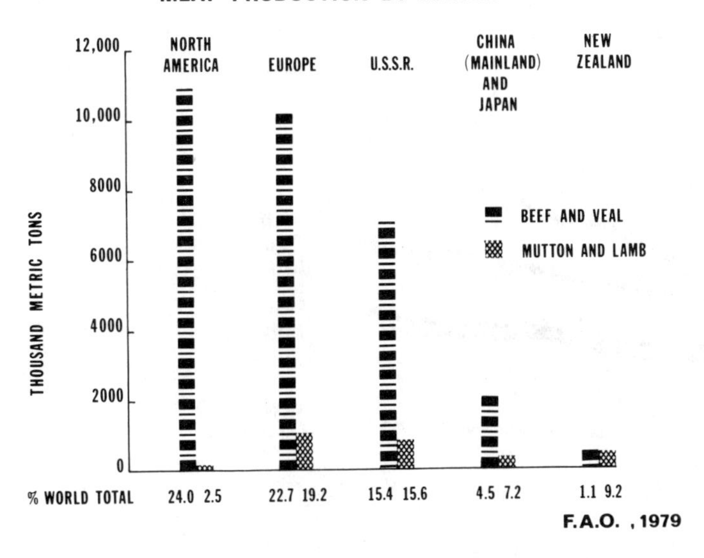

Fig. 1. Meat production in different temperate areas, as total production and as a percentage of world total, in 1979.

These are estimates of total livestock production by region and give no measure of individual animal performance, or productivity per hectare (ha), which are the problems to be considered at this conference, nor of how these may relate to quality of the diet. It has been suggested (McCloud 1974) that milk production per cow may be considered as an indicator of grassland productivity on a world and regional basis. Changes in the estimated levels of milk production per animal during the period 1950 to 1980 are summarized for temperate regions, and compared with world levels, in Fig. 2. The world average milk production per cow in 1950 was 1321 kg and increased to 1955 kg in 1979, an increase of 48 percent. By comparison, milk production per animal in Europe increased from 2030 to 3358 kg, an increase of 65 percent, and in North America from 2389 to 4794 kg, an increase of 101 percent during the same period. The more rapid increase in milk output per cow in the North American countries and Japan as compared with Europe undoubtedly reflects a shift to confinement systems of management and high levels of concentrate feeding. In western European countries, grass is the principal source of energy for dairy cattle, grazed herbage supplying approximately 50 percent and hay and silage 25 percent of the annual food intake, with concentrates supplying the remainder (Leaver 1976). In the United States it has been estimated that forages provide approximately 61 percent of the feed units utilized by dairy cattle (Barnes 1981). It is of interest to note, however, that the average milk production per cow in the Netherlands, on an intensive grass system, was 5094 kg in 1978, a level of production practically identical to that in the USA. These levels may be compared to an average output of 666 kg per cow in the developing countries of the world.

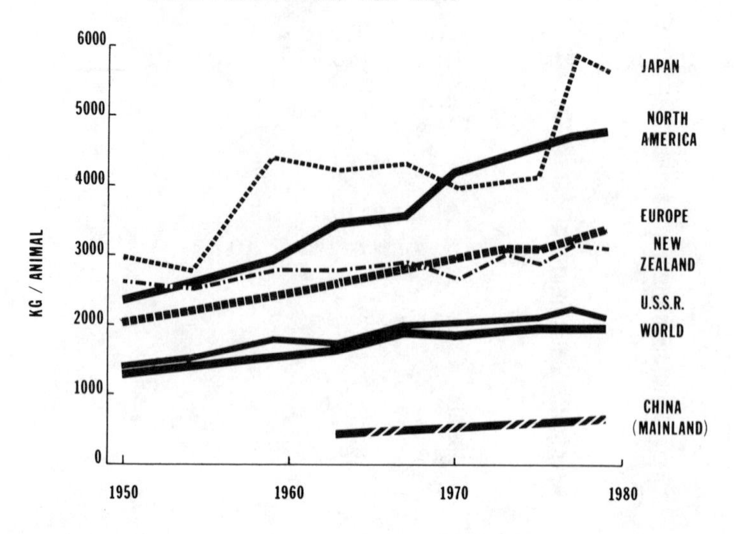

Fig. 2. Changes in estimated levels of milk production per cow in different temperate areas compared with world production during the period 1950 to 1980 (FAO Production Yearbooks).

A similar calculation can be made for meat production per animal, and data comparing mean levels of beef and veal, and mutton and lamb, output per head for the periods 1948-52 and 1978-79 in temperate areas are summarized in Fig. 3. It is apparent that, in general, increases in meat production per animal during the last 30 years are greater for beef animals than for sheep. It is also noteworthy that relative increases in beef production since 1950 have been greater in the European countries (82 percent) and the USSR (101 percent) than in North America (41 percent), where feed-lot fattening of beef cattle has been a common practice since the 1950's and 1960's. It might reasonably be suggested that, while the increase in livestock numbers per unit of land reflects an overall improvement in agronomic and management practices, the increase in milk and meat production per animal unit may result in part from an improvement in quality of the diet, as pasture, conserved forage and grain and other supplements, consumed by ruminant animals.

ANIMAL PRODUCTION SYSTEMS

It would be impracticable to review in detail the diverse sytems by which ruminant animals are produced on grasslands in temperate areas of the world. They range from extensive systems of sheep and beef cattle production on rangeland or hill country in the western and eastern United States and in parts of Europe, in which several hectares of land are required to support an animal unit, to the highly intensive use of improved pastures for milk production from dairy cattle as practised, for example, in western Europe and New Zealand. Rather, it may be of interest to consider briefly the utilization of grassland on the North American continent, and to relate levels and types of animal production attained to those in other temperate regions.

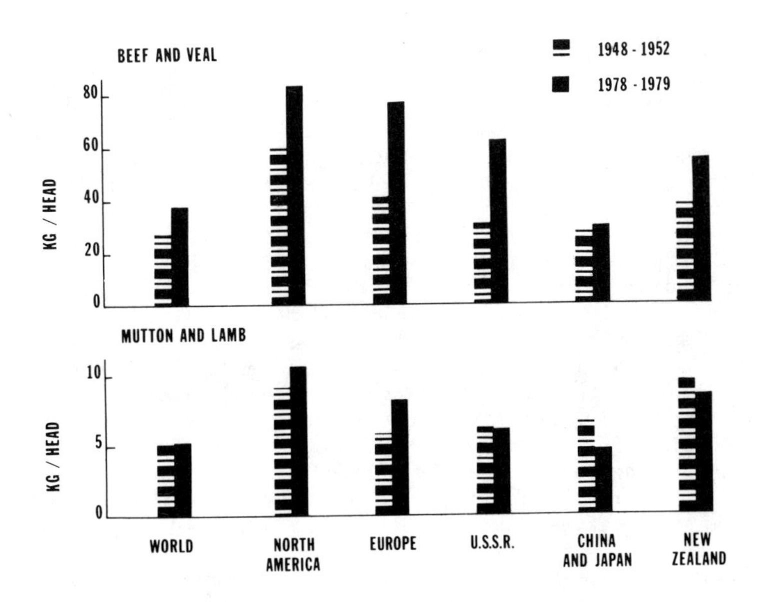

Fig. 3. Meat production per animal in temperate areas compared with world production. Source: FAO Production Yearbooks.

United States

Estimates of areas in rangeland and pasture use in the United States are summarized in Table 1 (Barnes 1981). It is apparent that estimates differ markedly with method of defining land use.

TABLE 1
Areas in pasture and forage in the United States (Barnes 1981).

Land resource	USDA† 1974	RPA†† 1977	NRI††† 1977
		Million ha	
Rangeland		332	168
Forest range		70	27
Pasture land and grassland pasture	242	45	54
Cropland pasture	34	34	12
Hay and silage	29		13
Total	305	481	274

† USDA (1978) Agricultural Statistics
†† Forest Service (1980) Resource Planning Act. An Assessment.
††† Soil Conservation Service (1977) National Resources Inventory with data for Alaska added.

The most extensive natural grasslands in Anglo-America occurred in an area east of the Rocky Mountains (Trewartha _et al_. 1967). The tall grass prairies, found in the more humid eastern region, extended south from the Prairie Provinces of Canada through the eastern parts of the Plains states to Texas; the short-grass steppe occurred through much of the semi-arid Great Plains between the 100th meridian and the Rocky Mountains. Zones of short bunchgrasses are found in restricted locations west of the Rocky Mountains. Much of this original grassland ecosystem has been modified drastically, first by extensive grazing and then by agricultural development. The prairie lands are now used intensively for the production of feed grains and livestock, and this use has extended into those sections of the semi-arid steppe land where rainfall is less limiting or where irrigation can be practised. Existing range is located mainly in the 17 states west of the Mississippi river and provides approximately 291 million ha of grazing for livestock, principally beef cattle and sheep. The remaining range area, some 28 million ha, is found east of the Mississippi and consists primarily of forest land grazing. Much of the pasture in the eastern United States is improved in type, seeded to cool season and warm season grasses and legumes, frequently subject to varying degrees of fertilization, management and conservation, and used in different forms to support both dairy and beef cattle production.

The changing role and potential of pasture and forages in meeting livestock needs in the United States has been reviewed (e.g. Hodgson 1977, Pimentel _et al_. 1980, Ward 1980, Barnes 1981). Beginning some 30 years ago, the availability of large supplies of cheap feed grains led to an intensification of cattle feeding and significantly modified traditional methods of pasture utilization. Pimentel _et al_. (1980) cite data by J.T. Reid that, in the period from 1960 to 1973, the use of hay as feed for cattle declined by more than 24 percent and the use of pasture by more than 32 percent. A summary of the relative proportions of forage to concentrates in the diets of ruminant livestock in the United States at the present time is given in Table 2. Pasture and forage provide nearly all the nutrient requirements for sheep, goats and beef cattle other than those being finished in feed lots. High quality hays and silage are the main sources of forage for dairy cattle, with pasture frequently filling a somewhat nebulous role as an exercise area and an unaccounted-for source of nutrients.

With anticipated constraints on availability of energy and land resources in the United States and changing demands for livestock products, it is probable that there will be at least a modified return to earlier methods of producing meat and milk, with increased emphasis on the use of pasture and forages (Forest Service 1980). Range and pasture use by beef animals, which is the main determinant of national grazing demand, is projected to increase steadily during the next 50 years. Quantitatively, total roughage requirements by the year 2030 are expected to increase by 38 percent, harvested roughage requirements by 5 percent, and all grazing by 54 percent. Within the grazing category, range grazing is projected to increase by 41 percent and non-range grazing (permanent pasture, cropland pasture, aftermath grazing such as crop residues in the field) by 57 percent.

TABLE 2
Use of feedstuffs by ruminants in the United States[†,††].
Source: Barnes (1981)

Types of animal	Proportion of ration		Proportion of total concentrate usage (%)	Proportion of total feed usage (%)
	Concentrates	Forage		
All livestock and poultry	37.5	62.5	100.0	100.0
All dairy cattle	38.8	61.2	16.6	16.0
Beef cattle on feed	72.4	27.6	20.7	10.7
Other beef cattle	4.2	95.6	5.1	46.2
All beef cattle	17.0	83.0	25.8	56.9
Sheep and goats	8.9	91.1	0.4	1.9

† Council for Agricultural Science and Technology (1980).
†† All feedstuffs measured on a corn-equivalent basis.

The total sheep population in the United States decreased from 27 million in 1964 to approximately 12 million in 1979 (USDA 1979). The decline in sheep numbers has been attributed to a lower demand for wool, problems in flock management, difficulties of predator control in range land, competing demands for use of range under Federal ownership, and a lesser demand for lamb by consumers. A similar drop in dairy cattle numbers has occurred, with a decline from approximately 16 million cows in 1964 to 11 million in 1978; milk production has remained fairly stable, however, due to an increase in output per cow from 3681 to 5109 kg over the same period. The beef cattle population tends to vary in a regular cyclical pattern and emphasis on pasture or forages as compared with grain feeding may change with phase of the cycle. In general, as summarized by Ward (1980), breeding herds and store operations tend to be located in areas of low forage costs, while feed lot finishing is concentrated in regions with suitable weather and cheap feed grains. This results, frequently, in the transportation of cattle over considerable distances. There has been a trend for centres of beef cattle production to move to the east; thus, Barnes (1981) reports that in 1978 the north-central and southern regions each maintained 40 percent of the cattle not on feed.

Europe and New Zealand

Less emphasis on grain feeding and greater reliance on pasture and forage crops is encountered in other temperate regions. In contrast to the United States, where beef cows outnumbered dairy cows by approximately 37 million to 11 million in 1979, Cunningham (1976) reported that, of the 80 million cattle in the European Economic Community (EEC), 25 million were either dairy or dual purpose cows, with some 6 million suckler cows. Much of the beef production in Europe is, therefore, from the dairy herd. Nearly half of the dietary energy for the dairy cow in the United Kingdom (UK) is derived from pasture, which is generally fertilized and intensively managed; systems of pasture utilization for dairy cattle have been described by Leaver (1976).

As in the United States, traditional methods of beef production in European countries have been described as a "leisurely process" (Craven & Kilkenny 1976), with cattle generally slaughtered from grass at the age of 2½-3 years. The process persists in certain areas; it is calculated that the proportion of grass in the crop areas used for beef production varies between 75 and 95 percent in different countries in the EEC, with a generally small contribution from root and succulent crops (Tayler 1976). There has, however, been an increasing trend to intensification of beef cattle production in recent years with the use of corn, corn silage and cereal-based systems, resulting in higher rates of liveweight gain and a reduction in the age and weight at which animals are slaughtered. Systems of beef production from the beef and dairy herd, utilizing different combinations of grassland and concentrate feeding, have been described by Baker (1975). Craven & Kilkenny (1976) estimated that 49 percent of steers and heifers in the UK were slaughtered between 12 and 24 months of age, and 45 percent when older; 51 percent were fattened in the equivalent of American feed lots in the period December to May, and 49 percent were slaughtered off grass between June and November. It is pointed out (Wilkinson 1976) that systems of producing beef from dairy type animals for slaughter at either 18 or 24 months provide farmers with the option of maximizing rate of individual animal growth on pasture or of maximizing output per ha of land. From the viewpoint of economics and efficiency of energy conversion, a feeding program based on maximum use of high quality pasture and conserved forage is considered more effective than one involving use of cereal grains and imported protein feeds.

At the opposite end of the spectrum from the contemporary American approach to animal production systems lies the situation in New Zealand where, for economic, topographical and climatic reasons, ruminant animal feeding is geared almost totally to the pasture nutrient cycle. Principal constraints to animal production, as defined by Rattray (1978) and Campbell & Bryant (1978), are "the attainable yield of pasture digestible dry matter" and the variation in pasture supply between years, between seasons of the year, and between geographic areas within the country. Farming systems range from intensive sheep and dairy production on relatively flat country, with a carrying capacity of 15 ewe equivalents per ha on ryegrass-clover pastures, to extensive sheep raising on tussock and sub-alpine grassland on high country in the South Island, with a carrying capacity of 0.75 ewe equivalents per ha (Brougham & Grant 1976).

TYPES OF PASTURE AND METHODS OF IMPROVEMENT

Natural grassland

Natural grassland areas in the temperate regions vary markedly in their botanical composition, depending largely on soil and climatic conditions; uniformly animal productivity from such areas tends to be low and subject to marked seasonal influences. For example, cattle management studies on sagebrush-bunchgrass range in the western United States typify the nutritional problems of livestock production on natural grasslands with limited precipitation (Raleigh 1970). Digestible nitrogen intake, in relation to the requirements of yearling steers for different rates of daily gain, became limiting in mid-June and digestible energy intake some two weeks later. Little or no animal gain was recorded after September 1. Mean digestibility coefficients of range forage declined from 62 percent for dry matter at the end of May to 48 percent at the beginning of September; apparent digestibility of nitrogen decreased from 65 to 25 percent.

Frequently, management systems on western range are highly complex, involving integration of desert, foothill and mountain areas with different plant communities of varying quality at different seasons of the year. The advantages of a diversified plant cover on seasonal ranges in meeting the annual nutrient requirements of grazing livestock have been indicated by Cook & Harris (1968); lactating animals were found to gain during the entire summer, while liveweight gains in calves and lambs declined markedly. Van Dyne *et al.* (1981) showed marked seasonal changes in the pattern of consumption of different plant groups by cattle, sheep, bison and pronghorn antelope grazing shortgrass prairie, indicating the possibility of complementary utilization of natural grassland by domesticated and indigenous animal species. Nutrient deficiencies on winter range are common, particularly with high intensity of use, and some form of supplementary feeding is generally necessary. It is, however, recognized that under typical western range management conditions it is not economical to supplement animals for maximum production.

Under heavy grazing pressure, selection by grazing animals may radically modify the botanical composition of natural grassland and lead to domination of the vegetative cover by undesirable species. A study by Bishop *et al.* (1975) of rangeland in semi-arid areas of Argentina indicated that, of forage species dominating sandhill pastures, many were totally unacceptable to grazing sheep. Sheep were found to select their diet from species making up less than 25 percent of all forage available; the primary nutritional constraint was an inadequate intake of energy, particularly during the winter months. A similar problem occurs on the upland grazing regions of western Europe (Hughes 1976, Hill Farming Research Organization 1979). Thus, Eadie (1973) commented that a major and general feature of sheep production on hill country in Scotland is a low level of individual sheep performance. Typical performance data were a stocking rate of 0.8 ewes per ha, a weaning percentage of 80 and a weaned lamb liveweight of 24 kg; the output per ha from such pasture was 15.3 kg weaned lamb per year, compared to an output of 612 kg from lowland pasture. Eadie concluded that the 40 fold difference in weaned lamb production was a function of a 5 fold difference in pasture production, a 3.5 fold difference in the efficiency with which the pasture was actually consumed, and a 2.5 fold difference in the efficiency with which the consumed herbage was converted to meat.

Large areas of natural grasslands occur in the USSR; estimates range from 375 to 421 million ha (Larin 1962, Morozov 1974), with much of the area in the south-east of the Russian Federation, Kazakhstan and Central Asia. Lack of moisture results in low yields of native grasses and low levels of animal productivity. China's natural grasslands occupy 37 percent of the total land area, and include temperate regions to the north and west of the isohyet of 400 mm rainfall, and subtropical regions in the south (Chia 1981). Grazing capacity of the winter pastures is described as low. In Japan, where a *Zoysia* type vegetation predominates in natural grasslands under continuous grazing, preliminary calculations (Okubo *et al.* 1977) showed low levels of animal gain and milk production. Similar results have been recorded in New Zealand and temperate areas of Australia. A comparison of the performance of sheep grazing native and improved pastures at different stocking rates in New South Wales (Langlands &

Bowles 1974) indicated that, while organic matter intakes were not different on the two types of vegetation, liveweights, wool production, digestibile organic matter, digestible nitrogen and nitrogen intakes were significantly lower on the native herbage. They attributed the differences mainly to the lower nutritive value of the diet selected from the natural pastures, particularly during winter.

In summary, the primary limiting nutritional features of natural grassland in temperate environments are: (1) a relatively short growing period when indigenous plants have a nutritive value equivalent to that of introduced species; (2) suppression of desirable species by selection and over-grazing, perhaps particularly with sheep; (3) the necessity, in certain areas, to under-utilize vegetation in spring and summer to maintain an adequate supply of winter feed, leading to accumulation of dead and mature herbage; (4) a relative deficit of protein, minerals and energy in the dominant plant species adapted to unfavourable soil and climatic environments.

Improved pastures

The principal objectives of pasture improvement are to increase production per ha and to prolong the period of availability of nutritive grazing so that "it becomes possible in many regions to depart from the primitive condition where animal physiology and reproductive behaviour follow the cycle of growth of the natural vegetation and to achieve a condition of reasonably uniform nutrition through the year" (Whyte *et al*. 1959). Greater or lesser attention may be paid to the improvement of nutritional quality *per se*, depending on: (1) the degree of development or sophistication of the improvement program; Cooper (1973), for example, pointed out that grasses and legumes are initially selected on the basis of survival and seasonal dry matter production rather than on nutritional composition; (2) the objectives of the grazing system.

Universally, the major problem encountered in improving pasture systems in temperate regions is to alleviate the marked seasonality of pasture production and the consequent effects on animal performance. The seasonal nature of pasture growth varies substantially in temperate countries; Hughes (1970) comments that areas in New Zealand have a 365 day growing period, whereas in Finland the period of pasture growth is approximately 125 days. Similarly, the grazing season in the United States ranges from some 150 days in the north-east to essentially year round grazing in the lower south. In the UK, 75 percent of total annual growth of hill pastures is produced within three months of the year. Ollerenshaw *et al*. (1976) have indicated the possibility of extension of the grazing season by selection of ecotypes of grasses and legumes adapted to low temperatures and low light intensities. In the eastern United States, where liveweight gains and carrying capacity of animals on pasture decline rapidly in late summer, management systems based on the complementary use of cool season and warm season grasses and legumes have been developed (Matches *et al*. 1975). While rates of gain on warm season grasses are generally not as high as on cool season species, their use permits high stocking rates to be maintained during periods when pasture growth is minimal. A major virtue of *Festuca arundinacea* in management systems is its ability to maintain relatively high digestibility in late autumn and early winter; this has led to the development of a variety of systems for stockpiling or baling this species for winter feeding of sheep and beef cattle on pasture. Other systems currently being examined in the United States to extend the grazing season are: (1) use of crop residues such as corn stover *in situ*; (2) introduction of fast-growing crops such as fodder root crops and brassicas, or annual grasses such as *Sorghum sudanense* or *Pennisetum americanum*, into permanent pastures by sod-seeding techniques.

Estimates of potential forage and animal productivity with increasing inputs of fertility and management are given in Table 3. The general principle is evident that, with improvement in fertility, renovation practices and grazing management, increases in herbage dry matter yield on both permanent pastures and arable land are accompanied by improvements in forage quality (digestibility) and animal production. The level of response to different inputs, in terms of both quantity and quality of herbage, varies markedly with soil characteristics, climatic conditions and nature of the existing vegetation. Hughes (1976) has summarized the results of studies in Wales with a *Molinia*-*Nardus* ecosystem on a peaty gley soil in Table 4. It is apparent that under conditions of poor soil fertility but adequate rainfall, a series of pasture renovation measures can result in major changes in pasture yield, organic matter digestibilty, protein and soluble carbohydrate concentrations in the herbage.

The responses to nitrogen may be noted. Use of nitrogenous fertilizers on perennial grass pastures has been the cornerstone of grassland improvement philosophy in western European countries since the 1960's, as it has on much of the permanent pasture in the southern and eastern United States. The price of nitrogenous fertilizers in the United States tripled in the ten year period between 1970 and 1980, with no commensurate increase in prices paid to farmers for animal products. This has led to a reappraisal of the role of legumes in pasture improvement. Relative input-output energy relationships for nitrogen fertilized grass pastures and grass-legume associations are summarized in Table 5. The advantages, in terms of animal production, of including legumes in pastures in temperate areas have been reviewed by Thomson (1979). For both sheep and cattle, rates of liveweight gain, herbage intake and efficiency of feed utilization are higher on grass-legume pastures than on grass alone; on the other hand, output and stock carrying capacity are generally lower. Thomson therefore suggests that the most appropriate strategies for utilization of the properties of legumes in grazing systems may lie in: (1) inclusion of the legume at a point in the animal production cycle when either quantity or quality (as in late season growth) of herbage are limiting; (2) provision at times of high nutrient demand, such as early stages of growth or early lactation in the dairy cow. Bula *et al*. (1977) state that at least 100 million ha of the world's humid temperate regions are used for cow-calf production at a stocking rate of two cow-calf units per ha; on the basis of typical responses in calf weight gains to the inclusion of clover in *Festuca arundinacea* pastures in grazing trials in Indiana, they calculated that simlar practices could increase liveweight meat production by 11 million metric tons.

TABLE 3
Estimates of annual forage production and animal productivity in the humid, temperate regions of the world. Source: Bula *et al*. (1977).

Forage production technology	Plant productivity		Animal productivity	
	Dry matter Tonnes/ha	Digestibility %	LW gain	Whole milk
			----Tonnes/ha----	
On permanent grazing land				
Present management	2	52	0.02	0.5
Improved fertility	4	54	0.08	1.6
Renovation, including interseeding with legumes	5	57	0.17	3.5
Improved grazing management	6	60	0.37	5.5
On arable land				
Present management	8	55	0.19	3.6
Adequate fertility	10	55	0.24	4.5
Improved varieties and pest control	12	57	0.44	8.1
Timely harvesting	14	60	0.86	12.8
Reduction of harvesting losses	15	65	1.56	20.7

TABLE 4
Yield and quality changes in *Molinia-Nardus* pastures in Wales 10 years after different treatments. Source: Hughes (1976).

Pasture treatment†	Yield Tonnes/ha	DOMD %	Crude protein %	Water soluble carbohydrate %
Original	1.1	37	7.1	3.6
Controlled grazing	2.3	49	15.6	8.9
Lime, basic slag	2.5	58	14.6	11.0
Lime, basic slag, 50 kg N/ha	3.4	66	17.3	12.7
Lime, basic slag, 250 kg N/ha	4.9	73	24.8	12.0
Dalaphon (Complete fertilizer and seed mixture)	4.9	73	18.6	14.3
Forage harvest (Complete fertilizer and seed mixture)	5.6	73	18.8	14.2
Rotavation (Complete fertilizer and seed mixture)	6.3	72	17.0	16.1
Plough and reseed, complete fertilizer				
Lolium perenne cv. S.23, 50 kg N/ha	6.1	75	18.0	16.7
Festuca rubra cv. S.59, 50 kg N/ha	6.3	70	21.4	13.5
Lolium perenne cv. S.23, 250 kg N/ha	10.2	72	18.9	13.1
Festuca rubra cv. S.59, 250 kg N/ha	6.9	75	24.4	11.5

† All treatments grazed rotationally. Stocking rate adjusted according to herbage on offer.

TABLE 5
Comparison of energy utilization between nitrogen fertilized grass pastures and grass-legume pastures (Gordon (1980).

	Grass + 450 kg N[†]	Grass + Clover[††]
Total energy input (MJ)	37,940	6,814
Output (kg DM/ha)	13,200	9,700
ME output (MJ)	145,200	106,700
ME output/MJ input	3.8	15.7

† Five applications/year; (kg) N, 450; P_2O_5, 90; K_2O, 190.
†† One application/year; (kg) N, 60; P_2O_5, 60; K_2O, 140.

The nutritive potential of major grass species used in the United States, with and without companion legumes, is summarized and compared with data from other temperate and tropical areas in Table 6, which includes results from a large number of grazing trials with certain common features. The features are daily liveweight gains and liveweight gains per ha of yearling steers (primarily of British beef breeding) grazing different pasture classes at low to moderate stocking intensity. The overriding effect of stocking rate on animal performance has been extensively discussed (e.g. Jones & Sandland 1974, Mott 1980); the present trials were selected to emphasize yield per animal as an expression of forage quality, with herbage available at the *ad libitum* level. This, as indicated by Mott, may underestimate animal yield per unit area. Average daily gain (ADG) of steers and steer gains per ha were calculated, in the main, at nitrogen fertilizer levels between 0 and 200 kg nitrogen per ha, although ranges for steer production per ha at fertilizer levels of up to 600 kg nitrogen are also included. Baker (1975) has cited target growth rates of 0.9 kg per day for yearling cross-bred steers finished on pasture at 20-24 months of age in the UK. Under experimental and good farm management conditions these growth rates are generally attained on either nitrogen fertilized grass or grass-legume pastures over the grazing season in western European countries with, as expected, significant seasonal effects. The lower growth rate and efficiency of energy utilization by cattle grazing pastures in autumn as compared with spring and summer has been indicated in a number of British studies (see also Armstrong 1982). Weight gains of 0.7 and 0.8 kg per day on improved pasture are regarded as reasonable in the USSR (Kutuzova 1972).

The generally poorer performance of animals grazing cool season perennial grass and grass-legume pastures in American grazing trials is therefore of considerable interest. The majority of such trials have been conducted in the northern and eastern regions of the United States with rather similar climatic conditions and lengths of grazing season to those obtained in Europe. The lower liveweight gains per ha may be attributed in part to a lesser use of nitrogen fertilizer and generally lower stocking rates. The lower average daily gain in steers, in comparison with European data, may relate to use of different grass species in the pastures. Digestibility of ryegrass, which is used extensively in Europe and New Zealand but mainly as winter pasture in the United States, is higher than that of other

TABLE 6
Estimates of steer performance on different pasture types over the total grazing season in the United States compared with the values from other temperate and tropical countries.

Country	Pasture type	Major species	Animal performance ADG, kg	kg/ha
U.S.A.[†]	Cool season perennial grass	*Dactylis glomerata*, *Festuca arundinacea*, *Bromus inermis*, *Poa pratensis*, *Phalaris arundinacea*	0.59 (0.38-0.93)	398 (216-708)
	Cool season grass-legume	Same grasses + *Trifolium repens*, *T. pratense*, *Lotus corniculatus*, *Medicago sativa*.	0.61 (0.42-0.90)	345 (207-611)
	Cool season annual grass	*Lolium multiflorum*, cereals	0.85 (0.70-1.15)	362 (253-579)
	Warm season perennial grass	*Cynodon dactylon*, *Paspalum notatum*, *Digitaria* spp., *Andropogon* spp., *Panicum virgatum*	0.47 (0.16-1.08)	493 (68-1043)
	Warm season annual grass	*Sorghum sudanense*, *S. sudanense* x *bicolor*, *Pennisetum americanum*	0.81 (0.47-0.97)	373 (217-525)
Puerto Rico	Tropical grass	*Panicum maximum*, *Pennisetum purpureum*, *Digitaria decumbens*, *Cynodon nlemfuensis*	0.56	1201
United Kingdom, Ireland, Australia (N.S.W.)[††]	Cool season perennial grass	*Lolium perenne*, *Dactylis glomerata*, *Agrostis-Festuca*	0.86 (0.75-1.13)	804 (560-1467)
	Cool season grass-legume	*Lolium perenne* + *Trifolium repens*	0.86 (0.73-1.07)	705 (500-940)
Africa, S. America, Australia[†††]	Tropical grass	*Panicum maximum*, *Pennisetum purpureum*, *Digitaria decumbens*, *Hyparrhenia* spp., *Chloris gayana*, *Paspalum notatum*	0.74 (0.39-1.27)	1093 (255-2027)

† Data from 28 reports of grazing trials; ranges of N fertilizer on grasses up to 400 kg N/ha; ADG's calculated for N fertilizer rates up to 200 kg N/ha.

†† Data from 15 reports of grazing trials; ranges of N fertilizer up to 600 kg N/ha; ADG's calculated for N fertilizer rates up to 200 kg N/ha.

††† Data from Smith (1970); Moore & Mott (1973); Stobbs (1976). N fertilizer rates up to 600 kg N/ha.

perennial grasses. Of the major species used for grazing in eastern and northern USA, both *Festuca arundinacea* and *Phalaris arundinacea* support generally low levels of individual animal performance, although their carrying capacity for livestock is high.

Perhaps surprisingly, comparison of individual liveweight gains on cool season perennial grasses as compared with grass-legume combinations in either the United States or other temperate areas indicates relatively minor differences, although the steer gains per ha on nitrogen fertilized grass pastures are generally higher. The nutritive characteristics of tropical pastures will be discussed in a subsequent paper (Norton 1982). It is sufficient to point out that, where warm season species such as *Cynodon dactylon* form the basis of grazing systems in the southern United States, average daily gain of cattle is lower than on cool season grasses. With high levels of nitrogen fertilization, adequate rainfall and an extended grazing season, the level of steer production per ha may be remarkably high (Burton 1972).

LIMITATIONS TO PRODUCTION

Nutritional limitations

The grazing animal exists in a highly dynamic situation in which its performance, in terms of growth, milk output or wool production, is determined not only by its changing nutrient requirements but by its physical environment and by the quantity and quality of herbage on offer. The purpose of this conference is primarily to emphasize quality aspects and it would be recognized that, under temperate conditions, the factor principally limiting animal response is the intake of digestible energy, with further consideration given to the efficiency of nutrient utilization. Additional limitations may be imposed by relative deficiencies or excesses in plants of specific minerals or the presence of metabolic inhibitors, resulting either in acute disease or, more commonly, in sub-optimal performance.

A simple model under which various factors affecting animal response may be examined is outlined in Fig. 4. A general relationship between productivity expressed as liveweight gain or milk output and intake of digestible energy by animals on pasture is well documented. A positive linear relationship between dry matter, organic matter or energy digestibility and level of intake for ruminants fed different forms of harvested forage and for grazing animals is also indicated (e.g. Hodgson *et al*. 1977, Voigtländer & Kühbauch 1978). Further, energy retention declines with concentration of metabolizable energy in the diet, even when metabolizable energy consumption is constant (Greenhalgh 1975). In terms of both intake and efficiency of energy utilization, therefore, the level of available energy in herbage is of major significance.

Factors influencing the digestible energy concentration of pasture plants fall into the principal categories of: (1) climate; (2) soil; (3) management.

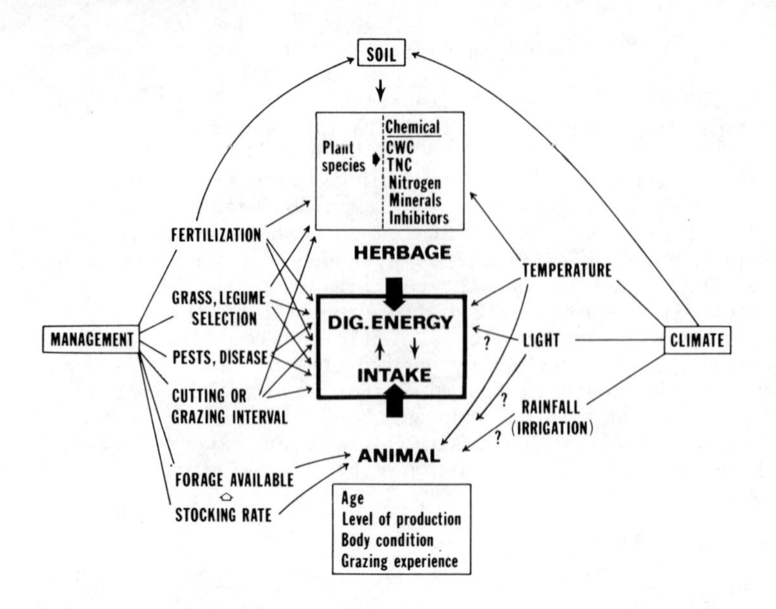

Fig. 4. Factors affecting the digestibility and intake of herbage by grazing animals.

Climate

Effects of temperature and light intensity on forage quality have been reviewed by Deinum & Dirven (1974) and will be discussed later in this symposium by Wilson (1982). In temperate regions a primary effect of increasing temperature is to depress concentrations of non-structural carbohydrates in the herbage. In the United States, an example of the consequences of climate on herbage digestibility is given in Fig. 5, a summary of data obtained with different classes of forages in Texas (Riewe 1981) and West Virginia. Digestibility is higher at the more northern latitudes. However, there appear to be climate x forage class interactions; thus, digestibility of the grasses is more severely depressed in southern conditions than is digestibility of the legumes, a factor which might account for a relatively greater animal response to legumes in a tropical or subtropical environment.

In terms of climate, the general principle emerges that "a better forage quality may be attained in the cooler parts of the year, provided temperature and radiation are high enough for grass growth" (Deinum & Dirven 1974). These authors indicate that regions and seasons where this principle apply include the northern European countries in spring and autumn, and the southern United States and southern USSR during the winter period. Direct effects of climatic factors such as rainfall, temperature and photoperiod on the animal may be expressed through effects on its endocrine status and on the distribution and length of time spent in grazing, and may result in decreased intake of pasture under extreme conditions.

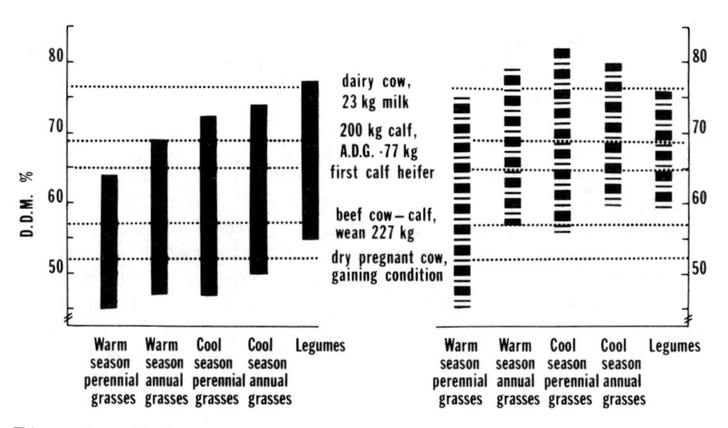

Fig. 5. Relative dry matter digestibilities of different classes of forage in Texas and West Virginia as related to nutrient requirements of cattle. Adapted from Riewe (1981).

Soils

Effects of regional variations in soils and soil fertility on the nutritional quality of pastures are difficult to quantify. There is considerable evidence that mineral limitations or imbalances in herbage limit animal productivity in both temperate and tropical environments (e.g. Butler & Jones 1973, McDowell 1976, Reid & Horvath 1980). Mineral imbalances are encountered both on native grassland under conditions of low soil fertility and, increasingly, on improved pastures subject to high inputs of fertilization and management. A point of major interest which arises from recent studies, particularly in Australia and the United States, is that deficiencies or imbalances of minerals such as calcium, phosphorus, sulphur, sodium, magnesium and certain of the micro-elements may alter ruminal function and digestibility and intake by the animal. While it might be expected that responses of this nature could be corrected by appropriate mineral supplementation, Australian research indicates that animals may react differently to specific minerals supplied in the diet or through the plant by soil amendment practices (e.g. Rees & Minson 1976).

Management

The principal nutritional constraints applied to the level and type of production of grazing ruminants will be those imposed through different forms of management. As pointed out by Greenhalgh (1975), maximum production per animal is obtained at low grazing intensity; most estimates of potential liveweight or milk production per ha have been predicated on maximum consumption of available herbage and this is usually achieved at the cost of reduced intake per animal. One solution is the provision of supplementary feeds to grazing animals; there is, however, considerable difference of opinion as to the form such supplementation should take.

A widely used approach to the improvement of animal performance on pasture has been the introduction of nutritionally superior grasses and legumes, superior in the sense of higher digestibility or intake characteristics. Application of selection techniques will be discussed elsewhere at this symposium (Bray 1982, Hacker 1982). In the United States, grazing trials with the *Cynodon dactylon* cultivar Coastcross-1, which is 12 percent more digestible than Coastal, have given a 34 percent higher average daily gain in yearly steers (Burton 1978).

It is probably unnecessary to point out that selection for increased digestibility or, for example, herbage concentrations of non-structural carbohydrates, lignin or cell wall components on the basis of greenhouse or small plot trials may have limited application in the grazing environment. Pasture evaluations in Wales (Evans *et al*. 1979, Davies & Morgan 1979) indicate that "apparently superior herbage quality of a species has not always resulted in significantly better animal performance from that species". Agronomic and quality characteristics are highly interrelated.

Of other factors less directly affecting quality, Burton (1978) has discussed progress in restructing the forage plant (dwarfing *Pennisetum americanum*, removing bloom on forage sorghum), changing the maturity (photoperiod responses) and increasing resistance to insects and disease. On the assumption that pasture production in the future will extend into areas now defined as marginal, i.e. with soil, climatic or topographical limitations, a particular need is seen in the area of selecting forage plants of reasonable quality adapted to acid and problem soils.

The effects of fertilization on herbage quality may be either indirect, through changes incurred in the botanical composition and yield of the pasture, or direct, through alterations in the concentration and structure of cell wall fractions, levels and ratios of non-structural carbohydrate and nitrogenous components, and concentrations of minerals and potentially inhibitory compounds such as nitrate and alkaloids. Use of high levels of nitrogen fertilization on improved grass pastures in countries in western Europe has led not only to markedly increased outputs of milk and beef per ha, but to an increased occurrence of metabolic problems such as hypomagnesaemic tetany, nitrate poisoning, and a somewhat nebulous category of "production diseases". The incidence of hypomagnaesemic tetany, in both its spring and winter forms, appears to have become more widespread in beef cow herds in the United States during the last 20 or 30 years. Effective preventive measures have been developed for dairy herds in western European countries; the condition has not obtained significant recognition in eastern Europe or the USSR. Presumably a greater dependence on grass-legume pastures in New Zealand and temperate areas of Australia has minimized the problem in those countries.

On intensively managed farms in the Netherlands, high levels of nitrate in herbage have been reported (Kemp & Geurink 1978). Russian workers (e.g. Kutuzova & Morozova 1980) state that "the content of nitrate and water soluble carbohydrate in grasses is the most important factor in determining the appropriate rates of nitrogen fertilizers for pastures". They further report a serious effect of nitrate on milk output and reproductive ability in dairy cows (Vorobev 1980). A problem of comparatively recent identification in the United

States has been that of impaired reproduction in beef cows grazing Festuca arundinacea pastures treated with high levels of nitrogen fertilizer. A seeming paradox has also arisen in that, while levels of nitrogenous compounds in fertilized spring pasture are generally much higher than the animal's requirements, milk production in dairy animals may be increased if protected protein is fed (Stobbs et al. 1977, Minson 1981a). These problems obviously require further attention.

A group of what have been termed "pastoral faux pas" or anti-quality factors in pasture herbage have resulted in considerable decreases in animal productivity and economic losses in a number of temperate countries. This subject has been reviewed (Matches 1973, Reid & Jung 1973, Voigtländer & Kühbauch 1978, Burns 1978) and will be further discussed later in the symposium by Hegarty (1982). Under western range conditions in the United States it has been estimated that 3-5 percent of cattle may be adversely affected by ingestion of toxic plants; losses result from death of grazing livestock, photosensitization, abortions, birth defects and decreased production. On sown pastures in the eastern regions the major problems are associated with the grazing of Festuca arundinacea, Phalaris arundinacea, forage sorghums, Sorghum sudanense, possibly Cynodon dactylon, and such legumes as Coronilla varia and Lespedeza cuneata. Apart from direct metabolic effects, a frequent consequence of the presence of inhibitors is a depression of intake of herbage by the grazing animals; poor steer performance on Festuca arundinacea, for example, has been associated with the presence of mycotoxins produced by fungi such as Epichloe typhina or Acremonium (Hoveland et al. 1980). The interactions of climate, soil and agronomic management with the production and effects of inhibitors in pasture plants are indeed complex and little understood at the present time.

The profound effects of grazing or cutting interval, level of forage available to the animal and grazing pressure on herbage quality and animal production in temperate regions have been reviewed (e.g. Morley 1978, Minson 1981b) and will be considered again at this conference. Herbage intake is directly affected by physiological status, level of production, body condition and grazing experience of the animal. These are animal rather than herbage quality characteristics but obviously are highly involved in determination of intake of digestible or metabolizable energy, and thereby level of performance of the grazing ruminant.

Other limitations

While it is theoretically possible to define most of the nutritional constraints influencing animal responses on pastures, and to devise management systems emphasizing either high individual animal output or high output of animal product per unit area, it should be recognized that the gap between theory and practice is frequently substantial and determined by factors unrelated to production principles. As stated by Brougham & Grant (1976): "Progress to a large degree has been determined by economic factors and their associated social effects. The result has been the adoption of farm management practices which, while not ensuring maximum productivity, have ensured economic survival."

If further increases in animal production from temperate grasslands are to be attained, presumably they will result either from reclamation and improvement of land which is presently regarded as marginal from the standpoint of soil, precipitation or topography, or from intensification of management systems on existing arable land. The susceptibility of both alternatives to political, economic and social constraints is apparent. Development of marginal lands in the upland areas of western European countries is presently supported by a range of governmental grants and subsidies, although the nature of returns to be expected from such investment is not yet evident. Improvement of grassland in New Zealand has been described as a cyclical process, geared to economic conditions. In the United States at the present time large areas of permanent pasture in the Appalachian region are reverting to native vegetation; factors involved include marginal returns in the beef cow-calf industry, small farm units, high costs of fertilizer and labour, inflationary gains in landholdings and a reluctance on the part of farmers to intensify management systems. The latter consideration - particularly the management skills involved in going to very high stocking rates and complete utilization of available herbage - has limited intensification of pasture use in many countries.

The world problem was defined by R.W. Phillips in 1970: "In 1960 there was available for each person in the world 1.20 acres of arable land (including land in tree crops) and 2.18 acres of permanent grasslands. Assuming that the amount of land available for food production remains unchanged, and the UN medium projections of population numbers in 2000 are realized, these amounts of land per person upon which to produce will drop to 0.6 acres of arable and tree-crop land and 1.09 acres of permanent grassland. In these circumstances each acre of land would be called upon to produce approximately twice as much to maintain present levels of production per person."

The area of permanent grassland in temperate regions has remained essentially constant during the last 25 years. Increases in domestic animal numbers in the period since 1970 have been only 50 percent of increases in the human population (Henzell 1981). Temperate regions presently produce some two-thirds of the world's ruminant livestock products on approximately 30 percent of the world's permanent grassland area. Considering the advances in forage technology, management and plant breeding which have occurred in recent years, a doubling of animal output in temperate regions appears technically feasible. The feature, defined in a comment by Whyte *et al.* (1959), "now depends on the creation by the agrarian sociologist and the political scientist of the economic and social environment or climate in which grass can be grown".

ACKNOWLEDGEMENTS

The authors are indebted to Ms. Melanie Norberg for assistance in preparation of this manuscript.

REFERENCES

Armstrong, D.G. (1982) Digestion and utilization of energy. In: Nutritional limits to animal production from pastures. Editor J.B. Hacker. Farnham Royal, U.K., Commonwealth Agricultural Bureaux, pp. 225-244.

Baker, H.K. (1975) Grassland systems for beef production from dairy bred and beef calves. Livestock Production Science 2, 121-136.

Barnes, R.F. (1981) Role of forage. In: Forage evaluation; concepts and techniques. Editors J.L. Wheeler and R.D. Mochrie. Melbourne, CSIRO and American Forage and Grassland Council, pp.1-19.

Bishop, J.P.; Froseth, J.A.; Verettoni, H.N.; Noller, C.H. (1975) Diet and performance of sheep on rangeland in semiarid Argentina. Journal of Range Management 28(1), 52-55.

Bray, R.A. (1982) Selecting and breeding better legumes. In: Nutritional limits to animal production from pastures. Editor J.B. Hacker. Farnham Royal, U.K., Commonwealth Agricultural Bureaux, pp. 287-303.

Brougham, R.W.; Grant, D.A. (1976) Hill country farming in New Zealand; an overview. In: Hill lands, Proceedings of an international symposium. Editors J. Luchok, J.D. Cawthon and M.J. Breslin. Morgantown, West Virginia University Books, pp. 18-23.

Bula, R.J.; Lechtenberg, V.L.; Holt, D.A. (1977) Potential of temperate zone cultivated forages. In: Potential of the world's forages for ruminant animal production. Morrilton, Arkansas, Winrock Report, pp.7-28.

Burns, J.C. (1978) Antiquality factors as related to forage quality. Journal of Dairy Science 61, 1809-1820.

Burton, G.W. (1972) Can the South become the worlds greatest grassland? Progressive Farmer 87(3), 22-24.

Burton, G.W. (1978) Advances in breeding a better quality forage plant. In: Advances in hay, silage and pasture quality. Raleigh, American Forage and Grassland Council Proceedings, pp. 96-102.

Butler, G.W.; Jones, D.I.H. (1973) Mineral biochemistry of herbage. In: Chemistry and biochemistry of herbage vol. 2. Editors G.W. Butler and R.W. Bailey. London and New York, Academic Press, pp. 127-162.

Campbell, A.G.; Bryant, A.M. (1978) Pasture constraints on dairy production. Proceedings of the Agronomy Society of New Zealand 8, 115-118.

Chia, Shen-Siu. (1981) China's grassland types: their locality and improvement. Proceedings of the 14th International Grassland Congress, Lexington, Kentucky, (In press).

Cook, C.W.; Harris, L.E. (1968) Nutritive value of seasonal ranges. Utah State University Agricultural Experiment Station Bulletin 472, 55 pp.

Cooper, J.P. (1973) Genetic variation in herbage constituents. In: Chemistry and biochemistry of herbage vol. 2. Editors G.W. Butler and R.W. Bailey. London and New York, Academic Press, pp. 379-417.

Council for Agricultural Science and Technology (1980) Food from animals; quantity, quality and safety. Report No. 82.

Coupland, R.T. (1979) Editor, Grassland ecosystems of the world; analysis of grasslands and their uses. Cambridge, Cambridge University Press, 401 pp.

Craven, J.A.; Kilkenny, J.B. (1976) The structure of the British cattle industry. In: Principles of cattle production. Editors H. Swan and W.H. Broster. London-Boston, Butterworths, pp. 1-43.

Cunningham, E.P. (1976) The structure of the cattle populations in the E.E.C. In: Optimisation of cattle breeding schemes. Editor P. McGloughlin. Luxembourg, Commission of the European Communities pp. 13-27.

Davies, D.A.; Morgan, T.E.H. (1979) Grazing evaluation of four grass species under upland conditions. Grass and Forage Science 34, 67-68.

Deinum, B.; Dirven, J.G.P. (1974) A model for the description of the effects of different environmental factors on the nutritive value of forages. Proceedings of the 12th International Grassland Congress, Moscow, vol. 1, part 1, 338-346.

Eadie, J. (1973) Sheep production systems development on the hills. Potassium Institute Ltd. Colloquium Proceedings No. 3, 131-137.

Evans, W.B.; Munro, J.M.M.; Scurlock, R.V. (1979) Comparative pasture and animal production from cocksfoot and perennial ryegrass varieties under grazing. Grass and Forage Science 34, 64-65.

Forest Service (1980) An assessment of the forest and range land situation in the United States. United States Department of Agriculture, Washington D.C., 631 pp.

Gordon, F.J. (1980) Grass and fresh grass products. Proceedings of the Nutrition Society 39, 249-256.

Greenhalgh, J.F.D. (1975) Factors limiting animal production from grazed pasture. Journal of the British Grassland Society 30, 153-160.

Hacker, J.B. (1982) Selecting and breeding better quality grasses. In: Nutritional limits to animal production from pastures. Editor J.B. Hacker. Farnham Royal, U.K., Commonwealth Agricultural Bureaux, pp. 305-326.

Hegarty, M.P. (1982) Deleterious factors in forages affecting animal production. In: Nutritional limits to animal production from pastures. Editor J.B. Hacker. Farnham Royal, U.K., Commonwealth Agricultural Bureaux, pp. 133-150.

Henzell, E.F. (1981) Contribution of forages to worldwide food production: now and in the future. Proceedings of the 14th International Grassland Congress, Lexington, Kentucky (In press).

Hill Farming Research Organisation (1979) Science and hill farming. Edinburgh, HFRO, 184 pp.

Hodgson, H.J. (1977) Gaps in knowledge and technology for finishing cattle on forages. Journal of Animal Science 44, 896-900.

Hodgson, J.; Rodriguez Capriles, J.F.; Fenlon, J.S. (1977) The influence of sward characteristics on the herbage intake of grazing calves. Journal of Agricultural Science, Cambridge 89, 743-750.

Hoveland, C.S.; Haaland, R.L.; King, C.C.; Anthony, W.B.; Clark, E.M.; McGuire, J.A.; Smith, L.A.; Grimes, H.W.; Holliman, J.L. (1980) Association of Epichloe typhina fungus and steer performance on tall fescue pasture. Agronomy Journal 72, 1064-1965.

Hughes, R. (1970) Factors involved in animal production from temperate pastures. Proceedings of the 11th International Grassland Congress, Surfers Paradise, Australia, A31-38.

Hughes, R. (1976) Hills and uplands in Britain - the limitations and the development of potential. In: Hill lands, Proceedings of an international symposium. Editors J. Luchok, J.D. Cawthon and M.J. Breslin. Morgantown, West Virginia University Books, pp. 1-8.

Jones, R.J.; Sandland, R.L. (1974) The relation between animal gain and stocking rate. Journal of Agricultural Science, Cambridge 83, 335-342.

Kemp, A.; Geurink, J.H. (1978) Grassland farming and minerals in cattle. Netherlands Journal of Agricultural Science 26, 161-169.

Kutuzova, A.A. (1972) Increasing protein production in hayfields and pastures. In: Methods for increasing plant protein production. Moscow, Kolos Publishers. Translated for the United States Department of Agriculture and the National Science Foundation, Washington, D.C. 290 pp.

Kutuzova, A.A.; Morozova, Z.V. (1980) Fertilizer application in cultivated pastures. In: Cultivated pastures in dairy management. New Delhi, Amerind Publishing Company. Translated for United States Department of Agriculture and National Science Foundation, Washington, D.C., pp. 63-123.

Langlands, J.P.; Bowles, J.E. (1974) Herbage intake and production of Merino sheep grazing native and improved pastures at different stocking rates. Australian Journal of Experimental Agriculture and Animal Husbandry 14, 307-315.

Larin, I.V. (1962) Pasture economy and meadow cultivation. Washington, National Science Foundation, Israel Program for Scientific Translations, 641 pp.

Leaver, J.D. (1976) Utilisation of grassland by dairy cows. In: Principles of cattle production. Editors H. Swan and W.H. Broster. London-Boston, Butterworths, pp.307-327.

McCloud, D.E.; Bula, R.J. (1973) Climatic factors in forage production. In: Forages, 3rd edn. Editors M.E. Heath, D.S. Metcalfe and R.F. Barnes. Ames, Iowa, Iowa State University Press, pp. 372-382.

McCloud, D.E. (1974) Man's impact on world grasslands. Proceedings of the 12th International Grassland Congress, Moscow, vol. 1, part 1, 62-75.

McDowell, L.R. (1976) Mineral deficiencies and toxicities and their effect on beef production in developing countries. In: Beef cattle production in developing countries. Editor A.J. Smith. Edinburgh, Centre for Tropical Veterinary Medicine, pp. 216-241.

Matches, A.G. (1973) Editor. Anti-quality components of forages. Madison, Wisconsin, Crop Science Society of America. Special Publication No. 4, 140 pp.

Matches, A.G.; Thompson, G.B.; Martz, F.A. (1975) Post-establishment harvesting and management systems of forages. Missouri Agricultural Experiment Station Journal Series No. 7371.

Minson, D.J. (1981a) The effects of feeding protected and unprotected casein on the milk production of cows grazing ryegrass. Journal of Agricultural Science, Cambridge 96, 239-241.

Minson, D.J. (1981b) Forage quality: assessing the plant-animal complex. Proceedings of the 14th International Grassland Congress, Lexington, Kentucky. (In press).

Moore, J.E.; Mott, G.O. (1973) Structural inhibitors of quality in tropical grasses. In: Antiquality components of forages. Editor A.G. Matches. Madison, Wisconsin, Crop Science Society of America Special Publication No. 4, pp.53-98.

Morley, F.H.W. (1978) Animal production studies on grassland. In: Measurement of grassland vegetation and animal production. Editor L. 't Mannetje. Farnham Royal, U.K., Commonwealth Agricultural Bureaux, pp. 103-162.

Morozov, P. (1974) Soviet agriculture. Proceedings of the 12th International Grassland Congress, Moscow, vol. 1, part 1, 43-61.

Mott, G.O. (1980) Measuring forage quantity and quality in grazing trials. Proceedings of the 37th Southern Pasture and Forage Crop Improvement Conference. Nashville, Science and Education Administration, United States Department of Agriculture, pp.3-9.

Norton, B.W. (1982) Differences between species in forage quality. In: Nutritional limits to animal production from pastures. Editor J.B. Hacker. Farnham Royal, U.K., Commonwealth Agricultural Bureaux, pp. 89-110.

Okubo, T.; Takahashi, S.; Akiyama, T. (1977) Modelling approach to animal production under grazing conditions in different types of grassland in Japan. World Review of Animal Production 13, 45-50.

Ollerenshaw, J.H.; Stewart, W.S.; Gallimore, J.F.; Baker, R.H. (1976) Extending the seasonality of growth of hill land pastures. In: Hill lands, Proceedings of an international symposium. Editors J. Luchok, J.D. Cawthon and M.J. Breslin. Morgantown, West Virginia University Books, pp. 583-586.

Pimentel, D.; Oltenacu, P.A.; Nesheim, M.C.; Krummel, J.; Allen, M.S.; Chick, S. (1980) The potential for grass-fed livestock: resource constraints. Science 207, 843-848.

Raleigh, R.J. (1970) Symposium on pasture methods for maximum production in beef cattle; manipulation of both livestock and forage management to give optimum production. Journal of Animal Science 30, 108-114.

Rattray, P.V. (1978) Pasture constraints to sheep production. Proceedings of the Agronomy Society of New Zealand 8, 103-108.

Rees, M.C.; Minson, D.J. (1976) Fertilizer calcium as a factor affecting the voluntary intake, digestibility and retention time of pangola grass (Digitaria decumbens) by sheep. British Journal of Nutrition 3, 179-187.

Reid, R.L.; Jung, G.A. (1973) Forage-animal stresses. In: Forages, 3rd edn. Editors M.E. Heath, D.S. Metcalfe and R.F. Barnes. Ames, Iowa State University Press, pp. 639-653.

Reid, R.L.; Horvath, D.J. (1980) Soil chemistry and mineral problems in farm livestock. A review. Animal Feed Science and Technology 5, 95-167.

Riewe, M.E. (1981) Expected animal response to certain grazing strategies. In: Forage evaluation: concepts and techniques. Editors J.L. Wheeler and R.D. Mochrie. Melbourne, CSIRO and American Forage and Grassland Council, pp. 341-355.

Smith, C.A. (1970) The feeding value of tropical grass pastures evaluated by cattle weight gains. Proceedings of the 11th International Grassland Congress, Surfers Paradise, Australia, 839-841.

Soil Conservation Service (1977) National resources inventory. United States Department of Agriculture, Washington, D.C.

Stobbs, T.H. (1976) Beef production from sown and planted pastures in the tropics. In: Beef cattle production in developing countries. Editor A.J. Smith. Edinburgh, Centre for Tropical Veterinary Medicine, pp. 164-183.

Stobbs, T.H.; Minson, D.J.; McLeod, M.N. (1977) The response of dairy cows grazing a nitrogen fertilized grass pasture to a supplement of protected casein. Journal of Agricultural Science, Cambridge 89, 137-141.

Tayler, J.C. (1976) Beef production in the E.E.C. and the co-ordination of research by the commission of the European communities. Livestock Production Science 3, 305-318.

Thomson, D.J. (1979) Effect of the proportion of legumes in the sward on animal output. Occasional symposium No. 10, British Grassland Society, University of York, pp. 101-109.

Trewartha, G.T.; Robinson, A.H.; Hammond, E.H. (1967) Physical elements of geography, 5th edn. New York, McGraw-Hill Book Company. 527 pp.

Trewartha, G.T.; Horn, L.H. (1980) An introduction to climate, 5th edn. New York, McGraw-Hill Book Company, 416 pp.

USDA (1978) Agricultural Statistics. Washington, D.C., U.S. Government Printing Office.

USDA (1979) Agricultural Statistics. Washington, D.C., U.S. Government Printing Office.

USDA (1981) Feed. Outlook and Situation. Economics and Statistics Service FdS-280, Washington, D.C. 29 pp.

Van Dyne, G.M.; Hanson, J.D.; Jump, R.C. (1981) Seasonal changes in botanical and chemical composition and digestibility of diets of large herbivores on shortgrass prairie. Proceedings of the 14th International Grassland Congress, Lexington, Kentucky (In press).

Voigtländer, G.; Kühbauch, W. (1978) Factors constraining animal production in grazing management. Proceedings of the 7th General Meeting European Grassland Federation, Gent, 4.3-4.27.

Vorobev, E.S. (1980) Nutrition of milk cows on pastures. In: Cultivated pastures in dairy management. New Delhi, Amerind Publishing Company. Translated for United States Department of Agriculture and National Science Foundation, Washington, D.C., pp. 142-199.

Ward, G.M. (1980) Energy, land and feed constraints on beef production in the 80's. Journal of Animal Science 51, 1051-1064.

Whyte, R.O.; Moir, T.R.G.; Cooper, J.P. (1959) Grasses in agriculture. Rome, F.A.O. Agricultural Studies No. 42. 417 pp.

Wilkinson, J.M. (1976) Beef from grass and forage crops. In: Principles of cattle production. Editors H. Swan and W.H. Broster. London-Boston, Butterworths, pp. 329-342.

Wilson, J.W. (1982) Environmental and nutritional factors affecting herbage quality. In: Nutritional limits to animal production from pastures. Editor J.B. Hacker. Farnham Royal, U.K., Commonwealth Agricultural Bureaux, pp. 111-131.

PROBLEMS OF ANIMAL PRODUCTION FROM MEDITERRANEAN PASTURES

W.G. ALLDEN

Waite Agricultural Research Institute, The University of Adelaide, Glen Osmond, South Australia, 5064.

ABSTRACT

Extensive cereal cropping and livestock production are features of mediterranean environments, but the use of land resources varies greatly between regions. Lack of integration between arable farming and livestock production has led to problems of soil fertility and overgrazing of communal rangelands in some areas, imposing serious constraints on animal production. This situation is compared with the integrated ley farming system which currently operates in southern Australia and South Africa.

Nutritional limitations to animal production from mediterranean annual-type pastures are closely related to the climatic environment. During the 4-6 months of summer drought the pasture herbage, although abundant, is of low nutritional value and livestock commonly lose weight. Pasture growth during the winter months is the key to animal productivity, the time of onset of the rainfall season, seed carryover, stocking rates, fertilizer use and choice of pasture legume being significant determinants of yield. The effects of the nutritional constraints associated with these factors are discussed in relation to the productivity of animals of different physiological status in intensive and non-intensive grazing systems.

INTRODUCTION

An appreciation of the problems of animal production from mediterranean type pastures calls for an understanding of the ecology of grazed pastures and of the nutritional and management factors which impair output of animal product, subjects which have been reviewed extensively by Rossiter (1966) and Purser (1981). In this paper I propose to cast my net wider and consider also the pasture-animal unit as a component of the agricultural holding or as an integral part of the national endeavour to produce food. This approach is consistent with land use in mediterranean environments, where extensive cereal cropping and grazing livestock are common elements of the agriculture.

FEATURES OF THE MEDITERRANEAN ENVIRONMENT

Location and climate

Lands with mediterranean type climates comprise less than 1 percent of the earth's total land surface, and are found in parts of 27 countries with a total population of about 300 million (Oram 1977). More than half of the lands lie around the Mediterranean sea; the remainder are located in regions of California, central Chile, south west Africa's coastal plain and in areas of south western and southern Australia. Fig. 1, from Aschmann (1973), which illustrates areas of the world with mediterranean climates, shows that the climate type is generally found between latitudes 30° and 40° on a continent's west face. The environment is characterized by hot dry summers and mild wet winters, the growth of crops and pastures being confined to the cool wet period. Aschmann (1973), using the effective rainfall index of Bailey (1958), sets the rainfall limits of the "true" mediterranean climate between 275 mm for coastal regions (350 mm inland) at the dry end of the scale and 900 mm at the wetter boundary, with a 4-9 months pasture growing season. At least 65 percent of the year's rainfall should fall during the winter half-year. Aschmann also suggests that for a climate to be classified as mediterranean the mean monthly winter temperatures should be less than 15°C with no more than three percent of the hours of the year below 0°C. This definition, based on the degree of winter cold, restricts the distribution of the mediterranean climate compared with the Köppen climate classification with its Cs or olive climate (Köppen & Geiger 1936) and the Ackerman (1941) variant. The distinction is important because winter cold may impair plant growth, and add to the problems of summer drought.

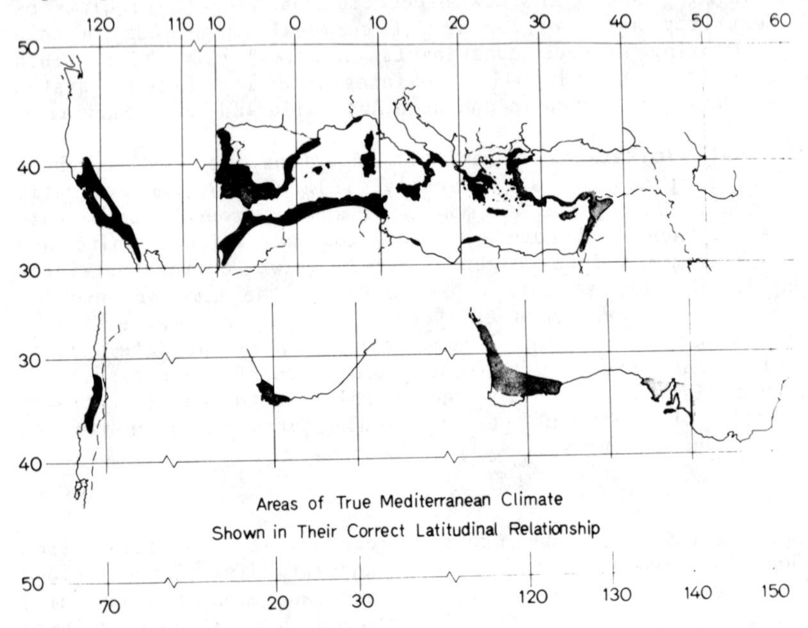

Fig. 1. Areas of mediterranean climate (Aschmann 1973)

Topography

Apart from Australia and parts of north Africa, both of which have moderate relief, the regions are topographically similar - coastal plains fringed by low hills which share the summer drought and a background of high mountain ranges which receive more precipitation in both summer and winter. The mountain ranges are valuable in providing grazing through transhumance, thereby alleviating the problem of summer drought experienced in the coastal fringe. The presence of river systems that originate in the mountains may provide a source of abundant water for irrigated pastures during the summer months, but this out-of-season feed for grazing livestock is not typical of mediterranean pastures per se. A notable example is the mediterranean environment of California where nearly 3 m ha is watered by the Sacramento and San Joaquin river systems. Areas of Chile and the Mediterranean basin are similarly favoured, but most of the water is used to grow crops for the human population, with only crop by-products available for livestock.

Land use

The use of land resources in mediterranean regions has been reviewed by Grigg (1974) who has outlined the evolution of traditional and "overseas" mediterranean agricultures. A major feature has been the great importance of both cereal growing and pastoralism at some time during the development of all regions. More particularly there have been significant differences in the ways land is used for cereal production and livestock.

For example, in much of the Mediterranean basin and Near East there is a farming system which for more than two millenia has been based on a balance between cropping, livestock, tree plantations and rangelands. There is a traditional dichotomy of ownership of arable land and livestock, with animals having infrequent access to the crop residues and fallows. Farmers give top priority to cereal growing and are reluctant to consider any alternative land use system. The main source of sustenance for grazing ruminants is from communal grazing on unimproved, unfertilized pastures, particularly in the regions of low rainfall. The rainfed cropping areas have serious problems associated with declining soil fertility and an inadequate feed supply for livestock. Pressures for arable land are encroaching on the more arid, pastoral regions, thereby restricting further the areas available for livestock (Carter 1978). Yet livestock numbers have increased dramatically since the early 1950's, leading to overgrazing and a growing shortfall in animal production targets (Oram 1977). Recent reports on developing countries in the Near East and Mediterranean regions have urged the adoption of the land use system currently practised in the mediterranean-type areas of Australia (Hardison & Fox 1973, Ford Foundation 1974, Oram 1975, Doolette 1977) but progress has been slow. (D.W. Puckridge and E.D. Carter, unpublished work, presented at International Congress on Dryland Farming, Adelaide, August, 1980).

In Australia, as in most of the "overseas" mediterranean lands, continuous cropping without livestock was also a feature of the early agriculture, but the practice proved so damaging to soil fertility and crop productivity that alternative measures were adopted to increase yields and to preserve the fragile landscapes. Pastures based on species of Mediterranean origin (subterranean clover (*Trifolium subterraneum*) and *Medicago* species) have now been integrated with the cereal operations into a stable agriculture which has supplanted the continuous cropping system and at the same time produced a significant improvement in cereal grain yields, and livestock production (Puckridge & Carter, *loc. cit*). South Africa is similarly placed, Chile less so.

There appear to be no serious ecological or edaphic constraints preventing pasture legumes filling much the same role in the Near East, Mediterranean basin and Chile, although problems of fertilizer use, tillage, fallowing and other technical and socio-economic aspects of integration need to be overcome (Carter 1966, Puckridge & Carter *loc. cit*).

In California there is little integration of livestock and grain production, because the availability of irrigation has allowed the development of a highly specialized agriculture.

Table 1, adapted from Carter (1974), compares land use, crop production and animal numbers in South Australia, (where livestock and crops have been integrated into a ley farming system for more than 50 years), and Algeria where there is little integration. The two regions are climatically similar, although winters on the Algerian plateau are more severe. Significant features of Table 1 are the great differences in the proportion of fallowed land (33 per cent Algeria vs 7 percent S. Australia) and pasture (< 1 percent vs 54 per cent), and nearly a fourfold difference in the use of phosphatic fertilizers. Land allocated to cereal production in South Australia is only 62 percent of the area cropped in Algeria, but total grain production is 15 percent higher because yields are almost double. Sheep and cattle numbers are nearly twice as great in South Australia. Carter (1974) estimated that pasture improvement and the integration of crops and

TABLE 1

Comparative agriculture statistics, Algeria and South Australia (adapted from Carter 1974).

	Algeria	South Australia
Distribution of land by rainfall ('000 ha)		
>600 mm	5081	1082
300-600 mm	13852	9644
<300 mm	218623	87674
Land use for agriculture ('000 ha) 1971/72	42449	65146
Rangeland	35345	59272
Cropping, vines, horticulture	8104	5874
of which Grain %	44%	38%
Fallow %	33%	7%
Pasture %	1%	54%
Superphosphate use (t P_2O_5)		
Crops	31360	60968) (384)+
Pastures	trace	59444)
Cereal production (1961-72 mean)		
Wheat and barley harvested ('000 ha)	2764	1727 (62)+
Yield kg/ha	616	1135 (184)+
Production ('000 t)	1702	1960 (115)+
Livestock		
Sheep ('000)	8367	18961 (227)+
Cattle ('000)	864	1239 (143)+

+ South Australia value as a percentage of Algerian value

livestock in Algeria had the potential to produce fodder for an additional 18 m ewe equivalents; in a subsequent survey of nine countries in the Near East and North Africa he indicated that ley farming practices could provide forage for 120 m ewe equivalents (Carter 1978). He also reported on the potential for large increases in stock numbers in Chile through improved pasture technology (Carter 1966).

The climatic and edaphic factors which limit the distribution of *T. subterraneum* and *Medicago spp.* are unlikely to be identical for Australia and the Mediterranean basin. Australian cultivars have been used with success in many regions, but there is evidence that the selection of local ecotypes adapted to local environments will prove to be more rewarding in some areas (Adem 1974, Saunders 1976, Carter 1974, 1978, Puckridge & Carter *loc. cit*). There is now a growing research program in the Near East and North Africa concerned with legumes for pastures, but it is hard to see how the problems of livestock-cropping integration in many of the overcropped and overgrazed lands can be overcome without radical changes to the customary land-use patterns.

The use of crop rotations including grain legumes for human consumption or fodder legumes for animals may provide a more acceptable method for improving soil fertility and is likely to provide more human food than the ley farming system. This is of greatest significance to the growing deficit between cereal production and consumption in North Africa and the Middle East (Oram 1977), yet it tends to be overshadowed by the desire to produce more meat.

In all regions economic factors are important, and the livestock enterprise has to compete for its share of resources on the basis of profitability. Even in Australia, where the ley farming system is firmly entrenched, depressed prices for animal products during the last 20 years have stimulated studies into ways whereby the cropping phase can be extended to allow the farmer maximum enterprise flexibility consistent with the maintenance of soil fertility (Wolfe *et al*. 1976, Downes 1976, Corbin 1976).

At the present time there is a clear distinction between the underlying factors which influence the nutritional problems of livestock in the grazing lands of the Mediterranean regions and those of the "overseas" lands, principally because livestock and pasture leys are rarely integrated into the arable farming system of the Mediterranean and Near East.

Livestock in the mediterranean lands

Regions with mediterranean climates are not separated by national boundaries but occupy restricted areas between deserts and temperate lands or the sea (Fig. 1) with a considerable movement of stock between climatic zones through nomadism and transhumance, as in the Mediterranean basin, parts of the western United States, and Chile. Oram (1975) gives an example where nine million nomads in the Near East own 75 percent of the sheep and goats and nearly 60 percent of the cattle and camels, while many farmers own only draught animals. In other countries there is commonly a stratification of the livestock industries, with breeding taking place in the dry rangelands and fattening or feed lotting in more favoured environments.

For these regions the assessment of livestock numbers and productivity within the climatic zone is extremely complex, and will not be attempted here. Sheep, goats and cattle are present in all regions in varying proportions which reflect social and customary preferences, the need for draught, and the suitability of land for a particular form of produciton.

The pastures and vegetation

The evolution of mediterranean floras has been reviewed by Raven (1973). The native vegetation, which differs between regions, is dominated by broad sclerophyll evergreen trees and shrubs, with diverse annual plants forming about half the total species. Extensive cropping and grazing throughout the mediterranean lands has so disturbed the natural ecosystems that the dominant components of the vegetation have been displaced by cool season annual species which were spread either by design or accident from other mediterranean climatic regions, especially from the Mediterranean basin and from the Cape Province of South Africa (Specht 1969). These pastures, referred to by Rossiter (1966) as the mediterranean annual-type pasture, comprise aggressive free-seeding grasses and herbs and include species of *Hordeum*, *Avena*, *Bromus*, *Lolium*, *Asphodelus*, *Homeria*, *Trifolium*, *Medicago*, *Melilotus*, *Oxalis*, *Echium*, *Rumex*, *Hypochoeris*, *Arctotheca*, *Erodium* etc. There is now a remarkably similar array of pasture plants represented in the grazing lands of all regions. Among the volunteers are those species sown at the present time, namely cultivars of annual ryegrass (*Lolium rigidum*), subterranean clover (*Trifolium subterraneum*), barrel medic (*Medicago truncatula*) and rose clover (*T. hirtum*). Phosphatic fertilizers, often augmented by sulphur and minor elements, are of great significance for legume-based leys in mediterranean lands; nitrogen is used less frequently, - e.g. on the Californian range (Hoglund *et al*. 1952, Jones 1960, Martin & Berry 1970). The mediterranean annual-type pasture does not have a longer growing season than the native species, but when adequately fertilized it produces more herbage of high quality during the rainfall season, and both animal performance and stock carrying capacity are greatly increased. Much of the native vegetation in the overseas lands is considered to be of low productivity; examples relating to South Africa and Californian annual grassland have been given by Haylett (1958) and Murphy *et al*. (1973).

Many of the introduced annuals have undesirable characteristics at some stage of the growth cycle. For example, the seeds of barley grass (*Hordeum leporinum*), the brome grasses (*Bromus spp*.) and geranium or crowsfoot (*Erodium spp*.) may penetrate the eyes and skin of sheep, impair growth and cause damage to the carcass. Indeed, even the sown pasture species present health hazards for livestock. The presence in clovers of pharmacologically active compounds such as oestrogenic isoflavones may cause ewe infertility (Braden & MacDonald 1970) and ingested annual ryegrass seedheads may cause deaths due to a nematode-bacterium disease complex of the plant (Price 1973). The subject of deleterious factors in forage production is discussed by Hegarty (1982).

I propose to confine my remarks to the constraints on animal production of the mediterranean annual-type pasture as defined by Rossiter (1966) and to consider the deficiencies within the broad context of the ley farming system. Documentation is taken mainly from Californian and Australian work, since few other sources are available.

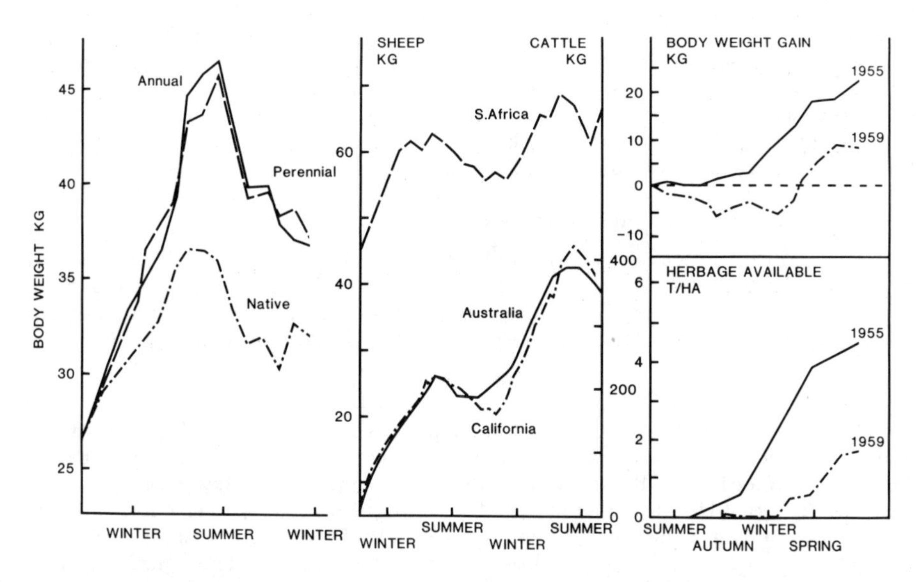

Fig. 2. (a) Growth of sheep on native, sown annual and sown perennial pastures.
(b) Seasonal changes in weight of sheep and cattle in Australia, South Africa and California.
(c) Weight changes of sheep in two different years (upper) in relation to winter pasture availability (lower).

NUTRITIONAL CONSTRAINTS TO ANIMAL PRODUCTION FROM PASTURE

Growth of animals

The most distinctive feature of animal growth on dryland pastures in mediterranean environments is the seasonal alternation between rapid liveweight gain during the moist winter and spring months and substantial liveweight losses during the hot dry summers. This switch-back progression as an animal grows from birth to maturity occurs in all regions for both sheep and cattle at even the most modest stocking rates, and has been quoted on sown annual and perennial pastures and on native pastures. Fig. 2(a) compares the growth of young sheep grazing a native Danthonia pasture, a subterranean clover, annual grass (Lolium rigidum) pasture, and a sub-clover perennial grass (Phalaris aquatica) pasture in a 500 mm rainfall environment in Australia (Neal Smith 1942). Stocking rates, although not identical, were comparable, being in the range 10-11.5/ha. All pastures received similar applications of superphosphate, but the native pasture was less responsive than the sown species, and animal gains were depressed during the winter. Nevertheless the growth trends from season to season were unchanged. Fig. 2(b) illustrates a similar seasonal pattern of growth for sheep in Australia (Allden 1968) and South Africa (Vosloo & Meissenheimer 1970) and for beef cattle in California (Guilbert & Hart 1946). The periods of low animal productivity during the summer may extend into the autumn and winter months if the onset of the rainfall season is delayed. As an example Fig. 2(c) illustrates variation in the gains of wether merino sheep (upper graph) and the amount of pasture available to them during the autumn-winter period (lower graph) in two contrasting years (Allden unpublished). In 1955 effective rainfall commenced in late summer providing adequate winter feed, whereas in 1959 production by similar stock on the same paddock was depressed by the lack of rain of any consequence until late winter.

Production and nutritional value of pasture

The variation in animal growth shown in Fig. 2 is closely related to the production of herbage, its nutritional value and its seasonal distribution, with animal health, adaptation, breeding and management as mediating factors. These latter factors will not be considered.

Pasture production

The general relationship between pasture production and effective rainfall needs no elaboration and has been discussed by Whittaker (1970) and Wheeler & Hutchinson (1973). During the dry summer there is no plant growth but plenty of dry senescent low quality herbage is available, being the carryover material from the growth in the previous spring. By late summer and autumn both the quality and quantity may be limiting, depending on the length of the summer drought and stocking rate. With the advent of the winter rainfall season the new season's herbage is of high quality but often limiting in the quantitative sense. Only during the spring months is pasture plentiful and of high quality, with growth rates of as much as 130 kg/ha/day (Silsbury *et al*. 1979).

Pasture growth during the winter months is the principal determinant of animal output (Fig. 2(c)), the time of onset of the rainfall season being the key to high productivity (Trumble & Cornish 1936). Autumn is the common period for lambing and calving, and if feed is scarce at this time perinatal mortalities may be high (Croker 1968) particularly when underfed livestock are subjected to unfavourable weather conditions (Obst & Day 1968, Alexander 1970).

When the livestock enterprise is integrated with cereal cropping the autumn is the time when the proportion of land available for grazing is at a minimum; pastures entering the cropping phase are under cultivation and reseeded pastures are not yet ready to be grazed. The farmer probably makes his long-term decisions about the number of animals to keep according to past management experiences during the late summer and autumn period.

Thus a major constraint to animal production on adequately fertilized pasture is the quantitative deficiency of feed associated with a late start to the pasture growing season, which exacerbates the effects of undernutrition resulting from low quality herbage during the summer.

Another factor influencing the productivity of annual pastures during the winter period and which deserves comment is plant density.

Plant density and animal productivity

The regeneration of mediterranean annual-type pastures is completely dependant on the seed carried over from previous seasons and there may be substantial variation in the amount in the ground at the commencement of the rainfall season. Poor seed set in the spring due to high stocking rates (Sharkey et al. 1964), or high seedling mortalities after a false start to the rainfall season in the late summer, are two common causes of reduced seed numbers, and the outcome is pastures of low plant density and poor winter production. Likewise the decision to sow pastures at either a high or a low seeding rate, or not to sow at all may have an important bearing on plant density after a cropping phase, thereby affecting the numbers of animals that can safely be carried in the season of establishment. Table 2 illustrates the weight changes of sheep grazing Lolium rigidum pastures sown at rates ranging from 4-1024 kg/ha and grazed at the same stocking rate. The most significant effect of seeding rate was on pasture yield during the winter months. Seeding rates of 16 kg/ha and less provided insufficient herbage to sustain animals, and mortalities of up to 100 percent were recorded in some treatments. At seeding rates of 32 kg/ha the plant population provided enough herbage to maintain the weight of the grazing animals during the winter months, whereas the sheep grazing the most dense pastures gained up to 12 kg. During the spring flush the animals on pastures of 32 kg/ha seeding rate quickly attained the same weight as those which had grown uninterrupted on pastures of high seeding rates.

TABLE 2

Influence of seeding rate on plant density, herbage yield and sheep liveweight gain during selected periods (adapted from Williams et al. 1980)

Seeding rate (kg/ha)	Initial Plant Density (plants/ m^2)	Pasture yield 7 weeks after sowing kg/ha	Liveweight change of sheep (kg) during winter 5/7-27/9	during spring 27/9-21/12	whole period	Sheep Mortality during experiment %	Cumulative yields of D.M. for whole experiment kg/ha
4	165	43	-18	*	*	100	3197
8	243	83	-18	+1	-17	60	5032
16	495	113	-16	+18	+2	40	6801
32	1126	237	0	+15	+15	0	9653
1024	35993	1620	+12	+4	+16	0	9851

* All sheep on this treatment died.

The principle is quite clear. Livestock numbers are completely dependent on pasture production, and in pastures which regenerate annually the amount of seed available at the time of the first reliable winter rains is the key to plant density, plant production and the number of stock that can be carried productively and safely.

Qualitative deficiencies

There are few records of the sequential changes in characters of the diet related to the seasonal fluctuations in animal growth on mediterranean pastures. Fig. 3 depicts the liveweight (upper graphs) of young Merino sheep grazing pastures based on either T. subterraneum or annual Medicago spp. in relation to herbage digestibility and nitrogen content of the diet (middle graphs) and the intake of digestive energy relative to the estimated maintenance requirement (M on lowest graphs). The studies were carried out at modest stocking rates, and pastures were not considered to be limiting in the quantitative sense.

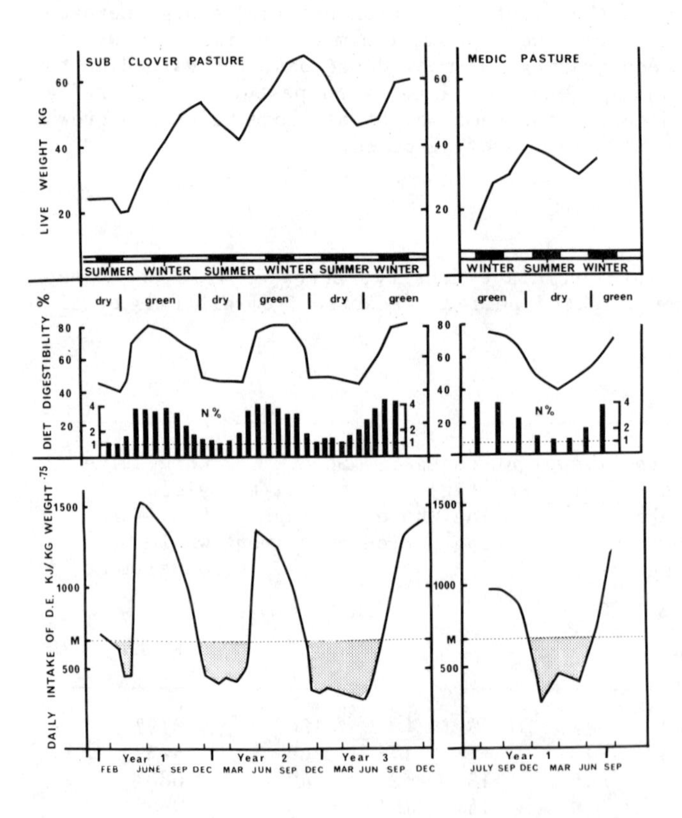

Fig. 3. Temporal changes in liveweight (top graphs), diet digestibility and N content (centre graphs) and daily intake of digestible energy relative to maintenance (M) (lower graphs) for sheep grazing pastures based on T. subterraneum (left) and annual Medicago spp. (right). Adapted from Hutchinson & Porter (1958) and Allden (1968).

Sheep growth followed the general pattern illustrated in Fig. 2, there being rapid weight gains during the period when pastures were green and growing, and substantial weight losses on the dry summer feed. Herbage digestibility and dietary nitrogen (centre graph) were closely related to each other and to animal performance. There was a rapid transition from herbage of low digestibility (ca. 45 percent) to high digestibility (ca. 80 percent) when the new seasons growth commenced in the autumn. Thereafter digestibility remained high until the late spring, when there was a sharp decline associated with the maturation and senescence of the annual pasture plants. Similar changes in digestibility with advancing maturity have been noted by Radcliffe & Cochrane (1970), Purser (1981) and many others, and are not confined to mediterranean environments.

Weight gains were closely related to the intake of digestible energy above maintenance (lower graph). The effects of low digestibility during the 4-5 months dry summer were compounded by a reduction in the amount of herbage eaten, there being up to a fourfold difference in the intake of useful energy between summer and winter. These low intakes of herbage in the presence of abundant feed are readily understandable from the results of studies which have shown that the intake of poorly digested feed is restricted by the slow rate of passage of herbage through the digestive tract.

Energy and protein

Digestible energy emerges as the significant deficiency of summer pasture. Purser (1981) has cited limited evidence to suggest that some annual pasture species are nutritionally superior to others during the dry summer, but the studies in this area are limited, and the data far from consistent. The seed of *T. subterraneum* is of a high digestibility but the pod (burr) is poorly digested (Wilson & Hindley 1968). Consequently the dry leaf and stem is nutritionally superior to the burr plus seed (Squires, personal communication). Likewise the pods of barrel medic (*Medicago truncatula*) are of little nutritional value (Vercoe & Pearce 1960; Denney *et al.* 1979). Pastures which dry off prematurely due to drought have a higher digestibility than those which mature in the presence of adequate rainfall (Pullman & Allden 1971), and desiccation, leaching by rainfall and exposure to the sun may seriously reduce the amount of standing feed and its digestibility (Guilbert & Mead 1931, Hart *et al.* 1932, Rossiter 1966).

Fig. 3 also illustrates that the nitrogen content of the mature herbage consumed by grazing animals during the summer ranged from 1.1-1.9 percent and it seems unlikely that intake would be impaired by nitrogen deficiencies except possibly at the lowest levels. Similar values have been observed by Cannon (1974) in a study of the integration of wheat and sheep production in Victoria. Weir & Torell (1967), in supplementary feeding studies on the Californian dry range, noted that the nitrogen content of the dry forage selected by sheep with oesophageal fistulae ranged between 0.8 percent and 1.2 percent whereas Theriez & Skouri (1970) working in central Tunisia observed values of about 1.3 percent nitrogen from equivalent samples during the summer months. The general picture which emerges is that mature pasture with a low legume content may be marginally deficient in nitrogen for the growth of young sheep (Allden 1959, Weir & Torell 1967, Davies *et al.* 1970). None of the protein values recorded in the field approaches the low levels (0.2-0.4 percent nitrogen) noted in low quality roughage studies with sheep in pens, where significant increases in intake accompanied supplementary nitrogen (Coombe & Tribe 1963, Round 1976). The nitrogen content of pasture during the late spring and dry summer months may also vary according to species and environment (Allden 1959, Rossiter 1966).

Rossiter (1966) also observed that the nitrogen contents of both grasses and capeweed (*Arctotheca calendula*) during the summer were higher in the presence of a legume than in its absence. It should be recognized that in ley farming systems one of the prime objects of the pasture phase in the intercropping period is to fix nitrogen for the cropping phase. Thus herbage with low nitrogen contents would suggest that the object of the pasture ley in building up soil nitrogen levels was not being achieved.

Livestock operations

The qualitative and quantitative deficiencies of the mediterranean annual-type climate set clear limitations to the potential of pastures for animal growth and define that part of the year to which the output effort must be directed.

Animal production is geared to fit in with the pasture growth cycles, lambing or calving being arranged to coincide with, or shortly precede, the onset of the winter rainfall season. This is not the most advantageous time in relation to prolificacy (Allden 1956), but it fits in well with the cereal growing operation. At high stocking rates a spring lambing may be preferable if the end-of-season rains are reliable (Davies 1962). In areas where the pasture growing season is of 4-6 months duration lambs are turned off at a marketable weight at the end of spring. Cattle may be sold as vealers or carried over into the following season, with a consequent loss in weight during the summer period (Wagnon *et al*. 1959). Poorly grown lambs, if carried into the summer drought period, commonly succumb to the effect of undernutrition and need special treatment to ensure survival (Allden & Anderson 1957). In this respect cereal stubbles play a useful role in providing better nutrition during the early summer on the cereal-livestock farm.

SIGNIFICANCE OF THE DEFICIENCIES OF PASTURE

The limitations of mediterranean pastures as an animal feed are not difficult to define. The problems which arise relate to decisions on the form of enterprise to adopt, the stocking rate, fertilizer use, choice of pasture legume, timing of the reproductive cycle and the development of economically sound strategies to deal with the seasonal and annual variation in climate. Before attempting to formulate remedial measures it is important to know the extent to which nutritional limitations depress potential productivity. In other words, how important are the seasonally recurrent periods of feast and famine within the context of the husbandry systems which have evolved? The extent to which undernutrition affects wool production or meat production is self evident. Wool growth is more or less proportional to the intake of useful energy (Ferguson 1972, Langlands & Donald 1977, Allden 1979a, Purser 1981) and more and better feed would result in more wool. Likewise, for meat animals plenty of good quality feed would ensure sustained growth, earlier turnoff, or heavier market weights. The situation is less clear for the reproducing animal and her progeny. A method that we have used at the Waite Institute to assess the overall effects of the deficiencies of mediterranean annual-type pastures in grazing systems is to include a treatment group which, in addition to grazing, is allowed an unlimited amount of high quality supplement throughout the year, thereby minimizing the effects of nutritional fluctuations. Some of the results are relevant to the topic of this paper.

Reproduction

The advantages of good feeding throughout the year were quite small when flocks of Merino breeding ewes grazing at low (L) and high (H) stocking rates of 7.5 and 10.0 ewes/ha respectively were compared with a group which in addition to paddock grazing was provided with unrestricted amounts of high quality feed at all times. Fig. 4 depicts the gross body weights (including wool and conceptus) of the experimental animals during a 5-year period, the numbers of lambs born and weaned, and the weaning weight of the lambs (Allden unpublished). The gross body weights of the animals that were well fed showed minor fluctuations (about a mean weight of 60 kg during the course of the experiment) mainly associated with pregnancy, lactation and the amount of fleece carried, but these were small when compared with the seasonal variations of up to 25 kg in the weights of animals which grazed pasture only.

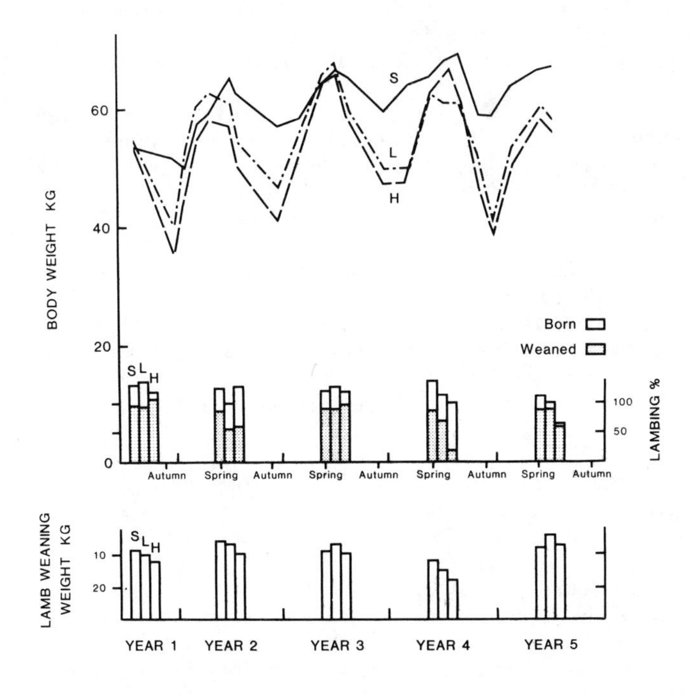

Fig. 4. Weight changes of ewes receiving continuous supplement (s) and those at low (L) and high (H) stocking rates (upper graph); lambs born and lambs weaned to ewes mated (upper histograms); weaning weights of lambs at 100 days (lower histograms).

For the full 5-year period the numbers of lambs born and weaned did not differ significantly between the unrestricted sheep and the low stocking rate group, the respective values being 126 vs. 116 percent born and 90 vs. 80 percent weaned; ewe mortality rates did not differ and weights of lambs at weaning were similar. In two years the high stocking rate flock was more precariously placed, and lamb survival and productivity were significantly reduced.

A feature of the graph is the remarkable capacity of the unsupplemented groups to make rapid gains during the spring, thereby compensating for the undernutrition experienced during the late summer and early winter months. In one year only was there a significant disparity in the weights of the supplemented and unsupplemented sheep at mating time in the late spring.

Growth of breeding flock replacements

Similar results in terms of growth were obtained with the ewe lambs retained to replace the cast-for-age ewes. During the period from birth to first mating at 19 months of age the growth paths of the three groups were strikingly different; those unrestricted by diet made rapid gains throughout (demonstrating the potential of better nutrition for early turnoff of the meat animal), whereas the young animals allowed access to pastures only at the low and high stocking rates showed appreciable weight losses during the summer. Fig. 5 depicts these weight changes (Allden unpublished). By the end of spring at the time of first mating (19 months) the sheep grazing on pasture had compensated for their earlier nutritional handicap. Good feed from birth, and plenty of it, had conferred no weight advantage.

Although the results do not provide data on the effects of these different growth paths in early postnatal life on subsequent productivity in the breeding flock, the evidence of Coop & Clark (1955), Giles (1968), and Allden (1979b) indicate that any effect on lifetime reproduction performance would be unlikely.

The results illustrate that summer undernutrition per se may have only small effects on the performance of Merino flocks at low stocking rates (although in this experiment the stocking rate was double the district average). However when the feed shortages are carried into the winter months either through overstocking, poor pasture establishment or through variation in seasonal growth the consequences may be serious.

In intensive production systems summer nutrition becomes more important. For example Geytenbeek (pers. comm.), studying the effects of stocking rate and frequency of lambing in prime lamb breeding flocks, observed a significant increase in total production when grazing flocks that lambed three times in two years were provided with high quality supplements throughout the year in addition to grazing. However the cost of providing better feed (1 kg/head/day) far outweighed the value of the additional returns. The flocks which depended solely on grazing produced only 75 percent of the annual meat output of those sheep unrestricted by diet (110 vs. 146 kg/ha).

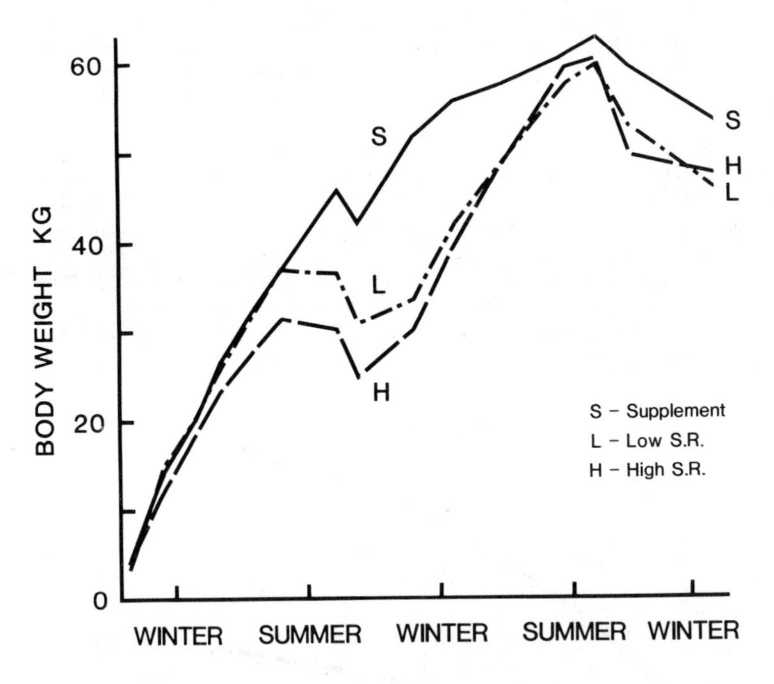

Fig. 5. Growth of ewe progeny from birth to first mating at 19 months for continuously supplemented group (S), and unsupplemented groups grazing at low (L) and high (H) stocking rates.

Thus it can be concluded that the qualitative deficiencies of pasture in mediterranean environments may not be of great significance to reproduction in non-intensive rearing systems, but they do represent a significant constraint to output when the full productive potential of prolific breeding animals is to be exploited.

Stocking rate

A further problem on mediterranean annual-type pasture is setting the optimum stocking rate, either at the on-farm or at the national level, and of devising economically sound strategies to minimize the effects of year to year variation in the climatic environment.

The growth of sheep depicted in Fig. 3 relates to animals grazing at moderate stocking rates with about 40-50 percent of the feed grown actually consumed in an average season. Much pasture research during the last 20 years, mainly with non-reproducing sheep, has shown that animal production from grassland may be greatly increased by ensuring that a high proportion of the material grown is consumed by the grazing animal, and this can only be achieved by increasing animal numbers (Carter & Day 1970). The consequence is an increased competition for feed, a decrease in production per head and an increase in output per unit area. However a point is eventually reached when the greater animal numbers can no longer derive sufficient feed from the pasture to meet their needs for survival, and the system becomes non-viable. Thus in mediterranean environments increasing the stocking rate may result in a deepening of the production troughs depicted in Fig. 3 and a prolongation of the period of stress into the winter months of pasture growth. The nutritional deficiencies associated with an increased stocking rate may reduce the output of the breeding flock through impaired reproduction, high lamb mortalities and metabolic disorders in the female (Davies 1964, 1968).

Reports from many regions of the Mediterranean basin, where livestock and cropping are not integrated, indicate that over-stocking of the communal lands is one of the greatest constraints to animal production (Oram 1975). In southern Australia, where the cereal and animal enterprises are commonly integrated, the evidence suggests that farmers set their stocking rates at a level well below the "optimum" determined in district experiments. Fig. 6 from Rossiter & Ozanne (1970) illustrates this point, and shows the relationship between rainfall and stocking rate observed in western Australia.

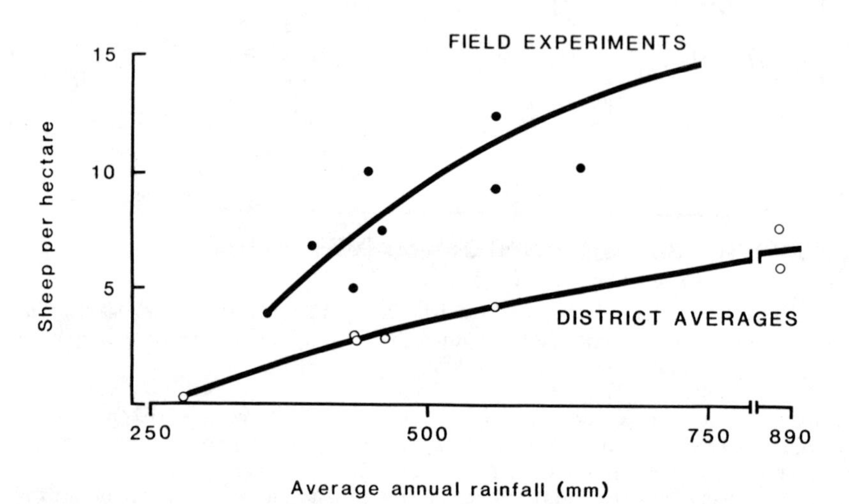

Fig. 6. Stocking rates on sown pastures in relation to rainfall. Field experiments compared with the district averages (Rossiter & Ozanne 1970).

Plant species and grazing management

Plant species might also be expected to play a role in determining annual herbage yield, but there has been limited research to assess the relative contribution and value of the component species in terms of yield, quality and seasonal production (reviewed by Rossiter 1966, Purser 1981). The presence of a legume is always important because of its capacity to fix nitrogen, whereas the companion grasses and forbs make varying contributions according to their capacity to utilize the nitrogen fixed by the legume and to adapt to seasonal and management variations.

Instability in botanical composition is a notable feature of the mediterranean annual-type pasture as Rossiter (1966) has recorded in his comprehensive review, there being dramatic changes from season to season and year to year in response to climate (drought, frosts, timing and duration of seasonal rains), fertilizer treatment, grazing management, seed dormancy and dispersion, competition from companion species, palatability, stocking rate, fodder conservation and insect pests. Thus the value of a species in annual self-regenerating pasture is not simply determined by its yield and nutritive value; the capacity to persist and sustain animal productivity during adverse seasonal conditions and at intensive levels of stocking is equally important. This factor has been commonly overlooked by agronomists. Suffice it to say that in many stocking rate experiments those pastures which support the greatest animal production/unit area consist of volunteer annual grasses and forbs which may have undesirable characteristics at some stage of their growth cycle, whereas the sown species commonly disappear under intensive grazing (e.g. Davies 1965, Davies & Humphries 1965, Rossiter 1966, Carter & Day 1970).

Table 3 illustrates the changes in botanical composition in response to stocking rate in an experiment in Western Australia. At the highest stocking rate Capeweed (*Arctotheca calendula*) and *Erodium* spp. were the principal components of a sward which sustained a highly productive breeding system. Yet these species are commonly considered to be undesirable pasture plants.

In the past much emphasis was placed on the role of sown perennial pasture species to stabilize botanical composition and provide out-of-season feed (Trumble & Davies 1931) but this potential has not been realized (Rossiter 1952, Donald 1970, Gibson 1976, 1977) other than in the wetter regions with a long grazing season. Even then the production pattern of perennials does not differ appreciably from that of annual species, and they are commonly heavily invaded by annuals after the season of establishment. In a few areas where soil water relations are favourable during the summer dryland lucerne (*Medicago sativa*) may provide out-of-season feed (Barrow & Pearson 1970) but taken as a whole the perennial pasture plants play an insignificant role in mediterranean pastures.

TABLE 3
Mean effect of stocking rate on botanical composition of a sown pasture at Kojonup, Western Australia (Davies 1966).

Stocking Rate ewes/ha	Percent (weight basis)				
	Sub. clover	Capeweed	*Erodium*	Misc. Forbs	Grasses
3.7	24	8	7	2	59
7.4	26	25	7	1	41
11.1	22	43	16	2	17

Thus, plant species do not appear to impose significant constraints to animal production unless the persistence of the legume component is affected (often an edaphic or fertilizer problem) or if they possess characters that influence animal health.

Finally, animals in mediterranean environments are kept at pasture throughout the year, a practice which confers a great economic advantage when compared with the housing, feeding and labour inputs that are necessary in animal production systems in northern Europe and North America. To sustain this advantage, management inputs in grazing systems need to be profitable. However, few studies into the effects of grazing management on animal production have shown any increase (often a decrease with higher costs) compared with continuous grazing (Heady 1961, 1975, Morley 1966, Arnold 1969, Wolloughby 1970, Myers 1972, Terblanche 1974), the rotational grazing of lucerne being a notable exception.

REFERENCES

Ackerman, E.A. (1941) The Köppen classification of climates in North America. Geographical Review 31, 105-111.

Adem, L. (1974) Etude du comportement des Medicago annuelles. These, Institut National Agronomique, Universitaire Alger.

Alexander, G. (1970) Thermogenesis in young lambs. In: Physiology of digestion and metabolism in the ruminant. Editor A.T. Phillipson. Newcastle-on-Tyne, Oriel Press, pp. 119-210.

Allden, W.G. (1956) Time of mating studies. II. The effect of time of mating on fat lamb and wool production in the lower south-east. South Australian Department of Agricultural Journal 59, 410-417.

Allden, W.G. (1959) The summer nutrition of weaner sheep: the relative roles of available energy and protein when fed as supplements to sheep grazing mature pasture herbage. Australian Journal of Agricultural Research 10, 219-236.

Allden, W.G. (1968) Undernutrition of the merino sheep and its sequelae. IV. Herbage consumption and utilization of feed for wool production following growth restrictions imposed at two stages of early post-natal life in a mediterranean environment. Australian Journal of Agricultural Research 19, 997-1007.

Allden, W.G. (1979a) Feed intake, diet composition and wool growth. In: Physiological and environmental limitations to wool growth. Editors J.L. Black and P.J. Reis. Armidale, Australia. University of New England Publishing Unit, pp.61-78.

Allden, W.G. (1979b) Undernutrition of the merino sheep and its sequelae. V. The influence of severe growth restriction during early post-natal life on reproduction and growth in later life. Australian Journal of Agricultural Research 30, 939-948.

Allden, W.G.; Anderson, R.A. (1957) "Unthriftiness" in weaner sheep. South Australian Department of Agriculture Journal 61, 69-86.

Arnold, G.W. (1969) Pasture management. Proceedings of the Australian Grasslands Conference, Perth, 1968 2, 189-211.

Aschmann, H. (1973) Distribution and peculiarity of mediterranean ecosystems. In: Mediterranean type ecosystems: origin and structure. Editors P. di Castri and H.A. Mooney. London, Chapman and Hall Ltd, pp. 11-19.

Bailey, H.P. (1958) A simple moisture index based upon a primary law of evaporation. Geografiska Annaler 40, 196-215.

Barrow, P.M.; Pearson, F.B. (1970) The mallee and mallee heaths. In: Australian grasslands. Editor R.M. Moore. Canberra, Australian National University Press, pp.219-227.

Braden, A.H.W.; Macdonald, I.W. (1970) Disorders of grazing animals due to plant constituents. In: Australian grasslands. Editor R.M. Moore. Canberra, Australian National University Press, pp. 381-391.

Cannon, D.J. (1974) Integrating sheep and wheat production: effect of substituting wheat production for pasture on the wool production of wethers. Australian Journal of Experimental Agriculture and Animal Husbandary 14, 454-460.

Carter, E.D. (1966) The pasture and livestock potential of Chile. The Rockefeller Foundation, Chile Agricultural Program, 54 pp.

Carter, E.D. (1974) The potential for increasing cereal and livestock production in Algeria. A report prepared for CIMMYT, Mexico and the Ministry of Agriculture and Agrarian reform, Algeria, 66 pp.

Carter, E.D. (1978) A review of the existing and potential role of legumes in farming systems of the Near East and North African region. A report to the International Centre for Agricultural Research in the Dry Areas (ICARDA), 120 pp.

Carter, E.D.; Day, H.R. (1970) Interrelationships of stocking rate and superphosphate rate on pasture as determinants of animal production. 1. Continuously grazed old pasture land. Australian Journal of Agricultural Research 21, 473-491.

Coombe, J.B.; Tribe, D.E. (1963) The effects of urea supplements on the utilization of straw plus molasses diets by sheep. Australian Journal of Agricultural Research 14, 70-92.

Coop, I.E.; Clark, V.R. (1955) The influence of method of rearing as hoggets on the lifetime productivity of sheep. New Zealand Journal of Science and Technology 37, 214-228.

Corbin, E.J. (1976) Strategies to extend the cropping phase in the southern New South Wales wheat belt. Australian Institute of Agricultural Science, National Conference, Canberra, pp.164-165.

Croker, K.P. (1968) Lamb mortality on agricultural research station. Western Australian Journal of Agriculture 9, 51-55.

Davies, H.L. (1962) Studies on time of lambing in relation to stocking in south-western Australia. Proceedings of the Australian Society of Animal Production 4, 113-120.

Davies, H.L. (1964) Lambing losses in south-western Australia. Proceeding of the Australian Society of Animal Production 5, 107-112.

Davies, H.L. (1965) Studies in nutrition and reproduction in sheep in south western Australia. Ph.D. thesis. Western Australian University.

Davies, H.L. (1968) Studies in pasture management for the breeding ewe. 1. Effects of ewe liveweights, evidence of pregnancy toxaemia, lamb birth weights, and neo-natal mortality. Australian Journal of Agricultural Research 19, 605-620.

Davies, H.L.; Humphries, A.W. (1965) Stocking rate and wool production at Kojonup. 1. Wether sheep. Western Australian Department of Agriculture Journal, Fourth Series 6, 409-413.

Davies, H.L.; Boundy, C.A.P.; Southey, I.N. (1970) Summer nutrition of weaner sheep in south western Australia. Proceeding of the Australian Society of Animal Production 8, 255-258.

Denney, G.D.; Hogan, J.P.; Lindsay, J.R. (1979) The digestion of barrel medic (Medicago truncatula) hay and seed pods by sheep. Australian Journal of Agricultural Research 30, 1177-1184.

Donald, C.M. (1970) Temperate pasture species. In: Australian grasslands. Editor R.M. Moore. Canberra, Australian National University Press, pp.303-320.

Doolette, J.B. (1977) The application of the Australian ley farming system in North Africa. In: Proceedings of an international symposium on rainfed agriculture in semi-arid regions. Editor G.H. Cannell. Riverside, California, p.589-608.

Downes, R.W. (1976) Towards better use of environmental resources. Australian Institute of Agricultural Science National Conference, Canberra, Australia pp.71-77.

Ferguson, K.A. (1972) The nutritional value of diets for wool growth. Proceedings of the Australian Society of Animal Production 9, 314-320.

Ford Foundation (1974) Regional workshop for sheep and forage production. The Arid Lands Development Program, Ford Foundation, Beirut, Lebanon. 8 pp.

Gibson, P.R. (1976) A comparison of annual and perennial based pastures for liveweight and wool production of sheep grazing on the lateritic podzolic soils of Kangaroo Island. Proceedings of the Australian Society of Animal Production 11, 325-328.

Gibson, P.R. (1977) Persistence of perennial ryegrass under grazing in a mediterranean-type environment of South Australia. Proceedings of the 13th International Grasslands Congress, Leipzig, 295-299.

Giles, J.R. (1968) The effects of different levels of nutrition from weaning to seventeen months of age on the lifetime production of merino ewes. Australian Journal of Experimental Agriculture and Animal Husbandry 8, 149-157.

Grigg, D.B. (1974) The Agricultural Systems of the World: An Evolutionary Approach. Cambridge, Cambridge University Press, 358 pp.

Guilbert, H.R.; Hart, G.H. (1946) Calfornia beef production. Californian University, Agricultural Experiment Station, Extension Service. Manual 2, 39 pp.

Guilbert, H.R.; Mead, S.W. (1931) The digestibility of bur clover as affected by exposure to sunlight and rain. Hilgardia 6, 1-12.

Hardison, W.A.; Fox, C.W. (1973) Present livestock situation in the Near East and the possibility of developing integrated crop-livestock farming system. Ford Foundation, Lebanon, 14 pp.

Hart, G.H.; Guilbert, H.R.; Goss, H. (1932) Seasonal changes in the chemical composition of range forage and their relation to nutrition of animals. Californian University, Agriculture Experiment Station, Bulletin. 543, 62 pp.

Haylett, D.G. (1958) Advances in crops and pastures in the Union of South Africa in the last twenty-five years. Empire Journal of Experimental Agriculture 26, 169-182.

Heady, H.F. (1961) Continuous vs. specialized grazing systems: A review and application to the California annual type. Journal of Range Management 14, 182-193.

Heady, H. (1975) Rangeland Management. New York, McGraw-Hill 460 pp.

Hegarty, M. (1982) Deleterious factors in forages affecting animal production. In: Nutritional limits to animal production from pastures. Editor J.B. Hacker. Farnham Royal, U.K., Commonwealth Agricultural Bureaux, pp. 133-150.

Hogland, O.K.; Miller, H.W.; Hafenrichter, A.L. (1952) Application of fertilizers to aid conservation on annual forage range. Journal of Range Management 5, 55-61.

Hutchinson, K.J.; Porter, R.B. (1958) Growth and wool production of merino hoggets related to grazing intake in a South Australian environment. Proceedings of the Australian Society of Animal Production 2, 33-41.

Jones, M.B. (1960) Responses of annual range to urea applied at various dates. Journal of Range Management 13, 188-192.

Köppen, W.; Geiger, R. (1936) Handbuch der Klimatologie, Band 1, Teil C, C 42-43. Berlin, Gebruder Borntrager.

Langlands, J.P.; Donald, G.E. (1977) Efficiency of wool production of grazing sheep. 4. Forage intake and its relationships to wool production. Australian Journal of Experimental Agriculture and Animal Husbandry 17, 247-315.

Martin, W.E.; Berry, L.J. (1970) Effects of nitrogenous fertilizers on California range as measured by weight gains of grazing cattle. Californian University Agricultural Experiment Station Bulletin. 846, 1-24.
Morley, F.H.W. (1966) The biology of grazing management. Proceedings of the Australian Society of Animal Production 6, 127-136.
Murphy, A.H.; Jones, M.B.; Clawson, J.W.; Street, J.E. (1973) Management of clovers on California annual grasslands. Californian University Agricultural Experiment Station, Extension Service. Circular. 564, 1-19.
Myers, L.F. (1972) Effects of grazing and grazing systems. In: Plants for sheep in Australia. Editors J.H. Leigh and J.C. Noble. Sydney, Angus and Robertson pp.183-192.
Neal Smith, C.A. (1942) The comparative grazing values of top-dressed natural and sown pastures in the middle south-east of South Australia. Journal of Agriculture of South Australia 45, 485-496.
Obst, J.M.; Day, H.R. (1968) The effect of inclement weather on mortality of Merino and Corriedale lambs on Kangaroo Island. Proceedings of the Australian Society of Animal Production 7, 239-242.
Oram, P.A. (1975) Livestock production and integration with crops in developing countries. In: Proceedings of the III World Conference on Animal Production. Editor R.L. Reid. Sydney, Sydney University Press, pp. 309-330.
Oram, P.A. (1977) Agriculture in the semi-arid regions: problems and opportunities. In: Proceedings of an International Symposium on Rainfed Agriculture in semi-arid regions. Editor G.H. Cannell. Riverside, California, pp. 2-59.
Price, P.C. (1973) Investigation of a nematode - bacterium disease complex affecting Wimmera ryegrass. Ph.D. thesis, Waite Agricultural Research Institute. The University of Adelaide, Adelaide, Australia, 164 pp..
Pullman, A.L.; Allden, W.G. (1971) Chemical curing of annual pastures in southern Australia for beef cattle and sheep. Australian Journal of Agricultural Research 22, 401-413.
Purser, D.B. (1981) Nutritional value of mediterranean pastures. In: World Animal Science B 1. Grazing Animals. Editor F.H.W. Morley. Amsterdam, Elsevier Scientific Publishing Company. pp. 159-180.
Radcliffe, J.C.; Cochrane, M.J. (1970) Digestibility and crude protein changes in ten maturing pasture species. Proceedings of the Australian Society of Animal Production 8, 531-536.
Raven, P.H. (1973) The evolution of mediterranean floras. In: Mediterranean type ecosystems: origin and structure. Editors F. di Castri and H.A. Mooney.
Rossiter, R.C. (1952) The effect of grazing on a perennial veldt grass -subterranean clover pasture. Australian Journal of Agricultural Research 3, 148-159.
Rossiter, R.C. (1966) Ecology of the mediterranean annual-type pasture. Advances in Agronomy 18, 1-56.
Rossiter, R.C.; Ozanne, P.G. (1970) South-western termperate forests, woodlands, and heaths. In: Australian grasslands. Editor R.M. Moore. Canberra, Australian National University Press, pp.199-218.
Round, M.H. (1976) Production response to urea supplements by ruminants consuming low quality forages in Australia and factors affecting response. Department of Agriculture and Fisheries, South Australia, Livestock Branch Technical Information Circular, No. 28, 22pp.
Saunders, D.A. (1976) Early management issues in establishing wheat-forage legume rotations. Proceedings 3rd Regional Wheat Workshop, Tunis. Tunisia 1975, pp.254-259. Centro Internacional de Mejoramiento de Maiz y Trigo.
Sharkey, M.J.; Davis, I.F.; Kenny, P.A. (1964) The effect of rate of stocking with sheep on the botanical composition of an annual pasture in southern Victoria. Australian Journal of Experimental Agriculture and Animal Husbandry 4, 34-38.
Silsbury, J.H.; Adem, L.; Baghurst, P.; Carter, E.D. (1979) A quantitative examination of the growth of swards of *Medicago truncatula* cv. Jemalong. Australian Journal of Agricultural Research 30, 53-63.
Specht, R.L. (1969) A comparison of the sclerophyllous vegetation characteristic of mediterranean type climates in France, California, and southern Australia. 1. Structure, morphology, and succession. Australian Journal of Botany 17, 277-292.
Terblanche, P.J.J. (1974) [Evaluation of a mixed pasture in a mediterranean region with merino ewes.] M.Sc. thesis. University of Stellenbosch, Stellenbosch, South Africa.
Theriez, M.; Skouri, M. (1970) Performance of sheep on a range in central Tunisia and relationship with diet. Proceedings of the 11th International Grasslands Congress, Surfers Paradise, Australia. 767-770.
Trumble, H.C.; Cornish, E.A. (1936) The influence of rainfall on the yield on a natural pasture. Journal of the Council of Scientific and Industrial Research Australia 4, 140-151.
Trumble, H.C.; Davies, J.G. (1931) The role of pasture species in regions of winter rainfall and summer drought. Journal of the Council of Scientific and Industrial Research, Australia 9, 19-28.
Vercoe, J.E.; Pearce, G.R. (1960) Digestibility of *Medicago tribuloides* (Barrel medic) pods. Journal of the Australian Institute of Agricultural Science, 26, 67-70.
Vosloo, L.P.; Meissenheimer, D.J.B. (1970) Wool production of merino wethers on cultivated pastures. Report on Technical Training. University of Stellenbosch, Stellenbosch, South Africa.

Wagnon, K.A.; Guilbert, H.R.; Hart, G.H. (1959) Beef cattle investigations on the San Joaquin Experimental Range. Californian University Agricultural Experiment Station, Bulletin. 765, 1-70.

Weir, W.C.; Torell, D.T. (1967) Supplemental feeding of sheep grazing on dry range. Californian University Agricultural Experiment Station, Bulletin. 832, 1-48.

Wheeler, J.L.; Hutchinson, K.J. (1973) Production and utilization of food for ruminants. In: Pastoral industries of Australia. Editors G. Alexander and O.B. Williams. Sydney, Sydney University Press, pp. 201-232.

Whittaker, R.H. (1970) Communities and ecosystems. Macmillan, London, U.K. 162 pp.

Williams, C.M.J.; Allden, W.G.; Geytenbeek, P.E. (1980) The influence of seeding rate on production from grazed and ungrazed annual ryegrass pastures. International Congress on Dryland Farming, Adelaide. Working papers 2, 76-78.

Willoughby, W.M. (1970) Grassland management. In: Australian grasslands. Editor R.M. Moore. Canberra, Australian National University Press, pp. 392-397.

Wilson, A.D.; Hindley, N.L. (1968) the value of the seed, pods and dry tops of subterranean clover (Trifolium subterraneum) in the summer nutrition of sheep. Australian Journal of Experimental Agriculture and Animal Husbandry 8, 168-171.

Wolfe, E.C.; Curll, M.L.; Fitzgerald, R.D.; Hall, D.C.; Southwood, O.R. (1976) Changing livestock production in the cereal zone of southern New South Wales. Australian Institute of Agricultural Science, National Conference, Canberra. pp. 209-210.

PROBLEMS OF ANIMAL PRODUCTION FROM TROPICAL PASTURES

L.'t MANNETJE

CSIRO, Division of Tropical Crops and Pastures, Cunningham Laboratory, St. Lucia, Brisbane, Queensland 4067, Australia

ABSTRACT

Tropical pastures are important but largely undeveloped resources for animal production. A brief description of native and improved pastures and pastures associated with crops is followed by comments on the main animal species that graze them. Levels of production are presented for beef cattle in arid, sub-humid and humid regions, for meat and fibre production of sheep and goats and milk production of cattle, sheep, goats and camels.

Tropical pastures do not meet the nutritional requirements of ruminants for maximum production. The main limitations are availability of green feed for at least half the year in seasonally dry regions and low nutritive value during most of the period of active pasture growth.

It is considered to be improbable that improving the genetic production potential of cattle will increase animal production, although there is a need for genetic improvement of sheep for fibre production, and for adaptation of cattle and sheep to tropical conditions.

INTRODUCTION

Livestock production in the tropics has four major economic roles: 1) to supply energy and proteins for indigenous populations, 2) to supply draught, 3) to produce fibres and hides and 4) to meet import demands by industrialized nations.

There are some 800 x 10^6 cattle, 500 x 10^6 sheep, 400 x 10^6 goats, 130 x 10^6 water buffaloes and 14 x 10^6 camels in the tropics (FAO 1978). These figures comprise 64 percent of the world's cattle, 51 percent of the sheep, 94 percent of the goats, practically all the buffaloes and nearly all the camels.

Of the main livestock producing countries with tropical climates only Australia is a major exporter. Most other countries are barely self-sufficient or have insufficient meat and milk production to feed their human population. This is evident from the level of human nutrition and particularly from the proportion of energy and protein derived from animals in developing countries. In developed countries the proportions of energy and proteins derived from animals for human consumption is about 34 and 59 percent respectively, but in developing countries they are only 8 and 21 percent (Reid & White 1980). In wealthy countries 75 g of animal protein are consumed per head per day as compared with 5 g in the poorest nations (Jasiorowski 1975).

In Asian countries animals graze on waste lands, common grazing areas, cut feed and by-products, but elsewhere most of the feed is obtained from grasslands, shrublands and forage crops.

Therefore there is a great need to increase and improve herbage-derived animal feed in tropical countries. Arable land is nearly all in use already and grazing lands, which can generally not be used for other economic purposes, are largely undeveloped and have a large potential for adding to the supply of human food.

This review consists of two parts. The first deals with the traditional role of animals and pastures in animal production and is therefore mainly descriptive. In the second part the limitations of tropical pastures for animal production in contrasting environments at different levels of nutrition are discussed.

TYPES OF PASTURES

Tropical pastures may be divided into 1) native pastures, 2) improved pastures and 3) pastures associated with crops.

Native pastures

These may be natural or derived; pasture management is restricted to grazing and burning. They may be subdivided and stock water may be provided. They may consist only of indigenous species, or also contain naturalized exotic species which have invaded naturally. In other words, they rely on natural conditions for pasture growth. Native tropical grasses generally contain low nutrient concentrations (French 1957). In some regions grazing animals receive mineral, energy and protein supplements.

Extensive areas of natural pastures in regions of high soil fertility and sufficient rainfall (> 600 mm) are now utilized for crop production. However, this has been offset by large areas of open woodlands with a grassy understorey now used as grazing lands. Tree density has often been reduced leading to greater herbage production (Walker *et al*. 1972, Van Niekerk *et al*. 1978, Gillard 1979). In Australia *Heteropogon contortus* is dominant in many such pastures.

Derived pastures occur where forests have been felled for either crops or pasture production. In the humid tropics rainforest has been removed in slash-and-burn agriculture. After a number of cycles with short intervals of fallow the soil fertility is too low for cropping. Regular grazing and burning prevent tree regrowth and inferior grasses such as *Imperata cylindrica* occupy the land (Whyte 1974; Falvey 1981). In eastern Queensland rainforest was removed for pasture production and indigenous and naturalized exotic species formed the basis of a dairy industry. As soil fertility declined, the pastures became less productive and were invaded by weeds (Bryan 1970). Large areas in Africa and Latin America have been cleared, but in many areas the grass cover has been replaced by dense inedible shrubs due to overgrazing and resultant lack of fuel for fires (Skovlin & Williamson 1978, Gonzalez Padilla 1980).

Improved pastures

These may be fully sown pastures consisting of grasses only or also including legumes, or native pastures oversown with legumes.

Grass-only pastures cannot persist at high levels of production without the use of nitrogen fertilizers (Mannetje & Shaw 1972, Henzell *et al*. 1975) except on soils of heavy texture with high organic matter contents (Coaldrake 1970). The dry matter yield response to nitrogen

TABLE 1
The main sown grasses and legumes adapted to the climatic zones of Troll (1966), (Henzell & Mannetje 1980).

Zone	Grasses	Legumes
Tropical rainy climates (V_1) and Tropical humid summer climates (V_2)	*Brachiaria decumbens*	*Centrosema pubescens*
	Brachiaria mutica	*Desmodium heterophyllum*
	Digitaria decumbens	*Leucaena leucocephala*
	Panicum maximum	*Pueraria phaseoloides*
	Pennisetum purpureum	*Stylosanthes guianensis* var. *guianensis*
	Setaria sphacelata	
Wet-and-dry-tropical climates (V_3)	*Cenchrus ciliaris*	*Leucaena leucocephala*
	Cenchrus setiger	*Macroptilium atropurpureum*
	Urochloa mosambicensis	*Stylosanthes hamata* cv. Verano
		Stylosanthes humilis
		Stylosanthes scabra
Tropical dry climates (V_4)	*Cenchrus ciliaris*	*Stylosanthes scabra*
	Panicum antidotale	
Steppe climates with short summer humidity (IV_3)	*Anthephora pubescens*	*Stylosanthes fruticosa*
	Dactyloctenium giganteum	*Stylosanthes viscosa*
	Cenchrus ciliaris	
	Eragrostis curvula	
Dry winter climates with long summer humidity (IV_4)	*Cenchrus ciliaris*	*Leucaena leucocephala*
	Chloris gayana	*Macroptilium atropurpureum*
	Digitaria decumbens	*Stylosanthes guianensis* var. *intermedia*
	Panicum maximum var. *trichoglume*	*Stylosanthes humilis*
	Setaria sphacelata	*Stylosanthes scabra*
Permanently humid climates with hot summers (IV_6 & IV_7)	*Digitaria decumbens*	*Desmodium uncinatum*
	Paspalum dilatatum	*Desmodium intortum*
	Pennisetum clandestinum	*Lotononis bainesii*
	Setaria sphacelata	*Macroptilium atropurpureum*
		Medicago sativa
		Neonotonia wightii
		Trifolium repens
		Trifolium semipilosum

depends on rainfall (Salette 1970, Mannetje & Shaw 1972). In dry regions the optimum application for most years may be no more than 150 kg N/ha/yr (Henzell et al. 1975). However, in humid regions responses up to 800 kg (Salette 1970) and with irrigation up to 1300 kg N/ha/yr (Rivera Brenes et al. 1961) have been recorded.

Sown grass-legume pastures are persistent and productive provided they consist of adapted species, are adequately fertilized and not overgrazed. Henzell & Mannetje (1980) listed some of the main grasses and legumes used for pasture establishment in climatic zones ranging from humid to arid (Table 1). New cultivars and species are continually being added. For example in Queensland a new cultivar of *Stylosanthes scabra* (Fitzroy) and one of *S. guianensis* (Graham) were released in 1980. The main emphasis in the past has been on providing new plants adapted to climatic and edaphic situations, which has been quite successful, except for arid and semi-arid regions (< 600 mm rainfall) and for legumes for heavy textured soils. The search for improved pasture species is continuing. For the wetter areas the main problems now are to provide cultivars with resistance to diseases, such as *Colletotrichum* sp. (Lenné et al. 1980) and with higher nutritive value.

The main legumes used for oversowing into native pastures are *Stylosanthes humilis* (Shaw 1961, Shaw & Mannetje 1970, Gillard 1979), *Macroptilium atropurpureum* (Partridge 1975, Tothill & Jones 1977) and *S. guianensis* (Chadhokar 1977, Bowen & Rickert 1979). *Stylosanthes* spp. have low phosphorus requirements (Jones 1974) and can form productive pastures with native grasses without fertilizers, although they will respond to phosphorus and potassium in dry matter and animal production (Shaw & Andrew 1979).

Pastures associated with crops

There are three situations in which pastures may be associated with crops - ley pastures in crop rotation, companion pastures with field crops, and companion pastures with tree crop plantations.

The main role of ley pastures is to cover and restore the soil after a cropping phase in order to combat erosion, rebuild fertility or control pests, diseases and weeds. Ideally, the pasture species used must be fast growing, established by seed and easily eradicated for the next crop. In the sub-tropics *Medicago sativa*, *Lablab purpureus* and *Sorghum almum* and in the wet tropics *Pueraria phaseoloides* and *Pennisetum purpureum* comply with these requirements.

Companion pastures with field crops serve the same purpose as ley pastures but they are sown with the crop and care must be taken that they do not reduce crop yield through competition. The best results have been obtained with *Stylosanthes humilis* and *S. guianensis* in dry-land crops of rice, sorghum or maize (Humphreys 1978).

Coconut, oil palm, rubber and fruit plantations normally have an understorey of weeds and native grasses. These need to be controlled by herbicides, slashing or grazing to reduce competition for the crop and in the case of coconuts to increase the recovery of nuts. Sown grasses and legumes provide a more productive understorey which can be used for cut-and-carry feed or for grazing, but competition for nutrients between the pasture and the crop must be considered (Santhirasegaram 1966). Additional fertilizers are normally required. The aim is to achieve a higher return from the combined forms of land use than from either form alone.

Pasture-tree crops combinations are most successful in coconut plantations because of the high light penetration (Ohler 1969, Steel & Humphreys 1974, Rika *et al.* 1981). In oil palm and rubber plantations the tree canopies only cover a small proportion of the ground for the first few years after planting, but this gradually increases until a practically closed canopy is obtained. This progressive reduction in light penetration leads to a reduction in pasture yield and the eventual replacement of the sown pasture by more shade tolerant weeds and grasses.

ANIMAL PRODUCTION

Type of animal

Cattle

There are numerous breeds belonging to or derived from crosses between *Bos indicus* and *B. taurus*. Payne (1970) has described their origin, distribution and main characteristics. In the absence of environmental constraints *B. taurus* breeds have a higher genetic potential than *B. indicus* breeds but they lack heat and tick tolerance (Vercoe & Frish 1980). Because of these factors there is an active and ongoing process of crossing and selecting between breeds of the two species (Turner 1975). *B. indicus* animals generally have a lower maintenance requirement per unit liveweight than those of *B. taurus* (Vercoe & Frish 1980), but there are few differences between cattle species in their ability to digest and utilize low quality roughage (Moran *et al.* 1979).

Although specialized breeds for draught or milk and meat production have been developed, in many countries the same animals are often used for all three purposes. In some areas, cattle are also important in providing blood for human consumption, for social security and prestige, as currency (bride price etc.), the production of dung for fuel and of fertilizer for crop, vegetable and fish production.

Buffaloes

The Asian buffaloes have been domesticated since early recorded history and have been introduced to other continents including South America and southern Europe; they are feral in northern Australia. The swamp buffaloes, from south-east Asia, are mainly used for draught and meat, having a very low milk yield, whereas there are various improved breeds of river buffaloes, from India, with high milk yields (Mason 1974).

Whilst there are reports that buffaloes have a superior digestive system compared to cattle (Chutikul 1975, Devendra 1980), recent research and critical analyses of research results have questioned this conclusion (Chalmers 1974, Ichhponani *et al.* 1977, Moran *et al.* 1979, Moran 1981).

Sheep and goats

The main products from both species in the tropics are wool (or hair), meat and milk. Sheep with a fine wool type (Merino) have been introduced to tropical Australia and southern Africa, whilst in other regions sheep breeds have coarse wool or hairy coats. Because goats are hardy and can thrive on poor quality diets (Devendra 1978), they are often important where overgrazing by cattle or sheep has resulted in a vegetation dominated by coarse herbage and shrubs.

Camels

Camels and dromedaries are not often considered, but they are the only source of meat, milk and draught in some arid regions. There are 9.6×10^6 dromedaries in Africa and 4.3×10^6 bactrian camels in Asia. In addition dromedaries were introduced into Australia (Knoess 1977).

Levels of production

The level of production of ruminants depends on the type of animal, its genetic potential, its health and the environment, but the overriding determinant is the intake of digestible energy, proteins, minerals and vitamins and the absence of harmful substances (Minson *et al*. 1976). Since feed intake is partitioned into requirements for body function, maintenance, work, heat and production and since half the energy intake is required for maintenance (Morrison 1956) it follows that feed requirements of an animal depends largely on body size. Thus, the availability and quality of feed sources determines to a large extent the type of livestock prevalent in a region.

Beef production

The amount of beef produced per 1000 head of cattle in the tropics averages 3 tonnes in Asia, 12 in Africa, 23 in Latin America and 33 in northern Australia (Mannetje 1978). This reflects partly the level of feeding and partly the main purpose for which animals are used. For example, in Asia many are mainly used for draught and in parts of India cattle are not slaughtered for religious reason. Production per 1000 ha of grazing land is lowest in northern Australia and Africa with 2 and 3 tonnes respectively, compared to 13 in Latin America and 23 in Asia (Mannetje 1978). The high figure for Asia is misleading, however, because most cattle are kept on waste lands and fed on by-products. Levels of beef production from grazing lands as a rule depend on the availability of moisture for pasture growth, although soil fertility also plays an important part. The drier the climate the lower the forage production and therefore the lower the carrying capacity and growth rate of grazing animals. Seasonality of rainfall distribution is the main cause of seasonality of production.

The following paragraphs apply only to animals on pasture under controlled grazing. This situation contrasts to uncontrolled grazing, such as on common grazing lands, which are frequently severely overgrazed. Overgrazing is a limitation to production which overrides all others. Also excluded are situations in which hand feeding is practised for example cut-and-carry systems or by-product feeding in the wet tropics, the feeding of conserved fodder or concentrates. Some of these aspects are covered by Mahadevan (1982).

Arid and semi-arid regions

The largest areas with tropical semi-desert, desert and tropical dry climates (Troll 1966) occur in northern Africa, Australia and Latin America. In Africa, areas receiving less than 250-380 mm of rain per year are considered suitable only for nomadic grazing (Chalmers 1976). Although Skerman (1975) was of the opinion that there is no future for nomadism due to pressure for cultivation and the need for greater efficiency of the livestock industries, Chalmers (1976) indicated that nomadic cattle herds in the Sudan were more productive than sedentary ones as shown below.

	Calving %	Deaths %	Gain in Herd numbers %
Nomadic herds	65	26	8
Sedentary herds	40	32	-35

There is a vital symbiotic link between the livestock industry and arable agriculture in semi-arid Africa. During the dry season livestock consume crop residues whilst manure from the grazing animals is beneficial to the subsequent crop (McCown *et al*. 1979).

In Australia there are some 5 million beef cattle in the tropical and sub-tropical arid and semi-arid regions (data estimated from BAE (1975) and Mannetje *et al*. (1976). The grazing management system is year-long set stocking on large properties with minimal internal fencing. In the Northern Territory average property size is 2500 km^2 carrying 6-8 animals per km^2 (Perry 1970). The main emphasis is on breeding. Branding percentages range between 40 and 60, but cattle are turned off at rates of 22 to 26 percent (BAE 1975) at the age of 3 to 4½ years with dressed carcass weights ranging from 225 to 360 kg (Beattie 1956). Most of the meat produced is exported as manufacturing beef. However, due to increasing costs of fuel and transport Squires (1978) does not consider that there is much future for the beef cattle industry in the arid zone of Australia.

The arid zones of Latin America (excluding Argentina and Chile, which are mostly in the temperate zone) comprise 2.2 x 10^6 km^2 and carry more than 23 x 10^6 head of cattle. The productivity of these animals is low. In the semi-arid north eastern part of Brazil there are some 13 x 10^6 head of cattle which produce only 14 kg of meat per head from an extraction rate of 10 percent per year (Gonzalez Padilla 1980).

Sub-humid regions

These are characterized by variable summer rainfall of 500-800 mm per annum. Cattle on native pastures may gain up to 100 kg liveweight during the wet season but lose up to half that during the following dry season and consequently age to slaughter is 3½ to 5 years. Coupled with low carrying capacity (3 to 10 ha per beast) liveweight gain per ha usually does not exceed 30 kg per annum (Shaw 1961, Norman 1965). Reproductive rate is low with average calving percentage below 60 (Alexander & Carraill 1973). However, where rainfall exceeds about 650 mm per annum pasture improvement is possible, which can lead to highly increased production. In Queensland 3.6 x 10^6 ha or only five percent of non-arid grazing lands are improved and these are used for dairy as well as beef production (Mannetje *et al*. 1976). The advantages of improved over native pastures for beef production lie in their higher dry matter production and nutritive value. Animal production can be increased from two to fifty fold, depending on the environment and the existing level of production (Stobbs 1976, McIvor *et al*. 1981). In addition, in south-east Queensland improved pastures produce four times the number of calves and five times the total weaning weight per ha (Mannetje & Coates 1976). Furthermore, age at slaughter (assuming a slaughter weight of 500 kg) can be reduced from 3½ years on native pasture to 2½ years on improved pasture (Fig.1). Efficiency of production can also be greatly increased by using improved pastures (Stobbs 1976, Mannetje et al. 1976). With the increasing demand for meat and milk we can expect an increase of the area of improved pastures. However, there are technical, sociological and financial problems inhibiting a rapid increase in the area of improved pastures, particularly in developing countries (Mannetje 1978).

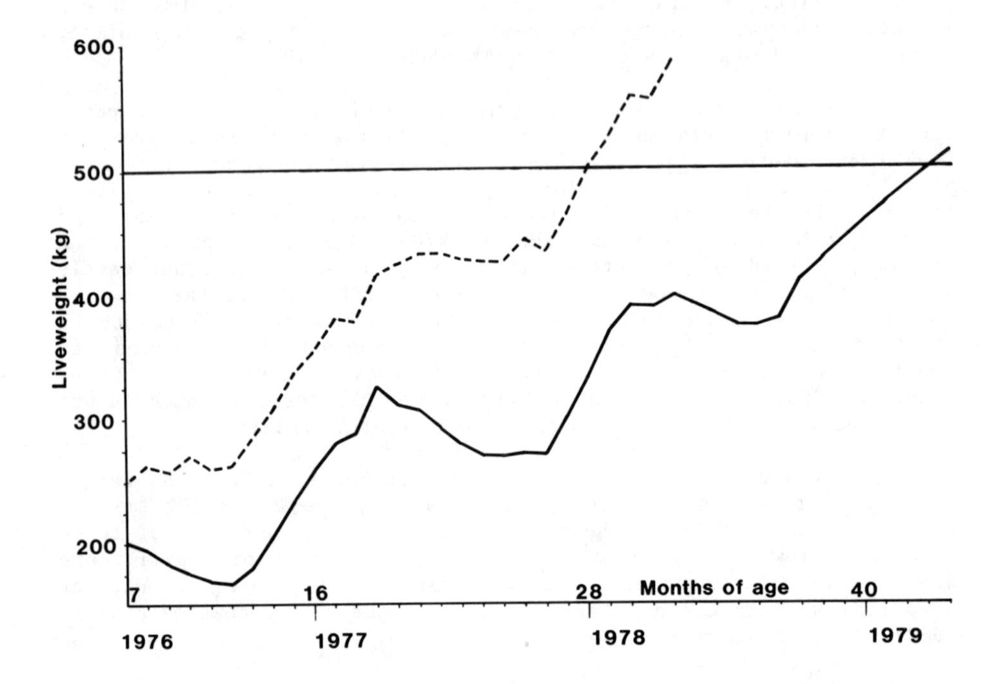

Fig. 1. Cumulative liveweight gain of steers grazing unimproved native pasture (———) at 0.27 steers/ha and Cenchrus ciliaris with Macroptilium atropurpureum (-------) fertilized with superphosphate at 1.1 steers/ha from weaning till 500 kg liveweight. (L.'t Mannetje, unpublished).

Humid regions

Cattle graze four types of pastures in humid regions - waste lands, degenerated crop lands, sown pastures and the herbaceous understorey of tree crop plantations. The first and second types are often covered in Imperata cylindrica. Magadan et al. (1974) recorded 100 kg/ha per year liveweight gain from an I. cylindrica pasture in the Philippines. Where this grass was replaced with Brachiaria mutica and Centrosema pubescens with phosphate fertilizer liveweight gain was 305 kg/ha per year. (See also Falvey 1981). Generally, grass-legume pastures adequately supplied with non-nitrogen fertilizer have yielded up to 500 kg/ha of liveweight gain (Vicente-Chandler et al. 1964, Bryan & Evans 1971, Mellor et al. 1973, Eng et al. 1978). With irrigation in northern Queensland a mixture of Centrosema pubescens and Brachiaria mutica gave a mean liveweight gain over three years of 890 kg/ha (Ebersohn, J.P. et al. unpublished).

Because of the well distributed high rainfall in equatorial regions, grass pastures fertilized with nitrogen have high dry matter yields and therefore high carrying capacities. Under conditions of nearly uninterrupted pasture growth liveweight gains per animal are also high. Using data from Puerto Rico (Vicente-Chandler et al. 1974), Dirven (1977) calculated the linear relation $y = 515 + 1.8X$, where y = liveweight gain (kg/ha/year) and x = kg N/ha/year. Up to nearly 700 kg N/ha/year was applied to grass pastures. A maximum liveweight gain of 1770 kg/ha/year commensurate with this equation was obtained. However, the equation implies that without the use of nitrogen liveweight gain would still be 515 kg/ha, which seems unlikely. Regions at higher latitudes have reduced pasture growth during the cool months, and therefore production figures are lower. Nevertheless, Evans (1970) reported annual liveweight gain of nearly 1300 kg/ha with 448 kg N/ha/year in southern coastal Queensland. With irrigation in north Queensland 2760 kg/ha liveweight gain was recorded using 673 kg N/ha at a stocking rate of 12.4 animals/ha (Ebersohn J.P. et al., unpublished). However, Blunt (1978), in the Kimberley region of Western Australia, obtained only a maximum of 1330 kg liveweight gain at 11.5 animals/ha with irrigation on nitrogen fertilized Digitaria decumbens. In this experiment there was no significant effect of nitrogen over the range of 310 to 800 kg/ha per year on pasture yield or liveweight gain.

Until recently very little information was available on animal production under tree crop plantations. In his review Ohler (1969) reported that carrying capacities under coconuts ranged from 0.5 to 3.0 animals/ha in various parts of the tropics. Rika et al. (1981) grazed Bos banteng steers and bulls on a sown grass-legume pasture under coconuts in Bali at 2.7, 3.6, 4.8 and 6.3 animals/ha. Over three years mean maximum liveweight gain was 135 kg/animal, which decreased by 8.2 kg per unit increase of stocking rate. The authors stated that stable pasture and animal production of about 550 kg/ha/year liveweight gain could be maintained at 5 animals/ha. Reynolds (1981) working in Western Samoa concluded that even in dry years 250-400 kg/ha liveweight gain could be produced at around 2.5 Hereford steers (200-260 kg) per hectare. With an estimated 4.5×10^6 ha of coconuts plantations in the world (Payne 1976) this could make a major contribution to animal production in developing countries.

In addition, oil palm plantations offer possibilities for grazing during the first five years after establishment and rubber plantations can often be grazed throughout the life of the plantation, albeit at low stocking rates due to reduced light penetration.

Meat production from sheep and goats

Thirty percent of domestic ruminants in South-East Asia are sheep and goats and they are an important source of meat for home consumption, particularly in Indonesia. Devendra (1979) indicated that the hot carcass weight of indigenous goats in Malaysia could be increased from 8.2 kg to 14.7 kg by better nutrition.

In the arid zones of Brazil, Mexico and Peru there are about 27×10^6 sheep and goats which produce a total of 47×10^6 kg of meat per year (Gonzalez Padilla 1980).

Milk production

High milk production requires a large intake of metabolizable energy (Dirven 1965, Stobbs 1971). Most tropical grasses have a low nutritive value, and the more intensive type of dairy production therefore relies more on improved pastures than does meat or fibre production. Nevertheless, maximum milk production per cow without concentrate feeding is 10-15 kg/day, compared to 20-25 kg on intensive temperate pastures. However, high stocking rates can be attained leading to milk production up to a maximum of 5000-8000 l/ha/yr in the wet tropics on grass-legume pastures (Stobbs 1971, Dirven 1977).

Goats and sheep are important milk producers in north Africa, the Indian sub-continent, the Middle East and Brazil as is indicated by the fact that 17 percent of total milk produced in Africa is from goats and sheep. Maximum production is about 2 kg/day, but in areas typical for goat husbandry it is often 300 g or less during a lactation period of 75 to 250 days (Gall 1975). Camels can produce as much milk as cows when fed better quality tropical feeds. Yields of 9 and 13 kg/day have been recorded from animals grazing irrigated pastures, but under less favourable conditions the average is about 6 kg/day (Knoess 1977).

Fibre production

In contrast to European sheep breeds, indigenous tropical breeds have coarse fleeces or a coat of short or long hairs. Greasy fleece weights range from 0.7 to 3.0 kg from tropical hairy breeds in India in contrast to 3.5 to 6.0 kg of fine wool from Merino in temperate Australia (Turner 1974). Although these differences are mainly genetic, environment and feeding also play an important role. For example, Merino sheep average a greasy fleece weight of 3.4 kg in dry tropical Queensland as against 6.0 kg or more in South Australia (Brown & Williams (1970). Nevertheless, because of the different carrying capacities wool production in tropical Queensland at 1.2 kg/ha is low compared to 20 kg/ha in more humid temperate zones of Australia.

LIMITATIONS TO ANIMAL PRODUCTION

Under the prevailing conditions in practically all densely populated or arid regions in developing countries overgrazing and lack of grazing lands are the main reasons for low animal production. This is not a simple technical problem, but it is associated with the strong link between man and his domestic animals. No amount of pasture or animal research will solve this problem. Only when the traditional cultural reliance has been replaced by more economically based considerations, coupled with a reduction in the human birth rate, will it be possible to solve the problems of animal production in these regions by technological means.

Leaving these considerations aside, animal production is limited by the feed if the animal's genetic potential for growth is not achieved; conversely it is limited genetically by the animal if the potential of the feed for animal production is not realized.

Beef production

It is assumed that the genetic potential for growth is equivalent to the highest liveweight gain recorded in pen feeding with the best quality feed, which invariablly means a mixture of herbaceous and

TABLE 2

Animal production from pastures providing different levels of animal nutrition in three contrasting climatic environments.

Locality and breed	Rain mm	Types of pasture	SR[1] b/ha	ADG[2] kg	Means over years	LWG[3] kg per beast	LWG[3] kg per ha	Reference
Malaysia	2400	Village conditions	-	0.16	17	58	-	Devendra & Lee 1975
Kedah-Kelantan		*P. maximum* + legumes	4	0.28	2	103	410	Eng *et al*. 1978
		P. maximum, *Digitaria* sp., *B. decumbens*) 300 kg N/ha	8	0.36	1	131	1048	Chen, C.P. *et al*., unpublished
South Johnstone, north Qld. *B. taurus* x *B. indicus*	3200	*P. maximum*	4.2	0.24	2	87	367	Grof & Harding 1970
		P. maximum + *C. pubescens*	4.2	0.30	2	108	452	
		P. maximum + 165 kg N/ha	4.2	0.38	2	139	585	
Mundubbera, south-east Qld. *B. taurus* and	720	Native pasture	0.27	0.31	10	113	30	Mannetje 1972, 1974 and unpublished
		C. ciliaris	0.74	0.28	10	102	76	
		C. ciliaris + *M. atropurpureum*	1.09	0.42	10	153	167	
B. taurus x *B. indicus*		*C. ciliaris* + 168 kg N/ha	1.09	0.42	10	153	167	

[1] Stocking rate [2] Average daily gain [3] Liveweight gain

concentrate feeds. The genetic potential depends on the breed of cattle. Preston & Willis (1970) listed a large number of pen feeding trials with large bovines of _B. taurus_ and _B. taurus_ x _B. indicus_ breeds and the maximum liveweight gain recorded over sustained periods is about 1.5 kg/day. For the smaller indigenous _B. indicus_ and _B. banteng_ breeds in South-East Asia, Devendra & Lee (1975) and Moran (1980) indicated a maximum growth rate of between 0.40 and 0.60 kg/day. For the purpose of this discussion the genetic potential for growth will be assumed to be 1.5 and 0.5 kg/day for the large and small bovines respectively.

However, in real grazing systems we cannot only consider the genetic potential of the animal. Liveweight gain is related to the type of animal, the type of pasture, the stocking rate and the environment. The maximum liveweight gain per animal obtained at very low stocking rates reflects the potential of the animal modified by the quality of the feed and the environment. However, at such low stocking rates the potential of the pasture to produce liveweight gain is underutilized. We must therefore consider liveweight gain at stocking rates which allow optimum animal production without deleterious effects to the pasture.

Table 2 presents animal production data at low and high nutritional levels pertaining in three contrasting situations - Malaysia, wet tropical Queensland and sub-tropical Queensland. No supplementary feeding was provided except for sodium and cobalt.

The rainfall in Malaysia is evenly distributed with only occasional short periods in which there would be little or no pasture growth. At South Johnstone rainfall is recorded in all months with 75 percent falling from January-June. In southern Queensland 68 percent of the 720 mm falls between 1 November and 1 May, but the rainfall is very unreliable with frequent long periods without any rain. In addition, there are frequent frosts between June and September. Active pasture growth is restricted to about 4-5 months of the year.

The data in Table 2 are annual means. The average daily gains are similar to those quoted for a wide range of tropical pastures by Whiteman (1980). In Malaysia the average daily gain under village conditions, that is grazing of wastelands and feeding of cut herbage, amounts to 32 percent of the assumed genetic potential, which can be increased to 56 percent on legume based and to 72 percent on nitrogen fertilized pasture. In both Queensland situations cattle grazing unimproved pastures or grass pastures without a source of nitrogen attain only 16 to 21 percent of their potential gain and on fully improved pastures this is still only 28 percent. This indicates that on a year-round basis tropical pastures are grossly deficient in providing the nutritional requirements of cattle, despite the fact that liveweight gain per animal can be increased 1.6 times and per ha up to 5.6 times by pasture improvement according to these data and even higher in other cases (McIvor _et al_. 1981).

The low average daily gains in relation to the potential is largely accounted for by seasonal differences. In Malaysia seasonal differences were small, but in north Queensland they were marked. In the experiment of Grof & Harding (1970) the average daily gain recorded between December and March on the nitrogen fertilized pasture was 0.64 kg compared to 0.28 kg during the remainder of the year. Seasonal differences in south-east Queensland were even larger. Mean data over ten years from the C. ciliaris/M. atropurpureum pasture indicated four distinct periods of liveweight gain (Mannetje, unpublished):

Days from start of growing season (average)	Average daily gain (kg)
1-200	0.70
201-280	0.15
281-337	-0.09
338-365	0.29

In an earlier analysis of data from this experiment it was shown that liveweight gain was asymptotically related to green material in the pasture (Mannetje 1974). At the asymptote average daily gain was 0.70 kg. This indicates that average daily gain below 0.70 kg is limited by green material in the pasture. However, at best the maximum mean average daily gain is only half the potential growth rate of these cattle. The explanation for this lies in the nutritive value of the green feed, not its quantity. Wilson & Minson (1980) have pointed out that this low nutritive value is partly due to the high growth rate when temperature and soil moisture are high. They therefore argued that less favourable growing conditions should result in lower dry matter yields, but higher nutritive values, provided feed is not limiting.

Because in south-east Queensland temperatures are less variable from year to year than is rainfall, the annual liveweight gain of steers grazing C. ciliaris and M. atropurpureum at 1.09 steers/ha was plotted against annual rainfall (Fig. 2). There was a positive linear relation with increasing rainfall up to 750 mm and a negative linear relation when rainfall exceeded 750 mm. Although a more detailed analysis is required to take into account rainfall distribution, there appears to be some support for the hypothesis that the most favourable growing conditions for pastures do not necessarily give the best liveweight gain. Since climate cannot be changed the only way to improve the growth rate of cattle on tropical pastures is to develop new cultivars (Mannetje & Ebersohn 1980), which Wilson & Minson (1980) defined as possessing "smaller rates of decline in DMD and crude protein, despite the rapid rate of growth and development under high temperatures".

Although maximum average daily gain on unsupplemented pastures is about half the genetic potential of large bovines, pasture improvement offers great scope for increasing animal production in the tropics (Table 2).

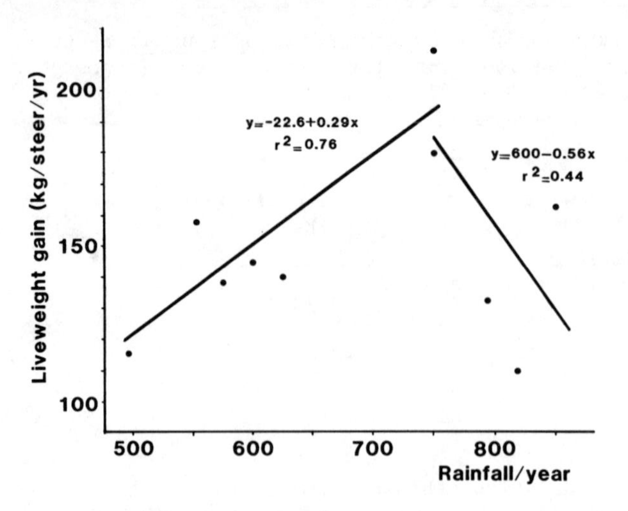

Fig. 2. Relation between liveweight gain/steer/year and total annual rainfall in a Cenchrus ciliaris and Macroptilium atropurpureum pasture grazed at 1.1 steers/ha (L.'t Mannetje, unpublished).

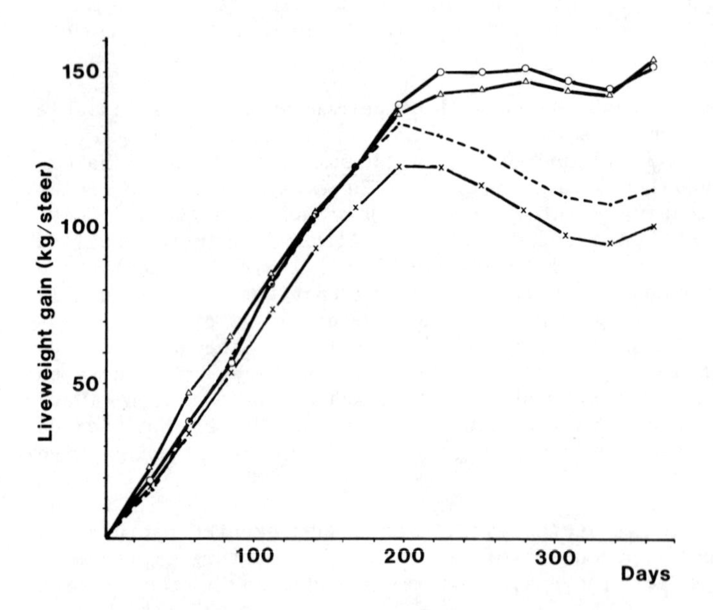

Fig. 3. Mean cumulative liveweight gain over 10 years of steers grazing unimproved native pasture at 0.27 steers/ha (..........), Cenchrus ciliaris at 0.74 steers/ha (x———x), C. ciliaris with Macroptilium atropurpureum at 1.1 steers/ha (o———o) and C. ciliaris with 168 kg N/ha at 1.1 steers/ha (Δ———Δ). All C. ciliaris pastures were fertilized with superphosphate (L.'t Mannetje, unpublished).

In south-east Queensland the deficiencies in nutritive value of native pasture and of unfertilized *C. ciliaris* pasture are mainly in winter and spring (Fig. 3). During the growing season average daily gain is similar on all pastures, but cattle on the *C. ciliaris* pastures with a legume or nitrogen fertilizer continue to gain weight for two to three months with only a slight loss towards the end of winter. This can only be attributed to the legume or nitrogen fertilizer. Pastures without a source of nitrogen have much lower dry matter yields in the growing season than those with a legume or nitrogen fertilizer. This is taken into account by adjusting the stocking rate.

The foregoing comments do not apply to arid regions where lack of soil moisture precludes pasture improvement. The main limitation to animal production in these regions is lack of feed over prolonged periods. Feed availability is determined by the interaction of rainfall and stocking rate (Burrows 1980).

Dairy production

Energy requirements for milk production are greater than for beef production. A dairy cow weighing 500 kg and producing 20 kg of milk per day requires 110 MJ net energy, but a steer weighing 300 kg and gaining 1 kg per day requires only 45 MJ (after McDonald *et al*. 1973).

If tropical pastures cannot supply the energy for the genetic potential of beef cattle it follows that they are even less capable of fulfilling the needs of dairy cows. This is borne out by the low milk yields of cattle even on good quality tropical pastures (Dirven 1977). Under extensive conditions total dry matter as well as its energy content are lacking, resulting in even lower yields.

Fibre production

Wool growth has a greater requirement for protein than for energy because wool fibres consist almost entirely of proteins. A sheep weighing 50 kg requires about 7 g N/day for maintenance and 1.3 g N/day to grow 3 kg of fibre per year. The daily energy requirement for the same sheep is 5000 kJ for maintenance and 230 for wool growth. Thus, 16 percent of nitrogen intake and only 4½ percent of energy intake are required for wool growth (after McDonald *et al*. 1973).

Sheep in the tropics are mostly restricted to semi-arid regions. Therefore, the fulfilment of their nutritional requirements for maximum production depends on pasture availability which is limited by rainfall and stocking rate. However, wool production per animal can be as high in semi-arid regions as in high rainfall regions (Wilson 1978). This is attributable to the continuation of wool growth at the expense of body growth when feed intake is below requirements (Ferguson 1962).

CONCLUSION

The genetic potential of ruminants for production is never reached on even the best improved pastures.

In climates with evenly distributed rainfall feed supply is limited mainly by stocking rate. In regions with pronounced dry seasons, feed supply is limited by the extent of the dry season and stocking rate. The main limitation of pastures for animal production in seasonally dry regions is deficiency of green feed for much of the year (Mannetje & Ebersohn 1980). Even when green feed supply is not limiting, intake of digestible energy is insufficient to achieve the production potential. This is most marked for milk production and least for fibre production. However, meat and milk production per unit area and liveweight gain and fibre production per animal during the growing season from improved tropical pastures are comparable to those from temperate pastures. In addition, tropical pastures can be grazed throughout the year, whereas temperate pasture in many regions are out of production for several months due to inclement weather.

On a world scale many resources are devoted to increasing the genetic production potential of cattle. It is questionable whether this in itself will lead to any increase in animal production because the level of nutrition is insufficient to meet the growth potential of existing animals. However, this does not apply to sheep, because there is a definite need for genetic improvement to increase the rate of food to fibre conversion (Turner 1974). There is also a need for genetic manipulation to improve adaptation to tropical conditions and parasites for both cattle and sheep, where non-tropical breeds are predominant.

The possibility of improving the nutritional value of tropical pasture species depends on scientific research to develop plants which combine some of the characteristics of the best temperate species and adaptation to tropical conditions. Whether this goal can be achieved is not predictable and any progress towards it will of necessity be slow and require a high input of research resources. However, because of the lack of pasture development in the tropics, it is undoubtedly more important to direct resources towards increasing the area of improved pastures as this will have a greater effect on producing more human food.

REFERENCES

Alexander, G.I.; Carraill, R.M. (1973) The beef cattle industry. In: The pastoral industries of Australia. Practice and technology of sheep and cattle production. Editors G. Alexander and O.B. Williams. Sydney, Sydney University Press, pp. 143-170.

Beattie, W.A. (1956) A survey of the beef cattle industry of Australia. CSIRO, Australia, Bulletin No. 278.

Blunt, C.G. (1978) Production from steers grazing nitrogen fertilized irrigated pangola grass in the Ord Valley. Tropical Grasslands 12, 90-96.

Bowen, E.J.; Rickert, K.G. (1979) Beef production from native pastures oversown to fine-stem stylo in the Burnett region of south-eastern Queensland. Australian Journal of Experimental Agriculture and Animal Husbandry 19, 140-149.

Brown, G.D.; Williams, O.B. (1970) Geographical distribution of the productivity of sheep in Australia. Journal of the Australian Institute of Agricultural Science 36, 182-198.

Bryan, W.W. (1970) Tropical and sub-tropical forests and heaths. In: Australian grasslands. Editor R.M. Moore. Canberra, Australian National University Press, pp. 101-111.

Bryan, W.W.; Evans, T.R. (1971) A comparison of beef production from nitrogen fertilized pangola grass and from a pangola grass-legume pasture. Tropical Grasslands 5, 89-98.

Bureau of Agricultural Economics (1975) The Australian beef cattle industry. Submissions to the Industries Assistance Commission Inquiry. Industry Economics Monograph No. 13. Canberra, Australian Government Publishing Service.

Burrows, W.H. (1980) Range management in the dry tropics with special reference to Queensland. Tropical Grasslands 14, 281-287.

Chadhokar, P.A. (1977) Establishment of stylo (*Stylosanthes guianensis*) in kunai (*Imperata cylindrica*) pastures and its effect on dry matter yield and animal production in the Markham Valley, Papua New Guinea. Tropical Grasslands 11, 263-272.

Chalmers, A.W. (1976) Advantages and disadvantages of nomadism with particular reference to the Republic of the Sudan. In: Beef cattle production in developing countries. Editor A.J. Smith. Edinburgh, University of Edinburgh, pp. 388-397.

Chalmers, M.I. (1974) Nutrition. In: The husbandry and health of the domestic buffalo. Editor W. Ross Cockrill. Rome, FAO, pp. 167-194.

Chutikul, K. (1975) Ruminant (buffalo) nutrition. In: The Asiatic water buffalo. Editors de Guzman, M.R.; Allo, A.V. Taiwan, Food and Fertilizer Technology Centre.

Coaldrake, J.E. (1970) The brigalow. In: Australian grasslands. Editor R.M. Moore. Canberra, Australian National University Press, pp. 123-140.

Devendra, C. (1978) The digestive efficiency of goats. World Review of Animal Production 14, 9-22.

Devendra, C. (1979) Goat and sheep production in the ASEAN region. World Animal Review 32, 33-41.

Devendra, C. (1980) The potential value of grasses and crop by-products for feeding buffaloes in Asia. In: Buffalo production for small farms. Editor W.H. Tetangco. Food and Fertilizer Technology Centre, Taiwan, Book Series No. 15.

Devendra, C.; Lee, K.C. (1975) Studies on Kedah-Kelantan cattle. 1. Effect of improved nutrition on growth. MARDI Research Bulletin 3, 68-86.

Dirven, J.G.P. (1965) Milk production on grassland in Surinam. Proceedings of the 9th International Grassland Congress, Sao Paulo, Brazil, 995-999.

Dirven, J.G.P. (1977) Beef and milk production from cultivated tropical pastures. A comparison with temperate pastures. Stikstof 20, 2-15.

Eng, P.K.; Mannetje, L.'t; Chen, C.P. (1978) Effects of phosphorus and stocking rate on pasture and animal production from a guinea grass-legume pasture in Johore, Malaysia. 2. Animal liveweight change. Tropical Grasslands 12, 198-207.

Evans, T.R. (1970) Some factors affecting beef production from subtropical pastures in the coastal lowlands of southeast Queensland. Proceedings of the 11th International Grassland Congress, Surfers Paradise, Australia, 803-807.

Falvey, J.L. (1981) Imperata cylindrica and animal production in south-east Asia: a review. Tropical Grasslands 15, 52-56.

FAO (1978) Production yearbook Vol. 32. Rome, Food and Agricultural Organization of the United Nations, 287 pp.

Ferguson, K.A. (1962) The efficiency of conversion of feed into wool. In: The simple fleece. Editor A. Barnard. Parkville, Australia, Melbourne University Press, pp. 145-154.

French, M.H. (1957) Nutritional value of tropical grasses and fodders. Herbage Abstracts 27, 1-9.

Gall, C. (1975) Milk production from sheep and goats. World Animal Production 13, 1-8.

Gillard, P. (1979) Improvement of native pasture with Townsville stylo in the dry tropics of sub-coastal northern Queensland. Australian Journal of Experimental Agriculture and Animal Husbandry 19, 325-336.

Gonzalez Padilla, E. (1980) Sistema bioeconomico de producion animal en las zonas aridas de America Latina In: Proceedings of the 4th World Conference of Animal Production, Buenos Aires, 3-33.

Grof, B.; Harding, W.A.T. (1970) Dry matter yields and animal production of guinea grass (Panicum maximum) on the humid tropical coast of north Queensland. Tropical Grasslands 4, 85-95.

Henzell, E.F.; Mannetje, L.'t (1980) Grassland and forage research in tropical and subtropical climates. In: Perspectives in world agriculture. Farnham Royal, U.K. Commonwealth Agricultural Bureaux, England pp. 485-532.

Henzell, E.F.; Peake, D.C.I.; Mannetje, L.'t; Stirk, G.B. (1975) Nitrogen response of pasture grasses on duplex soils formed from granite in southern Queensland. Australian Journal of Experimental Agriculture and Animal Husbandry 15, 498-507.

Humphreys, L.R. (1978) Tropical pastures and fodder crops. London, Longman, 135 pp.

Ichhponani, J.S.; Gill, R.S.; Makkar, G.S.; Ranjhan, S.K. (1977) Work done on buffalo nutrition in India - a review. Indian Journal of Dairy Science 30, 173-91.

Jasiorowski, H.A. (1975) Intensive systems of animal production. In: Proceedings of the 3rd World Conference on Animal Production, Melbourne. Editor R.L. Reid. Sydney, Sydney University Press, pp. 369-386.

Jones, R.K. (1974) A study of the phosphorus responses of a wide range of accessions from the genus Stylosanthes. Australian Journal of Agricultural Research 25, 847-862.

Knoess, K.H. (1977) The camel as a meat and milk animal. World Animal Review 22, 39-44.

Lenné, J.M.; Turner, J.W.; Cameron, D.F. (1980) Resistance to diseases and pests of tropical pasture plants. Tropical Grasslands 14, 146-152.

Magadan, P.B.; Javier, E.Q.; Madamba, J.C. (1974) Beef production on native (Imperata cylindrica (L.) Beauv.) and para grass (Brachiaria mutica (Forsk.) Stapf) pastures in the Philippines. Proceedings of the 12th International Grassland Congress, Moscow, 3, 883-892.

Mahadevan, P. (1982) Pastures and animal production. In: Nutritional limits to animal production from pastures. Editor J.B. Hacker. Farnham Royal, U.K., Commonwealth Agricultural Bureaux, pp. 1-17.

Mannetje, L.'t (1972) The effect of some management practices on pasture production. Tropical Grasslands 6, 260-263.

Mannetje, L.'t (1974) Relations between pasture attributes and liveweight gains on a subtropical pasture. Proceedings of the 12th International Grassland Congress, Moscow, 3, 299-304.
Mannetje, L.'t (1978) The role of improved pastures for beef production in the tropics. Tropical Grasslands 12, 1-9.
Mannetje, L.'t; Coates, D.B. (1976) Effects of pasture improvement on reproduction and pre-weaning growth of Hereford cattle in central sub-coastal Queensland. Proceedings of the Australian Society of Animal Production 11, 257-260.
Mannetje, L.'t; Durand, M.R.E.; Isbell, R.F.; Sturtz, J.D. (1976) Options for improved efficiency in the beef industry of non-arid tropical Australia. Tropical Grasslands 10, 151-164.
Mannetje, L.'t; Ebersohn, J.P. (1980) Relations between sward characteristics and animal production. Tropical Grasslands 14, 273-280.
Mannetje, L.'t; Shaw, N.H. (1972) Nitrogen fertilizer responses of a Heteropogon contortus and a Paspalum plicatulum pasture in relation to rainfall in central coastal Queensland. Australian Journal of Experimental Agriculture and Animal Husbandry 12, 28-35.
Mason, I.L. (1974) Species types and breeds. In: The husbandry and health of the domestic buffalo. Editor W. Ross Cockrill. Rome, FAO, pp. 1-47.
Mellor, W.; Hibberd, M.J.; Grof, B. (1973) Performance of Kennedy ruzi grass on the wet tropical coast of Queensland. Queensland Journal of Agricultural and Animal Sciences 30, 53-56.
Minson, D.J.; Stobbs, T.H.; Hegarty, M.P.; Playne, M.J. (1976) Measuring the nutritive value of pasture plants. In: Tropical pasture research - principles and methods. Editors N.H. Shaw and W.W. Bryan. Farnham Royal, U.K., Commonwealth Agricultural Bureaux, pp. 308-337.
Moran, J.B. (1980) The performance of Indonesian beef breeds under traditional and improved management systems. First Asian-Australian Animal Science Congress, Kuala Lumpur, Sept. 1980.
Moran, J.B. (1981) Aspects of nitrogen utilization in Asiatic water buffalo and zebu cattle. Journal of Agriculture Science, Cambridge (in press).
Moran, J.B.; Norton, B.W.; Nolan, J.V. (1979) The intake, digestibility and utilization of a low-quality roughage by Brahman cross, buffalo, banteng and shorthorn steers. Australian Journal of Agricultural Research 30, 333-340.
Morrison, F.B. (1956) Feeds and Feedings. Ithaca, New York: The Morrison Publishing Company, 22nd Edition, 1165 pp.
McDonald, P.; Edwards, R.A.; Greenhalgh, J.F.D. (1973) Animal Nutrition. 2nd Ed. Edinburgh, Oliver C. Boyd, 479 pp.
McIvor, J.G.; Jones, R.J.; Gardener, C.J.; Winter, W.H. (1981) The development of legume-based pastures for beef production in dry tropical areas of northern Australia. Proceedings of the 14th International Grassland Congress, Lexington, Kentucky, (in press).
Norman, M.J.T. (1965) Seasonal performance of beef cattle on native pasture at Katherine, N.T. Australian Journal of Experimental Agriculture and Animal Husbandry 5, 227-231.
Ohler, J.G. (1969) Cattle under coconuts. Tropical Abstracts 24, 639-645.
Partridge, I.J. (1975) The improvement of mission grass (Pennisetum polystachyon) in Fiji by topdressing superphosphate and oversowing a legume (Macroptilium atropurpureum). Tropical Grasslands 9, 45-51.
Payne, W.J.A. (1970) Cattle production in the tropics. Vol. 1: Breeds and breeding. Tropical Agriculture Series. London, Longman, 336 pp.
Payne, W.J.A. (1976) Systems of beef production in developing countries. In: Beef cattle production in developing countries. Editor A.J. Smith. Edinburgh, Centre for Tropical Veterinary Medicine, pp. 118-131.
Perry, R.A. (1970) Arid shrublands and grasslands. In: Australian grasslands. Editor R.M. Moore. Canberra, Australian National University Press, pp. 246-259.
Preston, T.R.; Willis, M.B. (1970) Intensive beef production. Oxford, Pergamon Press, 544 pp.
Reid, J.T.; White, O.D. (1980) The role of available animals. In: Proceedings of the 4th World Conference on Animal Production Buenos Aires, Vol 1. pp. 162-165.
Reynolds, S.G. (1981) Grazing trials under coconuts in western Samoa. Tropical Grasslands 15, 3-10.
Rika, I.K.; Nitis, I.M.; Humphreys, L.R. (1981) Effects of stocking rate on cattle growth, pasture production and coconut yield in Bali. Tropical Grasslands 15, (in press).
Rivera Brenes, L.; Torres, Mas, J.; Arroyo, J.A. (1961) Response of guinea, pangola and coastal bermuda grass to different nitrogen fertilization levels under irrigation in the Lajas Valley of Puerto Rico. Journal of Agriculture, University of Puerto Rico 45, 123-146.
Salette, J.E. (1970) Nitrogen use and intensive management of grasses in the wet tropics. Proceedings of the 11th International Grassland Congress, Surfers Paradise, Australia pp. 404-407.
Santhirasegaram, K. (1966) The effects of pasture on the yield of coconuts. Journal of the Agricultural Society, Trinidad, 66, 183-193.
Shaw, N.H. (1961) Increased beef production from Townsville lucerne (Stylosanthes sundaica Taub.) in the spear grass pastures of central coastal Queensland. Australian Journal of Experimental Agriculture and Animal Husbandry 1, 73-80.

Shaw, N.H.; Andrew, C.S. (1979) Superphosphate and stocking rate effects on a native pasture oversown with Stylosanthes humilis in central coastal Queensland. 4. Phosphate and potassium sufficiency. Australian Journal of Experimental Agriculture and Animal Husbandry 19, 426-436.

Shaw, N.H.; Mannetje, L.'t (1970) Studies on a speargrass pasture in central coastal Queensland - the effect of fertilizer, stocking rate, and oversowing with Stylosanthes humilis on beef production and botanical composition. Tropical Grasslands 4, 43-56.

Skerman, P.J. (1975) Nomadism overseas, and cattle management in northern Australia. In: Proceedings of the 3rd World Conference on Animal Production. Editor R.L. Reid. Sydney, Sydney University Press, pp. 264-267.

Skovlin, J.M.; Williamson, D.L. (1978) Bush control and associated tsetse fly problems of rangeland development on the coastal plain of East Africa. Proceedings of the 1st International Rangeland Congress, Denver, USA, 581-583.

Squires, V.R. (1978) The potential for animal production in the arid zone. Proceedings of the Australian Society of Animal Production 12, 75-85.

Steel, R.J.H.; Humphreys, L.R. (1974) Growth and phosphorus response of some pasture legumes sown under coconuts in Bali. Tropical Grasslands 8, 171-178.

Stobbs, T.H. (1971) Quality of pasture and forage crops for dairy production in the tropical regions of Australia. 1. Review of the literature. Tropical Grasslands 5, 159-170.

Stobbs, T.H. (1976) Beef production from sown and planted pastures in the tropics. In: Beef cattle production in developing countries. Editor A.J. Smith. Edinburgh, Centre for Tropical Veterinary Medicine, University of Edinburgh, pp. 164-183.

Tothill, J.C.; Jones, R.M. (1977) Stability in sown and oversown Siratro pastures. Tropical Grasslands 11, 55-65.

Troll, C. (1966) Seasonal climates of the earth. The seasonal course of natural phenomena in the different climatic zones of the earth. In: World maps of climatology. Editors E. Rodenwaldt and H.J. Justaz, Berlin, Springer-Verlag, pp. 19-28.

Turner, H.G. (1975) The tropical adaptation of beef cattle. World Animal Review 13, 16-21.

Turner, H.N. (1974) Some aspects of sheep breeding in the tropics. World Animal Review 10, 31-37.

Van Niekerk, J.P.; Bester, F.V.; Lombard, H.P. (1978) Control of brush encroachment of aerial herbicide spraying. Proceedings of the 1st International Rangeland Congress, Denver, USA, 659-663.

Vercoe, J.E.; Frish, J.E. (1980) Animal breeding and genetics with particular reference to beef cattle in the tropics. Proceedings of the 4th World Conference on Animal Production, Buenos Aires, 452-463.

Vicente-Chandler, J.; Abruna, F.; Caro-Costas, R.; Figarella, J.; Silva, S.; Pearson, R.W. (1974) Intensive grassland management in the humid tropics of Puerto Rico. Bulletin, University of Puerto Rico. Agricultural Experiment Station, No. 233. 164 pp.

Vicente-Chandler, J.; Caro-Costas, R., Pearson, R.W., Abruna, F., Figarella, J.; Silva, S. (1964) The intensive management of tropical forages in Puerto Rico. Bulletin, University of Puerto Rico. Agricultural Experiment Station, No. 187. 152 pp.

Walker, J.; Moore, R.M.; Robertson, J.A. (1972) Herbage response to tree and shrub thinning in Eucalyptus populnea shrub woodlands. Australian Journal of Agricultural Research 23, 405-410.

Whiteman, P.C. (1980) Tropical pastures science. New York, Oxford University Press, 392 pp.

Whyte, R.O. (1974) Tropical grazing lands: Communities and constituent species. The Hague, W. Junk. 222 pp.

Wilson, A.D. (1978) Future resource management in the arid rangelands of Australia. Journal of the Australian Institute of Agricultural Science 44, 157-165.

Wilson, J.R.; Minson, D.J. (1980) Prospects for improving the digestibility and intake of tropical grasses. Tropical Grasslands 14, 253-259.

PART 3

LIMITATIONS CAUSED BY CHEMICAL COMPOSITION AND DIGESTIBILITY

DIFFERENCES BETWEEN SPECIES IN FORAGE QUALITY

B.W. NORTON

Department of Agriculture, University of Queensland, St. Lucia, Queensland 4067, Australia.

ABSTRACT

The geographic origin of plant species commonly used as forages for animals is presented, and comparisons in quality made between temperate and tropical legumes and grasses. Morphological and anatomical differences are discussed, together with aspects of the biochemistry of carbon fixation (C_3 *versus* C_4) and protein metabolism in both legumes and grasses. The paper deals with the differences in chemical composition between plant species (protein, minerals, soluble carbohydrate, cell walls) and the effect these differences have on forage digestibility. The concentration, intracellular distribution and nature of the various chemical constituents is discussed together with the effects of plant maturity and fertilizer application on the various fractions. The implications of these species differences on productivity of ruminants are considered.

INTRODUCTION

Grazing animals throughout the world have access to a great diversity of plant species from which to select a diet. Within any sward forage quality varies not only with genus, species and cultivar, but also with different plant parts, stage of maturity, soil fertility and with local and seasonal conditions. Pasture quality may be defined in terms of the level of animal production sustained, and is affected by soil fertility, animal management, and plant species.

The value of sowing high quality species was first demonstrated in studies at Cockle Park, Northumberland in 1904 where the use of phosphatic fertilizers and white clover markedly improved the growth of grazing sheep. Since then, the quest for superior species and strains of grasses and legumes has led to major increases in animal productivity from pasture, although only 40 of the 10,000 grass species (Poaceae) and 30 of the 11,000 legume species (Papilionaceae) are presently used to any significant degree as sown pasture species. In the legume family Mimosaceae, only one species (*Leucaena leucocephala*) is potentially useful. The temperate grasses and legumes have largely been selected from the indigenous flora of the northern Eurasian and Mediterranean regions, with some 24 species (10 genera) of grasses of the sub-family Festucoideae and about 20 legume species, mainly from the genera *Trifolium*, *Medicago* and *Lotus* being important. The cultivated tropical species are native to eastern and southern Africa (10 grass and 5 legume genera) and from the central and southern American regions (4 grass and 7 legume genera). The grasses from these regions are mainly from sub-families Panicoideae and Chloridoideae, and the legumes are represented by many different genera.

This paper describes some of the plant factors that are responsible for differences in quality between legumes and grasses of temperate and tropical origin. Differences in the morphology, anatomy and biochemistry of pasture plants will be discussed together with the effects of plant species, stage of maturity and fertilizer on chemical components in plant cell contents and cell walls. These factors operate by their effects on voluntary intake, digestibility and efficiency of utilization of absorbed nutrients.

PLANT MORPHOLOGY

Morphological and physiological differences between species, such as growth habit, perenniality, proportions and distribution of leaf and stem and flowering behaviour, have significant effects on both the quantity and quality of forage available to grazing animals. Characteristics such as rate of tiller formation and growth in tufted grasses, stolon development in prostrate plants and deep rootedness and crown burial in legumes provide selective resistance to the effects of grazing and treading by grazing animals. The superiority of the ryegrasses, (*Lolium* spp.) is related to their tolerance of frequent grazing or cutting. The susceptibility of many tropical legumes to close grazing is associated with a twining habit, weak ability to root at nodes of trailing stems and delayed recovery from defoliation because grazing removes the youngest leaves and apical meristems.

Stobbs (1973) has suggested that the low animal production from tropical compared with temperate pastures may be caused by the more erect growth habit of most tropical grasses and legumes. The low density (kg DM/ha cm) of leaf in these pastures appears to restrict harvestability and intake of pasture by grazing animals. When leaf density of the pasture is increased by the use of plant growth retardants, forage quality is improved (Stobbs 1973). The effects of sward structure on animal performance are reviewed later in this symposium (Hodgson 1982).

As grasses flower and mature there is a decline in forage quality caused by the translocation of soluble carbohydrates from stem and leaves to the inflorescence, an increased content of lignified cell walls and a decrease in the ratio of leaf to stem. Flowering in legumes is not usually associated with large changes in nutritive value, despite leaf loss through senescence. The initiation of flowering in both grasses and legumes is often determined by day length and temperature regimes. Evans (1964) has grouped grasses into the temperate (sub-family Festucoideae) species which are either long day plants or indifferent to day length and the tropical grasses (sub-families Panicoideae and Chloridoideae) which are either short or intermediate day plants or are indifferent to day length. High night temperatures (12-18°C) inhibit flowering in long day (temperate) grasses irrespective of day length, but similar temperatures stimulate flowering in short day plants. Pasture plants in temperate and mediterranean environments are maintained in a vegetative (and high quality) state by environmental variables (vernalization and day length requirements). Tropical grasses grown in warm environments have higher growth rates and fewer environmental restraints to flowering and therefore usually progress to maturity and decline in quality rapidly. The relationship between environmental variables and forage quality will be discussed by Wilson (1982).

Leafiness in pasture plants is commonly associated with forage quality, because there is usually a positive correlation between leaf percentage in a given plant species and the protein and mineral composition and dry matter digestibility (Fagan & Jones 1924, Reid *et al.* 1959). However, differences *between* species in the proportions of leaf and stem are not necessarily indicative of differences in forage quality. It is often overlooked that at a young stage of growth, leaf in both temperate and tropical grasses is usually of lower digestibility than stem (Mowat *et al.* 1965, Hacker 1971). Minson *et al.* (1960a,b) have shown that despite higher leaf percentage, *Dactylis glomerata* is of lower digestibility than *Lolium perenne* due mainly to differences in leaf digestibility. Similarly Minson & Laredo (1972) have reported that, although there were large differences between tropical grass species in leaf percentage, these differences had no effect on either digestibility or voluntary intake of forage when fed to sheep. However, where comparisons were made between cultivars of *Panicum* spp., varieties with a high leaf percentage were consumed in greater quantities than those of similar digestibility with lower leaf contents (Minson 1971). The differences in intake between leaf and stem were more closely related to physical properties of the fraction (bulk density, surface area/g) than to differences in chemical composition. Ulyatt (1970) found that *Lolium perenne* was physically degraded more slowly than *L. perenne* x *L. multiflorum* in the rumen, despite similar structural carbohydrate contents, and these differences were reflected in animal performance. With tropical legumes, there are much larger differences in digestibility between leaf and stem than found with grasses, and the consumption of legume stems therefore results in greater penalties to animal production. Other aspects of plant morphology that influence forage quality have been discussed by Wilson & Minson (1980).

PLANT ANATOMY

It is only recently that variations between species in basic leaf anatomy, (which are related to biochemical and physiological differences between plants (Laetsch 1969, Walker & Crofts 1970)) have been related to their potential nutritive value for ruminants (Wilson & Minson 1980). Most tropical grasses have a C_4 pathway of photosynthesis (Hatch & Slack 1970) and are characterized by a specialized leaf anatomy (Kranz anatomy), the major distinguishing features being a radial arrangement of chlorenchyma cells (bundle sheath) around the vascular bundles, structural (granal or agranal) and size dimorphism of chloroplasts in the bundle sheath and surrounding mesophyll cells and the presence of more and larger mitochondria in the bundle sheath cells than found in mesophyll cells (Laetsch 1974).

Temperate grasses (sub-family Festucoideae) and both temperate and tropical legumes have C_3 pathways of photosynthesis and lack the above specialized features. Fig. 1 shows the cross sectional view of leaves from representative tropical and temperate legumes and grasses, and demonstrates the relatively high proportions of bundle sheath and vascular tissue and the low proportion of thin walled mesophyll cells in tropical grasses compared. The stems of tropical grasses also contain more vascular bundles than temperate grasses.

The bundle sheath cells vary in size and depth depending on species, have no intercellular air spaces and have thick suberized outer walls which render these leaves resistant to mechanical breakdown. The high tensile strength of tropical grass leaves has been associated with difficulties of harvesting leaf material by grazing cattle (Chacon 1976) and may be related to the longer rumination times recorded for animals grazing tropical pastures. The mesophyll cells in tropical grasses are more densely packed than those in temperate grasses (Carolin et al. 1973), and intercellular air spaces represent only 3-12 percent of leaf volume compared with 10-35 percent in temperate species. The lower surface area to weight ratio for both mesophyll and bundle sheath tissues restricts accessability of plant cells to microbial digestion in the rumen (Hanna et al. 1973), thereby decreasing the rate of digestion of the bundle sheath and enclosed vascular tissue (Akin & Burdick 1975). There is considerable variation in both the numbers and characteristics of vascular bundles in tropical grasses, and plant species or cultivars with high digestibility usually have either fewer vascular bundles in their leaf and stem tissue (Schank et al. 1973) or more rapid digestion of mesophyll and bundle sheath cells (Akin et al. 1974).

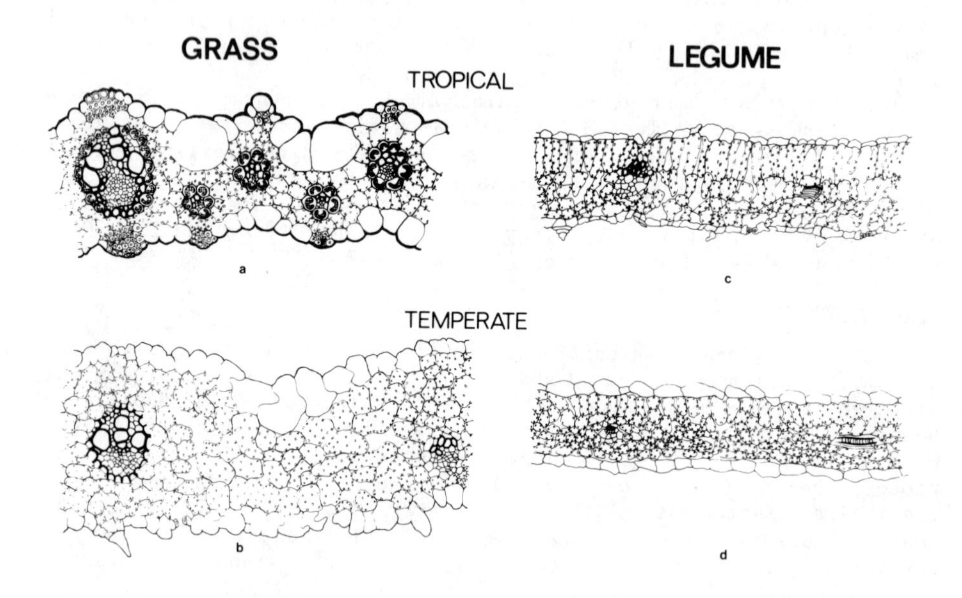

Fig.1. A cross sectional view of the leaves of tropical and temperate grasses and legumes (a. *Cenchrus ciliaris*, b. *Phalaris aquatica*, c. *Macroptilium atropurpureum*, d. *Trifolium repens*) (reproduced from Wilson & Minson 1980).

The high resistance offered to both mechanical and microbial degradation by the specialized leaf anatomy of tropical grasses may partly explain the longer retention time of tropical grass forage in the rumen, and the consequent lower voluntary intake by ruminants consuming these plants (Thornton & Minson 1973). The observation that the destruction of the physical properties of tropical grasses by grinding increases voluntary intake (Minson & Milford 1968) indicates that feed quality is determined by the physical properties of the forage in addition to its chemical composition.

PLANT BIOCHEMISTRY

The high potential yield (35-85 t/ha/year) of tropical grasses compared with temperate grasses (20-27 t/ha/year) (Roberts & Carbon

1969, Cooper 1970) is related to the high temperature at which they grow and to differences in the biochemical pathways through which carbon dioxide is fixed during photosynthesis. The major difference in the biochemical pathways for C_3 and C_4 plants that might influence quality is that in C_4 plants first photosynthetic products are oxaloacetic, malic and aspartic acids whereas in C_3 plant phosphoglyceric acid is produced.

The higher growth rates, nitrogen use efficiency and lower tissue protein contents of C_4 compared with C_3 plants have been related to differences in their pathways of carbon fixation (Brown 1978). RuDP carboxylase in C_3 plants represents about 50 percent of the soluble protein in the mesophyll cells, and its low activity is the major limiting step in photosynthetic carbon fixation. In C_4 plants RuDP carboxylase is restricted to the bundle sheath cells and concentrations are lower (20 percent soluble protein) than those found in mesophyll cells of C_3 species (Björkmann et al. 1976). Malic and aspartic acids, derived from PEP carboxylation, apparently increase intracellular CO_2 fixation through RuDP carboxylase. PEP carboxylase is found in relatively low concentrations in C_4 mesophyll cells but its higher activity results in high rates of CO_2 fixation per unit of cellular protein when compared to C_3 plants.

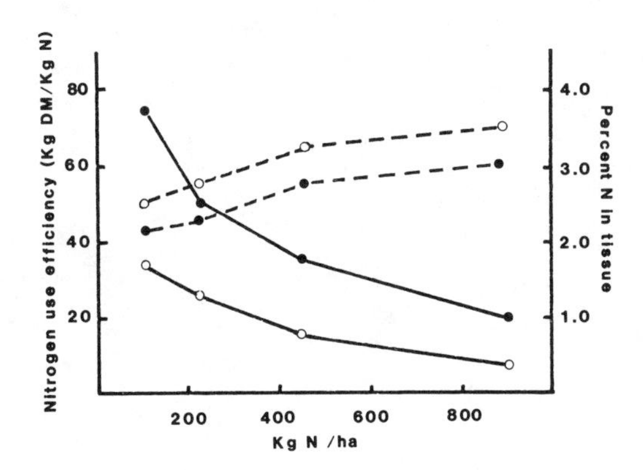

Fig. 2. The effects of N fertilizer on the efficiency of N-use (—) and tissue N contents (---) of C_3 (o) and C_4 (•) plants. (Adapted from Hallock et al. (1965)).

Fig. 2 shows the effects of nitrogen fertilizer application on the efficiency of nitrogen use (kg dry matter produced/kg N applied) and the nitrogen concentration in the tissues of C_3 and C_4 plants. It is clear that C_4 plants use nitrogen for dry matter accumulation with at least twice the efficiency of C_3 plants at all levels of fertilizer application. Higher efficiency in C_4 plants is associated with lower tissue nitrogen contents, but the rate at which concentration increases with fertilizer nitrogen is similar for C_3 and C_4 species. Colman & Lazenby (1970) have shown that at higher temperatures (23-35°C), the efficiency of nitrogen use by C_4 plants is higher than at low temperatures (13-24°C), but tissue nitrogen contents are also decreased by high temperatures. It may be concluded that the low protein contents often found in tropical grasses, even under nitrogen fertilized conditions, are an inherent characteristic of C_4 plant metabolism which is closely related to survival under conditions of low soil fertility.

CHEMICAL COMPOSITION OF PLANTS

High quality forages provide protein, minerals and energy in direct proportion to the animals requirement. Chemical composition, often used as an index of quality, has been expressed in many ways, ranging from the crude description provided by proximate analysis to the precise definition of specific compounds. The description of plant composition in terms of cell components and cell walls provides a separation of plant material into fractions of high (cell contents) and low (cell walls) availability to the ruminant animal (Van Soest 1967).

The chemical composition of plants varies not only within and between species, but is also affected by ontogenetic and environmental (light, temperature, water stress) factors. For the purposes of comparison, plant species have been grouped into tropical and temperate grasses and legumes, and the differences in the nature and concentrations of protein, minerals, non-structural carbohydrates (cell contents) and cell walls are presented. The effects of maturity and fertilizer application on these fractions, and the effects of these differences on forage quality will be discussed. Specific toxic factors will be reviewed by Hegarty (1982).

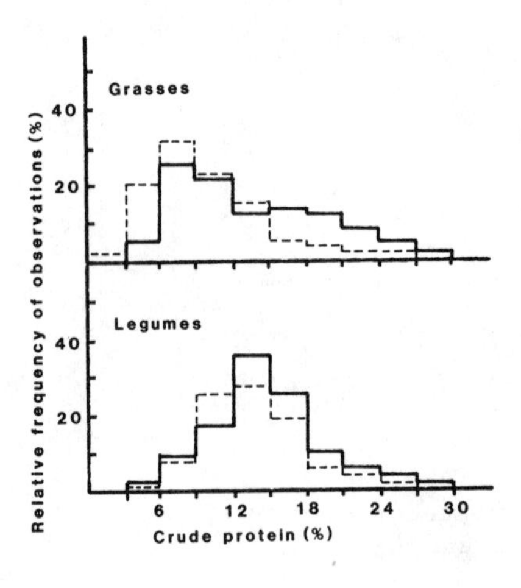

Fig. 3. The distribution of crude protein in tropical (---) and temperate (——) grasses and legumes (after Minson 1976).

PROTEIN CONTENT

Grasses usually contain less protein than legumes and this difference is shown in the frequency distribution of crude protein from 492 cuts of legume and 760 cuts of grass reported in the literature (Fig. 3a). Temperate and tropical legumes have a similar range in crude protein content, with few values falling below the 9 percent level considered minimal for ruminant requirements (Fig. 3b). Although the grasses covered a similar range to the legumes, 53 percent of all tropical grasses contained less than 9 percent crude protein compared with 32 percent of the temperate grasses. A minimum of 15 percent crude protein is required for lactation and growth and whilst most legumes and temperate grasses generally satisfy this requirement, less than 20 percent of the tropical grasses have crude protein contents above 15 percent. It is clear that where the protein requirement of ruminants is defined in terms of crude protein, animal production from well managed temperate pastures containing legumes will seldom be limited by protein deficiency (Johns 1955). However, the low protein content of tropical grasses poses a major limit to intensive forms of animal production, and the inclusion of tropical legumes in these pastures improves animal production by increasing protein availability for grazing stock. The crude protein content of grasses and legumes is significantly affected by the environmental conditions under which they are grown, and detailed aspects of these interactions have been reviewed by Lyttleton (1973).

Distribution and composition of plant proteins

The nitrogen in forage consists of soluble and insoluble proteins, amino acids, amides, ureides, nitrates and ammonia. The non protein component may represent up to 25 percent of total nitrogen, depending on the level of nitrogen fertilizer application and on the nutrient status of the plant (Hegarty & Peterson 1973). The effects of high levels of some of the non-protein amino acids and nitrates will be discussed by Hegarty (1982). The distribution of protein in the leaves of temperate grasses is shown in Table 1, and it is clear that chloroplast protein is the major protein found in the plant cell. Less is known about the distribution of these fractions in tropical grasses, although their different leaf anatomy and biochemical activity suggest that there will be lower concentrations of fraction I protein (RuDP carboxylase) in the bundle sheath cells than in mesophyll cells of temperate grasses. The amino acid composition of leaf proteins varies little between plant species (Lyttleton 1973), and does not change significantly with either increasing cell protein content after nitrogen fertilizer application (Goswami & Willcox 1969) or with declining protein content during maturity (Hodgson 1964). This constancy is not surprising since up to 50 percent of the cell protein is one enzyme (RuDP carboxylase). Leaf protein has high biological value, with relatively high lysine levels compared with seed proteins. Methionine is the first and isoleucine is the second amino acid in leaf proteins to limit animal growth (Brady 1976).

TABLE 1

The distribution of protein in the leaves of temperate grasses. (after Brady 1976).

Intracellular	Location	% of total protein
Chloroplast	membrane	30-45
	soluble	25-30
Mitochondria	membrane	3-4
	soluble	2-3
Cytoplasm	membrane	5-15
	soluble	15-20
Nucleus		1-2
Cell wall		1-2

Protein quality is usually considered to be less important in ruminants than in non-ruminants because extensive degradation of protein occurs in the rumen. However, the extent of dietary protein degradation varies with its source, and degradability in the rumen has been correlated with protein solubility (Buttery 1976). Aii & Stobbs (1980) have reported that stage of growth had little effect on protein solubility of 8 tropical grasses, but significant differences in protein solubility were found between grass and legume species and between different plant parts. These results are summarized in Table 2. Younger leaves in both legumes and grasses have higher proportions of soluble protein, and stem proteins are more soluble than leaf proteins. The low solubility of proteins in the leaves of *Desmodium* spp. has been related to a high tannin content in their leaves (McLeod 1974, Ford 1978) which leads to low rates of mineralization in soil (Vallis & Jones 1973) and possibly to higher weight gains of cattle grazing these legumes through protection of plant proteins from rumenal degradation (Jones 1972). The wide range of solubility found by Aii & Stobbs (1980) for different grasses and legumes suggests that forage proteins are protected to different extents from digestion in the rumen, and that forages with high levels of protein protection may be more productive than those with high proportions of readily degradable protein.

Plant maturity and protein content

In the vegetative stage of growth, protein levels in grasses are usually high, and it is only as the plant approaches maturity that low protein contents in both temperate and tropical grasses pose a major limitation to forage quality for grazing animals. With advancing maturity, the decline in protein content is slower in leaf than in stem, and protein content at maturity is determined by species differences in initial protein levels in vegetative tissue, the rate and extent of decline and the final proportions of leaf and stem in the mature plant. Few grasses maintain high protein levels throughout growth, the onset of flowering usually hastening the decline in protein content. *Pennisetum clandestinum* is a notable exception, in that protein levels remain at high levels throughout growth and at maturity (Milford & Haydock 1965, Colman & Holder 1968, Aii & Stobbs 1980). The inherently lower protein content of C_4 plants results in tropical grasses declining to lower levels at maturity than those found in temperate grasses, and where plants are severely water stressed, even lower protein contents are found. In the absence of water stress, the protein content of both temperate and tropical grasses declines at a similar rate after flowering (Lyttleton 1973).

The protein content of legumes declines only slowly with maturation, the older leaves showing the greatest decrease. The higher protein contents and their maintenance with maturity may be associated with the continuous supply of nitrogen available from rhizobial fixation, and variation between legume species in protein content also probably reflects the effectiveness of rhizobial nitrogen fixation under different environmental conditions.

TABLE 2

The solubility of protein in the leaves and stems of some tropical grasses and legumes (after Aii & Stobbs, 1980).

Species	Protein solubility (% total N)	
	Leaf	Stem
Grasses		
Setaria sphacelata cv. Kazungula	19.3	29.0
Digitaria decumbens cv. Pangola	24.4	22.7
Pennisetum clandestinum	24.0	66.0
Chloris gayana	29.7	48.2
Brachiaria mutica	33.5	53.0
Legumes		
Desmodium uncinatum	5.3	36.3
Desmodium intortum	7.6	15.9
Aeschynomene indica	21.0	48.5
Macroptilium atropurpureum cv. Siratro	40.8	52.9
Macrotyloma uniflorum	44.7	54.5

MINERAL CONTENT

The balance, concentration and nature of minerals in herbage varies with plant species, stage of growth and availability of minerals in the soil. Fertilizers may increase plant growth rate and mineral content, but species differences are still evident when optimal fertilizer requirements for growth are provided. The various factors affecting the mineral composition of herbage have been extensively reviewed (Whitehead 1966, Reid et al. 1970, Fleming 1973, Reid & Horvath 1980). The extent to which the relative proportions of elements in forages match microbial and animal requirements for growth will determine forage quality. A deficiency of any one of the seventeen elements considered essential for animals will limit digestion, absorption and utilization of all dietary components, as will toxic levels of minerals. These aspects are discussed in detail by Little (1982), and the following section only considers the range and species distribution of minerals that occur in tropical and temperate grasses and legumes.

The concencentration of minerals in legumes and grasses

A low content of minerals in plants may be caused by low availability in the soil, low genetic capacity for accumulation or by low requirements for growth. Alternatively, high and/or toxic levels may be the result of excessive availability in soil, genetic or physiological capacity for high rates of accumulation or be indicative of high requirements of growth. The wide range of mineral contents found in different plant species suggests all these factors are operative. Forage quality will only be affected when mineral contents of the forage fall outside the range required for optimal animal growth. Table 3 shows mean values for mineral content of tropical and temperate legumes and grasses (data taken from McDowell et al. 1974, ARC 1976, Minson 1977, 1982). Data in the table represent values from forages grown under a wide range of environmental conditions.

TABLE 3

Mean values for mineral content in the dry matter of tropical and temperate grasses and legumes. Values in parenthesis are the number of samples.

	Temperate		Tropical	
	Grass	Legume	Grass	Legume
Phosphorus (%)	0.33 (400)	0.36 (320)	0.22 (586)	0.26 (165)
Calcium (%)	0.59 (428)	1.86 (291)	0.40 (390)	1.21 (154)
Magnesium (%)	0.18 (335)	0.29 (193)	0.36 (280)	0.40 (48)
Sodium (%)	0.23 (318)	0.19 (121)	0.26 (192)	0.07 (40)
Copper (ppm)	6 (127)	12 (93)	15 (94)	10 (17)
Zinc (ppm)	32 (31)	55 (34)	36 (119)	42 (7)
Cobalt (ppm)	0.20 (111)	0.42 (21)	0.16 (45)	0.07 (3)

Phosphorus

Tropical grasses and legumes were generally lower in phosphorus than temperate species, although the range was similar (Fig. 4a). However the differences between legumes and grasses growing in the same climate were small. Few temperate legumes (13 percent) were below the required dietary level of 0.24 percent (ARC 1980), but 44 percent of temperate grasses were less than this value (Fig. 4b). In the tropical grasses, 63 percent were below requirement and 52 percent of the tropical legumes were also below 0.24 percent phosphorus (Fig. 4c). Little (1980) has reported that 0.12 percent phosphorus is adequate for growing stock. If this is the case few temperate species would be judged deficient in phosphorus, and only 7-13 percent of the tropical legumes and grasses would be below this value. Tropical legume species most often found to be deficient are *Stylosanthes humilis* (Little 1968, Fisher 1969), *S. scabra* cv. Fitzroy and *S. hamata* cv. Verano (McIvor 1979).

The phosphorus content of plant tissue declines with advancing maturity, and the rate and extent of decline varies with species. All species were low in phosphorus at a mature stage of growth despite phosphorus fertilizer application. Tropical grass species considered deficient for animals were most often grown under conditions of low soil fertility and were also at a mature stage of growth.

Calcium

Legumes had higher calcium contents than grasses and both tropical legumes and grasses contained less calcium than temperate legumes and grasses (Table 3, Figs. 4c and d). Where a dietary level of 0.43 percent calcium is required (Little 1982), 25 percent of the temperate grasses and 65 percent of the tropical grasses would be deficient for animal growth. Few temperate legumes were below this value and only 23 percent of the tropical legumes could be considered deficient. However, calcium in plant tissue may be rendered unavailable by formation of insoluble salts with oxalic acid. The relatively low calcium levels in tropical grasses and the high levels of oxalic acid (0.2-7.8 percent) found in some tropical species (Jones *et al*. 1970, Dijkshoorn 1973) suggest that calcium availability may limit forage quality.

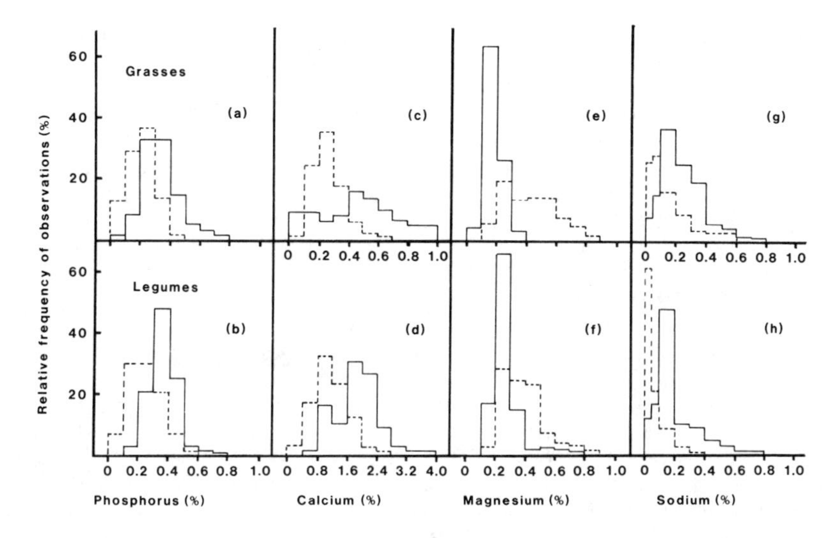

Fig. 4. The frequency distributions of concentrations of phosphorus, calcium, magnesium and sodium in tropical (---) and temperate (——) grasses and legumes.

The calcium content of herbage generally declines with maturity in both tropical and temperate grasses but a consistent pattern is not always seen in tropical species (Fleming 1973). The lower calcium contents of tropical compared with temperate legumes may be related to the higher temperature conditions during growth. Smith (1971) found that Medicago sativa grown at high temperatures contained less calcium than when grown at lower temperatures. Also, where rainfall is high, similar trends are seen in both calcium and phosphorus contents of forages (Daniel & Harper 1935), and calcium should be considered a mineral likely to be deficient in grasses grown under tropical conditions.

Magnesium

Legumes contained higher magnesium contents than grasses, and tropical legumes consistently had higher concentrations than temperate legumes (Fig. 4e and f). The magnesium content of temperate grasses was only half that found in tropical grasses (Table 4) and 35 percent of the temperate grasses contained less than dietary requirement (0.15 percent, Little 1982) whereas only 18 percent of the tropical grasses were below this value. In M. sativa high temperatures during growth increase magnesium concentration (Smith 1971) and the observed differences between tropical and temperate legumes in magnesium content may be an effect of environmental growth temperature. It is not known whether this mechanism operates also in grasses, although the total content of divalent cations (calcium and magnesium) is similar in tropical and temperate grasses. Magnesium deficiency appears to be more likely to occur in animals grazing temperate than tropical pastures.

Sodium

The sodium levels in forages varied from 0.005 to 2.13 percent, but the frequency distribution curves shown in Figs. 4g and h only include forages with 0.8 percent sodium or less. A significant proportion of both tropical grasses (27 percent) and tropical legumes (62 percent) contained less than 0.05 percent sodium, whereas few temperate species were below the level of 0.10 percent considered adequate for lactating cattle (Little 1982). Both temperate and tropical grasses may be categorized as high or low sodium accumulators; species in the former group usually show large varietal differences in sodium content (ap Griffiths & Walters 1966, Hacker 1974). Among the temperate grasses *Lolium perenne* has a high potential for sodium accumulation, whereas *L. multiflorum* and *Phleum pratense* have low potential. In a study of *Setaria* species and cultivars, Hacker (1974) found 15 with high and 27 low sodium accumulaters when grown on the same soil type. Low potential cultivars (low sodium content) increased potassium concentration during oxalic acid accumulation, whilst cultivars with high potential for sodium accumulation increased sodium content rather than potassium during oxalate accumulation. *Brachiaria* spp., *Panicum* spp. and varieties of *Chloris gayana* also have two distinct populations for high and low potential for sodium uptake, and there is evidence that sodium content of plants is under genetic control (see Hacker 1982). Andrew & Robins (1971) found a significant increase in the sodium content of some tropical grasses as the level of phosphorus fertilizer increased, although similar responses were not seen in tropical legumes, even where sodium was applied with the phosphorus (Andrew & Robins 1969). The low sodium contents of both tropical legumes and grasses may limit forage quality where animals are grazed for long periods on these pastures.

Trace elements

Relatively little information is available on the trace element content of forages, particularly for tropical legumes. Many soils, particularly in the tropics, are deficient in many of the trace elements required for plant growth, and also in cobalt and selenium which are not usually considered essential for plant growth but are needed for animal growth. Legumes contain more copper than grasses, although 65 percent of the tropical legumes and 67 percent of the temperate legumes contained less than the recommended dietary requirement (14 ppm, Little 1982). All temperate grasses were below this requirement, and a percentage of the tropical grasses contained less than 15 ppm copper.

Insufficient data are available for a comparison of tropical and temperate legumes for trace element concentration. Legumes were higher in zinc content than grasses, and their was little difference between temperate and tropical grasses. Few tropical and temperate legumes and grasses contained less than the required 20 ppm (Little 1982) and hence zinc deficiency in grazing animals.

The few values available for cobalt in tropical legumes are very low compared with other forages and with animal requirements (0.11 ppm, Little 1982). *Stylosanthes guianensis* cv. Schofield and *Macroptilium atropurpureum* cv. Siratro have recently been found to have low cobalt contents (0.005 to 0.048 ppm) (Winter *et al*. 1977, Nicol & Smith 1981), and responses to cobalt supplementation have been obtained in animals grazing *Panicum maximum* and *Digitaria decumbens*

(Mannetje *et al.* 1976, Norton & Hales 1976, Winter *et al.* 1977). Although more tropical grasses were deficient (46 percent less than requirement) than were temperate grasses, there is insufficient information on not only cobalt, but also the other trace elements, to ascribe species differences to high or low levels of these essential elements.

The effect of fertilizer on mineral content

The concentration of minerals in plant tissues increases when the rate of uptake by the roots exceeds tissue growth, and differences between plant species in mineral content are a consequence of both genetic and environmental processes. Where nutrient availability is limiting, fertilizer application improves forage quality by stimulating the growth of young digestible leaf and stem and by increasing plant mineral content. The effects of fertilizer on forage mineral content are complex, often causing either antagonistic or synergistic effects on elements other than those applied. This subject has been comprehensively reviewed by others (Fleming 1973, Reid & Horvath 1980) and will not be considered further.

NON-STRUCTURAL CARBOHYDRATES

Glucose, fructose and sucrose and the storage polysaccharides starch and fructosans are the major 'soluble' carbohydrates found in plant cells. Soluble carbohydrates accumulate in plant tissues when the rate of formation during photosynthesis exceeds the quantity required for growth and respiration (Brown & Blaser 1968). Accumulation will therefore depend on plant species and on those environmental factors (light, temperature, nutrient status) that affect plant growth. The water soluble sugars, glucose, fructose and sucrose, and fructosan are a readily available source of fermentable energy for microorganisms in the rumen, and forages with high levels of soluble carbohydrates are usually highly digestible. Starch has a variable solubility in water depending on its amylopectin content, and may be less easily digested in the rumen. Starch accumulated in the leaves and stems of legumes and tropical grasses is largely insoluble in water (Akazawa 1965, Smith 1972b). For silage making a minimum of 10 percent soluble carbohydrate is required for effective lactic acid formation, and the low concentration of soluble carbohydrates often found in tropical grasses leads to poor silage fermentation.

Species differences

Plant species may be grouped according to the nature of the storage carbohydrates. Legumes (tropical and temperate) and tropical grasses (C_4 plants) characteristically accumulate starch and sucrose. Legumes store starch in both leaves and stems but tropical grasses store high concentrations of starch in their leaves. Temperate grasses (C_3 plants) accumulate sucrose, and fructosans rather than starch, mainly in the stems. Long chain fructosans are characteristic of the tribe Aveneae, short chain fructosans are found in the tribe Hordeae whilst in the tribe Festuceae both long and short chain fructosans are stored (Smith 1968). There is no evidence to suggest that these differences affect forage quality.

When grown in the same cool environment temperate grasses and legumes have higher concentrations of soluble carbohydrates than do tropical grasses and legumes, but when grown in warm environments, soluble carbohydrate concentration is low in all groups (Smith 1972a,

Noble & Lowe 1974). In the sub-tropics soluble carbohydrate content of temperate grasses and legumes increases during winter, in contrast to tropical grasses and legumes in which concentrations remain low (Noble & Lowe 1974).

With increasing maturity, the soluble carbohydrate content of grasses increases with increased stem content. Different species reach peak carbohydrate contents at different stages of the growth period, and total levels of carbohydrates in tissues vary with species (Smith 1972b). Few consistent changes occur in the carbohydrate content of legumes during growth. The application of nitrogen fertilizer usually decreases carbohydrate content in grasses, due mainly to an increased requirement of reserves for tissue growth. Where nutrient deficiencies occur, photosynthesis exceeds growth and carbohydrates will accumulate in plant tissues.

STRUCTURAL CARBOHYDRATES

The nature and concentrations of structural carbohydrates in plant cell walls are major determinants of forage quality. Cell walls form 30 to 80 percent of plant dry matter and vary as a source of energy. Structural carbohydrates in higher plants are represented by the polysaccharides cellulose (β 1→4 linked D-glucose), hemicelluloses (β 1→4 linked xylopyranoses, β 1→4 linked D-glucose and D-mannose) and the pectic substances (α 1→4 linked D-galacturonic acids, galactans, arabans). The chemistry and synthesis of these polysaccharides in plants have been reviewed by Bailey (1973). Cell walls also contain tannins, protein, minerals and the phenolic polymer lignin. These minor cell wall components affect both cellulose and hemicellulose digestion in the rumen. The following discussion explores some of the major factors affecting the digestion of cell walls and their relationship to forage quality.

Species differences

Cell wall content

The cell wall content of leaves is usually lower than that of stem, with differences between leaf and stem being greater in legumes. Cell wall content of both legumes and grasses increases continuously during growth to maturity, although grasses have higher contents at maturity when compared with legumes. Since progress to maturity increases the stem content of forages, total cell wall content in plants at maturity will vary with the proportions of leaf and stem present. The high cell wall content of tropical grasses is related to higher proportions of vascular tissue associated with the specialized anatomy of these C_4 plants. However, Ford *et al.* (1979), in a study of 13 tropical and 11 temperate grasses, found a two-fold difference in leaf cell wall content of different species grown at the same temperature. With increased temperature, the cell wall content of tropical grasses decreased and that of temperate grasses increased, so that under higher temperature conditions there was little difference between the groups in mean cell wall content. A small decrease in cell wall content has been observed in both temperate and tropical grasses after nitrogen application (Waite 1970, Ford & Williams 1973), but applying nitrogen to tropical grasses does not appear to improve forage digestibilty (Minson 1973).

There appear to be significant differences between plant species in cell wall content, but meaningful comparisons between species are often confounded not only by morphological and anatomical differences between plants, but also by environmental factors which further modify plant cell wall content.

Crude fibre (ligno-cellulose) has been the most common fraction used to designate the structural carbohydrate content of herbage, although neither hemicellulose nor pectins are included in this fraction. Both tropical legumes and grasses are higher in crude fibre content than are temperate species, but the crude fibre content of temperate legumes and grasses are similar (Fig. 5). Mean values for the concentrations of crude fibre are: tropical grasses, 33.9 percent; tropical legumes, 30.3 percent; temperate grasses, 26.0 percent and temperate legumes, 25.3 percent.

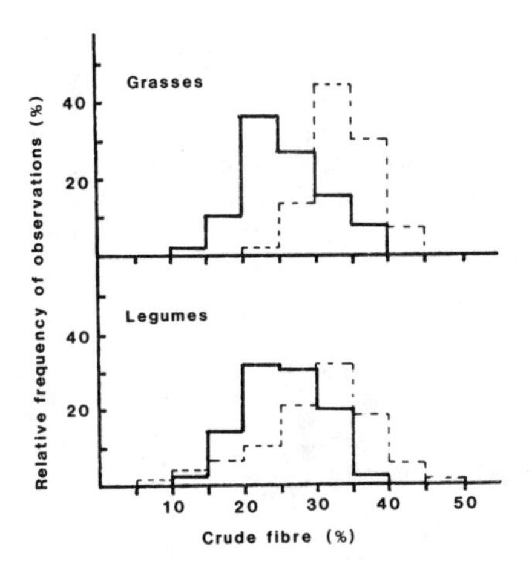

Fig. 5. The frequency distribution of crude fibre in tropical (---) and temperate (——) grasses and legumes.

Cell wall composition and digestiblity

Cellulose and hemicellulose are the major polysaccharides in the cell wall. The ratio of cellulose to hemicellulose is lower in temperate grasses (1:0.7→0.9) than in tropical grasses (1:1.0→1.2). Temperate and tropical grasses grown in the same warm environment have similar cellulose contents, but the tropical grasses have higher concentrations of hemicellulose than do temperate grasses. The rates of cellulose and hemicellulose breakdown in grasses are similar during digestion (Minson 1971, McLeod & Minson 1974, Ulyatt & Egan 1979) and although the extent of digestion varies with species, the ratio of cellulose to hemicellulose does not seem to affect cell wall digestibility.

Temperate legumes contain less cellulose and hemicellulose than do temperate grasses, and also have higher ratios of cellulose to hemicellulose (1:0.3→0.6) (Bailey 1973, Ulyatt & Egan 1979). Little information is available on the structural carbohydrates in the cell walls of different tropical legumes species although pectin concentration appears to be higher than in temperate legumes. Values from 14-20 percent have been found in *Stylosanthes* spp. and other tropical legumes (Hunter *et al*. 1970, Pangway & Richards 1971). Pectin concentration in grasses is usually less than 2 percent.

Lignin effects on cell wall digestion

Structural carbohydrates in the walls of sclerenchyma and some vascular cells are bound at random points by covalent and hydrogen bounds to the phenolic polymer lignin. Lignin is a heterogenous compound which is not digested either by ruminal microorganisms or by intestinal enzymes. By bonding to plant fibre it prevents swelling, thereby restricting entry of microbial digestive enzymes and consequently depressing fibre digestibility. The lignin content of the cell wall is the major determinant of the extent to which it can be digested. Tropical grasses and legumes tend to have higher lignin contents than temperate species (Harkin 1973), but where tropical and temperate grasses are grown under high temperature regimes, many temperate species have higher lignin contents in their leaves than do tropical grasses (Ford *et al*. 1979). Lignin is higher in stem than leaf, these differences being greater for legumes than grasses. As the plant matures, lignin content increases faster than cell wall content so that cell walls at maturity are more lignified and hence less digestible than at a young stage of growth. However lignin content of different grass and legume species is not an accurate guide to cell wall digestibility (Minson 1971, McLeod & Minson 1976). This is possibly caused by varying levels of protection of the cell wall to digestion. The different protective nature of lignin on cellulose digestion in temperate grasses and legumes is shown in Fig. 6.

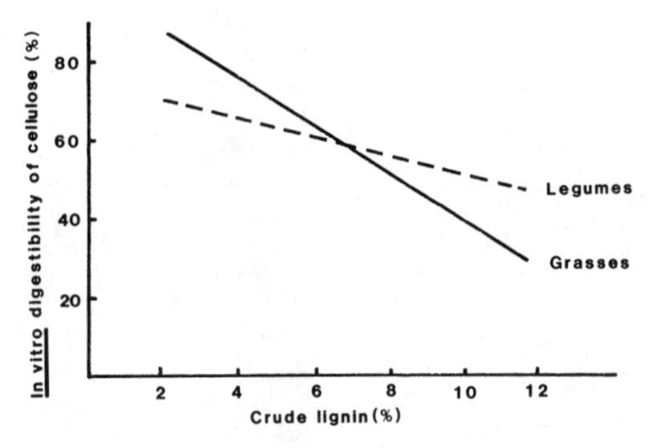

Fig. 6. The relationship between lignin content and cellulose digestibility in temperate legumes and grasses. (Adapted from Tomlin *et al*. 1965).

Silica effects on cell wall digestion

Minerals in the cell wall

In most forage plants, silica represents a small fraction of the minerals present depending on the soil type on which the plant was grown. McManus *et al*. (1977) have found silica present in plants as discrete intracellular inclusions of SiO_2 or sometimes included as a minor component of cell walls. Van Soest & Jones (1968) have found that for temperate forages ranging in silica content from 0.5 to 6.4 percent, silica content was inversely related to forage digestibility. However, Minson (1971) could demonstrate no effect of silica content (2.6-7.5 percent) on the digestibility or voluntary feed intake of six *Panicum* varieties, and this equivocal situation for the effects of silica on digestibility awaits resolution.

DIGESTIBILITY

Digestibility of forage dry matter by the ruminant is the summation of the digestibility of the components tissues as affected by morphology, anatomy and chemical composition.

Minson & Wilson (1980) have determined from a survey of the literature that the mean dry matter digestibility for tropical legumes was 56.6 percent compared with 60.7 percent for temperate legumes. Similarly temperate grasses had a significantly higher digestibility (68.2 percent) than did tropical grasses (55.4 percent). The frequency distributions of dry matter digestibility for tropical and temperate legumes and for tropical and temperate grasses are shown in Fig. 7. The variation shown is caused by differences between species, by varying stages of maturity within species and by widely differing environmental conditions during growth. Some temperate grasses retain their higher digestibility in early growth until flowering after which digestibility progressively declines (Minson *et al*. 1960b).

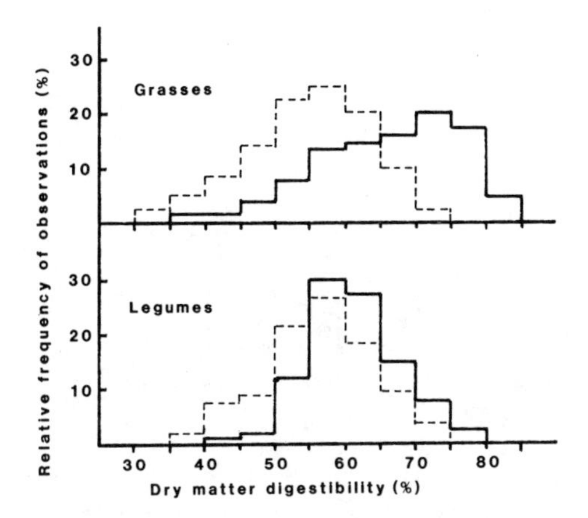

Fig. 7. The frequency distribution of dry matter digestibility in tropical (---) and temperate (——) legumes and grasses. (From Wilson & Minson 1980).

However digestibility in tropical grasses continuously declines during growth, and species differ markedly in digestibility at the early vegetative stage of growth and in the rate at which digestibility declines with maturity. Reid *et al*. (1973) have reported the relationships between *in vitro* dry matter digestibility and the age of regrowth for 42 tropical grasses and 11 tropical legumes. The rate of fall in digestibility with maturity is correlated with the digestibility of the immature forage (28 day regrowth) for both grasses and legumes (Fig. 8). Species with a high initial digestibility tended to decrease in digestibility at a much faster rate with maturity than those with low initial digestibilities. However, there were still large differences between species in the rate of fall in digestibility with maturity that could not be accounted for simply by differences in initial digestibility.

Forage species which maintain high digestibility for long periods during the growth season are of higher value for animal production than those which may have high digestibility at a young stage of growth but in which digestibility decreases rapidly. Most of the tropical legumes, with the exception of the *Desmodium* species, and many grass species from the genera *Brachiaria*, *Setaria* and *Digitaria* are characterized by relatively low rates of decline in digestibility in contrast to species from the genera *Panicum*, *Chloris* and

Hyparrhenia. It is clear that major differences occur between pasture plants in digestibility and overall quality for animal production, and the causes of these differences are those variable plant attributes (morphology, anatomy and chemical) discussed earlier. The procedures used for selecting forage species of high quality must therefore take into account not only digestibility but also the rate at which digestibility changes with increasing maturity.

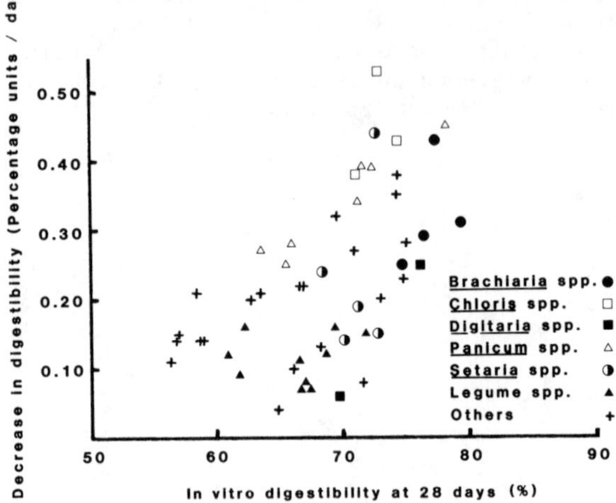

Fig. 8. The relationship between in vitro dry matter digestibility (28 days regrowth) and rate of decrease in digestibility for tropical grasses and legumes (from data of Reid et al. 1973).

ACKNOWLEDGEMENT

The author wishes to sincerely thank Dr D.J. Minson, for providing the collated data on the mineral composition of pasture species and for his interest, advice and encouragement during the preparation of this manuscript.

REFERENCES

ARC (1976) The nutrient requirements of farm livestock. No. 4. Composition of British feedstuffs. Technical review and tables. London, Agricultural Research Council, 710 pp.

ARC (1980) The nutrient requirements of ruminant livestock: Technical review by an Agricultural Research Council working party. Farnham Royal, U.K., Commonwealth Agricultural Bureaux, 351 pp.

Aii, T.; Stobbs, T.H. (1980) Solubility of the protein of tropical pasture species and the rate of its digestion in the rumen. Animal Feed Science and Technology 5, 183-192.

Akazawa, T. (1965) Starch, inulin, and other reserve polysaccharides. In: Plant biochemistry. Editors J. Bonner and J.E. Varner. New York and London, Academic Press, pp.258-297.

Akin, D.E.; Burdick, D. (1975) Percentage of tissue types in tropical and temperate grass leaf blades and degradation of tissues by rumen microorganisms. Crop Science 15, 661-668.

Akin, D.E.; Burdick, D.; Amos, H.E. (1974) Comparative degradation of coastal bermudagrass, coastcross-1 bermudagrass, and pensacola bahiagrass by rumen microorganisms revealed by scanning electron microscopy. Crop Science 14, 537-541.

Andrew, C.S.; Robins, M.F. (1969) The effect of phosphorus on the growth and chemical composition of some tropical pasture legumes. Nitrogen, calcium, magnesium, potassium and sodium contents. Australian Journal of Agricultural Research 20, 675-685.

Andrew, C.S.; Robins, M.F. (1971) The effect of phosphorus on the growth, chemical composition, and critical phosphorus percentages of some tropical pasture grasses. Australian Journal of Agricultural Research 22, 693-706.

Bailey, R.W. (1973) Structural carbohydrates. In: Chemistry and biochemistry of herbage. Editors G.W. Butler and R.W. Bailey. New York and London, Academic Press, Vol. 1, pp.157-211.

Beeson, K.C. (1978) Plants and foods of plant origin. In: Biochemistry and the environment. Distribution of trace elements related to the occurence of certain cancers, cardio vascular diseases and urolithiasis. Washington D.C., National Academy of Sciences Vol. III, pp.59-78.

Björkman, O.; Boynton, J.; Berry, J. (1976) Comparison of the heat stability of photosynthesis, chloroplast membrane reactions, photosynthetic enzymes, and soluble protein in leaves of heat adapted and cold adapted C_4 species. Carnegie Institution of Washington Yearbook 75, pp.400-407.

Brady, C.J. (1976) Plant proteins, their occurence quality and distribution. In: From plant to animal protein. Reviews in rural science II. Editors T.H. Sutherland, J.R. McWilliam and R.A. Leng. Armidale, University of New England Publishing Unit, pp.13-16.
Brown, R.H. (1978) A difference in N use efficiency in C_3 and C_4 plants and its implications in adaptation and evolution. Crop Science 18, 93-98.
Brown, R.H.; Blaser, R.E. (1968) Leaf area index in pasture growth. Herbage Abstracts 38, 1-9.
Buttery, P.J. (1976) Protein synthesis in the rumen. Its implications in the feeding of non-protein nitrogen to ruminants. In: Principles of cattle production. Editors H. Swan and W.H. Bróster. London, Butterworths, pp. 145-168.
Carolin, R.C.; Jacobs, S.W.L.; Usk, M. (1973) The structure of the cells of the mesophyll and parenchymatous bundle sheath of Gramineae. Botanical Journal of the Linnaean Society 66, 259-275.
Chacon, E.A. (1976) The effects of sward characteristics upon grazing behaviour, intake and animal production from tropical pastures. Ph.D. thesis. The University of Queensland.
Colman, R.L.; Holder, J.M. (1968) Effect of stocking rate on butterfat production of dairy cows grazing kikuyu grass pastures fertilized with nitrogen. Proceedings of the Australian Society of Animal Production 7, 129-133.
Colman, R.L.; Lazenby, A. (1970) Factors affecting the response of tropical and temperate grasses to fertilizer nitrogen. Proceedings of the 11th International Grassland Congress, Surfers Paradise, Australia, 392-397.
Cooper, R.L.; Lazenby, A. (1970) Factors affecting the response of tropical and temperate grasses to fertilizer nitrogen. Proceedings of the 11th International Grassland Congress, Surfers Paradise, Australia, 392-397.
Cooper, J.P. (1970) Potential production and energy conversion in temperate and tropical grasses. Herbage Abstracts 40, 1-15.
Daniel, H.A.; Harper, H.J. (1935) The relation between effective rainfall and total calcium and phosphorus in alfalfa and prairie hay. Journal of the American Society of Agronomy 27, 644-652.
Dijkshoorn, W. (1973) Organic acids, and their role in ion uptake. In: Chemistry and biochemistry of herbage. Editors G.W. Butler and R.W. Bailey. London and New York, Academic Press Vol. 2, pp. 163-188.
Evans, L.T. (1964) Reproduction. In: Grasses and grasslands. Editor C. Barnard. London, Macmillan & Co. Ltd, pp.126-153.
Fagan, T.W.; Jones, H.T. (1924) The nutritive value of grasses as shown by their chemical composition. Bulletin of the Welsh Plant Breeding Station, Series H, No.3, pp.85-130.
Fisher, M.J. (1969) The growth and development of Townsville lucerne (Stylosanthes humilis) in ungrazed swards at Katherine, N.T. Australian Journal of Experimental Agriculture & Animal Husbandry 9, 196-208.
Fleming, G.A. (1973) Mineral composition of herbage. In: Chemistry and biochemistry of herbage. Editors G.W. Butler and R.W. Bailey. London and New York, Academic Press Vol.1, pp.529-566.
Ford, C.W. (1978) In vitro digestibility and chemical composition of three tropical pasture legumes. Desmodium intortum cv. Greenleaf, D. tortuosum and Macroptilium atropurpureum cv. Siratro. Australian Journal of Agricultural Research 29, 963-974.
Ford, C.W.; Williams, W.J. (1973) In vitro digestibilty and carbohydrate composition of Digitaria decumbens and Setaria anceps grown at different levels of nitrogenous fertilizer. Australian Journal of Agricultural Research 24, 309-316.
Ford, C.W.; Morrison, I.M.; Wilson, J.R. (1979) Temperature effects on lignin, hemicellulose and cellulose in tropical and temperate grasses. Australian Journal of Agricultural Research 30, 621-634.
Goswami, A.K.; Willcox, J.S. (1969) Effect of applying increasing levels of nitrogen to ryegrass. 1. Composition of various nitrogenous fractions and free amino acids. Journal of the Science of Food and Agriculture 20, 592-599.
ap Griffiths, G.; Walters, R.J.K. (1966) The sodium and potassium content of some grass genera, species and varieties. Journal of Agricultural Science, Cambridge 67, 81-89.
Hacker, J.B. (1971) Digestibility of leaves, leaf sheaths and stems in setaria. Journal of the Australian Institute of Agricultural Science 37, 154-155.
Hacker, J.B. (1974) Variation in oxalate, major cations and dry matter digestibility of 47 introductions of the tropical grass setaria. Tropical Grasslands 8, 145-154.
Hacker, J.B. (1982) Selecting and breeding better quality grasses. In: Nutritional limits to animal production from pastures. Editor J.B. Hacker. Farnham Royal, U.K., Commonwealth Agricultural Bureaux, pp.305-326.
Hallock, D.L.; Brown, R.H.; Blaser, R.E. (1965) Relative yield and composition of Ky. 31 fescue and coastal bermuda grass at four nitrogen levels. Agronomy Journal 57, 539-542.
Hanna, W.W.; Monson, W.G.; Burton, G.W. (1973) Histological examination of fresh forage leaves after in vitro digestion. Crop Science 13, 98-102.
Harkin, J.M. (1973) Lignin. In: Chemistry and biochemistry of herbage. Editors G.W. Butler and R.W. Bailey. New York and London, Academic Press Vol. 1, pp.323-373.
Hatch, M.D.; Slack, C.R. (1970) In: Progress in phytochemistry. Editors L. Reinhold and Y. Lowschitz. London, Interscience, Vol.2, 35-106.
Hegarty, M.P. (1982) Deleterious factors in forages affecting animal production. In: Nutritional limits to animal production from pastures. Editor J.B. Hacker. Farnham Royal, U.K., Commonwealth Agricultural Bureaux, pp. 133-150.

Hegarty, M.P.; Peterson, P.J. (1973) Free amino acids, bound amino acids, amines and ureides. In: Chemistry and biochemistry of herbage. Editors G.W. Butler and R.W. Bailey. London and New York, Academic Press, Vol 1, pp.1-62.

Hodgson, H.C. (1964) The protein amino-acid composition and nitrogen distribution in two tropical grasses. Journal of the Science of Food and Agriculture 15, 721-724.

Hodgson, J. (1982) Influence of sward characteristics on diet selection and herbage intake by the grazing animal. In: Nutritional limits to animal production from pastures. Editor J.B. Hacker. Farnham Royal, U.K., Commonwealth Agricultural Bureaux, pp. 153-166.

Hunter, R.A.; McIntyre, B.L.; McIlroy, R.J. (1970) Water soluble carbohydrates of tropical pasture grasses and legumes. Journal of the Science of Food and Agriculture 21, 400-405.

Johns, A.T. (1955) Pasture quality and ruminant digestion. 1. Seasonal change in botanical and chemical composition of pasture. New Zealand Journal of Science and Technology A37, 301-311.

Jones, R.J. (1972) Stocking rate effects on three different pastures. Division of Tropical Pastures, CSIRO, Annual Report 1971-72, p.12.

Jones, R.J.; Seawright, A.A.; Little, D.A. (1970) Oxalate poisoning in animals grazing the tropical grass Setaria sphacelata. Journal of the Australian Institute of Agricultural Science 36, 41-43.

Laetsch, W.M. (1969) Relationship between chloroplast structure and photosynthetic carbon-fixation pathways. Science Progress, Oxford 57, 323-351.

Laetsch, W.M. (1974) The C_4 syndrome: A structural analysis. Annual Review of Plant Physiology 25, 27-52.

Laredo, M.A.; Minson, D.J. (1973) The voluntary intake, digestibility, and retention time by sheep of leaf and stem fractions of five grasses. Australian Journal of Agricultural Research 24, 875-888.

Little, D.A. (1968) Effect of dietary phosphate on the voluntary consumption of Townsville lucerne (Stylosanthes humilis) by cattle. Proceedings of the Australian Society of Animal Production 7, 376-380.

Little, D.A. (1980) Observations on the phosphorus requirement of cattle for growth. Research in Veterinary Science 28, 258-260.

Little, D.A. (1982) Utilization of minerals. In: Nutritional limits to animal production from pastures. Editor J.B. Hacker. Farnham Royal, U.K., Commonwealth Agricultural Bureaux, pp. 259-283.

Lyttleton, J.W. (1973) Proteins and nucleic acids. In: Chemistry and biochemistry of herbage. Editors G.W. Butler and R.W. Bailey. London and New York, Academic Press, Vol.1, pp.63-103.

McDowell, L.R.; Conrad, J.H.; Thomas, J.E.; Harris, L.E. (1974) Latin American table of feed composition. Gainsville, Florida, University of Florida, 509 pp.

McIvor, J.G. (1979) Seasonal changes in nitrogen and phosphorus concentrations and in in vitro digestibility of Stylosanthes species and Centrosema pubescens. Tropical Grasslands 13, 92-97.

McLeod, M.N. (1974) Plant tannins - their role in forage quality. Nutrition Abstracts and Reviews 44, 803-815.

McLeod, M.N.; Minson, D.J. (1974) Differences in carbohydrate fractions between Lolium perenne and two tropical grasses of similar dry matter digestibility. Journal of Agricultural Science, Cambridge 82, 449-454.

McLeod, M.N.; Minson, D.J. (1976) The analytical and biological accuracy of estimating the dry matter digestibility of different legume species. Animal Feed Science and Technology 1, 61-72.

McManus, W.A.; Robinson, V.N.E.; Grout, L.L. (1977) The physical distribution of mineral material on forage plant cell walls. Australian Journal of Agricultural Research 28, 651-662.

Mannetje, L. 't, Sidhu, A.S.; Murugaiah, M. (1976) Cobalt deficiency in cattle in Johore. Liveweight change and response to treatments. M.A.R.D.I. Research Bulletin 4, No.1, 90.

Milford, R.; Haydock, K.P. (1965) The nutritive value of protein in subtropical pasture species grown in south-east Queensland. Australian Journal of Experimental Agriculture and Animal Husbandry 5, 13-17.

Minson, D.J. (1971) Influence of lignin and silicon on a summative system for assessing the organic matter digestibility of Panicum. Australian Journal of Agricultural Research 22, 589-598.

Minson, D.J. (1971) The digestibility and voluntary intake of six varieties of Panicum. Australian Journal of Experimental Agriculture and Animal Husbandry 11, 18-24.

Minson, D.J. (1973) Effect of fertilizer nitrogen on digestibility and voluntary intake of Chloris gayana, Digitaria decumbens and Pennisetum clandestinum. Australian Journal of Experimental Agriculture and Animal Husbandry 13, 153-157.

Minson, D.J. (1976) Nutritional significance of protein in temperate and tropical pastures. In: From plant to animal protein. Reviews in Rural Science II. Editors T.M. Sutherland, J.R. McWilliam and R.A. Leng. Armidale, University of New England Publishing Unit, pp. 27-30.

Minson, D.J. (1977) The chemical composition and nutritive value of tropical legumes. In: Tropical forage legumes. Editor P.J. Skerman. Rome, FAO, pp. 186-194, 569-582.

Minson, D.J. (1982) The chemical composition and nutritive value of tropical grasses. In: Tropical grasses. Editor P.J. Skerman. Rome, FAO. (In press)

Minson, D.J.; Laredo, M.A. (1972) Influence of leafiness on voluntary intake of tropical grasses by sheep. Journal of the Australian Institute of Agricultural Science 38, 303-305.

Minson, D.J.; Milford, R. (1968) The nutritional value of four tropical grasses when fed as chaff and pellets to sheep. Australian Journal of Experimental Agriculture and Animal Husbandry 8, 270-276.

Minson, D.J.; Wilson, J.R. (1980) Comparative digestibility of tropical and temperate forage - a contrast between grasses and legumes. Journal of the Australian Institute of Agricultural Science 46, 247-249.

Minson, D.J.; Raymond, W.F.; Harris, C.E. (1960a) The digestibility of grass species and varieties. Proceedings of the 8th International Grassland Congress, Reading, 470-474.

Minson, D.J.; Raymond, W.F.; Harris, C.E. (1960b) Studies in the digestibility of herbage. 8. The digestibility of S37 cocksfoot, S23 ryegrass and S24 ryegrass. Journal of the British Grassland Society 15, 174-180.

Mowat, D.W.; Fulkerson, R.S.; Tossel, W.E.; Winch, J.E. (1965) The in vitro digestibility and protein content of leaf and stem portions of forages. Canadian Journal of Plant Science 45, 321-331.

Nicol, D.C.; Smith, L.D. (1981) Responses to cobalt therapy in weaner cattle in south-east Queensland. Australian Journal of Experimental Agriculture and Animal Husbandry 21, 27-31.

Noble, A.; Lowe, K.F. (1974) Alcohol soluble carbohydrates in various tropical and temperate pasture species. Tropical Grasslands 8, 179-188.

Norton, B.W.; Hales, J.W. (1976) A response of sheep to cobalt supplementation in south eastern Queensland. Proceedings of the Australian Society of Animal Production 11, 393-396.

Pangway, C.; Richards, G.N. (1971) Polysaccharides of tropical pasture herbage III. The distribution of the major polysaccharide components of Townsville lucerne (Stylosanthes humilis) during growth. Australian Journal of Chemistry 24, 1041-1048.

Reid, J.T.; Kennedy, W.T.; Turk, K.L.; Slack, S.; Trimberger, G.W.; Murphy, R.P. (1959) Symposium on forage evaluation. 1. What is forage quality from the animal 'stand point'? Agronomy Journal 51, 213-216.

Reid, R.L.; Horvath, D.J. (1980) Soil chemistry and mineral problems in farm livestock. A review. Animal Feed Science and Technology 5, 95-167.

Reid, R.L.; Post, A.J.; Jung, G.A. (1970) Mineral composition of forages. West Virginia University Bulletin 589T, 35 pp.

Reid, R.L.; Post, A.J.; Olsen, F.J.; Mugerwa, J.S. (1973) Studies on the nutritional quality of grasses and legumes in Uganda. 1. Applications of in vitro digestibility techniques to species and stage of growth effects. Tropical Agriculture, Trinidad 50, 1-15.

Roberts, F.J.; Carbon, B.A. (1969) Growth of tropical and temperate grasses and legumes under irrigation in south-west Australia. Tropical Grasslands 3, 109-116.

Schank, S.C.; Klock, M.A.; Moore, J.E. (1973) Laboratory evaluation of quality in subtropical grasses. II. Genetic variation among Hermarthrias in in vitro digestion and stem morphology. Agronomy Journal 65, 256-258.

Smith, D. (1968) Carbohydrates in grasses. IV. Influence of temperature on the sugar and fructosan composition of timothy plant parts at anthesis. Crop Science 8, 331-334.

Smith, D. (1971) Efficiency of water for extraction of total nonstructural carbohydrates from plant tissue. Journal of the Science of Food and Agriculture 22, 445-447.

Smith, D. (1972a) Total nonstructural carbohydrate concentrations in the herbage of several legumes and grasses at first flower. Agronomy Journal 64, 705-706.

Smith, D. (1972b) Carbohydrate reserve of grasses. In: The biology and utilization of grasses. New York and London, Academic Press pp.318-332.

Stobbs, T.H. (1973) The effect of plant structure on the intake of tropical pastures. 1. Variation in the bite size of grazing cattle. Australian Journal of Agricultural Research 24, 809-819.

Thornton, R.F.; Minson, D.J. (1973) The relationship between apparent retention time in the rumen, voluntary intake and apparent digestibility of legume and grass diets in sheep. Australian Journal of Agricultural Research 24, 889-898.

Tomlin, D.C.; Johnson, R.R.; Dehority, B.A. (1965) Relationship of lignification to in vitro cellulose digestibility of grasses and legumes. Journal of Animal Science 24, 161-165.

Ulyatt, M.J. (1970) Factors contributing to differences in the quality of short-rotation ryegrass, perennial ryegrass and white clover. Proceedings of the 11th International Grassland Congress, Surfers Paradise, Australia, 709-713.

Ulyatt, M.J.; Egan, A.R. (1979) Quantitative digestion of fresh herbage by sheep. V. The digestion of four herbages and prodiction of sites of digestion. Journal of Agricultural Science, Cambridge 92, 605-616.

Vallis, I.; Jones, R.J. (1973) Net mineralization of nitrogen in leaves and leaf litter of Desmodium intortum and Phaseolus atropurpureus mixed with soil. Soil Biology and Biochemistry 5, 391-398.

Van Soest, P.J. (1967) Development of a comprehensive system of feed analysis and its application to forages. Journal of Animal Science 26, 119-128.

Van Soest, P.J.; Jones, L.H.P. (1968) Effect of silica in forages upon digestibility. Journal of Dairy Science 51, 1644-1648.

Waite, R. (1970) The structural carbohydrates and the in vitro digestibility of a ryegrass and a cocksfoot at two levels of nitrogenous fertilizer. Journal of Agricultural Science, Cambridge 74, 457-462.

Walker, D.A.; Crofts, A.R. (1970) Photosynthesis. Annual Review of Biochemistry 39, 389-428.

Whitehead, D.C. (1966) Nutrient minerals in grassland herbage. Review Series 1/1966. Hurley, U.K., Commonwealth Bureau of Pastures and Field Crops, 83 pp.

Wilson, J.R. (1982) Environmental and nutritional factors affecting herbage quality. In: Nutritional limits to animal production from pastures. Editor J.B. Hacker. Farnham Royal, U.K., Commonwealth Agricultural Bureaux, pp. 111-131.

Wilson, J.R.; Minson, D.J. (1980) Prospects for improving the digestibility and intake of tropical grasses. Tropical Grasslands 14, 253-359.

Winter, W.H.; Siebert, B.D.; Kuchel, R.E. (1977) Cobalt and copper therapy of cattle grazing improved pastures in northern Cape York Peninsula. Australian Journal of Experimental Agriculture and Animal Husbandry 17, 10-15.

ENVIRONMENTAL AND NUTRITIONAL FACTORS AFFECTING HERBAGE QUALITY

J. R. WILSON

CSIRO, Division of Tropical Crops and Pastures, Cunningham Laboratory, St. Lucia, Queensland 4067, Australia.

ABSTRACT

This paper reviews the effects of environmental and nutritional factors on the digestibility and on some associated tissue characteristics of pasture herbage. High growth temperatures cause a substantial decrease in the digestibility of grasses and a similar but smaller effect on legumes. Variations in relative humidity or evaporative demand do not appear to influence herbage quality. Herbage grown under shade is of lower digestibility than herbage grown in full sunlight. Severe droughts are detrimental to herbage quality but soil water deficits of low to moderate severity may result in an improvement of herbage digestibility and nutritive value. Nitrogen fertilizer may have positive, nil or negative effects on herbage digestibility and these responses are discussed. Phosphorus, potassium, sulphur and calcium fertilizers generally have little influence on herbage digestibility, although several positive effects have been noted for sulphur and calcium.

INTRODUCTION

Climate and soil environment are prime determinants of the adaptation and potential growth of pasture species in any region. The inherent differences between species in nutritive quality and the limitations to animal production due to environmental effects on pasture productivity are considered in Chapters 2-5 (Allden 1982, Mannetje 1982, Norton 1982, Reid & Jung 1982). The present review concentrates on the manner in which particular climatic and nutritional factors influence the nutritive quality of pasture herbage in general, and hence may modify potential animal performance. Because digestible energy is a major limitation to animal production (Hardison 1966, Karue 1975) main emphasis is given to changes in dry matter digestibility and the factors such as cell wall content, lignification or plant morphology which may be responsible for these changes. Very few controlled experiments have examined environmental effects on herbage characteristics in relation to intake.

GENERAL CLIMATIC TRENDS

The changes in dry matter digestibility with season for several tropical and temperate herbage species are illustrated in Fig. 1.

These trends represent the general response of a wide range of species in sub-tropical and temperate regions, that is, dry matter digestibility highest in spring, then falling to a low value in mid-late summer, increasing somewhat in autumn and decreasing again in winter (e.g. Hacker & Minson 1972, Strickland 1973, Fletcher 1976, Langlands & Holmes 1978, Powell *et al*. 1978, Reed 1978, Andrews & Crofts 1979, Clark & Brougham 1979). The increase in autumn may be smaller or non-existent and the fall in winter more pronounced as one progresses from the wetter sub-tropics to drier areas or the monsoonal tropics, where the dry period is more defined, or regions where frosts are severe. In the humid tropics where seasonal variation in climate is small, the changes in dry matter digestibility seen in Fig. 1 are considerably damped (Chenost 1972, Caro-Costas et al. 1976). Voluntary intake of pasture is usually high in spring and much lower in late summer, autumn and winter (Chenost 1972, Reed 1978, Clark & Brougham 1979).

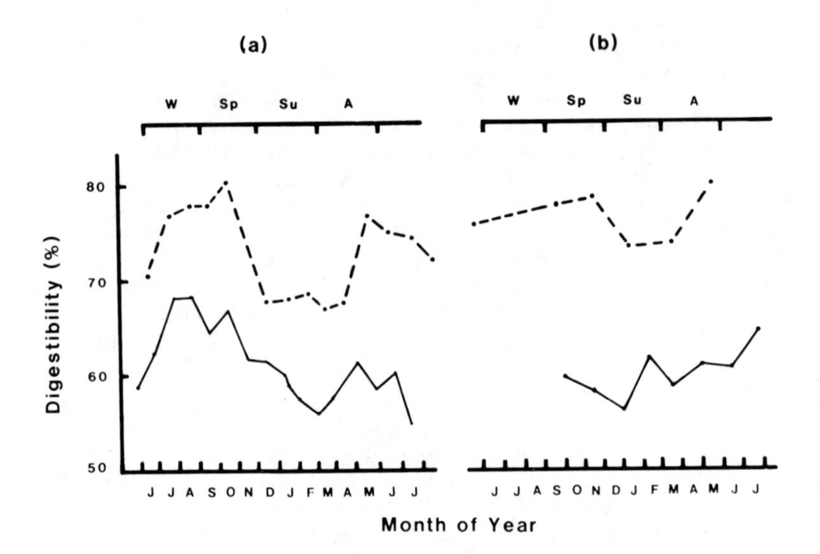

Fig. 1. Seasonal changes in digestibility for (a) Lolium perenne pasture in New Zealand (---Clark & Brougham 1979) and Setaria sphacelata in S.E. Qld., Australia (——Hacker & Minson 1972), and (b) Trifolium repens in Tasmania, Australia (---Michell 1973) and Macroptilium atropurpureum in N. Qld., Australia (——J.R. Wilson, unpubl. data 1981). W(winter), Sp(spring), Su(summer) and A(autumn).

Both the spring maximum and the summer minimum dry matter digestibility become progressively lower with change of region from high to low latitudes (Fig. 2).

These general climatic trends may be due to temperature, daylength or irradiance, climatic variables which are usually correlated. For example all increase in spring and decrease in late autumn, whilst rainfall and frosts are highly variable in occurence. Consequently, the interpretation of seasonal, and regional, differences in nutritive quality in relation to individual climatic variables is difficult. Experiments in which environmental factors are controlled are necessary to aid our understanding and the next section reviews the information available in this area.

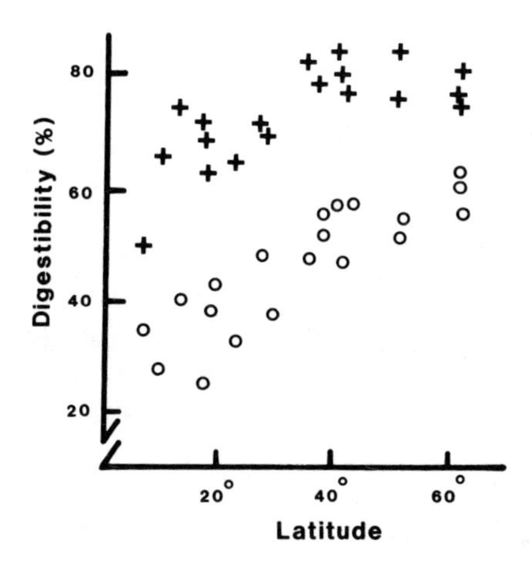

Fig. 2. Relationship between digestibility of perennial grasses and latitude for the earliest spring cutting (+) and for mature herbage (o). After Van Soest et al. 1978.

INDIVIDUAL ENVIRONMENTAL FACTORS

Light

Level of irradiance

Irradiance levels experienced by pastures vary widely between sites and seasons. For example total daily radiation may differ by as much as 100 percent between arid mediterranean and humid tropical environments or between winter and summer (Cooper & Tainton 1968). Tropical pasture species, particularly in the arid to semi-arid regions of the world, are generally subjected to higher levels of irradiance than temperate species (Cooper & Tainton 1968) and it has been suggested that the high insolation experienced by tropical species is a significant contributing factor to high lignification and low digestibility (French 1961, Sakurai 1963).

Contrary to this belief, the few experiments examining the effect of irradiance on herbage dry matter digestibility (Garza et al. 1965, Deinum et al. 1968, Masuda 1977, Wong 1978) indicate that herbage dry matter digestibility of both tropical and temperate grasses is decreased, not increased, by reducing the light level under which plants are grown. The decrease in dry matter digestibility of these grasses over a wide range of irradiances was mostly small (1-5 percentage units) except for the experiment of Wong (1978) in which green panic (Panicum maximum var. trichoglume) swards grown under 60 or 40 percent shade for 2-4 months were 10-12 percentage units lower in dry matter digestibility than swards grown in full sunlight (see Fig. 3). In the same experiment, Wong found no effect of shade on the dry matter digestibility of the legume siratro (Macroptilium atropurpureum); more work is needed to determine whether this lack of response is general for all legumes.

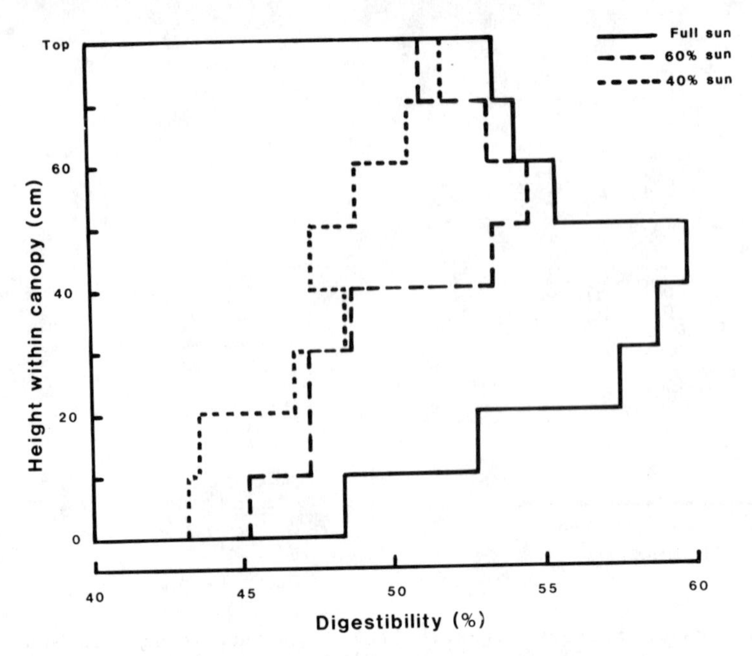

Fig. 3. Effect of light level during growth on the dry matter digestibility of an 8 week old green panic canopy divided into 10 cm strata (adapted from Wong 1978).

Shading lowers the soluble carbohydrate level in plants (Smith 1973) and usually there is an accompanying increase in cell wall content (Deinum 1966, Deinum & Dirven 1967, 1972, Hight et al. 1968, Deinum 1971). Lignification (Burton et al. 1959, Van Dyne & Heady 1965, Deinum et al. 1968, Hight et al. 1968) and silica content (Van Dyne & Heady 1965, Eriksen & Whitney 1977) also appear to be higher in shaded herbage, and the digestibility of the cell wall may be lower (Masuda 1977). The proportion of readily digested mesophyll tissue is decreased under shade relative to the less digestible epidermis (Evans 1964, Wilkinson & Beard 1975, Chabot & Chabot 1977). These characteristics of shaded plants coupled with accentuated stem elongation of grasses (Wong 1978) and reduced tillering (Auda et al. 1966, Masuda 1977) would all contribute to a lower herbage dry matter digestibility. Shade also increases the percentage moisture content of the tissues which may reduce herbage intake by animals (Burton et al. 1959). However, to offset these detrimental effects on herbage quality, plants under shade often have a higher crude protein content (Deinum 1966, Deinum et al. 1968, Wong 1978), reduced vascular development (see Wilkinson & Beard 1975) and a much thinner cuticle (Hull et al. 1975). There are also a few reports (Fujita 1942, Myhr & Saebo 1969) of lower cell wall or crude fibre content in shaded than in full-sun plants.

The only critical experiment found which involved animals was that of Hight et al. (1968) in which sheep fed perennial ryegrass (*Lolium perenne*) pasture shaded to 22 percent sunlight ate 9-15 percent less and had daily liveweight gains 38 percent lower than those fed on pasture grown in full sunlight. This supports comments often made (see Burton et al. 1959) that animals make poorer liveweight gains during cloudy, wet weather, although this could well be due to animal rather than plant factors. The weight of evidence from analysis of plant dry matter digestibility and composition indicates that shading (and not high levels of irradiance) reduces the nutritive value of herbage. Probably shade is not a major limitation for animal performance except perhaps at the level of shading experienced by pastures grown in conjunction with forestry or plantation crops.

Daylength

Daylengths which initiate flower and stem development in grasses will lead eventually to a sward of low digestibility and probably poor intake by animals. The few experimental studies of the effects of daylength on dry matter digestibility or cell wall content and lignification (Bowman & Law 1964, Allinson 1971, Faix 1974, Mannetje 1975) indicate that, excluding effects associated with flowering the effect of daylength on tissue dry matter digestibility is usually small and inconsistent. Table 1 illustrates data for tall fescue (*Festuca arundinacea*). Furthermore, the effect of daylength *per se* has not been separated in the various studies from differences in total daily radiant energy input which can lead to large treatment differences in plant yield as seen in Table 1.

TABLE 1

The effect of daylength on digestibility (DMD) of tall fescue (Allinson 1971).

	27/16°C		16/4°C	
	Yield (g/pot)	DMD (%)	Yield (g/pot)	DMD (%)
Long day (16 hours)	8.5	65.8	8.2	68.3
Short day (10 hours)	6.9	64.6	4.6	74.0

Soil moisture

Drought

Droughts of sufficient severity to stop growth and kill all above ground herbage undoubtedly limit animal production because of the low quality of dead herbage and insufficient available feed. This effect is accentuated as leaf is shed and only mature stem remains for consumption.

However, also of interest are limitations to herbage quality and hence animal production imposed by intermittent droughts of only light to moderate severity. Such droughts are experienced with varying frequency in most forage regions of the world. Some believe that these conditions are detrimental to forage quality (Woodman *et al*. 1931, French 1961, Sakurai 1963) while others maintain the opposite viewpoint (Dent & Aldrich 1963, Van Soest *et al*. 1978).

A survey of the literature on the effects of moisture stress on herbage dry matter digestibility is shown in Table 2. Most of these papers indicate that low soil moisture has either no effect or increases the digestibility of the pasture or of leaf or stem material. Moisture stress slows growth and delays stem development resulting in leafier swards of higher digestibility (Vough & Marten 1971, Wurster *et al*. 1971, Perry & Baltensperger 1979, Wilson 1981). This factor is perhaps of particular importance for tropical forage grasses in which stem develops rapidly under well-watered conditions, Wilson (1981) found droughted swards of *Panicum maximum* and *Cenchrus ciliaris* to be 8-11 units higher in dry matter digestibility than well-watered swards.

TABLE 2

Summary of published literature on effects of low soil moisture on herbage digestibility (DMD), cell wall (CWC) and lignin (lig.) content.

DMD (%)	CWC(%)	Lig.(%)	Plant material	Reference
+			G	Dent & Aldrich (1963)
0			G & L	Hiridiglou *et al*. (1966)
0/+	-	-	L	Vough & Marten (1971)
+	-	0	G	Wurster *et al*. (1971)
+			L	Snaydon (1972)
-			G	Hutton (1974)
+	-	0	G stem	Wilson & Ng (1975)
0/-	-	0	G leaf	
0/+			L	Pesant & Dionne (1976)
0			G	Taylor *et al*. (1976)
0			G	Spurway *et al*. (1976)
0			G	Silcock (1978)
0/+/-			G	Garwood *et al*. (1979)
0/+				Wilson (1981)
		-	L & M	Penfound (1931)
	+	+	M	Fujita (1942)
		-	M	Tiver (1942)
	-		G	Van Burg (1962)
	-		G	Lazenby & Rogers (1964)
	-		G	Deinum (1966)
		-	G	Koller & Clark (1965)
	-		G & L	Gifford & Jensen (1967)
	-	-	L	Jensen *et al*. (1967)
	-		G & L	Rahman *et al*. (1971)
	-	0/-	G	Wilson *et al*. (1980)
	0	+	L	

Increase (+), no effect (0) or decrease (-) with lower soil moisture level. Grass (G), legume (L), miscellaneous (M).

The ageing or maturation of tissue is not necessarily hastened by drought as some believe (e.g. Hoveland & Monson 1980), in fact there is evidence that in younger leaves ageing is delayed by water stress (Tiver 1942, Ludlow & Ng 1974) and the decline in nitrogen content (Wilson & Ng 1975) and dry matter digestibility (Wilson 1981) is slower than for leaves on plants kept well-watered. With low soil moisture the progress of plants to later stages of development may be markedly retarded (Calder & Macleod 1968).

Usually the changes in tissue composition associated with water stress are favorable rather than unfavorable to pasture quality. Increases in the concentration of nitrogen (Gifford & Jensen 1967, Wilson & Ng 1975), most minerals (Rahman *et al*. 1971, Gerakis *et al*. 1975) and soluble carbohydrates (Blaser *et al*. 1966, Ford & Wilson 1981) are usually recorded. Even cell wall and lignin content have frequently been reported to be lower in water-stressed herbage as is seen in Table 2, and Silcock (1978) reported a lower breaking strength for water-stressed compared with unstressed grass leaves. These latter responses are perhaps contrary to the general belief that plants grown under dry conditions have large amounts of mechanical supporting tissue with thick cell walls and high lignification (Shields 1950, Sinnott 1960, Wangermann 1961). Much of this information relates to xerophytic species ecologically adapted to arid regions. Most improved pasture species are mesophytes and their exposure to water stress does not necessarily lead to development of xerophytic characteristics (Barss 1930, Wangermann 1961, Hull *et al*. 1975, Wilson *et al*. 1980). Native pasture species may be regarded differently, in many arid regions of the world the naturally occurring plants are xerophytes with the tissue characteristics associated with low herbage digestibility (Newman 1973) and this imposes a severe limit on animal production for much of the year (see Romero & Siebert 1980).

Phosphorus is one element which is often at a low concentration in water-stressed herbage (Hawthorne 1956, Rahman *et al*. 1971, Pesant & Dionne 1976, Fisher 1980) and this could prove an important limitation to animal production in regions where soil phosphorus levels are low. Palatability problems due to increased alkaloid or hydrocyanic acid content may arise in some species when severely water-stressed (Hoveland & Monson 1980, Rouquette *et al*. 1980).

Overall, the information available indicates that unless the pasture is all old and dead, water-stressed herbage is likely to be of high quality. Provided low yield does not limit intake then animal daily liveweight gain could be better than average from this type of herbage. Observations of this nature have been made (Woodman *et al*. 1931) but critical animal data to demonstrate this are difficult to find. Results from a grazing trial in S.E. Queensland (L.'t Mannetje, personal communication)) indicated that in years of very low rainfall exceptionally high animal daily liveweight gains were obtained (see Wilson 1981). Launchbaugh (1957) assessing grazing studies on native rangeland in Kansas, USA commented that drought years often produced herbage of excellent quality and gave animal gains well above average. Improved animal production from droughted herbage could also result from better accessibility of leaf due to reduced stem development, and perhaps also from a higher herbage intake because of the lower water content of the tissue (Butterworth *et al*. 1961, Ketth & Ranawana 1971).

Relative humidity and evaporative demand

Minson & McLeod (1970) suggested that plant water deficits due to high evaporative demand may be a contributing factor to the lower dry matter digestibility of herbage grown in areas of high temperatures. The only experimental examination of relative humidity and herbage quality (Wilson *et al*. 1976) indicated no effect of humidity on dry matter digestibility. Generally, the effect of differences in humidity or evaporative demand on tissue characteristics which may be associated with herbage quality are relatively small (Cain & Potzger 1933, Fujita 1942, Wilson 1973, Morrison Baird & Webster 1978) and inconsistent (Wangermann 1961), with the exception of increased cuticle development which occurs under low humidity conditions (Baker 1974). Silica uptake is associated with active transpiration (Sangster & Parry 1971) but Johnston et al. (1967) were unable to detect any relationship between plant silica concentration and either current or long term differences in rainfall. Nor does there appear to much difference in the distribution frequency of silica concentrations between herbage of tropical and temperate species (Fig. 4) despite the generally higher evaporative demand experienced by tropical species.

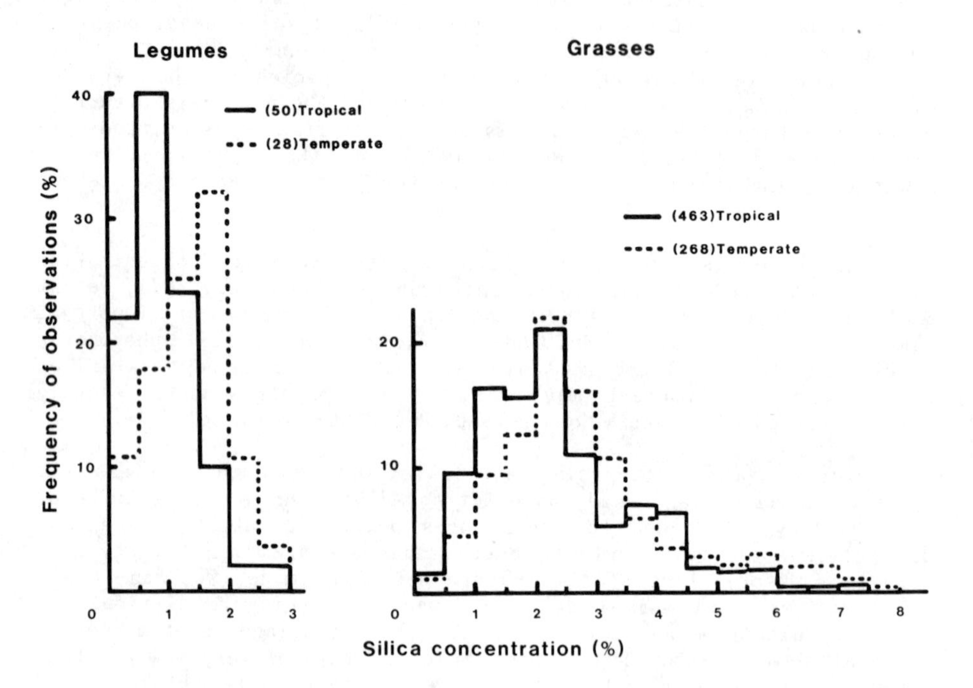

Fig. 4. Frequency distribution of silica concentrations in herbage of tropical and temperate grasses and legumes. Numbers of estimations shown in parentheses.

Excess water

Peterschmidt et al. (1979) reported that overwet conditions had no consistent effect on dry matter digestibility or crude protein in *Medicago sativa*. In grasses these conditions gave herbage of lower crude protein and higher cell wall content, but had no effect on lignin content (Pate & Snyder 1979).

Wind

The leaves of plants grown under high wind velocities often develop xeromorphic features of greater vascular development, more sclerenchyma and thicker cuticles (Venning 1949, Whitehead & Luti 1962, Grace & Russell 1977), although Russell & Grace (1978) found the opposite trend in their studies with *Lolium perenne*. The tendency towards increased xeromorphy could lead to lower tissue digestibility but no experimental measurements of wind effects on digestibility have been made.

Temperature

General temperature regime

The thermal adaptation of plants is a major determinant of the distribution of pasture species in any region (McWilliam 1978). Whilst the differences in nutritive value associated with species are outside the scope of this review, the generally lower dry matter digestibility of tropical compared with temperate pasture species, especially the grasses (Minson & McLeod 1970), and the extent to which temperature is a contributing factor, warrants consideration. Minson & McLeod (1970) selected herbage samples from different regions and determined a negative correlation between mean growth temperature and dry matter digestibility which indicated approximately one percentage unit decrease in dry matter digestibility per 1°C increase in growth temperature.

Many studies of the effects on herbage dry matter digestibility of controlled temperature regimes have been made and those relating to plant tops are summarized in Table 3. Similar data have also been collated for leaf and stem material and the mean effect of temperature on dry matter digestibility of all plant fractions is shown in Table 4. Clearly, high temperature has a strong detrimental effect on the dry matter digestibility of both tropical and temperate grasses but a much smaller effect on the dry matter digestibility of legumes. The data for tropical legumes are too few to determine whether the apparent beneficial effect of high temperature on leaf dry matter digestibility is real or just a consequence of the one species (siratro) for which data for leaves are available (J. R. Wilson-unpublished data). The small temperature response of the legumes may explain the overall small differences in dry matter digestibility between tropical and temperate legumes (Minson & Wilson 1980) and the small seasonal changes in dry matter digestibility often noted for many legumes (Minson & McLeod 1970, Jones 1973).

A detailed analysis of the effects of temperature on growth, morphology and development of grasses in relation to their nutritive quality has been presented by Dirven & Deinum (1977). Higher temperatures promote more rapid growth and stem development, particularly of tropical grasses (Dirven & Deinum 1977) and these effects are negatively associated with dry matter digestibility (Ivory *et al*. 1974, Deinum & Dirven 1973, 1975, 1976). Growth and stem development of temperate grasses is less likely to be accelerated by high temperature because of their lower optimum temperature for growth and because many species require a specific daylength for induction of flowering and culm elongation. Generally, in both groups of grasses the cell wall content of tops increases at high temperatures (Allinson 1971, Deinum and Dirven 1976, Kobayashi *et al*. 1977) either because of more rapid progress to advanced growth stages, or in the temperate grasses especially, because of a much lower accumulation of soluble carbohydrates (Smith 1973). Probably of more significance to nutritive quality is the fact that high temperatures decrease cell wall digestibility (Deinum & Dirven 1976, Moir *et al*. 1977). This effect may be seen clearly in the data of Table 5, and is probably due to the greater lignification which usually occurs at higher growth temperatures (Ehara & Tanaka 1961, Jensen *et al*. 1967, Vough & Marten 1971, Ford *et al*. 1979).

TABLE 3

Estimated change in digestibility of plant tops, percentage units per °C increase in growth temperature. Survey of studies using controlled temperature regimes.

Temperate grasses: Change in digestibility	Reference	Tropical grasses: Change in digestibility	Reference
-0.51	Deinum *et al*. 1968	-0.38, -0.58	Wilson & Ford 1971
-0.03 to -0.54	Smith 1970	-0.80	Deinum & Dirven 1973
-0.52, -1.08	Allinson 1971	-0.08 to -0.72	Wilson & Ford 1973
-0.44, -0.60	Wilson & Ford 1971	-0.38 to -1.02	Deinum & Dirven 1975
-0.12 to -1.01	Wilson & Ford 1973	0.13 to -0.67	Mannetje 1975
-0.65	Downes *et al*. 1974		
-0.41, -0.76	Deinum & Dirven 1975		
0.17 to -0.47	Smith 1975	-0.74 to -1.81	Deinum & Dirven 1976
-0.41, -0.63	Wilson *et al*. 1976	-0.41 to -1.17	Wilson *et al*. 1976
-0.61	Masuda 1977		
-0.70	Smith 1977		

Temperate legumes: Change in digestibility	Reference	Tropical legumes: Change in digestibility	Reference
-0.20	Garza *et al*. 1965	0.93 to -1.50	Mannetje 1975
-0.18	Smith 1969		
-0.31, -0.32	Vough & Marten 1971		
-0.56	Greenfield & Smith 1973	-0.15, -0.31	Wilson 1981 (unpublished data)
0.31 to -0.38	Faix 1974		
0 to -0.86	Mannetje 1975		

TABLE 4

Effect of temperature on dry matter digestibility (from Wilson and Minson 1980).

	Average change in DMD (% units) per °C increase in growth temperature			
	Grass		Legume	
	Tropical	Temperate	Tropical	Temperate
Tops	-0.60	-0.56	-0.28	-0.21
Leaf	-0.57	-0.64	+0.19	-0.09
Stem	-0.86	-0.76	-0.27	-0.22

TABLE 5

Effect of growth temperature on mean cell wall digestibility for five tropical grasses and two temperate grasses (adapted from Moir *et al*. 1977).

Indigestible cell wall as percentage of total cell wall content				
Day/night	Tropical grasses		Temperate grasses	
temperature	Tops	Stubble	Tops	Stubble
18/10 °C	30.4	28.1	27.7	22.6
25/17 °C	33.2	36.6	31.9	25.7
32/24 °C	39.1	40.1	37.4	31.1

High growth temperatures also hasten the maturation of individual tissues so that leaf of comparable age (Wilson *et al*. 1976) or stem of comparable size (Deinum & Dirven 1976) is of higher digestibility when grown at low temperatures (Fig. 5). Consequently it may be seen when water is not a limiting factor, that the rate of decline in dry matter digestibility of herbage with age is slower during the cooler autumn and winter than the hotter months in spring and summer (Johnson *et al*. 1973).

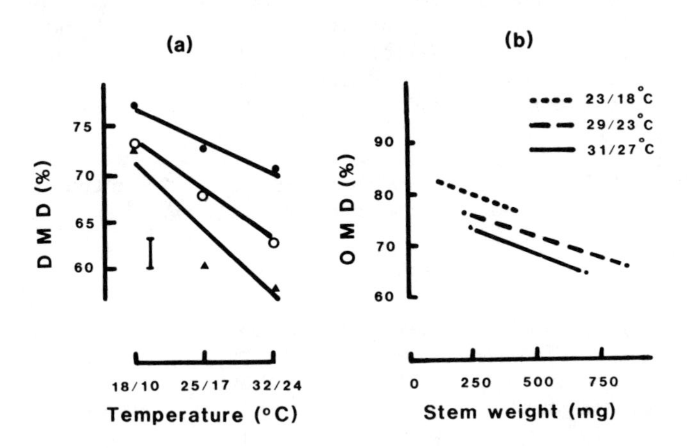

Fig. 5. Effect of temperature (day/night) on (a) dry matter digestibility (DMD) of leaf at 0(●), 8(o) 16 (▲) days after attainment of full expansion, average for six grasses (from Wilson *et al*. 1976), and (b) organic matter digestibility (OMD) of setaria stem in relation to stem size or development (from Deinum & Dirven 1977).

In some forages, for example *Lespedeza sericea*, high temperature may increase the accumulation of tannins, and thereby decrease forage digestibility (Hoveland & Monson 1980).

Minson & McLeod (1970) reported a 13 percentage unit difference in digestibility between temperate grasses grown in temperate climates and tropical grasses grown in tropical climates. Comparison of spring and summer temperatures of the United Kingdom and subtropical Queensland showed there to be an 8-16°C difference. If digestibility is depressed by 0.6 units per °C increase in temperature (Table 4) then 5-10 of the 13 percentage units difference reported by Minson & McLeod are attributable to the effect of temperature *per se*. A similar calculation for legumes shows that 2-4 of the 4 percentage units difference are attributable to temperature alone (Minson & Wilson 1980).

Frost

Frost causes significant damage to pastures in many of the sub-tropical areas of the world. Frosts may also occur well into tropical latitudes (Coleman 1964). The frosts in tropical areas may be mild but because most tropical pasture species have only a low frost tolerance (50 percent tissue death at -1 to -4°c) and show little or no capacity to harden during autumn (Ludlow 1980) these frosts cause considerable death of above ground herbage. Temperate pasture grasses can tolerate temperatures down to -25°C (Ludlow 1980) but nevertheless frost damage can still be significant in temperate pasture regions (Reid *et al*. 1967, Langlands & Holmes 1978, Beaty & Engel 1980). Frost brings about a rapid decline in nutritive quality of grasses (Milford 1960) and especially of tropical legumes, many of which shed their leaves after frosting (Milford 1967). Frost-killed leaves which remain on the plant often decline rapidly in dry matter digestibility (Wilson & Mannetje 1978) and nitrogen content (Muldoon & Pearson 1977). The freezing and thawing process leads to leaching and respiration losses of the more digestible tissue constituents (Ocumpaugh & Matches 1977, Phipps & Weller 1979). These degradation processes leading to lower nutritive quality are accelerated in regions where rain occurs after frosting (Prine & Burton 1956, Hart *et al*. 1969).

The lower quality of frost-killed herbage is clearly reflected in a marked decline of voluntary intake (Reid *et al*. 1967) and poorer animal performance (Utley *et al*. 1978).

NUTRITIONAL FACTORS

This section deals with fertilizer effects on the dry matter digestibility of herbage and not with the mineral composition of the plant. Often it is difficult to distinguish whether the effect results from a change in herbage composition or a correction of mineral deficiency or imbalance in the animal. The later topic is beyond the scope of this paper and is reviewed elsewhere (Reid & Jung 1974, Fleming 1977).

Nitrogen

Because of its importance for manipulating pasture grass production there is a voluminous literature on the influence of nitrogen supply on herbage quality. Intake of herbage is restricted when crude protein levels are less than *ca*. 7 percent (Milford & Minson 1965); this limitation occurs more frequently in tropical than temperate grasses and is rare for legumes. Increase in crude protein above this critical level usually does not lead to further improvement in intake (Horn *et al*. 1979). Nitrogen fertilizer may also increase the water content of the herbage (Wilman 1975) and some believe that this may be a factor limiting intake under some circumstances (Behaeghe & Carlier 1973).

The effect of nitrogen fertilizer on herbage dry matter digestibility is complex and variable (e.g. Kreuger & Scholl 1970, Minson 1973, Wilman 1975). A survey of 80 references from recent years indicates a roughly equal distribution of positive, nil or negative effects of nitrogen on dry matter digestibility of grasses. The ultimate effect on dry matter digestibility depends on the balance between the beneficial and the detrimental effects of nitrogen on sward development and tissue composition. The beneficial effects arise from the stimulated new growth of tissue which has high protein content and low cell wall and lignin content (Ford & Williams 1973, Whitney 1974) leading to higher dry matter digestibility of leaf and whole sward if harvested soon after nitrogen application (Whitney 1974, Gomide et al. 1969). These beneficial changes in young leaf may be more pronounced for tropical grasses where cell wall content varies inversely with crude protein level (Wilson 1973) than the temperate grasses in which higher crude protein is associated with lower soluble carbohydrate (i.e. an interchange of two readily digestible constituents) and does not change proportion of cell wall (Alberda 1965). However, with high nitrogen supply grasses may progress more rapidly to flowering (Wilson 1959, Donaldson and Rootman 1977) and mature growth stages with a higher ratio of flowering to vegative culms (Beliuchenko 1979) and greater stem development (Huguet & Gillet 1974, Wilman et al. 1976, Saibro et al. 1978). The accentuation of stem development will vary with species growth habit and propensity for flowering (Minson 1973). Unless managed correctly a high leaf area index will result with nitrogen fertilizer and this accelerates the rate of leaf senescence (Davies 1969, Wilson & Mannetje 1978) and the accumulation of dead material at the base of the sward (Beaty & Engel 1980). These changes eventually lead to higher cell wall or lignin content (Whitney 1974) and lower dry matter digestibility than for low nitrogen fertilized swards (Herrera 1977, Donaldson & Rootman 1977). Also, in tropical grasses especially, because of their rapid growth response to nitrogen, even high additions of nitrogen may have little effect on tissue crude protein level after long regrowth intervals (Henzell 1963) and hence are unlikely to greatly influence the cell wall content and thereby the dry matter digestibility.

Clearly, then, the effect of nitrogen fertilizer may strongly interact with growth interval after application and age or stage of growth of material when sampled. High nitrogen supply will increase or have no effect on the dry matter digestibility of young leaf material or vegetative regrowth but decrease the dry matter digestibility of old leaves and stemmy, mature swards (Gomide et al. 1969, Saibro et al. 1978, Wilson & Mannetje 1978, Horn et al. 1979). These same factors can lead to a similar interaction with season, especially for temperate grasses because in these species high nitrogen supply can decrease dry matter digestibility in late spring when stem development and flowering occurs and increase dry matter digestibility in late summer-autumn when the plants are vegetative and leafy (Kaiser et al. 1974).

Nitrogen fertilizer may also reduce pasture quality by inducing toxic levels of nitrate, oxalate and alkaloids; these aspects are discussed by Hegarty (1982).

Phosphorus

The overwhelming weight of evidence indicates that phosphorus fertilization has little effect on herbage dry matter digestibility (Reid & Jung 1965, Das *et al*. 1974, Van Adrichem & Tingle 1975, Pesant & Dionne 1976, Balasko 1977, Rehm *et al*. 1977) or cell wall content (Miller *et al*. 1961, Johnston *et al*. 1968). There is some indication however that higher phosphorus supply may improve herbage palatability (Reid & Jung 1965), and dry matter intake (Powell *et al*. 1978, Murray 1978).

Other nutrients

Potassium fertilizer appears to have little influence on dry matter digestibility or cell wall content of grasses (Miller *et al*. 1961, Reid & Jung 1965, Calder & MacLeod 1968, Rouquette *et al*. 1972, Hannaway & Reynolds 1979). Although Calder & MacLeod (1968) reported an increase in dry matter digestibility of lucerne with potassium fertilizer, and Fernandes *et al*. (1970) found that potassium fertilizer improved cellulose digestibility of tropical grasses.

Generally, other nutrient elements have been little studied with respect to their effect on pasture quality other than mineral content. Fertilizer sulphur increased the dry matter digestibility and intake of pangola grass (Rees *et al*. 1974), the improvement in nutritive value being greater than was achieved by feeding a sulphur supplement directly to the animal. However, in later work Goh & Kee (1978) and Rees & Minson (1978) found no effect of sulphur on dry matter digestibility of *Lolium perenne* and *Digitaria decumbens* respectively.

Calcium is important because of its application in superphosphate and its use in liming of pastures. It also has an important role in cell wall development. Rees & Minson (1976) found that calcium fertilization of pangola grass increased its dry matter digestibility and intake and concluded that this was because of changes in the structural composition of the grass. In contrast, Reid & Jung (1965) found no evidence for a change in herbage dry matter digestibility with liming, and Odom *et al*. 1980 reported that the calcium level in the herbage had no effect on the dry matter digestibility of three temperate grasses.

CONCLUSIONS

The most important environmental influence on nutritive quality is growth temperature. High growth temperatures accelerate stem development and the maturation processes in plants leading to increase in tissue cell wall content and lignification and decrease in herbage dry matter digestibility. Furthermore, the potential dry matter digestibility of newly formed tissue is lower for plants grown at high than at low temperature. These temperature influences are particularly evident in grasses and could lead to at least 5-10 percentage units difference in dry matter digestibility between species grown in tropical versus temperate climates. Beneficial effects of low temperature on herbage quality in cooler seasons of the year may not always be evident because slower growth at these times especially of tropical species usually means that even the youngest herbage on the pasture is relatively mature and of low dry matter digestibility. High irradiance and moderate soil water deficits or water stress are mostly beneficial rather than detrimental to the quality of grasses, although, severe droughts in which all available herbage is killed undoubtedly lower nutritive value and limit animal production. Generally, it seems that the nutritive quality of legumes is less influenced by environmental factors than that of the grasses but more information on the response of legumes is needed.

Fertilizer effects on herbage dry matter digestibility and associated cell wall and lignin characteristics are small and inconsistent. Nitrogen may have positive, nil or negative effects on these characteristics, often increasing dry matter digestibility in young tissue and early regrowth but decreasing dry matter digestibility of mature tissue. Phosphorus and potassium fertilization rarely seem to have much direct influence on herbage dry matter digestibility but may of course improve nutritive quality indirectly by stimulating rapid new growth of severely nutrient deficient plants or by alteration of botanical composition of the pasture. Several reports of improved digestibility from sulphur and calcium fertilization of herbage have been noted.

REFERENCES

Alberda, Th. (1965) The influence of temperature, light intensity and nitrate concentration on dry matter production and chemical composition of Lolium perenne L. Netherlands Journal of Agricultural Science 13, 335-360.

Allden, W.G. (1982) Problems of animal production from mediterranean pastures. In: Nutritional limits to animal production from pastures. Editor J.B. Hacker. Farnham Royal, U.K., Commonwealth Agricultural Bureaux. pp.45-65.

Allinson, D. W. (1971) Influence of photoperiod and thermoperiod on the IVDMD and cell wall components of tall fescue. Crop Science 11, 456-458.

Andrews, A. C.; Crofts, F. C. (1979) Hybrid Bermudagrass compared with kikuyu and common couch in coastal New South Wales. 2. Crude protein content, and estimated in vivo digestibility. Australian Journal of Experimental Agriculture and Animal Husbandry 19, 444-447.

Auda, H.; Blaser, R. E.; Brown, R. H. (1966) Tillering and carbohydrate contents of orchardgrass as influenced by environmental factors. Crop Science 6, 139-143.

Baker, E. A. (1974) The influence of environment on leaf wax development in Brassica oleracea var. gemmifera. New Phytologist 73, 955-966.

Balasko, J. A. (1977) Effects of N, P, and K fertilization on yield and quality of tall fescue forage in winter. Agronomy Journal 69, 425-428.

Barss, A. F. (1930) Effect of moisture supply on development of Pyrus communis. Botanical Gazette 90, 151-176.

Beaty, E. R.; Engel, J. L. (1980) Forage quality measurements and forage research-A review, critique and interpretation. Journal of Range Management 33, 49-54.

Behaege, T. J., Carlier, L. A. (1973) Influence of nitrogen levels on quality and yield of herbage under mowing and grazing conditions. Vaxtodling 28, 52-66.

Beliuchenko, I. S. (1979) Factors affecting the structure of pure grass pastures. 1. Influence of stalk types and soil fertility. Cuban Journal of Agricultural Science 13, 173-189.

Blaser, R. E.,; Brown, R. H.; Bryant, H. T. (1966) The relationships between carbohydrate accumulation and growth of grasses under different microclimates. Proceedings of the 10th International Grassland Congress, Helsinki 147-150.

Bowman, D. E.; Law, A. G. (1964) Effects of temperature and daylength on the development of lignin, cellulose and protein in Dactylis glomerata L. and Bromus inermis Leyss. Agronomy Journal 56, 177-179.

Burton, G. W.; Jackson, J. E.; Knox, F. E. (1959) Influence of light reduction on the production, persistence and chemical composition of coastal bermudagrass Cynodon dactylon. Agronomy Journal 51, 537-542.

Butterworth, M. H.; Groom, C. G.; Wilson, P. N. (1961) The intake of pangola grass (Digitaria decumbens, Stent.) under wet- and dry-season conditions in Trinidad. Journal of Agricultural Science, Cambridge 56, 407-409.

Cain, S. A.; Potzger, J. E. (1933) A comparison of leaf tissues of Gaylussacia baccata (Wang.) C. Koch. and Vaccinium vacillans Kalm. American Midland Naturalist 14, 97-112.

Calder, F. W.; MacLeod, L. B. (1968) In vitro digestibility of forage species as affected by fertilizer application, stage of development and harvest dates. Canadian Journal of Plant Science 48, 17-24.

Caro-Costas, R.; Vicente-Chandler, J.; Abruna, F. (1976) Comparison of heavily fertilized Congo, star and pangola grass pastures in the humid mountain region of Puerto Rico. Journal of Agriculture of University of Puerto Rico 60, 179-185.

Chabot, B. F.; Chabot, J. F. (1977) Effects of light and temperature on leaf anatomy and photosynthesis in Fragaria vesca. Oecologia 26, 363-377.

Chenost, M. (1972) Observations preliminaires sur les variations saisonnieres de la quantite d'aliment ingeree par les caprins en milieu tropical humide. Annales de Zootechnie 21, 113-120.

Clark, D. A.; Brougham, R. W. (1979) Feed intake of grazing Friesian bulls. Proceedings of the New Zealand Society of Animal Production 39, 265-274.
Coleman, R. G. (1964) Frosts and low night temperatures as limitations to pasture development in subtropical eastern Australia. CSIRO, Divison of Tropical Pastures Technical Paper No. 3.
Cooper, J. P.; Tainton, N. M. (1968) Light and temperature requirements for the growth of tropical and temperate grasses. Herbage Abstracts 38, 167-176.
Das, B.; Arora, S. K.; Luthra, Y. P. (1974) Fractionation of structural carbohydrates and in vitro digestibility of oat forage as influenced by nitrogen, phosphorus and stage of maturity. Zeitschrift fur Acker-und Pflanzenbau 139, 307-313.
Davies, I. (1969) The influence of management on tiller development and herbage growth. Welsh Plant Breeding Station Technical Bulletin No. 3. 121pp.
Deinum, B. (1966) Influence of some climatological factors on the chemical composition and feeding value of herbage. Proceedings of the 10th International Grassland Congress, Helsinki 415-418.
Deinum, B. (1971) Climate, nitrogen and grass. 3. Some effects of light intensity on nitrogen metabolism. Netherlands Journal of Agricultural Science 19, 184-188.
Deinum, B.; Dirven, J.G.P. (1967) Informative experiment on the influence of light intensity and temperature on dry-matter production and chemical composition of Brachiaria ruziziensis Germain et Evrard. Surinaamse Landbouw 15, 5-10.
Deinum, B.; Dirven, J.G.P. (1972) Influence of age, light intensity and temperature on the production and chemical composition of Congo grass (Brachiaria ruziziensis Germain et Evrard). Netherlands Journal of Agricultural Science 20, 125-132.
Deinum, B.; Dirven, J.G.P. (1973) Preliminary investigations on the digestibility of some tropical grasses grown under different temperature regimes. Surinaamse Landbouw 21, 121-126.
Deinum, B; Dirven, J.G.P. (1975) Climate, nitrogen and grass. 6. Comparison of yield and chemical composition of some tropical and temperate grass species grown at different temperatures. Netherlands Journal of Agricultural Science 23, 69-82.
Deinum, B.; Dirven, J.G.P. (1976) Climate nitrogen and grass. 7. Comparison of production and chemical composition of Brachiaria ruziziensis and Setaria sphacelata grown at different temperatures. Netherlands Journal of Agricultural Science 24, 67-78.
Deinum, B.; Van Soest, P. J.; Van Es, A.J.H. (1968) Climate, nitrogen and grass. 2. The influence of light intensity, temperature and nitrogen on vivo digestibility of grass and prediction of these effects from some chemical procedures. Netherlands Journal of Agricultural Science 16, 217-223.
Dirven, J.G.P.; Deinum, B. (1977) The effect of temperature on the digestibility of grasses. An analysis. Forage Research 3, 1-17.
Donaldson, C. H.; Rootman, G. T. (1977) Evaluation of Cenchrus ciliaris: 1. Effects of nitrogen level and cutting frequency on digestibility and voluntary intake. Proceedings of the Grassland Society of South Africa 12, 91-93.
Downes, R. W.; Christian, K. R.; Freer, M. (1974) Nutritive value of oats and sudan grass grown at controlled temperatures. Australian Journal of Agricultural Research 25, 89-97.
Dent, J. W.; Aldrich, D.T.A. (1963) The interrelationships between heading date, yield, chemical composition and digestibility in varieties of perennial ryegrass, timothy, cocksfoot and meadow fescue. National Institute of Agricultural Botany Journal 9, 261-281.
Ehara, K.; Tanaka, S. (1961) Effect of temperature on the growth behavior and chemical composition of the warm-and cool-season grasses. Proceedings of the Crop Science Society of Japan 29, 304-306.
Eriksen, F.; Whitney, A.S. (1977) Performance of tropical forage grasses and legumes under different light intensitites. Proceedings Regional Seminar on Pasture Research, Honiara, Solomon Islands, 180-190.
Evans, P.S. (1964) A comparison of some aspects of the anatomy and morphology of Italian ryegrass (Lolium multiflorum Lam.) and perennial ryegrass (L. perenne L.). New Zealand Journal of Botany 2, 120-130.
Faix, J. J. (1974) The effect of temperature and daylength on the quality of morphological components of three legumes. Ph.D. Thesis, Cornell University, U.S.A.
Fernandes, A.P.M.; Gomide, J.A.; Braga, J.M. (1970) Efeito da adubacao potassica sobre a producao e valor nutritivo de algumas gramineas forrageiras tropicais. Experientiae (Vicosa) 10, 187-208.
Fisher, M. J.(1980) The influence of water stress on nitrogen and phosphorus uptake and concentrations in Townsville style (Stylosanthes humilis). Australian Journal of Experimental Agriculture and Animal Husbandry 20, 175-180.
Fleming, G. A. (1977) Mineral disorders associated with grassland farming. Proceedings of an International Meeting on Animal Production from Temperate Grassland, Dublin, 88-95.
Fletcher, L. R. (1976) Effect of season and regrowth period on the in vitro digestibility of irrigated lucerne in Canterbury. New Zealand Journal of Experimental Agriculture 4, 469-471.
Ford, C.W.; Williams, W. T. (1973) In vitro digestibility and carbohydrate composition of Digitaria decumbens and Setaria anceps grown at different levels of nitrogen fertilizer. Australian Journal of Agricultural Research 24, 309-316.

Ford, C. W.; Wilson, J. R. (1981) Changes in levels of solutes during osmotic adjustment to water stress in leaves of four tropical pasture species. Australian Journal of Plant Physiology 8, 77-91.
Ford, C. W., Morrison, I. M.; Wilson, J. R. (1979) Temperature effects on lignin, hemicellulose and cellulose in tropical and temperate grasses. Australian Journal of Agricultural Research 30, 621-633.
French, M. H. (1961) Observations on the digestibility of pasture herbage. Turrialba 11, 78-84.
Fujita, T. (1942) Physiological studies on changes of membrane substances in higher plants. Kyushu University, Faculty of Agricultural Science Bulletin No. 10.
Garwood, E. A.; Tyson, K. C.; Sinclair, J. (1979) Use of water by six grass species 1. Dry-matter yields and response to irrigation. Journal of Agricultural Science, Cambridge 93, 13-24.
Garza, R. T.; Barnes, R. F.; Mott, G. O.; Rhykerd, C. L. (1965) Influence of light intensity, temperature and growing period on the growth, chemical composition and digestibility of Culver and Tanverde alflafa seedlings. Agronomy Journal 57, 417-420.
Gerakis, P. A.; Guerrero, F. P.; Williams, W. A. (1975) Growth, water relations and nutrition of three grassland annuals as affected by drought. Journal of Applied Ecology 12, 125-135.
Gifford, R. O.; Jensen, E. H. (1967) Some effects of soil moisture regimes and bulk density on forage quality in the greenhouse. Agronomy Journal 59, 75-77.
Goh, K. M.; Kee, K. K. (1978) Effects of nitrogen and sulphur fertilization on the digestibility and chemical composition of perennial ryegrass (Lolium perenne L.). Plant and Soil 50, 161-177.
Gomide, J. A.; Holler, C. H.; Mott, G. O.; Conrad, J. H.; Hill, D. L. (1969) Effect of plant age and nitrogen fertilization on the chemical composition and in vitro cellulose digestibility of tropical grasses. Agronomy Journal 61, 116-120.
Grace, J.; Russell, G. (1977) The effect of wind on grasses. III. Influence of continuous drought or wind on anatomy and water relations in Festuca arundinacea Schreb. Journal of Experimental Botany 28, 268-278.
Greenfield, P. L.; Smith, D. (1973) Influence of temperature change at bud on composition of alfalfa at first flower. Agronomy Journal 65, 871-874.
Hacker, J.B.; Minson, D.J. (1972) Varietal differences in in vitro dry matter digestibility in Setaria, and the effects of site, age, and season. Australian Journal of Agricultural Research 23, 959-967.
Hannaway, D. B; Reynolds, J. H. (1979) Seasonal changes in organic acids, water-soluble carbohydrates, and neutral detergent fiber in tall fescue forage as influenced by N and K fertilization. Agronomy Journal 71, 493-496.
Hardison, W. A. (1966) Chemical composition, nutrient content and potential milk-producing capacity of fresh tropical herbage. University of Philippines Dairy Training and Research Institute Technical Bulletin 1, 37pp.
Hart, R. H.; Monson, W. G.; Lowrey, R. S. (1969) Autumn-saved coastal bermudagrass (Cynodon dactylon (L.) Pers.): Effects of age and fertilization on quality. Agronomy Journal 61, 940-941.
Hawthorne, H. A. (1956) Phosphorus accumulation of plants as a function of moisture stress, Ph.D. thesis, University of California, U.S.A.
Hegarty, M.P. (1982) Deleterious factors in forages affecting animal production. In: Nutritional limits to animal production from pastures. Editor J.B. Hacker. Farnham Royal, U.K., Commonwealth Agricultural Bureaux. pp. 133-150.
Henzell, E. F. (1963) Nitrogen fertilizer responses of pasture grasses in South-eastern Queensland. Australian Journal of Experimental Agriculture and Animal Husbandry 3, 290-299.
Herrera, R. S. (1977) Nitrogen fertilization and age of regrowth in the chemical composition of Cynodon dactylon cv. Coast cross 1. Cuban Journal of Agricultural Science 11, 331-345.
Hight, G. K.; Sinclair, D. P.; Lancaster, R. J. (1968) Some effects of shading and of nitrogen fertilizer on the chemical composition of freeze-dried and oven-dried herbage, and on the nutritive value of oven-dried herbage fed to sheep. New Zealand Journal of Agricultural Research 11, 286-302.
Hiridiglou, M.; Dermine, P.; Hamilton, H. A. (1966) Chemical composition and in vitro digestibility of forage as affected by season in northern Ontario. Canadian Journal of Plant Science 46, 101-109.
Horn, F. P.; Telford, J. P.; McCroskey, J. E.; Stephens, D. F.; Whiteman, J. V.; Totusek, R. (1979) Relationship of animal performance and dry matter intake to chemical constituents of grazed forage. Journal of Animal Science 49, 1051-1058.
Hoveland, C. S.; Monson, W. G. (1980) Genetic and environmental effects on forage quality. In: Crop quality, storage and utilization. Madison, Wisconsin, American Society of Agronomy, 139-168.
Huquet, L.; Gillet, M. (1974) The influence of nitrogen fertilizer and autumn management on the quality of green forages. Vaxtodling 29, 100-110.
Hull, H. M.; Morton, H. L., Wharrie, J. R. (1975) Environmental influences on cuticle development and resultant foliar penetration. Botanical Review 41, 421-452.
Hutton, J. B. (1974) The effect of irrigation on forage yield, dairy cow production and intake under intensive grazing conditions in New Zealand. Proceedings of the 19th International Dairy Congress, New Delhi. Vol IE, 73-74.

Ivory, D. A.; Stobbs, T.H.; McLeod, H.N.; Whiteman, P.C. (1974) Effect of day and night temperatures on estimated dry matter digestibility of Cenchrus ciliaris and Pennisetum clandestinum. Journal of the Australian Institute of Agricultural Science 40, 156-158.

Jensen, E. H.; Massengale, M. A.; Chilcote, D. O. (1967) Environmental effects on growth and quality of alfalfa. University of Nevada, Western Region Research Publication T9, 1-36.

Johnson, W. L.; Guerrero, J.; Pezo, D. (1973) Cell-wall constituents and in vitro digestibility of Napier grass (Pennisetum purpureum). Journal of Animal Science 37, 1255-1261.

Johnston, A.; Bezau, L. M.; Smoliak, S. (1967) Variation in silica content of range grasses. Canadian Journal of Plant Science 47, 65-71.

Johnston, A.; Bezau, L. M.; Smith, A. D.; Lutwick, L. E. (1968) Nutritive value and digestibility of fertilized rough fescue. Canadian Journal of Plant Science 48, 351-355.

Jones, R. J. (1973) The effect of cutting management on the yield, chemical composition and in vitro digestibility of Trifolium semipilosum grown with Paspalum dilalatum in a subtropical environment. Tropical Grasslands 7, 277-284.

Kaiser, C. J., Matches, A. G.; Martz, F. A.; Mott, G. O. (1974) Seasonal trend of in vitro dry matter digestibility and animal performance from grazed tall fescue (Festuca arundinacea Schreb.) pastures. Proceedings of the 12th International Grassland Congress, Moscow 1, 294-305.

Karue, C. N. (1975) The nutritive value of herbage in semi-arid lands of E. Africa. II. Seasonal influence on the nutritive value of Themeda triandra. East Africa Agriculture and Forestry Journal 40, 372-387.

Ketth, J. M.; Ranawana, S.S.E. (1971) Kikuyu grass: Pennisetum clandestinum, Hochst ex Chiov and its value in the montane region of Ceylon. 2. Nutritive value and animal production aspects. Tropical Agriculturist 127, 93-103.

Kobayashi, T.; Nishimura, S.; Tanaka, S. (1977) Comparative growth responses of seven tropical and subtropical grasses to various controlled temperatures. Science Bulletin Faculty of Agriculture, Kyushu University 32, 93-99.

Koller, H. R., Clark, N. A. (1965) Effect of plant density and moisture supply on the forage quality of Sudan grass (Sorghum sudanense (Piper) Staph). Agronomy Journal 57, 591-593.

Kreuger, C. R.; Scholl, J. M. (1970) Performance of bromegrass, orchardgrass and reed canary grass grown at five nitrogen levels and with alfalfa. University of Wisconsin, College of Agriculture and Life Sciences Research Report 69.

Langlands, J. P.; Holmes, C. R. (1978) The nutrition of ruminants grazing native and improved pastures. 1. Seasonal variation in the diet selected by grazing sheep and cattle. Australian Journal of Agricultural Research 29, 863-874.

Launchbaugh, J.L. (1957) The effect of stocking rate on cattle gains and on native short grass vegetation in west-central Kansas. Kansas Agricultural Experiment Station Bulletin 394.

Lazenby, A.; Rogers, H. H. (1964) Selection criteria in grass breeding. III. Chemical composition. Journal of Agricultural Science, Cambridge 63, 323-333.

Ludlow, M. M. (1980). Stress physiology of tropical pasture plants. Tropical Grasslands, 14, 136-145.

Ludlow, M. M.; Ng, T. T. (1974) Water stress suspends leaf ageing. Plant Science Letters 3, 235-240.

Mannetje, L.'t (1975) Effect of daylength and temperature on introduced legumes and grasses for the tropics and subtropics of coastal Australia. 2. N-concentration, estimated digestibility and leafiness. Australian Journal of Experimental Agriculture and Animal Husbandry 15, 256-263.

Mannetje, L. 't (1982) Problems of animal production from tropical pastures. In: Nutritional limits to animal production from pastures. Editor J.B. Hacker. Farnham Royal, U.K., Commonwealth Agricultural Bureaux. pp. 67-85.

Masuda, Y. (1977) Comparisons of the in vitro dry matter digestibility of forage oats grown under different temperatures and light intensities. Journal of the Faculty of Agriculture, Kyushu University 21, 17-24.

McWilliam, J. R. (1978) Responses of pasture plants to temperature. In: Plant relations in pastures. Editor J.R. Wilson. Melbourne, CSIRO, pp.17-34.

Michell, P. J. (1973) Digestibility and voluntary intake measurements on regrowths of six Tasmanian pasture species. Australian Journal of Experimental Agriculture and Animal Husbandry 13, 158-164.

Milford, R. (1960) Criteria for expressing nutritional values of subtropical grasses. Australian Journal of Agricultural Research 11, 121-137.

Milford, R. (1967) Nutritive values and chemical composition of seven tropical legumes and lucerne grown in subtropical south-eastern Queensland. Australian Journal of Experimental Agriculture and Animal Husbandry 7, 540-545.

Milford, R., Minson, D. J. (1965) Intake of tropical pasture species. Proceedings of the 9th International Grassland Congress, Sao Paulo, 815-822.

Miller, W. J.; Donker, J. D.; Adams, W. E.; Stelly, M. (1961) Effect of nitrogen fertilization on the crude fiber content and crude fiber-nitrogen relationship of coastal and common bermudagrass. Agronomy Journal 53, 173-174.

Minson, D. J. (1973) Effect of fertilizer nitrogen on digestibility and voluntary intake of Chloris gayana, Digitaria decumbens and Pennisetum clandestinum. Australian Journal of Experimental Agriculture and Animal Husbandry 13, 153-157.

Minson, D. J.; McLeod, M. N. (1970) The digestibility of temperate and tropical grasses. Proceedings of the 11th International Grassland Congress, Surfers Paradise, Australia, 719-722.

Minson, D.J.; Wilson, J.R. (1980) Comparative digestibility of tropical and temperate forage - a contrast between grasses and legumes. Journal of the Australian Institute of Agricultural Science 46, 247-249.

Moir, K. W.; Wilson, J. R.; Blight, G. W. (1977) The in vitro digested cell wall and fermentation characteristics of grasses as affected by temperature and humidity during their growth. Journal of Agricultural Science, Cambridge 88, 217-222.

Morrison Baird, L. A.; Webster, B. D. (1978) Relative humidity as a factor in the structure and biochemistry of plants. Hortscience 13, 556-558.

Muldoon, D. K.; Pearson, C. J. (1977) Hybrid pennisetum in a warm temperature climate: regrowth and stand-over forage production. Australian Journal of Experimental Agriculture and Animal Husbandry 17, 277-283.

Murray, R. M. (1978) Metabolism and nutrition of beef cattle in the tropics. In: Beef cattle production in the tropics. Editors R.M. Murray and K.W. Entwhistle. Townsville, Australia, James Cook University Press, pp.247-298.

Myhr, K.; Saebo, S. (1969) The effects of shade on growth, development and chemical composition in some grass species. State Experiment Station Fureneset, Norway, Report 14, 297-315.

Newman, D.M.R. (1973) The influence of rainfall on the nutritive value of a semi-arid mulga pasture in south-west Queensland. Tropical Grasslands 7, 143-147.

Norton, B.W. (1982) Differences between species in forage quality. In: Nutritional limits to animal production from pastures. Editor J.B. Hacker. Farnham Royal, U.K., Commonwealth Agricultural Bureaux. pp.89-100.

Ocumpaugh, W. R; Matches, A. G. (1977) Autumn-winter yield and quality of tall fescue. Agronomy Journal 69, 639-643.

Odom, J. W.; Haaland, R. L.; Hoveland, C. S.; Anthony, W. B. (1980) Forage quality response of tall fescue, orchardgrass and phalaris to soil fertility level. Agronomy Journal 72, 401-402.

Pate, F. M.; Synder, G. H. (1979) Effect of high water table in organic soil on yield and quality of forage grasses-Lysimeter study. Proceedings Soil and Crop Science Society of Florida 38, 72-75.

Penfound, W. T. (1931) Plant anatomy as conditioned by light intensity and soil moisture. American Journal of Botany 18, 558-572.

Perry, L. J.; Baltensperger, D. D (1979) Leaf and stem yields and forage quality of three N-fertilized warm-season grasses. Agronomy Journal 71, 355-358.

Pesant, A. R.; Dionne, J. L. (1976) Effets de la fertilisation et des regimes hydriques sur le rendement, l'utilization d'eau et la composition chimique de la luzerne et du trefle Ladino. Canadian Journal of Plant Science 56, 293-302.

Peterschmidt, N. A.; Delaney, R. H.; Greene, M. C. (1979) Effects of overirrigation on growth and quality of alfalfa. Agronomy Journal 71, 752-754.

Phipps, R. H.; Weller, R. F. (1979) The development of plant components and their effects on the composition of fresh and ensiled forage maize. 1. The accumulation of dry matter, chemical composition and nutritive value of fresh maize. Journal of Agricultural Science, Cambridge 92, 471-483.

Powell, K.; Reid, R. L.; Balasko, J. A. (1978) Performance of lambs on perennial ryegrass, smooth bromegrass, orchardgrass and tall fescue pastures. II. Mineral utilization, in vitro digestibility and chemical composition of herbage. Journal of Animal Science 46, 1503-1514.

Prine, G. M.; Burton, G. W. (1956) The effect of nitrogen rate and clipping frequency upon the yield, protein content and certain morphological characteristics of coastal bermudagrass (Cynodon dactylon (L) Pers.). Agronomy Journal 48, 296-301.

Rahman, A. A. Abdel; Shalaby, A. F.; Monayeri, M. O. El (1971) Effect of moisture stress on metabolic products and ions accumulation. Plant and Soil 34, 65-90.

Reed, K. F. M. (1978) The effect of season of growth on the feeding value of pasture. Journal of the British Grassland Society 33, 227-234.

Rees, M. C.; Minson, D. J. (1976) Fertilizer calcium as a factor affecting the voluntary intake, digestibility and retention time of pangola grass (Digitaria decumbens) by sheep. British Journal of Nutrition 36, 179-187.

Rees, M. C.; Minson, D. J. (1978) Fertilizer sulphur as a factor affecting voluntary intake, digestibility and retention time of pangola grass (Digatara decumbens) by sheep. British Journal of Nutrition 39, 5-11.

Rees, M. C.; Minson, D. J.; Smith, F. W. (1974) The effect of supplementary and fertilizer sulphur on voluntary intake, digestibility, retention time in the rumen, and site of digestion of pangola grass in sheep. Journal of Agricultural Science, Cambridge 82, 419-422.

Rehm, G. W.; Sorensen, R. C.; Moline, W. J. (1977) Time and rate of fertilization on seeded warm-season and bluegrass pastures. II. Quality and nutrient content. Agronomy Journal 69, 955-961.

Reid, R. L.; Jung, G. A. (1965) Influences of fertilizer treatment on the intake, digestibility and palatability of tall fescue hay. Journal of Animal Science 24, 615-625.

Reid, R. L.; Jung, G. A. (1974) Effects of elements other than nitrogen on the nutritive value of forage In: Forage fertilization. Editor D.A. Mays. Madison, American Society of Agronomy, 395-435.

Reid, R.L.; Jung, G.A. (1982) Problems of animal production from temperate pastures. In: Nutritional limits to animal production from pastures. Editor J.B. Hacker. Farnham Royal, U.K., Commonwealth Agricultural Bureaux. pp.21-43.

Reid, R. L.; Jung, G. A.; Kinsey C. M. (1967) Nutritive value of nitrogen-fertilized orchardgrass pasture at different periods of the year. Agronomy Journal 59, 519-525.

Romero, A.; Siebert, B. D. (1980) Seasonal variations of nitrogen and digestible energy intake of cattle on tropical pasture. Australian Journal of Agricultural Research 31, 393-400.

Rouquette, F. M.; Holt, E. C.; Ellis, W. C. (1972) Effect of N, P, and K fertilizer and stages of maturity on chemical composition of fiber in Kleingrass (Panicum coloratum L.) Agronomy Journal 64, 456-459.

Rouquette, F. M., Keisling, T. C.; Camp, B. J.; Smith, K. L. (1980) Characteristics of the occurrence and some factors associated with reduced palatability of pearl millet. Agronomy Journal 72, 173-174.

Russell, G.; Grace, J. (1978) The effect of wind on grasses. IV. Some influences of drought or wind on Lolium perenne. Journal of Experimental Botany 29, 245-255.

Saibro, J. C.; Hoveland, C. S.; Williams, J. C. (1978) Forage yield and quality of phalaris as affected by N fertilization and defoliation regimes. Agronomy Journal 70, 497-500.

Sakurai, M. (1963) Histological study on the decomposition of pasture grass tissue by livestock digestion. Grasslands Division of Kaoto-Tosan Agricultural Experiment Station Research Report 15.

Sangster, A. G.; Parry, D. W. (1971) Silica deposition in the grass leaf in relation to transpiration and the effect of dinitrophenol. Annals of Botany 35, 667-677.

Shields, L. M. (1950) Leaf xeromorphy as related to physiological and structural influences. Botanical Review 16, 399-447.

Silcock, R. G. (1978) Transpiration and water use efficiency as affected by leaf characteristics of Festuca species. Ph.D. Thesis, University College of Wales, U.K.

Sinnott, E. W. (1960) Plant Morphogenesis. London, McGraw-Hill Book Co. Inc.

Smith, A. E. (1977) Influence of temperature on tall fescue forage quality and culm base carbohydrates. Agronony Journal 69, 745-747.

Smith, D. (1969) Influence of temperature on the yield and chemical composition of 'vernal' alfalfa at first flower. Agronomy Journal 61, 470-472.

Smith, D. (1970) Influence of cool and warm temperatures and temperature reversal at inflorescence emergence on yield and chemical composition of timothy and brome grass at anthesis. Proceedings of the 11th International Grassland Congress, Surfers Paradise, Australia, 510-514.

Smith, D. (1973) The non-structural carbohydrates. In: Chemistry and biochemistry of herbage. Editors G.W. Butler and R.W. Bailey. London, Academic Press, vol. 1, pp.106-155.

Smith, D. (1975) Influence of temperature on growth of Froker oats for forage. II. Concentrations and yields of chemical constituents. Canadian Journal of Plant Science 55, 897-901.

Snaydon, R. W. (1972) The effect of total water supply and frequency of application upon lucerne. II. Chemical composition. Australian Journal of Agricultural Research 23, 253-256.

Spurway, R. A.; Hedges, D. A.; Wheeler, J. L. (1976) The quality and quantity of forage oats sown at intervals during autumn: effects of nitrogen and supplementary nitrogen. Australian Journal of Experimental Agriculture and Animal Husbandry 16, 555-563.

Strickland, R. W. (1973) Dry matter production, digestibility and mineral content of Eragrostis superba Peyr. and E. curvula (Schrad.) Nees at Samford, south eastern Queensland. Tropical Grasslands 7, 233-241.

Taylor, A. O.; Haslemore, R. M.; McLeod, M. N. (1976) Potential of new summer grasses in Northland. III. Laboratory assessments of forage quality. New Zealand Journal of Agricultural Research 19, 483-488.

Tiver, N. S. (1942) Studies of the flax plant. 1. Physiology of growth, stem anatomy and fibre development in fibre flax. Australian Journal of Experimental Biology and Medical Science 20, 149-160.

Utley, P. R.; McCormick, W. C.; Lowerey, R. S. (1978) Weathered grass for wintering brood cows. University of Georgia, College of Agriculture Experiment Stations, Research Report 293, 11pp.

Van Adrichem, M.C.J.; Tingle, J. N. (1975) Effects of nitrogen and phosphorus on the yield and chemical composition of meadow foxtail. Canadian Journal of Plant Science 55, 949-954.

Van Burg, P. F. J. (1962) Interne stikstofbalans, produktie van droge stof en veroudering van gras. Verslag Landbouwkundig Onderzoek Nederland 68-12, 1-131.

Van Dyne, G. M.; Heady, H. F. (1965) Dietary chemical composition of cattle and sheep grazing in common on a dry annual range. Journal of Range Management 18, 78-86.

Van Soest, P.J.; Mertens, D.R.; Deinum, B. (1978) Preharvest factors influencing quality of conserved forage. Journal of Animal Science 47, 712-720.

Venning, F. D. (1949) Stimulation by wind motion of collenchyma formation in celery petioles. Botanical Gazette 110, 511-514.

Vough, L R.; Marten, G. C. (1971) Influence of soil moisture and ambient temperature on yield and quality of alflafa forage. Agronomy Journal 63, 40-42.

Wangermann, E. (1961) The effect of water supply and humidity on growth and development. In: Handbuch der Pflanzenphsiologie. Editor W. Ruhland. Berlin, Springer Verlag. Vol XVI, pp.618-633.

Whitehead, F. H.; Luti R. (1962) Experimental studies of the effect of wind on plant growth and anatomy. 1. Zea mays. New Phytologist 61, 56-58.

Whitney, A. S. (1974) Growth of kikuyu grass (Pennisetum clandestinum) under clipping. Effects of nitrogen fertilization, cutting interval, and season on yields and forage characteristics. Agronomy Journal 66, 281-287.

Wilkinson, J. F.; Beard, J. B. (1975) Anatomical response of 'Merion' Kentucky bluegrass and 'Pennlawn' red fescue at reduced light intentsities. Crop Science 16, 189-194.

Wilman, D. (1975) Nitrogen and Italian reygrass. 1. Growth up to 14 weeks: dry matter yield and digestibility. Journal of the British Grassland Society 30, 141-147.

Wilman, D.; Ojuederie, B. M.; Asare, E. O. (1976) Nitrogen and Italian ryegrass. 3. Growth up to 14 weeks: yield, proportions, digestibilities and nitrogen contents of crop fractions, and tiller populations. Journal of the British Grassland Society 31, 73-79.

Wilson, J. R. (1959) The influence of time of tiller origin and nitrogen level on the floral initiation and ear emergence of four pasture grasses. New Zealand Journal of Agricultural Research 2, 915-932.

Wilson, J. R. (1973) The influence of aerial environment, nitrogen supply, and ontogenetical changes on the chemical composition and digestibility of Panicum maximum Jacq. var. trichoglume Eyles. Australian Journal of Agricultural Research 24, 543-556.

Wilson, J. R. (1981) The effects of water stress on herbage quality. Proceedings of the 14th International Grassland Congress, Lexington, U.S.A. (in press)

Wilson, J. R; Ford, C. W. (1971) Temperature influences on the growth, digestibility, and carbohydrate composition of two tropical grasses, Panicum maximum var. trichoglume and Setaria sphacelata and two cultivars of the temperate grass, Lolium perenne. Australian Journal of Agricultural Research 22, 563-571.

Wilson, J. R.; Ford, C. W. (1973) Temperature influences on the in vitro digestibility and soluble carbohydrate accumulation of tropical and temperate grasses. Australian Journal of Agricultural Research 24, 187-198.

Wilson, J. R.; Mannetje, L.'t (1978) Senescence, digestibility and carbohydrate content of buffel grass and green panic leaves in swards. Australian Journal of Agricultural Research 29, 503-516.

Wilson, J.R.; Minson, D.J. (1980) Prospects for improving the digestibility and intake of tropical grasses. Tropical Grasslands 14, 253-259.

Wilson, J.R.; Ng T. T. (1975) Influences of water stress on parameters associated with herbage quality of Panicum maximum var. trichoglume. Australian Journal of Agricultural Research 26, 127-136.

Wilson, J. R; Taylor, A. O.; Dolby, G. R. (1976) Temperature and atmospheric humidity effects on cell wall content and dry matter digestibility of some tropical and temperate grasses. New Zealand Journal of Agricultural Research 19, 41-46.

Wilson, J. R.; Ludlow, M. M.; Fisher, M. J.; Schulze, E. D. (1980), Adaptation to water stress of the leaf water relations of four tropical forage species. Australian Journal of Plant Physiology 7, 207-220.

Wong, C. C. (1978) The influence of shading and defoliation on growth and forage quality of green panic and siratro in pure and mixed swards. M. Agr. Sci. Thesis, University of Queensland, Australia.

Woodman, H. E.; Norman, D. B.; French, M. H. (1931) Nutritive value of pasture. VII. The influence of the intensity of grazing on the yield, composition and nutritive value of pasture herbage. Journal of Agricultural Science, Cambridge 21, 267-323.

Wurster, M. J.; Ross, J. G.; Kamstra, L. D., Bullis, S. S. (1971) Effect of droughty soil on digestibility criteria in three cool season forage grasses. Proceedings of the South Dakota Academy of Science 50, 90-94.

DELETERIOUS FACTORS IN FORAGES AFFECTING ANIMAL PRODUCTION

M.P. HEGARTY

CSIRO, Division of Tropical Crops and Pastures, Cunningham Laboratory, St. Lucia, Queensland 4067, Australia.

ABSTRACT

Reduced productivity is the commonest indication that livestock grazing certain forages are suffering from a diet-dependent disorder. Deleterious or toxic substances in the forage or associated with a forage are important causes of such disorders. Some factors seriously affect the health of the animals while others cause reduced production without visible signs of ill-health.

Livestock on rangelands in many countries are exposed to a wide variety of poisonous plants. Environmental conditions and management practices can exacerbate the problems. Grasses and legumes used in sown pastures also contain deleterious substances. Some general information from the literature is presented to show the estimated costs to the livestock industry caused by poisonous plants.

Examples of disorders associated with the following classes of plant constituents are presented - proteins (bloat), amino acids (mimosine toxicity), alkaloids (pyrrolizidine alkaloid toxicities, and problems with alkaloids in improved pasture grasses and legumes), nitrate and nitro-compounds (nitrate toxicity and miserotoxin), organic acids (oxalic acid and fluoro-acids), cyanoglycosides (cyanide poisoning and goitre), toxic sulphur compounds (goitre and anaemia), heterocyclic compounds (infertility caused by phytooestrogens), mycotoxins (facial eczema, lupinosis, and several staggers syndromes).

Some disorders have a complex of causes which involve soil, plant, weather and microbial factors and are very difficult to predict. Prevention and control of dietary dependent disorders is also difficult because of the wide range of substances involved, the differing susceptibilities of animals and the influence of management. Methods aimed at preventing poisoning which are being investigated include new management practices, manipulation of rumen metabolism to give enhanced destruction of toxins in the rumen, manipulation of liver metabolism for certain hepatotoxins, protective immunization and protection by oral administration.

There have been a number of successful plant breeding programs in both grasses and legumes, which have resulted in production of non-toxic cultivars or cultivars of low toxicity.

INTRODUCTION

Many factors contribute to disorders observed in livestock ingesting forage. Reid (1973), in a comprehensive review of dietary-dependent disorders, classified these factors into seven groups according to their causes (1) nutritional mismatching; (2) mineral poisoning; (3) poisoning by naturally occuring substances; (4) disturbances of ruminant physiology or metabolism; (5) interactions of nutrition and noxious agents; (6) poisoning by chemicals introduced by man; and (7) physical damage. Although the groups frequently overlap, the classification is particularly useful in demonstrating the wide variety of disorders that can occur. This review is concerned principally with disorders caused by poisonous substances in or associated with the forage and with disturbances in physiology and metabolism produced by some of these substances. Some aspects of disorders produced by nutritional mismatching and mineral toxicity are presented later in this symposium (Little 1982).

The commonest indicator of the presence of a diet-dependent disorder is a reduction in productivity (weight gains, reduced production of milk, meat or wool). In some disorders this reduction takes place without seriously affecting the health of the animals but in others there are dramatic acute and chronic effects which put the animals' health at risk.

Some of the most useful and widely grazed sown pasture legumes and grasses contain substances that under certain conditions cause undesirable effects in livestock. High nutritive value is often associated with potential toxicity and these risks are accepted in taking advantage of the productive capacity of the plants. When the conditions leading to toxicity are understood the plants may often be used to advantage without serious risk.

Livestock grazing the ranges throughout the world are in a dangerous environment. Many of the species regularly grazed may be toxic. Patterns of land use and climatic conditions may restrict animals to a diet of which toxic species form the major part. For example, serious losses of sheep in the inland south east of New South Wales and Victoria were shown to be caused by a combination of copper poisoning and pyrrolizidine alkaloidosis caused by ingestion of *Heliotropium europaeum* - a free seeding weed which provides practically the only green material after summer rainstorms in a predominatly winter rainfall area (Bull *et al*. 1956). Numerous examples of seasonal toxicity problems in arid ranges and the mountain ranges of the western United States have been reported by Keeler (1975).

ECONOMIC SIGNIFICANCE OF SOME DISORDERS

The limitations to animal production imposed by deleterious substances in forages in terms of increased costs to the livestock industries are considerable but it is difficult to express them in monetary terms. The following information is intended to draw attention to the major disorders caused by plant toxins and to give some idea of their extent and cost to the industries. Because adequate systems for recording stock losses caused by poisonous plants are not widely used these figures are estimates only and were made some years ago. Despite their obvious limitations they are useful guides.

In the United States the Agricultural Research Service (1965) estimated that the annual monetary losses from plant poisonings for the period 1951-60 was $17 million for cattle and $6 million for sheep. James (1978a), using a mortality figure of 3-5 percent estimated that the loss amounted to $51 million annually of which $23 million was lost in the western states. The National Academy of Science (1968) calculated that 9 percent of the nutritionally sick animals in the western United States were suffering from plant poisoning. This figure is in addition to the 5 percent mortality loss but does not include various associated losses.

Information on losses associated with improved pasture is also scanty. The annual economic loss from bloat in the United States has been estimated (Agricultural Research Service 1965) as $105 million with a mortality rate of 0.5 percent. Similar figures for the New Zealand dairy cattle industry are NZ$5 million with a mortality rate in excess of 1 percent (Clarke & Reid 1970). Steffert (1970) estimated that in New Zealand facial eczema was responsible for the loss of 7500 dairy cows in a season.

In Western Australia, Lightfoot (1974) estimated that at that time about 1 million ewes of the Australia flock failed to lamb because of ingestion of oestrogenic cultivars of Trifolium subterraneum. However management methods have now been established which minimize the effects in existing pastures.

Estimates of annual losses of sheep from annual ryegrass (Lolium rigidum) toxicity rarely exceed 1000 in Western Australia and South Australia. Losses from lupinosis in Western Australia from 1970-1975 varied from 1000 to 3000 sheep a year depending on the season (C.C.J. Culvenor, personal communication).

ROLE OF SECONDARY PLANT METABOLITES

Most of the potentially toxic substances occurring in forages may be classed as secondary plant products, that is compounds not essential for the basic biochemical reactions required to sustain the growth and development of the plants. Although many secondary compounds are potentially toxic to vertebrates it does not necessarily follow that these compounds have evolved because of the benefits such toxicity could confer on its possessor. To explain the role of secondary compounds, Culvenor (1970) has suggested that those substances may protect the plants from vertebrate and invertebrate predators and sees this protective role as a logical evolutionary development. Kingsbury (1978) and Bell (1980) believe that these compounds may protect the plants from a wide range of predators such as insects and nematodes and in some cases may act as fungicides or even as phytotoxins which discourage competing plant species.

It may be quite fortuitous that these compounds are toxic to man or livestock because these have not exercised the environmental pressures which have led to the selection of plants containing these compounds, but these compounds have evolved in response to selectionary pressures exercised by insects and fungi.

THE NATURE OF DELETERIOUS SUBSTANCES AND THEIR EFFECTS ON LIVESTOCK

The list of plants that has been proven to contain substances which may be deleterious to livestock or that have been suspected of being toxic is a long one. A bibliography of plant poisonings in animals over the period 1960-1972 is available (Crane 1973) as well as detailed tables of phytotoxins (Duke 1977). Plant toxins have been the subject of comprehensive and current reviews by Matches (1973), Committee of Food Protection, N.R.C. (1973), Keeler (1975), Burns

(1978), McDonald (1981), and by contributors to a joint U.S.-Australian symposium on poisonous plants (Keeler *et al*. 1978).

No attempt will be made here to provide a catalogue of all the plant constituents that are toxic to livestock. Duke (1977) has published detailed tables of phytotoxins and Butler & Bailey (1973) have discussed the occurrence, chemistry and biochemistry of the constituents of herbage.

In this chapter information is presented on the more important classes of plant toxins and ways of predicting and controlling losses in productivity caused by them are discussed. The examples have been chosen to illustrate basic principles rather than the full range of pharmacological effects produced by various toxins.

Nitrogenous compounds

Proteins

The principal disorder caused by plant proteins is bloat. Bloat is an important disorder in beef and dairy cattle grazing succulent pastures in temperate regions. Many pasture species have been implicated as causes of bloat but the most serious problems are usually associated with grazing *Medicago sativa* (lucerne), *Trifolium repens* (white clover) and *T. pratense* (red clover). Under certain conditions a very stable foam is formed in the reticulo-rumen of affected animals. This interferes with the normal eructation of the gaseous products of ruminal fermentation, resulting in a dangerous rise in intraruminal pressure and gastric distention. This distention interferes with respiration and causes circulatory disorders and other effects which can result in death. The physiological aspects of bloat have been the subject of a number of recent reviews (Clarke & Reid 1970, Leng & McWilliam 1973, Reid *et al*. 1975).

The formation of a stable foam is caused by an extremely complex interaction of plant, animal and microbial factors as well as involving a delicate balance between foam producing and foam inhibiting factors. Secondary plant products such as saponins, cyanoglucosides and amines have at various times been suspected of causing bloat but it now seems unlikely that they play a major role as causative agents although they may sometimes modify sensitivity or cause secondary effects.

Soluble leaf proteins appear to play the major role in persistent foam formation (Mangan 1959), but there is disagreement on which particular fraction is principally responsible. McArthur & Miltmore (1969) considered that the 18S (Fraction I) protein was responsible for the foams produced by *M. sativa*, but Jones & Lyttleton (1972) have produced stable foams *in vitro* from Fraction I and Fraction II protein fractions from *T. pratense*. However the properties of rumen foams cannot be explained in terms of a single component being responsible for the foam characteristics (Jones & Lyttleton 1972) and other naturally occurring surfactants such as polygalacturonic acid (pectic acid) and salivary macromolecules can increase the persistence of leaf protein foams (Jones *et al*. 1978) *in vitro*. There are a number of factors which can inhibit the formation of stable foam and recent interest has been concentrated on the possible role of flavanol polymers (flavolans), a term synonymous with condensed tannins, which occur in many legumes, in preventing foam formation by reacting with the soluble proteins and precipitating them. Jones & Lyttleton (1971) demonstrated a negative association between the occurrence of bloat and the presence of protein precipitants in 11 species of temperate legumes. Bloat rarely occurs in animals grazing tropical legumes in Australia. This is related to the presence of protein precipitants in some species, and in others to low levels of soluble (Jones & Lyttleton 1971).

Animals vary greatly in their susceptibility to bloat but the factors determining susceptibility have yet to be defined (Reid et al. 1975). Microbial factors also are very important in the development of bloat (Walker 1973). Leng (1973) has put forward a complex hypothesis which attempts to integrate plant, animal and microbial factors but consideration of it is outside the scope of this review. Methods for controlling bloat have been only partially effective. These include pasture management measures such as increasing the proportion of grass in the pasture and supplementary feeding of hay. The use of antifoaming agents is a logical approach. Spraying bloat pastures with oils has been successful in some areas but synthetic antifoaming agents hold the most promise (Bartley 1967). A slow-release antibloat capsule containing a surfactant which gives effective control for up to 30 days has been developed (Laby 1975, 1980) and has greatly reduced bloat in experimental beef cattle.

Amino acids

More than 300 non-protein amino acids occur in higher plants (Hegarty & Peterson 1973). Many of these have growth inhibitory effects on microorganisms and laboratory animals (Rosenthal & Bell 1979) but only a small number have been implicated as toxic to humans and livestock. The most important toxic amino acids in forages are mimosine in *Leucaena leucocephala*, indospicine in *Indigofera* species, and the selenium - containing amino acids involved in "alkali disease" and 'blind staggers' in livestock grazing selenium - containing vegetation (for review see Olson 1978).

Mimosine

Mimosine occurs in high concentrations in the leaves (8-10 percent of the dry weight) and seeds (3-5 percent) of *Leucaena leucocephala*, a tropical shrub legume of increasing importance as a pasture legume for the tropics and sub-tropics (Jones 1979). In Australia and New Guinea prolonged ingestion of *L. leucocephala* by cattle has caused low liveweight gains, hair loss and goitre in adult cattle (Holmes 1976, Jones et al. 1976). Goitrous offspring have been produced by sheep and cattle fed during pregnancy on diets containing a high proportion of *L. leucocephala*. Mimosine causes loss of hair and a wide range of physiological effects in non ruminants but is extensively metabolized in the rumen to 3-hydroxy-4(IH) - pyridone (3-dihydroxy pyridine, DHP) (Hegarty et al. 1964) and excreted in the urine. The goitres produced in ruminants ingesting this legume (Jones et al. 1978) are caused by DHP which is a potent goitrogen with thiouracil-type activity such that its effects cannot be overcome by iodine supplementation (Hegarty et al. 1979).

Goitre has not been reported outside Australia and New Guinea in ruminants ingesting *L. leucocephala*, even though the foliage contains similar concentrations of mimosine. Jones (1981) has compared the effects of feeding diets containing up to 100 percent *L. leucocephala* to goats in Hawaii and in north Queensland. In Queensland, the diets were highly toxic and caused a rapid fall in serum thyroxine levels and produced goitres. In Hawaii goats showed no clinical signs of toxicity. Thryoxine levels were normal and thyroids were not enlarged. The urine of the animals in Queensland contained high concentrations of DHP while that from the Hawaiian goats had very low concentrations. On the basis of these results and other studies of metabolism of mimosine *in vitro* Jones (1981) has suggested that ruminants in Hawaii ingesting *L. leucocephala* have very low concentrations of DHP circulating in their blood and has suggested that the rumen microorganisms necessary to degrade DHP do not occur in Australia. Further work is at present in progress to test this hypothesis which may lead to a 'biological' solution to the problem of toxicity in cattle grazing *L. leucocephala* in Australia and elsewhere. Progress in the breeding of low-mimosine types of *L. leucocephala* is discussed later in this symposium (Bray 1982). Jones (1979) has suggested that diets containing less than 30 percent *L. leucocephala* would not be expected to cause toxicity problems in livestock. In spite of the problems mentioned above *L. leucocephala* is regarded by other workers as a particularly useful browse legume in the dry tropics and sub-tropics.

Indospicine

Indospicine (L-2-amidino-6 amino hexanoic acid) was originally isolated from *Indigofera spicata*, a potentially useful tropical pasture legume (Hegarty & Pound 1970). It has not been detected outside the genus *Indigofera*. It is a potent hepatotoxin and also causes serious problems in reproduction in cattle. Other species of *Indigofera* are potentially useful pasture legumes for certain applications and soil types in Australia, and biological and chemical screening of the large number of species for indospicine has shown that there are a number of species which appear to be free of indospicine (E.A. Bell, personal communication). On some of these species rats show weight gains comparable with those obtained on lucerne diets (Lambourne 1979). Further testing in ruminants is needed to determine the suitability of these species as pasture legumes.

Nitrate and nitro-compounds

Nitrate occurs normally in many pasture species such as maize, lucerne and *Pennisetum clandestinum*, and may accumulate to high concentrations under such conditions as high levels of nitrogen fertilization, low light intensity and drought. Most cases of nitrate poisoning in the field occur when animals that have been starved for a period ingest large amounts of potentially toxic plants (containing more than 1.5 percent nitrate, expressed as KNO_3). Ingestion of sublethal doses can cause abortion, depression of lactation and digestive disturbances. When such forage is ingested by ruminants, the relatively non-toxic nitrate is reduced to nitrite which can cause fatal methaemoglobinaemia if it is produced more rapidly that it can be reduced to ammonia by the ruminal microorganisms. Dietary factors may influence the selection of ruminal populations with different capacities of nitrate or nitrite reduction (Allison 1978).

Glucose derivatives of 3-nitropropanoic acid occur in the foliage of a number of species of the legumes *Indigofera*, *Astragalus* and in *Coronilla varia* (crown vetch). Miserotoxin (3-nitro-1-propyl-β-D-glucopyranoside) is the acutely poisonous constituent of a number of

Astragalus species (Stermitz & Yost 1978). The glucosides are hydrolyzed in the rumen to the 5-nitro compounds. 3-nitro propanol is the more toxic as it is more rapidly absorbed (Williams & Jones 1978). These compounds occur in only a few potentially useful pasture legumes but careful chemical and biological screening will be required in assessing whether these species may be used under conditions where they can form a major part of the animals' diet (Williams 1980).

Alkaloids

There is a variety of groups within the alkaloid class, but from the point of view of deleterious effects of livestock the more important ones are the pyrrolizidines, the quinolizidines, diterpenes, tropanes and steroidal sub classes (Everist 1974). Alkaloids are uncommon in grasses, being known in only 21 species out of a total number of about 8,000 (Culvenor 1973), but the β-phenylethylamine derivatives and tryptamine derivaties found in some Phalaris species and perloline which occurs in Festuca and Lolium species are implicated in various disorders of ruminants.

Pyrrolizidine alkaloids

Pyrrolizidine group alkaloids are potent hepatotoxins and are found in species of Amsinckia, Crotalaria, Senecio, Heliotropium and other genera which are common in range areas. In Australia, the most important disorder, called toxaemic jaundice, is due to the consumption of Heliotropium europeum. Approximately 100 million sheep are exposed to the plant over 50,000 km^2 in the south-eastern states. The alkaloids, notably lasiocarpine, heliotrine and their N-oxides, cause severe chronic effects which can result in substantial mortalities after several heavy exposures. Less severe exposures can cause a substantial reduction in the useful life of sheep (Culvenor 1978). The chronic liver damage can lead to photosensitization and to the accumulation in the liver of abnormally large amounts of copper. When the affected animals are stressed, the copper is released into the blood causing acute intravascular haemolysis and death (Bull et al. 1956).

Under Australian conditions it is largely impossible to control weeds such as H. europeum and various prophylactic measures have been investigated to enable sheep to graze the plants without severe poisoning. The measures are (1) enhanced destruction of alkaloids in the rumen by the use of inhibitors of methanogenesis, since metabolic hydrogen plays a role in the detoxification process and there is a competitive relation between this process and methanogenesis (Lanigan et al. 1978). Iodoform prophylaxis may be a useful protective measure in the field for sheep exposed to the plant for one season, but further work is needed to determine the extent of the protection and to find less toxic inhibitors; (2) manipulation of liver metabolism to decrease the extent of formation of toxic metabolites (pyrroles) in the liver by enzyme inhibition and to enhance the reactions which destroy the biological activity of the compounds (enzyme induction). Although of considerable theoretical importance it seems unlikely that this approach will be successful for heliotrope poisoning; (3) protective immunization by coupling the alkaloid to a protein to form an antigen. This approach has not been particularly promising because of difficulties in preparing suitable alkaloid-protein conjugates which will bind the alkaloids irreversibly (Culvenor 1978). Some of these methods may be applicable to other plant toxins.

Alkaloids in Phalaris species and Festuca arundinacea

Extensive studies have been carried out in the USA and Australia on the alkaloids in *Phalaris arundinacea* and *P. aquatica* to assess their contribution to dietary dependent disorders reported with these species. The results of the research in the USA have been summarized by Marten (1973) and Burns(1978) and only the main findings are presented here.

In the USA the disorder associated with *P. arundinacea* consists of reduced liveweight gains and milk production and general symptoms of unthriftiness. So far nine alkaloids have been characterized: five indole derivatives, three derivatives of β carboline and one phenolic alkaloid. Total alkaloid concentration was negatively correlated with palatibility in grazing trials with sheep (Simons & Marten 1971). In other experiments (Marten *et al*. 1976), genotypes with high alkaloid concentrations or genotypes containing carboline-tryptamine alkaloids depressed intake and caused greater frequency of diarrhoea. Burns (1978) has drawn attention to recent research in which diverse genotypes free of tryptamine - β - carboline alkaloids and with a wide range of gramine concentrations have been identified (Hovin & Martin 1975, Coulman *et al*. 1976). There is clearly potential for breeding of cultivars free of deleterious constituents.

In Australia, three syndromes have been recognized in animals grazing *P. aquatica*. These are peracute, acute and chronic. In the peracute or "sudden death" syndrome sheep collapse and die with symptoms of heart failure. In the acute syndrome there is evidence of nervous disorder; the chronic condition is known as "phalaris staggers", a slowly progressive and usually fatal neurological disorder. The peracute and acute syndromes are caused by various dimethyl tryptamine alkaloids that can be present in the plant in high concentrations (Gallagher *et al*. 1966) but the toxin causing the staggers has not yet been identified. Introduction of heavy cobalt pellets into the rumen has been effective in preventing phalaris staggers in those areas known to be deficient in the element. Few cases of *P. aquatica* toxicity have been reported outside Australia and in the USA cattle graze the grass without any untoward effects. This could be due to adequacy of cobalt supplies, differences in susceptibility of sheep and cattle or lower concentrations of the alkaloids (Burns 1978).

Cattle grazing *Festuca arundinacea* (tall fescue) pastures during the summer months in the USA suffer "summer toxicosis" a syndrome characterized by decreased forage intake, reduced weight gain and increased respiration. Milk production in dairy cows is depressed. In addition to this syndrome, cattle ingesting the grass during cold weather often show symptoms of a disease called "fescue foot". The poor performance parallels the accumulation of the alkaloid perloline and related compounds in the grass. Perloline (a normal constituent of *Lolium* species) inhibits ruminal cellulose digestion, fatty acid production and growth of celluloytic bacteria, ultimately decreasing energy and nutrient available to the animal (Bush & Buckner 1973). However, toxicity symptoms have been reported in cows and sheep fed on experimental material selected for low perloline concentration and it seems likely that factors other than perloline are involved and may even have been increased in selection for low perloline content (Hemken *et al*. 1979). Alkaloids produced by fungi parasitic to the grass may also be implicated in fescue toxicosis (Bacon *et al*. 1975).

Cyanoglycosides

Cyanoglycosides, which are hydrolyzed to free hydrogen cyanide when ingested, occur in a large number of rangeland weeds in many countries and in such well known sown pasture species as *Trifolium repens*, *Cynodon dactylon*, *C. aethiopicus* and various *Sorghum* species. Losses due to acute cyanide toxicity usually occur under one or more of the following conditions: (1) when the animals are hungry; (2) when the plants are young and actively growing or have been subjected to a check in growth; (3) when the animals have been stressed. High levels of application of nitrogen fertilizers increase the concentrations of cyanoglycosides in the forage. Cyanide interferes with the capacity of the blood to carry oxygen, respiration in all tissues is inhibited and death may occur from respiratory paralysis.

Prevention of acute cyanide poisoning can best be accomplished by recognizing the environmental conditions which cause accumulation of the cyanoglycosides and by using suitable management practices. In ruminants, hydrogen cyanide is metabolized to thiocyanate via oxidation by sulphur donors in the rumen and principally in the liver, and then excreted. Thiocyanate is goitrogenic because it inhibits the trapping of iodide by the thyroid gland. Thiocyanate-type goitres have been observed in sheep grazing *T. repens* (Butler *et al*. 1957) and *C. aethiopicus* (Rudert & Oliver 1978). However, these goitrogenic effects are readily overcome by iodine supplementation. A significant proportion of the sulphur ingested by animals grazing *Sorghum* species may be used to detoxify the hydrogen cyanide. It is possible that low live weight gains sometimes observed in animals grazing *Sorghum* pastures with appreciable concentrations of cyanoglycosides may be due to an induced sulphur deficiency which can be corrected by giving sulphur supplements (Wheeler *et al*. 1975).

Heterocyclic compounds

Oestrogenic isoflavones and coumestans

Although there are many oestrogenic substances occuring in plants, the principal pasture plants which cause oestrogenic effects are *Trifolium subterraneum*, *T. pratense*, *Medicago sativa* and *M. truncatula*. The oestrogens in the *Trifolium* species are usually isoflavones, while the *Medicago* species contain coumestans. For recent reviews on phytooestrogens see Cox & Braden (1974), Cox (1978).

The main isoflavones are formononetin, genistein, diadzein and biochanin A. and they may occur in high concentrations in highly oestrogenic cultivars. For example, a typical analysis of the Yarloop cultivar of *Trifolium subterraneum* is formononetin 1.5, genistein 2.8 and biochanin A 0.5 percent of dry matter (Cox 1978). The isoflavones usually occur as the glucosides in the plant and are readily hydrolysed by plant enzymes when the fresh material is crushed. The isoflavones (and the coumestans) are relatively weak oestrogens but their activity can be modified by metabolism in the rumen. Formononetin, believed to be mainly responsible for the oestrogenicity of *T. subterraneum* and *T. pratense* in sheep, is only weakly active, but in the rumen is converted to equol which is the active agent. In contrast, diadzein and biochanin A, which are oestrogenic when given parenterally, are rendered almost inactive in the rumen (Shutt & Braden 1968). The coumestans have not been responsible for extensive long term fertility problems like those that have been associated with the isoflavones (Bickoff 1968) but they may be produced in high concentrations when certain normally non-oestrogenic plants of *T. repens* are infected with fungi. These infected plants can produce oestrogenic effects in sheep (Wong & Latch 1971). Cattle grazing oestrogenic pastures do not appear to suffer the severe infertility problems that affect sheep.

Infertility in sheep caused by ingestion of *T. subterraneum* in Australia manifests itself in two forms. The chronic form involves severe infertility together with maternal dystocia, post-natal death of lambs and other symptoms. The symptoms are progressive and cumulative and usually lead to permanent infertility. This form is now seen only rarely as a result of management practices designed to control the disorder, but reproductive wastage caused by *T. subterraneum* is still widespread (Lightfoot 1974) and costly to the sheep industry.

A "temporary infertility" may occur in ewes grazing highly oestrogenic pastures around the time of mating (Morley *et al*. 1966). Fertility is restored when the sheep move to other pastures; the sheep suffer no other ill effects. This problem can partially be overcome by avoiding mating sheep when clover is dominant in the pasture.

It is now possible to overcome many of the losses previously experienced by a combination of the following methods (1) changes in husbandry practices associated with joining and lambing, (2) methods of pasture management involving the replacement of undesirable cultivars, and dilution of oestrogenic cultivars in the pasture and by application of fertilizers, particularly, phosphate because the concentration of formononetin increases markedly when the soil is deficient in phosphate. New cultivars with low formononetin content have been developed (Francis 1975) for use under different climatic conditions but the highly oestrogenic cultivar Yarloop is proving difficult to eradicate and replace.

Because of the importance of the problem in Australia considerable effort has gone into protective measures, such as immunization against phytooestrogens and modification of ruminal metabolism. Preliminary tests in sheep with isoflavones coupled to various proteins have given encouraging results (Cox 1978) but more research is needed to determine the practical feasibility of this approach.

Toxic sulphur compounds

Although not strictly a forage or pasture species *Brassica oleracea* (kale) is now grown extensively in the British Isles as a fodder for cattle and sheep. Ruminants grazing the plant may develop a haemolytic anaemia characterized by haemoglobinuria, loss of appetite and growth retardation.

The cause of the anaemia is S-methylcysteine sulphoxide, an amino acid derivative which occurs in a number of *Brassica* species. The active substance is dimethyl disulphide which is formed from S-methyl cysteine sulphoxide in the rumen (Smith *et al*. 1978). Restriction of the intake of the plant is the only effective way of preventing the disorder.

Kale also contains glucosinolates (thioglucosides) which can cause goitre (Van Etten & Tookely 1978) in sheep and lambs (Paxman & Hill 1974).

Organic acids

Oxalic acid (oxalate) poisoning is an important animal disorder in many countries. Oxalic acid often occurs in high concentrations (as high as 10-14 percent of the dry matter) in rangeland weeds particularly those belonging to the genus *Rumex* or genera of the families Chenopodiaceae and Oxalidaceae (James 1978b). Many oxalate-containing plants are palatable to livestock and dramatic stock losses have been reported in animals eating *Oxalis pescaprae* in Australia and *Halogeton glomeratus* in the USA.

While many species of tropical pasture grasses contain low concentrations of soluble oxalate (< 1 percent) (Jones & Ford 1972), *Cenchrus ciliaris* contains 1-2 percent and much higher concentrations (3-5 percent) have been reported in grasses in the *Setaria sphacelata* complex (Jones & Ford 1972, Hacker 1974). However, there have been only occasional reports of deaths of cattle grazing this grass (Jones *et al*. 1970). Horses grazing these pastures are not able to metabolize the ingested oxalate and develop a disorder called osteodystrophia fibrosa which is a degenerative condition of the bone associated with an induced calcium and phosphorus imbalance. *C. ciliaris* and *Panicum maximum* have also caused this disorder in horses (Walthall & McKenzie 1976).

High levels of soluble oxalate are also undesirable because oxalate has a low energy value. Hacker (1974) concluded that it was unlikely that breeding for low oxalate concentrations in *S. sphacelata* could be achieved without sacrifices in yield and possibly digestibility.

Ruminants are able to metabolize oxalate and the rumen microflora can adapt so that increasing quantities of oxalate are degraded. Hence increased tolerance to free oxalate was acquired by cattle and sheep by gradually increasing the quantity of oxalate-containing material fed (James *et al*. 1967, Allison 1978). When the ability of the rumen microbes to degrade oxalate is exceeded large amounts of oxalate may be absorbed and death may result from hypocalcaemia in acute cases and from uremia caused by kidney damage in chronic cases (James 1972).

The other organic acid implicated in toxicity to livestock is fluoroacetic acid which occurs in some plants which are ingested by animals grazing on rangelands. These include species of *Gastrolobium*, *Oxylobium*, *Acacia* and *Dichapetalum*. McEwan (1978) has reviewed the organofluorine compounds in plants. Fluoride ingested in amounts greater than the optimum level may have harmful effects. Plants rarely accumulate fluoride to toxic levels and fluoride toxicosis usually results from contamination of feeds, water or vegetation near certain industrial operations (Shupe *et al*. 1978).

DISORDERS CAUSED BY MYCOTOXINS

A number of important disorders associated with sown pastures are caused by mycotoxins. These substances are either metabolites produced by the infecting agent (usually a fungus) or by the plant as a result of infection. The mycotoxins belong to various classes of chemical compounds and in a number of instances have not been fully characterized. Some of the more important mycotoxicoses that can adversely affect animal production are discussed here.

Facial eczema

Facial eczema is an important endemic disease of cattle and sheep in New Zealand (see Reid 1973) and also occurs in Australia and South Africa. It is caused by a hepatotoxin, sporidesmin, which is produced by a fungus *Pithomyces chartarum* that grows on plant litter and in warm humid weather multiplies very rapidly. Damaging amounts of spores are likely to be ingested by livestock when grazing pressure is high and animals are forced to graze close to the ground. When eaten the spores cause severe damage to the liver, especially in that part which produces and excretes bile. The chlorophyll in the plant material is metabolized in the rumen to phylloerythrin which is normally excreted with the bile. However, when the liver is affected by sporidesmin it is unable to excrete bile and so bile containing high concentrations of phylloerythrin circulates in the blood stream. Direct sunlight activates the phylloerythrin in the blood beneath the skin in exposed areas to a toxic compound. Areas of skin exposed to sunlight, particularly the face, ears, eyelids and lips, develop severe dermatitis. Affected sheep rapidly lose condition and may die.

In New Zealand, periods when outbreaks are likely to occur can be predicted on the combined basis of spore counts and weather conditions (Percival & Thornton 1958). Management practices that give some measure of control include reducing the consumption of pasture by feeding hay or crops, lax grazing over the whole farm, or spraying with certain fungicides. Drenching with 50 percent zinc sulphate solution has a marked prophylactic effect but can reduce appetite (Smith *et al*. 1978).

Pithomyces chartarum is probably involved in causing photosensitivity in young cattle grazing *Brachiaria decumbens* and *B. brizantha* in South America (Nobre & Andrade 1976, Camargo *et al*. 1976).

Some other pasture plants or weeds, for example *Lupinus angustifolius*, *Panicum miliaceum*, *Trifolium hydridum* and *Hypericum perforatum* contain hepatotoxins or phytodynamic agents which can cause photensitization.

Annual ryegrass toxicity

A dual infection of the seed heads of *Lolium rigidum* (annual or Wimmera ryegrass) by a nematode *Anguina sp*. and a bacterium *Cornynebacterium nathay* occurs in certain areas of Western and South Australia. Nematode larvae climb the young plants and eventually bore into the developing seeds taking the bacterium with them. The seed does not develop properly and becomes a gall, containing nematode or bacterium but rarely both. Sheep grazing the grass at the seeding stage develop a severe neurological disease (staggers and convulsions) which can rapidly become fatal. In addition to causing a large number of deaths the disease may cause pregnant animals to abort. This disease has drastically reduced the areas available for grazing on some properties.

Ryegrass toxicity is associated with the bacterial galls but it is not clear whether it is of bacterial origin or produced by the grass in response to infection (Lanigan *et al*. 1976). Preliminary evidence (Culvenor *et al*. 1978) indicates that it is not a protein but resembles lipopolysaccharide - peptide toxins. Even though the toxin has not been identified, some success has been achieved in reducing the losses by careful daily monitoring of the pastures for the development of infection and removal of stock to a safe area at the first sign of the disease (Trotman 1978). Nematicides also provide a means of controlling the disease but are expensive. The drug librium has been tested as an antidote in the field with moderate success and other drugs of this type may prove to be useful.

Ryegrass staggers

Ryegrass staggers is a neuromuscular disorder that affects sheep and less frequently cattle, grazing sown pastures in which there is a predominance of *Lolium perenne* (perennial ryegrass). It has been reported in New Zealand, Australia, USA and England (for brief review see Mortimer 1978). Morbidity can be high in both sheep and cattle but mortality is usually low. Loss of production is not usually serious except in severe outbreaks and the animals usually recover slowly. The results from Australia (Cockrum *et al*. 1979) and New Zealand (Mortimer 1978, di Menna & Mantle, 1978) indicate that the disorder is caused by ingestion of soil containing tremorgenic (neurotoxic) metabolites of *Penicillium* and *Aspergillus* species. In Australia the principal tremorgenic species found in the soil and in the faeces of affected animals are *Penicillium janthinellum* and *P. paxilli*. These fungal species and their active metabolites differ from those reported from New Zealand (di Menna & Mantle, 1978) but the overall effects are very similar.

In New Zealand, avoidance of grazing short ryegrass pastures and use of forage crops are obvious measures for controlling the disorder. Prior stimulation of liver microsomal enzymes that detoxify the tremorgens with a chlorine-containing fungicide has given a high degree of protection to sheep for a period of several weeks, but the residual effects of the fungicide make it unacceptable for general use (Mortimer 1978).

Paspalum staggers

The symptoms of paspalum staggers closely resemble those of ryegrass staggers. This disorder occurs widely in cattle and sheep ingesting *Paspalum dilatatum* infected by ergot (*Claviceps paspali*) (Cysewski 1973). Sclerotia of *C. paspali* also contain non-alkaloidal mycotoxic tremorgens, structurally allied to the indolic tremorgens in *Penicillium* species which produce the neurotoxic symptoms (Mantle *et al*. 1977).

Lupinosis

Lupinosis is a disorder associated with the use of *Lupinus* spp. as fodder crops (Gardiner 1976). In parts of Australia where lupins are grown extensively for seed, sheep graze the stubble and are affected. It is a mycotoxicosis caused by hepatotoxins present in the fungus *Phomopsis leptostromeformis* which infects the plant (Van Warmelo *et al*. 1970). Sheep lose appetite and eventually collapse. The main toxic metabolite of the fungus has been isolated but its final structure has not been determined (Culvenor *et. al*. 1978). The development of toxicity cannot be predicted and the stubble may remain toxic for long periods. Successful use of the plant as forage depends on recognition of the early symptoms of the disorder and removal of the sheep to other areas.

CONCLUSIONS

Grazing animals ingest a wide variety of chemical compounds that can adversely affect productivity. Many of these substances are detoxified in the rumen and in body tissues, especially in the liver; however problems may arise when the intake of a toxin exceeds the animals' capacity to detoxify it. Furthermore there are some substances that cannot readily be detoxified and others that are themselves harmless but are converted into toxic derivatives in the digestive tract or in the liver. In some disorders such as bloat and mycotoxicoses the cause is complex and may involve a combination of chemical, biological and environmental factors.

With sown pasture species in which the disorder is caused by simple chemical factors management practices can often control and considerably reduce losses in productivity. In rangelands, however, prevention of production losses resulting from toxicity is difficult. Where the deleterious chemical factors are under simple genetic control the breeding of non-toxic cultivars or cultivars with low toxicity has been successful; for example the breeding of non-toxic cultivars of *Trifolium subterraneum*. In contrast, disorders which have a complex cause, for example mycotoxicoses are controlled through predication and early diagnosis of the problem and the use of management practices and treatments such as drenches or controlled-release capsules.

Attempts to protect animals against toxins by immunization have so far not progressed beyond the preliminary testing stage. Considerably more research is needed to overcome the problem associated with making suitable toxin-protein conjugates. Attempts to develop other prophylactic measures based on the administration to animals of chemicals which cause enhanced destruction of toxins in the rumen or increased inactivation of hepatotoxins in the liver have so far given only limited protection. However further research is warranted to develop chemicals which do not themselves have undesirable side effects. The recent observations of Jones (1981) on possible differences in ruminal metabolism of mimosine indicate the importance of understanding the mechanisms by which many toxins are metabolized in the rumen.

Screening for possible deleterious chemical factors should be carried out at an early stage in the evaluation of potentially useful new pasture species. Rapid chemical methods are now available for a wide range of plant constituents (for a brief review see Jones & Hegarty 1981) and these should be complemented with feeding experiments with rodents and ruminants and detailed histological studies on a range of tissues.

REFERENCES

Agricultural Research Service (1965) Losses in Agriculture. Agricultural Handbook No. 291, Agricultural Research Service, U.S. Department of Agriculture, Washington, D.C.

Allison, M.J. (1978) The role of ruminal microbes in the metabolism of toxic constituents from plants. In: Effects of poisonous plants on livestock. Editors R.F. Keeler, K.R. Van Kampen and L.F. James. New York, Academic Press, pp. 101-118.

Bacon, C.W.; Porter, J.K.; Robbins, J.D. (1975) Toxicity and occurrence of Balansia on grass from toxic fescue pastures. Applied Microbiology 29, 553-556.

Bartley, E.E. (1967) Progress in bloat prevention. Agricultural Science Review 5, 5-13.

Bell, E.A. (1980) The possible significance of secondary compounds in plants. In: Encyclopedia of plant physiology, new series. Vol. 8 Secondary plant products. Editors E. A. Bell and B. V. Charlwood. Berlin, Springer-Verlag, pp.11-12.

Bickoff, E.M. (1968) Oestrogenic constituents of forage plants. Commonwealth Bureau of Pastures Field Crops Rev. Ser. 1.

Bray, R.A. (1982) Selecting and breeding better legumes. In: Nutritional limits to animal production from pastures. Editor J.B. Hacker. Farnham Royal, U.K., Commonwealth Agricultural Bureaux, pp. 287-303.

Bull, L.B.; Albison, H.E.; Edgar, G.; Dick, A.T. (1956) Toxaemic jaundice of sheep: phytogenous copper poisoning, heliotrope poisoning and hepatogenous copper poisoning. Australian Veterinary Journal 32, 229-236.

Burns, J.C. (1978) Antiquality factors as related to forage quality. Journal Dairy Science 61, 1809-1820.

Bush, L.; Buckner, R.C. (1973) Tall fescue toxicity. In: Antiquality components of forages. Editor A. G. Matches. Crop Science Society of America Special Publication No. 14. Madison, Crop Science Society of America, pp. 99-110.

Butler, G.W.; Bailey, R.W. (1973) Editors. Chemistry and biochemistry of herbage, Volume 1. New York, Academic Press, 639 pp.

Butler, G.W.; Flux, D.S.; Peterson, G.B.; Wright, E.W.; Glenday, A.C.; Johnson, J.R. (1957) Goitrogenic effect of white clover. New Zealand Journal of Science and Technology 38A, 798-802.

Camargo, W.V.A.; Nazario, W.; Fernandes, N.S.; Amaral, R.E.M. (1976) Photosensitivity in calves. Probable involvement of the fungus Pithomyces chartarum in the development of the disease. Biologico 42, 259-61.

Clarke, R.T.J.; Reid, C.W.S. (1970) Legume bloat. In: Physiology of digestion and metabolism in the ruminant. Editor A.T. Phillipson. Newcastle-upon-Tyne, Oriel Press, pp. 596-606.

Cockrum, P.A.; Culvenor, C.C.J.; Edgar, J.A.; Payne, A.L. (1979) Chemically different tremorgenic mycotoxins in isolates from Penicillium paxilli from Australia and North America. Journal of Natural Products 42, 534-536.

Committee of Food Protection, N.R.C. (1973) Toxicants occurring naturally in foods. National Academy of Sciences, Washington, D.C. 624 pp.

Coulman, B.E.; Woods, D.L.; Clark, K.W. (1976) Identification of low alkaloid genotypes of reed canary grass. Canadian Journal of Plant Science 56, 830-845.

Cox, R.I. (1978) Plant oestrogens affecting livestock in Australia. In: Effects of poisonous plants on livestock. Editors R. F. Keeler, K. Van Kampen and L. F. James. New York, Academic Press, pp.451-463.

Cox, R.I.; Braden, A.W. (1974) The metabolism and physiological effects of phytooestrogens. Proceedings of the Australian Society of Animal Production 10, 122-129.

Crane, T.D. (1973) Plant poisoning in animals - a bibliography. Veterinary Bulletin 43, 165-177 and 231-249.

Culvenor, C.C.J. (1970) Toxic plants - a reevaluation. Search 1, 103-110.

Culvenor, C.C.J. (1973) Alkaloids. In: Chemistry and biochemistry of herbage. Editors G. W. Butler and R. W. Bailey. New York, Academic Press, Vol. 1, pp.375-439.

Culvenor, C.C.J. (1978) Prevention of pyrrolizidine alkaloid poisoning -animal adaptation or plant control? In: Effects of poisonous plants on livestock. Editors R. F. Keeler, K. R. Van Kampen and L. F. James. New York, Academic Press, pp.189-200.

Culvenor, C.C.J.; Frahn, J.L.; Jago, M.V.; Lanegan, G.W. (1978) The toxic of Lolium rigidum (annual rye grass) seed heads associated with nematode - bacterium infection. In: Effect of poisonous plants on livestock. Editors R.F. Keeler, K.R. Van Kampen and R.F. James. New York, Academic Press, pp.349-352.

Culvenor, C.C.J.; Smith, L.W.; Frahn; J.L.; Cockrum, P.A. (1978) Lupinosis: Chemical properties of phomopsin A, the main toxic metabolite of Phomopsis leptostromiformis. In: Effects of poisonous plants to livestock. Editors R. F. Keeler, K. R. Van Kampen and L. F. James. New York, Academic Press, pp. 565-570.

Cysewski, S.J. (1973) Paspalum staggers and tremorgen intoxication animals. Journal of American Veterinary Medical Association 163(11), 1291-1292.

Duke, J.A. (1977) Phytotoxin tables. C.R.C. Critical Reviews in Toxicology, 5, 189-237.

Everist, S.L. (1974) Poisonous plants of Australia. Sydney, Angus and Robertson, 684 pp.

Francis, C. M. (1975) Trikkala: A new safe clover for wet areas. Journal of Agriculture Western Australia, Series 4, 16, 2-4.

Gallagher, C.H.; Koch, J. H.; Hoffman, H. (1966) Diseases of sheep due to ingestion of Phalaris tuberosa. Australian Veterinary Journal 42, 279-284.

Gardiner, M.R. (1976) Lupinosis. Advances in Veterinary Science 11, 85-138.

Hacker, J.B. (1974) Variation in oxalate, major cations, and dry matter digestibility of 47 introductions of the tropical grass setaria. Tropical Grasslands 8, 145-154.

Hegarty, M. P.; Pound, A.W. (1970) Indospicine, a hepatotoxic amino acid from Indigofera spicata: Isolation, structure and biological studies. Australian Journal of Biological Sciences 23, 831-842.

Hegarty, M.P.; Peterson, P.J. (1973) Free amino acids, bound amino acids, amines and ureides. In: Chemistry and biochemistry of herbage. Editors G.W. Butler and R.W. Bailey. New York, Academic Press, Vol. 1, pp.1-62.

Hegarty, M.P.; Schinckel, P.G.; Court, R.D. (1964) Reaction of sheep to the consumption of Leucaena glauca and to its toxic principle mimosine. Australian Journal of Agricultural Research 15, 153-167.

Hegarty, M.P.; Lee, C.P.; Christie, G.S.; Court, R.D.; Haydock, K.P. (1979) The goitrogen 3-hydroxy-4(IH)-pyridone, a ruminal metabolite from Leucaena leucocephala: Effects in mice and rats. Australian Journal of Biological Sciences 32, 27-40.

Hemken, R.W.; Bull, L.S.; Bolling, J.A.; Kane, E.; Bush, L.P.; Buckner, R.C. (1979) Summer fescue toxicosis in lactating dairy cows and sheep-fed experimental strains of rye grass tall fescue hybrids. Journal of Animal Science 49, 641-646.

Hovin, A.W.; Marten, G.C. (1975) Distribution of specific alkaloids in reed canarygrass cultivars. Crop Science 15, 705-707.

Holmes, J.H.G. (1976) Growth of Brahman cross heifers grazing Leucaena. Proceedings of the Australian Society of Animal Production 11, 453-456.

James, L.F. (1972) Oxalate toxicosis. Clinical Toxicology 5, 231-243.

James, L.F. (1978a) Overview of poisonous plant problems in the United States. In: Effects of poisonous plants on livestock. Editors R.F. Keeler, K.R. Van Kampen and L.F. James. New York, Academic Press, pp. 3-5.

James, L.F. (1978b) Oxalate poisoning in livestock. In: Effects of poisonous plants on livestock. Editors R.F. Keeler, K.R. Van Kampen and L.F. James. Yew York, Academic Press, pp. 139-146.

James, L.F.; Street, J.C.; Butcher, J.E. (1967) In vitro degradation of oxalate and of cellulose by rumen ingesta from sheep fed Halogeton glomeratus. Journal of Animal Science 26, 1438-1444.

Jones, R.J. (1979) The value of Leucaena leucocephala as a feed for ruminants in the tropics. World Animal Review 31, 13-23.

Jones, R.J. (1981) Does ruminal metabolism of mimosine explain the absence of Leucaena toxicity in Hawaii? Australian Veterinary Journal 57, 55-56.

Jones, R.J.; Ford, C.W. (1972) The soluble oxalate content of some tropical pasture grasses grown in South East Queensland. Tropical Grasslands, 6, 201-204.

Jones, R.J.; Hegarty, M.P. (1981) Screening plants for possible toxic effects on livestock. In: Forage evaluation: concepts and technique . Editors J.L. Wheeler and R.D. Mochrie. Lexington, The American Forage and Grassland Co cil and Melbourne, CSIRO. pp. 237-247.

Jones, R.J.; Seawright, A.A.; Little, D.A. (1970) Oxalate poisoning in animals grazing the tropical grass Setaria sphacelata. Journal of the Australian Institute of Agricultural Science 36, 41-43.

Jones, R.J.; Blunt, C.G.; Holmes, J.H.G. (1976) Enlarged thyroid glands in cattle grazing leucaena pastures. Tropical Grasslands 10, 113-116.

Jones, R.J.; Blunt, C.G.; Nurnberg, B.I. (1978) Toxicity of Leucaena leucocephala: the effect of iodine and mineral supplements on penned steers fed a sole diet of leucaena. Australian Veterinary Journal 54, 387-392.

Jones, W.T.; Lyttleton, J.W. (1971) Bloat in cattle XXXIV. A survey of legumes that do and do not produce bloat. New Zealand Journal of Agricultural Research 14, 101-107.

Jones, W.T.; Lyttleton, J.W. (1972) Bloat in cattle XXXVI. Further studies on the foaming properties of soluble leaf proteins. New Zealand Journal of Agricultural Research 15, 267-278.

Jones, W.T.; Lyttleton, J.W.; Mangan, J.L. (1978) Interaction between fraction 1 leaf protein and other surfactants involved in the bloat syndrome 1 foam stabilizing materials. New Zealand Journal of Agricultural Research 21, 401-407.

Keeler, R.F. (1975) Toxins and teratogens in higher plants. Lloydia 38, 56-86.

Keeler, R.F.; Van Kampen, K.R.; James, L.F. (1978) Editors. Effects of poisonous plants on livestock. New York, Academic Press, 600pp.

Kingsbury, J.M. (1978) Ecology of poisoning. In: Effects of poisonous plants on livestock. Editors R. F. Keeler, K. R. Van Kampen and L. F. James. New York, Academic Press, pp. 81-91.

Laby, R.H. (1975) Surface active agents in the rumen. In: Digestion and metabolism in the ruminant. Editors I. W. McDonald and A. C. I. Warner. Armidale, Australia, University of New England Publishing Unit, pp. 537-550.

Laby, R.H. (1980) Modern technological developments. Proceedings of the Australian Society of Animal Production, 13, 6-10.

Lambourne, L.J. (1979) Biological evaluation of tropical forage legumes. Australian CSIRO, Division of Tropical Crops and Pastures, Annual Report 1978/79, p. 81.

Lanigan, G.W.; Payne, A.L.; Frahn, J.L. (1976) Origin of toxicity in parasitised annual rye grass (Lolium rigidum). Australian Veterinary Journal 52, 244-246.

Lanigan, G.W.; Payne, A.L.; Peterson, J.E. (1978) Antimethanogenic drugs and Heliotropum europaeum poisoning in penned sheep. Australian Journal of Agricultural Research 29, 1281-1292.

Leng, R.A. (1973) Ruminal fermentation and bloat: The possible role of protozoa in the development of bloat. In: Bloat. Editors R. A. Leng and J. R. McWilliam. Reviews in Rural Science No. 1. Armidale, Australia, University of New England, pp. 57-62.

Leng, R.A.; McWilliam, J.R. (1973) Editors. Bloat. Reviews in Rural Science No. 1. Armidale, Australia, University of New England, 103 pp.

Lightfoot, R.J. (1974) A look at recommendations for the control of infertility due to clover disease in sheep. Proceeding of the Australian Society of Animal Production 10, 113-121.

Little, D.A. (1982) Utilization of minerals. In: Nutritional limits to animal production from pastures. Editor J.B. Hacker. Farnham Royal, U.K., Commonwealth Agricultural Bureaux, pp. 259-283.

McArthur, J.M.; Miltmore, J.E. (1969) Bloat investigations. Studies on soluble proteins and nucleic acids in bloating and non-bloating forages. Canadian Journal of Animal Science 49, 69-75.

McDonald, I.W. (1981) Detrimental substances in plants consumed by grazing animals. In: World animal science. B1. Grazing animals. Editor F. W. Morley. Amsterdam, Elsevier, pp. 349-360.

McEwan, T. (1978) Organo-fluorine compounds in plants. In: Effects of poisonous plants on livestock. Editors R. F. Keeler, K. R. Van Kampen, L. F. James. New York, Academic Press. pp. 147-160.

Mangan, J.L. (1959). Bloat in cattle XI. Foaming properties of proteins, saponins and rumen liquor. New Zealand Journal of Agricultural Research 2, 47-61.

Mantle, P.G.; Mortimer, P.H.; White, E.P. (1977) Mycotoxic hemogens of Claviceps paspali and Penicillium cyclopium: a comparative study of effects on sheep and cattle in relation to natural staggers syndromes. Research in Veterinary Science 24, 49-56.

Marten, G.C. (1973) Alkaloids in reed canary grass. In: Antiquality components of forages. Editor A. G. Matches. Crop Science Society of America Special Publication No. 4. Madison, Crop Science Society of America, pp. 15-30.

Marten, G.C.; Jordan, R.M.; Hovin, A.W. (1976) Biological significance of reed canary grass alkaloids and associated palatibility variation to grazing cattle and sheep. Agronomy Journal 68, 909-914.

Matches, A.G. (1973) Antiquality components of forages. Crop Science Society of America Special Publication No. 4. Madison, Crop Science Society of America, 140 pp.

Menna, M.E. di; Mantle, P.G. (1978) The role of penicillia in rye grass staggers. Research in Veterinary Science 24, 347-351.

Morley, F.H.W.; Axelsen, A.; Bennet, D. (1966) Recovery of normal fertility after grazing on oestrogenic red clover. Australian Veterinary Journal 42, 204-206.

Mortimer, P.H. (1978) Perennial rye grass staggers in New Zealand. In: Effects of poisonous plants on livestock. Editors R. F. Keeler, K. R. Van Kampen, L. F. James. New York, Academic Press. pp. 353-361.

National Academy of Science (1968) Prenatal and postnatal mortality in cattle. National Academy of Science Publication No. 1685, Washington D.C.

Nobre, D.; Andrade, S.O. (1976) Relationship between photosensitivity in young cattle and the grass Brachiaria decumbens Stapf. Biologico 42, 248-258.

Olson, S.E. (1978) Selenium plants as a cause of livestock poisoning. In: Effects of poisonous plants on livestock. Editors R. F. Keeler, K. R. Van Kampen, L. F. James. New York, Academic Press, pp.121-134.

Paxman, P.J.; Hill, R. (1974) The goitrogenicity of kale and its relation to thiocyanate content. Journal of the Science of Food and Agriculture 25, 329-337.

Percival, J.C.; Thornton, R.H. (1958) Relationship between the presence of fungal spores and a test for hepatotoxic grass. Nature, London, 182, 1095-1096.

Reid, C.S.W. (1973) Limitations to the productivity of the herbage-fed ruminant that arise from the diet. In: Chemistry and biochemistry of herbage. Editors G. W. Butler and R. W. Bailey. New York, Academic Press, Vol. 1, pp.215-262.

Reid, C.S.W.; Clarke, R.T.J.; Cockrem, F.R.M.; Jones, W.T.; McIntosh, J.T.; Wright, D.E. (1975) Physiological and genetic aspects of pasture (legume) bloat. In: Digestion and metabolism in the ruminant. Editors I. W. McDonald and A. C. I. Warner. Armidale, Australia, University of New England Publishing Unit, pp. 524-536.

Rosenthal, G.A.; Bell, E.A. (1979) Naturally occurring non protein amino acids. In: Herbivores - their interaction with secondary plant metabolites. Editors G.A. Rosenthal and D. H. Janzen. New York, Academic Press, pp.353-385.

Rudert, C.P.; Oliver, J. (1978) The effect of fertilizer nitrogen on the content of hydrocyanic acid, nitrate and some minerals in star grass (Cynodon aethiopicus Clayton and Harlan) Rhodesian Journal of Agricultural Research 16, 23-29.

Shupe, J.L.; Peterson, H.B.; Olson, A.E.; Miller, G.W. (1978) Inorganic toxicants and poisonous plants. In: Effects of poisonous plants on livestock. Editors R. F. Keeler, K. R. Van Kampen, L. F. James. New York, Academic Press Inc., pp. 35-46.

Shutt, D.A.; Braden, A.W.H. (1968) The significance of equol in relation to the oestrogenic responses in sheep ingesting clover with a high formononetin content. Australian Journal of Agricultural Research 19, 545-553.

Simons, A.B.; Martin, G.C. (1971) Relationship of indole alkaloids to palatability of Phalaris arundinacea L. Agronomy Journal 63, 915-919.
Smith, B.L.; Coe, B.D.; Embling, P.P. (1978) Protective effect of zinc sulphate in a natural facial eczema outbreak in dairy cows. New Zealand Veterinary Journal 26, 314-315.
Steffert, I.J. (1970) Facial exczema: The extent of damage in dairy cows. Dairy Farming Annual 95-101.
Stermitz, F.R.; Yost, G.S. (1978) Analysis and characterisation of nitro compounds from Astragalus species. In: Effects of poisonous plants on livestock. Editors R. F. Keeler, K. R. Van Kampen, L. F. James. New York, Academic Press, pp:379-390.
Trotman, C.H. (1978) Controlling annual rye grass toxicity. Journal of Agriculture Western Australia 19, 84-89.
Van Etten, C.H.; Tookey, H.L. (1978) Glucosinolates in cruciferous plants. In: Effects of poisonous plants on livestock. Editors R. F. Keeler, K. R. Van Kampen and L. F. James. New York, Academic Press, pp. 507-520.
Van Warmelo, K.T.; Marasas, W.F.O.; Adelaar, T.F.; Kelleriman, T.S.; Van Rensburg, I.B.J.; Minne, J.A. (1970) Experimental evidence that lupinosis of sheep is a mycotoxicosis caused by the fungus, Phomopsis leptostromiformis (Kuhn) Bubok. Journal of South African Veterinary Medical Association 41, 235-247.
Walker, D.J. (1973) The rumen microbes and bloat. In: Bloat. Editors R. A. Leng and J. R. McWilliam. Armidale, Australia, University of New England, Publishing Unit, pp.49-52.
Walthall, J.C.; McKenzie, R.A. (1976) Osteodystrophia fibrosa in horses at pasture in Queensland: Field and laboratory observations. Australian Veterinary Journal 52, 11-16.
Wheeler, J.L.; Hedges, D.A.; Till, A.R. (1975) A possible effect of cyanogenic glucoside in sorghum on animal requirements for sulphur. Journal of Agricultural Science, Cambridge 84, 377-379.
Williams, M.C. (1980) Toxicological investigations on Astragalus hamosus and Astragalus sesameus. Australian Journal of Experimental Agriculture and Animal Husbandry 20, 162-165.
Williams, M.C.; James, L.F. (1978) Livestock poisoning from nitro-bearing Astragalus. In: Effects of poisonous plants on livestock. Editors R. F. Keeler, K. R. Van Kampen and L. F. James. New York, Academic Press, pp. 379-390.
Wong, E.; Latch, G.C.M. (1971) Effect of fungal diseases on phenolic contents of white clover. New Zealand Journal of Agricultural Research 14, 633-638.

PART 4

LIMITATIONS TO INTAKE

INFLUENCE OF SWARD CHARACTERISTICS ON DIET SELECTION AND HERBAGE INTAKE BY THE GRAZING ANIMAL

J. HODGSON

Hill Farming Research Organisation, Bush Estate, Penicuik, Midlothian, EH26 OPY, Scotland.

ABSTRACT

The botanical and morphological composition of vegetation can exert a marked effect upon the selection drive of grazing animals, and variations in sward structure and the distribution of components within the sward canopy influence the opportunity for selection. The factors influencing the choice of grazing site and the choice between alternative plants or plant components within sites are complex and difficult to rationalize with present knowledge. The green leaf content and nutrient concentration of the diet are almost invariably greater than those of the sward as a whole, but in some circumstances this may be due as much to the stratified distribution of plant components within the sward as to active selection by the animal.

The mass and structure of vegetation in the sward canopy affect ingestive behaviour and hence herbage intake. The vertical distribution of foliage exerts the major influence on ingestive behaviour in sown temperate swards, whereas in tropical swards variables associated with leaf density and leaf/stem ratio are of dominant importance. It is not clear to what extent this contrast simply reflects conditions at opposite ends of a continuous spectrum of response.

A better understanding of the principles and implications of the control of ingestive behaviour and selective grazing requires a greater degree of experimental manipulation of both swards and animals, and a more detailed definition of sward structure and the distribution of grazing activity, than has normally been the case in the past.

INTRODUCTION

Restricted nutrient intake is probably the major factor limiting production from grazing animals World-wide. Of the variables influencing nutrient intake, the digestibility and metabolisability of the diet consumed may each vary by a factor of about two, whereas the herbage intake of grazing animals may vary by a factor of at least four, even under relatively unrestricted conditions (Agricultural Research Council 1980, Hodgson & Grant 1981, Minson 1982). This indicates the importance of an understanding of the factors influencing the herbage intake of grazing animals, and of the means of controlling it to advantage.

The influence of the nutritive value and internal structure of ingested herbage on voluntary herbage consumption by housed animals is discussed by Minson (1982). In this paper attention is concentrated upon the influence of characteristics of the sward canopy upon (a) the composition of the diet selected and (b) the amount of herbage eaten by animals under grazing conditions.

DIET SELECTION

Preference and selection

The World literature on the botanical composition of the diets of grazing and browsing ungulates is extensive, particularly for agriculturally important animals (Van Dyne *et al*. 1980). However, the conclusions to be drawn about the factors influencing the associations between sward and diet composition are very limited, except in the most general terms. This is partly a reflection of the fact that many studies have been specific to particular locations, and partly because descriptions of vegetation characteristics have frequently been inadequate.

The characteristics of plants thought to influence the preferences exhibited by grazing and browsing animals have been discussed by Heady (1964), Arnold & Hill (1972) and many others. It is usually possible to explain the avoidance of individual plant species or parts in terms of the presence of particular chemical constituents or surface characteristics (Arnold 1964a, Marten *et al*. 1973, Heady 1975). There may be less certainty about the features contributing to the attractiveness of plants or parts of plants with a high preference rating, and an explanation in terms of the absence of undesirable characteristics hardly seems to be satisfactory. Most of the evidence on the factors influencing preferences for different plant species and morphological components is derived by inference from uncontrolled field studies, and care is therefore needed in its application. The reader is referred to reviews by Arnold & Hill (1972) and Arnold (1981) for critical assessment of the principles involved in the determination and interpretation of dietary preferences. Arnold (1981) states categorically that the plant components measured in conventional proximate analyses cannot influence preference directly because they are not chemical entities active at the molecular level. He also suggests that, because of the complexity of the matrix of plant and animal factors influencing preference, any attempt to identify "palatability factors" in plant species is likely to be wasted effort. This is not to suggest that some success has not been achieved in identifying chemical agents which influence preference (Arnold 1981), but successes have been very few.

The diets of grazing animals consistently contain more leaf and less stem, and more live and less dead material, than the average of the vegetation to which they have access (Van Dyne *et al*. 1980, Arnold 1981). This reflects in part the preferential use of areas containing green herbage (Low *et al*. 1981) and of plant communities in an active phase of growth (Hunter 1962). Selection for green leaf within a sward canopy is well recognized, and can be extreme (Chacon & Stobbs 1976, Hendricksen & Minson 1980) (Table 1). The reasons for these behaviour patterns are not so easy to explain however, though they may be associated with differences in the structural strengths of leaf and stem and of young and mature leaf tissue, and in the surface characteristics and turgidity of live and dead tissue (Heady 1975). In many intensively managed temperate swards, the composition of the diet appears to be the natural consequence of a largely unselective grazing habit superimposed on a stratified distribution of plant tissue. This may be true of mixed grass/clover swards (Fig. 1) as well as on simple grass swards (Hodgson & Ollerenshaw 1969). Thus, it should not necessarily be assumed that the existence of a difference between the composition of the sward and that of the herbage selected from it is indicative of the deliberate exercise of choice by the animal.

TABLE 1

An example of selection for green leaf by cattle grazing a _Setaria sphacelata_ cv. Kazungula sward (from Chacon & Stobbs, 1976).

	Days from start of grazing 1	15
Herbage mass (kg DM/ha)	7300	3600
Proportion of green leaf in:		
Sward	0.33	0.08
Diet	0.96	0.52

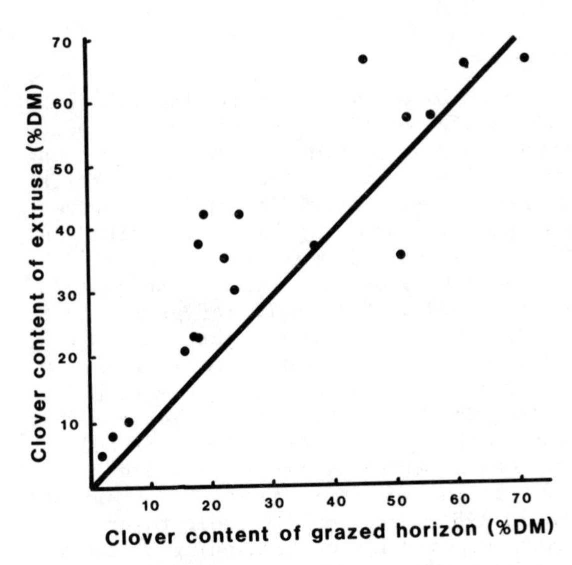

Fig. 1. The relationship between the proportion of clover in the top horizons of the sward grazed by sheep, and the proportion in the herbage consumed; grass and clover plants were defoliated to the same height (J.A. Milne, J. Hodgson, W. Senter & G.T. Barthram, unpublished data).

Diet selection may be viewed as a function of the _preferences_ which would be exhibited between the individual components of a sward if choice were unlimited, modified by the degree to which the characteristics of the vegetation canopy influence the _opportunity_ for selection (Hodgson 1979). Complete freedom of choice is seldom possible except in the artificial conditions of a cafeteria trial, and isolation of the separate effects of preference and opportunity under field conditions is extremely difficult. The concept is a useful one, however, particularly for consideration of the interactions between the effects of sward composition and structure on diet selection and of differences in diet selection between ruminant species. This concept of diet selection does not differ in any material sense from the views of other authorities (Heady 1975, Arnold 1981), though there are differences in terminology.

Listings of the relative "preferences" exhibited for different plant species or varieties by grazing animals have been given by many authors (see Arnold 1981), and are frequently described in terms of the ratio of the proportion of a component in the diet to the proportion in the sward, or some analogous expression. This is strictly a measure of selection as described above, and is best described as the selection ratio; the ranking may be specific to the conditions in which the measurements were made (Heady 1975). Controlled studies of the cafeteria type on small plots may yield more direct information on preferences, but the results are not necessarily indicative of differences in potential intake, or of differences in selection from mixed swards (Arnold 1981). Thus, with the exception of observations on plant species which are particularly high or low on the preference or selection rankings (e.g. Leigh & Mulham 1966a,b, Hodgson & Grant 1981), the results of such studies are likely to be equivocal unless they are pursued over a range of sward conditions and seasons of the year.

Selection and sward characteristics

Selection is likely to be modified, sometimes substantially, by the relative proportions of the different plant species or components and their distribution in space. Arnold (1964b) defined the grazing process as "movement in a horizontal plane and selection in a vertical plane". This seems an unnecessarily restrictive definition, however, and the concept of a two-phase process involving "site selection" and "bite selection" (Milne et al. 1979) seems more appropriate. Site selection can occur on a scale varying from a few meters (Keogh 1973) to several kilometers (Low et al. 1981). The choice of grazing site may be influenced by the distribution of green areas (Low et al. 1981), differences in the constituent species and stage of growth of different communities (Hunter 1962), the patchy distribution of dung and urine, and soil contamination (Marsh & Campling 1970, Keogh 1973). The distribution of grazing activity may also be affected by variations in ground slope, aspect and micro-topography (Arnold 1981), and can be manipulated by the siting of fence lines, watering and feeding points, and shade (Arnold 1981). It would be idle to pretend, however, that we have more than a superficial understanding of the way in which these effects might interact (Arnold & Dudzinski 1978).

Bite selection is essentially the choice of individual bites of herbage from the vegetation at a chosen site, and is the aspect most directly relevant to this paper. The opportunity for bite selection reflects the intimacy of admixture of the different plant components and their distribution in both the horizontal and vertical planes. The structure of the vegetation canopy may also be important, the open structure of most annual grasslands and tussock communities allowing relatively easy access to all levels within the sward (Leigh & Mulham 1966a,b, Hodgson & Grant 1980), in contrast to the closed canopy typical of many temperate short-grass communities (Hodgson & Ollerenshaw 1969).

Selection will depend upon the preference contrasts between alternative components of the sward, as well as their distribution within the canopy. For example, sheep may graze relatively indiscriminately at the surface of intensively managed *Lolium perenne/Trifolium repens* swards (Fig. 1), but selection for *T. repens* growing close to the base of indigenous *Agrostis-Festuca* swards can be much more extreme (Hodgson & Grant 1980). The choice of grazing or browse in plant communities containing shrubs and trees as well as herbaceous plants may be regarded as either site or bite selection, depending upon circumstances.

Animal effects

Both preference and selection vary between and within animal species. Differences in preference may reflect innate or induced differences in sensitivity to plant characteristics (Arnold 1981). Selection differences reflect differences in the size and shape of the body and mouth parts and in grazing strategy (Bell 1969), which would be expected to influence both site selection and bite selection. It is seldom possible to differentiate between these effects from published evidence, except in the most general terms, or indeed to determine cause and effect in many cases. Most of the comparative evidence on diet composition in different animal species (Van Dyne *et al.* 1980) relates to species grazing different plant communities, or at least different strata within the same community as a result of differences in size or agility, and there are few direct comparisons of diet selection by animals confined to the same community. In our own studies on indigenous hill plant communities in Scotland, sheep usually graze deeper within the sward canopy than do cattle and select a diet containing higher proportions of live leaf and of broad-leaved plants (Hodgson & Grant 1980).

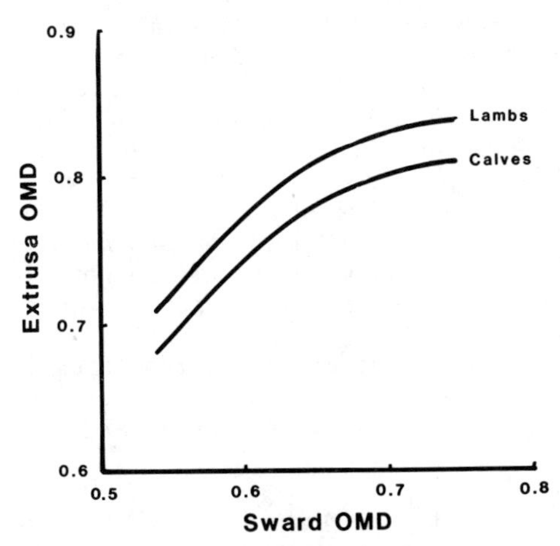

Fig. 2. The relationship between the digestibility of the sward and that of the diet selected by grazing lambs and calves (from Jamieson & Hodgson 1979b).

The diets of grazing sheep often contain higher concentrations of green leaf and of digestible nutrients than do those of cattle (Langlands & Sanson 1976, Jamieson & Hodgson 1979b, Fig. 2) and it is tempting to suggest that sheep are the more "efficient" selectors. However, this begs the question (vide Arnold 1981) about the grazing strategies of the two species, and the degree to which differences in diet composition can be taken to reflect differences in selection capability rather than differences in what might be termed selection drive. It is clear that in their selective grazing behaviour sheep and cattle may respond to different characteristics of the sward (Langlands & Sanson 1976, T.D.A. Forbes 1981, unpublished work). In extreme cases there may simply be species or size differences in the physical ability of animals to reach some components of particularly tall or impenetrable swards.

These concepts can be generalized into the views of the complementary grazing strategies and tactics of wild ungulates occupying complex areas of vegetation like the African savannahs (Jarman & Sinclair 1980). However, neither the specific nor the general case has been tested experimentally. Evidence on differences in diet selection between animals of the same species differing in age or physiological state is also sparse. The differences observed appear to be relatively small and unimportant, except that diet selection by young animals may be less consistent than that by older animals (Hodgson & Jamieson 1981).

HERBAGE INTAKE

Herbage intake and ingestive behaviour

Variations in the nutrient content or internal structure of plant material exert a strong influence upon the amount consumed by ruminants. These effects, which can readily be investigated with housed animals, are discussed by Minson (1982). The influence upon voluntary herbage intake of variables like the digestibility of the diet may be similar for grazing animals (e.g. Hodgson et al. 1977) and housed animals fed on all-forage diets (Minson 1982). However, the herbage intake of grazing animals may in addition be affected by non-nutritional characteristics of the sward associated primarily with variations in the mass of herbage and its distribution within the foliage canopy (Hodgson 1977). Evidence on the influence of these characteristics has usually been taken as presumptive evidence that intake may be restricted by limitations to the adaptability of ingestive behaviour as well as by the chemostatic controls and limitations to digestive tract capacity postulated for housed animals (Hodgson 1977). However, increases in intake with increasing herbage mass, for example, might be explained either by greater ease of prehension and ingestion of herbage (Arnold & Dudzinski 1966) or by greater opportunity for selection (Hamilton et al. 1973) resulting in a higher nutrient concentration in the diet. There was no conclusive evidence on the importance of limitations to ingestive behaviour until the work of Chacon & Stobbs (1976).

Herbage intake over 24 h, or some shorter time interval (I) can be considered as the product of the time spent grazing (GT) and the rate of herbage consumption per unit of grazing time (RI); the rate of herbage consumption is itself the product of the amount of food ingested per bite (IB) and the rate of biting during grazing (RB) (Allden & Whittaker 1970). Thus:

$$I = IB \times RB \times GT$$

Variations in sward conditions would be expected to exert direct effects upon intake per bite and rate of biting; modifications to the time spent grazing, and possibly also to the rate of biting, would be seen as compensating animal responses.

This view of ingestive behaviour is somewhat mechanistic, and the expression of continuously variable parameters as single means or totals may be too simplistic. However, the concept is useful because it provides a convenient basis for considering the influence of sward characteristics on herbage intake.

The variation in intake per bite is usually substantially greater than variations in either biting rate or grazing time (Stobbs 1973b, Hodgson 1981). Thus, although increases in biting rate often do occur when intake per bite is depressed, they are seldom of a magnitude to avoid some reduction in the rate of herbage intake (IB x RB) (Hodgson 1981). Increases in grazing time may initially be enough to offset a declining rate of intake on continuously stocked swards, but compensating changes are limited (Arnold 1964b). Under strip-grazing management (Jamieson & Hodgson 1979a) or where swards are progressively grazed down (Chacon & Stobbs 1976) declines in the rate of intake and in grazing time may reinforce one another as sward conditions become more difficult. Stobbs (1973a), working with tropical swards, suggested that the number of grazing bites taken by cows during a 24 h period (RB x GT) rarely exceeds 36,000, a value also approached by cattle and sheep on some temperate swards (Jamieson & Hodgson 1979a); this may set an effective upper limit to the compensation for reduced intake per bite.

Herbage intake and gross sward characteristics

Herbage intake usually increases at a progressively decreasing rate with increase in either herbage mass or sward height, though there is marked variation in the pattern of response observed in individual studies (Hodgson 1977). This variation may reflect in part the confounding effect of concomitant changes in the nutritive value of the herbage consumed. It is very likely, for instance, that in some of the more extreme examples of an asymptotic relationship between herbage mass and herbage intake (Hodgson 1977) there was a progressive decline in herbage digestibility with increasing mass which would tend to limit artificially the response to variation in mass alone.

The bulk density of herbage within the sward (weight per unit volume) would be expected to exert an influence upon intake per bite, and hence upon daily herbage intake, but this effect has not often been examined in grazing studies. Stobbs (1975) demonstrated significant relationships between leaf bulk density and intake per bite in cattle grazing Setaria sphacelata swards, but the effects of leaf density and leaf/stem ratio could not be separated.

The pattern of response to variations in sward height may also depend upon the way in which height is measured. Variations in mass and in surface height are often closely correlated, and tend to influence intake in the same way. If measurements are made of the extended height of leaves or tillers, however, the relationship may be quadratic, with intake declining on either side of an optimum extended height. This observation is usually interpreted in terms of the increasing difficulty of prehending and ingesting both excessively long and very short leaves from the sward. The phase of increasing intake or rate of intake with increasing height is usually apparent in temperate swards (Allden & Whittaker 1970, Fig. 3) though the declining phase may be observed on particularly long herbage (Waite *et al*. 1950), and intake may be maximized at an extended sward height of 40-45 cm (Hodgson *et al*. 1977). The declining phase of intake with increasing height is more common in tall-growing tropical swards, and probably is demonstrated most dramatically in the case of the trailing tropical legumes (Stobbs & Hutton 1974). In temperate swards the rate of biting tends to fall progressively with increasing sward height as the ratio of manipulative to ingestive jaw movements increases (Chambers *et al*. 1981). However, the decline in rate of biting is not rapid enough to offset the advantages of increases in intake per bite with increasing sward height (Hodgson 1981).

Herbage intake and sward structure

It is frequently difficult to disentangle the independent effects of variations in gross sward characteristics such as herbage mass, sward height and bulk density upon herbage intake, and to separate them from the effects of variations in the nutritive value of the herbage consumed. Also, observations on gross characteristics like these almost certainly provide an inadequate description of the vegetation as perceived by the grazing animal. Descriptions of the distribution of plant components within the vegetation canopy, and particularly their association with short-term measurements of ingestive behaviour within specified sward strata (Stobbs 1973b), provide a means of rationalizing some of the inconsistent intake responses to variations in herbage mass or sward height noted earlier, and of improving the objectivity of studies on herbage intake and diet selection. Examination of the results of studies involving this approach indicates interesting and often substantial differences in the conclusions drawn for temperate and for tropical swards. It is not clear at present whether this reflects fundamental differences between temperate and tropical forage plants, or responses measured at opposite ends of a common spectrum of response to variations in sward structure. The latter seems the more likely, but for the present it is simpler to consider the two sets of conditions separately.

In intensively managed *Lolium perenne* swards, daily herbage intake or the short-term rate of intake increases progressively with increase in the height of the grazed surface, and appears to be insensitive to variations in the bulk density of herbage within the grazed horizon (Hodgson 1981, Fig. 4). This is consistent with the evidence that intake is higher on relatively erect spring swards than on relatively short summer or autumn re-growths at equivalent levels of herbage mass and digestibility (Hodgson *et al*. 1977, Jamieson & Hodgson 1979a), suggesting that management or selection for erect growth habit could have an important influence on intake potential. The absence of a response to variation in herbage bulk density is not easy to understand, but it may simply reflect the overriding influence of sward height in circumstances where it is difficult to achieve independent variation in height and density. A probable explanation for the sward height effect was provided recently by Barthram (1980), who showed that grazing sheep seldom penetrated into the horizons containing pseudostem or dead material, even when herbage intake was severely limited in consequence, and that there was a direct relationship between the surface height of the sward and the depth of the grazed horizon (Fig. 5). Thus the beneficial effects of breeding

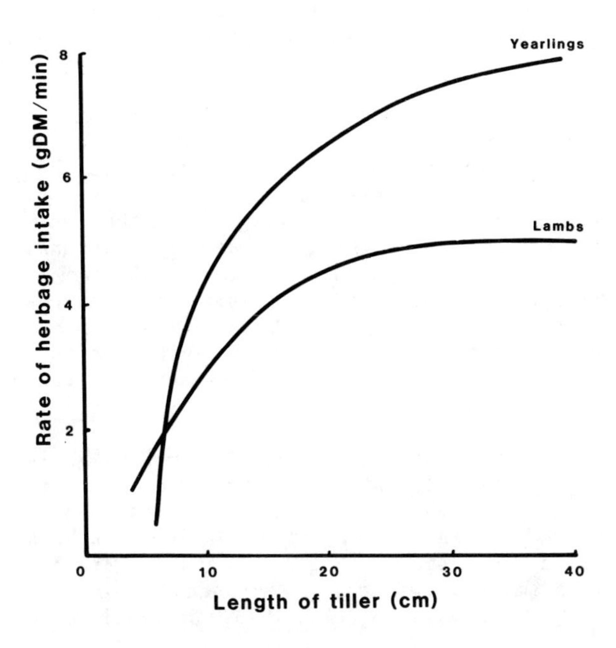

Fig. 3. The relationship between tiller length (cm) and rate of herbage intake (g DM/min) in lambs and yearling sheep (from Allden & Whittaker 1970).

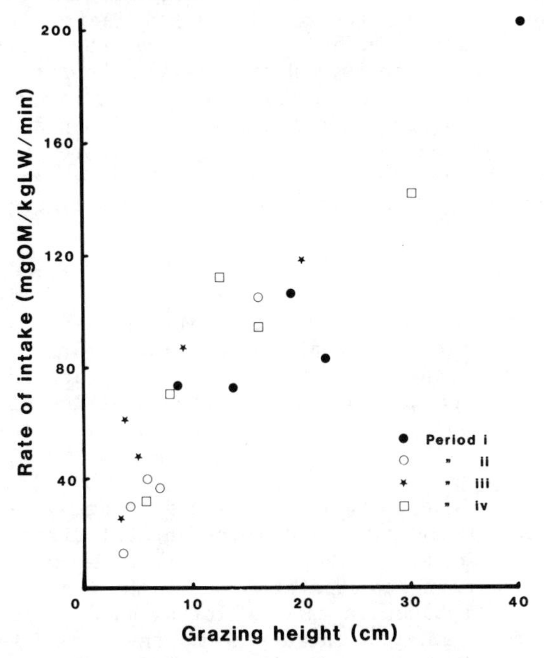

Fig. 4. The relationships between the rate of herbage intake by grazing calves (g OM/kg LW/min) and the height of the grazed horizon (cm) (from Hodgson 1981 with permission of Agricultural Institute, Eire).

for erect growth might be limited if there were a correlated increase in pseudostem height or (in view of the influence on selective grazing and the nutrient concentration of the diet) in stem/leaf ratio or leaf fibre content (Minson 1982).

Where variation in herbage bulk density is small, changes in intake per bite must reflect changes in bite volume. The results of Barthram (1980) indicate the possibility of change in bite depth, but there is little information on bite area and the sward factors affecting it. Cows usually sweep an area of herbage into the mouth with the tongue, whereas sheep are capable of removing individual leaves from a plant. Cattle are also capable of fine resolution in removing individual components from the sward in some circumstances (Stobbs 1973a, Hodgson & Grant 1980). The depth of a bite is not necessarily limited by the dimensions of the buccal cavity, because both cattle and sheep frequently grip leaves or tillers and tear them off before drawing them into their mouths.

Sward surface height may be substantially greater and the bulk density of the herbage substantially lower in tropical than in temperate swards (Stobbs 1973b). In these circumstances variations in height apparently have a smaller influence and variations in bulk density a greater influence on intake (Fig. 6) and particular importance is attached to parameters reflecting leaf density or leaf to stem ratio (Stobbs 1973b, Chacon & Stobbs 1976, Hendrickson & Minson 1980).

The nature of the responses observed has usually been complex, but there is a clear inference that efforts in sward management and in plant breeding and selection should be towards increased leaf bulk density and reduced stem to leaf ratio in order to improve levels of herbage intake from tropical swards (Stobbs & Hutton 1974). However, there is also the associated question of access within the sward canopy. The different patterns of selective grazing within closed and open canopies have already been mentioned. Breeding for increased leaf density will not necessarily be beneficial if it reduces access within the sward canopy particularly if senescent material accumulates at the surface.

Influence of selective grazing

Selective grazing, whatever its basis, almost invariably results in some improvement in the nutrient content of the diet (Arnold 1964a). The degree to which this can be equated with an increase in total nutrient intake, however, depends largely upon the magnitude of the penalty to the short-term rate of nutrient intake which selection may incur.

Though it would be logical to expect that selective grazing would tend to limit intake per bite and the rate of herbage intake, direct evidence on these effects is sparse. Stobbs (1973b) associated a reduction in intake per bite amongst grazing cows with greater selectivity in swards of S. sphacelata and Chloris gayana at progressively later stages of maturity. Increases in the rate of biting by cows with increasing sward maturity in another trial (Stobbs 1974) were interpreted as an attempt to compensate for reduced intake per bite. A similar pattern of response has been observed in recent studies on indigenous hill plant communities at this Institute, intake per bite tending to fall and rate of biting to increase in both sheep

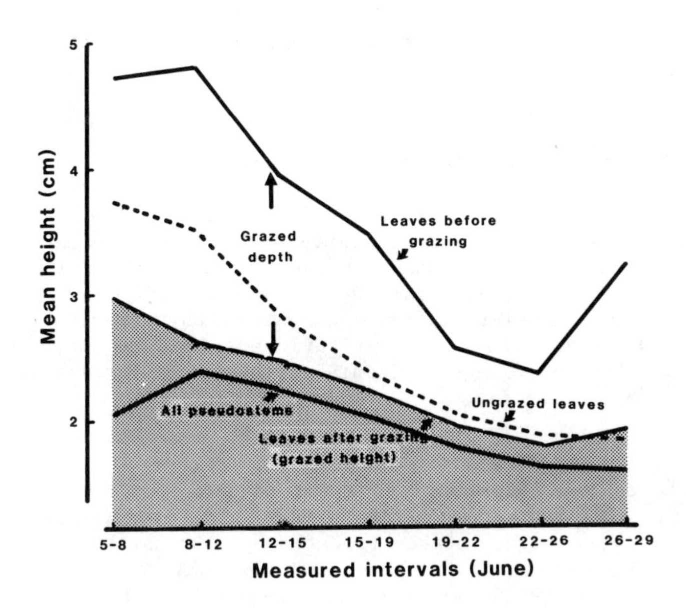

Fig. 5. The relationships between sward surface height, pseudostem height and the depth of the grazed horizon (all cm) in a sheep-grazed sward (from Barthram 1980).

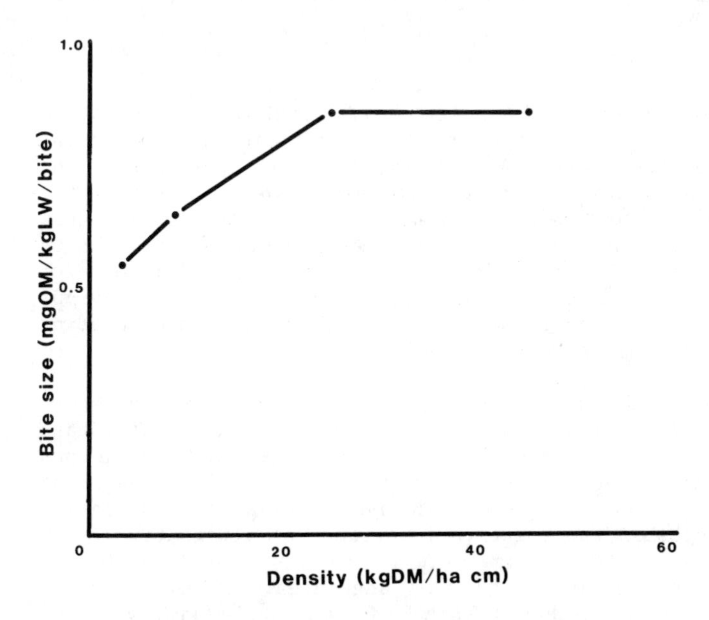

Fig. 6. The relationship between the bulk density of herbage in the surface horizons of a sward (kg DM/ha cm) and intake per bite (mg OM/kg LW/bite) in grazing cattle (from Stobbs 1975).

and cattle with declining green herbage mass and green/dead ratio (T.D.A. Forbes 1981, unpublished work).

Animal effects

Sheep are probably better able to maintain intake on very short swards than are cattle (J.C. Arosteguy 1981, unpublished), and young sheep appear to maintain intakes better than older sheep (Allden & Whittaker 1970). With these exceptions, patterns of response in ingestive behaviour to variations in sward characteristics appear to be very similar in cattle and sheep (Jamieson & Hodgson 1979b) and between animals of the same species differing in age or physiological state (Arnold 1975). However, there is some evidence that response patterns may be unstable in young ruminants with little grazing experience (Hodgson & Jamieson 1981).

Appraisal

The current state of knowledge on the impact of sward structure and the distribution of plant components within the sward upon ingestive behaviour and herbage intake illustrates the practical problems of the work involved. It is difficult to achieve enough independent variation in the sward variables of interest to achieve a clear indication of the magnitude of their separate effects, and particularly of the way in which their relative importance changes as sward conditions change. The different response patterns to variations in sward height, density and leafiness in temperate and tropical swards are tantalizing examples of the scope for manipulation of ingestive behaviour and herbage intake by plant selection and sward management strategies, but at present we can only surmise the reasons for these differences.

CONCLUSIONS

A proper understanding of the mechanisms controlling ingestive behaviour and selective grazing, and their implications to grazing management, requires a greater degree of experimental manipulation and a more detailed definition of grazing activity, in relation to the distribution of plant components within the sward canopy, than has normally been the case in the past. Much of the existing evidence on diet selection is descriptive, and specific to particular sets of circumstances. The information from studies on ingestive behaviour and herbage intake is frequently more useful because some manipulation of sward conditions has often been involved.

Despite these limitations, many of the sward characteristics likely to influence ingestive behaviour, and hence the nutrient intake of grazing animals, can be described in terms relevant to decisions about priorities for plant selection and for grassland management (Stobbs & Hutton 1974, Hodgson 1980). The need now is for comparative information on changes in the relative importance of alternative sward characteristics across the range of combinations likely to be encountered in practice and with the ungulate species of major agricultural importance. Although such information is likely to be obtained principally in studies limited to particular geographical areas and systems of grassland management, the value of a generalized body of theory cannot be over-emphasized.

ACKNOWLEDGEMENTS

I am grateful to many colleagues for discussions on the background to the views expressed in this paper. In particular, I wish to acknowledge the contribution made by the late Dr Harry Stobbs, who played such an active role in developing the concepts and techniques which have made it possible to examine in detail the relationships between sward characteristics and ingestive behaviour.

REFERENCES

Agricultural Research Council (1980) The nutrient requirements of ruminant livestock. Technical review by an Agricultural Research Council working party. Farnham Royal, U.K., Commonwealth Agricultural Bureaux, 351 pp.

Allden, W.G.; Whittaker, I.A.McD. (1970) The determinants of herbage intake by grazing sheep: The interrelationship of factors influencing herbage intake and availability. Australian Journal of Agricultural Research 21, 755-766.

Arnold, G.W. (1964a) Some principles in the investigation of selective grazing. Proceedings of Australian Society of Animal Production 5, 258-271.

Arnold, G.W. (1964b) Factors within plant associations affecting the behaviour and performance of grazing animals. In: Grazing in terrestrial and marine environments. Editor D.J. Crisp. British Ecological Society Symposium No. 4. pp. 133-154.

Arnold, G.W. (1975) Herbage intake and grazing behaviour in ewes of four breeds at different physiological states. Australian Journal of Agricultural Research 26, 1017-1024.

Arnold, G.W. (1981) Grazing behaviour. In: Grazing animals. Editor F.H.W. Morley. World animal science, B1. Amsterdam, Elsevier, pp. 79-104.

Arnold, G.W.; Dudzinski, M.L. (1966) The behavioural responses controlling the food intake of grazing sheep. Proceedings of the 10th International Grassland Congress, Helsinki, 367-370.

Arnold, G.W.; Dudzinski, M.L. (1978) The ethology of free-ranging domestic animals. Amsterdam, Elsevier, 198 pp.

Arnold, G.W.; Hill, J.L. (1972) Chemical factors affecting selection of food plants by ruminants. In: Phytochemical ecology. Editor J.B. Harborne. Proceedings of the Phytochemical Society Symposium, 1971. London, Academic Press, No. 8 pp. 71-101.

Barthram, G.T. (1980) Sward structure and the depth of the grazed horizon. Proceedings of the British Grassland Society Winter Meeting 1980. Grass and Forage Science 36, 130-131.

Bell, R.H.V. (1969) The use of the herb layer by grazing ungulates in the Serengeti National Park, Tanzania. Ph.D. Thesis, University of Manchester.

Chacon, E.; Stobbs, T.H. (1976) Influence of progressive defoliation of a grass sward on the eating behaviour of cattle. Australian Journal of Agricultural Research 27, 709-727.

Chambers, A.R.M.; Hodgson, J.; Milne, J.A. (1981) The development and use of equipment for the automatic recording of ingestive behaviour in sheep and cattle. Grass and Forage Science 36, 97-105.

Hamilton, B.A.; Hutchinson, K.J.; Annis, P.C.; Donnelly, J.B. (1973) Relationships between the diet selected by grazing sheep and the herbage on offer. Australian Journal of Agricultural Research 24, 271-277.

Heady, H.F. (1964) Palatability of herbage and animal preference. Journal of Range Management 17, 76-82.

Heady, H.F. (1975) Rangeland Management. New York, McGraw-Hill, 460 pp.

Hendricksen, R.; Minson, D.J. (1980) The feed intake and grazing behaviour of cattle grazing a crop of Lablab purpureus cv. Rongai. Journal of Agricultural Science, Cambridge 95, 547-554.

Hodgson, J. (1977) Factors limiting herbage intake by the grazing animal. Proceedings of the International Meeting of Animal Production from Temperate Grassland, Dublin 70-75.

Hodgson, J. (1979) Nomenclature and definitions in grazing studies. Grass and Forage Science 34, 11-18.

Hodgson, J. (1980) Testing and improvement of pasture species. In: World animal science, B1. Grazing animals. Editor F.H.W. Morley. Amsterdam, Elsevier, pp.309-317.

Hodgson, J. (1981) The influence of variations in the surface characteristics of the sward upon the short-term rate of herbage intake by calves and lambs. Grass and Forage Science 36, 49-57.

Hodgson, J.; Grant, Sheila A. (1980) Grazing animals and forage resources in the hills and uplands. In: The effective use of forage and animal resources in the hills and uplands. Editor J. Frame. Proceedings of Occasional Symposium No. 12, The British Grassland Society, Edinburgh, 1980, pp. 41-57.

Hodgson, J.; Jamieson, W.S. (1981) Variation in herbage mass and digestibility, and the grazing behaviour and herbage intake of adult cattle and weaned calves. Grass and Forage Science 36, 39-48.

Hodgson, J.; Ollerenshaw, J.H. (1969) The frequency and severity of defoliation of individual tillers in set-stocked swards. Journal of the British Grassland Society 24, 226-234.

Hodgson, J.; Rodriguez Capriles, J.M.; Fenlon, J.S. (1977) The influence of sward characteristics on the herbage intake of grazing calves. Journal of Agricultural Science, Cambridge 89, 743-750.

Hunter, R.F. (1962) Hill sheep and their pasture: A study of sheep-grazing in south-east Scotland. Journal of Ecology 50, 651-680.

Jamieson, W.S.; Hodgson, J. (1979a) The effects of daily herbage allowance and sward characteristics upon the ingestive behaviour and herbage intake of calves under strip-grazing management. Grass and Forage Science 34, 261-271.

Jamieson, W.S.; Hodgson, J. (1979b) The effects of variation in sward characteristics upon the ingestive behaviour and herbage intake of calves and lambs under a continuous stocking management. Grass and Forage Science 34, 273-282.

Jarman, P.J.; Sinclair, A.R.E. (1980) Feeding strategy and the pattern of resource partitioning in ungulates. In: Serengeti: dynamics of an ecosystem. Editors A.R.E. Sinclair and M. Morton-Griffiths. Chicago, Chicago University Press, pp.130-163.

Keogh, R.G. (1973) Herbage growth, and the grazing pattern of sheep set-stocked, on a ryegrass-dominant, staggers-prone pasture during summer. New Zealand Journal of Experimental Agriculture 1, 51-54.

Langlands, J.P.; Sanson, J. (1976) Factors affecting the nutritive value of the diet and the composition of rumen fluid of grazing sheep and cattle. Australian Journal of Agricultural Research 27, 691-707.

Leigh, J.H.; Mulham, W.E. (1966a) Selection of diet by sheep grazing semi-arid pastures on the Riverine Plain. 1. A bladder saltbush (Atriplex vesicaria) - cotton bush (Kochia aphylla) community. Australian Journal of Experimental Agriculture and Animal Husbandry 6, 460-467.

Leigh, J.H.; Mulham, W.E. (1966b) Selection of diet by sheep grazing semi-arid pastures on the Riverine Plain. 2. A cotton bush (Kochia aphylla) - grassland (Stipa variabilis - Danthonia caespitosa) community. Australian Journal of Experimental Agriculture and Animal Husbandry 6, 468-474.

Low, W.A.; Dudzinski, M.L.; Muller, W.J. (1981) The influence of forage and climatic conditions on range community preference of shorthorn cattle in central Australia. Journal of Applied Ecology 18, 11-26.

Marsh, R.; Campling, R.C. (1970) Fouling of pastures by dung. Herbage Abstracts 40, 123-130.

Marten, G.C.; Barnes, R.F.; Simons, A.B.; Wooding, F.J. (1973) Alkaloids and palatability of Phalaris arundinacea L. grown in diverse environments. Agronomy Journal 65, 199-201.

Milne, J.A.; Bagley, L.; Grant, Sheila, A. (1979) Effects of season and level of grazing on the utilisation of heather by sheep. 2. Diet selection and intake. Grass and Forage Science 34, 45-53.

Minson, D.J. (1982) Effects of chemical and physical composition of herbage eaten upon intake. In: Nutritional limits to animal production from pastures. Editor J.B. Hacker. Farnham Royal, U.K., Commonwealth Agricultural Bureaux. pp. 167-182.

Stobbs, T.H. (1973a) The effect of plant structure on the intake of tropical pastures. I. Variation in the bite size of grazing cattle. Australian Journal of Agricultural Research 24, 809-819.

Stobbs, T.H. (1973b) The effect of plant structure on the intake of tropical pastures. II. Differences in sward structure, nutritive value, and bite size of animals grazing Setaria anceps and Chloris gayana at various stages of growth. Australian Journal of Agricultural Research 24, 821-829.

Stobbs, T.H. (1974) Rate of biting by Jersey cows as influenced by the yield and maturity of pasture swards. Tropical Grasslands 8, 81-86.

Stobbs, T.H. (1975) The effect of plant structure on the intake of tropical pasture. III. Influence of fertiliser nitrogen on the size of bite harvested by Jersey cows grazing Setaria anceps cv. Kazungula swards. Australian Journal of Agricultural Research 26, 997-1007.

Stobbs, T.H.; Hutton, E.M. (1974) Variations in canopy structures of tropical pastures and their effects on the grazing behaviour of cattle. Proceedings of the 12th International Grassland Congress, Moscow, Section V, 680-687.

Van Dyne, G.M.; Brockington, N.R.; Szocs, Z.; Duek, J.; Ribic, C.A. (1980) 4. Large herbivore subsystem. In: Grasslands, systems analysis and man. Editors A.I. Bremeyer and G.M. Van Dyne. International biological programme 19. Cambridge, Cambridge University Press, pp. 269-537.

Waite, R.; Holmes, W.; Campbell, Jean I.; Fergusson, D.L. (1950) Studies in grazing management. II. The amount and chemical composition of herbage eaten by dairy cattle under close-folding and rotational methods of grazing. Journal of Agricultural Science, Cambridge 40, 392-402.

EFFECTS OF CHEMICAL AND PHYSICAL COMPOSITION OF HERBAGE EATEN UPON INTAKE

D.J. MINSON

CSIRO, Division of Tropical Crops and Pastures, Cunningham Laboratory, St Lucia, Brisbane, Queensland 4067, Australia

ABSTRACT

The voluntary intake of forage has been positively related to the digestibility of the dry matter and energy. This relation is empirical and not mechanistic in origin; the main factors controlling intake are the proportion of indigestible residue in the feed, the transit time of this residue through the rumen and the size of the rumen. Feeds vary in the time required for them to be broken down to particles sufficiently small to leave the rumen. These differences lead to different relationships between intake and digestibility for chopped _versus_ pelleted forages, leaf _versus_ stem, legumes _versus_ grasses, grass with and without calcium fertilizer, pasture species and cultivars and temperate _versus_ tropical grasses.

The chemical factors in pastures that influence intake are considered under two headings; (1) chemical factors related to the quantity and composition of the fibre and (2) nutrients that are essential for the microbial population of the rumen and for the host animal. It is shown that intake is correlated with a wide range of cell wall and cell content fractions but that there are large differences in intake between leaf and stem and between pasture species that cannot be accounted for by differences in chemical composition. Intake of pastures will be depressed below the level set by physical factors if there is a deficiency of protein, sulphur, sodium, phosphorus, cobalt or selenium.

It is concluded that intake is controlled by many unrelated physical and chemical characteristics of the pasture and that intake can rarely be predicted accurately from a single laboratory analysis. Improvement in prediction of intake will only be achieved once the many physical factors controlling intake have been identified and their relative importance determined.

INTRODUCTION

The quantity of herbage eaten by the grazing animal depends on three factors: (1) the availability of suitable herbage, (2) the physical and chemical composition of the herbage and (3) the nutrient requirements of the animal. This paper will consider the effects of physical and chemical composition of the herbage on intake; factors 1 and 3 are reviewed by Hodgson (1982) and Weston (1982) respectively. The effect of animal type and method of feeding on herbage intake has been previously reviewed (Minson 1981).

For more than 50 years it has been known that a relation exists between the bulk of food and the quantity voluntarily eaten (intake) by ruminants (Balch & Campling 1962). This relation appears to be associated with the filling effect in the gut, especially in the reticulo-rumen (rumen) and hence is related to such factors as the digestibility and rate of passage of the food through the rumen (Balch & Campling 1962).

In 1961 Blaxter et al. elegantly demonstrated the way herbage intake is related to digestibility, rate of passage and rumen fill. Using poor, medium and good quality hays they showed that low intake was associated with low digestibility and that the fall in intake was inversely related to the transit time of stained particles through the rumen. From the faecal excretion curves of stained particles they estimated that the dry matter contained in the digestive tract was similar for all three hays and that intake of hay is mainly controlled by rumen capacity. More recently it has been shown that the intake of herbage is not limited by the capacity of the small and large intestines to transport bulk (Grovum & Phillips 1978). The positive relationship between intake and digestibility has also been observed with calves and dairy cattle grazing temperate swards (Hodgson 1977) with no evidence of the curvilinear response reported for pen studies by Conrad (1964) and Balch & Campling (1969).

If it could be shown that intake was always closely related to digestibility then it would be possible to estimate intake from digestibility estimated by in vitro techniques based on rumen fluid (Tilley & Terry 1963) or cellulase (McLeod & Minson 1978). Many studies with both sheep and cattle have confirmed the existence of a correlation between intake and digestibility but deviations from the general trend have also been found. These deviations from the general relation will be described and the possible causes of these deviations discussed.

PHYSICAL FACTORS

Evidence is accumulating that the prime physical factor in a plant which influences intake is the rate at which it is broken down to particles small enough to leave the rumen. This evidence will be discussed in the following section. The effect of water content of the herbage on the quantity of dry matter eaten will also be considered.

Chopped vs. pelleted forages

The intake of a forage is usually increased when it is ground and pelleted (Minson 1963, Campling & Milne 1972, Jarrige et al. 1973, Wilkins 1973). The increase in intake is not constant, but depends on the quality of the unpelleted forage; large increases occur with forages that are eaten in small quantities and give poor rates of growth while little improvement occurs when good quality forages are pelleted (Minson 1963). Although pelleting increases intake of the forage, this rise is not associated with a rise in digestibility. In most studies digestibility has been depressed by pelleting (Minson 1963, Greenhalgh & Wainman 1972) leading to different intake/digestibility relationships for chopped and pelleted forages (Fig. 1). This difference in intake is associated with a faster rate of passage of the pelleted forage through the rumen and this permits more feed to be consumed (Campling & Freer 1966, Minson 1967, Laredo & Minson 1975a). Most feed particles greater than 1 mm (sieve hole size) are prevented from leaving the rumen through the reticulo-omasal orifice (Poppi et al. 1980) and are regurgitated and reduced in size by rumination. Finely ground forages contain few particles larger than 1 mm and animals fed these forages ruminate very little (Balch 1971). Thus the difference between chopped and pelleted forages in the relation between intake and digestibility appears to be associated with the reduction in the need for the forage to be broken down by rumination and in consequence the faster passage of forage through the rumen. Other aspects of pelleting forages are considered in greater detail by Wilkins (1982).

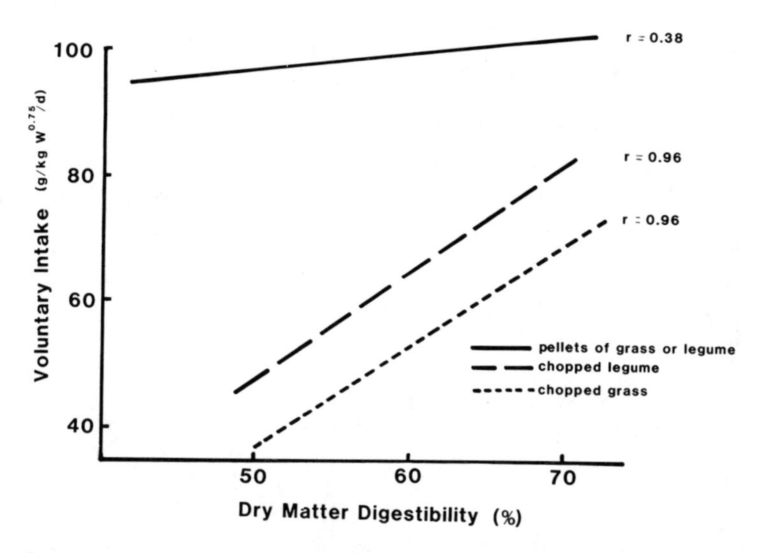

Fig. 1. Effect of pelleting on the relation between intake and digestibility (Data of Heaney et al. 1963).

Leaf vs. stem

Physical differences also exist between different parts of the same plant. It was generally believed that leaf and stem of the same digestibility were eaten in similar quantities. However, studies with separated leaf and stem fractions have shown leaf to be eaten in greater quantities than stem of similar dry matter digestibility (Fig. 2). The higher intake of leaf has been found with sheep fed tropical grasses (Laredo & Minson 1973, 1975a), a temperate grass (Laredo & Minson 1975b) and a tropical legume (Hendricksen et al. 1981). Cattle also eat more leaf than stem of tropical grasses (Poppi et al. 1981a) and of a tropical legume (Hendricksen et al. 1981). The mean difference in intake between leaf and stem fractions in 30 comparisons was 42 percent with a difference of only one percent in digestibility between the two fractions (Table 1).

The higher intake of the leaf fraction was associated with the shorter time the leaf fraction was retained in the rumen compared with the stem fraction. The mean retention time in the rumen of leaf and stem of 26 forages was 24.6 and 33.3 hours respectively (Laredo & Minson 1973, 1975a, Poppi et al. 1981a, Hendricksen et al. 1981). The weight of dry matter in the rumen was similar for animals fed leaf or stem fractions (Laredo & Minson 1973, Poppi et al. 1981a, Hendricksen et al. 1981) so the shorter time the leaf fraction stayed in the rumen permitted more feed to be consumed. The mechanism controlling the difference in intake between leaf and stem fraction was therefore similar to that operating with pelleted and chopped forages.

TABLE 1

Mean voluntary intake and digestibility of the dry matter of separated leaf and stem fractions of grasses and a legume fed to sheep and cattle.

Pasture species	Animal species	Number of samples	Mean Intake $g/kg^{0.75}/d$ Leaf	Stem	Diff.	Digestibility % Leaf	Stem	Diff.	Reference
Digitaria decumbens	sheep	3	58	40	18	53	54	-1	Laredo & Minson 1973
"	"	1	44	34	10	57	51	6	Laredo & Minson 1975a
"	"	2	29	25	4	50	52	-2	Poppi *et al*. 1981a
"	cattle	2	27	19	8	52	54	-2	" " "
Chloris gayana	sheep	3	57	45	12	53	58	-5	Laredo & Minson 1973
"	"	1	36	29	7	40	42	-2	Laredo & Minson 1975a
"	"	2	28	21	7	51	49	2	Poppi *et al*. 1981a
"	cattle	2	29	22	7	55	54	1	" " "
Setaria sphacelata	sheep	3	59	32	27	56	59	-3	Laredo & Minson 1973
"	"	1	41	27	14	46	42	4	Laredo & Minson 1975a
Panicum maximum	"	3	64	47	17	51	56	-5	Laredo & Minson 1973
Pennisetum clandestinum	"	3	50	35	15	51	52	-1	" " "
Lolium perenne	"	2	74	62	12	67	65	2	Laredo & Minson 1975a
Lablab purpureus	"	1	91	53	38	56	49	6	Hendricksen *et al*.1981
"	cattle	1	92	52	40	55	55	0	" " "
Mean			51	36	15	53	54	-1	

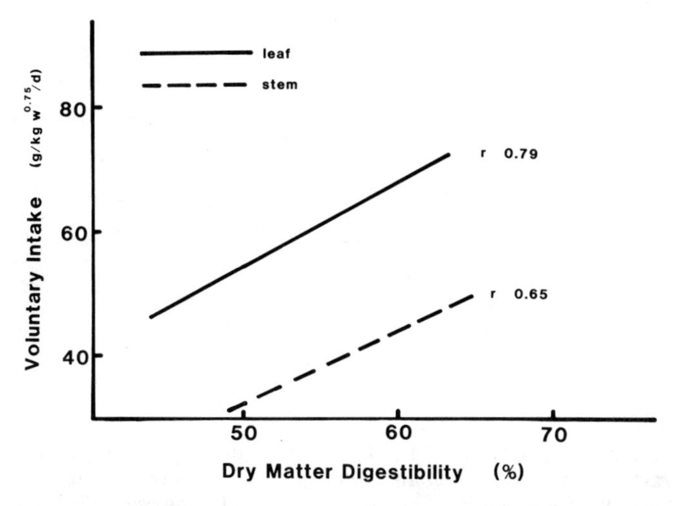

Fig. 2. The relationship between intake and digestibility of leaf and stem fractions (Data of Laredo & Minson 1973).

There are many physical differences between leaf and stem fractions and any of these might be the cause of the shorter retention time of the leaf fraction. The leaf fraction has a larger surface area per unit weight than the stem fraction (Laredo & Minson 1973) and this might allow more rapid digestion, but studies of rate of digestion *in vitro* showed no difference in rate of digestion (Laredo & Minson 1973, 1975b). This result is not in agreement with the suggestion of Crampton *et al*. (1960a) that rate of digestion could be used to predict forage intake. However, in their work intake differences were obtained from feeds differing in species and stage of growth.

As a result of studies with pelleted and chopped forages Baumgardt (1970) suggested that high bulk density of forages leads to high voluntary intake. However, although stem has a bulk density three times higher than that of the leaf fraction, the leaf fraction is eaten in greater quantities, indicating that density *per se* is not a factor controlling intake of herbage (Laredo & Minson 1973).

The most probable reason for the longer retention time of the stem fraction in the rumen is the higher proportion of large particles in masticated stem than in masticated leaf (Table 2) because of the greater resistance of stem to physical breakdown. It has been found that the energy required to grind 1 g samples in a hammer mill fitted with a 1 mm screen is 74 percent higher for stem than leaf fractions (Laredo & Minson 1973, 1975b). The large particles of the stem fraction tended to stay longer in the rumen than the large particles of the leaf fraction (Poppi *et al*. 1981b).

TABLE 2

Effect of mastication by cattle and sheep on the proportion of large particles (> 1.18 mm screen) in swallowed samples of four forages (Poppi *et al*. 1981b)

	Large particles (g/g)		
	Leaf	Stem	Difference
Chopped forage	0.85	0.86	0.01 NS
Masticated by cattle	0.58	0.76	0.18**
Masticated by sheep	0.56	0.69	0.13**

Legumes vs. grasses

Crampton (1957) was probably the first to recognize that legumes are eaten in greater quantities than grasses of similar energy digestibility (Table 3). Subsequent studies showed the intake of *Medicago sativa* (Figs. 1 and 3, Demarquilly & Jarrige 1974) and *Trifolium pratense* (Demarquilly & Weiss 1970) to be higher than that of chopped grasses of the same digestibility. Tropical legumes also are eaten in greater quantities than tropical grasses of the same digestibility (Table 4) although the differences are not always so pronounced. In another study with eight grasses and six legumes sheep ate 28 percent more legume than grass, a difference attributed to the 17 percent shorter time legumes were retained in the rumen and the higher packing density of legume in the rumen (Thornton & Minson 1973).

Fertilizer

Fertilizer calcium has been found to increase both the intake and the digestibility of pasture, but the relation between intake and digestibility is different for calcium fertilized and unfertilized pasture (Fig. 4). This difference was associated with the shorter time calcium fertilized grass was retained in the rumen (Rees & Minson 1976).

Pasture species and cultivars

Differences have been reported between pasture species in the relationship between intake and digestibility (Table 4, Osbourn *et al*. 1966, Corbett 1969, Walters 1971, 1974). The most comprehensive data on temperate species were published by Demarquilly & Weiss (1970) and this is shown in graphic form in Fig. 3.

In comparisons of diploid and tetraploid *Lolium perenne* Osbourn *et al*. (1966) found that the tetraploid was eaten in smaller quantities than the diploid and suggested that physical characteristics of the pasture might be causing this difference in intake. Very large differences in intake were found in a study of six varieties of *Panicum* of similar digestibility (Fig. 5). When compared at the same digestibility the intake of *P. maximum* cv. Hamil was 50 percent higher than that of *P. coloratum* cv. Kabulabula. This

TABLE 3

Intake of dry matter and digestibility of the energy of legumes and grasses by sheep.

Forages	Intake ($g/W^{0.75}/d$)	Digestibility %
Study 1.		
Medicago lupulina	79	63
Trifolium pratense	85	67
Bromus sp.	57	60
Phleum pratense	45	61
Study 2.		
Trifolium pratense		
Early bloom	78	55
Late bloom	74	53
Phleum pratense		
Early bloom	53	58
Late bloom	55	50

Derived from Crampton 1957, Crampton *et al*. 1960b.

TABLE 4

Estimated intake of dry matter of 7 tropical grasses and 2 tropical legumes at the same dry matter digestibility (Milford & Minson 1966).

Species	Number of observations	Estimated Intake at 55% digestibility ($g/kg\ W^{0.73}$)
Grasses		
Setaria sphacelata cv. Nandi	11	37
Digitaria decumbens cv. Pangola	15	38
Chloris gayana (C.P.I. 16144)	7	40
Pennisetum clandestinum	16	52
Sorghum almum	33	58
Chloris gayana cv. Callide	9	58
Cenchrus ciliaris cv. Molopo	22	63
Legumes		
Macroptilium atropurpureum cv. Siratro	12	69
Neonotonia wightii cv. Cooper	5	82

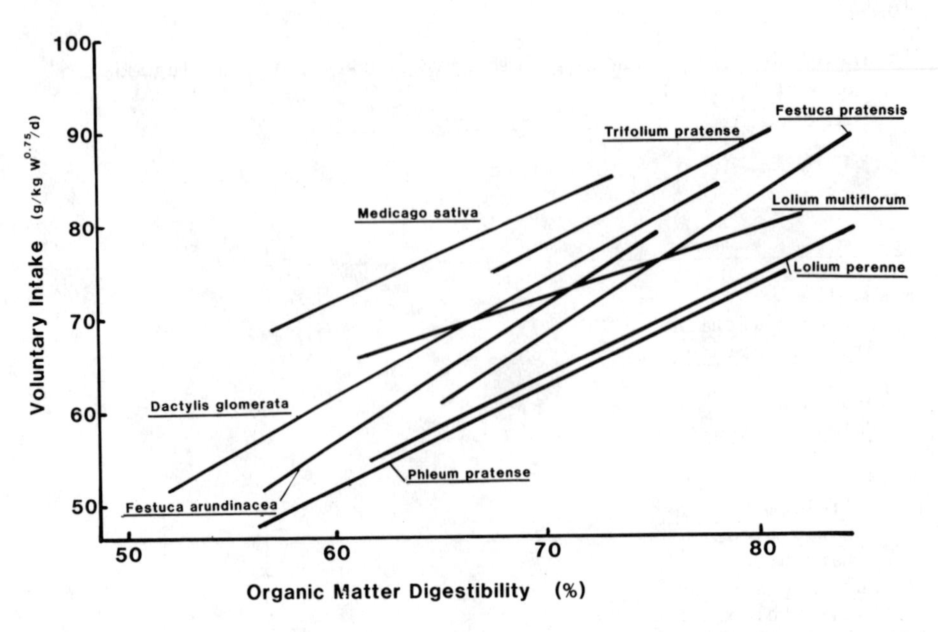

Fig. 3. Relationship between intake and digestibility of organic matter for 6 grasses and 2 legumes (Data of Demarquilly & Weiss 1970).

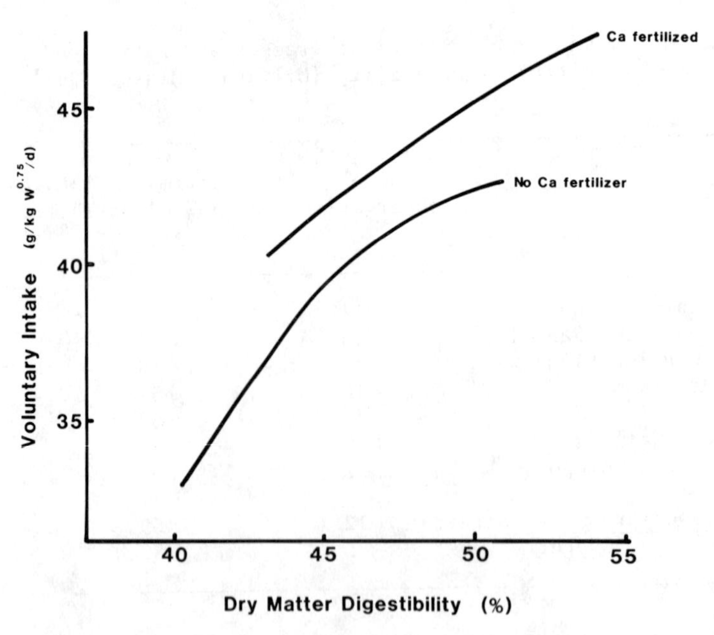

Fig. 4. Effect of fertilizer calcium on the relationship of intake with digestibility (After Rees & Minson).

difference between varieties was closely related to differences in leafiness (Minson & Laredo 1972).

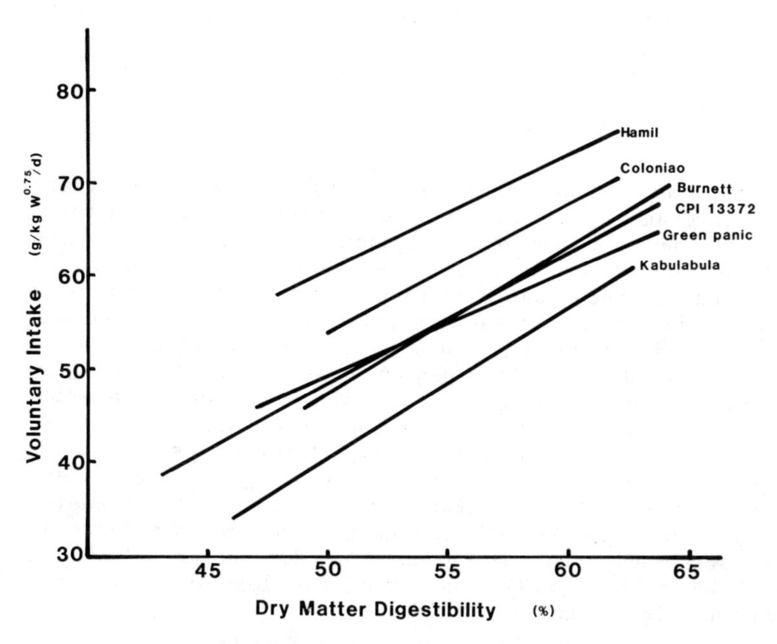

Fig. 5. Relation between intake and digestibility for six varieties of *Panicum* (Data from Minson 1971).

Temperate vs. tropical grasses

The voluntary intake of tropical grasses is usually less than that of temperate grasses grown for the same time and this is associated with a lower digestibility of the dry matter. The relationship between voluntary intake and digestibility of a wide range of temperate and tropical grasses has been compared by Minson (1980). At digestibilities around 60 percent tropical grasses are immature and leafy whereas temperate grasses are mature and stemmy. In consequence, at this level of digestibility, intake of tropical grass is about 20 percent higher than that of temperate grasses.

Water content of the herbage

No discussion of bulk factors limiting herbage intake would be complete without considering the effect of high water content on the intake of herbage dry matter. When animals are fed fresh pasture the daily intake of "wet matter" appears to be constant, so the higher the water content, the lower the daily intake of dry matter (Davies 1962). This study was made over the range of 8-18 percent dry matter. In a comparison of fresh and dried grass fed to daily cows, intake was decreased by 0.337 kg dry matter for each one percent fall in dry matter below 18.1 percent dry matter (Verite & Journet 1970). Studies with *Trifolium alexandrinum* and *Medicago sativa* have indicated that intake was depressed by water even when the fresh herbage contained 26 percent dry matter (Yoelao *et al*. 1970). However, factors other than excess water may have caused this apparent depression since in this study drying increased the digestibility coefficients of all nutrients by 6-20 percent.

CHEMICAL FACTORS

The chemical factors in pasture plants that may influence intake of the pasture by animals may be divided into three classes; (1) chemical fractions that are related to the quantity and composition of the fibre in the plant (2) chemical fractions that are essential nutrients for the microbial population of the rumen and for the host animal and (3) toxic factors.

Fibre

As pasture plants mature there is usually an increase in the proportion of fibre and a reduction in the protein and non-structural carbohydrates of the cell contents. Associated with these changes digestibility and intake by animals are reduced. The existence of this general trend has been confirmed for 14 grass and legume samples (Donefer *et al*. 1963) as significant correlations between intake and dry matter soluble in water ($r = 0.86$), potassium phthalate ($r = 0.81$), hydrochloric acid ($r = 0.86$), pepsin ($r = 0.87$), cellulase ($r = 0.85$) and cellulase plus pepsin ($r = 0.87$). Van Soest (1965) reported significant correlations between intake and neutral detergent fibre ($r = -0.65$), protein ($r = 0.54$) and acid detergent fibre ($r = -0.53$) but not with lignin ($r = -0.13$) in a study of 82 samples of grass and legume. In another study with 35 grass and 14 legume samples Donefer *et al*. (1966) found intake of grass and legume samples was significantly correlated with the quantity of dry matter soluble in acid pepsin (PSDM) ($r = 0.94$) but for the legumes alone the correlation was not significant ($r = 0.53$).

TABLE 5

Regressions relating intake of dry matter by sheep (y) to the chemical composition of pasture (x).

Chemical component	Number of samples	Regression	r	RSD
Crude fibre[1]	58 Mixed	$y = 106.1-1.69x$	-0.79	±7.6
MADF[1]	58 Mixed	$y = 120.7-1.87x$	-0.84	±6.6
MADF[2]	60 Mixed	$y = 118.6-1.80x$	-0.82	±7.0
NDF[3]	56 Grasses & legumes	$y = 95-0.73x$	-0.88	-
NDF[4]	83 Grasses & legumes	$y = 110.4-1716/(100-x)$	-0.65	-
NDF[4]	38 Grasses & legumes	$y = 92.4-1177/(100-x)$	-	-
PSDM[5]	57 Panicums	$y = 24.0+1.94x$	0.67	±7.7

MADF = modified acid detergent fibre. NDF = Neutral detergent fibre.
PSDM = Pepsin soluble dry matter.

1 Wilson *et al*., 1966
2 Clancy & Wilson 1966
3 Osbourn *et al*. 1974
4 Van Soest 1965
5 Minson & Haydock 1971

Regression equations have been published relating intake of pasture by sheep to various chemical fractions (Table 5). Although all these regressions are significant the chemical composition fails to account for all the differences in intake between pasture samples. Some of this residual variation is caused by errors in estimating intake _in vivo_ and in analysis, but most of it is caused by true differences in intake of different pasture species or varieties with the same chemical composition. In the case of pepsin soluble dry matter (PSDM, Minson & Haydock 1971) large differences were found between regressions for six different cultivars of _Panicum_; when PSDM was 14 percent, intake of the cultivars Kabulabula and Hamil was 40 and 62 $g/wt^{0.75}/d$ respectively. Differences in intake regressions have also been found between leaf and stem fractions of grasses; the intake of the leaf fraction was significantly higher than that of the stem fraction when compared at the same levels of neutral detergent fibre, acid detergent fibre and lignin (Laredo & Minson 1973).

Essential nutrients

Protein

The intake of pasture by animals will only be limited by the level of fibre and its physical composition if protein, vitamins and minerals are available in sufficient quantity. When the crude protein content of the pasture falls below 6-8 percent appetite is depressed and pasture intake by the animal will be less than might be expected from a consideration of the physical composition of the pasture (Blaxter & Wilson 1963, Milford & Minson 1966, Minson & Milford 1967). Feeding a protein supplement or urea increased by 16-82 percent the intake by cattle of nitrogen deficient hay and silage (Morris 1958, 1966). In other studies nitrogenous fertilizer has increased the intake of nitrogen deficient hay (Chapman & Kretschmer 1964, Minson 1967). Adequate levels of dietary sulphur are required for the conversion of urea into microbial protein and where the pasture is deficient in sulphur then feeding urea has little effect on intake (Kennedy & Siebert 1973). The depressing effect on intake of a deficiency of protein appears to be caused by factors other than rumen distension since grinding and pelleting protein deficient herbage has little effect on intake (Minson 1967). The depression of intake appears to be caused by a deficiency of circulating amino acids, since intake of a protein deficient diet can be increased if casein (but not urea) is infused into the duodenum (Egan 1965).

Sulphur

Sulphur is required in the formation of bacterial protein in the rumen so any deficiency of sulphur is likely to lead to a protein deficiency and reduced intake. Feeding sulphur supplements has increased the intake of grasses with less than 0.10 percent sulphur by 21 to 49 percent (Playne 1969a, Kennedy & Siebert 1972, 1973, Rees _et al_. 1974). This rise only occurs when nitrogen levels in the diet are adequate for the utilization of the sulphur by the microbial population in the rumen (Kennedy & Siebert 1973). Subsequent work showed that the intake response was related to the sulphur content of the grass and that no increase in intake by sheep would be expected when the grass contained more than 0.17 percent sulphur (Rees & Minson 1978).

Sodium

Many pasture plants are deficient in sodium and this will lead to a reduction in intake if other sources of sodium are not available. Where pastures are deficient feeding a supplement containing sodium has increased intake by 7 to 28 percent (Joyce & Brunswick 1975, Minson 1980).

Phosphorus

Pasture may contain less phosphorus than is required by the animal. This deficiency can be caused by pasture species, advanced maturity or low levels of soil phosphorus. When animals fed phosphorus deficient pastures are supplemented with phosphorus the intake has been increased by 15 and 25 percent for sheep and cattle respectively (Playne 1969b, Little 1968).

Trace elements

The intake of pasture may also be limited by deficiencies of trace elements. The intake of sheep eating cobalt deficient pasture has been increased 60 percent by oral administration of cobalt (Marston *et al*. 1938). Feeding selenium supplements has increased intake of lambs grazing selenium deficient pastures in New Zealand (McLean *et al*. 1962).

Toxic factors

If herbage contains elements in toxic quantities or toxic organic compounds these will cause ill health, loss of appetite and sometimes death. The quantities of different elements that result in toxicity have been considered by Underwood (1966) and the toxic organic compounds by Hegarty (1982).

FUTURE DEVELOPMENTS

The most reliable method of estimating pasture intake (as a pasture characteristic) is to measure it with the required animal species and under defined environmental and physiological conditions. Thus intake is the integrated product of all the many chemical, physical and physiological factors that control intake and their interactions. We all accept this fact so why attempt to evaluate it in the laboratory? There are two different reasons why laboratory methods have been developed for assessing the intake of pasture.

(1) When animals fail to achieve the desired level of production samples of the pasture being selectively eaten are often analysed in an attempt to identify the cause of the problem.

(2) In plant breeding and selection programs large numbers of samples have to be evaluated in terms of potential intake and insufficient material is available for intake measurements with animals.

In this paper I have described some of the many unrelated pasture factors that control intake; the existence of more than one factor automatically means that the intake of a pasture can rarely be accurately predicted from the results of a single laboratory analysis. This is very well illustrated by the way deficiencies of many different elements depress pasture intake.

The first priority in pasture analysis should be the determination of the essential mineral elements and protein. This is particularly important in areas with heavily leached soils which have not received fertilizer or when dealing with a new pasture cultivar. Until relatively recently a complete elemental analysis was expensive and very time consuming. This has been changed by the introduction of the direct reading emission spectroscopic method of analysis using a briquetting technique (Johnson & Simons 1972). The concentration of 21 elements may be estimated within 24 hours of receipt of the pasture sample.

Where mineral and protein contents are above the levels likely to cause deficiencies then intake will be controlled by physical factors in the diet. Many of these physical attributes are related to the stage of growth of the pasture and the concentration of fibre. Thus analysis for any fibre fraction or pepsin soluble dry matter usually provides a rough estimate of the potential intake of a pasture (Table 5). Large errors are attached to all these estimates of intake and care must be exercised when a regression is used to determine differences in intake between leaf and stem fractions, chopped and pelleted forages or even between cultivars at the same stage of growth. All too often regression equations give misleading estimates.

In conclusion it is suggested that current laboratory methods of analysis may provide guides as to the potential intake of pastures but for accurate evaluation there is no satisfactory alternative to measurements made with the animal. Only animals are capable of responding to all the physical characters of the pasture that control intake most of which have not yet been quantified in the laboratory. Improvement in prediction from laboratory analysis will only be achieved once all these physical attributes of the pasture have been identified and their relative importance determined.

REFERENCES

Balch, C.C. (1971) Proposal to use time spent chewing as an index of the extent to which diets for ruminants possess the physical property of fibrousness characteristics of roughages. British Journal of Nutrition 26, 383-392.

Balch, C.C.; Campling, R.C. (1962) Regulation of voluntary food intake in ruminants. Nutrition Abstracts and Reviews 32, 669-686.

Balch, C.C.; Campling, R.C. (1969) Voluntary intake of food. In: Handbuch der Tierernahrung Part 1. Editors W. Lenkeit, K. Breirem and E. Grasemann. Hamburg, Verlag Paul Pavey, pp. 554-579.

Baumgardt, B.R. (1970) Regulation of feed intake and energy balance. In: Physiology and Digestion and Metabolism in the Ruminant. Editor A.T. Phillipson. Newcastle-upon-Tyne, Oriel Press. pp.235-253.

Blaxter, K.L.; Wainman, F.W.; Wilson R.S. (1961) The regulation of food intake by sheep. Animal Production 3, 51-61.

Blaxter, K.L.; Wilson, R.S. (1963) The assessment of a crop husbandry technique in terms of animal production. Animal Production 5, 27-42.

Campling, R.C.; Freer, M. (1966) Factors affecting the voluntary intake of food by cows. 6. A preliminary experiment with ground, pelleted hay. British Journal of Nutrition 17, 263-272.

Campling, R.C.; Milne, J.A. (1972) The nutritive value of processed roughages for milking cows. Proceedings of the British Society of Animal Production 1, 53-60.

Chapman, H.L.; Kretschmer, A.E. (1964) Effect of nitrogen fertilization on digestibility and feeding value of pangola grass hay. Proceedings of the Soil and Crop Science Society of Florida 24, 176-183.

Clancy, M.J.; Wilson, R.K. (1966) Development and application of a new chemical method for predicting the digestibility and intake of herbage samples. Proceedings of the 10th International Grassland Congress, Helsinki 445-453.

Conrad, H.R.; Pratt, A.D.; Hibbs, J.W. (1964) Regulation of feed intake in dairy cows. 1. Changes in importance of physical and physiological factors with increasing digestibility. Journal of Dairy Science 47, 54-62.

Corbett, J.C. (1969) The nutritional value of grassland herbage In: International encyclopaedia of food and nutrition Volume 17: Nutrition of animals of agricultural importance Part 2. Editor D. Cuthbertson. Oxford & New York Permagon Press. pp. 593-644.

Crampton, E.W. (1957) Interrelations between digestible nutrient and energy content, voluntary dry matter intake, and the overall feeding value of forages. Journal of Animal Science 16, 546-552.
Crampton, E.W.; Donefer, E.; Lloyd, L.E. (1960a) A nutritive value index for forages. Proceedings of the 8th International Grassland Congress, Reading 462-466.
Crampton, E.W.; Donefer, E.; Lloyd, L.E. (1960b) A nutritive value index for forages. Journal of Animal Science 19, 538-544.
Davies, H. Lloyd (1962) Intake studies in sheep involving high fluid intake. Proceedings of the Australian Society of Animal Production 4, 167-171.
Demarquilly, C.; Jarrige, R. (1974) The comparative nutritive value of grasses and legumes. Vaxtodling 28, 33-41.
Demarquilly, C.; Weiss, Ph. (1970) Tableaux de la valeur alimentaires des fourrages. Ministere de l'Agriculture Institute National de la Recherche Agronomique, Servier d'Experimentation et d'Information I.N.R.A., Versailles. 65 pp.
Donefer, E.; Crampton, E.W.; Lloyd, L.E. (1966) The prediction of digestible energy intake potential (NVI) of forages using a simple, in vitro technique. Proceedings of the 10th International Grassland Congress, Helsinki, 442-445.
Donefer, E.; Niemann, P.J.; Crampton, E.W.; Lloyd, L.E. (1963) Dry matter disappearance by enzyme and aqueous solutions to predict the nutritive value of forages. Journal of Dairy Science 46, 965-970.
Egan, A.R. (1965) Nutritional status and intake regulation in sheep. 2. The influence of sustained duodenal infusions of casein or urea upon voluntary intake of low protein roughages by sheep. Australian Journal of Agricultural Research 16, 451-462.
Greenhalgh, J.F.D.; Wainman, F.W. (1972) The nutritive value of processed roughages for fattening cattle and sheep. Proceedings of the British Society of Animal Production 1, 61-72.
Grovum, W.L.; Phillips, G.D. (1978) Factors affecting the voluntary intake of food by sheep. 1. The role of distension, flow rate of digesta and propulsive motility in the intestines. British Journal of Nutrition 40, 323-335.
Heaney, D.P.; Pigden, W.J.; Minson, D.J.; Pritchard, G.I. (1963) Effect of pelleting on energy intake of sheep from forages cut at three stages of maturity. Journal of Animal Science 22, 752-757.
Hegarty, M.P. (1982) Deleterious factors in forages affecting animal production. In: Nutritional limits to animal production from pastures. Editor J.B. Hacker. Farnham Royal, U.K., Commonwealth Agricultural Bureaux. pp. 133-150.
Hendricksen, R.E.; Poppi, D.P.; Minson, D.J. (1981) The voluntary intake, digestibility and retention time by cattle and sheep of leaf and stem fractions of a tropical legume (Lablab purpureus). Australian Journal of Agricultural Research 32, 389-398.
Hodgson, J. (1977) Factors limiting herbage intake by the grazing animal. In: Proceedings International Meeting on Animal Production from Temperate Grassland. Editor B. Gilsenan, Dublin, An Foras Taluntais. pp. 70-75.
Hodgson, J. (1982) Influence of sward characteristics on diet selection and herbage intake by the grazing animal. In: Nutritional limits to animal production from pastures. Editor J.B. Hacker. Farnham Royal, U.K., Commonwealth Agricultural Bureaux, pp.153-166.
Kennedy, P.M.; Siebert, B.D. (1972) The utilization of spear grass (Heteropogon contortus). 2. The influence of sulphur on energy intake and rumen and blood parameters in cattle and sheep. Australian Journal of Agricultural Research 23, 45-56.
Kennedy, P.M.; Siebert, B.D. (1973) The utilization of spear grass (Heteropogon contortus). 3. The effect of the level of dietary sulphur on the utilization of spear grass by sheep. Australian Journal of Agricultural Research 24, 143-152.
Jarrige, R.; Demarquilly, C.; Journet, M.; Beranger, C. (1973) The nutritive value of processed dehydrated forage with special reference to the influence of physical form and particle size. Proceedings of the 1st International Green Crop Drying Congress, Oxford, 99-118.
Johnson, A.D.; Simons, J.G. (1972) Direct reading emmission spectroscopic analysis of plant tissue using a briquetting technique. Communications in Soil Science and Plant Analysis 3, 1-9.
Joyce, J.P.; Brunswick, L.C.F. (1975) Sodium supplementation of sheep and cattle fed lucerne. New Zealand Journal of Experimental Agriculture 3, 299-304.
Laredo, M.A.; Minson, D.J. (1973) The voluntary intake, digestibility and retention time by sheep of leaf and stem fractions of five grasses. Australian Journal of Agricultural Research 24, 875-888.
Laredo, M.A.; Minson, D.J. (1975a) The effect of pelleting on the voluntary intake and digestibility of leaf and stem fractions of three grasses. British Journal of Nutrition 33, 159-170.
Laredo, M.A.; Minson, D.J. (1975b) The voluntary intake and digestibility by sheep of leaf and stem of Lolium perenne. Journal of the British Grassland Society 30, 73-77.
Little, D.A. (1968) Effect of dietary phosphorus on the voluntary consumption of Townsville lucerne (Stylosanthes humilis) by cattle. Proceedings of the Australian Society of Animal Production 7, 376-380.
Marston, H.R.; Thomas, R.G.; Murnane, D.; Lines, E.W.L.; McDonald, I.W.; Moore, H.O.; Bull, L.B. (1938) Studies on coast disease of sheep in South Australia. Council of Scientific and Industrial Research, Bulletin 113.

McLean, J.W.; Thompson, G.G.; Iverson, C.E.; Jagusch, K.T.; Lawson, B.M. (1962) Sheep production and health on pure species pasture. Proceedings of the New Zealand Grassland Association 24, 57-70.

McLeod, M.N.; Minson, D.J. (1978) The accuracy of the pepsin-cellulase technique for estimating the dry matter digestibility in vivo of grasses and legumes. Animal Feed Science and Technology 3, 277-287.

Milford, R.; Minson, D.J. (1966) Intake of tropical pasture species. Proceedings of the 9th International Grassland Congress, São Paulo, 814-822.

Minson, D.J. (1963) The effect of pelleting and wafering on the feeding value of roughages - a review. Journal of the British Grassland Society 18, 39-44.

Minson, D.J. (1967) The voluntary intake and digestibility in sheep, of chopped and pelleted Digitaria decumbens (pangola grass) following a late application of fertilizer nitrogen. British Journal of Nutrition 21, 587-597.

Minson, D.J. (1971) The digestibility and voluntary intake of six varieties of Panicum. Australian Journal of Experimental Agriculture and Animal Husbandry 11, 18-25.

Minson, D.J. (1980) Nutritional differences between tropical and temperate pastures. In: Grazing animals. Editor F.H.W. Morley. Amsterdam, Elsevier Scientific Publishing Company. pp. 143-157.

Minson, D.J. (1981) The measurement of digestibility and voluntary intake of forages with confined animals. In: Forage Evaluation: Concepts and Techniques. Editors J.L. Wheeler and R.D. Mochrie. CSIRO, Melbourne. pp. 159-174.

Minson, D.J.; Haydock, K.P. (1971) The value of pepsin dry matter solubility for estimating the voluntary intake and digestibility of six Panicum varieties. Australian Journal of Experimental Agriculture and Animal Husbandry 11, 181-185.

Minson, D.J.; Laredo, M.A. (1972) Influence of leafiness on voluntary intake of tropical grasses by sheep. Journal of the Australian Institute of Agricultural Science 38, 303-305.

Minson, D.J.; Milford, R. (1967) The voluntary intake and digestibility of diets containing different proportions of legume and mature pangola grass. Australian Journal of Experimental Agriculture and Animal Husbandry 7, 546-551.

Morris, J.G. (1958) Drought feeding studies with cattle and sheep. 1. The use of native grass hay (bush hay) as the basal component of a drought fodder for cattle. Queensland Journal of Agricultural Science 15, 161-180.

Morris, J.G. (1966) Supplementation of ruminants with protein and non-protein nitrogen under Northern Australian conditions. Journal of the Australian Institute of Agricultural Science 32, 178-189.

Osbourn, D.F.; Thomson, D.J.; Terry, R.A. (1966) The relationship between voluntary intake and digestibility of forage crops using sheep. Proceedings of the 10th International Grassland Congress, Helsinki 363-366.

Osbourn, D.F.; Terry, R.A.; Outen, G.E.; Cammell, S.B. (1974) The significance of a determination of cell walls as the rational basis for the nutritive evaluation of forages. Proceedings of the 12th International Grassland Congress, Moscow 4, 514-519.

Playne, M.J. (1969a) Effect of sodium sulphate and gluten supplements on the intake and digestibility of a mixture of spear grass and Townsville lucerne hay by sheep. Australian Journal of Experimental Agriculture and Animal Husbandry 9, 393-399.

Playne, M.J. (1969b) The effect of dicalcium phosphate supplements on the intake and digestibility of Townsville lucerne and spear grass by sheep. Australian Journal of Experimental Agriculture and Animal Husbandry 9, 192-195.

Poppi, D.P.; Minson, D.J.; Ternouth, J.H. (1981a) Studies of cattle and sheep eating leaf and stem fractions of grasses. 1. The voluntary intake, digestibility and retention time in the reticulo-rumen. Australian Journal of Agricultural Research, 32, 99-108.

Poppi, D.P.; Minson, D.J.; Ternouth, J.H. (1981b) Studies of cattle and sheep eating leaf and stem fractions of grasses. 3. The retention time in the rumen of large feed particles. Australian Journal of Agricultural Research 32, 123-137.

Poppi, D.P.; Norton, B.W.; Minson, D.J.; Hendricksen, R.E. (1980) The validity of the critical size theory for particles leaving the rumen. Journal of Agricultural Science, Cambridge 94, 275-280.

Rees, M.C.; Minson, D.J. (1976) Fertilizer calcium as a factor affecting the voluntary intake, digestibility and retention time of pangola grass (Digitaria decumbens) by sheep. British Journal of Nutrition 36, 179-187.

Rees, M.C.; Minson, D.J. (1978) Fertilizer sulphur as a factor affecting voluntary intake, digestibility and retention time of pangola grass (Digitaria decumbens) by sheep. British Journal of Nutrition 39, 5-11.

Rees, M.C.; Minson, D.J.; Smith, F.W. (1974) The effect of supplementary and fertilizer sulphur on voluntary intake, digestibility, retention time in the rumen, and site of digestion of pangola grass in sheep. Journal of Agricultural Science, Cambridge 82, 419-422.

Thornton, R.F.; Minson, D.J. (1973) The relationship between apparent retention time in the rumen, voluntary intake, and apparent digestibility of legumes and grass diets in sheep. Australian Journal of Agricultural Research 24, 889-898.

Tilley, J.M.A.; Terry, R.A. (1963) A two stage technique for in vitro digestion of forage crops. Journal of the British Grassland Society 18, 104-111.

Underwood, E.J. (1966) The mineral nutrition of livestock. Food and Agriculture Organization of the United Nations and Commonwealth Agricultural Bureaux, 237pp.

Van Soest, P.J. (1965) Symposium on factors influencing the voluntary intake of herbage by ruminants. Voluntary intake in relation to chemical composition and digestibility. Journal of Animal Science 24, 834-843.

Verite, R.; Journet, M. (1970) Influence, de la teneur en eau et de la deshydration de l'herbage sur sa valeur alimentaire pour les vaches laitieres. Annales de Zootechnie 19, 255-268.

Walters, R.J.K. (1971) Variation in the relationship between in vitro digestibility and voluntary dry matter intake of different grass varieties. Journal of Agricultural Science, Cambridge 75, 243-252.

Walters, R.J.K. (1974) Variation between grass species and varieties in voluntary intake. Vaxtodling 29, 184-192.

Weston, R.H. (1982) Animal factors affecting feed intake. In: Nutritional limits to animal production from pastures. Editor J.B. Hacker. Farnham Royal, U.K., Commonwealth Agricultural Bureaux, pp. 183-198.

Wilkins, R.J. (1973) The effects of processing on the nutritive value of dehydrated forages. Proceedings of the 1st International Green Crop Drying Congress, Oxford, 119-134.

Wilkins, R.J. (1982) Improving forage quality by processing. In: Nutritional limits to animal production from pasture. Editor J.B. Hacker. Farnham Royal, U.K. Commonwealth Agricultural Bureaux, pp. 389-408.

Wilson, R.K.; Sillane, T.A.; Clancy, M.J. (1966) The influence of fibre content on herbage intakes by ruminants. Irish Journal of Agricultural Research 5, 142-143.

Yoelao, K.; Jackson, M.G.; Saran, Ishwar (1970) The effect of wilting berseem and lucerne herbage on voluntary dry matter intake by buffalo heifers. Journal of Agricultural Science, Cambridge 74, 47-51.

ANIMAL FACTORS AFFECTING FEED INTAKE

R.H. WESTON

CSIRO Division of Animal Production, P.O. Box 239, Blacktown, NSW 2148, Australia

ABSTRACT

A broad outline is given of data relating to the effect of genotype, physiological state and climate on the voluntary consumption of feed by ruminants.

Part of the variation in the capacity of ruminants to consume feed has a genetic basis, the magnitude of which is difficult to establish. Animals with higher potential for feed consumption exhibit enhanced tissue metabolism as indicated by higher basal metabolism and maintenance energy requirement.

Physiological state influences the level of consumption of both herbage and herbage + concentrate diets. Increases occur during lactation and with enhanced growth potential as in young animals. Decreases generally prevail with increased body fatness, during late pregnancy and in disease states. The extent of the feed consumption change is variable and cannot be predicted accurately.

Climate may influence feed consumption appreciably. Day-length affects the grazing pattern during the day and limited data suggest enhanced feed intake with longer day-length. Conditions militating against the animal's maintenance of thermal balance modulate feeding. High ambient temperature, solar radiation and relative humidity are associated with decreased feed consumption whereas cold conditions increase consumption provided severe cold stress is not established.

Herbage intake regulation with change in genotype, physiological state and climate appears to involve interplay between the animal's metabolism and transactions in the rumen. It is suggested that key roles in this regulation may be played by the rumen digesta load and the animal's energy deficit, the latter being the difference between the capacity of the animal to use energy and the energy it receives from absorbed nutrients.

INTRODUCTION

The overall regulation of voluntary feed consumption in ruminants embraces many physiological processes and several factors may act as important determinants of the level achieved. This paper provides a broad outline of knowledge relating to the influence of genotype, physiological state and climate on voluntary feed consumption, various other aspects having been considered earlier in this symposium by Hodgson (1982) and Minson (1982).

The effects of genotype, physiological state and climate on voluntary feed consumption are mediated via the animal's metabolism and the direction of change in consumption is generally independent of diet. Thus factors enhancing voluntary feed consumption do so with low quality roughage diets as well as with high quality diets such as ground and pelleted mixtures of roughages and concentrates. Hence the same direction of change prevails for diets limited in intake by the rate of organic matter removal from the rumen as for diets limited in intake by the animal's capacity to use energy.

Organic matter is removed from the rumen by absorption and eructation and by transfer to the omasum. The significance of various processes in contributing to this removal was considered in detail by Weston (1979a). Fig. 1 shows some of these processes and provides a link between factors relating to the rumen transactions and those of metabolic and physiological moment "external" to the rumen. It is assumed that the central nervous system controls feeding according to feedback signals relating to satiety on the one hand and to energy need on the other. Baile & Forbes (1974) describe a range of stimuli involved in this inhibition and facilitation of feeding and these are not discussed here. It is pertinent that when voluntary feed consumption is limited by transactions in the rumen, the animal fails to achieve its potential to use energy and a state of energy deficit prevails. In this context the energy deficit is defined as the difference between the animal's potential to use energy, as modulated by hormone status and availability of essential nutrients (Fig. 1), and the amount of energy available for metabolism (energy supply).

Difficulties exist in comparing voluntary feed consumption data from various sources due to inadequate description of experimental conditions and particularly to problems with the description of body size. Animals with the same recorded body weight (BW) can vary widely in gut fill, fat content and, with sheep, fleece weight, hence body weight can be an imprecise scaler with respect to body size. The literature abounds with power functions to which body weight should be raised to minimize variations arising from body size differences. This will not be discussed here but it is noteworthy that the general interspecific scaler $BW^{0.75}$ is not applicable to comparisons between adult sheep and cattle, $BW^{0.9}$ being more appropriate (e.g. Graham 1972, Ternouth *et al*. 1979). Again $BW^{0.75}$ is inferior to $BW^{1.0}$ in comparing breeds of sheep (Blaxter *et al*. 1966a) and with various cattle comparisons (Frisch & Vercoe 1977). As discussed later, the use of fasting heat production as a scaler between and within species reduces variation but such data are not generally available. Accordingly, unless specific data indicate to the contrary, little would appear to be gained with comparisons of adults within species by use of other than the simple $BW^{1.0}$ scaler.

GENOTYPE AND VOLUNTARY FEED CONSUMPTION

Ruminants vary in their capacity to consume feed but it is difficult to estimate reliably the extent of genetic variation. However, an heritability estimate of 0.42 ± 0.10 has been calculated for net energy consumption by lactating cows (Miller *et al*. 1972).

Under optimal conditions of the diet and environment, feed intake should be determined by the animal's genetic potential to use energy; hence feed intake differences between animals should reflect the difference in potential. Baumgardt & Peterson (1971) and Clancy *et al*. (1977) measured a maximum daily feed intake equivalent to about 1090 kJ digestible energy per unit $BW^{0.75}$ for their growing crossbred lambs. By comparison, the corresponding values for Border Leicester x Merino lambs of comparable maturity in the studies of Weston (1974, 1979b) may be calculated to be a least 15 percent higher. As conditions should have been near optimal in both sets of studies, the intake difference probably indicates a genetic difference between the two types of lamb in capacity to use energy.

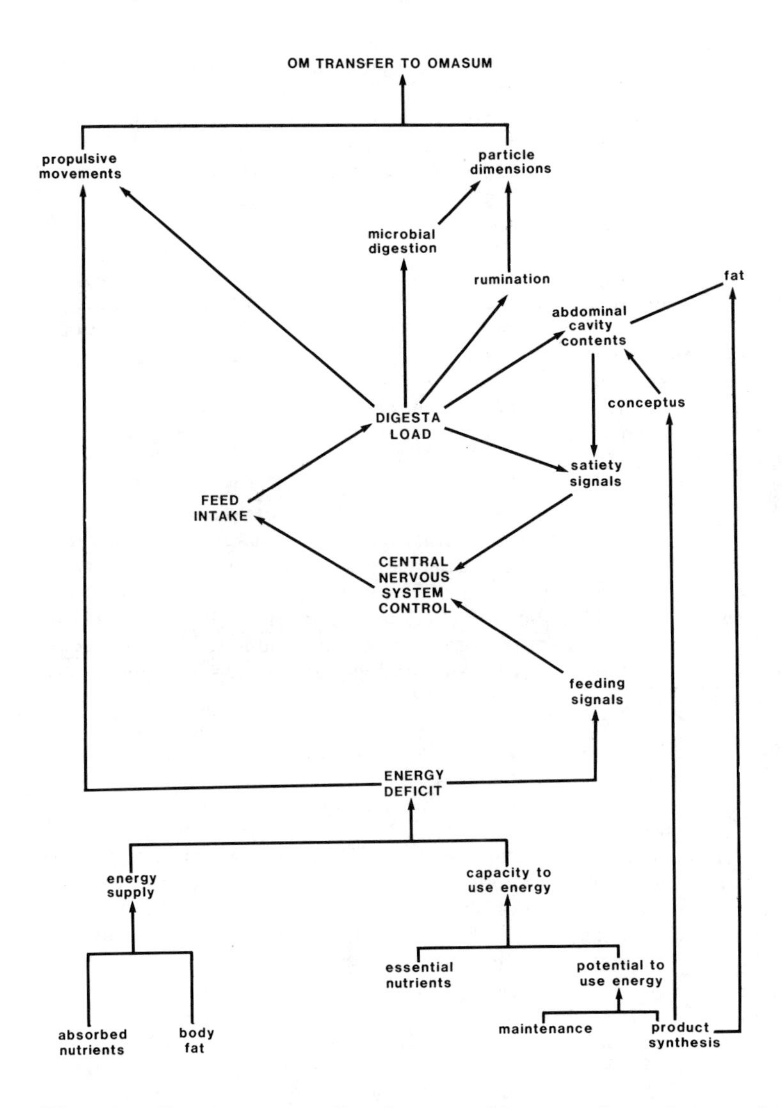

Fig. 1. Factors contributing to the transfer of organic matter (OM) from the rumen to the omasum.

Most studies of voluntary feed intake differences between species, breeds and strains are deficient with respect to the selection and preparation of the experimental animals, and the data may not reliably indicate population means. In the study of Blaxter et al. (1966a) with six breeds of sheep, the coefficient of variation (CV) for voluntary feed consumption per unit body weight was 4.8 percent, with the maximum difference between breeds being 15 percent. At pasture no consistent differences can be seen between high and low wool producing Merinos or between Merino, Border Leicester and Dorset

Horn sheep (e.g. Langlands 1968, Langlands & Hamilton 1969, Arnold 1975); variation shown between other breeds needs to be confirmed. As discussed by Frisch & Vercoe (1977) many studies indicate a significant voluntary consumption advantage of *Bos taurus* cattle over *Bos indicus* cattle under conditions of minimal environmental stress; further, the crosses of the two types have values intermediate to those of the parents.

Lack of an appropriate reference weight precludes reliable estimation of differences between species in voluntary consumption and the possible existence of breed differences within species renders the interpretation of comparisons difficult. However, at present there appears to be no convincing evidence of a general difference in voluntary consumption per unit body size between sheep, goats, cattle and buffalo.

Voluntary consumption differences within and between species are clearly related to the animal's metabolic activity. Thus Blaxter *et al.* (1966a, b) showed that a general proportionality exists between voluntary consumption and maintenance energy requirement both within breeds of sheep and between sheep and cattle. Further, voluntary feed consumption is proportional to fasting heat production within breeds of cattle, between breeds of cattle, and between cattle and buffalo (Frisch & Vercoe 1977, Vercoe & Frisch 1977).

Voluntary feed consumption is affected by genotype interactions with type of diet and various components of the environment. Interaction with heat load is indicated by differences in voluntary feed consumption per unit body weight between Brahman and Shorthorn cattle allowed to grow at 8° and 27°C (Johnson *et al.* 1958). Again, feed consumption is more affected in *Bos taurus* than in *Bos indicus* cattle by stresses associated with heat and parasites as discussed by Frisch & Vercoe 1978.

PHYSIOLOGICAL STATE AND VOLUNTARY FEED CONSUMPTION

Age

Allden (1979) summarized various data relating voluntary feed consumption to body weight for ruminants fed high quality feed from weaning onwards. With Merino sheep, voluntary feed consumption in absolute terms increased progressively until some 30-40 percent of mature body weight was achieved, after which it remained steady or decreased slightly. However, when voluntary feed consumption was expressed per unit $BW^{0.75}$, a steady decrease was shown after the maximum at about 35 percent of mature body weight was reached; the voluntary feed consumption at maturity was about 50 percent of the maximum attained. This relationship between voluntary feed consumption per unit $BW^{0.75}$ and body weight (percent of mature weight) was found to apply to both sheep and cattle fed high quality roughage plus concentrate diets. The point of maximum intake relative to body size needs to be confirmed to ensure that a deficiency of essential amino acids at the tissue level, as discussed by Weston (1979a), was not limiting voluntary consumption when the animals were young. Comparable data with roughage-fed animals do not appear to be available.

Langlands (1968) and Langlands & Hamilton (1969) observed higher intakes at pasture with younger animals. Thus in the latter study with four breeds of sheep, mean voluntary feed consumption per unit

$BW^{1.0}$ at 8 months was 22 percent higher than at 20 months, and 27 percent higher than at 32 months. With roughage diets fed indoors weaner lambs have been found to consume some 19-45 percent more feed per unit $BW^{1.0}$ (Egan & Doyle 1980, Weston 1980). These higher intakes are accompanied by larger quantities of digesta in the rumen, and at pasture enhanced rumen digesta weight per unit body weight has been observed; the younger animals also perform more rumination bites per day but do not spend more time ruminating (Weston 1980 and unpublished data).

Although the decrease in voluntary consumption with advancing age is consistent with change in fasting metabolism, a strict proportionality may not apply in the roughage-fed animal (Weston 1980).

Body composition

The adult rat grows to a particular body size or fat content which it maintains by varying voluntary consumption to compensate for change in energy needs arising from change in environmental conditions. Unequivocal demonstration of such a set point in ruminants appears lacking, although the study of Paquay *et al*. (1979) shows a tendency in sheep towards a constant high body weight accompanied by a proportionately lower voluntary consumption.

No clear quantitative relationship exists between voluntary consumption and body composition. In the study of Donnelly *et al*. (1974a) grazing Merino sheep with 19 percent fat consumed no less feed that their counterparts with 11 percent fat. Further, Paquay *et al*. (1979) found voluntary consumption to be almost unchanged over 16 weeks, during which time the body weight of their mature sheep advanced from 75 kg to 95 kg. Similarly, cattle when fat (610 kg) were found to consume no less straw than when thin (433 kg) (Bines *et al*. 1969). However, pasture intake decreased by 10 percent in cattle of higher fat content (15 kg as compared with 10 kg internal fat) (Tayler 1959) and fat cattle consumed 19 percent less of a hay diet and 23 percent less of concentrates + hay (Bines *et al*. 1969). Again, Foot (1972) found fat ewes consumed 26 percent less dried grass than ewes with 10 percent less fat. During lactation, voluntary feed composition tends to be inversely related to fat content. In dairy cattle higher fat content has been found to be associated with lower voluntary feed consumption in early lactation (-6 percent) (Lodge *et al*. 1975), and comparisons of lactating ewes in different body condition show 5 to 15 percent lower feed consumption in fatter animals with the extent of decline not clearly related to diet or level of body fat (Foot & Russel 1979, Cowan *et al*. 1980).

Higher fat content has been shown to be associated with a reduction in the quantity of digesta in the rumen or the alimentary tract as a whole (Tayler 1959, Bines *et al*. 1969, Cowan *et al*. 1980). This association has led to the suggestion that increasing abdominal fat physically reduces potential rumen capacity and accordingly feed intake. However, as discussed by Forbes (1980), metabolic changes may accompany increase in body fat content and these in turn may affect voluntary consumption. Thus increase in the size of the adipocytes as the animal fattens may be accompanied by a reduction in their capacity to synthesize triglycerides. Further, with increased adipocyte size, fatty acids released during the normal turnover of triglycerides may more readily escape from the cell. This enhanced "leakage" of fatty acids could influence metabolic receptors associated with feeding.

Pregnancy

During pregnancy the volume and nutrient demand of the conceptus progressively increase and the dam's endocrine status changes. These major physical and metabolic changes affect voluntary feed consumption as discussed by Forbes (1970, 1971).

Feed intake in monogastric animals generally increases during pregnancy to match the enhanced nutrient demand. However, with cattle and sheep, a decline is often shown in late pregnancy. A marked decrease is usually observed in the last few days (e.g. Reid & Hinks 1962).

The voluntary feed consumption response in late pregnancy is variable. In the grazing situation changes ranging from -3 to +26 percent have been recorded (Arnold & Dudzinski 1967, Dulphy *et al.* 1980). With studies indoors no changes were observed by Hadjipieris & Holmes (1966), decreases of 8-10 percent were found by Broster *et al.* (1964), Campling (1966) and Lamberth (1969) and larger decreases (16-25 percent) were obtained by Owen *et al.* (1980). The rate of change in voluntary consumption in late pregnancy is also variable. Examination of studies with 10 diets shows that during the last 5-6 weeks of pregnancy voluntary consumption declined by 7-12 percent per week in 3 cases and by 1-3 percent per week in a further 5; zero change and increase of about two percent per week prevailed in the other two. As discussed by Forbes (1970, 1971), voluntary feed consumption decrease in late pregnancy is not confined to diets limited in intake by physical factors, decline being equally shown with diets based on concentrates.

Why feed intake fails to increase in late pregnancy to match the increasing energy demand is not established. Additional digestible energy is not available from the feed at this time (e.g. Weston 1979c). Deficiency of amino acids at the tissue level does not limit voluntary feed consumption nor does ruminating activity or the propulsive capacity of the alimentary tract; it could be significant that propulsive activity appears to increase in late pregnancy, the extent of change appearing to be independent of number of foetuses or level of feeding (Graham & Williams 1962, Campling 1966, Weston 1979c, unpublished data). The upward displacement of the ventral wall of the rumen in late pregnancy has been shown to be associated with a reduction in rumen digesta volume and voluntary feed consumption (Forbes 1970, 1971) and accordingly it has been suggested that pressure on the rumen, or abdominal wall distension (Forbes 1980) might modulate voluntary consumption. However, a rumen volume decrease (30 percent) at constant feed intake has been observed in late pregnancy (Weston 1979c), so the relation between reduced digesta volume and reduced voluntary consumption is not clear. Metabolic changes inducing a limitation in capacity to use energy could possibly be involved in control of voluntary consumption (Reid & Hinks 1962). Forbes (1971) has discussed the possible significance of increased oestrogen secretion and his data may indicate that oestradiol reduces the sheep's capacity to use energy.

Lactation

The ruminant, like the monogastric animal, achieves its highest voluntary consumption during lactation, the maximum energy intake by highly productive dairy cows being some 70 percent above that of growing steers (Baumgardt 1970). Voluntary consumption increases promptly after parturition, reaching a maximum after a variable time, thence remaining fairly steady, or declining slowly.

Comparisons of lactating and non-pregnant, non-lactating animals have shown voluntary consumption of fresh pasture herbage to be increased by 50 percent in lactating dairy cows (Hutton 1963) and by 26 to 43 percent in ewes during early-mid lactation (Cook *et al*. 1961, Arnold & Dudzinski 1967, Arnold 1975). Indoors, increases of 14 percent (roughage plus concentrate mixture) and 8 percent (concentrate diet) were observed with dairy cows (Campling 1966) and corresponding changes with ewes were 7 to 30 percent with four low-medium quality roughages, 20 to 48 percent with roughage plus concentrate mixtures, and 104 percent with immature *Cryptostemma* sp. forage (Reid & Hinks 1962, Lloyd Davies 1962, Hadjipieris & Holmes 1966, Owen *et al*. 1980). Examination of studies with 9 diets shows that ewes with twins consumed 10 (S.E. ± 3) percent more feed than their counterparts with singles.

Peak milk yield in dairy cows, and accordingly maximum energy demand, generally occurs 5 to 8 weeks after parturition but the voluntary feed consumption peak tends to occur later (Bines 1979). Increasing the proportion of concentrates in mixtures of roughages and concentrates tends to reduce the time lag between peak milk yield and peak voluntary feed consumption (Bines 1976). Also rate of voluntary feed consumption increase after parturition was faster with a ground and pelleted mixture than with similar diets fed unprocessed (Journet & Remond 1976). Thus maximum voluntary feed consumption may be achieved more rapidly with diets having low resistance to removal from the rumen; this could be relevant with pastures of differing quality.

During lactation more time is spent in both eating and ruminating and rate of eating increases (Campling 1966, Arnold & Dudzinski 1967, Arnold 1975, Dulphy *et al*. 1980). More frequent feeding has been reported with ewes (Forbes 1980) but not with cows (Campling 1966). Rumen capacity, digesta load and tissue weight have been shown to be lower in early than late lactation (Hutton *et al*. 1964, Tulloh 1966, Cowan *et al*. 1980). Comparisons of lactating and non-pregnant non-lactating ewes show increased rumen digesta load and rumen tissue weight; this accompanies the enhanced voluntary feed consumption and the independently controlled enhanced rumen propulsive activity (Weston 1979c, unpublished data). The various physiological factors possibly involved in the lag between peak yield and peak voluntary feed consumption have been discussed by Bines (1976, 1979). The rumen may not be able to attain its maximum capacity until the fat in the abdominal cavity that accrued during pregnancy is removed. Again, time may be needed for the rumen to hypertrophy to the degree consistent with maximum capacity, or for the rumen and other body tissues to become adapted to enhanced metabolic activity. Further, some endocrine change could be involved, but no such evidence is available.

Disease

Infectious, parasitic and metabolic diseases usually result in decreased feed intake and casual observations indicate that localized diseases of the mouth and feet may physically hinder feed acquisition.

A decrease in voluntary feed consumption with helminth infestation is generally observed, as discussed by Steel and Symons

(1979) and Weston (1979a); this effect occurs with different species of helminths and with infection at different sites, including the abomasum (e.g. Trichostrongylus axei), the small intestine (e.g. Strongyloides papillosus), the large intestine (e.g. Oesophagostomum spp.) and the bile ducts (Fasciola hepatica). The magnitude of the change can vary appreciably. The general complexity of the voluntary feed consumption response is indicated by Steel & Symons (1979) for a study in which T. colubriformis was given continuously for 24 weeks to simulate natural infection. It was shown that (1) the extent of voluntary feed consumption decline was related to the number of infective larvae ingested, (2) a threshold level existed for level of infection below which consumption was not significantly depressed, and (3) consumption returned to normal following the depression, possibly due to the development of resistance. Companion studies with O. circumcincta showed similar trends but with the effect on voluntary feed consumption being smaller. Similar long term dosing with Fasciola hepatica (Sykes et al. 1980) showed a 15 percent reduction in voluntary feed consumption independent of dose level, a reduction that continued to be maintained over about 25 weeks; however, consumption may be severely depressed with large doses of this helminth (Hawkins & Morris 1978). In general, little is known about the basis for depressed feed intake with different types of helminth infection, although many physiological processes have been shown to be impaired (Steel & Symons 1979, Sykes et al. 1980). However, Symons & Hennessy (1981) have shown that the high level of cholecystokinin in the plasma of sheep infected with T. colubriformis could be involved in depressing feed intake.

Reduced voluntary feed consumption has been shown to occur with metabolic diseases such as acetonaemia, pregnancy toxaemia and hypomagnesaemia, and the ectoparasite Boophilus microplus is also associated with feed intake decline in cattle (e.g. Seebeck et al. 1971).

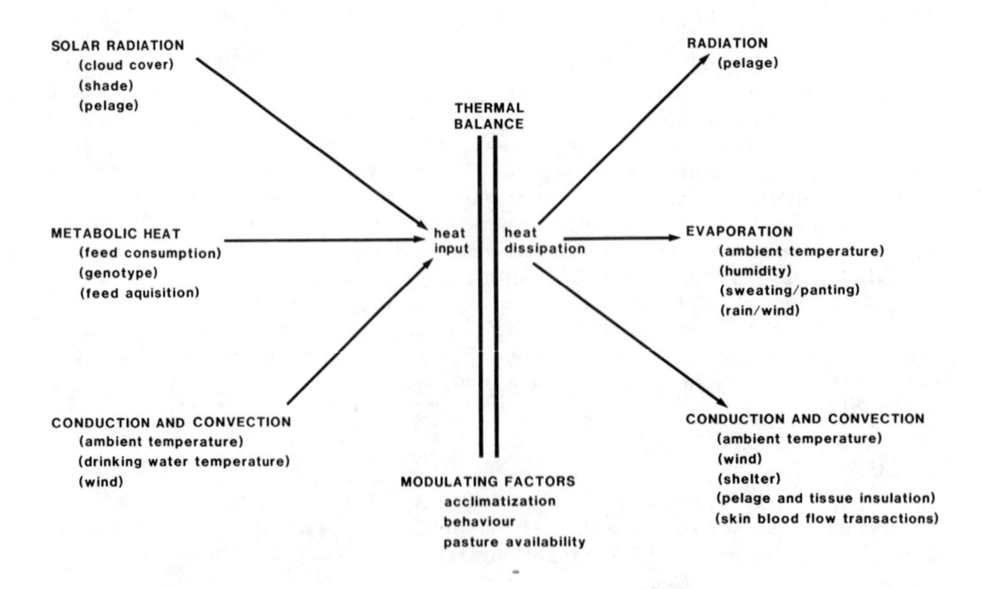

Fig. 2. Factors influencing thermal balance.

CLIMATE AND VOLUNTARY FEED CONSUMPTION

The effects of the various components of climate on voluntary feed consumption have not been studied comprehensively and much of the relevant data is derived from studies having other primary objectives. Thermal stress and daylength are of particular relevance.

Thermal stress

The effects of thermal stress generally relate to the maintenance of thermal balance, as indicated in studies with cattle showing onset of depression in voluntary consumption to coincide with increase in body temperature (Worstell & Brody 1953). As indicated in Fig. 2, the maintenance of thermal balance involves many factors which contribute to heat input and heat dissipation. Accordingly, the response in voluntary consumption to any particular climatic component varies with the concomitant significance of other relevant factors.

Heat input components

Solar radiation contributes to heat input in accordance with latitude, cloud cover, available shade, and the animal's pelage (hair or wool). In studies conducted indoors, increase in radiation levels from 15 to 570 watt/m^2 decreased voluntary consumption in Holstein cows by 24 percent at an ambient temperature of 27°C, by 20 percent at 24°C, and by only 9 percent at 10°C (Brody _et al_. 1954). High levels of radiation may affect cattle deleteriously; lack of shade during very hot days has been shown to reduce grazing during the cooler conditions prevailing at night (Larkin 1954). Pattern of grazing may be affected, with day grazing decreased in hot weather (e.g. Cowan 1975) and night grazing decreased in cold weather (Hutchinson & McRae 1969). The physical properties of the pelage affect the degree of reflection of solar radiation and accordingly heat input. Although shorn sheep are more affected by radiation than their fully fleeced counterparts, effects on their consumption of feed do not appear to have been studied.

Metabolic heat input depends mainly on the diet and the animal can readily adjust its thermal balance by an appropriate change in food consumption. The diet and metabolic state of the animal affect metabolic heat through influence on voluntary feed consumption. Contrasting situations are seen with (A) the non-pregnant non-lactating animal fed low quality roughage, and (B) the lactating animal fed high quality roughage. Heat production with sheep in A could be about 50 watt compared with about 100 watt in B, hence the upper and lower critical temperatures would be elevated in A relative to B. Calculations using average values for relevant parameters show that, following shearing, the lower critical temperatures could be 28°C in A and 14°C in B. Such differences must affect the level of thermal load needed to evoke a voluntary feed consumption response, but definitive data do not appear to be available. High energy expenditure in feed acquisition under difficult grazing conditions could possibly increase metabolic heat by some 10 watt.

Conduction and convection must contribute to feed intake limitation under very hot windy conditions, but no specific data are available. However, cooling of drinking water has been shown to increase consumption of feed under conditions of high heat load (Ittner _et al_. 1951).

Heat dissipation components

Ambient temperature markedly affects heat dissipation, but as many factors affect thermal balance, the onset of change in voluntary consumption of feed with increasing thermal stress does not consistently occur within narrow ranges of temperatures.

In studies of a few weeks' duration, decline in the voluntary consumption of feed by lactating cows with increasing heat load occurred at 24°-27°C or 31°C with Holsteins, 27°-30°C with Jerseys, and 30°C with Brown Swiss (Ragsdale *et al*. 1948, 1951, Wayman *et al*. 1962). Depressed voluntary feed consumption has also been shown in Holsteins for high day temperature (38°C) coupled with moderate night temperature (21°C) (Brody *et al*. 1955). With non-pregnant non-lactating *Bos taurus* cattle the onset of decline in consumption has been shown to range between 27° and >35°C (Johnson & Yeck 1964, Gengler *et al*. 1970, Martz *et al*. 1971, Olbrick *et al*. 1973, Lippke 1975). In long term studies with growing Shorthorn cattle, voluntary consumption was higher at 10° than at 27°C (Johnson *et al*. 1958). Low ambient temperature is often associated with increased consumption at least over periods of a few weeks. Thus increase of about 30 percent was shown at 9° compared with 18°C (Olbrick *et al*. 1973) and changes ranging from 0 to +40 percent were shown in comparisons between intakes at 10°C and those at 5°, -1°, -13° and -15°C (Ragsdale *et al*. 1949, 1950). However, no clear relationship has been shown between voluntary feed consumption response and temperature difference. No data appear to be available on low temperature effect on voluntary consumption over a long term but McDowell *et al*. (1976) found with dairy cows that voluntary consumption in the first 50 days of lactation was 14 percent higher in winter (mean 7°C) than in summer (mean 31°C).

Temperature stress in sheep has received scant study. Acclimatized Merino sheep tolerate high temperatures well. Thus lactating ewes maintained at 49°C without shade, consumed no less roughage diet than their counterparts provided with shade at 37°C (R.G. Stephenson *et al*. personal communication). Decreases in voluntary consumption of 14 to 50 percent, depending on diet, were found under heat load (21°-32°C) for sheep having access to feed for only part of the day (Bhattacharya & Hussain 1974) and in the study of Moose *et al*. (1969) voluntary consumption by lambs was not affected at 27°C but was lower at 29°C than at 5°C. Cold stress following shearing decreases voluntary consumption (e.g. Donnelly *et al*. 1974b), but in the lamb, voluntary consumption at 0°C did not differ from that at 23°C (Moose *et al*. 1969).

Humidity, by affecting rate of heat loss, can modify voluntary consumption of feed (Ragsdale *et al*. 1953, Johnson *et al*. 1963). With cattle maintained below 27°C, relative humidity has little effect on voluntary consumption. However, above 27°C, voluntary consumption decreased more at 70 percent than at 40 percent humidity, and the difference widened at higher temperatures. At 32°C voluntary consumption was normal at low relative humidity (20 percent) but declined appreciably with the relative humidity at 40 percent.

The pelage and tissue insulation markedly influences heat dissipation by conduction and convection. When insulation is reduced by shearing, consumption of feed increases, provided that the conditions do not cause cold stress. For mean ambient temperature of about 25°C, Minson & Ternouth (1971) observed little increase in voluntary consumption of 3 hays by sheep following shearing. Near 13°C, consumption increased by 5 to 13 percent with 4 different hays and with a mean of about 6°C larger increases (35-60 percent) were obtained with *Medicago sativa* hay and pasture (A.W.F. Davey, personal communication, Wheeler *et al*. 1963, Weston 1970, Minson & Ternouth 1971). The availability of shelter may affect intake as Winfield *et al*. (1968) found that pregnant ewes kept in exposed conditions outdoors (mean 5°C) consumed some 20 percent more feed than their counterparts housed indoors (mean 7°C). The properties of the pelage and the capacities to sweat, pant and vary blood flow to the periphery relate largely to genotype and form the basis of genotype x thermal stress interactions with voluntary consumption.

Factors modulating voluntary feed consumption responses

Factors such as acclimatization, animal behaviour and pasture availability may modulate the effect of thermal stress on voluntary consumption, but few definitive data are available. In studies with Holstein cows, Brody *et al*. (1955) found that voluntary consumption decreased by 50 percent in the first 8 days of exposure to heat load but the decrease was only 10 percent during the period 17-24 days of exposure. Breeds of sheep seem to differ in their use of shelter (Arnold & Dudzinski 1978) and this type of behaviour difference may affect voluntary consumption.

Temperature stress is associated with changes in rumen function. In the cold, rumen propulsive activity increases (Kennedy *et al*. 1976) and rumen digesta load increases in association with greater consumption of feed (A.W.F. Davey, personal communication). There appears to be no enhancing effect of cold *per se* on rumination although time spent ruminating tends to increase due to higher feed intake (Weston 1977 and unpublished data). With heat load, propulsive activity probably declines (Lippke *et al*. 1975).

Daylength

Daylength affects the pattern of grazing (Dulphy *et al*. 1980) and possibly affects feed intake. Observations at various research centres in the U.K., as listed by Sykes *et al*. (1980), together with the studies of Milne *et al*. (1978), show voluntary food consumption to increase from the winter to the summer months. In some instances the precise change cannot be assessed due to possible changes with time in body size or diet composition. However, under controlled conditions of artificial lighting, feed consumption was found to be about 10 percent higher with longer daylength in growing lambs (Forbes *et al*. 1975, 1979). Feed consumption response to daylength could possibly be related to genotype.

REGULATION OF VOLUNTARY FEED CONSUMPTION WITH CHANGE IN METABOLIC STATE

Regulation of voluntary consumption of feed is considered here in the context of herbage feeding, a situation in which the rate of organic matter removal from the rumen generally limits consumption. As discussed earlier, an energy deficit prevails under these conditions, its magnitude being equal to the animal's capacity to use energy minus the energy available to the tissues (Fig. 1).

The data reviewed here indicate that voluntary consumption and rate of organic matter transfer from the rumen may vary with change in the energy deficit. Accordingly, it is necessary in considering regulation of voluntary consumption to link energy metabolism with the rumen transactions. It would appear that the energy deficit and the rumen digesta load may play key roles in this interplay between the metabolism and the rumen. Thus a change in energy deficit may be considered to change the intensity of signals to the central nervous system that facilitate feeding. Again, alteration in digesta load will affect the intensity of signals that inhibit feeding. Accordingly, when metabolic state changes, the animal may achieve satiety by modifying its digesta load and the resulting inhibitory stimulus, to offset the changed feeding stimulus that relates to the energy deficit. Alteration in digesta load would be expected to change rate of removal of organic matter, not only by affecting the rate of digestion, which would change in relation to substrate availability, but also by influencing rumination and propulsive activity. Thus change in organic matter removal rate due to modulation of digesta load would permit a different voluntary feed consumption to be achieved.

It seems possible that the increases in voluntary consumption due to lactation, moderate cold exposure and higher growth potential as in young animals, could be modulated in this way. These responses are generally associated with enhanced energy deficit, digesta load, rumination and propulsive activity. The depressed voluntary consumption with increasing fatness could be affected similarly. Reduced energy deficit would accompany increased fatness if, as has been suggested, the potential rate of triglyceride synthesis decreases and/or fatty acid "leakage" from the adipocytes increases under such conditions (Fig. 1); the reduced digesta load observed in the fatter animal would be consistent with reduced energy deficit. Alternatively, large amounts of fat in the abdominal cavity could prevent the rumen achieving its potential capacity, and thereby limit digesta load.

During late pregnancy energy requirement is increased and processes relating to organic matter removal appear to be normal. Nevertheless, digesta load is clearly reduced. On the one hand this lower load may be due to reduced voluntary consumption arising from a reduced energy deficit, in turn due to an impairment in the capacity to use energy. On the other hand, the load may be reduced by competition for space in the abdominal cavity (Fig. 1) and possibly result in reduced voluntary consumption. Critical studies are required to clarify the situation.

Insufficient information is available to indicate how voluntary consumption is regulated in response to genotype, disease and daylength. Similar considerations apply with heat stress, but here the reduced energy deficit and propulsive activity would be relevant, as would the known existence of thermoreceptors in the central nervous system. Again, factors other than the maintenance of thermal balance could be involved as normal voluntary consumption may not be restored for some time after heat stress is removed (Wayman *et al*. 1970).

REFERENCES

Allden, W.G. (1979) Feed intake, diet composition and wool growth. In: Physiological and environmental limitations to wool growth. Editors J.L. Black and P.J. Reis. Armidale, Australia, University of New England Publishing Unit, pp. 61-78.

Arnold, G.W. (1975) Herbage intake and grazing behaviour in ewes of four breeds at different physiological states. Australian Journal of Agricultural Research 26, 1017-1024.

Arnold, G.W.; Dudzinski, M.L. (1967) Studies on the diet of the grazing animal. II. The effect of physiological status in ewes and pasture availability on herbage intake. Australian Journal of Agricultural Research 18, 349-359.

Arnold, G.W.; Dudzinski, M.L. (1978) Ethology of free ranging domestic animals. Amsterdam, Elsevier Scientific Publishing Company, pp. 84-86.

Baile, C.A.; Forbes, J.M. (1974) Control of feed intake and regulation of energy balance in ruminants. Physiological Reviews 54, 160-214.

Baumgardt, B.R. (1970) Control of feed intake in the regulation of energy balance. In: Physiology of digestion and metabolism in the ruminant. Editor A.T. Phillipson. Newcastle-upon-Tyne, U.K., Oriel Press, pp. 235-253.

Baumgardt, B.R.; Peterson, A.D. (1971) Regulation of food intake in ruminants. 8. Caloric density of diets for young growing lambs. Journal of Dairy Science 54, 1191-1194.

Bhattacharya, A.N.; Hussain, F. (1974) Intake and utilization of nutrients in sheep fed different levels of roughage under heat stress. Journal of Animal Science 38, 877-886.

Bines, J.A. (1976) Regulation of food intake in dairy cows in relation to milk production. Livestock Production Science 3, 115-128.

Bines, J.A. (1979) Voluntary food intake. In: Feeding strategy for the high-yielding dairy cow. Editors W.H. Broster and H. Swan. London, Granada, pp. 23-48.

Bines, J.A.; Suzuki, S.; Balch, C.C. (1969) The quantitative significance of long-term regulation of feed intake in the cow. British Journal of Nutrition 23, 695-704.

Blaxter, K.L.; Clapperton, J.L.; Wainman, F.W. (1966a) The extent of differences between six British breeds of sheep in their metabolism, feed intake and utilization, and resistance to climatic stress. British Journal of Nutrition 20, 283-294.

Blaxter, K.L.; Wainman, F.W.; Davidson, J.L. (1966b) The voluntary intake of food by sheep and cattle in relation to their energy requirements for maintenance. Animal Production 8, 75-83.

Brody, S.; Ragsdale, A.C.; Thompson, H.J.; Worstell, D.M. (1954) The thermal effects of radiation intensity (light) on milk production, feed and water consumption, and body weight in Holstein, Jersey and Brahman cows at air temperatures 40°, 70° and 80°F. University of Missouri Agricultural Experiment Station, Research Bulletin, 556.

Brody, S.; Ragsdale, A.C.; Yeck, R.G.; Worstell, D.M. (1955) Milk production, feed and water consumption, and body weight of Jersey and Holstein cows in relation to several diurnal temperature rhythms. University of Missouri Agricultural Experiment Station, Research Bulletin, 578.

Broster, W.H.; Tuck, V.J.; Balch, C.C. (1964) Experiments on the nutrition of the dairy heifer. V. Nutrition in late pregnancy. Journal of Agricultural Science, Cambridge 63, 51-58.

Campling, R.C. (1966) A preliminary study of the effect of pregnancy and of lactation on the voluntary intake of food by cows. British Journal of Nutrition 20, 25-39.

Clancy, M.; Bull, L.S.; Wangsness, P.J.; Baumgardt, B.R. (1977) Digestible energy intake of complete diets by wethers and lactating ewes. Journal of Animal Science 42, 960-969.

Cook, C.W.; Mattox, J.E.; Harris, L.E. (1961) Comparative daily consumption and digestibility of summer range forage by wet and dry ewes. Journal of Animal Science 20, 866-870.

Cowan, R.T. (1975) Grazing time and pattern of grazing of Friesian cows on a tropical grass-legume pasture. Australian Journal of Experimental Agriculture and Animal Husbandry 15, 32-37.

Cowan, R.T.; Robinson, J.J.; McDonald, I.; Smart, R. (1980) Effects of body fatness at lambing and diet in lactation on body tissue loss, feed intake and milk yield of ewes in early lactation. Journal of Agricultural Science, Cambridge 95, 497-514.

Donnelly, J.R.; Davidson, J.L.; Freer, M. (1974a) Effect of body condition on the intake of food by mature sheep. Australian Journal of Agricultural Research 25, 813-823.

Donnelly, J.B.; Lynch, J.J.; Webster, M.E.D. (1974b) Climatic adaptation in recently shorn Merino sheep. International Journal of Biometeorology 18, 233-247.

Dulphy, J.P.; Remond, B.; Theriez, M. (1980) Ingestive behaviour and related activities in ruminants. In: Digestive physiology and metabolism in ruminants. Editors Y. Ruckebusch and P. Thivend. Lancaster, U.K., M.T.P. Press, pp. 103-122.

Egan, J.K.; Doyle, P.T. (1980) The comparative intake and digestion of herbage diets by weaner and mature sheep. Proceedings of the Australian Society of Animal Production 13, 475.

Foot, J.Z. (1972) A note on the effect of body condition on the voluntary intake of dried grass wafers by Scottish Blackface ewes. Animal Production 14, 131-134.

Foot, J.Z.; Russel, A.J.F. (1979) The relationship in ewes between voluntary food intake during pregnancy and forage intake during lactation and after weaning. Animal Production 28, 25-39.

Forbes, J.M. (1970) The voluntary food intake of pregnant and lactating ruminants: A review. The British Veterinary Journal 126, 1-11.

Forbes, J.M. (1971) Physiological changes affecting voluntary food intake in ruminants. Proceedings of the Nutrition Society 30, 135-142.

Forbes, J.M. (1980) Hormones and metabolites in the control of food intake. In: Digestive physiology and metabolism in ruminants. Editors Y. Ruckebusch and P. Thivend. Lancaster, U.K., M.T.P. Press, pp. 145-160.

Forbes, J.M.; Driver, P.M.; El Shahat, A.A.; Boaz, T.G.; Scanes, C.G. (1975) The effect of daylength and level of feeding on serum prolactin in growing lambs. Journal of Endocrinology 64, 549-552.

Forbes, J.M.; El Shahat, A.A.; Jones, R.; Duncan, J.G.S.; Boaz, T.G. (1979) The effect of daylength on the growth of lambs. 1. Comparisons of sex, level of feeding, shearing and breed of sire. Animal Production 29, 33-42.

Frisch, J.E.; Vercoe, J.E. (1977) Food intake, eating rate, weight gains and metabolic rate and efficiency of feed utilization in Bos taurus and Bos indicus crossbred cattle. Animal Production 25, 343-358.

Frisch, J.E.; Vercoe, J.E. (1978) Utilizing breed differences in growth of cattle in the tropics. World Animal Review 25, 8-12.

Graham, N.McC. (1972) Units of metabolic body size for comparisons amongst adult sheep and cattle. Proceedings of the Australian Society of Animal Production 9, 352-355.

Graham, N.McC.; Williams, A.J. (1962) The effects of pregnancy on the passage of food through the digestive tract of sheep. Australian Journal of Agricultural Research 13, 894-900.

Hadjipieris, G.; Holmes, W. (1966) Studies on feed intake and feed utilization by sheep. 1. The voluntary feed intake of dry, pregnant and lactating ewes. Journal of Agricultural Science, Cambridge 66, 217-223.

Hawkins, C.D.; Morris, R.S. (1978) Depression of productivity in sheep infected with Fasciola hepatica. Veterinary Parasitology 4, 341-351.

Hodgson, J. (1982) Influence of sward characteristics on diet selection and herbage intake by the grazing animal. In: Nutritional limits to animal production from pastures. Editor J.B. Hacker. Farnham Royal, U.K., Commonwealth Agricultural Bureaux, pp. 153-166.

Hutchinson, K.J.; McRae, B.H. (1969) Some factors associated with the behaviour and survival of newly shorn sheep. Australian Journal of Agricultural Research 20, 513-521.

Hutton, J.B. (1963) The effect of lactation on intake in the dairy cow. Proceedings of the New Zealand Society of Animal Production 23, 39-52.

Hutton, J.B.; Hughes, J.W.; Newth, R.P.; Watanabe, K. (1964) The voluntary intake of the lactating dairy cow and its relation to digestion. Proceedings of the New Zealand Society of Animal Production 24, 29-42.

Ittner, N.R.; Kelly, C.F.; Guilbert, H.R. (1951) Water consumption of Hereford and Brahman cattle and the effect of cooled drinking water in a hot climate. Journal of Animal Science 10, 742-751.

Johnson, H.D.; Yeck, R.G. (1964) Age and temperature effects on T.D.N., water consumption and balance of dairy calves and heifers exposed to environmental temperatures of 35 and 95°F. University of Missouri Agricultural Experiment Station, Research Bulletin, 865.

Johnson, H.D.; Ragsdale, A.C.; Yeck, R.G. (1958) Effects of constant environmental temperatures of 50° and 80°F on the feed and water consumption of Brahman, Santa Gertrudis, and Shorthorn calves during growth. University of Missouri Agricultural Experiment Station, Research Bulletin, 683.

Johnson, H.D.; Ragsdale, A.C.; Berry, I.L.; Shanklin, M.D. (1963) Temperature-humidity effects including influence of acclimation in feed and water consumption of Holstein cattle. University of Missouri Agricultural Experiment Station, Research Bulletin, 846.

Journet, M.; Remond, B. (1976) Physiological factors affecting the voluntary intake of feed by cows: A review. Livestock Production Science 3, 129-146.

Kennedy, P.M.; Christopherson, R.J.; Milligan, L.P. (1976) The effect of cold exposure of sheep on digestion, rumen turnover time and efficiency of microbial synthesis. British Journal of Nutrition 36, 231-242.

Lamberth, J.L. (1969) The effect of pregnancy in heifers on voluntary intake, total rumen contents, digestibility and rate of passage. Australian Journal of Experimental Agriculture and Animal Husbandry 9, 493-496.

Langlands, J.P. (1968) The feed intake of grazing sheep different in age, breed, previous nutrition, and liveweight. Journal of Agricultural Science, Cambridge 71, 167-172.

Langlands, J.P.; Hamilton, B.A. (1969) Efficiency of wool production of grazing sheep. 2. Differences between breeds and strains varying in age. Australian Journal of Experimental Agriculture and Animal Husbandry 9, 254-257.

Larkin, R.M. (1954) Observations on the grazing behaviour of beef cattle in tropical Queensland. Queensland Journal of Agricultural Science 11, 115-141.

Lippke, H. (1975) Digestibility and volatile fatty acids in steers and wethers at 21 and 32°C ambient temperature. Journal of Dairy Science 58, 1860-1864.

Lloyd Davies, H. (1962) Intake studies in sheep involving high fluid intake. Proceedings of the Australian Society of Animal Production 4, 167-171.

Lodge, G.A.; Fisher, L.J.; Lessard, J.R. (1975) Influence of prepartum feed intake on performance of cows fed ad libitum during pregnancy. Journal of Dairy Science 58, 696-702.

McDowell, R.E.; Hoover, N.W.; Camoens, J.K. (1976) Effect of climate on performance of Holsteins in first lactation. Journal of Dairy Science 59, 965-973.

Martz, F.A.; Mishra, M.; Campbell, J.R.; Daniels, L.B.; Hildebrand, E. (1971) Relation of ambient temperature and time post-feeding on ruminal, arterial and venous volatile fatty acids, and lactic acid in Holstein steers. Journal of Dairy Science 54, 520-525.

Miller, R.H.; Hoover, N.W.; Smith, J.W.; Creegan, M.E. (1972) Feed consumption differences among lactating cows. Journal of Dairy Science 55, 454-459.

Milne, J.A.; Macrae, J.C.; Spence, A.M.; Wilson, S. (1978) A comparison of the voluntary intake and digestion of a range of forages at different times of the year by the sheep and the red deer (Cervus elaphus). British Journal of Nutrition 40, 347-357.

Minson, D.J. (1982) Effects of chemical and physical composition of herbage eaten upon intake. In: Nutritional limits to animal production from pastures. Editor J.B. Hacker. Farnham Royal, U.K., Commonwealth Agricultural Bureaux, pp. 167-182.

Minson, D.J.; Ternouth, J.H. (1971) The expected and observed changes in the intake of three hays by sheep after shearing. British Journal of Nutrition 26, 31-39.

Moose, M.G.; Ross, C.V.; Pfander, W.H. (1969) Nutritional and environmental relationships with lambs. Journal of Animal Science 29, 619-627.

Olbrich, S.E.; Martz, F.A.; Hildebrand, E.S. (1973) Ambient temperature and ration effects on nutritional and physiological parameters of heat and cold tolerant cattle. Journal of Animal Science 37, 574-580.

Owen, J.B.; Lee, R.F.; Lerman, P.M.; Miller, E.L. (1980) The effect of reproductive state of ewes on their voluntary intake of diets varying in straw content. Journal of Agricultural Science, Cambridge 94, 637-644.

Paquay, R.; Doize, F.; Bouchat, J.C. (1979) Long term control of voluntary food intake in sheep. Annales de Recherches Veterinaires 10, 223-225.

Ragsdale, A.C.; Brody, S.H.; Thompson, H.J.; Worstell, D.M. (1948) Influence of temperature, 50° to 105°F, on milk production and feed consumption in dairy cattle. University of Missouri Agricultural Experiment Station, Research Bulletin, 425.

Ragsdale, A.C.; Worstell, D.M.; Thompson, H.J.; Brody, S. (1949) Influence of temperature, 50° to 0°F and 50° to 95°F, on milk production, feed and water consumption and body weight in Jersey and Holstein cows. University of Missouri, Agricultural Experiment Station, Research Bulletin, 449.

Ragsdale, A.C.; Thompson, H.J.; Worstell, D.M.; Brody, S. (1950) Milk production and feed and water consumption responses of Brahman, Jersey, and Holstein cows to changes in temperature. University of Missouri Agricultural Experiment Station, Research Bulletin, 460.

Ragsdale, A.C.; Thompson, H.J.; Worstell, D.M.; Brody, S. (1951) Influence of increasing of temperature, 40° to 105°F on milk production in Brown Swiss cows, and on feed and water consumption and body weight in Brown Swiss and Brahman cows and heifers. University of Missouri Agricultural Experiment Station, Research Bulletin, 471.

Ragsdale, A.C.; Thompson, H.J.; Worstell, D.M.; Brody, S. (1953) The effect of humidity on milk production and composition, feed and water consumption, and body weight in cattle. University of Missouri Agricultural Experiment Station, Research Bulletin, 521.

Reid, R.L.; Hinks, N.T. (1962) Studies on the carbohydrate metabolism of sheep. XVII. Feed requirements and voluntary feed intake in late pregnancy, with particular reference to prevention of hypoglycaemia and hyperketonaemia. Australian Journal of Agricultural Research 13, 1092-1111.

Seebeck, R.M.; Springell, P.H.; O'Kelly, J.C. (1971) Alterations in host metabolism by the specific and anorectic effects of the cattle tick (Boophilus microplus). I. Food intake and body weight growth. Australian Journal of Biological Sciences 24, 373-380.

Steel, J.W.; Symons, L.E.A. (1979) Current ideas on the mechanisms by which gastro-intestinal helminths influence the rate of wool growth. In: Physiological and environmental limitations to wool growth. Editors J.L. Black and P.J. Reis. Armidale, Australia, University of New England Publishing Unit, pp. 311-320.

Sykes, A.R.; Coop, R.L.; Rushton, B. (1980) Chronic subclinical fascioliasis in sheep: effects of food intake, food utilisation and blood constituents. Research in Veterinary Science 28, 63-70.

Symons, L.E.A.; Hennessy, D.R. (1981) Cholecystokinin and anorexia in sheep infected by the intestinal nematode Trichostrongylus colubriformis. International Journal of Parasitology 11, 55-58.

Tayler, J.C. (1959) A relationship between weight of internal fat, "fill" and the herbage intake of grazing cattle. Nature, London 184, 2021-2022.

Ternouth, J.H.; Poppi, D.P.; Minson, D.J. (1979) The voluntary food intake, ruminal retention time and digestibility of two tropical grasses fed to cattle and sheep. Proceedings of the Nutrition Society of Australia 4, 152.

Tulloh, N.M. (1966) Physical studies of the alimentary tract of grazing cattle. IV. Dimensions of the tract in lactating and non-lactating cows. New Zealand Journal of Agricultural Research 9, 999-1008.

Vercoe, J.E.; Frisch, J.E. (1977) The importance of voluntary food intake and metabolic rate to production in different genotypes of cattle in different environments. Animal Breeding Papers, Third International Congress of the Society for the Advancement of Breeding Researchers in Asia and Oceania, Canberra, pp. 1C-42-1C-45.

Wayman, O.; Johnson, H.D.; Merilan, C.P.; Berry, I.L. (1962) Effect of ad libitum or force-feeding of two rations on lactating dairy cows subject to temperature stress. Journal of Dairy Science 45, 1472-1478.

Weston, R.H. (1970) Voluntary consumption of low quality roughage by sheep during cold exposure. Australian Journal of Experimental Agriculture and Animal Husbandry 10, 679-684.
Weston, R.H. (1974) Factors limiting the intake of feed by sheep. VIII. The roughage requirement of the ruminant lamb fed concentrate diets based on wheat. Australian Journal of Agricultural Research 25, 349-362.
Weston, R.H. (1977) Metabolic state and roughage consumption in sheep. Proceedings of the Nutrition Society of Australia 2, 88.
Weston, R.H. (1979a) Feed intake regulation in the sheep. In: Physiological and environmental limitations to wool growth. Editors J.L. Black and P.J. Reis. Armidale, Australia, University of New England Publishing Unit, pp. 163-177.
Weston, R.H. (1979b) Factors limiting the intake of feed by sheep. IX. Further studies of the roughage requirement of the ruminant lamb fed on concentrate diets based on wheat. Australian Journal of Agricultural Research 30, 533-541.
Weston, R.H. (1979c) Digestion during pregnancy and lactation in sheep. Annales de Recherches Veterinaires 10, 442-444.
Weston, R.H. (1980) Roughage intake and digestion comparisons with lambs and adult sheep. Proceedings of the Nutrition Society of Australia 5, 191.
Wheeler, J.L.; Reardon, T.F.; Lambourne, J.L. (1963) The effect of pasture availability and shearing stress on herbage intake of grazing sheep. Australian Journal of Agricultural Research 14, 364-372.
Winfield, C.J.; Brown, W.; Lucas, I.A.M. (1968) Some effects of compulsory exposure over winter on in-lamb Welsh mountain ewes. Animal Production 10, 451-463.
Worstell, D.M.; Brody, S. (1953) Comparative physiological reactions of European and Indian cattle to changing temperature. University of Missouri Agricultural Experiment Station, Research Bulletin, 515.

PART 5

LIMITATIONS TO DIGESTION AND UTILIZATION

MICROBIAL BREAKDOWN OF FEED IN THE DIGESTIVE TRACT

D. E. AKIN

USDA, ARS, Richard B. Russell Agricultural Research Center, Athens, GA, 30613, USA.

ABSTRACT

The microbial population in the rumen is made up of a diverse collection of species consisting primarily of bacteria and protozoa. Although the types of microorganisms in the rumen are relatively stable due to consistent environmental conditions in the rumen, the proportions of the various species can be influenced by the type of feed in the diet, by the quality of forage, and by the ruminant species. The nutrients provided to the microorganisms by the forage diet can vary extensively in their availability for microbial use. Much of the energy in forage diets is in the form of structural carbohydrates in the plant cell wall, which is degraded primarily by the bacteria. Often bacteria adhere closely to the plant cell wall and degrade the structural carbohydrates apparently by cell-bound carbohydrases. Differences in the predominate morphotypes of adhering bacteria exist for different forages, but no consistent relationship to digestibility is found when a range of forages is examined. However, tissues that are more rapidly digested are often degraded by unattached but nearby bacteria, suggesting that the carbohydrates in these cell walls are less rigidly complexed and more readily available to cell-free carbohydrases than those in slowly digested tissues. Of the rumen protozoa, *Epidinium ecaudatum* has been shown to degrade and ingest plant cell walls of particular forages and to show a preference for forages with more easily digested cell walls. Differences in the anatomy of forage leaf blades exist and potentially can influence the digestion of cell walls by rumen microorganisms. Additionally, similar tissue types of different forages vary in the ease of digestion indicating that the inherent cell wall structure markedly influences digestibility. Lignin and phenolic acids are present in forage tissues and appear to bind to cell wall-type carbohydrates, especially those of the hemicellulose, and prevent microbial utilization of these potential substrates. ρ-Coumaric acid is toxic to fibre-digesting bacteria. Living tissues that are slowly degraded are often found to give a chlorine-sulphite positive reaction suggesting that syringyl lignin could influence the rate of digestion of living tissues. The digestion of forages results from a complex interaction of rumen microorganisms and forage cell wall structure. Continued research is essential to identify and characterize factors inherent in the structurally intact plant cell wall that limit the availability of nutrients for maximum use of forages.

MICROORGANISMS PRESENT IN THE RUMEN

Since the discovery of the rumen protozoa by Gruby and Delafond in 1843 (Hungate 1966), research has continuously provided new information on the microbes inhabiting the rumen, and considerable information is available on the processes involved in the fermentation of feedstuffs by rumen microorganisms. It has been established that the predominant microbes that inhabit the rumen are bacteria and protozoa. The numbers of bacteria in the rumen are estimated to be 10^{10} to 10^{11}/ml while protozoal numbers are estimated at 10^5 to 10^6/ml (Hungate 1975).

Bacteria

Since the rumen functions as a storage compartment in which plant materials undergo fermentation by microorganisms, species capable of degrading plant fibre (i.e., cell walls consisting of cellulose, hemicellulose, pectin) comprise an important part of the microbial population. Rumen bacteria are the more important degraders of the cellulosic materials that ruminants ingest as part of their diet (Bryant 1973). Hungate (1966) in his classic text, *The Rumen and Its Microbes*, described several species of cellulolytic rumen bacteria including *Bacteroides succinogenes*, *Ruminococcus albus* and *R. flavefaciens*, *Clostridium longisporum* and *C. lochheadii*, *Cillobacterium cellulosolvens*, *Cellulomonas fimi*, and *Butyrivibrio fibrisolvens*. The most important cellulolytic species, based on numbers in the rumen and ability to attack cellulose, are *B. succinogenes*, *R. albus*, and *R. flavefaciens* (Bryant 1973). Although *B. fibrisolvens* often is not considered to contribute markedly to cellulose digestion in the rumen (Bryant 1973, Hungate 1966), Kistner (1965) in studies in South Africa reported that *Butyrivibrio* species were the major cellulolytic bacteria in sheep fed a low-protein *Eragrostis tef* hay.

In addition to cellulose, hemicellulose is another major component of forage cell walls. *Ruminococcus* spp. are hemicellulolytic as well as cellulolytic. Although *B. succinogenes* has been reported to degrade hemicellulose, it cannot utilize xylan, which comprises a large part of the hemicellulose (Dehority & Scott 1967). *B. fibrisolvens*, *Bacteroides*, and *Eubacterium* are important non-cellulolytic bacteria that attack hemicellulose (Hungate 1966, Dehority & Scott 1967). Pectins are attacked by rumen treponemes, *B. succinogenes*, *B. ruminicola*, *B. fibrisolvens*, *Succinivibrio dextrinosolvens*, and *Lachnospira multiparus*, which appears to be one of the most active pectinolytic rumen species (Clarke *et al*. 1969, Bryant 1977, Wojciechowicz & Ziolecki 1979).

A polymer not present in cell walls but important in many feeds, especially grains, is starch. According to Hungate (1966), among the prevalent amylolytic species are a few of the cellulolytic ones including *B. succinogenes* and the non-cellulolytic species *Streptococcus bovis*, *Bacteroides amylophilus*, *B. ruminicola*, *Succinimonas amylolytica*, and *Selenomonas ruminantium*.

Reviews of rumen microbiology have stressed the diversity of bacteria involved in the fermentation of feed (Hungate 1966, Bryant 1977). In addition to those attacking components of fibre and starch, other microorganisms are prevalent and active in fermenting simple carbohydrates such as xylose, glucose, fructose, galactose, sucrose (e.g., *Eubacterium ruminantium*, *Fusobacterium necrophorum*, *Treponema* and *Borrelia* spp., *Megasphaera elsdenii*), lactate (e.g., *Veillonella alcalescens*, *Selenomonas lactilytica*), and glycerol (e.g., *Anaerovibrio lipolytica*, *F. necrophorum*). The "large" rumen bacteria such as Oscillospira, Lampropedia, and Quin's and Eadie's ovals are often present in large masses on the cuticular surfaces of green leaves, but their exact function as related to the cuticle is unknown (Clarke 1979b). Specific populations of bacteria have been shown by electron microscopy to associate with the rumen epithelium in sheep (Bauchop et al. 1975) and cattle (Cheng et al. 1979b). Cheng et al. (1979b) suggested that the existence of this taxonomically distinct population may be important for scavenging oxygen (since about 20 percent of the strains were facultatively anaerobic), recycling dead tissue, and digesting urea to provide a source of ammonia for other microorganisms.

Protozoa

Although flagellates are present in the rumen (Clarke 1979a), the majority of rumen protozoa are ciliates, which have been divided into the holotrichs (Family Isotrichidae) and the entodiniomorphs (Family Ophryoscolecidae of the Order Entodiniomorphida) (Hungate 1966, Kudo 1966, Bryant 1977). Hungate (1966) stated that the holotrichs in the rumen consisted of Isotricha prostoma, I. intestinalis, and Dasytricha ruminantium. Holotrichs have complete body ciliation, move quickly, and rapidly assimilate soluble sugars with extensive amounts stored as amylopectin (Hungate 1966, Kudo 1971, Bryant 1977, Clarke 1979a). Other research (Orpin & Letcher 1978) indicates that both Isotricha species have a chemotactic response to sucrose, glucose, and fructose, and that an attachment mechanism could give this genus an advantage over others in assimilating soluble carbohydrates.

The entodiniomorphs lack complete body ciliation but have bands of syncilia for movement and food ingestion. They are a more diverse group than the holotrichs and are represented in the rumen by many genera including Ophryoscolex, Entodinium, Epidinium, Diplodinium, Polyplastron among others (Hungate 1966, Kudo 1971). The entodiniomorphs are especially important in the ingestion and digestion of starch and data suggest that entodiniomorphs can digest cellulose (Hungate 1966). Epidinium ecaudatum has been shown to possess xylanase (Bailey et al. 1962), to associate in large numbers with alfalfa stems (Bauchop & Clarke 1976), and to digest and ingest cell walls of certain forage material (Amos & Akin 1978, Akin & Amos 1979).

Other microorganisms

Although the above mentioned bacterial and protozoal species are prominent in the rumen, other types of microorganisms have also been reported. Lund (1974) reported the presence of nine fungal species, of which several could reproduce under in vitro anaerobic conditions. Orpin (1975, 1977) reported that the rumen flagellates Piromonas communis and Neocallimastix frontalis are zoospores of phycomyceteous fungi inhabiting the rumen. Bauchop (1979b) reported that large numbers of phycomyceteous zoospores attached to fibrous fragments of plant material within the rumen.

Microorganisms of lesser or undefined importance in the rumen include mycoplasma (Robinson & Hungate 1973), cellulolytic actinomycetes (Hungate 1946, Maluszynska & Janota-Bassalik 1974), an unidentified filamentous microbe that attacks particular lignified tissue (Akin 1980b), and various types of bacteriophages (Paynter et al. 1969).

Hungate (1960) suggested that the more abundant rumen bacteria were known, but better selective methods in the future would reveal additional microorganisms. While this has been true, it is apparent that under certain feeding regimes or conditions, other types of microbes can be present or shifts in major types can result.

VARIATIONS IN MICROBIAL POPULATIONS

Concentrate versus Roughage

Concentrate diets high in grain offer a readily available form of rapidly fermentable carbohydrates to the rumen microbial populations. In general, the feeding of concentrates results in higher numbers of bacteria in the rumen (Bryant & Burkey 1953, Latham *et al*. 1974). The numbers of cellulolytic bacteria have been reported to decline with the feeding of grain (Grubb & Dehority 1975, Henning *et al*. 1980, Slyter 1976) but numbers of certain cellulolytic types, notably the cocci, are relatively constant in concentrate and roughage rations (Bryant & Burkey 1953, Latham *et al*. 1971). Often large numbers of homofermentative lactic acid-producing bacteria such as lactobacilli and sometimes *S. bovis* are prevalent on concentrate diets (Slyter *et al*. 1970), and selenomonads, streptococci, and peptostreptococci have been reported to increase with the addition of grain (Latham *et al*. 1971). Protozoal numbers have been found to increase with grain (Grubb & Dehority 1975) but others found fewer holotrichs (Latham *et al*. 1971).

Forages

Just as variations exist in the microbial types with variation in the concentrate to roughage ratio, variations occur in the microbial populations with different quality roughages, and the fibre-digesting species are more important in feed degradation. In comparing wheat straw with alfalfa hay, Bryant & Burkey (1953) reported that the proportion of cellulolytic rods was greater on wheat straw and less on *Medicago sativa* than the cellulolytic cocci. Kistner (1965), in comparing *M. sativa* and *Eragrostis tef* hay, concluded that *R. albus* and *R. flavefaciens* are more prevalent on *M. sativa* while *B. succinogenes* or *B. fibrisolvens* are the predominant cellulolytic bacteria on low-quality roughage. Based upon electron microscopic observation of populations attacking various fibres and substrates within nylon bags, Dinsdale *et al*. (1978) concluded that grass blades appear to be attacked primarily by cocci while the residual plant cell walls as well as cotton fibre select a rod-like microorganism. Data from our laboratory (Akin 1979), based on electron microscopic examination of leaf blades of various grasses all incubated with rumen fluid from a single source, show that microorganisms resembling ruminococci are prevalent and even predominate with some forage species. However, bacteria resembling *B. succinogenes* are prevalent on other grasses, and in particular, predominate on cell walls of *Festuca arundinacea*; these bacteria are more prevalent on other grasses as *in vitro* digestion proceeds. Variations in major types of bacteria that adhere to and degrade cell walls do not relate consistently to differences in cell wall digestibility (Akin 1980a). We found that the protozoan *Epidinium ecaudatum* associates with cool- but not warm-season forages and degrades about 11 percent of the dry matter in *Dactylis glomerata* (Amos & Akin 1978).

Influence of ruminant species

Cipolloni *et al.* (1951) reviewed digestibility values from the literature and concluded that digestibility of forages could vary with ruminant species. Indeed there are data indicating the microbial types and species vary with the host animal. However, research has also indicated that microbial types can vary considerably among animals of the same species, even on similar diets (Latham *et al.* 1974, Grubb & Dehority 1975). Hungate (1966) reviewed research and reported that many microbial species are common to sheep and cattle rumens but that greater numbers of selenomonads, Quin's oval, and *Veillonella alcalescens* are present in sheep. Pant & Roy (1970) reported higher numbers of rumen bacteria in buffalo than in zebu cattle, but the numbers of protozoa and the cellulolytic activity are similar. Other research on zebu and buffalo cows indicated higher total bacterial counts in the zebu before feeding, but lower counts at 3 and 6 hours after feeding (Panjarathinam & Laxminarayana 1974). Playne (1978) showed that cattle digest dry matter, cellulose and hemicellulose of low quality tropical hay more efficiently than do sheep. He suggested that better utilization of nutrients occurs with bovine rumen microbes and perhaps higher microbial activity results. It is interesting that with *Lolium perenne* diets in the United Kingdom (Dinsdale *et al.* 1978), diets of *Eragrostis tef* hay and *Medicago sativa* in South Africa (Kistner 1965), and diets of wood pulp and other additives in Japan (Takechi *et al.* 1978), few or no *B. succinogenes* were found in sheep rumens. Host-specificity has been suggested by researchers for some protozoal species (see discussion of Dehority 1978), but Dehority (1978) studied 24 species and found no evidence of host specificity for sheep or cattle based upon cross-inoculation, although variation did occur with diets. Panjarathinam & Laxminarayana (1974) found significant differences in protozoal populations between zebu and buffalo cows on different diets.

Digestion other than in the rumen

High numbers of bacteria have been isolated from the ileum and caecum of ruminants (Allison *et al.* 1975) but little is reported about the particular species. Hungate *et al.* (1959) reported that all but about 4 percent of the fermentation occurs in the rumen. Considerably fewer cellulolytic bacteria have been found in the caecum than in the rumen (Kern *et al.* 1974). Ulyatt & Egan (1979) reported that 6-26 percent and 3-13 percent of the digestible hemicellulose and cellulose, respectively, is digested after the rumen; Hoover (1978) reported higher post-ruminal digestibility values for hemicellulose (30 to 40 percent) and cellulose (18 to 27 percent). Further, other reports (reviewed by Phillipson 1977) indicated about 10 percent and sometimes more of the cellulose is digested in the caecum but the proteolytic activity is stronger in the caecum than the rumen. Infusion of purified wood cellulose into the caecum results in a 33 percent cellulose digestibility (Warner *et al.* 1972).

REQUIREMENTS FOR MICROBIAL GROWTH

Rumen environment

The rumen has been extensively studied and as a result, the environmental conditions are generally listed as follows (Bryant 1977):

temperature - 39°C (range of 38° to 42°C).
oxidation-reduction potential - -250 to -450 mv
pH - 6 to 7.

The pH of the rumen has been studied by numerous researchers partly because it can vary substantially with diets such as concentrates from the 6 to 7 range and effect significant variations in the rumen microbial populations. Slyter *et al.* (1966), using a continuous-culture system, found that a pH below 6 results in marked decreases in the production of volatile fatty acids and methane and a high percentage of bacteria not identified with the major rumen bacteria found at pH 6.7. Using pure cultures of selected bacteria, Russell *et al.* (1979) reported decreased bacterial growth as the pH is lowered from 6.7 to 4.7. Overfeeding with grain may reduce the pH below 5 and this results in changes in the populations in the rumen and lower intestinal tract; protozoa also decrease in number with decreased pH (Allison *et al.* 1975).

Nutrients

Major nutrient requirements for rumen microorganisms are well known. Minerals are important in the rumen system for such functions as maintenance of desirable pH, redox potential, and osmotic pressure. Further, many enzymes require minute amounts of metal ions as co-factors. Optimal and toxic concentrations of minerals for cellulose and fibre digestion have been reported (Burroughs *et al.* 1951, Hubbert *et al.* 1958, Martinez & Church 1970). However, different minerals vary in availability (Playne *et al.* 1978) and also vary in availability depending on forage species, age (Powell *et al.* 1978), and the chemical form of the mineral (Ward *et al.* 1979).

In reviewing requirements for cellulolytic rumen bacteria, Bryant (1973, 1977) indicated that ammonia is essential as the main nitrogen source for most strains of the predominant bacterial species, but apparently some bacteria can use the amide-nitrogen from amino acids or peptides. Rumen protozoa apparently can use ammonia, free amino acids, and bacterial protein (Bryant 1973, Coleman & Sandford 1980). Although requirements vary for individual species, required vitamins include biotin, ρ-aminobenzoic or folic acid, thiamine, pyridoxine, and pantothenic acid; volatile fatty acids are also required by some species (Hungate 1966). Caldwell and Bryant (1966) developed a semisynthetic medium (medium 10) that supports growth of bacterial populations similar to that on rumen fluid medium. These researchers replaced rumen fluid with hemin, trypticase, yeast extract and a volatile fatty acid mixture of acetic, propionic, butyric, isobutyric, n-valeric, isovaleric, and DL-α-methylbutyric acids.

The carbohydrates, which are used as sources of carbon and energy for microbial growth, vary in availability within the feed. The amount of microbial material produced depends in part on the amount of carbohydrate fermented and the amount of adenosine triphosphate (ATP) made available through individual reactions (Bryant 1977). In a review of rumen microbial growth efficiency, Hespell (1979) reported that nutritional limits of the diet and decreased growth rates are responsible for low cell yields, and that maxima theoretical yields of Y_{ATP} of 26 be attained. He further suggested that *in vivo* growth of rumen bacteria in the fluid phase could be limited by carbohydrate availability. Carbohydrates in forage include cellulose, hemicellulose, pectins, starch and soluble sugars.

Many of the fibre-digesting bacteria, such as the ruminococci, degrade more than one type of structural carbohydrate. *B. succinogenes* has been shown in some studies to be the most active cellulolytic species (Halliwell and Bryant 1963, Dehority and Scott 1967); this microbe degraded but did not utilize xylan (Dehority 1973). Further research using pure cultures indicated that synergism among species occurs, and enhanced digestion of the structural carbohydrates in plant fibre can result from combinations of degrading and non-degrading bacteria (Dehority 1973). The type and maturity of the forage material exerts a marked influence on the availability of the structural carbohydrates for rumen microbial fermentation; lignin is thought to bind to the carbohydrates and prevent microbial utilization (Van Soest 1973).

Starch is more rapidly digested than structural carbohydrates and resembles in feed value the soluble sugars; protozoa are particularly prominent in the utilization of starch (Hungate 1966). Our work has suggested that localization of starch within certain cells (parenchyma bundle sheaths of warm-season grasses) can influence the availability of starch to rumen microorganisms (Akin & Burdick 1977).

Soluble carbohydrates are the most rapidly metabolized (Hungate 1966) and completely utilized (Van Soest 1967) of the carbohydrate types. Numerous bacteria and protozoa can utilize the soluble carbohydrates and different rates of utilization have been found for the various monosaccharides (Hungate 1966). Research using non-cellulolytic rumen bacteria grown on soluble sugars indicated that catabolite regulatory mechanisms occur with and that preferences exist for individual sugars by the bacteria (Russell & Baldwin 1978). In general, the soluble carbohydrates are considered to be almost totally utilized, i.e., average true digestibility of 98 percent (Van Soest 1967). Cool-season forages generally contain more soluble carbohydrates than warm-season species (Wilson & Ford 1973).

Additives which affect microbial growth or forage digestion

Considerable research has been carried out to overcome dietary deficiencies and to increase rumen microbial growth and forage digestibility by including additives such as additional sources of energy, minerals, and nitrogen in the diet. The addition of limiting nutrients within certain bounds has shown positive results. For example, increased cellulose digestion is found by addition of 0.25 percent glucose, whereas 0.5 percent depresses digestion (Arias *et al.* 1951).

Of the minerals necessary for microbial growth, additional sulphur has been shown to enhance cellulose and lignocellulose digestion (Bull & Vandersall 1973). Since minerals are not available from all forages equally (Playne *et al.* 1978), supplementation of the minerals to optimal available levels can result in increased microbial activity for low quality forages. Care is necessary to avoid exceeding toxic levels. Direct supplementation of animals is the most feasible means of ensuring adequate levels of particular minerals in the diet (Hunter *et al.* 1978, Wheeler *et al.* 1980).

Supplementation of a low-protein *Eragrostis tef* hay with urea and urea plus branched-chain volatile fatty acids increases the proportion of ruminococci to butyrivibrios while the addition of fatty acids results in higher numbers of cellulolytic bacteria (Van Glyswyk 1970). Voluntary hay intake and percentage cellulose and hemicellulose digestibility were increased with urea. Non-protein nitrogen supplements could be useful if the ruminal ammonia is less that 50 mg ammonia-nitrogen/l in the rumen fluid (Satter & Slyter 1974). Additionally, for some forage species, increases in crude protein and *in vitro* digestibility have been found with increased nitrogen fertilization (Taliaferro *et al*. 1975).

The previous discussion shows the importance of research and suggests possible applications into the factors influencing microbial growth for maximum fibre digestion. However, degradation of forage cell walls is also influenced to a large extent by factors inherent in the plants.

ATTACK ON FORAGE CELL WALLS BY RUMEN MICROORGANISMS

Attachment to forage cell walls by rumen bacteria

Studies involving transmission electron microscopy (TEM) have demonstrated the direct adherence to plant cell walls by particular morphotypes of rumen bacteria during the process of fibre degradation (Akin 1976). In inocula from fistulated steers, the predominant adhering bacteria are of two types (Akin 1976) 1) an encapsulated coccus with characteristics identical to those of *Ruminococcus* (Hungate 1966, Latham *et al*. 1978a) and 2) an irregularly shaped bacterium characteristic of *B. succinogenes* (Hungate 1966, Costerton *et al*. 1974, Latham *et al*. 1978b). Both types of bacteria can degrade the entire cell wall of digestible tissues in forages. Forages of various digestibilities were examined by TEM for possible variations in the proportion of the fibre-digesting bacteria. Table 1 shows the percentage of the various morphotypes of rumen bacteria adhering to and degrading cell walls of various grass leaves incubated *in vitro* with sub-samples of the same inoculum.

TABLE 1.

Percentage morphological types of rumen bacteria attached to cell walls of forage leaf blades after 12 hours *in vitro* digestion (Akin 1979).

Forage[a]	Morphological types of rumen bacteria			
	Encapsulated coccus	Irregularly-shaped bacterium	Regularly shaped bacteria	Others
Cynodon dactylon cv. Coastal	36	34	22	8
Cynodon cv. Coastcross-1	20	64	12	4
Paspalum notatum cv. Pensacola	12	66	15	8
Festuca arundinacea cv. Kentucky-31	9	82	8	0
Festuca arundinacea cv. Kenhy	63	15	35	11
Phleum pratense cv. Clair	35	37	23	5
Dactylis glomerata cv. Boone	52	28	19	1

[a]Ranked in increasing digestibility based on SEM evaluation of cell wall degradation.

No consistent relationship exists among forages for the proportion of attached bacteria and cell wall digestibility. For example, *Cynodon dactylon* and *Phleum pratense*, which vary markedly in digestibility, are virtually identical in the percentage of adhering morphotypes after digestion for 12 hours. Certain forages, such as *Paspalum notatum* and *Festuca arundinacea* have a high proportion of the irregularly-shaped bacterium attached to the cell walls; *F. arundinacea* consistently has a high proportion of this morphotype (Akin 1979).

Work in this laboratory has been conducted with the entire microbial population in an effort to understand fibre digestion in the normal ecological niches of the various microorganisms. However, other important and necessary research has been conducted on fibre digestion using pure cultures of rumen bacteria. Extensive maceration of tissues resulting from attack on the intercellular substances by the pectinolytic bacterium *Lachnospira multiparus* has been shown to expose other portions of the cell walls for subsequent attack by cellulolytic bacteria (Cheng *et al*. 1979a). Microbial features involved in the attachment of *R. albus* to fibre have been revealed (Leatherwood 1973, Patterson *et al*. 1975), and extracellular carbohydrates which can facilitate the adherence of bacteria to substrates have been shown for many rumen microorganisms (Costerton *et al*. 1974, Cheng & Costerton 1975). In co-cultures of *R. flavefaciens* and *B. succinogenes*, *R. flavefaciens* predominates on the epidermis, phloem, and sclerenchyma cell walls whereas *B. succinogenes* predominates on the mesophyll (Latham *et al*. 1978b). Other studies involving whole rumen populations (Akin 1980a) have not shown a consistent preference of a particular morphotype to adhere to specific tissue types when several grasses were studied, but trends within a single grass are found, especially for the sclerenchyma cell walls. Dinsdale *et al*. (1978) studied the attachment of rumen bacteria to various fibres and concluded that cell walls of grass leaves are degraded mainly by cocci, whereas cotton fibre and the resistant tissues from faecal residue elicit a rod-like population. However, data obtained from digestion studies with several warm- and cool-season species (Table 1) are not in agreement with this conclusion, but rather indicate that the irregularly-shaped bacterium resembling *B. succinogenes* predominates on certain forages. Although it is not known what effect the selection for a particular microbe, if possible, might have in *in vivo* fibre digestion, it has been shown that *B. succinogenes* is a more active cellulolytic bacterium than the ruminococci (Halliwell & Bryant 1963, Dehority & Scott 1967). Increased cellulolytic activity has been shown *in vitro* with rumen populations from *F. arundinacea* having a higher proportion of *B. succinogenes*-like bacteria in comparison with other populations (Akin 1980a).

Degradation without bacterial attachment

The attachment of rumen bacteria to plant cell walls often plays an important role in the process of forage digestion. However, many species of rumen bacteria are capable of producing extracellular enzymes active against cellulose and hemicellulose (Gill & King 1957, Dehority 1968, 1973, Smith *et al*. 1973). Research on the production of carboxymethylcellulase, xylanase, and pectin lyase by *R. flavefaciens* indicates that these enzymes are primarily cell-associated during exponential growth but accumulate in the supernatant during the stationary phase (Pettipher & Latham 1979).

There is conflicting information in the literature concerning the activity of cell-free cellulases in rumen fluid supernatant (Gill & King 1957, Krishnamurti & Kitts 1969). Further, it has been suggested that inhibitors in rumen fluid can influence the activity of these cell-free enzymes (Krishnamurti & Kitts 1969, Francis *et al*. 1978).

The less rigid cell walls of tissues such as the phloem and mesophyll may be degraded without the direct adherence of rumen bacteria and apparently by enzymes free from the bacteria, but bacteria are always near the degraded zones (Akin & Amos 1975). Additionally, in highly digestible grasses, other tissues such as the parenchyma bundle sheath and inner part of the epidermis may be degraded by nearby but unattached bacteria (Akin 1980a). These observations suggest that the manner of degradation varies with specific cell walls. Data on the comparison of the number of culturable bacteria on hemicellulose-type carbohydrates from populations adapted to two grasses of different digestibilities show 75 percent more total bacteria, 105 percent more pectinolytic, 82 percent more xylanolytic, and 54 percent more cellobiolytic bacteria from populations adapted to *Dactylis glomerata* compared to populations adapted to *Cynodon* (Akin 1980a). The number of xylanolytic bacteria is significantly higher ($P \leq .05$) although less total hemicellulose and xylan are present initially in *D. glomerata* fibre (F. E. Barton, 1979, personal communication). These data suggest that cell wall structure, and, in particular, the hemicellulose component, can affect the digestibility of these two grasses. Indeed, cell wall organization with particular emphasis on the hemicelluloses has been implicated by others as a factor influencing cell wall degradation (Bailey & Jones 1971, Dehority 1973).

Attack on forage cell walls by rumen protozoa

Clarke (1979a) reported that the rumen protozoa are capable of using all the major plant constituents, including structural polysaccharides, but the majority of carbohydrates used are the soluble sugars and starch. The total contribution to cellulose digestion by protozoa appears small in comparison with the bacteria (Hungate 1975). Frequent references do occur in relation to the association of *Epidinium ecaudatum* with forage fibre. High numbers of *E. ecaudatum* have been shown to associate with damaged regions of *Medicago sativa* stems and *Trifolium repens* (Bauchop & Clarke 1976) and to ingest large amounts of cell walls of *M. sativa* stems (Bauchop 1979a). We found that a spined protozoan identified as *E. ecaudatum* form *caudatum* varies in its association with forage grasses of different digestibilities (Amos & Akin 1978). Table 2 presents data on the protozoal association with six grasses at three fermentation times using subsamples of the same inoculum. Virtually all of the protozoa fit the description for *E. ecaudatum* form *caudatum* (Hungate 1966). The protozoa preferentially associate with the more digestible, cool-season grasses. Follow-up studies using rumen fluid with and without bacteria having fibre-digesting activity (rumen fluid plus 1.6 mg/ml streptomycin) show that these protozoa could degrade *in vitro* about 11 percent of the *D. glomerata*, but only about 3.7 percent of the *C. dactylon* (Amos & Akin 1978).

TABLE 2.

Numbers of protozoa found within degraded zones of mesophyll tissue of warm- and cool-season grasses (Amos & Akin 1978).

Forage	Fermentation time (hour)		
	4	10	22
Cynodon dactylon cv. Coastal	0	0	0.4
Paspalum notatum cv. Pensacola	0	2.3	0
Digitaria decumbens cv. Pangola	0	2.8	0.3
Festuca arundinacea cv. Kentucky-31	18.7	10.5	4.0
Phleum pratense cv. Clair	15.7	41.8	3.3
Dactylis glomerata cv. Boone	TNC[a]	71.0	6.5

[a]Too numerous to count.

TEM of protozoal attack on *D. glomerata* cell walls show definitively that *E. ecaudatum* degrades cell walls via extracellular enzymes and ingests the partially degraded cell walls including the mesophyll, parenchyma bundle sheath, and epidermis (Akin & Amos 1979). Research had previously indicated that extracts from this protozoan have hemicellulase and xylanase activity, but the protozoan cannot degrade cellulose (Bailey *et al.* 1962). These data suggest that the availability of the hemicellulose components in the cool-season grasses is such that the extracellular enzymes of this protozoan can degrade the tissues and ingest the cell walls; however, in the warm-season grasses examined, the components are more rigidly held within the plant cell wall and apparently are unavailable to the protozoa. This agrees with anaerobic culture studies which show that a higher number of hemicellulolytic bacteria is elicited with *D. glomerata* than with *C. dactylon* (Akin 1980a). Delignification of the cell walls of *C. dactylon* with sodium hydroxide (Spencer & Akin 1980) and potassium permanganate (Barton & Akin 1977) results in increased association of *E. ecaudatum* with the treated leaf blades compared with untreated ones.

FORAGE CHARACTERISTICS AFFECTING MICROBIAL DEGRADATION OF FEED BY RUMEN MICROORGANISMS

Plant anatomy and ultrastructure

Tissues present in the leaf blades of all grasses include vascular tissue (divided into phloem and xylem cells), parenchyma bundle sheath(s) surrounding the vascular tissue, sclerenchyma patches connecting the vascular bundles to the epidermises, single-layered abaxial and adaxial epidermal cells covered by a protective cuticle, and mesophyll cells between the vascular bundles and epidermal layers (Metcalfe 1960). However, variation exists in the anatomy of grasses, including the number and kind of sheaths and the cellular arrangements of the mesophyll, and this variation has been used for grouping basic grass types (Brown 1958). Estimation of the tissue types, using a morphometric technique (Akin & Burdick 1975), has shown considerable variation in the amounts of tissue types among forage species

(Table 3). Although variations in the anatomy can occur with different blades from plants of a species (Wilson 1976, Akin *et al.* 1977), data in Table 3 show marked differences in tissue types among several species of forage grasses. In particular, the greater amount of vascular tissue due to the prominent parenchyma bundle sheaths and the lower mesophyll content are evident for warm-season grasses when compared to cool-season species.

Of interest to researchers of forage quality is the effect that variation in anatomy might have on digestibility. In general, the following trend exists for tissue digestion among the grasses (Hanna *et al.* 1973; Akin & Burdick 1975):

mesophyll > epidermis > sclerenchyma > lignified vascular tissue.
phloem parenchyma sheath

It is apparent from data in Table 3 that, if the specific tissues from different grasses are degraded at the same rate, certain grasses would have a faster rate of digestion because of the presence of more of the easily digested mesophyll; this is especially true for cool-season grasses. Anatomical information could provide insight into proposed modifications of forages for improved quality, especially in genera such as *Panicum* which have been shown to vary in plant structure (Brown & Brown 1975).

Although anatomical characteristics do influence the relative rate of digestion, other factors inherent in the cell wall structure also influence the ability of rumen bacteria to degrade forage fibre as has been referred to previously in this report. The parenchyma bundle sheath, a prominent structure in warm-season grasses, but less prominent in cool-season species, can be used to show the differences in digestibility of specific tissue types among forages (Table 4). In addition to indicating differential digestion of similar tissues among grass species, these data also show differential digestion of specific tissues within cultivars of the same species. Coastcross-1 is about 12 percent (6.6 percentage units) more digestible than *Cynodon dactylon* cv. Coastal (Lowrey *et al.* 1968), and *Festuca arundinacea* cv. Kenhy is about 6 percent more digestible than cv. Kentucky-31 (Buckner *et al.* 1967). The improved digestibilities of these cultivars within each species are reflected in greater digestibilities of their parenchyma bundle sheaths. Among grasses, differences in digestibility of specific tissues have also been demonstrated for mesophyll and epidermis (Akin & Burdick 1975).

Knowledge of specific tissue digestibility may be important in breeding programs for increased forage yield. For example, breeding for the more efficient C_4 pathway for carbon dioxide fixation could improve forage yield (Downton & Tregunna 1968). However, C_4 plants have well-developed parenchyma bundle sheaths that occupy a high cross-sectional area of the leaf (Table 3) and often are more slowly degraded than other tissues, especially in certain warm-season grasses (Table 4).

TABLE 3.
Percentage of tissue types in cross-sections of warm- and cool-season leaf blades from 4 to 8-week-old plants (Akin & Burdick 1975 and unpublished).

	Tissue Type							
	Total vascular tissue	Lignified vascular tissue	Parenchyma bundle sheaths	Phloem	Adaxial epidermis	Abaxial epidermis	Sclereu-chyma	Meso-phyll
Warm-Season								
Cynodon dactylon cv. Coastal	37	5	28	4	16	10	10	27
Cynodon cv. Coastcross-1	36	5	27	4	20	10	11	23
Paspalum notatum cv. Pensacola	18	5	11	2	17	8	5	52
P. dilatatum	14	2	10	2	25	10	7	44
Digitaria decumbens cv. Pangola	13	2	10	1	36	12	6	33
D. sanguinalis	13	3	9	1	32	17	3	35
Bothriochloa caucasica	23	7	13	3	29	13	4	31
Average	22	4	16	2	25	11	7	35
Cool-Season								
Bromus inermis	20	10	7	3	10	9	8	53
Dactylis glomerata cv. Boone	22	10	9	3	11	9	5	53
Phleum pratense cv. Clair	13	7	5	1	15	11	6	55
Poa pratensis	11	6	4	1	9	8	7	65
Festuca arundinacea cv. Kentucky-31	13	6	6	1	14	8	5	60
F. arundinacea cv. Kenhy	11	5	4	2	13	7	7	62
Lolium multiflorum	11	4	6	1	17	17	2	53
Average	14	7	6	1	13	10	6	57

[a]Amounts of tissue estimated by morphometric technique.

TABLE 4.

Percentage degradation of parenchyma bundle sheaths in grass leaf blades after 12 hours *in vitro* digestion (Akin & Burdick 1975).

Forage	Percent parenchyma bundle sheath		
	<25% degraded	25-75% degraded	>75% degraded
Cynodon dactylon cv. Coastal	56	35	9
Cynodon cv. Coastcross-1	25	71	4
Paspalum notatum cv. Pensacola	0	13	87
Festuca arundinacea cv. Kentucky-31	14	43	43
F. arundinacea cv. Kenhy	0	64	36
Dactylis glomerata cv. Boone	0	3	97

There are other specialized structures that limit the degradation of leaf blades by rumen microorganisms. The cuticle, which is a chemically complex, waxy layer providing protection to the plant (Esau 1965), prevents rumen microbial entry into the leaf blade (Monson *et al*. 1972), and generally is not degraded by rumen microorganisms (Hanna *et al*. 1973, Akin *et al*. 1974, Powell *et al*. 1974). However, studies of lines of *Pennisetum americanum* indicated that differences exist in the ability of the cuticle to crack under stress (Hanna & Akin 1978). Cracks in the cuticle could allow penetration by rumen bacteria into the leaves and result in improved digestibility (Hanna *et al*. 1974). Guard cells are often less available to microbial digestion than other, less specialized epidermal cells (Akin unpublished observations). Also tissues containing lignin are largely unavailable for microbial degradation (Baker & Harriss 1947, Drapala *et al*. 1947, Akin & Burdick 1975). Another structural factor probably related to the rate of digestion is that of the mesophyll arrangement in leaves (Hanna *et al*. 1973). Cool-season grasses usually have loose, irregularly arranged mesophyll cells with air spaces within the tissue, whereas the warm-season grasses have a more densely packed mesophyll with cells often radially arranged outside the vascular tissue (Brown 1958, Metcalfe 1960).

Whereas the preceeding discussion has emphasized leaf blades of grasses, studies of grass stems have also shown variation in tissue digestibility. For example, the percentage of vascular bundles in cross sections of stems of *Hemarthria altissima* has been related to *in vitro* organic matter digestibility (Schank *et al*. 1973). It is well known that, with increased maturity, stem digestibility declines more than that of leaf blades (Pritchard *et al*. 1963), and the tissues that limit digestibility in *Cynodon* stems have been identified (Hanna *et al*. 1976, Akin *et al*. 1977). In addition to the highly lignified and undigested vascular, sclerenchyma, and epidermal cells, portions of the parenchyma tissue become less digestible and this contributes to a 15 percentage unit decline in *in vitro* digestibility with increased maturity (Akin *et al*. 1977). At maturity not only the vascular structures but also the surrounding tissues become lignified and would be expected to contribute to reduced stem digestibility (Pigden 1953).

Chemical factors

Silica

Reports on the effect of silica (silicon dioxide) on forage digestibility are not consistent. Silica has been reported to decrease the digestibility of structural carbohydrates, possibly functioning physiologically in place of lignin (Van Soest 1968), and *in vitro* studies have indicated a decrease in digestibility of approximately one percent for each percentage increase in silica (Smith *et al*. 1971). In contrast, *in vivo* studies (Minson 1971) have shown that silica is not a factor in controlling the digestibility of structural carbohydrates in *Panicum* spp. Phytoliths or opaline silica bodies are present primarily in the epidermal cells (Parry & Smithson 1958) and silicon has been found in guard (Gallaher *et al*. 1973) and bulliform cells (Sangster & Parry 1969). Leaf surfaces containing phytoliths resist rumen microbial digestion (Brazle *et al*. 1979). However, research using TEM equipped with an energy-dispersive X-ray spectrometer indicated 1 to 2 percent silica in mesophyll cells of *Lolium multiflorum*, but no interference with digestion (Dinsdale *et al*. 1979).

Lignin

The association of lignin with cell walls is well known to limit their degradation by rumen microorganisms (Drapala *et al*. 1947, Van Soest 1973). Lignin is a polymer of phenylpropanoid units (Sarkanen & Ludwig 1971) and can vary in type based upon the predominate monomeric alcohol, i.e., coniferyl or sinapyl, used in forming the polymer (Vance *et al*. 1980). Histochemical techniques are available which identify specific lignin types in plant tissues. Acid phloroglucinol (Weisner test) indicates the presence of cinnamaldehyde groups whereas chlorine-sulphite and the Mäule test indicate syringyl groups within the lignin (Vance *et al*. 1980).

Lignified forage tissues vary in their response to the histochemical tests for lignin. Acid phloroglucinol-lignin is present primarily in the inner bundle sheath and xylem cell walls of leaf blades and the epidermis, sclerenchyma ring, and vascular tissue of stems; chlorine-sulphite-lignin is located in the sclerenchyma of blades and the more mature parenchyma cells in the stem (Stafford 1962, Akin & Burdick 1975, Akin *et al*. 1977). Acid phloroglucinol-positive tissues totally resist attack by rumen bacteria, but sclerenchyma tissues are less resistant to attack by rumen microorganisms as indicated by degradation of the sclerenchyma periphery by the rumen bacteria that predominate in fibre digestion (Akin *et al*. 1974). A filamentous bacterium isolated from rumen fluid degrades chlorine-sulphite-positive tissues more extensively than acid phloroglucinol-positive tissues (Akin 1980b). These observations indicate that the different types of lignified tissues have different abilities to resist microbial degradation and that lignified tissues in leaves having syringyl units (as indicated by chlorine sulphite) are less resistant than the lignified tissues having predominately cinnamaldehyde units.

Research from our laboratory (Akin & Burdick in press) has shown that some living tissue (i.e., those containing cytoplasm and organelles for metabolic activity) in leaf blades give a positive response for lignin with chlorine-sulphite, but not with acid phloroglucinol, and are only slowly or partially digested. These tissues include the parenchyma bundle sheath of *Cynodon* spp., and some of the mesophyll, parenchyma sheaths, and abaxial epidermal tissues of *F. arundinacea* samples. These observations indicate that, under certain conditions, lignin can be deposited in living cell walls, perhaps as a result of extreme crosslinking of the phenolic acid-complexes. Reaction only with chlorine-sulphite indicated that syringyl units were involved (Vance *et al*. 1980). Further, it is this type of lignin that occurs with advanced maturity in parenchyma cells of *Cynodon* stems and results in the inhibition of cell wall digestibility (Akin *et al*. 1977).

In addition to variation in response to microbial attack by the different histochemical types of lignified tissue, variation exists with treatment of intact lignified tissues by delignifying agents and by the detergent reagents of Van Soest (Van Soest 1963, Van Soest & Wine 1967). Extraction with the delignifying agent potassium permanganate results in a greater disruption of sclerenchyma than the lignified vascular tissue (Barton & Akin 1977). Further, whereas extraction with neutral detergent reagent causes little change in intact cell wall structure as observed by SEM, extraction with acid detergent reagent removes easily digested tissues and fragments the sclerenchyma tissues into individual cells, but causes less destruction of the lignified vascular tissue (Akin *et al*. 1975). Determination of the acid detergent fibre (ADF), which consists of lignin associated with cellulose and other structural polysaccharides (Colburn & Evans 1967, Bailey & Ulyatt 1970), has been suggested as the best chemical method for estimating *in vivo* dry matter digestibility (Rohweder *et al*. 1978). However, for tropical grasses, and especially *Cynodon* spp., the values for ADF are not closely related to digestibility (Moore 1977). Optimum conditions for determining ADF for each forage should be known to provide the best estimates of digestibility (McLeod & Minson 1972). Since the botanical make-up of ADF varies with grasses, it is possible that anatomical comparisons of ADF and the residue from digestion could provide insight into the value of ADF in interpreting and predicting forage quality.

Treatment of highly lignified, poor quality forages with alkali has mitigated some of the effects of lignin and resulted in substantial increases in digestibility. For example, alkali treatment of *C. dactylon* cv. Coastal increases the *in vitro* dry matter digestibility by 5 to 46 percent, depending upon the concentration of alkali and quality of the forage (Spencer & Amos 1977). A possible explanation offered for the improvement in digestibility of cell walls treated with alkali is the breaking of lignin-carbohydrate bonds, thus making the structural carbohydrates more readily available to the rumen microorganisms (Klopfenstein 1978). Electron microscopic comparisons of treated (10 percent KOH) with untreated *C. dactylon* leaves reveal that alkali causes distortion of lignified and unlignified cell walls and fragments tissues into individual cells (Spencer & Akin 1980). The tissues that are slowly degraded by adhering bacteria in untreated grasses are rapidly degraded and often, apparently, by cell-free enzymes from the bacteria as indicated by unattached but nearby bacteria. Further, alkali treatment effects a separation of xylem tissue into individual cells and results in the degradation of sclerenchyma tissue, especially the middle lamella between cells, by rumen bacteria. Certain protozoa also are able to attack the tissues in *C. dactylon* after alkali treatment (Spencer & Akin 1980). Therefore, it appears that alkali treatment increases forage digestibility by making the structural polysaccharides in both lignified and unlignified tissues more available to the rumen microorganisms. It is possible that alkali saponifies the ester bonds between lignin-carbohydrate or phenolic acid-carbohydrate complexes (Hartley & Jones 1977) in plant cell walls resulting in improved digestibility of lignified and unlignified cell walls.

Phenolic acids associated with forage cell walls

Phenolic compounds are important components of plant cell walls and are present in lignified and unlignified tissues (Harris & Hartley 1976, Hartley & Buchan 1979). Complexes consisting of phenolic acids (ρ-coumaric and ferulic) and carbohydrates (glucans and xylans) have been isolated intact from forage cell walls (Hartley 1972, Morrison 1974). Phenolic acid-carbohydrate complexes have been isolated by treatment with cellulase (Hartley *et al*. 1974) and have been found free in rumen fluid (Gaillard & Richards 1975). The binding, particularly of residues of xylose and arabinose to lignin or phenolic acids, has been reported to limit the rumen microbial degradation of these carbohydrates in forage cell walls (Gordon *et al*. 1977, Francis *et al*. 1978, Morrison 1979). Therefore, carbohydrates complexed to phenolic acids appear to be unavailable for microbial fermentation.

The addition of isolated lignin to microbial fermentation has been reported to have no influence on substrate digestion (Kamstra *et al*. 1958, Han *et al*. 1975). As a result of their data, it is concluded that lignin *per se* is not inhibitory to rumen microorganisms, but reduces fibre digestibility by complexing with cell wall components and rendering them unavailable for microbial fermentation as discussed previously. While the complexing does inhibit carbohydrate use, the possibility exists that the phenolic acids are also toxic to rumen bacteria. Studies in our laboratory (Akin 1980, unpublished data) have shown that ρ-coumaric acid inhibits growth of cellulolytic and xylanolytic bacteria, cellulose degradation, and motility of the entodiniomorph protozoa. Only slight and variable effects are found with feruluc acid and no toxic effect is found with sinapic acid. These data suggest that phenolic acids, expecially ρ-coumaric acid, if they become free of the carbohydrate complexes, can exert a toxic effect on fibre-digesting bacteria and thereby reduce forage digestibility. Research using ^{13}C nuclear magnetic resonance spectroscopy (Himmelsbach & Barton 1980) and gas chromatography (Snook 1980, personal communication) have shown that grasses differ in their amounts of ρ-coumaric and ferulic acids with the more digestible cool-season forages having comparatively less of the ρ-coumaric acid (complexed to carbohydrates or within the lignin) than warm-season species. To date, research has indicated that phenolic acids are found in cell walls primarily as complexes with carbohydrates (Hartley 1972, Snook 1980, personal communication). Further research is necessary to more fully assess the role of phenolic acids and their carbohydrate complexes for a potential role in limiting forage digestion by exerting a toxic effect on rumen microorganisms.

CONCLUSION

The digestion of forages results from a complex interaction of various microorganisms and forage cell wall structure. Although research has provided valuable information on forage quality and utilization, continued research is essential to make optimal use of specific forages in the forage-livestock interaction. Of particular importance is the understanding of inherent characteristics within the structurally intact cell walls of forages that limit the availability of nutrients for maximum production of energy and protein for the ruminant and ultimately for mankind.

REFERENCES

Akin, D. E. (1976) Ultrastructure of rumen bacterial attachment to forage cell walls. Applied and Environmental Microbiology 31, 562-568.

Akin, D. E. (1979) Microscopic evaluation of forage digestion by rumen microorganisms - a review. Journal of Animal Science 48, 701-710.

Akin, D. E. (1980a) Evaluation by electron microscopy and anaerobic culture of types of rumen bacteria associated with digestion of forage cell walls. Applied and Environmental Microbiology 39, 242-252.

Akin, D. E. (1980b) Attack on lignified grass cell walls by a facultatively anaerobic bacterium. Applied and Environmental Microbiology 40, 809-820.

Akin, D. E.; Amos, H. E. (1975) Rumen bacterial degradation of forage cell walls investigated by electron microscopy. Applied Microbiology 29, 692-701.

Akin, D. E.; Amos, H. E. (1979) Mode of attack on orchardgrass leaf blades by rumen protozoa. Applied and Environmental Microbiology 37, 332-338.

Akin, D. E.; Barton, F. E., II; Burdick, D. (1975) Scanning electron microscopy of Coastal bermuda and Kentucky-31 tall fescue extracted with neutral and acid detergents. Journal of Agricultural and Food Chemistry 23, 924-927.

Akin, D. E.; Burdick, D. (1975) Percentage of tissue types in tropical and temperate grass leaf blades and degradation of tissues by rumen microorganisms. Crop Science 15, 661-668.

Akin, D. E.; Burdick, D. (1977) Rumen microbial degradation of starch-containing bundle sheath cells in warm-season grasses. Crop Science 17, 529-533.

Akin, D. E.; Burdick, D.; Michaels, G. E. (1974) Rumen bacterial interrelationships with plant tissue during degradation revealed by transmission electron microscopy. Applied Microbiology 27, 1149-1156.

Akin, D. E.; Robinson, E. L.; Barton, F. E., II; Himmelsbach, D. S. (1977) Changes with maturity in anatomy, histochemistry, chemistry, and tissue digestibility of bermudagrass plant parts. Journal of Agricultural and Food Chemistry 25, 179-186.

Allison, M. J.; Robinson, I. M.; Dougherty, R. W.; Bucklin, J. A. (1975) Grain overload in cattle and sheep: Changes in microbial populations in the cecum and rumen. American Journal of Veterinary Research 36, 181-185.

Amos, H. E.; Akin, D. E. (1978) Rumen protozoal degradation of structurally intact forage tissues. Applied and Environmental Microbiology 36, 513-522.

Arias, C.; Burroughs, W.; Gerlaugh, P.; Bethke, R. M. (1951) The influence of different amounts and sources of energy upon in vitro urea utilization by rumen microorganisms. Journal of Animal Science 10, 683-692.

Bailey, R. W. (1973) Structural carbohydrates. In: Chemistry and biochemistry of herbage. Vol. 1. Editors G.W. Butler and R.W. Bailey. New York, Academic Press. pp. 157-211.

Bailey, R. W.; Jones, D. I. H. (1971) Pasture quality and ruminant nutrition. III. Hydrolysis of ryegrass structural carbohydrates with carbohydrases in relation to rumen digestion. New Zealand Journal of Agricultural Research 14, 847-857.

Bailey, R. W.; Ulyatt, M. J. (1970) Pasture quality and ruminant nutrition: II. Carbohydrate and lignin composition of detergent-extracted residues from pasture grasses and legumes. New Zealand Journal of Agricultural Research 13, 591-604.

Bailey, R. W.; Clarke, R. T. J.; Wright, D. E. (1962) Carbohydrases of the rumen ciliate Epidinium ecaudatum (Crawley). Biochemistry Journal 83, 517-523.

Baker, F.; Harriss, S. T. (1947) Microbial digestion in the rumen (and caecum), with special reference to the decomposition of structural cellulose. Nutrition Abstracts and Reviews 17, 3-12.

Barton, F. E., II; Akin, D. E. (1977) Digestibility of delignified forage cell walls. Journal of Agricultural and Food Chemistry 25, 1299-1303.

Bauchop, T. (1979a) The rumen ciliate Epidinium in primary degradation of plant tissues. Applied and Environmental Microbiology 37, 1217-1223.

Bauchop, T. (1979b) Rumen anaerobic fungi of cattle and sheep. Applied and Environmental Microbiology 38, 148-158.

Bauchop, T.; Clarke, R. T. J. (1976) Attachment of the ciliate Epidinium Crawley to plant fragments in the sheep rumen. Applied and Environmental Microbiology 32, 417-422.

Bauchop, T.; Clarke, R. T. J.; Newhook, J. C. (1975) Scanning electron microscope study of bacteria associated with the rumen epithelium of sheep. Applied and Environmental Microbiology 30, 668-675.

Brazle, F. K.; Harbers, L. H.; Owensby, C. E. (1979) Structural inhibitors of big and little bluestem digestion observed by scanning electron microscopy. Journal of Animal Science 48, 1457-1463.

Brown, R. H.; Brown, W. V. (1975) Photosynthetic characteristics of Panicum milioides, a species with reduced photorespiration. Crop Science 15, 681-685.

Brown, W. V. (1958) Leaf anatomy in grass systematics. Botanical Gazette 119, 170-178.

Bryant, M. P. (1973) Nutritional requirements of the predominant rumen cellulolytic bacteria. Federation Proceedings 32, 1809-1813.

Bryant, M. P. (1977) Microbiology of the rumen. In: Duke's physiology of domestic animals, 9th edn. Editor M.J. Swenson London, Cornell University Press, Ltd. pp. 287-304.
Bryant, M. P.; Burkey, L. A. (1953) Cultural methods and some characteristics of some of the more numerous groups of bacteria in the bovine rumen. Journal of Dairy Science 36, 205-217.
Buckner, R. C.; Todd, J. R.; Burrus, P. B., II; Barnes, R. F. (1967) Chemical composition, palatability, and digestibility of ryegrass-tall fescue hybrids, 'Kenwell', and 'Kentucky 31' tall fescue varieties. Agronomy Journal 59, 345-349.
Bull, L. S.; Vandersall, J. H. (1973) Sulfur source for in vitro cellulose digestion and in vivo ration utilization, nitrogen metabolism, and sulfur balance. Journal of Dairy Science 56, 106-112.
Burroughs, W.; Latona, A.; DePaul, P.; Gerlaugh, P.; Bethke, R. M. (1951) Mineral influences upon urea utilization and cellulose digestion by rumen microorganisms using the artificial rumen technique. Journal of Animal Science 10, 693-705.
Caldwell, D. R.; Bryant, M. P. (1966) Medium without rumen fluid for nonselective enumeration and isolation of rumen bacteria. Applied Microbiology 14, 794-801.
Cheng, K.-J.; Costerton, J. W. (1975) Ultrastructure of cell envelopes of bacteria of the bovine rumen. Applied Microbiology 29, 841-849.
Cheng, K.-J.; Dinsdale, D.; Stewart, C. S. (1979a) Maceration of clover and grass leaves by Lachnospira multiparus. Applied and Environmental Microbiology 38, 723-729.
Cheng, K.-J.; McCowan, R. P.; Costerton, J. W. (1979b) Adherent epithelial bacteria in ruminants and their roles in digestive tract function. The American Journal of Clinical Nutrition 32, 139-148.
Cipolloni, M. A.; Schneider, B. H.; Lucas, H. L.; Pavlech, H. M. (1951) Significance of the differences in digestibility of feeds by cattle and sheep. Journal of Animal Science 10, 337-343.
Clarke, R. T. J. (1979a) Protozoa in the rumen ecosystem. In: Microbial ecology of the gut. Editors R.T.J. Clarke and T. Bauchop. New York, Academic Press, Inc. pp. 251-275.
Clarke, R. T. J. (1979b) Niche in pasture-fed ruminants for the large rumen bacteria Oscillospira, Lampropedia, and Quin's and Eadie's ovals. Applied and Environmental Microbiology 37, 654-657.
Clarke, R. T. J.; Bailey, R. W.; Gaillard, B. D. E. (1969) Growth of rumen bacteria on plant cell wall polysaccharides. Journal of General Microbiology 56, 79-86.
Colburn, M. W.; Evans, J. L. (1967) Chemical composition of the cell wall constituent and acid detergent fibre fractions of forages. Journal of Dairy Science 50, 1130-1135.
Coleman, G. S.; Sandford, D. C. (1980) The uptake and metabolism of bacteria, amino acids, glucose and starch by the spined and spineless forms of the rumen ciliate Entodinium caudatum. Journal of General Microbiology 117, 411-418.
Costerton, J. W.; Damgaard, H. N.; Cheng, K.-J. (1974) Cell envelope morphology of rumen bacteria. Journal of Bacteriology 118, 1132-1143.
Dehority, B. A. (1968) Mechanism of isolated hemicellulose and xylan degradation by cellulolytic rumen bacteria. Applied Microbiology 16, 781-786.
Dehority, B. A. (1973) Hemicellulose degradation by rumen bacteria. Federation Proceedings 32, 1819-1825.
Dehority, B. A. (1978) Specificity of rumen ciliate protozoa in cattle and sheep. Journal of Protozoology 25, 509-513.
Dehority, B. A.; Scott, H. W. (1967) Extent of cellulose and hemicellulose digestion in various forages by pure cultures of rumen bacteria. Journal of Dairy Science 50, 1136-1141.
Dinsdale, D.; Gordon, A. H.; George, S. (1979) Silica in the mesophyll cell walls of Italian rye grass (Lolium multiflorum Lam. cv. RvP). Annals of Botany 44, 73-77.
Dinsdale, D.; Morris, E. J.; Bacon, J. S. D. (1978) Electron microscopy of the microbial populations present and their modes of attack on various cellulosic substrates undergoing digestion in the sheep rumen. Applied and Environmental Microbiology 36, 160-168.
Downton, W. J. S.; Tregunna, E. B. (1968) Carbon dioxide compensation - its relation to photosynthetic carboxylation reactions, systematics of the Gramineae, and leaf anatomy. Canadian Journal of Botany 46, 207-215.
Drapala, W. J.; Raymond, L. C.; Crampton, E. W. (1947) Pasture studies XXVII. The effects of maturity of the plant and its lignification and subsequent digestibility by animals as indicated by methods of plant histology. Science in Agriculture 27, 36-41.
Esau, K. (1965) Plant anatomy. New York, John Wiley and Sons, Inc. 767 pp.
Francis, G. L.; Gawthorne, J. M.; Storer, G. B. (1978) Factors affecting the activity of cellulases isolated from the rumen digesta of sheep. Applied and Environmental Microbiology 36, 643-649.
Gaillard, B. D. E.; Richards, G. N. (1975) Presence of soluble lignin-carbohydrate complexes in the bovine rumen. Carbohydrate Research 42, 135-145.
Gallaher, R. N.; Perkins, H. F.; Stormer, J.; Carlton, W. M. (1973) Electron probe microanalysis of silicon in grass leaves. Communications in Soil Science and Plant Analysis 4, 67-75.
Gill, J. W.; King, K. W. (1957) Characteristics of free rumen cellulases. Journal of Agricultural and Food Chemistry 5, 363-367.

Gordon, A. H.; Hay, A. J.; Dinsdale, D.; Bacon, J. S. D. (1977) Polysaccharides and associated components of mesophyll cell-walls prepared from grasses. Carbohydrate Research 57, 235-248.

Grubb, J. A.; Dehority, B. A. (1975) Effects of an abrupt change in ration from all roughage to high concentrate upon rumen microbial numbers in sheep. Applied Microbiology 30, 404-412.

Halliwell, G.; Bryant, M. P. (1963) The cellulolytic activity of pure strains of bacteria from the rumen of cattle. Journal of General Microbiology 32, 441-448.

Han, Y. W.; Lee, J. S.; Anderson, A. W. (1975) Chemical composition and digestibility of ryegrass straw. Journal of Agricultural and Food Chemistry 23, 928-931.

Hanna, W. W.; Akin, D. E. (1978) Microscopic observations on cuticle from trichomeless, tr, and normal, Tr, pearl millet. Crop Science 18, 904-905.

Hanna, W. W.; Monson, W. G.; Burton, G. W. (1973) Histological examination of fresh forage leaves after in vitro digestion. Crop Science 13, 98-102.

Hanna, W. W.; Monson, W. G.; Burton, G. W. (1974) Leaf surface effects on in vitro digestion and transpiration in isogenic lines of sorghum and pearl millet. Crop Science 14, 837-838.

Hanna, W. W.; Monson, W. G.; Burton, G. W. (1976) Histological and in vitro digestion study of 1-and 4-week stems and leaves from high and low quality bermudagrass genotypes. Agronomy Journal 68, 219-222.

Harris, P. J.; Hartley, R. D. (1976) Detection of the bound ferulic acid in cell walls of the Gramineae by ultraviolet fluorescence microscopy. Nature, London 259, 508-510.

Hartley, R. D. (1972) p-Coumaric and ferulic acid components of cell walls of ryegrass and their relationships with lignin and digestibility. Journal of the Science of Food and Agriculture 23, 1347-1354.

Hartley, R. D.; Buchan, H. (1979) High performance liquid chromatography of phenolic acids and aldehydes derived from plants or from the decomposition of organic matter in soil. Journal of Chromatography 180, 139-143.

Hartley, R. D.; Jones, E. C. (1977) Phenolic components and degradability of cell walls of grass and legume species. Phytochemistry 16, 1531-1534.

Hartley, R. D.; Jones, E. C.; Fenlon, J. S. (1974) Prediction of the digestibility of forages by treatment of their cell walls with cellulolytic enzymes. Journal of the Science of Food and Agriculture 25, 947-954.

Henning, P. A.; Linden, Y. Van Der; Mattheyse, M. E.; Nauhaus, W. K.; Schwartz, H. M. (1980) Factors affecting the intake and digestion of roughage by sheep fed maize straw supplemented with maize grain. Journal of Agricultural Science, Cambridge 94, 565-573.

Hespell, R. B. (1979) Efficiency of growth by ruminal bacteria. Federation Proceedings 38, 2707-2712.

Himmelsbach, D. S.; Barton, F. E., II (1980) ^{13}C Nuclear magnetic resonance of grass lignins. Journal of Agricultural and Food Chemistry 28, 1203-1208.

Hoover, W. H. (1978) Digestion and absorption in the hindgut of ruminants. Journal of Animal Science 46, 1789-1799.

Hubbert, F., Jr.; Cheng, E.; Burroughs, W. (1958) Mineral requirement of rumen microorganisms for cellulose digestion in vitro. Journal of Animal Science 17, 559-568.

Hungate, R. E. (1946) Studies on cellulose fermentation. II. An anaerobic cellulose-decomposing actinomycete, Micromonospora propionici, N. Sp. Journal of Bacteriology 51, 51-56.

Hungate, R. E. (1960) Symposium: Selected topics in microbial ecology. Bacteriological Reviews 24, 353-364.

Hungate, R. E. (1966) The rumen and its microbes. New York, Academic Press, Inc., 533 pp.

Hungate, R. E. (1975) The rumen microbial ecosystem. In: Annual review of ecology and systematics. Vol. 6. Editor R.F. Johnston. Palo Alto, Annual Reviews, Inc. pp. 39-66.

Hungate, R. E.; Phillips, G. D.; McGregor, A.; Hungate, D. P.; Buechner, H. K. (1959) Microbial fermentation in certain mammals. Science 130, 1192-1194.

Hunter, R. A.; Miller, C. P.; Siebert, B. D. (1978) The effect of supplementation or fertilizer application on the utilization by sheep of Stylosanthes guianensis grown on sulphur deficient soils. Australian Journal of Experimental Agriculture and Animal Husbandry 18, 391-395.

Kamstra, L. D.; Moxon, A. L.; Bentley, O. G. (1958) The effect of stage of maturity and lignification on the digestion of cellulose in forage plants by rumen microorganisms in vitro. Journal of Animal Science 17, 199-208.

Kern, D. L.; Slyter, L. L.; Leffel, E. C.; Weaver, J. M.; Oltjen, R. R. (1974) Ponies vs. steers: Microbial and chemical characteristics of intestinal ingesta. Journal of Animal Science 38, 559-564.

Kistner, A. (1965) Possible factors influencing the balance of different species of cellulolytic bacteria in the rumen. In: Physiology of digestion in the ruminant. Editor R.W. Dougherty. Washington D.C., Butterworth. pp. 419-432.

Klopfenstein, T. (1978) Chemical treatment of crop residues. Journal of Animal Science 46, 841-848.

Krishnamurti, C. R.; Kitts, W. D. (1969) Preparation and properties of cellulases from rumen microorganisms. Canadian Journal of Microbiology 15, 1373-1379.

Kudo, R. R. (1966) Protozoology, 5th edn. Springfield, Charles C. Thomas. 1174 pp.
Latham, M. J.; Brooker, B. E.; Pettipher, G. L.; Harris, P. J. (1978a) Ruminococcus flavefaciens cell coat and adhesion to cotton cellulose and to cell walls in leaves of perennial ryegrass (Lolium perenne). Applied and Environmental Microbiology 35, 156-165.
Latham, M. J.; Brooker, B. E.; Pettipher, G. L.; Harris, P. J. (1978b) Adhesion of Bacteroides succinogenes in pure culture and in the presence of Ruminococcus flavefaciens to cell walls in leaves of perennial ryegrass (Lolium perenne). Applied and Environmental Microbiology 35, 1166-1173.
Latham, M. J.; Sharpe, M. E.; Sutton, J. D. (1971) The microbial flora of the rumen of cows fed hay and high cereal rations and its relationship to the rumen fermentation. Journal of Applied Bacteriology 34, 425-434.
Latham, M. J.; Sutton, J. D.; Sharpe, M. E. (1974) Fermentation and microorganisms in the rumen and the content of fat in the milk of cows given low roughage rations. Journal of Dairy Science 57, 803-810.
Leatherwood, J. M. (1973) Cellulose degradation by Ruminococcus. Federation Proceedings 32, 1814-1818.
Lowrey, R. S.; Burton, G. W.; Johnson, J. C., Jr.; Marchant, W. H.; McCormick, W. C. (1968) In vivo studies with Coastcross 1 and other bermudas. Georgia Agricultural Experiment Stations Research Bulletin 55, 5-22.
Lund, A. (1974) Yeast and molds in the bovine rumen. Journal of General Microbiology 81, 453-462.
Maluszynska, G. M.; Janota-Bassalik, L. (1974) A cellulolytic rumen bacterium, Micromonospora ruminantium sp. nov. Journal of General Microbiology 82, 57-65.
Martinez, A.; Church, D. C. (1970) Effect of various mineral elements on in vitro rumen cellulose digestion. Journal of Animal Science 31, 982-990.
McLeod, M. N.; Minson, D. J. (1972) The effect of method of determination of acid-detergent fibre on its relationship with the digestibility of grasses. Journal of the British Grassland Society 27, 23-27.
Metcalfe, C. R. (1960) Anatomy of the monocotyledons. I. Gramineae. London, Oxford University Press. 731 pp.
Minson, D.J. (1971) Influence of lignin and silicon on a summative system for assessing the organic matter digestibility of Panicum. Australian Journal of Agricultural Research 22, 589-598.
Monson, W.G.,; Powell, J.B.; Burton, G.W. (1972) Digestion of fresh forage in rumen fluid. Agronomy Journal 64, 231-233.
Moore, J. E. (1977) Southern forages in the new hay standards. In: Proceedings of the 34th Southern Pasture and Forage Crop Improvment Conference, New Orleans; U.S. Department of Agriculture, SEA-AR, p. 129-135.
Morrison, I. M. (1974) Structural investigations on the lignin-carbohydrate complexes of Lolium perenne. Biochemistry Journal 139, 197-204.
Morrison, I. M. (1979) Changes in the cell wall components of laboratory silages and the effect of various additives on these changes. Journal of Agricultural Science, Cambridge 93, 581-586.
Orpin, C. G. (1975) Studies on the rumen flagellate Neocallimastix frontalis. Journal of General Microbiology 91, 249-262.
Orpin, C. G. (1977) The rumen flagellate Piromonas communis: Its life-history and invasion of plant material in the rumen. Journal of General Microbiology 99, 107-117.
Orpin, C. G.; Letcher, A. J. (1978) Some factors controlling the attachment of the rumen holotrich protozoa Isotricha intestinalis and I. prostoma to plant particles in vitro. Journal of General Microbiology 106, 33-40.
Panjarathinam, R.; Laxminarayana, H. (1974) Studies of rumen microflora in cows and buffaloes under different feeding regimes: Total and viable bacterial counts and protozoal counts. Indian Journal of Animal Science 44, 737-741.
Pant, H. C.; Roy, A. (1970) Studies on the rumen microbial activity of buffalo and zebu cattle. Indian Journal of Animal Science 40, 600-609.
Parry, D. W.; Smithson, F. (1958). Techniques for studying opaline silica in grass leaves. Annals of Botany 22, 543-550.
Patterson, H.; Irvin, R.; Costerton, J. W.; Cheng, K. J. (1975) Ultrastructure and adhesion properties of Ruminococcus albus. Journal of Bacteriology 122, 278-287.
Paynter, M. J. B.; Ewert, D. L.; Chalupa, W. (1969) Some morphological types of bacteriophages in bovine rumen contents. Applied Microbiology 18, 942-943.
Pettipher, G. L.; Latham, M. J. (1979) Production of enzymes degrading plant cell walls and fermentation of cellobiose by Ruminococcus flavefaciens in batch and continuous culture. Journal of General Microbiology 110, 29-38.
Phillipson, A. T. (1977) Ruminant digestion. In: Duke's physiology of domestic animals, 9th edn. Editor M.J. Swenson. London, Cornell University Press, Ltd. pp. 250-286.
Pigden, W. J. (1953) The relation of lignin, cellulose, protein, starch and ether extract to the "curing" of range grasses. Canadian Journal of Agricultural Science 33, 364-378.
Playne, M. J. (1978) Differences between cattle and sheep in their digestion and relative intake of a mature tropical grass hay. Animal Feed Science and Technology 3, 41-49.

Playne, M. J.; Echevarria, M. G.; Megarrity, R. G. (1978) Release of nitrogen, sulphur, phosphorus, calcium, magnesium, potassium, and sodium from four tropical hays during their digestion in nylon bags in the rumen. Journal of the Science of Food and Agriculture 29, 520-526.

Powell, J. B.; Monson, W. G.; Chatterton, N. J. (1974) Isolation of cuticle and vascular bundles of leaves by maceration in rumen fluid. Stain Technology 49, 29-33.

Powell, K.; Reid, R. L.; Balasko, J. A. (1978) Performance of lambs on perennial ryegrass, smooth bromegrass, orchardgrass, and tall fescue pastures. II. Mineral utilization, in vitro digestibility and chemical composition of herbage. Journal of Animal Science 46, 1503-1514.

Pritchard, G. I.; Folkins, L. P.; Pigden, W. J. (1963) The in vitro digestibility of whole grasses and their parts at progressive stages of maturity. Canadian Journal of Plant Science 43, 79-87.

Robinson, J. P.; Hungate, R. E. (1973) Acholeplasma bactoclasticum sp. n., an anaerobic mycoplasma from the bovine rumen. International Journal of Systematic Bacteriology 23, 171-181.

Rohweder, D. A.; Barnes, R. F.; Jorgensen, N. (1978) Proposed hay grading standards based on laboratory analyses for evaluating quality. Journal of Animal Science 47, 747-759.

Russell, J. B.; Baldwin, R. L. (1978) Substrate preferences in rumen bacteria: Evidence of catabolite regulatory mechanisms. Applied and Environmental Microbiology 36, 319-329.

Russell, J. B.; Sharp, W. M.; Baldwin, R. L. (1979) The effect of pH on maximum bacterial growth rate and its possible role as a determinant of bacterial competition in the rumen. Journal of Animal Science 48, 251-255.

Sangster, A. G.; Parry, D. W. (1969) Some factors in relation to bulliform cell silicification in the grass leaf. Annals of Botany 33, 315-323.

Sarkanen, K. V.; Ludwig, C. H. (1971) Definition and nomenclature. In: Lignins: Occurrence, formation, structure, and reactions. Editors K.V. Sarkanen and C.H. Ludwig. New York, Wiley-Interscience. pp. 1-18.

Satter, L. D.; Slyter, L. L. (1974) Effect of ammonia concentration on rumen microbial protein production in vitro. British Journal of Nutrition 32, 199-208.

Schank, S. C.; Klock, M. A.; Moore, J. E. (1973) Laboratory evaluation of quality in subtropical grasses: II. Genetic variation among Hemarthrias in in vitro digestion and stem morphology. Agronomy Journal 65, 256-258.

Slyter, L. L. (1976) Influence of acidosis on rumen function. Journal of Animal Science 43, 910-929.

Slyter, L. L.; Bryant, M. P.; Wolin, M. J. (1966) Effect of pH on population and fermentation in a continuously cultured rumen ecosystem. Applied Microbiology 14, 573-578.

Slyter, L. L.; Oltjen, R. R.; Kern, D. L.; Blank, F. C. (1970) Influence of type and level of grain and diethylstilbestrol on the rumen microbial populations of steers fed all-concentrate diets. Journal of Animal Science 996-1002.

Smith, G. S.; Nelson, A. B.; Boggino, E. J. A. (1971) Digestibility of forages in vitro as affected by content of "silica". Journal of Animal Science 33, 466-471.

Smith, W. R.; Yu, I.; Hungate, R. E. (1973) Factors affecting cellulolysis by Ruminococcus albus. Journal of Bacteriology 114, 729-737.

Spencer, R. R.; Akin, D. E. (1980) Rumen microbial degradation of potassium hydroxide-treated Coastal bermudagrass leaf blades examined by electron microscopy. Journal of Animal Science 51, 1189-1196.

Spencer, R. R.; Amos, H. E. (1977) In vitro digestibility of chemically treated Coastal bermudagrass. Journal of Animal Science 45, 126-131.

Stafford, H. A. (1962) Histochemical and biochemical differences between lignin-like materials in Phleum pratense L. Plant Physiology 37, 643-649.

Takechi, T.; Shibata, F.; Kurihata, Y. (1978) Microbiological and chemical characteristics of the rumen ingesta in sheep fed on purified diets. Japanese Journal of Ecology 28, 85-96.

Taliaferro, C. M.; Horn, F. P.; Tucker, B. B.; Totusek, R.; Morrison, R. D. (1975) Performance of three warm-season perennial grasses and a native range mixture as influenced by N and P fertilization. Agronomy Journal 67, 289-292.

Ulyatt, M. J.; Egan, A. R. (1979) Quantitative digestion of fresh herbage by sheep. V. The digestion of four herbages and prediction of sites of digestion. Journal of Agricultural Science, Cambridge 92, 605-616.

Vance, C. P.; Kirk, T. K.; Sherwood, R. T. (1980) Lignification as a mechanism of disease resistance. Annual Review of Phytopathology 18, 259-288.

Van Gylswyk, N. O. (1970) The effect of supplementing a low-protein hay on the cellulolytic bacteria in the rumen of sheep and on the digestibility of cellulose and hemicellulose. Journal of Agricultural Science, Cambridge 74, 169-180.

Van Soest, P. J. (1963) The use of detergents in the analysis of fibrous feeds. II. A rapid method for the determination of fibre and lignin. Journal of the Association of Official Analytical Chemists 46, 829-835.

Van Soest, P. J. (1967) Development of a comprehensive system of feed analyses and its application to forages. Journal of Animal Science 26, 119-128.

Van Soest, P. J. (1968) Structural and chemical characteristics which limit the nutritive value of forages. In: Forages: Economics/Quality, ASA Special Publication No. 13. Madison, Wisconsin, American Society of Agronomy, pp. 63-76.

Van Soest, P. J. (1973) The uniformity and nutritive availability of cellulose. Federation Proceedings 32, 1804-1808.

Van Soest, P. J.; Wine, R. H. (1967) Use of detergents in the analysis of fibrous feeds. IV. Determination of plant cell-wall constituents. Journal of the Association of Official Analytical Chemists 50, 50-55.

Ward, G.; Harbers, L. H.; Blaha, J. J. (1979) Calcium-containing crystals in alfalfa: Their fate in cattle. Journal of Dairy Science 62, 715-722.

Warner, R. L.; Mitchell, G. E., Jr.; Little, C. O. (1972) Post-ruminal digestion of cellulose in wethers and steers. Journal of Animal Science 34, 161-165.

Wheeler, J. L.; Hedges, D. A.; Archer, K. A.; Hamilton, B. A. (1980) Effect of nitrogen, sulphur, and phosphorus fertilizer on the production, mineral content and cyanide potential of forage sorghum. Australian Journal of Experimental Agriculture and Animal Husbandry 20, 330-338.

Wilson, J. R. (1976) Variation of leaf characteristics with level of insertion on a grass tiller. II. Anatomy. Australian Journal of Agricultural Research 27, 355-364.

Wilson, J. R.; Ford, C. W. (1973) Temperature influences on the in vitro digestibility and soluble carbohydrate accumulation of tropical and temperate grasses. Australian Journal of Agricultural Research 24, 187-198.

Wojciechowicz, M.; Ziolecki, A. (1979) Pectinolytic enzymes of large rumen treponemes. Applied and Environmental Microbiology 37, 136-142.

DIGESTION AND UTILIZATION OF ENERGY

D.G. ARMSTRONG

Department of Agricultural Biochemistry and Nutrition, The University, Newcastle upon Tyne, U.K. NE1 7RU.

ABSTRACT

Brief attention is given to forage composition with particular reference to differences between temperate and tropical plants. The importance of digestibility as the major determinant of the net energy value/unit dry matter of forages as well as being important in determining voluntary feed intake is emphasized. With reference to the need to consider protein/energy inter-relationships in discussing efficiency of energy utilization, sites of digestion of energy within the digestive tract are important in determining amino acid uptake/unit intake of metabolizable energy. Differences between forages in efficiencies of utilization of metabolizable energy for growth and fattening have been identified. At least part of the effect may be accounted for by differences in amounts of protein relative to fat deposited, with lower kf values reflecting increased deposition of energy as protein rather than as fat. The possible contribution to some of the differencies noted of (i) amino acid uptake relative to total metabolizable energy uptake, (ii) energy costs associated with absorption and metabolism in the gut wall, (iii) variation in proportions of the volatile fatty acids, and (iv) disturbance in carbohydrate metabolism within the body, are discussed.

It is concluded that further knowledge of digestion and of cellular metabolism within the body, linked to whole body energy metabolism studies, is required before it will be possible to account fully for differences in efficiency of utilization of metabolizable energy between different forages.

INTRODUCTION

A major determinant of the nutritive value of forages fed fresh or in conserved form is their capacity to supply useful (net) energy to the animal, firstly to meet maintenance needs, and in excess to meet the requirements for growth and fattening, pregnancy and lactation. The distinctive features of the ruminant animal's digestive tract and the processes that take place therein greatly complicate the relationship between feed energy intake and the amount of useful energy the animal can derive from it for a number of reasons.

Firstly, extensive fermentation of plant carbohydrates, the major energy-yielding constituents of forages, means that while enabling the ruminant animal to derive useful energy from cellulose and hemicellulose, the resulting end-products of digestion, e.g. the volatile fatty acids (VFA), lactic acid, carbon dioxide and methane are numerous and, with particular reference to the volatile fatty acids are varied in their proportions one to another; in addition, the metabolism of the three major volatile fatty acids differ significantly.

The proteins and other nitrogenous constituents of forages undergo very extensive transformation in the reticulo-rumen, thus making direct assessment of their contribution to amino acid, and hence energy supply to the host animal impracticable. Ingested lipids, the third category of energy-yielding nutrient, undergo lipolysis and biohydrogenation of unsaturated fatty acids within the reticulo-rumen but in considering the contribution of lipids to the energy status of the host animal, allowance must also be made for the synthesis of microbial lipids within the reticulo-rumen. It is the objective of this paper to examine, with reference to forages as sources of energy for ruminant livestock, the significance of these various events in the light of knowledge currently available.

MAJOR ENERGY-YIELDING CONSTITUENTS OF GRASSES AND LEGUMES.

The structural carbohydrates comprising cellulose and hemicelluloses are the major energy-yielding substrates present in forages and have been extensively reviewed by Bailey (1973). The non-structural carbohydrates comprise the sugars, glucose, fructose and sucrose and the polysaccharides starch and fructosan. Smith (1973) points out that fructosan is predominant in grasses and legumes of temperate origin and tends to accumulate in the stem; in tropical species starch predominates and accumulates in the leaves. Sugars occur only in low concentrations in forage dry matter of most forage species.

The lipid levels in leaves range from 3-10 percent of dry weight, the older the plant the lower the level of lipid present (Hawke 1973). The fatty acids present are generally rich in linolenic acid (60-75 percent by weight of total fatty acids) with linoleic and palmitic acids the next most abundant (each 6-20 percent of total fatty acids). In a detailed study of the lipids in perennial ryegrass (Lolium perenne) (Body & Hansen 1978), palmitic acid at 15.7 percent by weight of total fatty acid, was the major component of the 25.3 percent n-saturated fatty acids present with 56.5 percent of $C_{18.3}$ and 11.9 percent of $C_{18.2}$ comprising the major n-unsaturated fatty acids present (74.6 percent by weight of total fatty acid). There were some 7.2 percent of the total fatty acids in perennial ryegrass present as above-C_{18} comprising primarily n-saturated fatty acids ranging in chain length from C_{19} to C_{34}; as to be expected those with an even number of C atoms were predominant.

Of the nitrogenous constituents, proteins comprise 75-85 percent of the total nitrogen present (Lyttleton 1973). The 'soluble nitrogen' fraction prepared by extracting the plant tissue with 70-80% v/v ethanol is made up of free amino acids and then amides together with small amounts of other components including ureides, low molecular weight peptides and inorganic nitrogen constituents (Hegarty & Petersen 1973).

DIGESTIBILITY OF GROSS (CHEMICAL) ENERGY IN FORAGES

Two of the most important factors that govern the quality of a forage are level of voluntary intake and digestibility. The first mentioned, reviewed by Tyrrell & Moe (1975), is the subject of a separate paper at this symposium (Minson 1982) and little further reference will be made to the subject herewith, except to emphasize that as the digestibility of a forage rises voluntary intake generally increases.

It is well established that the major determinant of energy value/unit of forage dry matter is its digestibility (Armstrong 1964) and that as a forage matures so its digestibility declines. The relationships between chemical composition and physical structure on the one hand and digestibility on the other varies to such an extent between legumes and grasses, and within each group, between temperate and tropical grown forages that caution has to be exercised in attempting to predict digestibility of organic matter from chemical composition of the dry matter.

Not surprisingly, in view of the low lipid content of forages, there is a very close relationship between digestibility of organic matter and of energy. Minson & Milford (1966) reported that the mean energy value of the digestible organic matter in a considerable number of tropical grasses was 18.24 k J/g; Minson (1980) extended their range of grasses and included legumes in a further study and reported an overall mean value from 143 samples of tropical grasses and legumes to be 18.49 k J/g.

The conclusion of Minson (1980) that the mean energy value of the digestible organic matter in tropical forages is very similar to that found in temperate grasses is well supported by the mean value of 18.41±0.54 k J/g reported for roughages by Swift (1957), and a mean value of 18.57±0.13 k J/g found in experiments with sheep fed varying mixtures of hay and maize by Blaxter & Wainman (1964).

It is noteworthy that Terry *et al*. (1974) observed a relatively high degree of accuracy in predicting *in vivo* values for digestible energy (DE) content of forage dry matter by use of the two stage *in vitro* digestion method of Tilley & Terry (1963) and the addition to the relationship of a value for the crude protein (N x 6.25) content of the forage dry matter. Thus DE (MJ/kg DM) = 0.1233 x CP + 0.1705 D - 0.285 (±0.334) where DE = digestible energy content, CP = (N x 6.25)% DM, and D is the content of digestible organic matter on a dry matter basis determined *in vitro*. Some 94 percent of the variation in digestible energy was accounted for by this equation; the residual standard deviation of the multiple regression coefficient, expressed as a percentage of the mean digestible energy value, was ± 2.60 percent.

'D' values are generally related to *in vivo* values determined with sheep. Playne (1978a) has shown that dry matter digestibility coefficients for low-sulphur tropical grass hay may be as much as 15 percentage units higher for cattle than for sheep and considers that at least part of the difference is explainable in terms of the greater amounts of sulphur recycled in the saliva of cattle compared to sheep. In an examination of 17 sets of data relating to digestibility of dry matter, derived from 10 experiments reported in the literature in which both cattle and sheep were fed the same hay, some of temperate and some of tropical origin and all having a dry matter digestibility of less than 60 percent, Playne (1978b) again noted the superiority in terms of digestibility of cattle fed poor quality forages. The data fitted the linear equation $y = 0.67x + 20.3$, $r = 0.843$, where x and y represent the digestibilities of hay dry matter for sheep and cattle respectively.

The importance of sulphur level to energy utilization in sheep is well illustrated by the findings of Rees *et al*. (1980). These workers studied energy utilization of the grass, *Digitaria pentzii* by sheep when the crop was grown on sulphur-deficient land (0.11 percent S in feed dry matter) or on sulphur-fertilized land (0.17 percent S in feed dry matter); both feeds were similar in contents of nitrogen (1.2 percent) and a wide range of other components. For either of the two levels of dry matter fed, digestibility of energy was some 10-12 percentage units lower on the low sulphur crop and net availability of metabolizable energy for maintenance (k_m) was 24 percent lower; values for kf could not be determined since intakes were too low.

As already indicated, overall digestibility of energy is the major factor determining the net energy value/unit of forage dry matter and also exerts a very significant positive influence on voluntary dry matter intake. However, in any discussion of overall efficiency of energy utilization in feeds and possible differences between feeds it is necessary to consider energy-nitrogen interrelationships. One such relationship is the contribution of energy uptake as amino acids to total energy uptake from the gut. By virtue of the nature of the digestive processes in the ruminant animal, it is necessary to consider the sites of digestion of energy within the digestive tract in order to establish the contribution of amino acid uptake to total energy uptake.

SITES OF ENERGY DIGESTION

It is now understood that fermentation of the carbohydrate constituents of the feed in the reticulo-rumen is closely linked to the synthesis of microbial biomass therein. It is this biomass subsequently entering the abomasum and small intestine, which, under the influence of host-enzyme attack (as distinct from microbial enzyme attack in the reticulo-rumen) provides the ruminant with its major supply of amino acids. The source of nitrogen for the microbes comes in part from the protein of the feed and its associated non-protein nitrogenous constituents and in part from the protein and urea in saliva; under certain conditions (Armstrong 1980a) urea also enters the rumen across the wall in appreciable amounts. Feed proteins which escape fermentation in the rumen contribute to amino acid nitrogen uptake from the small intestine. Since the subject is one for a separate paper at this symposium (Hogan 1982), only those aspects that relate to overall energy utilization will be referred to here. That the influence of microbial fermentation in the rumen has very important consequences to the host animal is well illustrated by the studies of Egan (1974); in a study of 17 herbages fed to sheep he noted that a 12-fold range in dietary crude protein concentration was condensed to a 4-fold range of protein concentration in the organic matter digested by the animal.

Under normal circumstances the amount of microbial biomass yielded is proportional to the amount of energy-yielding substrates fermented in the rumen (McMeniman _et al_. 1976, Armstrong 1976, 1980b). It thus follows that if a portion of the energy-yielding substrates, such as cellulose and hemicellulose, escape fermentation within the rumen, then synthesis of microbial biomass will be curtailed and amino acid supply to the host animal from microbial protein reduced. It can be seen from the data shown in Table 1 that it is only when forages are dried artificially and fed ground and pelleted that this happens to any appreciable extent and then the carbohydrates escaping fermentation tend to be digested in the caecum and colon producing further quantities of volatile fatty acids. There is no evidence that the host animal can benefit, in terms of amino acid supply, from

TABLE 1

Sites of disappearance of digestible energy and proportions of digestible cellulose and digestible hemicellulose which disappear in the caecum and colon in sheep fed forages.

Forage	Proportion of digestible energy disappearing:			Propn. disappearing in caecum and colon of digestible:		Reference
	prior to sm.intestine	in small intestine	in caecum and colon	cellulose	hemicellulose	
Lolium perenne cv. S24 (fresh)	0.61	0.24	0.15	0.13	0.12	Ulyatt & MacRae 1974
Trifolium repens (fresh)	0.60	0.37	0.09	0.12	0.15	
L. perenne, spring cut (fresh, frozen)	0.60	0.29	0.12	0.10	0.27	Beever *et al*. 1978
L. perenne, autumn cut (fresh, frozen)	0.56	0.40	0.04	0.00	0.05	
T. pratense (fresh, frozen)	0.63	0.33	0.05	0.00	ND	Beever & Thomson1981
T. pratense (dried, wafered)	0.48	0.38	0.11	0.06	ND	
L. perenne (dried, chopped)	0.58	0.36	0.06	0.05	0.09	Beever *et al*. 1972
L. perenne (dried, ground and pelleted)	0.52	0.32	0.16	0.30	0.34	
Medicago sativa (dried, chopped)	0.39	0.34	0.28	0.18	0.31	Thomson *et al*. 1972
Medicago sativa (dried, ground and pelleted)	0.23	0.51	0.26	0.27	0.41	

microbial synthesis within the caecum (Elliot & Little 1977). However, as Beever, Osbourn, Cammell & Terry (personal communication, 1981) have clearly demonstrated in studies with dried forages fed either chopped or ground and pelleted, the reduced rumen fermentation occurring on pelleted feeds is associated with increased passage of feed protein into the small intestine in amounts that more than compensate for the reduced quantities of microbial protein entering therein (see Table 2). It can also be seen from Table 2 that volatile fatty acid production in the rumen was reduced by grinding and pelleting of the dried forages, although, in agreement with the conclusions of Osbourn et al. (1976), volatile fatty acid proportions were but little affected by the processing treatment. As these last-mentioned workers have indicated, when ground and pelleted legumes are fed the increased amounts of structural carbohydrates escaping fermentation in the rumen appear to undergo effective fermentation in the caecum and therefore overall digestibility is not affected; with dried grasses however digestion of residual structural carbohydrates in the caecum does not appear to compensate entirely for their reduced digestion in the rumen and thus depressions in overall digestibility can occur, particularly at high levels of intake.

TABLE 2.

Mean quantities of protein consumed and microbial and undegraded food protein flowing to the small intestine of sheep fed two dried forages each chopped and ground and pelleted, together with values relating to determined VFA production in the rumen (data of Beever, Osbourn, Cammell & Terry, personal communication).

	Lolium multiflorum		*Phleum pratense*	
	chopped	ground and pelleted	chopped	ground and pelleted
1. Amino acids (g/d)				
i) consumed	121.5	115.8	136.9	125.9
ii) entering small intestine:				
total	139.8	160.7	147.7	159.3
microbial	87.8	80.6	87.0	74.7
undegraded feed	36.5	62.5	44.0	67.1
2. Volatile fatty acids				
i) VFA production in rumen (m/d)	7.67	5.66	6.23	5.28
ii) C_2/C_3 ratio(molar basis)	3.08	2.97	3.87	3.63
iii) Total rumen VFA energy (MJ/d)	8.79	6.62	6.83	5.81
iv) Total VFA energy as proportion of energy disappearing in rumen	0.92	0.96	0.92	0.86

Clearly two factors must be taken into account when considering protein/energy interrelationships in digestion for the animal fed ground and pelleted forage in contrast to chopped forage. The reduced amount of microbial protein entering the small intestine is more than compensated for by an increased amount of undegraded protein entering therein; therefore total amino acid uptake is increased. Since some additional feed protein escapes fermentation in the rumen, then even if there were complete compensatory fermentation of residual carbohydrate in the caecum for the reduced amount of carbohydrate fermentation occurring in the rumen, the total amount of volatile fatty acid energy absorbed will be reduced; as already mentioned in many ground and pelleted forages digestion of structural carbohydrates in the caecum does not compensate fully for their reduced fermentation in the rumen. That energy from volatile fatty acid is reduced by pelleting has also been shown by Osbourn et al. (1976).

Another consideration is the extent to which soluble carbohydrates in forages may escape fermentation within the rumen and subsequently be digested within the small intestine under the influence of host-secreted carbohydrases possibly to be absorbed as hexoses. The subject has been reviewed by Armstrong & Smithard (1979) and is certainly an important aspect in ruminants fed diets high in maize or sorghum. With all-forage diets however the soluble carbohydrates are virtually completely fermented within the rumen and only very small amounts of α linked glucose polymer, probably of protozoal origin, enter the small intestine (Beever et al. 1972, Thomson et al. 1972, Ulyatt & MacRae 1974).

The flow of dietary long chain fatty acids into the small intestine is certainly augmented by lipids of microbial origin. In studies with sheep fed dried cocksfoot (Dactylis glomerata) either chopped or ground and pelleted (Outen et al. 1974) amounts entering the small intestine were some 75 and 97 percent greater than the amounts ingested. Knight et al. (1979), in experiments with sheep fed a hay/concentrate diet, noted that when they used an efficient method for extracting lipids from the feed, increases at the small intestine were only of the order of 25 percent - much lower than values previously obtained by them. Outen et al. (1974) estimated that long chain fatty acid uptake in the small intestine of sheep fed dried forage amounted to 23 and 17 percent of the total energy absorbed therefrom.

Table 3 shows, for a range of temperate forages, the amounts of amino acid absorbed from the small intestine in g/MJ of metabolizable energy absorbed. With one or two notable exceptions the values for grasses are lower than those for legumes; high values are to be seen for dried timothy (Phleum pratense) fed ground and pelleted (Beever et al. personal communication 1981) and for short rotation ryegrass fed by MacRae & Ulyatt (1974). These last-mentioned workers also reported a low value for white clover (Trifolium repens). Excluding the data for silages in Table 3, the mean value for the contribution of protein metabolizable energy to total metabolizable energy for the legumes, 18.9±1.03 percent, is significantly greater ($P < 0.01$) than the corresponding value for grasses, 15.4±0.41 percent.

With forage conserved as silage it is noticeable (see Table 3) that the use of formaldehyde in the ensiling process has had an appreciable effect in enhancing amino acid uptake from the small intestine, the result of reduced degradation of feed protein within

TABLE 3

Amino acid absorbed from small intestine g/MJ of metabolizable energy consumed by ruminants fed a range of temperate forages.

Forage	Digestibility of gross energy (%)	g Amino acid absorbed/ MJ of ME	Contribution of protein ME to total ME (%)	Reference
(a) Fed fresh:				
Lolium perenne	76.4	8.92	14.9	MacRae & Ulyatt 1974
L. perenne x *L. multiflorum*	77.7	10.96	18.3	
Trifolium repens	78.1	8.45	14.2	
L. perenne	-	8.81	14.7	Beever *et al.* 1980
L. perenne	-	8.43	14.1	
T. repens	-	12.45	20.8	
T. repens	-	13.81	23.1	
(b) Fed fresh, frozen:				
L. perenne, spring cut	70.2	9.00	15.1	Beever *et al.* 1978
L. perenne, autumn cut	66.2	10.43	17.5	
T. pratense	68.4	11.06	18.5	Beever & Thomson 1981
(c) Dried forages:				
early cut dried grass,				
chopped	80.6	8.83	14.8	Coelho da Silva *et al.* 1972b
ground and pelleted	74.4	7.95	13.3	
medium cut dried grass,				
chopped	70.8	8.44	14.1	Coelho da Silva *et al.* 1972b
ground and pelleted	71.2	8.99	15.0	
Medicago sativa, chopped	56.5	10.84	18.1	Coelho da Silva *et al.* 1972a
Medicago sativa, ground & pelleted	57.8	11.46	19.2	
L. multiflorum,				
chopped	76.6	8.17	13.7	Beever personal communication
ground and pelleted	71.6	9.65	16.1	
Phleum pratense, chopped	72.5	9.55	16.0	
Phleum pratense, ground & pelleted	66.9	10.83	18.1	
T. pratense				Beever & Thomson 1981
ground and pelleted	66.0	11.17	18.7	
(d) Silages:				
Lolium (control)	73.2	7.72	12.9	Thomson personal communication
Lolium (HCHO treated)	69.6	10.73	18.0	
Lolium (HCHO treated + urea)	70.3	10.68	17.9	

the rumen (Thomson, Beever, Lonsdale, Haines, Cammell & Austin 1981, personal communication).

UTILIZATION OF METABOLIZABLE ENERGY

There are numerous data in the literature indicating significant differences in the abilities of forages to produce animal products (Joyce & Newth 1967, Ulyatt 1971, Joyce & Brunswick 1975, Gibb & Treacher 1976, Thomson 1979). In many instances the effect is due in part to differences in voluntary feed intake as well as to differences in the value of unit dry matter. Ulyatt (1973) uses the term herbage feeding value to define the overall effect in terms of animal production response and the term nutritive value to define response/unit of intake. In the present context attention will be focussed on nutritive value although this in no way implies that the characteristics of a forage that determine voluntary intake are not of equal or possibly greater significance.

In terms of nutritive value the decline, with maturity of forage, in efficiency of utilization of metabolizable energy of forages for growth and fattening (kf) is well documented (Armstrong 1964, ARC 1980). A general equation for first growth forages that describes this relationship is $kf = 1.32\ q_m - 0.318\ (\pm 0.087)$ (ARC 1980), where q_m is the ratio of ME to GE/unit feed dry matter forage; thus for a q_m value of 0.40, kf = 0.211 and for $q_m = 0.60$, kf becomes 0.474.

It should be noted that most of the values used in obtaining the above equation relate to studies involving relatively mature sheep. Graham (1980) in a study of energy retention in young and mature sheep ranging in age from 2 months of age to 6 years, and fed a mixed forage/concentrate ration, observed that kf values increased from 0.32±0.02 at 2 months of age to 0.55±0.02 at 10 months of age and showed little change thereafter; the contribution to energy balance of protein at 2 months was some 50 percent, this value falling to 13 percent at 10 months. The increase in kf value noted by Graham would be in keeping with the lower efficiency for protein synthesis (0.44) compared to fat synthesis (0.74) reported by Pullar & Webster (1977). Support for this view is to be seen in the work of Ayala (1974) on Aberdeen Angus cattle referred to by Blaxter (1980a). kf values for feed were negatively correlated with the proportion of protein energy in the total energy gain. Thus kf values and proportions of protein energy in total energy gain were:

	kf value		proportion of protein energy in total energy gain	
	Abderdeen Angus	Holstein	Abderdeen Angus	Holstein
Bulls	0.414	0.379	0.27	0.45
Steers	0.483	0.407	0.20	0.40
Heifers	0.653	0.450	0.13	0.28

Clearly this aspect is an important one and warrants further study. Graham (1980) observed little change with age in efficiency of utilization of metabolizable energy for maintenance, k_m values ranged from 73-80 percent.

The higher kf values for spring *versus* autumn cut grass of similar dry matter digestibility has been shown by Corbett *et al.* (1966) and Blaxter *et al.* (1971) although no such effect was reported for lucerne (Joyce & Brunswick 1975). Rattray & Joyce (1974), Gibb & Treacher (1976), Thomson (1979) and Tyrell *et al.* (1981) have shown that legumes or mixtures of legumes and grasses have higher kf values than do grasses of equal or even higher digestibility. With reference to dried forages Osbourn *et al.* (1976) and Thomson & Cammell (1979) have observed that, although k_m values are little affected, kf values are significantly increased when the forage is fed in the ground and pelleted form rather than in the chopped form. This is reflected in the predictive equation for kf values for pelleted forages given by ARC (1980):

$$kf = 0.024q_m + 0.465 \ (\pm 0.030)$$

Hence at $q_m = 0.40$, kf = 0.475 and at $q_m = 0.60$, kf = 0.479. At low q_m values the kf values are appreciably higher than the corresponding ones for chopped forage referred to earlier. Table 4 shows a selection of kf values relating to the various comparisons referred to above. Although most of the evidence has been obtained using temperate crops there would seem little reason not to believe that similar differences exist in tropical species.

TABLE 4

Differences in utilization of metabolizable energy surplus to maintenance (kf) for liveweight gain in sheep.

Forage	ME (MJ/kg DM)	kf	NE (MJ/kg DM)	Reference
(a) *Spring vs. autumn cut grass*:				
early cut	9.79	0.435	4.26	Corbett
late cut	9.67	0.325	3.14	*et al.* 1966
early cut	11.88	0.452	5.37	Blaxter
late cut	10.79	0.328	3.54	*et al.* 1971
(b) *Legumes vs. grasses*:				
Lolium perenne	12.18	0.33	4.02	Rattray &
Trifolium repens	11.54	0.51	5.86	Joyce 1974
(c) *Chopped vs. ground & pelleted*:				
Medicago sativa, dried				
chopped	8.70	0.28	2.44	Thomson &
ground & pelleted	8.80	0.53	4.66	Cammell 1979

At the present time it does not seem possible, to this author at least, to account for these important differences in kf values. However, recent research has thrown light on a number of factors, some or all of which may contribute at least in part to some of them and these will now be considered.

PROPORTION OF METABOLIZABLE ENERGY INTAKE THAT IS ABSORBED AS PROTEIN

Reference has already been made to data in the literature, some of which is shown in Table 3, which indicate that in general, the digestion of legumes results in increased quantities of amino acid absorbed from the small intestine/unit of metabolizable energy intake compared with grasses. The superiority of white clover over perennial ryegrass in this capacity is well illustrated by some recent excellent studies with 5-month-old calves fed on fresh perennial ryegrass or white clover indoors (Beever *et al.* 1980), or allowed to graze these pastures (Ulyatt *et al.* 1980). Some of the results of both studies are given in Table 5 (also see Table 3). Noticeable features to be seen in Table 5 are the much smaller proportion of totally digested organic matter that apparently is digested in the rumen on the white clover and the associated very significantly increased flow of non ammonia N (NAN) into the small intestine which occurs on the clover diet. No data are yet available to allow partition of the NAN flows into the small intestine between microbial nitrogen and protein of feed origin which has passed through the rumen intact. It would seem likely that the last mentioned is considerably increased on the white clover diet. Despite changes in organic matter digestion methane production was constant for all diets (mean 2.23 mole/kg DOMI) except for the first growth of perennial ryegrass which was significantly lower at 1.87 mole/kg DOMI. The ratio of ME/DE varied but slightly over the range 0.78-0.81.

To what extent the increased uptake of amino acid/unit intake of metabolizable energy, associated with a shift in site of organic matter digestion, contributes to the enhanced voluntary intake and net energy value of clover compared to grass is difficult to ascertain.

In an attempt to assess the possible significance of these differences in relation to nutritive value data relating to protein and energy, requirements taken from ARC (1980) have been used. Protein requirements are expressed in terms of rumen degradable protein (RDP) and undegraded protein (UDP) and to convert these into equivalent daily amounts of absorbed amino acid (g), the rumen degradable protein requirements have been multiplied by 0.56 (0.80 x 0.70) while the undegraded protein requirements have been multiplied by 0.7. The summation of the two gives the requirement in terms of daily requirement for absorbed amino acids. Using the energy requirements which are listed in terms of MJ of metabolizable energy per day, one can then calculate the weight of amino acid to be absorbed/MJ of metabolizable energy. A selection of values so calculated is given in Table 6, from which it can be seen that, with the possible exception of the 100 kg calf gaining 0.75-1.0 kg/d, the quantities of absorbed amino acids required/MJ metabolizable energy intake can be met from the values for grasses shown in Table 3.

On the basis of the criteria above it would appear that differences in nutritive value between clover (legumes) and grasses do not lie in their capacity to supply the amounts of amino acids required for tissue protein synthesis *per se*. However, the majority of amino acids can act as gluconeogenic substrates and it is possible that in this role the increased amounts of amino acids taken up from the small intestine play an important part in enhancing the efficiency of utilization of metabolizable energy above maintenance. This aspect is dealt with more fully later in this paper.

TABLE 5
Data relating to non ammonia nitrogen flow into the small intestine of 5-month-old calves fed fresh *Lolium perenne* or *Trifolium repens* (data of Beever *et al*. 1980 and Ulyatt *et al*. 1980).

Forage	N content of feed OM (g/kg)	OM digestibility (%)	OMADR[1] / OMAD	NAN[2] flow (g) /kg DOMI indoors	grazing
L. perenne					
primary growth	37.3			32.3	33.7
trimmed primary growth	26.3	82	0.71	30.1	32.7
regrowth	24.4			32.1	37.8
T. repens					
primary growth	42.1	74	0.56	43.7	48.2
regrowth	48.7	81	0.48	44.8	48.8

1 - OMADR = OM apparently digested in the rumen;
OMAD = total OM apparently digested.
2 - NAN = non-ammonia N;
DOMI = digestible organic matter intake.

TABLE 6
Requirements in terms of amino acid absorbed (g) per MJ of metabolizable energy intake based on values for protein and energy requirements given in ARC (1980).

Animal	ME/GE	Liveweight (kg)	Level of production	
			Liveweight gain(kg/d)	
			0.2	0.3
Castrate lambs	0.6	20	6.42	-
		30	4.96	-
		40	4.32	-
	0.7	20	7.10	-
		30	5.41	5.52
		40	4.73	4.54
			Liveweight gain(kg/d)	
			0.75	1.0
Young cattle	0.6	100	8.66	9.75
		200	5.82	6.36
	0.7	100	9.55	10.43
		200	6.36	6.90
			Milk yield (kg/d)	
			30	40
Friesian cow yielding milk containing 36.8 g fat/kg	0.6	600	6.50	6.75
	0.7	600	6.89	7.10

It must be emphasized that the conclusion that amino acid supply in amounts for tissue protein synthesis is not likely to be an important factor must be treated with very considerable caution. No account has been taken of individual amino acid supply, the accuracy of the rumen degradable protein and undegradable protein values have yet to be established (they include for example a factor for efficiency of utilization of amino acid nitrogen within the body of 0.75) and furthermore no allowance is made in the values given in Table 3 relating to amino acid uptake/MJ of metabolizable energy for that portion which is of endogenous origin arising in the gastric secretions.

With reference to the well recognized and established superiority of spring vs. autumn cut herbage, Ribeiro et al. (1981) studied various aspects of digestion in sheep fed spring and autumn harvested dried grasses at approximately 1.5 x maintenance. These workers observed that volatile fatty acid production rates were very similar for the two grasses but in the sheep fed spring cut grass there was a 25 percent increase in total nitrogen entering the small intestine/unit intake of metabolizable energy and that this was associated with a 17 percent increase in flow of free amino acid nitrogen in the portal veins of these sheep. The high urinary nitrogen output on the autumn cut grass was considered to reflect the extensive degradation of crude protein to ammonia in the rumen. Ribeiro and his colleagues concluded that the increased flow of organic matter into the small intestine with the spring grass could be accounted for as increased microbial nitrogen flow therein, and suggested that the higher content of water soluble carbohydrate in the spring grass improved efficiency of microbial nitrogen capture. They suggested that the additional amino acid nitrogen made available to sheep on the spring grass may have improved utilization of energy-yielding substrate by providing extra gluconeogenic precursors required to support fat synthesis.

In contrast to the findings of Ribeiro et al., Beever et al. (1978) fed fresh (frozen) spring and autumn cut grasses to sheep and observed that on the autumn grass, containing a higher content of nitrogen and lower content of water-soluble carbohydrate, the actual amount of amino acid absorbed from the small intestine was similar in both grasses. However, they observed a considerably more efficient fermentation within the rumen on the spring cut associated with a higher yield of VFA/mole substrate fermented and total volatile fatty acid productions of 5.14 (spring) and 3.90 (autumn) moles/24 h; the C_2/C_3 ratio widened from 2.56:1 (spring) to 3.15:1 (autumn). As a result of the increased yield of volatile fatty acids on the spring cut grass, the autumn cut material had a higher proportion of metabolizable energy as absorbed amino acid (see Table 3). These workers concluded that a contributory factor to the superiority of spring cut grass lies in the increased yield of energy available for metabolism within the animal and furthermore that increased proteolysis in the rumen of autumn grass fed animals may be a contributory factor to its reduced yield of volatile fatty acids.

With reference to higher kf values for ground and pelleted over chopped lucerne, Thomson & Cammell (1979) attributed the cause to change in site of digestion related to increased amounts of amino acids absorbed from the small intestine/unit metabolizable energy intake together with an increased apparent absorption of methionine on the pelleted diet. They emphasized that the differences in energy costs of eating and ruminating between the two physical forms was small. Evidence relating to energy costs associated with forage digestion will now be considered.

THE ENERGY COST OF DIGESTION AND ABSORPTION

Studies at the Rowett Research Institute with sheep fed forage diets have provided quantitative data on the energy costs of eating and ruminating (Osuji *et al.* 1975) and those associated with the fermentation process and associated metabolism in the wall of the digestive tract (Webster *et al.* 1975). As Webster (1980) points out the contribution to the heat increment of food (HIF), i.e. ME-NE, of the energy costs of ingestion of feed and its rumination is small, varying from about 3 kJ/MJ of metabolizable energy for pelleted feeds which are eaten rapidly and with virtually no subsequent rumination to about 40 kJ/MJ of metabolizable energy for fresh herbage on which an animal might spend up to 9 h/d eating and ruminating. However, the energy cost associated with absorption and metabolism in the tissues of the gut wall makes an appreciable contribution to HIF. In sheep fed either chopped hay or fresh herbage, then above maintenance, energy costs associated with absorption and metabolism in the digestive tract are each estimated at 26 percent of HIF (Webster 1980); on forage diets some 40-45 percent of the total HIF can be accounted for in the energy costs associated with ingestion of feed, its digestion and the subsequent metabolic activities in the wall of the tract associated with the absorption of end-products.

One point brought out very clearly by Webster (1980) is that the contribution of the energy costs occurring in the gut wall both in absolute terms and as a proportion of the total HIF increases markedly above maintenance and is clearly a contributory factor to the decline in net availability of metabolizable energy that occurs as metabolizable energy intake rises from below maintenance (k_m) to above maintenance (kf). Thus with the fresh herbage referred to above, the relationship between heat increment of the food, below and above maintenance, and energy costs related to absorption and metabolism in the digestive tract were:

Heat increment of food kJ/MJ of ME	Energy costs kJ/MJ of ME
352 below maintenance	32 (9%)
612 above maintenance	156 (24.8%)

The tissues of the alimentary tract are very active metabolically and it has been suggested (Webster *et al.* 1975) that the very appreciable heat loss associated with gut tissue metabolism may well reflect an increased rate of cell division and cell growth for rumen mucosa associated with the ingestion of forage feeds. Fell & Weekes (1975) showed that a 3-fold increase in feed intake doubled the mass of rumen epithelium within about 7 days and also induced increases in weight of abomasum and small intestine.

TABLE 7

The contribution of various factors to overall heat increment of feeding in sheep (see MacRae & Lobley 1980).

Forage	Total HIF	Energy cost of:		
	(kJ/MJ of ME)	Eating	Fermentation	Digestive tract metabolism
			(kJ/MJ of ME)	
Dried grass:				
chopped	634	22	92	196
ground and pelleted	557	4	92	162
Dried lucerne:				
chopped	689	22	60	243
ground and pelleted	557	4	60	243

MacRae & Lobley (1980) further analyzed the data of Webster *et al.* (1975) with particular reference to the processing of dried forages (see Table 7). They showed that whereas with dried grass and dried lucerne, grinding and pelleting lowered the heat increment of the feed overall, the contribution of digestive tract metabolism, although slightly lower on the grass diet, remained the same on the pelleted lucerne. They concluded that differences in gut metabolism could not account for the higher kf values associated with grinding and pelleting of dried forages.

INFLUENCE OF THE VOLATILE FATTY ACIDS

Data relating to the significance or otherwise of the VFA to the variation in kJ values have been summarized by Blaxter (1980b) and ARC (1980). Of particular interest are the proportions of C_2 and C_4 to C_3 acids arising from fermentation in the rumen. The elegant studies of Ørskov *et al.* (1979) with sheep sustained entirely by liquid nutrients infused into various regions of their digestive tract appear to indicate that, within the range of volatile fatty acid proportions found in the rumens of animals ingesting natural feeds, variation in proportions of individual volatile fatty acids has little influence on the utilization of metabolizable energy.

In considering these various studies one or two points are worthy of attention. Firstly, in the studies by Armstrong *et al.* (see ARC 1980 for references) dried forage was used as the basal ration whereas in many of those yielding contradictory evidence (see Ørskov *et al.* 1979 for references) animals were fed appreciable amounts of concentrates. Recently Tyrell *et al.* (1979) observed that the type of basal diet fed markedly affects the partial efficiency of utilization of acetate for body tissue synthesis in mature cattle. Dry, non-pregnant cattle were fed either an all forage diet or one containing a major proportion as concentrates; on the former efficiency of utilization of the metabolizable energy of acetic acid infused into the rumen was 0.27, on the latter it was 0.69. Maize comprised over half the concentrate diet and may well have contributed starch into the small intestine, thus potentially increasing the supply of glucose or gluconeogenic substrate uptake from the region of the gut (Armstrong & Smithard 1979).

Secondly, it has always been stressed (Armstrong 1965, Annison & Armstrong 1970) that, for efficient utilization of acetate and butyrate, the body tissues need an adequate supply of gluconeogenic substrates to supply the necessary amounts of NADPH and glycerol for fat synthesis and possibly to suppress endogenous acetate production. In the sheep glucose formation from gluconeogenic amino acids is less than in other species (Lindsay 1979, 1980). Nevertheless, non-essential amino acids surplus to requirements for tissue protein synthesis may have an important role to play in sparing glucose, rather than contributing directly to its synthesis (Lindsay 1979). In reviewing the evidence relating to the heat increment of feed, MacRae & Lobley (1980) have emphasized the possible significance of gluconeogenic precursors other than propionate in markedly affecting the efficiency of utilization of acetate.

Finally, as pointed out by Webster (1980), the sheep sustained entirely by liquid infusions in the experiments of Ørskov et al. (1979) had abnormally undeveloped gut tissues. The structural development of the rumen can be divided into that of mucosal tissue and muscular tissue. The studies of Harrison et al. (1960) showed that muscular development of the rumen is largely governed by the bulky nature of the diet offered. Sander et al. (1959) showed that mucosal papillary development is primarily stimulated by the end products of fermentation, particularly the volatile fatty acids . Webster (1980) refers to the observations of Sakata & Tamate (1979) that the mean mitotic index following administration of C_2, C_3 and C_4 acids into the rumen were 8.5; 2.2 and 6.5 respectively, although it must be noted that in these last mentioned studies the observations were made following single intraruminal injections of the acids, and therefore the mitotic indices may relate to non-steady state conditions. However, it may well be that in the presence of solid digesta within the rumen, acetate and butyrate may have a stimulatory effect on the energy costs associated with overall metabolism of the gut wall of which the rumen forms a considerable part.

It is clear from the foregoing that much yet remains to be learnt about the significance of metabolism of the volatile fatty acids and the role they play in contributing to the heat increment of forage feeds, and in particular whether they have any part to play in the decline in kf value that occurs as a forage matures and thus becomes more fibrous.

EXCESS AMMONIA PRODUCTION IN THE RUMEN AND CARBOHYDRATE METABOLISM

A decrease in plasma glucose concentration in sheep fed twice/day when the crude protein of the feed was raised from 10.7 to 22.1 percent by the addition of urea (Leonard et al. 1977), or when lambs were given a urea-supplemented ration rather than a soyabean-supplemented one (Prior et al. 1972) has been suggested as indicating an impairment in gluconeogenesis. No such effect was noted when the urea was fed little and often (Prior et al. 1976). In the studies of Leonard et al. (1977) glucose space and metabolic clearance rate were not significantly reduced. However, in a glucose-kinetics study with cattle given a single, subtoxic dose of urea into the rumen, Spires & Clark (1979) observed a decrease in rate of glucose utilization. Prior et al. (1970) observed that in lambs fed a protein-free, purified diet containing a high level of urea (4.2 percent), in contrast to animals fed an isonitrogenous amount of soyabean protein, liver NAD was elevated while NADH and NADPH concentrations were decreased thereby greatly enhancing the ratios of the oxidized to reduced co-enzymes. Liver NADP-isocitrate dehydrogenase activities were 57 percent lower in the urea-fed animals.

Although the above effects were associated with urea feeding, the *in vitro* studies of Weekes et al. (1979) with isolated sheep liver cells suggest that detoxification of ammonia in the liver does reduce glucose formation at least from propionic acid (but not from alanine - Weekes, personal communication); in these studies the competition for energy (ATP) between glucose formation and urea formation could only account for about one-third of the inhibitory effect of ammonia on gluconeogenesis from propionate. Thus an additional factor that might have a role to play in affecting overall efficiency of utilization, over and above that associated with the energy costs of urea synthesis and its excretion may arise if the nitrogen containing substances in the forage consumed lead to a very rapid rise in rumen ammonia levels. This aspect may be of little consequence, but if it has any significance it is likely to be greater with silages made without additives in which there is rapid degradation of crude protein to ammonia in the rumen.

CONCLUDING REMARKS

Although a major factor in the lowered values of metabolizable energy for growth and fattening as compared to values for maintenance is the change in magnitude of the energy costs associated with absorption and metabolism in the gut wall and particularly the rumen, it is not yet possible to explain the cause of the decline in metabolizable energy values for growth with increasing maturity of the forage or the differences in the values between legumes and grasses, between spring and autumn cut forage and between chopped and ground and pelleted forages. Part of the effect may be related to differences in amounts of protein relative to fat deposited, with lower kf values reflecting increased deposition of energy as protein rather than as fat. It would also seem probable from biochemical considerations that the ratio of gluconeogenic to non-gluconeogenic substrates produced in digestion will be shown to be of significance in at least contributing to these differences. In this connection, Archer (1980) has emphasized that the most consistent factor emerging from studies demonstrating the superiority of clovers over grasses and of spring over autumn grass is the apparent relationship between level of water-soluble carbohydrates and animal performance through alteration in volatile fatty acid proportions in favour of propionic acid.

However, it seems certain that in the near future the increasing integration of biochemical studies in the whole animal and at the cellular level, carried out in conjunction with studies of overall energy metabolism will do much to clarify the situation.

ACKNOWLEDGEMENTS

The author is indebted to Drs. Beever and Thomson and their colleagues at the Grassland Institute for being allowed access to material being prepared for publication.

REFERENCES

ARC (1980) The nutrient requirements of ruminant livestock. Technical Review by an Agricultural Research Council Working Party. Farnham Royal, U.K., Commonwealth Agricultural Bureaux. 351 pp.

Annison, E.F.; Armstrong, D.G. (1970) Volatile fatty acid metabolism and energy supply. In: Physiology of digestion and metabolism of the ruminant. Editor A.T. Phillipson. Newcastle upon Tyne, Oriel Press, pp. 422-437.

Archer, K. (1980) Low productivity of lambs on improved pasture. In: Recent advances in animal nutrition. Editor D.J. Farrell. Armidale N.S.W.,. University of New England Publishing Unit, pp. 20-27.

Armstrong, D.G. (1964) Evaluation of artificually dried grass as a source of energy for sheep. II: The energy value of cocksfoot, timothy and two strains of ryegrass at varying stages of maturity. Journal of Agricultural Science, Cambridge 62, 399-416.

Armstrong, D.G. (1965) Carbohydrate metabolism in ruminants and energy supply. In: Physiology of digestion in the ruminant. Editor R.W. Dougherty. Washington, Butterworths, pp. 272-288.

Armstrong, D.G. (1976) Protein digestion and absorption in simple-stomached and ruminant animals. Ubers. Tiernahrung 4, 1-24.

Armstrong, D.G. (1980a) Some aspects of nitrogen digestion in the ruminant and their practical significance. North of Scotland College of Agriculture, Aberdeen. The 2nd Tom Miller Memorial Lecture. pp.1-16.

Armstrong, D.G. (1980b) Net efficiencies (in vivo) of microbial N synthesis in ruminant livestock. In: Proceedings of the 3rd EAAP symposium on protein metabolism and nutrition. Volume 2. Editors H.J. Oslage and K. Rohr. EAAP Publication No. 27, Braunschweig, pp. 400-413.

Armstrong, D.G.; Smithard, R.S. (1979) The fate of carbohydrates in the small and large intestines of the ruminant. Proceedings of the Nutrition Society 38, 283-293.

Ayala, H. (1974) Ph.D. Thesis, Cornell University. Ann Arbor, University Microfilms.

Bailey, R.W. (1973) Structural carbohydrates. In: Chemistry and biochemistry of herbage, Volume 1. Editors G.W. Butler and R.W. Bailey. London, Academic Press, pp.157-211.

Beever, D.E.; Coelha da Silva, J.F.; Prescott, J.H.D.; Armstrong, D.G. (1972) The effect in sheep of physical form and stage of growth on the sites of digestion of a dried grass. 1. Sites of digestion of organic matter, energy and carbohydrate. British Journal of Nutrition 28, 347-356.

Beever, D.E.; Terry, R.A.; Cammell, S.B.; Wallace, A.S. (1978) The digestion of spring and autumn harvested perennial ryegrass by sheep. Journal of Agricultural Science, Cambridge 90, 463-470.

Beever, D.E.; Thomson, D.J. (1981) The effect of drying and processing red clover on the digestion of the energy and nitrogen moieties in the alimentary tract of sheep. Grass and Forage Science (in press).

Beever, D.E.; Ulyatt, M.J.; Thomson, D.J.; Cammell, S.B.; Austin, A.R.; Spooner, M.C. (1980) Nutrient supply from fresh grass and clover fed to housed cattle. Proceedings of the Nutrition Society 39, 66A.

Blaxter, K.L. (1980a) The efficiency of energy utilization by beef cattle. Annales Zootechnie 29, 145-155.

Blaxter, K.L. (1980b) Feeds as sources of energy for ruminant animals. Massey-Ferguson Papers, Massey-Ferguson (UK) Ltd., pp. 7-60.

Blaxter, K.L.; Wainman, F.W. (1964) The utilization of the energy of different rations by sheep and cattle for maintenance and for fattening. Journal of Agricultural Science, Cambridge 63, 113-128.

Blaxter, K.L.; Wainman, F.W.; Dewey, P.J.S.; Davidson, J.; Dennerley, H.; Gunn, J.B. (1971) The effects of nitrogenous fertiliser on the nutritive value of artificially dried grass. Journal of Agricultural Science, Cambridge 76, 307-319.

Body, D.R.; Hansen, R.P. (1978) The occurrence of C_{13} to C_{31} branched-chain fatty acids in the faeces of sheep fed ryegrass and of C_{12} to C_{34} normal acids in both faeces and the ryegrass. Journal of the Science and Food and Agriculture 29, 107-114.

Coelho da Silva, J.F.; Seeley, R.G.; Beever, D.E.; Prescott, J.H.D.; Armstrong, D.G. (1972a) The effect in sheep of physical form and stage of growth on the sites of digestion of a dried grass. 2. Sites of nitrogen digestion. British Journal of Nutrition 28, 357-371.

Coelho da Silva, J.F.; Seeley, R.C.; Thomson, D.J.; Beever, D.E.; Armstrong, D.G. (1972b) The effect in sheep of physical form on the sites of digestion of a dried lucerne diet. 2. Sites of nitrogen digestion. British Journal of Nutrition 28, 43-61.

Corbett, J.L.; Langlands, J.P.; Mcdonald, J.; Pullar, J.D. (1966) Composition by direct animal calorimetry of the net energy values of an early and a late season growth of herbage. Animal Production 8, 13-27.

Egan, A.R. (1974) Protein-energy relationships in the digestion products of sheep fed on herbage diets differing in digestibility and nitrogen content. Australian Journal of Agricultural Research 25, 613-630.

Elliott, R.; Little, D.A. (1977) Fate of cyst(e)ine synthesised by microbial activity in the ruminant caecum. Australian Journal of Biological Science 30, 203-206.

Engels, E.A.N.; Malan, A. (1975) The energy intake and excretion of sheep grazing oat pasture. Agroanimalia 7, 81-86.

Fell, B.F.; Weekes, T.E.C. (1975) Food intake as a mediator of adaptation in the rumen epithelium. In: Digestion and metabolism in the ruminant. Editors I.W. MacDonald and A.I.D. Warner. Armidale, Australia, University of New England Publishing Unit, pp. 101-118.

Gibb, M.J.; Treacher, T.T. (1976) The effect of herbage allowance on herbage intake and performance of lambs grazing perennial ryegrass and red clover swards. Journal of Agricultural Science, Cambridge 86, 355-365.

Graham, N.McC. (1980) Variation in energy and nitrogen utilization by sheep between weaning and maturity. Australian Journal of Agricultural Research 31, 335-345.

Harrison, N.H.; Warner, R.G.; Sander, E.G.; Loosli, J.K. (1960) Changes in the tissue and volume of the stomachs of calves following the removal of dry feed or consumption of inert bulk. Journal of Dairy Science 43, 1301-1312.

Hawke, J.C. (1973) Lipids. In: Chemistry and biochemistry of herbage, Volume 1. Editors G.W. Butler and R.W. Bailey. London, Academic Press. pp.213-264.

Hegarty, M.P.; Petersen, P.J. (1973) Free amino acids, bound amino acids, amides and ureides. In: Chemistry and biochemistry of herbage, Volume 1. Editors G.W. Butler and R.W. Bailey. London, Academic Press, pp.2-62.

Hogan, J.P. (1982) Digestion and utilization of protein. In: Nutritional limits to animal production from pastures. Editor J.B. Hacker. Farnham Royal, U.K. Commonwealth Agricultural Bureaux, pp. 245-257.

Joyce, J.P.; Brunswick, L.C.F. (1975) Effects of stage of growth, season and conservation method on the nutritive value of lucerne. Proceedings of the New Zealand Society of Animal Production 35, 152-158.

Joyce, J.P.; Newth, R.P. (1967) Use of the comparative slaughter technique to estimate the nutritive value of pasture for hoggets. Proceedings of the New Zealand Society of Animal Production 27, 166-180.

Knight, R.; Sutton, J.D.; Storry, J.E.; Brumby, P.E. (1979) Rumen microbial synthesis of long chain fatty acids. Proceedings of the Nutrition Society 38, 4A.

Leonard, M.C.; Buttery, P.J.; Lewis, D. (1977) The effects on glucose metabolism of feeding a high-urea diet to sheep. British Journal of Nutrition 38, 455-462.

Lindsay, D.B. (1979) Is gluconeogenesis from amino acids important in ruminants. In: Protein metabolism in the ruminant. Editor P.J. Buttery. London, Agricultural Research Council, pp.7.1-7.9.

Lindsay, D.B. (1980) Amino acids as energy sources. Proceedings of the Nutrition Society 39, 53-59.

Lyttleton, J.W. (1973) Proteins and nucleic acids. In: Chemistry and biochemistry of herbage, Volume 1. Editors G.W. Butler and R.W. Bailey. London, Academic Press, pp. 63-105.

MacRae, J.C.; Lobley, G.E. (1980) Factors influencing energy losses during metabolism-thermal losses in ruminants. In EAAP Publication No. 31. Munich (in press).

MacRae, J.C.; Ulyatt, M.J. (1974) Quantitative digestion of fresh herbage by sheep II. The sites of digestion of some nitrogenous constituents. Journal of Agricultural Science, Cambridge 82, 309-319.

McMeniman, N.P.; Ben-Ghedalia, D.; Armstrong, D.G. (1976) Nitrogen-energy interactions in rumen fermentation. In: Protein metabolism and nutrition. Editors D.J.A. Cole, K.N. Boorman, P.J. Buttery, D. Lewis, R. Neale and H. Swan. EAAP Publication No. 16, London, Butterworths, pp.217-229.

Minson, D.J. (1980) Relationships of conventional and preferred fractions to determine energy values. In: Standardization of analytical methodology for feeds. Editors W.J. Pigden, C.C. Balch and M. Graham. Ottawa, International Development Research Centre, pp.72-78.

Minson, D.J. (1982) Effects of physical and chemical composition of herbage eaten upon intake. In: Nutritional limits to animal production from pastures. Farnham Royal, U.K., Commonwealth Agricultural Bureaux, pp. 167-182.

Minson, D.J.; Milford, R. (1966) The energy values and nutritive value indices of Digitaria decumbens, Sorghum almum and Phaseolus atropurpureus. Australian Journal of Agricultural Research 17, 411-423.

Ørskov, E.R.; Grubb, D.A.; Smith, J.S.; Webster, A.J.F.; Corrigal, W. (1979) Efficiency of utilisation of volatile fatty acids for maintenance and energy retention in sheep. British Journal of Nutrition 41, 541-551.

Osbourn, D.E.; Beever, D.E.; Thomson, D.J. (1976) The influence of physical processing on the intake, digestion and utilization of dried herbage. Proceedings of the Nutrition Society 35, 191-200.

Osuji, P.O.; Gordon, J.G.; Webster, A.J.F. (1975) Energy exchanges associated with eating and rumination in sheep given grass diets of different physical forms. British Journal of Nutrition 34, 59-71.

Outen, G.E.; Beever, D.E.; Osbourn, D.F. (1974) Digestion and absorption of lipids by sheep fed chopped and ground dried grass. Journal of the Science of Food and Agriculture 25, 981-987.

Playne, M.J. (1978a) Differences between cattle and sheep in their digestion and relative intake of a mature tropical grass hay. Animal Feed Science and Technology 3, 41-49.

Playne, M.J. (1978b) Estimation of the digestibility of low quality hays by cattle from measurements made with sheep. Animal Feed Science and Technology 3, 51-55.

Prior, R.L. (1976) Effects of dietary soy or urea nitrogen and feeding frequency on nitrogen metabolism, glucose metabolism and urinary metabolite excretion in sheep. Journal of Animal Science 42, 160-167.

Prior, R.L.; Clifford, A.J.; Hogue, D.E.; Visek, W.J. (1970) Enzymes and metabolites of intermediary metabolism in urea-fed sheep. Journal of Nutrition 100, 438-444.

Prior, R.L.; Milner, J.A.; Visek, W.J. (1972) Carbohydrates and amino acid metabolism in lambs fed purified diets containing urea or isolated soy protein. Journal of Nutrition 102, 1223-1231.

Pullar, J.D.; Webster, A.J.F. (1977) The energy cost of fat and protein deposition in the rat. British Journal of Nutrition 37, 355-363.

Rattray, P.V.; Joyce, J.P. (1974) Nutritive value of white clover and perennial ryegrass. 4. Utilisation of dietary energy. New Zealand Journal of Agricultural Research 17, 401-408.

Rees, M.C.; Graham, N.McC.; Searle, T.W. (1980) Net energy value of the grass Digitaria pentzii grown with and without sulphur fertilizer. Proceedings of the Australian Society of Animal Production 13, 466.

Ribeiro, J.M. de C.R.; MacRae, J.C.; Webster, A.J.F. (1981) An attempt to explain differences in the nutritive value of spring and autumn harvested dried grass. Proceedings of the Nutrition Society 40, 12A.

Sakata, T.; Tamate, H. (1979) Rumen epithelium cell proliferation accelerated by propionate and acetate. Journal of Dairy Science 62, 49-52.

Sander, E.G.; Warner, R.G.; Harrison, N.H.; Loosli, J.K. (1959) The stimulatory effect of sodium butyrate and sodium propionate on the development of rumen mucosa in the young calf. Journal of Dairy Science 42, 1600-1605.

Smith, D. (1973) The non structural carbohydrates. In: Chemistry and biochemistry of herbage, Volume 1. Editors G.W. Butler and R.W. Bailey. London, Academic Press, pp. 106-156.

Spires, H.R.; Clark, J.H. (1979) Effect of intraruminal urea administration on glucose metabolism in dairy steers. Journal of Nutrition 109, 1438-1447.

Swift, R.W. (1957) The calorific value of TDN. Journal of Animal Science 16, 753-756.

Terry, R.A.; Osbourn, D.F.; Cammell, S.B.; Fenlon, J.S. (1974) In vitro digestibility and the estimation of energy in herbage. Växtodling 28, 19-28.

Thomson, D.J. (1979) Effect of the proportion of legumes in the sward on annual output. British Grassland Society, Occasional Symposium No. 10, pp.101-109.

Thomson, D.J.; Cammell, S.B. (1979) The utilization of chopped and pelleted lucerne (Medicago sativa) by growing lambs. British Journal of Nutrition 41, 297-310.

Thomson, D.J.; Beever, D.E.; Coelho da Silva, J.F.; Armstrong, D.G. (1972) The effect in sheep of physical form on the sites of digestion of a lucerne diet 1. Sites of organic matter, energy and carbohydrate digestion. British Journal of Nutrition 28, 31-41.

Tilley, J.M.A.; Terry, R.A. (1963) A two-stage technique for the in vitro digestion of forages. Journal of the British Grassland Society 18, 104-111.

Tyrell, H.F.; Moe, P.W. (1975) Effect of intake on digestive efficiency. Journal of Dairy Science 58, 1151-1163.

Tyrell, H.F.; Reynolds, P.J.; Moe, P.W. (1979) Effect of diet on partial efficiency of acetate use for body tissue synthesis by mature cattle. Journal of Animal Science 48, 598-605.

Tyrell, H.F.; Thomson, D.J.; Waldo, D.R.; Goering, H.K.; Haarland, G.L. (1981) Utilization of energy and nitrogen by yearling Holstein cattle fed direct cut alfalfa or orchard grass ensiled with formic acid plus formaldehyde. Journal of Animal Science, (in press).

Ulyatt, M.J. (1971) Studies on the causes of differences in pasture quality between perennial ryegrass, short-rotation ryegrass and white clover. New Zealand Journal of Agricultural Research 14, 352-467.

Ulyatt, M.J. (1973) The feeding value of herbage. In: Chemistry and biochemistry of herbage, Volume 3. Editors G.W. Butler and R.W. Bailey. London, Academic Press, pp.131-178.

Ulyatt, M.J.; MacRae, J.C. (1974) Quantitative digestion of fresh herbage by sheep. 1. The sites of digestion of organic matter, energy, readily fermentable carbohydrate, structural carbohydrate and lipid. Journal of Agricultural Science, Cambridge. 82, 295-307.

Ulyatt, M.J.; Beever, D.E.; Thomson, D.J.; Evans, R.T.; Haines, M.J. (1980) Measurement of nutrient supply at pasture. Proceedings of the Nutrition Society 39, 67A.

Webster, A.J.F. (1980) Energy costs of digestion and metabolism in the gut. In: digestive physiology and metabolism in ruminants. Editors Y. Ruckebusch and P. Thivend. Lancaster, UK, MTP Press Ltd., pp.469-484.

Webster, A.J.F.; Osuji, P.O.; White, F.; Ingram, J.F. (1975) The influence of food intake on portal blood flow and heat production in the digestive tract of the sheep. British Journal of Nutrition 34, 125-139.

Weekes, T.E.C.; Richardson, R.L.; Geddes, N. (1979) The effect of ammonia on gluconeogenesis by isolated sheep liver cells. Proceedings of the Nutrition Society 38, 3A.

DIGESTION AND UTILIZATION OF PROTEINS

J.P. HOGAN

CSIRO, Division of Animal Production, Blacktown, NSW 2148, Australia.

ABSTRACT

Protein metabolism depends particularly on the supply of essential amino acids from the small intestine. Supply involves protein intake, the relative rates of protein breakdown and microbial protein synthesis in the rumen and the extent of release of amino acids in the small intestine. Protein metabolism in both the rumen and the tissues depends on interactions between energy and amino acids.

Both protein and energy intakes are highest when ruminants are fed immature forages and decline with advancing forage maturity. In the rumen the fate of a particular molecule in a forage depends on three factors - whether the molecule is fermented or not, whether metabolites from fermentation are absorbed or resynthesized into cells and whether those cells are destroyed and refermented or pass on to the small intestine. The mechanisms controlling these factors are discussed.

Quantitative estimates are presented of the amounts of protein that pass from the stomach to the intestines with forages of varying maturity and of the fate of that protein in the small intestine. Estimates are also made of the extent of amino acid wastage caused by inefficiencies of functioning of the rumen and small intestine.

Amino acid utilization depends on the following factors - whether the amino acid is required for maintenance or is available for production; what forms of protein storage receive priority for nutrients; how the mixture of absorbed amino acids differs from that in products; whether amino acids or energy first limit production. These factors are discussed and an estimate is made of the loss of productivity caused by amino acid wastage in the digestive tract.

INTRODUCTION

The protein value of a diet is an expression of the capacity of the diet to provide adequate ammonia and "essential" and "non-essential" amino acids to support protein synthesis at the rate permitted by the energy supply. The interaction between energy and protein occurs both during microbial fermentation in the rumen and large intestine and during amino acid metabolism in the tissues. As reviewed by Smith (1975) and Armstrong & Hutton (1975) dietary crude protein degraded by microbial enzymes in the rumen to ammonia and amino acids is partly reconverted to microbial proteins. Energy for this process is derived as high energy phosphates (ATP) released during the fermentation to volatile fatty acids of carbohydrates and the glycerol portion of fat. Amino acids subsequently released are absorbed from the small intestine and incorporated into proteins in the liver and peripheral tissues; the energy for the process is again supplied as high energy phosphates produced during catabolism of metabolites derived from carbohydrates, fats and proteins, i.e. volatile fatty acids, long chain fatty acids, glucose, ketones and other amino acids.

An understanding of the utilization of protein by the ruminant thus requires parallel consideration of energy transactions. In this review protein will be considered as crude protein (C.P.) i.e. N x 6.25, while dietary energy will be expressed as digestible organic matter (DOM). With forages, digestible organic matter has a relatively fixed energy content and provides an indication of the energy that becomes available both to the rumen microbes and to the tissues of the animal. In this review, data will be presented from a number of collaborative studies (e.g. Weston & Hogan 1971). In that work, twenty-three temperate grasses and clovers, harvested at varying stages of maturity and dried with a minimum application of heat, were fed at about 90 percent of *ad libitum* levels to 44 kg Merino wethers and quantitative studies of digestion were made.

FATE OF SUBSTRATE IN THE RUMEN

Many events occur simultaneously in the rumen and the fate of a particular component of a forage depends on three major factors (Fig. 1) - whether or not the molecule is fermented, whether metabolites from fermentation are absorbed or resynthesized into cells and whether those cells are destroyed and fermented or pass on to the small intestine. Each factor has many determinants. Fermentation depends on the amount and chemical and physical form of energy and protein substrates in relation to enzyme supply and time available for the reactions. Synthesis of microbial cells, and thus of all enzymes in the rumen, depends particularly on the relative availability of energy and other nutrients, while the fate of the microbial cell depends on the level of predation by other microbes and on time spent before the microbes pass from the rumen.

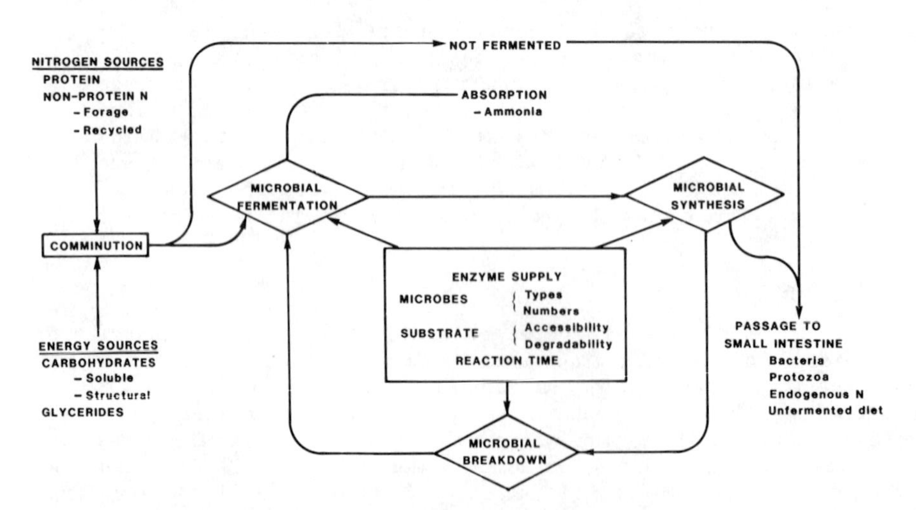

Fig. 1. Factors affecting protein metabolism in the rumen.

Variations in substrate ingested

Protein metabolism in the rumen is greatly influenced by the amount, chemical and physical form and distribution of nitrogen and energy substrates. Immature plants contain high levels of crude protein of which more than 80 percent is true protein, found mainly in the cytoplasm and especially the chloroplasts. Although half the leaf protein may be present in insoluble form associated with lipids in plant membranes (Lyttleton 1973), much occurs in solution. Most of the non-protein nitrogen present in the leaf is also soluble. This fraction increases under adverse growing conditions such as cold dull weather, particularly following the application of nitrogenous fertilizer; with such plants an associated fall in the levels of soluble carbohydrate probably has a substantial effect on nitrogen metabolism in the rumen. With advancing plant maturity, changes occur in the distrubition of protein and solubility of nitrogen within the cell (Lyttleton 1973) but the metabolic consequences have not been directly studied. During ensiling, plant proteins are extensively broken down to amino acids and only about one-third of the nitrogen in silage is present as protein (Siddons et al. 1979, Thomas et al. 1980). As indicated later, part of the forage protein also escapes attack by the anaerobic microbes in the rumen.

Further nitrogen is added to the rumen in sloughed rumen epithelial cells, salivary protein and urea recycled from the blood in saliva and across the rumen wall. Epithelial cells probably contribute only small amounts of nitrogen, but saliva, which can be secreted at perhaps 20 l/day in forage-fed sheep and ten times that amount in cattle, could supply 1-8 g N/day in sheep and 6-28 g N/day in cattle; of this half could be present in urea and most of the remainder in protein. At low rumen ammonia concentrations, urea appears to pass readily across the wall from the blood. However, this process is limited and ceases at rumen ammonia levels of about 0.2 g N/l (Kennedy & Milligan 1980) which are observed at nitrogen intakes of about 20 g/d. It has been suggested that facultatively aerobic bacteria attached to the rumen wall produce urease which facilitates the passage of urea through the wall (Cheng & Costerton 1979). The suppression of urease production with increasing levels of rumen ammonia then inhibits urea transport.

As immature forages contain low levels of cell wall constituents and are readily fermentable, they form a rich source of both protein and energy for the rumen microbes. With increasing maturity, rising cell wall levels accompanied by falling protein levels greatly alter the composition of the feed. The inverse relationship between cell wall and protein is most apparent with temperate grasses and clovers (Fig. 2a); in tropical species it is less distinct because of high cell wall levels in relatively immature plants (Laredo & Minson 1973). With advancing plant maturity, alterations in the physical and chemical properties of cell walls cause a decline in organic matter digestibility. In consequence the intakes of organic matter, digestible organic matter and protein fall (Fig. 2b). In this figure the upper limits to the intake of digestible organic matter and protein would be respectively 1100 g and 400 g/d and the lower limits 300 and 30 g/d.

As a relatively constant proportion of the organic matter digestion in the whole tract occurs in the rumen, a decrease in organic matter digestibility indicates that less energy becomes available not only to the tissues of the animal but also to the rumen microbes. This might be thought to create an excess of protein relative to energy for the microbes, but as plants mature, protein content declines more rapidly than organic matter digestibility, and the ratio of digestible organic matter to crude protein actually rises (Fig. 2c). Forage diets with ratios of digestible organic matter to crude protein less than 2:1 would probably not be encountered, but mature tropical grasses frequently have ratios greater than 20:1.

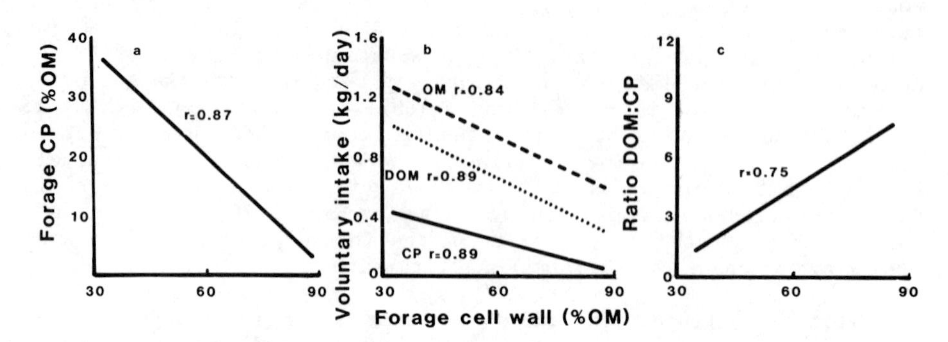

Fig. 2. With increasing levels of forage cell wall a decline occurs in the level of crude protein (CP) (Fig. 2a) and in the intakes of organic matter (OM) of digestible OM (DOM) and of CP (Fig. 2b), while the ratio of DOM to CP increases (Fig. 2c).

Comminution

During eating, forage is ground between the molars and some cell wells are ruptured. The degree of rupture, a reflection of the length of time spent by feed in the mouth, is probably determined by the time necessary for the animal to form a bolus suitable for swallowing. It is thus affected by factors determining bite size (Stobbs 1973) as well as those affecting the physical strength of the cell such as the species and stage of maturity of plants. This process is subsequently augmented by additional chewing during rumination. With immature forages, many cells are ruptured during chewing (Reid *et al*. 1962), but comparable data are not available for more mature plant material. While the rupture of plant cells makes substrate more accessible to microbial enzymes, it may also cause the release and mixing of intracellular components such as tannins which can denature proteins and render them less fermentable by microbes.

Microbial fermentation

As microbial enzymes are almost entirely intracellular, proteolysis requires contact between cell and substrate (see discussion by Akin 1982). The fate of protein on arrival in the rumen is thus influenced by the availability of plant fragments small enough to be ingested by protozoa, by the extent of release of soluble plant proteins to be metabolized by bacteria in the rumen liquor and by the numbers of bacteria available to attach themselves to plant particles. The initial rate of proteolysis then depends largely on the extent of rupture of plant cells during comminution. The proportion of "attached" bacteria free to attach themselves to newly arrived particles may have a significant effect. This is seen with *in vitro* studies in which the intial rate of ammonia production is more rapid if the rumen liquor inoculum is taken from an animal fasted overnight than from an animal fed a few hours previously (J.A. Hemsley, personal communication).

The initial rate of proteolysis is nutritionally important because proteins available for microbial attack are often readily transportable from the rumen. Increased passage of protein from the stomach can be induced by increasing the rate of digesta flow (Harrison & McAllan 1979), as well as by reducing accessibility of protein to enzymes by treatment with formaldehyde (Ferguson et al. 1967) or tannins (McLeod 1974).

Proteolytic activity comprises the breakdown of proteins to peptides and then to amino acids which may be incorporated into microbial protein, discharged into the rumen or deaminated with the release of ammonia and usually a volatile fatty acid. Such activity, which is associated with many species of rumen microbes, may be a necessary adjunct to specialized activities such as starch or fibre fermentation. Proteolytic activity in rumen liquor is highest with the dense microbial population associated with immature forage diets but declines with decreasing bacterial numbers as pastures mature. Rumen dysfunction may follow abrupt changes from low quality diets to lush pasture, but the enzyme supply should be adequate within about ten days when adaptation of microbes to a dietary change is complete.

Uptake and absorption of metabolites

Amino acids and ammonia released in the rumen may be either re-incorporated into microbial cells or absorbed: microbial cell synthesis, as indicated earlier, is energy dependent and the efficiency of conversion of dietary to microbial nitrogen depends on the rate of release to ATP compared with that of amino acids and ammonia during substrate fermentation. Although amino acids may reach significant levels for a short time after the consumption of lush pasture (Mangan et al. 1959) utilization and deamination ensure that levels are usually extremely low. Rumen microbes are well adapted, though, to obtaining any required amino acids from solutions of low concentration. By contrast, the microbes require higher concentrations of ammonia, a minimal level to maintain normal rumen function with forage diets being about 0.02 g N/1 (R.H. Weston, personal communication, Satter & Slyter 1974) and with concentrate diets perhaps ten times higher.

Although ammonia levels reflect protein intake (Fig. 3a) they are regulated by the relative availability of protein and energy (Weston & Hogan 1968) levels of 0.02 g N/1 or less being associated with diets with a digestible organic matter to crude protein ratio of 10:1 or more (Fig. 3b). As this ratio decreases towards 3:1, ammonia levels increase gradually, but rise sharply at ratios less than 3:1. With the latter protein, the source of potential ammonia release, displaces carbohydrate which is the main source of energy for ammonia uptake by the microbes. Ammonia levels are affected by small changes in the relative proportions of soluble nitrogen and carbohydrate with such diets.

Ammonia appears to cross membranes in the non-ionized form and absorption across the rumen wall to the blood, lymph and peritoneal fluid is favoured by alkaline pH (Chalmers et al. 1976). While alkaline conditions may occur following the intake of toxic amounts of urea, with forage diets high, levels of ammonia are associated with much greater levels of volatile fatty acids, and rumen pH is frequently below 6.0. Ammonia absorption is then dependent on the concentration in the rumen, but is relatively slow, perhaps one third of the amount in the rumen being absorbed per hour, even when at high concentrations. Uptake of ammonia by the microbes should take precedence over absorption, and hence ammonia absorption should not limit nitrogen metabolism in the rumen. High concentrations of rumen ammonia, while indicative of wastage of protein, are not likely to be harmful to the rumen microbes (Satter & Slyter 1974).

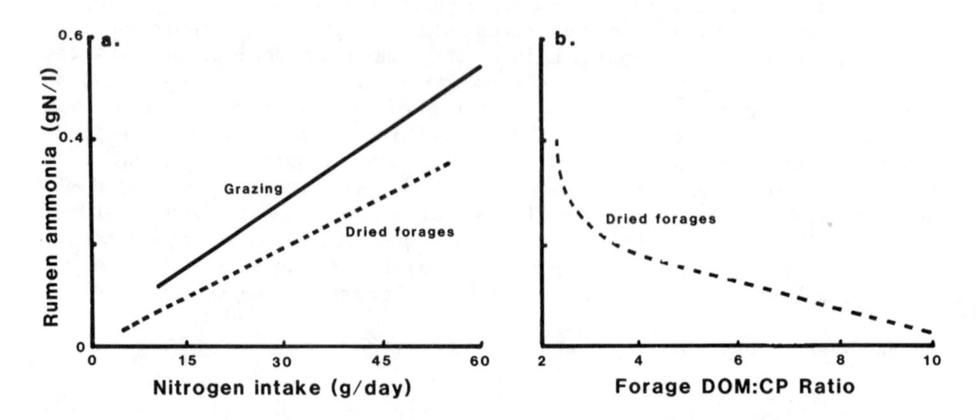

Fig. 3. The influence on rumen ammonia level of nitrogen intake and of the ratio of digestible organic matter to crude protein (DOM:CP) in the forage.

Cell synthesis and death

Microbial cells store protein and energy during development but at maturity use both protein and energy only for maintenance. However the breakdown of substrate continues; the efficiency of conversion of plant material to microbial cells which thus depends on the proportion of dividing, rapidly developing cells in the rumen, is undoubtedly higher when there is a high rate of microbial cell turnover (Harrison & McAllan 1979, John *et al*. 1980). With forage diets, the time spent by bacteria in the rumen is influenced by the relative proportions that travel at the same rate as water and those that move attached to plant fragments. Protozoa do not travel freely from the rumen but sequester themselves. Their contribution to protein reaching the abomasum may be less than half that expected from the numbers present in the rumen (Harrison & McAllan 1980); this factor substantially influences protein flow with diets such as sugar cane (Preston & Leng 1980) but probably has relatively minor effects with other forage diets.

Duration of residence time for bacterial cells in the rumen influences not only cell synthesis but also cell destruction by lysis, by anaerobic fungi, by bacteriophage and by ingestion by protozoa. The total destruction may amount to 30 percent of the nitrogen in the microbial "pool" (Nolan & Stachiw 1979). Individual protozoa could ingest about 3000 bacteria per day (Coleman 1975) but as protozoal numbers generally vary between 10^4 and 10^5/ml while bacterial numbers range between 10^{10} and 10^{11}/ml (Prins & Clarke 1980), at most only about 3 percent of the bacteria would be destroyed by protozoa. The relative importance of the other destructive agents is not known.

Protein passage from the stomach

The most reliable quantitative estimates of protein metabolism in the rumen are based on the flow of organic nitrogen from the abomasum. Allowance must be made for nitrogen of endogenous origin added in the abomasum; this was found by Harrop (1974) to range in sheep from 0.5 to 2.6 g N/day of which 60-70 percent was in the form of protein and 10 percent in mucin.

With the diets shown in Fig. 2, the gain or loss of nitrogen in the stomach show a general relationship with the levels of ammonia in the rumen (Fig. 4a) and with the ratio of digestible organic matter to crude protein in the diet (Fig. 4b).

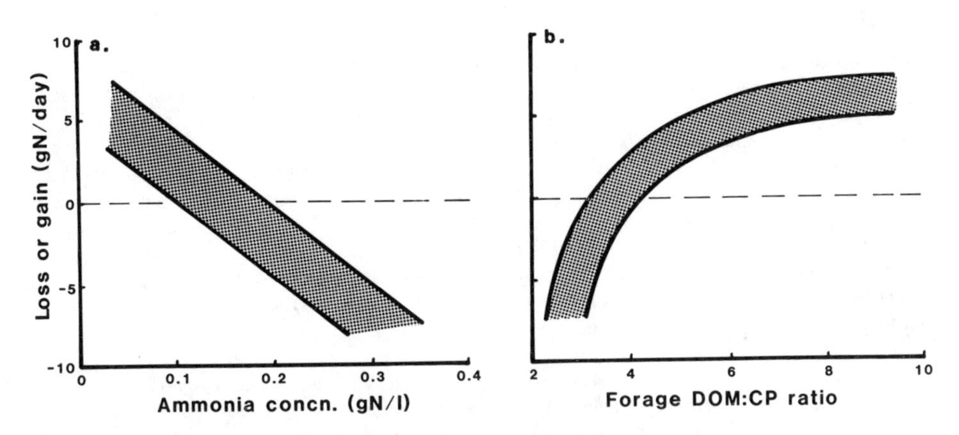

Fig. 4. The loss or gain of nitrogen in the stomach in relation to (a) ammonia concentration and (b) the ratio of digestible organic matter to crude protein (DOM:CP) in the forage.

For diets with digestible organic matter to crude protein ratios between about 2:1 and 10:1 the amount of crude protein entering the small intestine per unit crude protein intake (y) is related to the digestible organic matter to crude protein ratio of the diet (x) (Weston & Hogan 1973) as $y = 0.33 + 0.18\ x$. As indicated earlier with diets with ratios above 10:1, nitrogen limitation to microbial protein synthesis is likely, while with ratios between 2:1 and 3:1, "y" value is sensitive to small changes in soluble protein and soluble carbohydrate levels in the forage and to fermentation conditions in the rumen.

The above relationship modified by the inclusion of additional data from forage diets has been expressed as:

CP entering intestine = 360 g/kg CP intake + 160 g CP/kg DOM intake + 6, assuming an addition of 6 g/d crude protein in the abomasum (Hogan & Weston 1981).

This model over-estimates by up to 25 percent values for four diets based on lucerne (*Medicago sativa*) for unknown reasons and by more than 15 percent for pasture silages, probably because microbial protein synthesis is less with silages than with fresh forages (Thomas *et al*. 1980). However it underestimates protein flow by approximately

25 percent in lambs exposed to cold conditions (Kennedy & Milligan 1978) and by 10 percent in pregnant ewes (R.H. Weston, personal communication). The model suggests that for particular conditions a fixed proportion of dietary protein escapes degradation and that microbial protein synthesis is in fixed proportion to digestible organic matter intake. Similar relationships have been developed by Vérité et al. (1979) for diets of herbage and concentrates and by J.L. Corbett and F.S. Pickering (personal communication) for grazed forages. However the component assigned to unfermented dietary protein varied between 41 and 10 percent while protein synthesis ranged between 124 and 188 g/kg digestible organic matter. With all relationships the proportion of the digesta protein of microbial origin is highest with mature forages.

DIGESTION IN THE SMALL AND LARGE INTESTINES

True proteins comprise about 80 percent of the crude protein analysed from duodenal digesta derived from a wide range of diets, much of the remaining nitrogen being present in nucleic acids (Smith 1975). The amino acid composition of proteins in bacteria and protozoa derived from different dietary regimes, while reasonably constant, varies appreciably in nutritionally important areas. For instance with ten species of rumen bacteria, Purser and Buechler (1966) observed a two-fold range in the levels of lysine and methionine. Significantly, though, less variability is seen in the amino acid composition of digesta that pass from the abomasum with different forage diets (Hogan & Weston 1981).

Estimated net absorption of individual amino acids (the difference between the amounts passing into the small intestine from the stomach and out to the large intestine) have varied between about 55 and 80 percent (Armstrong & Hutton 1975). The sulphur amino acids cyst(e)ine and methionine appear to be absorbed less readily than the rest. The amino acid mixture that passes in the blood from the small intestine to the liver is probably less variable in composition because it includes amino acids derived from the appreciable quantitites of protein, largely in epithelial cells, discharged into the small intestine (McDougall 1966) and digested there.

When allowance is made for the effects of endogenous protein discharged into the small intestine true losses appear to be equivalent to about 70 percent of the amino acids or 545 g/kg crude protein passing out of the abomasum (Lindsay et al. 1980), though values for diets containing protein concentrates appear higher while those for diets containing tannins are probably lower (Egan & Ulyatt 1980). The remaining 30 percent of amino acids pass into the large intestine, are deaminated there, and play no further direct part in protein metabolism.

The amounts of amino acids derived by sheep offered the diets of Fig. 2 at ad libitum levels, calculated on the basis of 545 g/kg crude protein leaving the stomach, ranged from 35 to 180 g/d (Fig. 5). However, these animals, through excessive protein destruction in the rumen with some diets and inefficiencies in the release and absorption of amino acids in the small intestine with all diets, effectively wasted 20-110 g of amino acids per day.

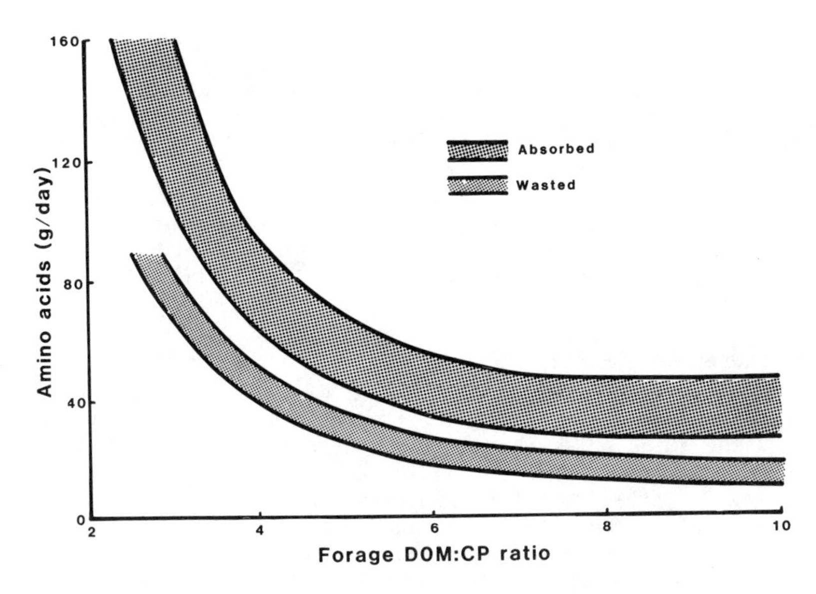

Fig. 5. Estimates of the amounts of amino acids absorbed from the small intestine with sheep fed forage diets varying in the ratio of digestible organic matter to crude protein (DOM:CP) and of amino acids wasted through deamination in the rumen and large intestine.

PROTEIN UTILIZATION

Utilization of proteins begins with the transfer of the products of protein digestion, ammonia, non-essential amino acids and essential amino acids across the wall of the digestive tract. Although the essential amino acids pass quantitatively to the liver, many of the non-essential amino acids are extensively metabolized in the wall of the small intestine and must be synthesized as required from appropriate carbon skeletons and ammonia. The ammonia is derived both from the digestive tract and from deamination of amino acids, and with ruminants there is no evidence that any aspects of function are limited by deficiency of "non-essential" amino acids. Surplus ammonia is converted to urea. Amino acid utilization in the tissues depends on whether the amino acid is required for maintenance or is available for production, what forms of protein storage receive priority, how the composition of the available amino acid mixture differs from that of products, and whether amino acids or energy first limit production.

Maintenance or production

Maintenance requirements for amino acids, based on the amount of nitrogen excreted in the urine after a 96 hour fast and on faecal nitrogen derived from epithelial cells and enzymes discharged into the digestive tract, are about 12-13 g/d for an adult sheep and 90 g/d for a 500 kg steer. The individual amino acids are probably required in proportions similar to those in muscle protein.

Priority for protein storage

Some processes seem to have priority for nutrients over others -the loss of body weight that occurs, even in the well fed dairy cow, in early lactation suggests that nutrients are diverted into milk in preference to body tissues; in the undernourished pregnant female, foetal growth generally takes precedence over storage in maternal tissues, while the depression in wool growth observed during periods of rapid protein synthesis both in the ewe during late pregnancy and lactation and in the rapidly growing lamb suggests that wool growth has a relatively low priority for limiting nutrients (Corbett 1979). However wool growth continues in the sheep receiving sub-maintenance amounts of dietary energy, presumably because, when insufficient energy is available for other forms of production, the low energy process of wool growth is able to make use of any available amino acids.

Amino acids in digesta and products

The efficiency of protein synthesis is reduced by differences that exist in the proportions of essential amino acids between digesta and products. The amino acids that differ most are the limiting amino acids, that is, the amino acids the supply of which first limits the rate of protein synthesis. The limiting amino acids for milk production are likely to be methionine and leucine, for muscle protein synthesis, histidine, lysine and methionine, and for wool production cystine and methionine, methionine being partly convertible to cystine.

Nutrient limits

The adequacy of available amino acids to permit the most efficient use of energy has been assessed in three ways, by theoretical calculation, by establishing optimal relationships between amino acid and energy supply and by observing production responses to the provision of supplementary amino acids.

Theoretical calculations on the adequacy of amino acid supply are made along the following general lines: a calculation is made of the amount of protein that can be stored for a given supply of net energy. Information pertinent to such a calculation is presented by ARC (1980). From data on the efficiency of conversion of absorbed amino acids to stored protein, a calculation is made of the amounts of amino acids that need to be absorbed for that level of protein synthesis. Efficiency for meat and milk production is regarded as 0.75 (ARC 1980) and for wool growth 0.13 (Hogan et al. 1979). The quantities of amino acids absorbed from the small intestine are then measured or calculated as described earlier. From such calculations it has been claimed that with forage-fed ruminants amino acid supply is not likely to be the first limiting nutrient for growth and milk production though it is for wool growth (Weston & Hogan 1973); a contrary view on the adequacy of amino acid supply for growth in the 15-25 kg ruminant lamb has been advanced by Black et al. (1976).

Optimal relationships between amino acids and energy have been established for the milk-fed lamb (Black & Griffiths 1975) but not for forage-fed animals. However, a useful alternative calculation has been made using the ratio of intestinally digested crude protein to digestible organic matter (DCPi:DOM). Thus with growing lambs fed a roughage and concentrate diet (Weston 1971, Faichney & Weston 1971) no increase was observed in either voluntary food intake or growth rate when this ratio exceeded 0.17:1. A similar value probably applies to the forage fed animal, and is met by diets in which the digestible organic matter to total crude protein ratio is less than about 10:1.

However the DCPi: DOM ratio cannot reach the desired level with diets which supply the small intestine mainly with protein synthesized in the rumen because, as indicated earlier, such synthesis has an efficiency of only about 0.16 kg/kg digestible organic matter, corresponding to a DCPi:DOM ratio of 0.12-0.13:1. Hence diets with digestible organic matter to crude protein ratios greater than 10:1, which supply mainly microbial protein, are potentially protein deficient.

Reports of production responses to amino acid supplementation are largely associated with wool growth (Reis 1979), probably through the supply of additional methionine. Lactation responses to additional amino acids have at times been observed with dairy cows fed pasture and fresh and ensiled forages (Stobbs _et al_. 1977, Rogers _et al_. 1979, 1980). Interpretation is difficult because the mammary gland draws essential amino acids from tissue catabolism as well as from the small intestine. With silages at least the response may be associated with an augmented methionine supply (Gill & Ulyatt 1979). The efficiency of conversion of supplementary protein to milk production is low; improved milk production may be related to increased production of lactose from amino acids via glucose, or simply to improved energy supply, because amino acids are used as sources of net energy for production with about 30 percent greater efficiency if absorbed from the small intestine than if deaminated and absorbed as volatile fatty acids from the rumen (Blaxter & Martin 1962). Enhanced growth has followed protein supplementation of beef cattle fed forage diets such as sugar cane (Preston & Leng 1980) or mature spear grass (_Heteropogon contortus_) (J.A. Lindsay, personal communication) in which the digestible organic matter to crude protein ratio exceeded 20:1. While such diets were undoubtedly protein deficient, the growth response could also reflect improved energy utilization.

General agreement thus exists from the three separate assessments that amino acids are rarely likely to be the first nutrients limiting productive processes other than wool growth with forage diets with digestible organic matter to crude protein ratio less than 10:1. This being so, what penalty is paid for wastage of amino acids caused by inefficiency of the digestive tract? This is difficult to estimate because it is not possible to assign priorities given by the animal for the use of nutrients. However, with the data of Fig. 5, the calculated wastage of net energy from this source ranged from 0.047 to 0.47 MJ/day, capable of increasing body growth by about 10 percent in the sheep involved. Alternatively, wool growth could have risen by 20 to 100 percent, and would have been limited with many diets only by the genetic potential of the sheep.

CONCLUSIONS

The ruminant digestive tract undoubtedly functions efficiently in improving the amino acid supply to animals fed roughages of moderate and low quality, though improvement is limited when the nitrogen content of the diet is very low. With high protein forages, the digestive tract is much less efficient in metabolizing protein and the amino acids absorbed from the small intestine may represent a disappointing part of the amount potentially available. For physiological processes other than wool growth, the amino acid supply appears adequate for the efficient use of available energy, though if all possible amino acids become available to the animal, an increase of perhaps 10 percent in growth rate and 20-100 percent in wool growth rate could be feasible.

REFERENCES

ARC (1980) The nutrient requirement of ruminant livestock. Technical Review by an Agricultural Research Council Working Party. Farnham Royal, U.K., Commonwealth Agricultural Bureaux. 351 pp.

Akin, D.E. (1982) Microbial breakdown of feed in the digestive tract. In: Nutritional limits to animal production from pastures. Editor J.B. Hacker. Farnham Royal, U.K., Commonwealth Agricultural Bureaux, pp. 201-223.

Armstrong, D.G.; Hutton, K. (1975) Fate of nitrogenous compounds entering the small intestine. In: Digestion and metabolism in the ruminant. Editors I.W. McDonald and A.C.I. Warner. Armidale, Australia, University of New England Publishing Unit, pp. 432-447.

Black, J.L.; Faichney, G.J.; Graham, N.McC. (1976) Future role of computer simulation in research and its application to ruminant protein nutrition. In: Protein metabolism and nutrition. Editors D.J.A. Cole et al. London, Butterworths, pp. 477-491.

Black, J.L.; Griffiths, D.A. (1975) Effect of liveweight and energy intake on nitrogen balance and total N requirements of lambs. British Journal of Nutrition 33, 399-413.

Blaxter, K.L.; Martin, A.K. (1962) The utilization of protein as a source of energy in fattening sheep. British Journal of Nutrition 16, 397-407.

Chalmers, M.I.; Grant, I.; White, F. (1976) Nitrogen passage through the wall of the ruminant digestive tract. In: Protein metabolism and nutrition. Editors D.J.A. Cole, K.N. Boorman, P.J. Buttery, D. Lewis, R.J. Neale and H. Swan. London, Butterworths, pp.159-179.

Cheng, K-J.; Costerton, J.W. (1979) Adherent rumen bacteria - their role in the digestion of plant material urea and epithelial cells. In: Digestive physiology and metabolism in ruminants. Editors Y. Ruckebusch and P. Thivend. Lancaster, U.K., MTP Press, pp. 227-250.

Coleman, G.S. (1975) The interrelationship between rumen ciliate protozoa and bacteria. In: Digestion and metabolism in the ruminant. Editors I.W. McDonald and A.C.I. Warner. Armidale, Australia, University of New England Publishing Unit, pp.149-164.

Corbett, J.L. (1979) Variation in wool growth with physiological state. In: Physiological and environmental limitations to wool growth. Editors J.L. Black and P.J. Reis. Armidale, Australia, University of New England Publishing Unit, pp. 79-98.

Egan, A.R.; Ulyatt, M. (1980) Quantitative digestion of fresh herbage by sheep. VI. Utilization of nitrogen in five herbages. Journal of Agricultural Science, Cambridge 94, 47-56.

Faichney, G.J.; Weston, R.H. (1971) Digestion by ruminant lambs of a diet containing formaldehyde-treated casein. Australian Journal of Agricultural Research 22, 461-468.

Ferguson, K.A.; Hemsley, J.A.; Reis, P.J. (1967) Nutrition and wool growth. The effect of protecting dietary protein from microbial degradation in the rumen. Australian Journal of Science 30, 215-217.

Gill, M.; Ulyatt, M.J. (1979) The metabolism of methionine in the silage-fed sheep. British Journal of Nutrition 41, 605-609.

Harrison, D.G.; McAllan, A.B. (1980) Factors affecting microbial growth yields in the reticulo-rumen. In: Digestive physiology and metabolism in ruminants. Editors Y. Ruckebusch and P. Thivend. Lancaster, U.K., MTP Press, pp.205-226.

Harrop, C.J.F. (1974) Nitrogen metabolism in the ovine stomach. 4. Nitrogenous components of the abomasal secretions. Journal of Agricultural Science, Cambridge 83, 249-257.

Hogan, J.P.; Elliott, N.M.; Hughes, A.D. (1979) Maximum wool growth rates expected from Australian Merino genotypes. In: Physiological and environmental limitations to wool growth. Editors J.L. Black and P.J. Reis. Armidale, Australia, University of New England Publishing Unit, pp.43-59.

Hogan, J.P.; Weston, R.H. (1981) Laboratory methods for protein evaluation. In: Forage evaluation: concepts and techniques. Editors J.L. Wheeler and R.D. Mochrie. Melbourne, CSIRO and American Forage and Grassland Council, pp.75-88.

John, A.; Ulyatt, M.J.; Jones, W.T.; Shelton, I.D. (1980) Factors influencing nitrogen flow from the rumen. Proceedings of the New Zealand Society of Animal Production 40, 226.

Kennedy, P.M.; Milligan, L.P. (1978) Effects of cold exposure on digestion, microbial synthesis and nitrogen transformations in sheep. British Journal of Nutrition 39, 105-117.

Kennedy, P.M.; Milligan, L.P. (1980) The degradation and utilization of endogenous urea in the gastro-intestinal tract of ruminants: A review. Canadian Journal of Animal Science 60, 205-221.

Laredo, M.A.; Minson, D.J. (1973) The voluntary intake, digestibility and retention time by sheep of leaf and stem fractions of five grasses. Australian Journal of Agricultural Research 24, 875-88.

Lindsay, J.R.; Hogan, J.P.; Donnelly, J.B. (1980) The digestion of protein from forage diets in the small intestine of the sheep. Australian Journal of Agricultural Research 31, 589-600.

Lyttleton, J.W. (1973) Proteins and nucleic acids. In: Chemistry and biochemistry of herbage. Vol 1. Editors G.W. Butler and R.W. Bailey. New York, Academic Press, pp.62-103.

Mangan, J.L.; Johns, A.T.; Bailey, R.W. (1959) Bloat in cattle. XIII. The effect of orally administered penicillin on the fermentation and foaming properties of rumen contents. New Zealand Journal of Agricultural Research 2, 342-354.

McDougall, E.I. (1966) Proteins of the Succus entericus from the jejunum of the sheep. Biochemistry Journal 100, 19-26.
McLeod, M.N. (1974) Plant tannins - their role in forage quality. Nutrition Abstracts and Reviews 44, 803-815.
Nolan, J.V.; Stachiw, S. (1979) Fermentation and nitrogen dynamics in Merino sheep given a low-quality-roughage diet. British Journal of Nutrition 42, 63-79.
Preston, T.R.; Leng, R.E. (1980) Utilization of tropical feeds by ruminants. In: Digestive physiology and metabolism in ruminants. Editors Y. Ruckebusch and P. Thivend. Lancaster, U.K., MTP Press, pp.621-640.
Prins, R.A.; Clarke, R.T.J. (1980) Microbial ecology of the rumen. In: Digestive physiology and metabolism in ruminants. Editors Y. Ruckebusch and P. Thivend. Lancaster U.K., MTP Press, pp.179-204.
Purser, D.B.; Buechler, S.M. (1966) Amino acid composition of rumen organisms. Journal of Dairy Science 49, 81-84.
Reid, C.S.W.; Lyttleton, J.W.; Mangan, J.L. (1962) Bloat in Cattle. XXIV. A method of measuring the effectiveness of chewing in the release of plant cell contents from ingested feed. New Zealand Journal of Agricultural Research 5, 237-248.
Reis, P.J. (1979) Effects of amino acids on the growth and properties of wool. In: Physiological and environmental limitations to wool growth. Editors J.L. Black and P.J. Reis. Armidale, Australia, University of New England Publishing Unit, pp.223-242.
Rogers, G.L.; Bryant, A.M.; McLeay, L.M. (1979) Silage and dairy cow production. III. Abomasal infusions of casein, methionine and glucose and milk yield and composition. New Zealand Journal of Agricultural Research 22, 533-541.
Rogers, G.L.; Porter, R.H.D.; Clarke, T.; Stewart, J.A. (1980) Effect of protected casein supplements on pasture intake, milk yield and composition of cows in early lactation. Australian Journal of Agricultural Research 31, 1147-1152.
Satter, L.D.; Slyter, L.L. (1974) Effect of ammonia concentration on rumen microbial protein production in vitro. British Journal of Nutrition 32, 199-208.
Siddons, R.C.; Evans, R.T.; Beever, D.E. (1979) The effect of formaldehyde treatment before ensiling on the digestion of wilted grass silage by sheep. British Journal of Nutrition 42, 535-545.
Smith, R.H. (1975) Nitrogen metabolism in the rumen and the composition and nutritive value of nitrogen compounds entering the duodenum. In: Digestion and metabolism in the ruminant. Editors I.W. McDonald and A.C.I. Warner. Armidale, Australia, University of New England Publishing Unit, pp. 399-415.
Stobbs, T.H. (1973) The effect of plant structure on the intake of tropical pastures. I. Variation in the bite size of grazing cattle. Australian Journal of Agricultural Research 24, 809-819.
Stobbs, T.H.; Minson, D.J.; McLeod, M.N. (1977) The response of dairy cows grazing on nitrogen fertilized grass pasture, to a supplement of protected casein. Journal of Agricultural Science, Cambridge 89, 137-141.
Thomas, P.C.; Chamberlain, D.G.; Kelly, N.C.; Wait, M.K. (1980) The nutritive value of silages. Digestion of nitrogenous constituents in sheep receiving diets of grass silage and grass silage and barley. British Journal of Nutrition 43, 469-479.
Vérité, R.; Journet, M.; Jarrige, R. (1979) A new system for the protein feeding of ruminants: the PDI system. Livestock Production Science 6, 349-367.
Weston, R.H. (1971) Factors limiting the intake of feed by sheep. V. Feed intake and the productive performance of the ruminant lamb in relation to the quantity of crude protein digested in the intestines. Australian Journal of Agricultural Research 22, 307-320.
Weston, R.H.; Hogan, J.P. (1968) Rumen ammonia in relation to characteristics of the diet and parameters of nitrogen metabolism. Proceedings of the Australian Society of Animal Production 7, 359-363.
Weston, R.H.; Hogan, J.P. (1971) Digestion of pasture plants by sheep. V. Studies with subterranean and berseem clovers. Australian Journal of Agricultural Research 22, 139-157.
Weston, R.H.; Hogan, J.P. (1973) Nutrition of herbage-fed ruminants. In: Pastoral industries of Australia. Editors G. Alexander and O.B. Williams. Sydney, Sydney University Press, pp. 233-268.

UTILIZATION OF MINERALS

D.A. LITTLE

Division of Tropical Crops and Pastures, Cunningham Laboratory, St. Lucia, Queensland 4067, Australia.

ABSTRACT

Diagnosis of dietary mineral deficiency depends largely on chemical analysis of pasture which should be sampled in a manner that recognizes grazing selectivity, and ideally with oesophageally-fistulated animals. Positive animal response to the mineral given as a supplement is the critical confirmation of diagnosis, but failure to respond does not necessarily imply adequacy.

Herbage mineral concentrations vary widely among species, stages of growth and plant parts, all of which are important with respect to selective grazing. Fertilizer application also has marked effects. The chemical forms in which minerals occur in plant tissue seem to be of relatively minor consequence in relation to their utilization by ruminants, because major changes in chemical form occur both within the rumen because of microbial activity, and during subsequent digestion. Examples of reduced mineral utilization following rumen microbial activity include magnesium, which, when bound to bacterial cell walls may survive subsequent digestion, and copper, when transformed to the sulphide or thiomolybdate forms. In some situations, calcium is sequestered in sites within the plant that are resistant to microbial attack.

The ability of a plant to supply a mineral is dependent on concentration within the plant and true absorption of the mineral within the animal. Data on apparent absorption, which does not take into account the excretion of the faecal endogenous component, are frequently published. However, for many minerals these are highly variable and may be misleading.

Absorption differs for different minerals, and for some minerals there are plant genetic and maturity effects on mineral absorption. Other factors which influence mineral requirement and utilization include physiological status, dietary mineral deficiency and genetic effects. Soil ingestion and parasitism also exert pronounced effects, and all of these factors need to be taken into account in mineral utilization studies.

Examination of the potential of a herbage to supply a particular mineral should be made using animals whose requirement is substantial, and a realistic assessment is unlikely if that mineral is fed greatly in excess of requirement. Interaction between dietary minerals affects their utilization, the most widely recognized example being that between copper, molybdenum and sulphur. For pasture-fed animals ratio of calcium to phosphorus in the diet, interactions between calcium and phosphorus, and interference with their utilization by magnesium, zinc or aluminium are unlikely to be of consequence. However, the marked influence of potassium on the absorption of magnesium is of great importance in the aetiology of hypomagnesaemia.

Copper and zinc can each interfere with the absorption of the other, mainly through competition for binding sites in the intestinal mucosa, and moderate levels of dietary iron markedly depress copper absorption. This finding is likely to be of great importance in grazing animals, due to the ingestion of soil.

The potential problem of metal pollutants is briefly considered, in that both cadmium and lead adversely affect the metabolism of copper and zinc.

INTRODUCTION

The essential dietary minerals are classified either as macro- or trace-element nutrients depending upon the quantity required by the animal. The macro-elements are mainly utilized either for structural purposes, (e.g. calcium, phosphorus, sulphur) or in the maintenence of acid-base balance, (e.g. sodium, potassium and chlorine), as well as making vital contributions to energy transfer, nerve impulse transmission and enzyme activation (e.g. potassium, calcium, magnesium). The trace minerals function mainly as enzyme co-factors (e.g. manganese, copper), or by contributing structurally or functionally to the activities of enzymes, (e.g. zinc, molybdenum, selenium), hormones (iodine) or vitamins, (e.g. cobalt). Usually they are present only in trace quantities in pasture, and due to technical problems of measurement in some cases it is only recently that their function within the animal has been determined. Some minerals, when in excess, cause toxicity problems. This has occurred notably in the cases of selenium and fluorine, the toxic roles of which were recognized well before their essentiality was demonstrated. Problems of toxicity were discussed earlier in this symposium by M.P. Hegarty (1982).

Several essential minerals are rarely, if ever, the source of deficiency problems in grazing ruminants. These include potassium, and iron, because grazing animals consume appreciable amounts of soil (Healy 1973). Similar comments apply to those elements for which essentiality has only recently been demonstrated, using highly purified diets and isolated environments: these include chromium, fluorine, silicon, vanadium, tin and nickel (see review by Schwarz 1974), and these will not be dealt with further here.

The major purpose of this review is to examine the potential of pasture plants as sources of minerals for grazing animals, including aspects of diagnosis of deficiency and factors that influence mineral absorption and metabolism by the animal. The discussion will be restricted very largely to cattle and sheep.

MINERAL DEFICIENCY, DIETARY REQUIREMENTS AND DIAGNOSIS OF DEFICIENCIES

The biological functions of essential minerals take place within a range of optimum concentrations in the tissues, and homeostatic mechanisms attempt to maintain these concentrations within fairly narrow limits. The initial consequence of dietary deficiency of an essential mineral is to render the maintenance of optimum tissue concentrations more difficult. Concentrations below optimum cause the development of a biochemical lesion and the attendant impairment of physiological function(s) appropriate to the mineral in question. Appetite is usually impaired, and this also reduces the rate of animal production. Specific clinical symptoms can aid diagnosis, but by the time these become apparent the deficiency state is usually well developed, as with the skeletal abnormalities of aphosphorosis and the tetany of hypomagnesaemia. By contrast, the clinical picture may be so non-specific as to be of no diagnostic assistance, as in the wasting disease of classical cobalt deficiency. Mineral deficiency in the field is further complicated in that several nutrients are often involved (Morrison & Whitehair 1956, Underwood 1966), and intercurrent disease problems of infectious or other aetiology usually elevate dietary requirements (Morrison & Whitehair 1956). In the assessment of nutritional problems in the field the chemical analysis of animal tissues and fluids, and plant materials is very useful in determining aetiology, but the critical diagnostic test is a positive response to correction of the suspected deficiency.

TABLE 1
Net and dietary mineral requirements of cattle for growth (200 kg liveweight; 0.5 kg/d) and lactation (500 kg liveweight; 10 kg/d).

	Growth		Lactation	
	Net§	Dietary	Net§	Dietary
		(% in DM)		
Calcium	0.29	0.43	0.22	0.32
Phosphorus	0.19	0.24*	0.17	0.30
Magnesium	0.03	0.15	0.03	0.18
Sodium	0.06	0.07	0.09	0.10
		(ppm in DM)		
Zinc	3.6-6.0	12-20	5.4-7.5	18-25
Copper	0.3-0.6	8-14	0.3-0.6	10-14
Manganese	0.1-0.2	10-20	0.1-0.2	10-20
Cobalt +		0.11		0.11
Selenium +		0.03-0.05		0.03-0.05
Iodine†		0.5		0.5

§ Net requirements tabulated from dietary concentration figures and coefficients of absorption (ARC 1980).
* 0.12 percent phosphorus has recently been shown to be adequate (Little 1980).
+ These levels regarded as probably marginal.
† Probably adequate for all cattle and sheep in the absence of dietary goitrogens.

The dietary mineral requirements are based on the quantities contained in the tissues or secretions produced in various physiological states, and the inevitable or endogenous losses from the body. The sum of these factors yields the physiological requirement at the tissue level, known as the net requirement (ARC 1980). This figure is divided by the appropriate absorption coefficient to give the actual dietary requirement. In the field, however, the quantity of feed consumed is not known, and requirements are therefore frequently expressed as concentrations in the dry matter, after assuming a level of dry matter intake. Table 1 presents such figures for two moderate levels of production, taken from the ARC (1980), as a basis against which to compare figures discussed subsequently.

The comparison of pasture mineral concentrations with dietary requirements is more meaningful when account is taken of the method of pasture sampling. Grazing animals are able to select a higher quality diet than that of the average pasture on offer (e.g. Little 1969), and samples for chemical analysis should be as similar as possible to the ingested herbage. For this purpose, the use of experimental animals fitted with oesophageal fistulae is ideal. In general it may be assumed that the quality of diet selected by grazing animals will be higher than that indicated by pasture samples obtained manually, and the actual difference will depend on the precise method of pasture sampling adopted.

The other approach to the diagnosis of deficiency states involves the assessment of animal status with regard to the mineral in question. Commonly the analysis is of samples of blood or plasma, which is quite suitable in the case of magnesium, for which no labile body store of significant magnitude exists, but is unsuitable for minerals such as phosphorus. Many factors influence plasma inorganic phosphorus concentrations, and measurements of plasma phosphorus can be misleading for the assessment of phosphorus status (Little & McMeniman 1973, McMeniman & Little 1974). Although blood inorganic phosphorus concentrations less than 2 mg/100 ml in ruminants generally indicate deficiency, higher levels cannot be interpreted with any confidence (Gartner _et al_. 1980). Wide examination has been made of the skeletal stores and the effects on them of variations in dietary minerals and other constituents (e.g. Hill 1962, Sykes _et al_. 1973, Field _et al_. 1975), and a rib biopsy technique suitable for serial sampling of cattle and sheep has been developed (Little 1972). Subsequent work has improved the sensitivity of the method by adding measurements of bone thickness and osteoid content to the earlier density and mineral concentration determinations (Little, unpublished data). It is noteworthy that deficiencies of protein and mineral may produce similar skeletal changes via their respective influences on bone matrix formation and mineral supply (Sykes & Field 1972, Sykes _et al_. 1973).

In the case of sodium, the quantity normally present in the rumen contents is a significant body store, and plasma determinations are unsuited to the diagnosis of deficiency, because the animal maintains normal concentration quite rigorously. Homeostasis is mediated by the adrenal secretion of aldosterone which efficiently reduces the quantities of sodium in saliva and urine; that in saliva is replaced by potassium. The resultant change in salivary Na:K ratio is a very sensitive index of sodium status (Kemp _et al_. 1972, Committee on Mineral Nutrition 1973, Murphy & Gartner 1974), provided the potentially confounding influence of stress is avoided (Post 1965). A useful recent review of this subject was presented by Morris (1980).

Methods of analysis of specific tissues for individual trace elements or their metabolites are well established, but the examination of blood has the great advantage of convenience for field use. It is therefore reassuring to note the conclusion of Underwood (1971) that "...estimations of whole blood or plasma trace element concentrations have exceedingly wide applicability and represent by far the most valuable diagnostic criteria". Nevertheless, the analysis of samples of milk for iodine deficiency, and of liver for copper, provide the most sensitive assessments of these problems (Committee on Mineral Nutrition 1973).

MINERAL CONCENTRATIONS IN PLANTS

Great variability exists in the concentration of mineral nutrients in plant species. This was amply demonstrated by Whitehead (1966), for mainly temperate species, and by McDowell et al. (1977) and Minson (1977, 1982) for tropical species. These surveys indicate that the potential incidence and severity of deficiency problems is much greater in tropical areas, and that the minerals most likely to be widely involved are phosphorus, sodium, magnesium and copper. Following correction of gross deficiencies, trace element problems are likely to become apparent in many areas. The actual concentrations of minerals in plant material are influenced by many soil and other factors (reviewed by Fleming 1973, Reid & Horvath 1980). In any given field situation it is obviously important to be able to identify those aspects of the botanical components of pasture most sensitive to manipulation, so that potential deficiencies may be efficiently rectified. It is well recognized that within a given environment, mineral concentrations vary widely among different plants, and the occurrence of this inter- and intraspecific variation is discussed elsewhere in this volume by Norton (1982) and Hacker (1982) respectively.

The ranges of concentrations of several minerals in plants are shown in Table 2. Clearly the physiological requirements of the animal at the tissue level (Table 1) may not be met in some circumstances even if the minerals were totally available for utilization. The necessity for a coefficient to be included to account for incomplete absorption gives a useful impression of magnitude of the dietary deficit that may occur in particular situations.

TABLE 2
The range of concentrations in plant material of various minerals (taken from Whitehead 1966, Minson 1977, 1982).

% in DM		ppm in DM	
Ca	0.04 - 6.0	Zn	1 - 120
P	0.02 - 0.71	Cu	1.1 - 100
Mg	0.03 - 1.00	Mn	9 - 2400
Na	0.001 - 2.12	Co	0.016 - 4.7
		I	0.09 - 5
		Se	< 0.01 - 4000
		Mo	0.01 - 156

Stage of growth

The changes in mineral contents that occur with advancing plant maturity are largely related to alterations in proportions of leaf and stem, flowering and the setting of seed. Very variable effects have been observed, but in general terms, calcium levels remain relatively constant, and those of phosphorus, nitrogen, sulphur and most trace elements decline with increasing age (Whitehead 1966, Fleming 1973). Extensive observations made on tropical species over many years were reported by Bisschop (1965) from South Africa, who showed that concentrations of phosphorus, sodium, magnesium, potassium and chlorine were all at their peak following active growth in the wet season, and declined rapidly with advancing maturity. Phosphorus and sodium levels were uniformly low and magnesium marginal for cattle nutrition, but calcium concentrations were adequate, and remained relatively constant throughout the year, even increasing slightly through the dry season. Essentially similar results were obtained in northern Zimbabwe by Jones (1963), who noted decreased levels of sodium and magnesium, and marked increases in calcium levels with advancing maturity. Data from northern Australia (Davies *et al*. 1938, Norman 1963, Robinson & Sageman 1967, Shaw 1978) and Brazil (de Sousa 1978) demonstrate similar patterns of change.

Effect of fertilizer application

The application of a given nutrient usually increases the concentration of that nutrient in the plant (Whitehead 1966). However, in some circumstances its application may increase plant yield without increasing concentration. Such responses are clearly of great significance to the grazing animal in terms of correcting dietary inadequacies. Fertilizer may also have an effect on plant concentrations of minerals not present in the fertilizer. Whether these concentrations increase or decrease depends largely on the soil status with respect to the minerals in question (Fleming 1973); the more rapid growth stimulated by the application of one element may produce a dilution effect on a second element under conditions of marginal soil availability, or luxury uptake if sufficient is available. These effects occur with both nitrogenous and phosphatic fertilizers, and are generally more marked with the former. The application of a high rate of nitrogen often increases plant magnesium content (Reid *et al*. 1974), and the liming of acid soils has produced increases in plant calcium, magnesium and phosphorus concentrations; the resultant changes in soil pH also frequently result in increases in plant copper, zinc and molybdenum (Reid & Jung 1974). Andrew & Robins (1969) found that phosphate applied to tropical legumes increased plant nitrogen and magnesium, and reduced potassium. Application of fertilizer potassium may result in a reduction in plant magnesium, calcium and sodium (Grunes *et al*. 1970, Fleming 1973). Slightly increased levels of potassium followed the application of magnesium to a series of temperate grasses and legumes (Reid *et al*. 1978). Obviously the very variable effects recorded in the literature indicate that any given pasture situation requires empirical evaluation. The potential changes in botanical composition of pasture induced by fertilizer application should also be taken into account in such assessments.

Form and distribution of minerals in plant tissue

Specific data on mineral distribution within the plant are surprisingly sparse, but of obvious importance in relation to selective grazing and herbage sampling for nutritive evaluation. One of the few studies was made by Fleming (1973) on *Dactylis glomerata*, *Lolium perenne* and *Trifolium pratense*. He found the concentrations of calcium, phosphorus, magnesium, copper, molybdenum and zinc to be much higher in leaf than stem of all three (except for molybdenum in *T. pratense*, where the reverse was true). There were accumulations of plant phosphorus in the heads of all three, and of zinc in the grass heads. In *T. pratense*, manganese and cobalt concentrations in leaf greatly exceeded those in stem, in contrast to the situation in grasses, where the differences were not marked. Other studies reviewed by Fleming (1973) also indicated that mineral concentrations in leaf generally exceeded those in stem in other temperate grasses. Concentration of phosphorus is almost invariably higher in leaves of *Stylosanthes* than in stems, and that in inflorescence exceeds that of leaves (McIvor 1979).

The availability of minerals for absorption is likely to be influenced by the chemical form in which they occur. Potassium, sodium, chlorine, fluorine and iodine occur mainly in ionic form, and are therefore readily available (Butler & Jones 1973), and even where potassium and sodium are not present as ions, they occur in readily exchangeable form. Bremner (1970) and Bremner & Knight (1970) found a large proportion of manganese, copper and zinc in *Lolium perenne*, to be readily soluble, but some zinc and manganese appeared to be in insoluble form associated with the fibre matrix. However, after rumen digestion the majority of copper and zinc, and probably of manganese, seemed to be associated with the microbial fraction. Most zinc and manganese were liberated in the abomasum, and a gradual increase in solubility of copper occurred with passage down the gut (Bremner 1970). This work illustrates the great changes that can occur in the chemical forms of minerals from the plant to the digesta. Both sulphur and selenium are protein-bound, associated with the sulphur-containing amino acids; some sulphur occurs as sulphate and sulphate-ester linked compounds such as Coenzyme A, and these compounds also are subject to substantial chemical modification in the rumen.

Plant phosphorus exists in many states of combination, although the greater proportion is present as inorganic phosphate (Butler & Jones 1973). Phosphorus is recognized as a very mobile nutrient within the plant, and much is metabolically active in the form of ester linked phosphates. Phytic acid occurs in seeds but not in vegetative material, and while phosphorus in this form is relatively unavailable to non-ruminant species (Taylor 1979), ruminants are able to utilize it quite efficiently, because phytate is readily broken down in the rumen by microbial activity (Reid *et al*. 1947).

About 50 percent of plant magnesium is water soluble (Butler & Jones 1973), and some 10 percent is organically combined in the porphyrin complex of chlorophyll (Wilkinson 1976). Chloroplasts released during mastication or subsequent digestion, if not ruptured, are usually rapidly ingested by entodiniomorphid protozoa (Mangan & West 1977). In any event, thorough digestion and utilization of chloroplast protein occurs, so that the associated magnesium is likely to be at least initially available. Under certain circumstances magnesium is firmly adsorbed onto bacterial cell walls (Fitt *et al*. 1972), and significant quantities of magnesium may be complexed in non-ionic form by certain organic acids and lignin (Molloy & Richards 1971). The possible significance of these phenomena in relation to availability is discussed below.

Within the plant some cobalt is ionic and some complexed with other compounds, while molybdenum is present as an enzyme component, but is probably translocated in the gut as $MoO_4^=$ (Butler & Jones 1973). Calcium is present in plants in various soluble, partially soluble (phosphate) and insoluble (oxalate) salts, and is also bound or adsorbed to protein and pectin (Butler & Jones 1973). Gallaher (1975) has reviewed the occurrence of calcium oxalate crystals in plant materials. A large proportion of plant calcium was found by Molloy & Richards (1971) to be complexed in non-ionic form by pectin, lignin and various organic acids. Jones (1978) found calcium to be the most likely of the metal ions to be bound by carboxyl and phenolic hydroxyl groups of the organic compounds of cell walls, and possibly also by the silica that is closely associated with these compounds in secondarily thickened cell walls. Calcium oxalate is normally broken down in the rumen, providing the microorganisms have become adapted (R.J.W. Gartner, B.J. Blaney & R.A. McKenzie, personal communication). However, in *Medicago sativa* up to 33 percent of the calcium may be in the form of calcium oxalate crystals within indigestible cells of the vascular bundle sheath (Ward *et al*. 1979), and this fraction of the calcium was unavailable to the animal. This finding illustrates the need for plant anatomical studies to be conducted in association with nutritional evaluation.

MINERAL ABSORPTION

Sites and mechanisms

Calcium

The rumen wall is relatively impermeable to calcium (Dobson & Phillipson 1968), and the major site for the absorption of calcium is the small intestine (Braithwaite 1976, Dillon & Scott 1979). However, in sheep fed fresh grass or clover, Grace *et al*. (1974) observed significant absorption of calcium anterior to the pylorus. Stevenson & Unsworth (1978) obtained evidence that absorption of calcium anterior to the duodenum may occur in sheep, and Bertoni *et al*. (1976) observed that some calcium was absorbed from this region in lactating, but not dry cows.

Schlachter *et al*. (1960) used the small intestine of the rat to show that most of the calcium transferred to the serosal surface was in ionized form, indicating that the process is an active cation transport system. They found also that the transport was greater in young than in old animals, and in pregnant rather than non-pregnant animals, and this demonstrated facultative changes in the rate of calcium transport in relation to the requirements of the animal. Similar observations have been made in cows (van't Klooster 1976) and sheep (Braithwaite 1974, 1975).

Mucosal phosphatases are involved in intestinal calcium transport (Krawitt *et al*. 1973), but the active transport system for calcium absorption depends very largely upon a calcium-binding protein in intestinal mucosa. This has been identified in a number of species, including chickens (Wasserman & Taylor 1966) and cattle (Fullmer & Wasserman 1973). The formation of this protein in the intestine is induced by 1,25-dihydroxy vitamin D which in turn is produced by the kidney under the control of parathyroid hormone secreted in response to low dietary calcium levels (Peacock 1976). Calcitonin apparently has no direct effect on calcium absorption (Peacock 1976).

Phosphorus

There is negligible absorption of phosphate from the reticulo-rumen (Dobson & Phillipson 1968), and again the major absorptive site is the small intestine; some phosphate absorption occurs from the omasum (Engelhardt & Hauffe 1975). In contrast with calcium, however, no mechanism has been described for the control of phosphorus absorption, and Braithwaite (1976) considered absorption to be by simple diffusion, since its rate was directly related to the concentration of phosphorus in the lumen of the small intestine. However, Wasserman & Taylor (1973) have demonstrated that the phosphorus absorptive process is saturable, and its transport by the wall of the small intestine is stimulated by 1,25-dihydroxy vitamin D (Chen *et al*. 1974). The extent of phosphorus absorption is influenced by the animal's phosphorus status, in that the short-term rate of faecal excretion of an oral dose of ^{32}P by cattle well supplied with phosphorus was twice that of cattle of similar body weight, but which had been fed a low-phosphorus diet (Little, unpublished data) indicating the probable saturation of the absorption mechanism in the former group. Thewis *et al*. (1978) measured considerable secretion of total and phospholipid phosphorus into the duodenum of sheep, and absorption of similar magnitude from the rest of the small intestine, from which they postulated the operation of an enterohepatic circulation of phospholipid. Their results were claimed not to support the notion of a saturable phosphorus absorption mechanism, but different mechanisms may well be involved.

Sodium, potassium, chloride

Dobson (1959) demonstrated the active transport of sodium across the rumen wall, a phenomenon now known as the sodium pump. Chloride also moves across the rumen wall against a concentration gradient (Stevens 1964). Jackson & Smyth (1971) observed the active absorption of sodium from both the small intestine and colon in dogs and man. Potassium diffuses freely across the rumen (Dobson & Phillipson 1968) and intestinal (Phillips & Code 1967) walls according to concentration gradients, and recent work in lambs has shown the major site of absorption of both sodium and potassium to be the small intestine, both before and after the development of rumen function (Dillon & Scott 1979). Grace *et al*. (1974) found the small intestine to be the major site for potassium absorption, and the large intestine that for sodium. Engelhardt & Hauffe (1975) observed up to 30 percent of the sodium and less than 10 percent of potassium that entered the omasum to be absorbed in the organ.

Magnesium

Any significant net passage of magnesium across the wall of the reticulo-rumen is unlikely (Dobson & Phillipson 1968). Although several workers concluded that most magnesium is absorbed from the small intestine (Stewart & Moodie 1956, Field 1961, Care & van't Klooster 1965), a number of later reports demonstrate substantial absorption anterior to the pylorus (Rogers & van't Klooster 1969, Ben Ghedalia *et al*. 1975, Horn & Smith 1978, Dillon & Scott 1979). The work of Fitt *et al*. (1979), in which supplements of magnesium to a low-magnesium diet were variously given via the mouth, rumen, omasum, abomasum and duodenum, strongly indicated that the major site of magnesium absorption is the omasum. Fitt *et al*. (1972) demonstrated that a portion of dietary magnesium could be bound quite strongly to bacterial cell walls in the rumen, and that the strength of this binding could increase during the passage of the cell wall residues down the tract. This phenomenon clearly constitutes a potentially significant reduction in the amount of magnesium available to the animal.

Copper, zinc and manganese

Copper, like calcium, binds to a duodenal protein (probably metallothionein, Bremner & Marshall 1974, Saylor et al. 1980) thought to be involved in its absorption (Sass-Korstak 1965). The same situation, and probably the same protein, seems to apply also to zinc absorption. High levels of copper interfere with the absorption of zinc (Van Campen 1969), and high zinc with copper absorption (Van Campen & Scaife 1967), indicating competition between the same ions for binding sites. The small intestine is the major site for the absorption of copper (O'Dell & Campbell 1970), zinc (Pate et al. 1970) and manganese (Miller et al. 1972). Some zinc may be absorbed from the abomasum also (Miller & Cragle 1965).

Measurement

The conventional digestibility type approach of measuring mineral intake and output has been used frequently in the evaluation of the mineral supply from herbage. This yields a measure of absorption (frequently termed availability), or more correctly apparent absorption, because the faecal mineral includes not only the unabsorbed fraction of that consumed, but also that secreted into the gut and not reabsorbed. The latter is termed the endogenous faecal fraction, and its presence ensures that the figure for apparent absorption is less than that for true absorption. The faecal endogenous fraction may be estimated in some cases by labelling the animal with the appropriate isotope, thus allowing measurements of true absorption to be made. A further observation often made is that of mineral retention, following measurement of urinary mineral excretion. A very useful review of errors associated with mineral balance measurements was made by Duncan (1958). The most frequently encountered of these determinations in the literature is apparent absorption. Unfortunately there is a plethora of factors both of plant and animal origin which influence these measurements, and they must be taken into account if a realistic interpretation of such data is to be made.

Of the macrominerals, sodium and potassium are very largely excreted in the urine; apparent absorption values for these nutrients therefore are usually high, providing reasonable approximations to true absorption. On the other hand, calcium, phosphorus and magnesium are excreted mainly in the faeces. This produces a usually large faecal endogenous component, with resultant low values for apparent absorption. Apparent absorption figures for trace minerals also are frequently very low in other than very young animals, because excess quantities are often excreted in the bile (e.g. manganese, Hall & Symonds 1981). Hartmans (1971) noted mean apparent absorption percentages of 3, 5 and -0.7 respectively for zinc, manganese and copper in the cows involved in his work cited in Table 3. The very wide range encountered in such observations on macro minerals is illustrated in Table 3, which presents figures for diets composed entirely or substantially of roughage.

TABLE 3

Apparent absorption (%) of macro-minerals in roughage or mainly roughage diets.

Ca	P	Mg	Na	K	S	Reference
			Lactating cows			
18-25	34-38	16-22	84-87	88-89	-	1
16-47	10-46	7-33	66-92	80-95	64-82	2
19*	26	26	77	88	59	3
-	-	0-37	-	-	-	4
			Wethers			
-26 to 8	-7 to 20	8-38	60-79	91-95	-	5
22-31	7-15	31-41	61-90	89-96	45-66	6
15-46	-28 to 36	23-48	-	87-93	51-73	7

* No range available.

[1] Rogers & van't Klooster (1969)
[2] Kemp & Guerink(1978)
[3] Hartmans (1971)
[4] Hutton *et al*. (1965)
[5] Pfeffer *et al*.(1970)
[6] Joyce & Rattray (1970a,b)
[7] Powell *et al*. (1978)

Playne (1976) provided an extensive review of measurements of the availability of phosphorus in ruminants. The figures he cited for true absorption determined by isotopic procedures ranged from 15-95 percent for sheep and 36-84 percent for cattle (excluding milk diets). The vast majority of the diets included concentrates and/or phosphorus supplements, but hay alone, mostly *Medicago sativa*, was used in five studies with sheep. They yielded a range of 55-95 percent true absorption, demonstrating the substantial variability that can attend such determinations. The following discussion examines several factors that are known to influence the degree of absorption of various dietary minerals by animals. No consideration is given here to the utilization of minerals from inorganic or other sources used in the feed formulation industry or as supplements. However, this topic has been thoroughly covered both for macro-minerals (Peeler 1972) and the trace elements (Ammerman & Miller 1972).

Effect of plant maturity and curing on mineral absorption

Reid *et al*. (1974) observed the apparent absorption and retention of magnesium in *Dactylis glomerata* regrowth to be much higher than it was in first growth herbage. Subsequent work from the same group (Powell *et al*. 1978) confirmed this observation for magnesium in several grasses, but maturation was found to be associated with decreased apparent absorption and retention of calcium, phosphorus, potassium and sulphur. These changes were unrelated to differences in intake, since, in several instances, the animals apparently consumed greater quantities of the mature herbage. In contrast, Gueguen & Demarquilly (1965) concluded that the true absorption of phosphorus *increased* with plant maturity, but the changes for calcium and magnesium were less clear cut. However, these results must be questioned since 'true absorptions' were calculated using assumed values for the faecal endogenous fractions which appear to be gross overestimates (ARC 1980), hence resulting in a substantial overestimate of true absorption.

Comparable observations have also been made for trace minerals. Absorption of copper is believed to be higher from mature than from young herbage, and from herbage made into hay rather than grazed fresh (Hartmans & Bosman 1970). This hypothesis was based on much higher liver copper concentrations recorded from cattle consuming the mature or hayed material. Lamand et al. (1977) compared the absorption of minerals from grass frozen after harvest with that from the same material made into hay after being rained on about ten times over five weeks after cutting. The low quality hay exhibited lowered apparent absorption and retention of calcium, phosphorus, copper and zinc, but not of magnesium or manganese. Although this result for copper apparently conflicts with the conclusion of Hartmans & Bosman (1970), variations in feed constituents not mentioned (e.g. sulphur and molybdenum, see below) may well have been responsible. Nevertheless, the treatment of this hay crop was rather extreme, and lesser effects might normally be expected.

Variation among plant species in mineral absorption

Differences between legumes and grasses, and between species and strains of grass have been recorded in apparent mineral absorption by ruminants. For example, Patil & Jones (1970) concluded that the availability of cobalt differed between two varieties of *Phleum pratense*, since despite similar plant cobalt concentrations, a response to cobalt supplementation was obtained in animals fed only one of them. Similarly, application of magnesium fertilizer may increase or decrease absorption of magnesium from different grass species (Reid et al. 1978). Apparent absorption of magnesium from legumes does not respond to magnesium fertilizer application, with the possible exception of *Medicago sativa* (Reid et al. 1978).

Physiological status and mineral absorption

The physiological status of the animal has a fundamental influence on the utilization of dietary minerals. Thus the requirements of growth, pregnancy and lactation increase the quantity of minerals required at the tissue level, which in turn causes increases in the true absorption of minerals. For example, Gütte et al. (1961) observed the true absorption of phosphorus by sows to increase from 31 to 62 percent between mid-pregnancy and the end of lactation. The demand of lactation for zinc promotes an increased absorptive ability in cows (Stake et al. 1975), and, of course, the maximum retention of calcium is much higher in young than old sheep, to cater for the requirements of growth (Braithwaite 1975).

The depletion of body mineral reserves caused by a period of dietary inadequacy also invokes homeostatic influences, which involve increased absorption and/or decreased endogenous excretion. Dietary deficiency of phosphorus has been shown in sheep to lead to greatly reduced metabolic faecal excretion of phosphorus (Young et al. 1966a) and to increased capacity for phosphorus absorption in sheep (Young et al. 1966b, Lee et al. 1979), and cattle (Little unpublished data). The occurrence of calcium depletion during pregnancy in sheep resulted in the animals being able to absorb during lactation twice the amount of calcium absorbed by non-depleted lactating sheep (Sykes & Dingwall 1975). Depletion of zinc and manganese also give rise to elevated absorptions in rats (Schwarz & Kirchgessner 1980).

The development of rumen function may be associated with profound effects upon utilization of trace elements. For example, Suttle (1975b) observed the availability of copper to milk-fed lambs to decrease from 71 to 47 percent just before weaning, and to 11 percent two weeks after weaning. Sheep in which rumen function was well established had copper-availability levels of 4-8 percent. The reduction in copper availability as sheep age appears to be due to production of sulphide by the rumen microflora; this allows the formation of much less available compounds such as cupric thiomolybdate and copper sulphide (Suttle 1974). Similarly, net absorption of zinc decreased from over 50 percent in very young calves to 12 percent in mature cows (Miller & Cragle 1965). Suttle (1975a) noted that the propensity of trace elements to form complexes with anionic ligands contributes greatly to this situation.

All of these observations indicate that, not only should care be exercised in selecting the types of animals to be used in evaluating feeds as sources of minerals, but also that the value of the published reports would be enhanced by including detailed descriptions of the type of animal used. The experimental use of a given group of animals will provide valid comparisons of feeds in the short term, but there is no point in attempting to assess the potential of a herbage to supply a particular mineral by using animals whose current requirement for it is minimal. Indeed, Suttle (1975a) suggested that the experimental use of animals marginally deficient in a particular mineral offered many advantages in trace element work, and this should apply equally to the macro-minerals.

Level of mineral intake and absorption

In work involving quite high rates of mineral intake, Lueker and Lofgreen (1961) showed that the amount of both calcium and phosphorus absorbed was directly related to the level of mineral intake. Similarly, Braithwaite (1975) found that the degree of absorption of calcium was related to calcium intake until an absorption maximum was reached, and also that apparent absorption of phosphorus was directly related to its level of intake. In later work, Braithwaite (1979) demonstrated a constant absorption coefficient for calcium in sheep fed between 40 and 100 mg calcium/day/kg body weight, but above this level, absorption increased in proportion to intake. He proposed that at lower calcium intakes the increasing diffusion of calcium across the gut wall was offset by corresponding decreases in active absorption until this became negligible.

These findings are at variance with the conclusion of the ARC (1980) that "... for a given requirement ..... the amount of calcium absorbed from the diet is independent of dietary intake", which implies that the coefficient of calcium absorption must vary inversely with the level of dietary calcium intake. As noted above, the calcium and phosphorus intakes employed by Lueker & Lofgreen (1961) and Braithwaite (1975, 1979) were in most cases greatly in excess of currently accepted requirements, and caution is needed before extrapolating to a grazing environment the results of studies employing formulated rations of such a nature.

It is becoming established that the proportion of a mineral absorbed increases as the amount of it present in the diet decreases, as noted in the previous section, and there is a clear need for further experimental work to define the extent of these influences so that dietary requirements may be more accurately established. For example, in demonstrating the overestimation of dietary phosphorus requirements for cattle growth made by the ARC (1965) and the NRC (1976), it was pointed out by Little (1980) that the phosphorus absorption data available to these groups were almost certainly inappropriately low for his experimental conditions, and that situation apparently has not greatly changed more recently (ARC 1980). Indeed, it is probable that both sheep (Young *et al*. 1966a,b) and cattle (Little 1980) adaptively modify both absorption and endogenous excretion of phosphorus, depending on their phosphorus status.

Differences between breeds in mineral absorption

Genetic differences occur in mineral metabolism between and within breeds. Miller (1975) noted the greater incidence of parturient paresis in Jerseys compared with other breeds, and there are genetic differences among sheep in the incidence of both deficiency and toxicosis with copper, and several other minerals (Wiener 1971). By using monozygotic twin cattle, Field & Suttle (1970, 1979) were able to demonstrate genetically-based differences in various aspects of mineral metabolism, including the utilization of magnesium, urinary excretion of calcium, and phosphorus excretion in both urine and faeces. The results of Manston & Vagg (1970) also suggested a genetic component controlling the urinary excretion of phosphorus by cows.

Soil ingestion and mineral absorption

Mineral absorption is influenced by the ingestion of soil; Grace & Healy (1974) found increased retention of calcium, magnesium and phosphorus following soil ingestion, and Healy (1974) pointed out that ingested soil may constitute a useful and signficant source of copper, manganese, selenium and iodine. He also demonstrated that soil consumption made moderate contributions to the animal's intake of macro-elements, and caused massive increases in ingested iron (Healy 1973). On the debit side, however, the ingestion of soil provoked a reduction of some 50 percent in the absorption of copper (Suttle *et al*. 1975). These observations have a most important bearing on the grazing situation, because grazing animals may ingest substantial quantities of soil (Healy 1973).

Effect of parasites on nutrient absorption

Pathological affections of the gut may impair mineral absorption; for example, *Trichostrongylus colubriformis* infection produced negative or substantially reduced balances of calcium and phosphorus in young sheep, while greatly reducing the bodyweight gain realized by non-infected lambs at the same level of intake (Sykes & Coop 1976). A most important observation to emerge from this work is that the infected lambs gained no body calcium or phosphorus over 14 weeks despite a bodyweight increase of some 9 kg, whereas a body weight gain of 15 kg by pairfed non-infected animals was associated with increases in body calcium and phosphorus of over 50 percent. Helminth infestation apparently reduced mineral utilization to a marked extent.

DIETARY INTERACTIONS

Absorption or utilization of most dietary minerals is affected by other minerals or organic constituents of the diet. The list of interactions in Table 4 is not claimed to be exhaustive, but illustrates the need to consider interactions when assessing field

TABLE 4

Essential mineral nutrients and those minerals (or organic nutrients in parenthesis) which may exert an influence upon their absorption, retention or metabolism.

Mineral	Interacting nutrients
Ca	P[1,2,3] Zn[4] Si[5] Na[6]
P	Ca[1,7] Zn[4] Na[8,9] (Digestible energy[10])
Mg	Ca[1] K[11,12,13,14,15] P[2,16] Na[15] Fe[2] (N[17]) (fatty acids[18])
S	Mg[19]
Na	Ca[7] K[13]
Zn	Cu[20,21] Mn[22] Ca[23] P[23,24] Cd[25,26,27] Pb[27]
Cu	Mo[28,29] S[28] Zn[30,31] Fe[2,32,33] Ca[32] Cd[25,26,27] Pb[27]
Mn	Ca[7,33] P[34] Zn[22]
Fe	Ca[2] P[2,32] Cd[25]
Mo	S[35,36,37] Cu[37]
Se	S[38]

[1] Chicco *et al.* (1973)
[2] Standish *et al.* (1971)
[3] Nel & Moir (1974)
[4] Thompson *et al.* (1959)
[5] Jones (1978)
[6] Timet *et al.* (1978)
[7] Hartmans (1971)
[8] Harrison & Harrison (1963)
[9] McHardy & Parsons (1956)
[10] McMeniman & Little (1974)
[11] Fontenot *et al.* (1973)
[12] Field & Suttle (1979)
[13] Fitt & Hutton (1974)
[14] McGregor & Armstrong (1979)
[15] Care *et al.* (1967)
[16] Smith & McAllan (1966)
[17] Reid *et al.* (1974)
[18] Kemp *et al.* (1966)
[19] Reid *et al.* (1979)
[20] Van Campen (1969)
[21] Becker & Hoekstra (1971)
[22] Schwarz & Kirchgessner (1980)
[23] Heth *et al.* (1966)
[24] Greger & Snedeker (1980)
[25] Hill *et al.* (1963)
[26] Mills & Dalgarno (1972)
[27] Petering (1974)
[28] Suttle (1975a)
[29] Kovalsky *et al.* (1974)
[30] Van Campen & Scaife (1967)
[31] Kincaid *et al.* (1976)
[32] Yano *et al.* (1978)
[33] Humphries *et al.* (1981)
[34] Lassiter *et al.* (1970)
[35] Suttle & Grace (1978)
[36] Grace & Suttle (1979)
[37] Mason *et al.* (1978)
[38] Pope *et al.* (1979)

nutrition situations. It should be noted also that interactions may involve more than two constituents.

Interations involving macro-minerals

Calcium and phosphorus

It is widely believed that the ratio of calcium to phosphorus in the diet of ruminants should be between 1:1 and 2:1. As calcium concentration in plants is almost invariably higher than phosphorus concentration (McDonald 1968), the deleterious effects which may be associated with ratios less than 1:1 (Wise et al. 1963) are restricted to hand feeding situations such as the feeding of grain. However Lueker & Lofgreen (1961) could detect no effect on calcium or phosphorus absorption of ratios from 0.8:1 to 6.0:1.

On the other hand, reports in the literature of the effects of wide calcium:phosphorus ratios greater than 2:1 are conflicting; in an earlier review, Little (1970) concluded that such ratios were probably of no consequence with respect to fertility, and possibly only marginally so for growth. It is salutary that ten years later, the ARC (1980) noted that ".. it is not possible to state the optimal ratio of calcium:phosphorus for animal performance or whether such a ratio actually exists".

A high level of dietary phosphorus associated with a calcium:phosphorus ratio less than 1:1 increased faecal calcium excretion, but when the ratio was greater than 1:1 a similar level of phosphorus reduced faecal calcium (Chicco et al. 1973). By contrast, an earlier experiment in the same laboratory (Standish et al. 1971) showed the apparent absorption of calcium to be significantly decreased by high dietary phosphorus at a constant ratio of 1.1:1, illustrating the problems of interpreting such studies. Hartmans (1971) demonstrated significant increases in apparent absorption and retention of phosphorus and sodium following the provision of a calcium supplement which changed the calcium:phosphorus ratio from 1.2 to 2.7. In that work the dietary phosphorus level was more than adequate at 0.41 percent and it has been suggested (e.g. ARC 1980) that a wide ratio is deleterious when the diet is deficient in phosphorus. This combination is of frequent occurrence in the tropics (Little 1970), but under these circumstances the extent to which the response to a phosphorus supplement is due to the correction of an inappropriate ratio rather than the direct correction of deficiency is obviously open to debate. Interaction between those minerals in terms of absorption occurs, in that calcium absorption is increased by low dietary phosphorus (Friedlander et al. 1977), and phosphorus transport in the duodenum is increased by calcium (Chen et al. 1974).

Thompson et al. (1959) reported the true absorption and retention of calcium to be inhibited by zinc, which had a similar but less marked effect upon phosphorus. The levels of dietary zinc employed, however, in the order of g/kg, would be judged potentially toxic by current standards (ARC 1980), and it seems most unlikely that such problems would occur at pasture. It is reassuring to note also that high levels of aluminium added to the diet had no detectable effect on calcium or phosphorus metabolism (Thompson et al. 1959). The experiments of Reid et al. (1979), in which herbage magnesium concentration was increased from 0.20 to 0.32 by application of fertilizer, likewise showed no influence of magnesium upon apparent absorption or retention of calcium or phosphorus.

The absorption of phosphate from the small intestine of the rat was shown to be directly related to the concentration of sodium in the digesta (McHardy & Parsons 1956, Harrison & Harrison 1963), but I know of no comparable observations in ruminants. Quite recently, however, Timet *et al*. (1978) used an *in vitro* technique to demonstrate that sodium concentration has a similar influence on calcium absorption in the bovine rumen, omasum and abomasum; this mechanism could be very important at times of high demand.

Magnesium

Table 4 shows magnesium to be greatly affected by interactions, and the extremely widespread problem of bovine hypomagnesaemia has prompted much experimental work. The high levels of plant potassium combined with lowered plant magnesium that occur following potash application (e.g. Reid & Jung 1974) frequently precipitate the clinical appearance of the disorder. Care *et al*. (1967) observed the true absorption of magnesium to be decreased by potassium and increased by sodium. Newton *et al*. (1972) showed that increasing dietary potassium from 0.6 to 4.9 percent provoked a reduction in apparent absorption of magnesium from *ca*. 50 to 26 percent, and that the change was basically decreased absorption rather than increased faecal endogenous excretion; this effect was confirmed by Field & Suttle (1979). Fontenot *et al*. (1973) concluded that high dietary nitrogen did not interfere with magnesium absorption or retention to a degree significant in the development of hypomagnesaemia, and suggested that nitrogen-induced increases in plant potassium might explain some observed problems. This is in contrast to the work of several others, for example, Reid *et al*. (1974), who applied fertilizer nitrogen and increased plant nitrogen from 2 percent to over 3 percent. This was associated with a marked reduction in both apparent absorption and retention of magnesium, despite a concurrent increase in plant magnesium from 0.13 to 0.20 percent, and little change in potassium. Kemp *et al*. (1966) related this association of high plant nitrogen and magnesium absorption and plasma concentration to the greater production of fatty acids at higher nitrogen levels in the rumen, leading to the possible formation of insoluble magnesium-soaps and consequently reduced absorption. The high-potassium diet used by McGregor & Armstrong (1979) reduced net magnesium absorption before the small intestine from 44 to 20 percent. This was associated with greatly increased absorption of potassium from the same area, and net retention of sodium was also greatly reduced. Fitt & Hutton (1974) postulated that the deleterious effect of potassium on magnesium utilization may operate in part via stimulated microbial activity in the rumen providing increased potential for the binding of magnesium to bacterial cell walls (Fitt *et al*. 1972).

Sulphur

The ruminant requirement for sulphur is almost entirely related to protein and associated sulphur-amino-acid requirement (McDonald 1968). The current recommendation for the dietary nitrogen:sulphur ratio is 14:1 (ARC 1980). The rumen microbial population degrades organic and inorganic sources of both nitrogen and sulphur, and uses them to synthesize microbial protein, and the degradability of dietary nitrogen and sulphur sources is clearly of significance for such synthesis. It is pertinent therefore that Reid *et al*. (1979) observed the application of magnesium fertilizer to increase the apparent absorption of plant sulphur. Pope *et al*. (1979) presented evidence that sulphur and selenium are closely related in metabolism, in that increasing dietary sulphur exerted a significant influence on urinary selenium excretion.

Interactions involving trace elements

Copper and molybdenum

The antagonism between dietary copper, molybdenum and sulphur, is probably the most widely known and thoroughly researched. The total interaction is very complex and operates both in the gut and systemically; space does not permit a thorough consideration of all aspects here. Briefly, the availability of dietary copper is reduced by increasing molybdenum concentrations in the presence of adequate levels of sulphur. The clinical appearance of copper toxicity and deficiency were originally associated respectively with very low and very high levels of dietary molybdenum, and these observations initially prompted the many investigations that have been conducted. Suttle & McLauchlan (1976) produced equations relating dietary copper availability to molybdenum and sulphur concentrations, which indicated a major independent effect of sulphur and a lesser sulphur-dependent effect of molybdenum. This is consistent with the intra-ruminal complexing of copper as the insoluble sulphide and poorly-available thiomolybdate, the former compound being implicated in the large depression in copper availability mentioned earlier in connection with the development of rumen function.

An excellent review of this subject was given by Suttle (1975a), and a very useful summary of the practical implications by the ARC (1980). Apart from the antagonism just described, another type of interaction has been identified. The enzyme xanthine oxidase, which occurs in milk, contains both copper and molybdenum, and its activity is determined by the ratio between copper- and molybdenum-containing isoenzymes. Kovalsky *et al*. (1974) found that varying amounts of copper and molybdenum in the diet directly influenced the amounts of those minerals in the enzyme. This phenomenon may allow the development of a sensitive diagnostic field test for copper and molybdenum deficiency.

Copper and zinc

The mutual antagonism between copper and zinc for absorption sites was mentioned earlier. Interference with zinc absorption by calcium observed in monogastrics is mainly due to phytate complexing (Becker & Hoekstra 1971) and may therefore be unimportant for ruminants, but such an effect nevertheless has been demonstrated in the absence of phytic acid, where phosphate level also was important (Heth *et al*. 1966). Recently, however, Kincaid (1979) noted that large amounts of calcium as the carbonate did not influence absorption of zinc supplied as several inorganic salts to lactating cows. Yano *et al*. (1978) also found no influence of a diet containing 1.2 percent calcium on zinc or magnesium, but increased faecal excretion and decreased retention of both copper and iron were observed. Schwarz & Kirchgessner (1980) described an interaction between zinc and manganese that operates at the intermediary level in that a deficiency of either has an influence on the retention of both, but neither has any effect on the absorption of the other from the gut.

Standish *et al*. (1971) obtained a significantly decreased apparent availability of copper when the iron content of the diet was increased from 0.1 to 1 g/kg, a modest level in the light of the data of Healy (1973). Further recent work (Humphries et al. 1981) has confirmed that a relatively minor increase in dietary iron to almost 1 g/kg for cattle rapidly reduced liver plasma copper to concentrations indicative of deficiency. This finding has considerable practical significance as large quantities of iron may be consumed with soil, as noted earlier.

Other minerals

Interactions also occur between essential and non-essential minerals; the problem is mainly associated with industrial pollution. High levels of cadmium adversely affect liver and blood copper and caeruloplasmin activity; Mills & Dalgarno (1972) in an experiment to examine the effects of feed cadmium levels commonly found in proximity to some industrial complexes, obtained marked deleterious effects at a concentration of only 12.3 ppm cadmium. Liver stores of copper, and to a lesser extent zinc were substantially reduced. According to Petering (1974) copper and iron are both involved in the ferroxidase activity of caeruloplasmin, and since both cadmium and lead inhibit the formation of caeruloplasmin, at least part of their toxicity relates to interference with the metabolism of copper, zinc and probably iron. Further study on such interactions is necessary due to the increasing incidence of industrial pollution.

CONCLUSION

In the routine evaluation of feed materials as sources of minerals for animals, greater attention should be paid to the limitations of measurements of apparent absorption. Further research is required on rates of faecal endogenous excretion of macro-minerals, so that the degree of influence exerted by the various factors involved may be more accurately defined.

The animals used in evaluation work should have a substantial physiological requirement for the minerals being studied, mediated by growth, lactation or by being marginally deficient in the element in question. This should ensure that the full potential of the feed to supply the minerals will be assessed.

The activity of rumen microflora in the main appears to render the chemical form in which minerals occur in plants of minimal consequence to their utilization by the animal, except in the cases of magnesium, copper and molybdenum. The marked influence on mineral utilization associated with the ingestion of soil and the incidence of gastro-intestinal parasitism is of great potential importance in the grazing environment, and warrants much more research work.

REFERENCES

ARC (1965) Nutrient requirements of farm livestock. No.2 Ruminants. London, Agricultural Research Council, 264 pp.

ARC (1980) The nutrient requirements of ruminant livestock. Technical review by an Agricultural Research Council working party. Farnham Royal, U.K., Commonwealth Agricultural Bureaux, 351 pp.

Ammerman, C.B.; Miller, S.M. (1972) Biological availability of minor mineral ions : a review. Journal of Animal Science 35, 681-694.

Andrew, C.S.; Robins, M.F. (1969) The effect of phosphorus on the growth and chemical composition of some tropical pasture legumes. II Nitrogen, calcium, magnesium, potassium and sodium contents. Australian Journal of Agricultural Research 20, 675-685

Becker, W.M.; Hoekstra, W.G. (1971) The intestinal absorption of zinc. In: Intestinal absorption of metal ions, trace elements and radionuclides. Editors S.C. Skoryna and D. Waldron-Edward. Oxford, Pergamon Press. pp. 229-256.

Ben-Ghedalia, D.; Tagari, H.; Zamwel, S.; Bondi, A. (1975) Solubility and net exchange of calcium, magnesium and phosphorus in digesta flowing along the gut of the sheep. British Journal of Nutrition 33, 87-94.

Bertoni, G.; Watson, M.J.; Savage, G.P.; Armstrong, D.G. (1976) The movements of minerals in the digestive tract of dry and lactating Jersey cows. 1. Net movements of Ca, P, Mg, Na, K and Cl. Zootecnica e Nutrizione Animale 2, 107-118.
Bisschop, J.H.R. (1965) Feeding phosphates to cattle. South African Department of Agricultural Technical Services Bulletin No. 365.
Braithwaite, G.D. (1974) The effect of changes of dietary calcium concentration on calcium metabolism in sheep. British Journal of Nutrition 31, 319-331.
Braithwaite, G.D. (1975) Studies on the absorption and retention of calcium and phosphorus by young and mature Ca-deficient sheep. British Journal of Nutrition 34, 311-324.
Braithwaite, G.D. (1976) Calcium and phosphorus metabolism in ruminants with special reference to parturient paresis. Journal of Dairy Research 43, 501-520.
Braithwaite, G.D. (1979) The effect of dietary intake of calcium and phosphorus on their absorption and retention by mature Ca-replete sheep. Journal of Agricultural Science, Cambridge 92, 337-342.
Bremner, I. (1970) The nature of trace element binding in herbage and gut contents. In: Trace element metabolism in animals. Editor C.F. Mills. Edinburgh, Livingston, pp.366-369.
Bremner, I.; Knight, A.H. (1970) The complexes of zinc, copper and manganese present in ryegrass. British Journal of Nutrition 24, 279-289.
Bremner, I.; Marshall, R.B. (1974) Hepatic copper- and zinc-binding proteins in ruminants. 1. Distribution of Cu and Zn among soluble proteins of livers of varying Cu and Zn content. British Journal of Nutrition 32, 283-291.
Butler, G.W.; Jones, D.I.H. (1973) Mineral biochemistry of herbage. In: 'Chemistry and biochemistry of herbage'. Editors G.W. Butler and R.W. Bailey. London, Academic Press. Vol 2, pp. 127-162.
Care, A.D.; van't Klooster, A.T.H. (1965) In vivo transport of magnesium and other cations across the wall of the gastro-intestinal tract of sheep. Journal of Physiology, London 177, 174-191.
Care, A.D.; Vowles, L.E.; Mann, S.O.; Ross, D.B. (1967) Factors affecting magnesium absorption in relation to the aetiology of acute hypomagnesaemia. Journal of Agricultural Science, Cambridge 68, 195-204.
Chen, T.C.; Castillo, L.; Korycka-Dahl, M.; DeLuca, H.F. (1974) Role of vitamin D metabolites in phosphate transport of rat intestine. Journal of Nutrition 104, 1056-1060.
Chicco, C.F.; Ammerman, C.B.; Feaster, J.P.; Dunavant, B.G. (1973) Nutritional interrelationships of dietary calcium, phosphorus and magnesium in sheep. Journal of Animal Science 36, 986-993.
Committee on Mineral Nutrition (1973) Tracing and treating mineral disorders in dairy cattle. Wageningen, Centre for Agricultural Publishing and Documentation, 61 pp.
Davies, J.G.; Scott, A.E.; Kennedy, J.F. (1938) The yield and composition of a Mitchell grass pasture for a period of twelve months. Journal of the Council for Scientific and Industrial Research, Australia 11, 127-139.
De Sousa, J.C. (1978) Interrelationships among mineral levels in soil, forage and animal tissues on ranches in northern Matto Grosso, Brazil. Ph.D. Thesis, University of Florida.
Dillon, J.; Scott, D. (1979) Digesta flow and mineral absorption in lambs before and after weaning. Journal of Agricultural Science, Cambridge 92, 289-297.
Dobson, A. (1959) Active transport through the epithelium of the reticulo-rumen sac. Journal of Physiology, London 146, 235-251.
Dobson, A.; Phillipson, A.T. (1968) Absorption from the ruminant forestomach. In: Handbook of Physiology, Section 6, Volume V. Editors C.F. Code and W. Heidel. Washington, D.C., American Physiological Society, pp.2761-2774.
Duncan, D.L. (1958) The interpretation of studies of calcium and phosphorus balance in ruminants. Nutrition Abstracts and Reviews 28, 695-715.
Engelhardt, W.V.; Hauffe, R. (1975) Funktionen des Blättermagens bei kleinen Hauswiederkäuern. IV. Resorption und Sekretion von Elektrolyten. Zentralblatt für Veterinärmedizin A22: 363-375.
Field, A.C. (1961) Studies on magnesium in ruminant nutrition. 3. Distribution of ^{28}Mg in the gastro-intestinal tract and tissues of sheep. British Journal of Nutrition 15, 349-359.
Field, A.C.; Suttle, N.F. (1970) Mineral excretion by three pairs of monozygotic cattle twins. Proceedings of the Nutrition Society 29, 34-35A.
Field, A.C.; Suttle, N.F. (1979) Effect of high potassium and low magnesium intakes on the mineral metabolism of monozygotic twin cows. Journal of Comparative Pathology 89, 431-438.
Field, A.C.; Suttle, N.F.; Nisbet, the late D.I. (1975) Effects of diets low in calcium and phosphorus on the development of growing lambs. Journal of Agricultural Science, Cambridge 85, 435-442.
Fitt, T.J.; Hutton, K. (1974) Effect of potassium ions on the uptake of magnesium by isolated cell walls of rumen bacteria. Proceedings of Nutrition Society 33, 107A-108A.
Fitt, T.J., Hutton, K., Thompson, A. and Armstrong, D.G. (1972) Binding of magnesium ions by isolated cell walls of rumen bacteria and the possible relation to hypomagnesaemia. Proceedings of the Nutrition Society 31, 100A.
Fitt, T.J.; Hutton, K.; Armstrong, D.G. (1979) Site of absorption of magnesium from the ovine digestive tract. Proceedings of the Nutrition Society 38, 65A.
Fleming, G.A. (1973) Mineral composition of herbage. In: Chemistry and biochemistry of herbage. Editors G.W. Butler and R.W Bailey. London, Academic Press. Vol 1, pp.529-566.

Fontenot, J.P.; Wise, M.B.; Webb, K.E. (1973) Interrelationships of potassium, nitrogen and magnesium in ruminants. Federation Proceedings 32, 1925-1928.
Friedlander, E.J.; Henry, H.L.; Norman, A.W. (1977) Studies on the mode of action of calciferol. Journal of Biological Chemistry 252, 8677-8683.
Fullmer, C.S.; Wasserman, R.H. (1973) Bovine intestinal calcium-binding proteins purification and some properties. Biochimica et Biophysica Acta 317, 172-186.
Gallaher, R.N. (1975) The occurrence of calcium in plant tissue as crystals of calcium oxalate. Communications in Soil Science and Plant Analysis 6, 315-330.
Gartner, R.J.W.; McLean, R.W.; Little, D.A.; Winks, L. (1980) Mineral deficiencies limiting producton of ruminants grazing tropical pastures in Australia. Tropical Grasslands 14, 266-272.
Grace, N.D.; Healy, W.B. (1974) Effect of ingestion of soil on faecal losses and retention of Mg, Ca, P, K and Na in sheep fed two levels of dried grass. New Zealand Journal of Agricultural Research 17, 73-78.
Grace, N.D.; Suttle, N.F. (1979) Some effects of sulphur intake on molybdenum metabolism in sheep. British Journal of Nutrition 41, 125-136.
Grace, N.D.; Ulyatt, M.J.; MacRae, J.C. (1974) Quantitative digestion of fresh herbage by sheep. III The movement of Mg, Ca, P, K and Na in the digestive tract. Journal of Agricultural Science, Cambridge 82, 321-330.
Greger, J.L.; Snedeker, S.M. (1980) Effect of dietary protein and phosphorus levels on the utilization of zinc, copper and manganese by adult males. Journal of Nutrition 110, 2243-2253.
Grunes, D.L.; Stout, P.R.; Brownell, J.R. (1970) Grass tetany of ruminants. Advances in Agronomy 22, 331-374.
Gueguen, L.; Demarquilly, C. (1965) Influence of the vegetative cycle and the growth stage on the mineral value of some herbage plants for adult sheep. Proceedings of the 9th International Grassland Congress, Sao Paulo, 745-754.
Gütte, J.O.; Lantzsch, H.J.; Molnár, S.; Lenkeit, W. (1961) Veränderungen der intestinalen Phosphor Absorption and Exkretion in Verlauf der Gravidität und Laktation des Schweines bei konstanter Ernährung. Zeitschrift für Tierphysiologie, Tierernährung und Futtermittelkunde 16, 75-90.
Hacker, J.B. (1982) Selecting and breeding better quality grasses. In: Nutritional limits to animal production from pastures. Editor J.B. Hacker. Farnham Royal, U.K., Commonwealth Agricultural Bureaux, pp. 305-326.
Hall, E.D.; Symonds, H.W. (1981) The maximum capacity of the bovine liver to excrete manganese in bile, and the effects of a manganese load on the rate of excretion of copper, iron and zinc in bile. British Journal of Nutrition 45, 605-611.
Harrison, H.E.; Harrison, H.C. (1963) Sodium, potassium and intestinal transport of glucose, l-tyrosine, phosphate and calcium. American Journal of Physiology 205, 107-111.
Hartmans, J. (1971) Effects of calcium on resorption and excretion of major and some minor elements in cattle. Proceedings of the 8th Colloquium International Potash Institute 207-211.
Hartmans, J.; Bosman, M.S.M. (1970) Differences in the copper status of grazing and housed cattle and their biochemical backgrounds. In: Trace element metabolism in animals. Editor C.F. Mills. Edinburgh, E. and S. Livingston, pp. 362-366.
Healy, W.B. (1973) Nutritional apsects of soil ingestion by grazing animals. In: Chemistry and biochemistry of herbage. Editors G.W. Butler and R.W. Bailey. London, Academic Press, Vol. 1, pp.567-588.
Healy, W.B. (1974) Ingested soil as a source of elements to grazing animals. In: Trace element metabolism in animals - 2. Editors W.G. Hoekstra, J.W. Suttie, H.E. Ganther, W. Mertz. Baltimore, University Park Press, pp.448-450.
Hegarty, M.P. (1982) Deleterious factors in forage affecting animal production. In: Nutritional limits to animal production from pastures. Editor J.B. Hacker. Farnham Royal, U.K., Commonwealth Agricultural Bureaux, pp. 133-150.
Heth, D.A.; Becker, W.M.; Hoekstra, W.G. (1966) Effect of Ca, P and Zn on ^{65}Zn absorption and turnover in rats fed semipurified diets. Journal of Nutrition 88, 331-337.
Hill, R. (1962) The provision and metabolism of calcium and phosphorus in ruminants. World Review of Nutrition and Dietetics 3, 129-148.
Horn, J.P.; Smith, R.H. (1978) Absorption of magnesium by the young steer. British Journal of Nutrition 40, 473-484.
Humphries, W.R.; Young, B.W.; Phillippo, M.; Bremner, I. (1981) The effects of iron and molybdenum on copper metabolism in cattle. Proceedings of the Nutrition Society 40, 68A.
Hutton, J.B.; Jury, K.E.; Davies, E.B. (1965) Studies of the nutritive value of New Zealand dairy pastures. IV The intake and utilization of magnesium in pasture herbage by lactating dairy cattle. New Zealand Journal of Agricultural Research 8, 479-496.
Jackson, M.J.; Smyth, D.H. (1971) Intestinal absorption of sodium and potassium. In: Intestinal absorption of metal ions, trace elements and radionuclides. Editors S.C. Skoryna and D. Waldron-Edward. Oxford, Pergamon Press, pp.137-150.
Jones, D.I.H. (1963) The mineral content of six grasses from a Hyparrhenia-dominant grassland in northern Rhodesia. Rhodesian Journal of Agricultural Research 1, 35-38.
Jones, L.H.P. (1978) Mineral components of plant cell walls. American Journal of Clinical Nutrition 31, 594-598.

Joyce, J.P.; Rattray, P.V. (1970a) Nutritive value of white clover and perennial ryegrass. II Intake and utilisation of sulphur, potassium and sodium. New Zealand Journal of Agricultural Research 13, 792-799.

Joyce, J.P.; Rattray, P.V. (1970b) Nutritive value of white clover and perennial ryegrass. III. Intake and utilisation of calcium, phosphorus and magnesium. New Zealand Journal of Agricultural Research 13, 800-807.

Kemp, A.; Guerink, J.H. (1978) Grassland farming and minerals in cattle. Netherlands Journal of Agricultural Science 26, 161-169.

Kemp, A.; Deijs, W.B.; Kluvers, E. (1966) Influence of higher fatty acids on the availability of magnesium in milking cows. Netherlands Journal of Agricultural Science 14, 290-295.

Kemp, A.; Korzeniowski, A.; Geurink, J.H. (1972) Method for estimating the sodium status of cows. Zeitschrift für Tierphysiologie, Tierernährung und Futtermittelkunde 29, 257-263.

Kincaid, R.L. (1979) Biological availability of zinc from inorganic sources with excess dietary calcium. Journal of Dairy Science 62, 1081-1985.

Kincaid, R.L.; Miller, W.J.; Gentry, R.P.; Neathery, M.W.; Hampton, D.L. (1976) Intracellular distribution of zinc and zinc-65 in calves receiving high but non-toxic amounts of zinc. Journal of Dairy Science 59, 552-555.

Kovalsky, V.V.; Vorotnitskaya, I.E.; Tsoi, G.G. (1974) Adaptive changes of the milk xanthine oxidase and its isoenzymes during molybdenum and copper action. In: Trace element metabolism in animals - 2. Editors W.G. Hoekstra, J.W. Suttie, H.E. Ganther and W. Mertz. Baltimore, University Park Press, pp.161-170.

Krawitt, E.L.; Stubbert, P.A.; Ennis, P.H. (1973) Calcium absorption and brush border phosphatases following dietary calcium restriction. American Journal of Physiology 224, 548-551.

Lamand, M.; Amboulou, D.; Rayssiguier, Y. (1977) Effect of quality of forage on availability of trace elements and some major elements. Annales Recherche Veterinaire 8, 303-306.

Lassiter, J.W.; Morton, J.D.; Miller, W.J. (1970) Influence of manganese on skeletal development in the sheep and rat. In: Trace element metabolism in animals. Editor C.F. Mills. Edinburgh, E. & S. Livingstone, pp. 130-132.

Lee, D.B.N.; Brautbar, N.; Walling, N.W.; Silis, V.; Coburn, J.W.; Kleeman, C.R. (1979) Effect of phosphorus depletion on intestinal calcium and phosphorus absorption. American Journal of Physiology 236, E451-E457.

Little, D.A. (1969) Oesophageal fistulation of cattle. Queensland Veterinary Proceedings 1969-1971, 41-42.

Little, D.A. (1970) Factors of importance in the phosphorus nutrition of beef cattle in northern Australia. Australian Veterinary Journal 46, 241-248.

Little, D.A. (1972) Bone biopsy in cattle and sheep for studies of phosphorus status. Australian Veterinary Journal 48, 668-670.

Little, D.A. (1980) Observations on the phosphorus requirement of cattle for growth. Research in Veterinary Science 28, 258-260.

Little, D.A.; McMeniman, N.P. (1973) Variation in bone composition of grazing sheep in south-western Queensland, related to lactation and type of country. Australian Journal of Experimental Agriculture and Animal Husbandry 13, 229-233.

Lueker, C.E.; Lofgreen, G.P. (1961) Effects of intake and calcium to phosphorus ratio on absorption of these elements by sheep. Journal of Nutrition 74, 233-239.

Mangan, J.L.; West, J. (1977) Ruminal digestion of chloroplasts and the protection of protein by glutaraldehyde treatment. Journal of Agricultural Science, Cambridge 89, 3-15.

Manston, R.; Vagg, M.T. (1970) Urinary phosphate excretion in the dairy cow. Journal of Agricultural Science, Cambridge 75, 161-167.

Mason, J.; Lamand, M.; Tressol, J.C.; Lab, C. (1978) The influence of dietary sulphur, molybdate and copper on the absorption, excretion and plasma fraction levels of ^{99}Mo in sheep. Annales Recherche Veterinaire 9, 577-586.

McDonald, I.W. (1968) The nutrition of grazing ruminants. Nutrition Abstracts and Reviews 38, 381-400.

McDowell, L.R.; Conrad, J.H.; Thomas, J.E.; Harris, L.E.; Fick, K.R. (1977) Nutritional composition of Latin American forages. Tropical Animal Production 2, 273-279.

McGregor, R.C.; Armstrong, D.G. (1979) The effect of increasing potassium intake on absorption of magnesium by sheep. Proceedings of the Nutrition Society. 38, 66A.

McHardy, G.J.R.; Parsons, D.S. (1956) The absorption of inorganic phosphate from the small intestine of the rat. Quarterly Journal of Experimental Physiology and Cognate Medical Sciences. 41, 398-409.

McIvor, J.G. (1979) Seasonal changes in nitrogen and phosphorus concentrations and in vitro digestibility of Stylosanthes species and Centrosema pubescens. Tropical Grasslands 13, 92-97.

McMeniman, N.P.; Little, D.A. (1974) Studies on the supplementary feeding of sheep consuming mulga (Acacia aneura). 1. The provision of phosphorus and molasses supplements under grazing conditions. Australian Journal of Experimental Agriculture and Animal Husbandry 14, 316-321.

Miller, J.K. and Cragle, R.G. (1965) Gastro intestinal sites of absorption and endogenous secretion of zinc in dairy cattle. Journal of Dairy Science 48, 370-373.

Miller, W.J.; Neathery, M.W.; Gentry, R.P.; Blackmon, D.M.; Lassiter; J.W.; Pate, F.M. (1972) Distribution and turnover rates of radio-active manganese in various tissues after duodenal dosing in Holstein calves fed a practical-type diet. Journal of Animal Science 34, 460-464.

Mills, C.F.; Dalgarno, A.C. (1972) Copper and zinc status of ewes and lambs receiving increased dietary concentrations of cadmium. Nature, London 239, 171-173.

Minson, D.J. (1977) The chemical composition and nutritive value of tropical legumes. In: Tropical forage legumes. Editor P.J. Skerman. Rome, FAO, pp.186-194.

Minson, D.J. (1982) The chemical composition and nutritive value of tropical grasses. In: Tropical grasses. Editor P.J. Skerman. Rome, FAO, (in press)

Molloy, L.F.; Richards, E.L. (1971) Complexing of Ca and Mg by the organic constituents of Yorkshire fog (Holcus lanatus). II. Complexing of Ca^{2+} and Mg^{2+} by cell wall fractions and organic acids. Journal of the Science of Food and Agriculture 22, 397-402.

Morris, J.G. (1980) Assessment of sodium requirements of grazing beef cattle : a review. Journal of Animal Science 50, 145-152.

Morrison, S.H.; Whitehair, C.K. (1956) Nutrition. In: Diseases of cattle. Editors M.G. Fincher, W.J. Gibbons, K. Mayer and S.E. Park. Illinois, American Veterinary Publications. pp.45-69.

Murphy, C.W.; Gartner, R.J.W. (1974) Sodium levels in the saliva and faeces of cattle on normal and sodium-deficient diets. Australian Veterinary Journal 50, 280-281.

NRC (1976) Nutrient requirements of domestic animals No. 4. Nutrient requirements in beef cattle. Washington, D.C., National Research Council, National Academy of Sciences, 56 pp.

Nel, J.W.; Moir, R.J. (1974) The effect of ruminal and duodenal application of different levels of calcium and phosphorus to sheep on semi-purified diets. South African Journal of Animal Science 4, 1-20.

Newton, G.L.; Fontenot, J.P.; Tucker, R.E.; Polan, C.E. (1972) Effects of high dietary potassium intake on the metabolism of magnesium by sheep. Journal of Animal Science 35, 440-445.

Norman, M.J.T. (1963) The pattern of dry matter and nutrient content changes in native pastures at Katherine, N.T. Australian Journal of Experimental Agriculture and Animal Husbandry 3, 119-124.

Norton, B.W. (1982) Differences between species in forage quality. In: Nutritional limits to animal production from pastures. Editor J.B. Hacker. Farnham Royal, U.K., Commonwealth Agricultural Bureaux, pp. 89-100.

O'Dell, B.L.; Campbell, B.J. (1970) Trace elements: metabolism and metabolic function. In: Comprehensive biochemistry. Editors M. Florkin and E.H. Stotz. Amsterdam, Elsevier. Vol. 21 pp. 179-266.

Pate, F.M.; Miller, W.J.; Blackmon, D.M.; Gentry, R.P. (1970) ^{65}Zn absorption rate following single duodenal dosing in calves fed zinc-deficient or control diets. Journal of Nutrition 100, 1259-1266.

Patil, B.D.; Jones, D.I.H. (1970) The mineral status of some temperate herbage varieties in relation to animal performance. Proceedings of the 11th International Grassland Congress, Surfers Paradise, Australia 726-730.

Peacock, M. (1976) Parathyroid hormone and calcitonin. In: Calcium phosphate and magnesium metabolism: chemical physiology and diagnostic procedures. Editor B.E.C. Nordin. Edinburgh, Churchill Livingstone pp.405-443.

Peeler, H.T. (1972) Biological availability of nutrients in feeds: availability of major mineral ions. Journal of Animal Science 35, 695-712.

Petering, H.G. (1974) The effect of cadmium and lead on copper and zinc metabolism. In: Trace element metabolism in animals - 2. Editors W.G. Hoekstra, J.W. Suttie, H.E. Ganther, W. Mertz. Baltimore, University Park Press. pp.311-325.

Pfeffer, E.; Thompson, A.; Armstrong, D.G. (1970) Studies on intestinal digestion in the sheep. 3. Net movement of certain inorganic elements in the digestive tract on rations containing different proportions of hay and rolled barley. British Journal of Nutrition 24, 197-204.

Phillips, S.F.; Code, C.F. (1967) Sorption of potassium in the small and the large intestine. American Journal of Physiology 211, 607-613.

Playne, M.J. (1976) Availability of phosphorus in feedstuffs for utilization by ruminants. In: Prospects for improving efficiency of phosphorus utilization. Reviews in Rural Science No. III. Editor G.J. Blair. Armidale, Australia, University of New England, pp. 155-164.

Pope, A.L.; Moir, R.J.; Somers, M.; Underwood, E.J.; White, C.L. (1979) The effect of sulphur on ^{75}Se absorption and retention in sheep. Journal of Nutrition 109, 1448-1455.

Post, T.B. (1965) Changes in levels of salivary sodium and potassium associated with the mustering of beef cattle. Australian Journal of Biological Sciences 18, 1235-1239.

Powell, K.; Reid, R.L.; Balasko, J.A. (1978) Performance of lambs in perennial ryegrass, smooth bromegrass, orchard grass and tall fescue pastures. II. Mineral utilization, in vitro digestibility and chemical composition of herbage. Journal of Animal Science 46, 1503-1514.

Reid, R.L.; Horvath, D.J. (1980) Soil chemistry and mineral problems in farm livestock. A review. Animal Feed Science and Technology 5, 95-167.

Reid, R.L.; Jung, G.A. (1974) Effects of elements other than nitrogen on the nutritive value of forage. In: Forage fertilization. Editor D. Mays. Madison, Amercian Society of Agronomy. pp. 395-435.

Reid, R.L.; Franklin, M.C.; Hallsworth, E.G. (1947) The utilization of phytate phosphorus by sheep. Australian Veterinary Journal 23, 136-140.

Reid, R.L.; Daniel, K.; Bubar, J.D. (1974) Mineral relationships in sheep and goats maintained on orchard grass fertilized with different levels of nitrogen, or nitrogen with micro-elements, over a five year period. Proceedings of the 12th International Grassland Congress, Moscow 3(1), 426-437.
Reid, R.L.; Jung, G.A.; Roemig, I.J.; Kocher, R.E. (1978) Mineral utilization by lambs and guinea pigs fed Mg fertilized grass. Agronomy Journal 70, 9-14.
Reid, R.L.; Jung, G.A.; Wolf, C.H.; Kocher, R.E. (1979) Effects of magnesium fertilization on mineral utilization and nutritional quality of alfalfa for lambs. Journal of Animal Science 48, 1191-1201.
Robinson, D.W.; Sageman, R. (1967) The nutritive value of some pasture species in north-western Australia during the late dry season. Australian Journal of Experimental Agriculture and Animal Husbandry 7, 533-539.
Rogers, P.A.M.; van't Klooster, A.T.H. (1969) Observations on the digestion and absorption of food along the gastro-intestinal tract of fistulated cows. 3. The fate of Na, K, Ca, Mg and P in the digesta. Mededelingen Landbouwhogeschol, Wageningen 69, 26-39.
Sass-Korstak, A. (1965) Copper metabolism. Advances in Clinical Chemistry 8, 1-65.
Saylor, W.W.; Morrow, F.D.; Leach, R.M. (1980) Copper- and zinc-binding proteins in sheep liver and intestine: effects of dietary levels of the metals. Journal of Nutrition 110, 460-468.
Schlachter, D.; Dowdle, E.G.; Schenker, H. (1960) Active transport of calcium by the small intestine of the rat. American Journal of Physiology 198, 263-268.
Schwarz, K. (1974) New essential trace elements (Sn, V, F, Si): progress report and outlook. In: Trace element metabolism in animals - 2. Editors W.G. Hoekstra, J.W. Suttie, H.E. Gouther, W. Mertz. Baltimore, University Park Press, pp. 355-380.
Schwarz, F.J.; Kirchgessner, M. (1980) Experimentelle Untersuchungen zur Interaktion zwischen den Spurenelementen Zink und Mangan. Zietschrift für Teirphysiologie, Tierernährung und Futtermittelkunde 42, 272-282.
Shaw, N.H. (1978) Superphosphate and stocking rate effects on a native pasture oversown with Stylosanthes humilis in central coastal Queensland. 1. Pasture production. Australian Journal of Experimental Agriculture and Animal Husbandry 18, 788-799.
Smith, R.H.; McAllan, A.B. (1966) Binding of magnesium and calcium in the contents of the small intestine of the calf. British Journal of Nutrition 20, 703-718.
Stake, P.E.; Miller, W.J.; Neathery, N.W.; Gentry, R.P. (1975) Zinc-65 absorption and tissue distribution in two- and six-month old Holstein calves and lactating cows. Journal of Dairy Science 58, 78-81.
Standish, J.F.; Ammerman, C.B.; Palmer, A.Z.; Simpson, C.F. (1971) Influence of dietary iron and phosphorus on performance, tissue mineral composition and mineral absorption in steers. Journal of Animal Science 33, 171-178.
Stevens, C.F. (1964) Transport of sodium and chloride by the isolated rumen epithelium. American Journal of Physiology 206, 1099-1105.
Stevenson, H.M.; Unsworth, E.F. (1978) Studies on the absorption of calcium, phosphorus, magnesium, copper and zinc by sheep fed on roughage-cereal diets. British Journal of Nutrition 40, 491-496.
Stewart, J.; Moodie, E.W. (1956) The absorption of magnesium from the alimentary tract of sheep. Journal of Comparative Pathology 66, 10-21.
Suttle, N.F. (1974) Recent studies of the copper-molybdenum antagonism. Proceedings of the Nutrition Society 33, 299-305.
Suttle, N.F. (1975a) Trace element interactions in animals. In: Trace elements in soil-plant-animal systems. Editors D.J.D. Nicholas and A.R. Egan. New York, Academic Press, pp. 271-289.
Suttle, N.F. (1975b) Changes in the availability of dietary copper to young lambs associated with age and weaning. Journal of Agricultural Science, Cambridge 84, 255-261.
Suttle, N.F.; McLauchlan, M. (1976) Predicting the effects of dietary molybdenum and sulphur on the availability of copper to ruminants. Proceedings of the Nutrition Society 35, 22A-23A.
Suttle, N.F.; Alloway, B.J.; Thornton, I. (1975) An effect of soil ingestion on the utilization of dietary copper by sheep. Journal of Agricultural Science, Cambridge 84, 249-254.
Suttle, N.F.; Grace, N.D. (1978) A demonstration of marked recycling of molybdenum via the gastro-intestinal tract of sheep at low sulphur intakes. Proceedings of the Nutritional Society 37, 68A.
Sykes, R.; Coop, R.L. (1976) Intake and utilization of food by growing lambs with parasitic damage to the small intestine caused by daily dosing with Trichostrongylus colubriformis larvae. Journal of Agricultural Science, Cambridge 86, 507-515.
Sykes, A.R.; Dingwall, R.A. (1975) Calcium absorption during lactation in sheep with demineralized skeletons. Journal of Agricultural Science, Cambridge 84, 245-248.
Sykes, A.R.; Field, A.C. (1972) Effects of dietary deficiencies of energy, protein and calcium on the pregnant ewe. 1. Body composition and mineral content of the ewes. Journal of Agricultural Science, Cambridge 78, 109-117.
Sykes, A.R.; Nisbet, the late D.I.; Field, A.C. (1973) Effects of dietary deficiencies of energy, protein and calcium on the pregnant ewe. V. Chemical analyses and histological examination of some individual bones. Journal of Agricultural Science, Cambridge 81, 433-440.

Taylor, T.G. (1979) Availability of phosphorus in animal feeds. Proceedings of the 13th Nutrition Conference for Feed Manufacturers, University of Nottingham, 23-33.

Thewis, A.; Francois, E.; Thielemans, M.F. (1978) Etude quantitative de l'absorption et de la sécrétion du phosphore total et du phosphore phospholipidique dans le tube digestif du Mouton. Annales de Biologie Animale Biochemie Biophysique 18, 1181-1195.

Thompson, A.; Hansard, S.L.; Bell, M.C. (1959) The influence of aluminium and zinc upon the absorption and retention of calcium and phosphorus in lambs. Journal of Animal Science 18, 187-197.

Timet, D.; Emanović, D.; Herak, M.; Kraljević, P.; Mitin, V. (1978) The effect of sodium concentration in the contents on gastric absorption of calcium in cattle. Veterinarski Arhiv 48 (Suppl) 537-538.

Underwood, E.J. (1966) The mineral nutrition of livestock. Farnham Royal, U.K., Commonwealth Agricultural Bureaux, 237 pp.

Underwood, E.J. (1971) Trace elements in human and animal nutrition. 3rd Edition. New York, Academic Press, 543 pp.

Van Campen, D.R. (1969) Copper interference with the intestinal absorption of Zinc-65 by rats. Journal of Nutrition 97, 104-108.

Van Campen, D.R.; Scaife, P.V. (1967) Zinc interference with copper absorption in rats. Journal of Nutrition 91, 473-476.

van't Klooster, A. Th. (1976) Adaptation of calcium absorption from the small intestine of dairy cows to changes in the dietary calcium intake and at the onset of lactation. Zeitschrift für Teirphysiologie, Tierernährung und Futtermittelkunde 37, 169-182.

Ward, G.; Harbers, L.H.; Blaha, J.J. (1979) Calcium containing crystals in alfalfa: their fate in cattle. Journal of Dairy Science 62, 715-722.

Wasserman, R.H.; Taylor, A.N. (1966) Vitamin-D_3-induced calcium-binding in chick intestinal mucosa. Science 152, 791-793.

Wasserman, R.H.; Taylor, A.N. (1973) Intestinal absorption of phosphate in the chick: effect of vitamin D_3 and other parameters. Journal of Nutrition 103, 586-599.

Whitehead, D.C. (1966) Nutrient minerals in grassland herbage. Commonweath Bureau of Pastures and Field Crops. Mimeograph Publication No. 1/1966.

Wiener, G. (1971) Genetic variation in mineral metabolism of ruminants. Proceedings of the Nutrition Society 30, 91-101.

Wilkinson, R. (1976) Absorption of calcium, phosphorus and magnesium. In: Calcium, phosphate and magnesium metabolism: clinical physiology and diagnostic procedures. Editor B.E.C. Nordin, Edinburgh, Churchill Livingstone, pp. 36-112.

Wise, M.B.; Ordoveza, A.L.; Barrick, E.R. (1963) Influence of variation in dietary calcium:phosphorus ratio on performance and blood constituents of calves. Journal of Nutrition 79, 79-84.

Yano, H.; Nokata, M.; Kawashima, R. (1978) Effects of supplemental calcium carbonate on the metabolism of iron, copper, zinc and manganese in sheep. Japanese Journal of Zootechnical Science 49, 625-631.

Young, V.R.; Lofgreen, G.P.; Luick, J.R. (1966a) The effects of phosphorus depletion, and of calcium and phosphorus intake, on the endogenous excretion of these elements by sheep. British Journal of Nutrition 20, 795-805.

Young, V.R.; Richards, W.P.C.; Lofgreen, G.P.; Luick, J.R. (1966b) Phosphorus depletion in sheep and the ratio of calcium to phosphorus in the diet with reference to calcium and phosphorus absorption. British Journal of Nutrition 20, 783-794.

PART 6

OVERCOMING LIMITATIONS TO ANIMAL PRODUCTION

SELECTING AND BREEDING BETTER LEGUMES

R.A. BRAY

CSIRO, Division of Tropical Crops and Pastures, Davies Laboratory, Townsville, Queensland 4814, Australia.

ABSTRACT

Selecting for quality in legumes is not often an important objective, since legumes generally are of high quality. However by increasing yield of legumes the quality of the overall pasture is increased.

In selecting better legumes, different criteria are used at different stages in the development of a species. In early development stages, selection is for general adaptation and yield. As a species (or variety) becomes more widely grown, its limitations become recognized, and may be remedied by breeding. Disease resistance and some antiquality factors are examples. Finally, in the fine-tuning stage the breeder may select for such characters as seed production, specific environmental adaptation, and "quality".

Some examples are given of selection against negative quality factors. Successful programs have been mainly on relatively simple biological systems, often governed by a single gene. Programs that as yet are unsuccessful (such as selection against bloat) have often dealt with complex systems, involving interaction of many factors.

Breeding for digestibility and nitrogen content is briefly considered, although most legumes are already high in these attributes.

It is suggested that factors that limit the possibilities for improving quality include the complexity of quality characteristics, the existence of undesirable correlations, and the lengthy programs needed.

Finally, an attempt is made to suggest where the major problems lie in quality breeding in legumes, and what is the likelihood of solving them.

INTRODUCTION

The importance of legumes as components of natural rangelands and sown pasture depends on their ability to fix atmospheric nitrogen by means of the *Rhizobium* symbiosis. Legumes thus have a high nitrogen content, and are generally rich in minerals and highly nutritious (Davies *et al*. 1967). Legumes may be cut for hay, grazed as a pure sward, or be a component of a mixed pasture. In the latter situation they play a dual role in increasing animal production: not only do they provide protein to the grazing animal, but they also provide a source of nitrogen for the associate grass.

What then would make a "better" legume? The answer must surely be "an improvement in any character which is limiting animal production". What these limiting characters are depends on the particular species, its history as a pasture legume, and the purpose for which it is being used.

At a Symposium such as this, with an audience composed (I suspect) mainly of animal scientists, one might reasonably expect a paper on "Selecting and breeding better legumes" to concentrate mainly on the prospects of selecting legumes for *higher quality*. Indeed the Organising Committee has asked me to do this. However, I feel that the question of improving quality in legumes needs to be put into proper perspective. Legumes grown in mixed pastures serve to increase the quality of those pastures, and therefore any increase in legume yield or persistence will be of value. Hence, much selection work has the general objective of improving *pasture* quality rather than *legume* quality.

In order to determine where the greatest advances in legume improvement are to be made, and where selection for quality fits into the overall scheme, I would like to look first at the general process of developing new and better legumes.

STAGES OF PASTURE LEGUME IMPROVEMENT

In general, selection of better legumes is a progressive process, moving through three broad stages. It is instructive to look at each of these stages, with examples of the sort of characters that might be important in each stage.

Preliminary stage

Some agricultural systems use legumes that have been cultivated for thousands of years. There is evidence that lucerne (*Medicago sativa*) was used as a forage over 3,000 years ago (Bolton *et al.* 1972). It is now probably the most widely sown pasture legume in the world, and used over vast areas.

In contrast to this, in some situations no suitable legume is available. The primary requirement is then to find a legume that is productive and persistent. A good example of this is the breeding of *Macroptilium atropurpureum* cv. Siratro by E.M. Hutton (Hutton 1962). When intensive pasture research in the Australian sub-tropics began in the mid-1950's there were literally no suitable legumes available. Although *Centrosema pubescens* was being used in the wet tropics, and *Stylosanthes humilis* was beginning to make its mark in the semi-arid tropics, the range of species on which experimental data were available was still very limited (Shaw & Whiteman 1977). The annual *M. lathyroides* had grown well on a range of soil types, mixed well with grasses, and was highly acceptable to animals, but still had serious deficiencies, such as poor regeneration from seed under grazing and susceptibility to root-knot nematode (Hutton 1962). Hutton therefore set out to breed, within this genus, a strongly perennial, disease resistant legume that would grow well over a wide range of climates and soils in northern Australia. Breeding proceeded from crosses involving Mexican introductions, combining good vigour with stoloniferous development. During the breeding program selection was mainly for yield, adaptation, and tolerance to grazing - all basic prerequisites for success.

Another example of this early stage is the selection of *Stylosanthes* cultivars in north Queensland. The annual *S. humilis* was an accidental introduction into Australia during the nineteenth century, and spread naturally over considerable areas, becoming known as a valuable pasture legume. The first attempt at improvement in this species was by D.F. Cameron, who selected cultivars on the basis of maturity (flowering time) and yield, to provide a range of adapted types. The promise of *S. humilis* (as well as its environmental limitations) led to the evaluation of other species in the genus, with selection for persistence, yield and adaptation. This has produced a number of cultivars of *S. guianensis*, *S. hamata*, and *S. scabra*, suitable for many tropical areas (Edye & Grof 1981).

The requirements for passage out of this preliminary stage of development are persistence and productivity under grazing, with all the implications of these generalized criteria: satisfactory yield, adaptation to a range of environments, satisfactory establishment and nodulation, and ready acceptance by animals. All legumes in commercial use must have fulfilled these requirements at some point during their evaluation and selection.

Development stage

The next phase of legume improvment is the process of discovering the limitations of the plant, and proceeding to overcome these. The history of *Medicago sativa* in the USA provides a useful example.

Almost all the *M. sativa* grown in the south-western and central states traces back to Spanish germplasm, introduced *via* Peru and Mexico. For many years various regional strains (known as "commons") provided the only sources of seed. Some of these regional strains were considerably changed from the original introductions by natural selection, but none were really hardy enough for the northern states. However, introductions from central Europe, planted in these cold areas, led to the development of winter-hardy varieties such as Grimm (selected from a farmer's field in Minnesota). The popularization of Grimm probably completed the preliminary phase referred to above, and enabled the next stage of development to proceed.

Since the mid-1920's much of the emphasis has been on breeding for disease and insect resistance, as a means of increasing persistence and yield. As specific pests and diseases became important, the responsible organisms recognized, and sources of resistance identified, plant breeders rapidly produced resistant varieties. In the USA, between 1950 and 1980, registration descriptions of 95 cultivars of *M. sativa* were published in Crop Science. It is possible, from these descriptions, to establish their selection criteria. Obviously, most cultivars have been selected for more than one character, but in general terms, of the 95, 65 have been selected for disease or pest resistance, and 25 for persistence or winterhardiness. (I cannot resist the observation that none has been specifically selected for quality. Indeed, few descriptions even consider it - some varieties are "dark green", others "leafy". No variety is claimed to be more nutritious.)

Table 1 summarizes the characteristics of those registered alfalfa varieties, and also of some other important North American legume cultivars. Selection in all cases has been based largely on yield and adaptation, rather than quality.

Success in improving disease resistance in lucerne is not restricted to North America, as is shown by successful selection for

TABLE 1

Selection criteria for new varieties of legumes registered in the USA during the period 1950-1980.

Species	No. of cultivars	Criterion					
		Disease and/or pest resistence	Persistence, winterhardiness	Forage yield, recovery after cutting	Seed yield	Adaptation, maturity	Other
Medicago sativa	95	68	25	37	20	3	18
Trifolium repens	4	1	4	3			
T. pratense	7	7		3			
Lespedeza cuneata	9	5		3	1	3	1
Onobrychis viciifolia	3			3			
Lotus corniculatus	4	1	1	4	2		

resistance to *Verticillium* wilt in Europe (e.g. cvv. Sabilt and Verneuil) and *Phytophthora* in Australian (Rogers *et al*. 1978, Bray & Irwin 1978, Irwin *et al*. 1980). Lucerne, with its wide availability of germplasm, seemingly inexhaustible genetic variation, and ease of genetic manipulation is without doubt the star in the disease resistance breeding field. However it must be remembered that it is also the most widely grown forage legume, and certainly has had more scientific effort expended on it than on all other forage legumes put together.

Many other legumes have also been selected for resistance to diseases or pests that have come to be recognized as important. In Western Australia, resistance of subterranean clover (*Trifolium subterraneum*) to clover scorch (*Kabatiella*) has become an important objective. In the last few years it has become apparent that anthracnose (*Colletotrichum gloeosporioides*) is threatening the success of *Stylosanthes* in much of the tropical country to which it is adapted. This has led to the selection of resistant cultivars (*S. scabra* cv. Seca and *S. guianensis* cv. Graham) for areas where the disease is important. Recently, rust (*Uromyces appendiculatus*) has become widespread on siratro in Australia. The extent to which this disease affects animal production and plant persistence is not yet known, but its occurrence serves to illustrate the possibility of new disease problems arising as a species becomes widely grown.

In addition to selection for disease and pest resistance, this development phase often includes selection against "negative" quality factors. These include oestrogens and toxic compounds. Frequently their importance only becomes recognized once a plant has been widely grown, as was the case with the oestrogen problem in *T. subterraneum*. I will not consider these factors here, but will treat them in detail in a later section.

The characters that are selected for in this development stage generally become essential criteria for any future selections, and screening for them becomes routine.

Fine tuning stage

We have now arrived at a stage where we have available high yielding, persistent, palatable legumes. What remains to be done? Williams (1981) suggests that in *Trifolium repens* what remains is to select characters involving compatibilities with other organisms: complementarity or compatibility with grasses, quality for the grazing animal, compatibility with *Rhizobium*, and compatibility with insect pollinators.

In general these requirements also apply to most other species of legumes, to a greater or lesser degree, depending on the particular plant.

i) Compatibility with grass. In *T. repens*, varieties are often classified according to leaf size. Although small leafed varieties tend to be more persistent under heavy grazing, this is not a simple relationship. The relative persistence of varieties depends on several factors including management, fertilizer use and the incidence of fungal disease. Larger leaved varieties with long petioles are less likely to suffer from competition from grass whose growth has been stimulated by nitrogen fertilizer. There are few other legumes where the characters necessary for successful combination with grass can be so readily stipulated.

ii) Quality for the grazing animal. Characters considered here may be thought of as "positive" factors. Breeders are concerned with increasing the levels of desirable nutrient qualities. I shall return to this aspect in a later section.

iii) Compatibility with Rhizobium. Most attempts to increase the nitrogen-fixing potential of legumes have concentrated on the selection of superior strains of Rhizobium. However, there is ample evidence that plant genetic factors are involved, and that selection for improved nodulation (and hopefully subsequent growth) should be possible in M. sativa (eg. Brockwell & Hely 1966, Leach 1968), T. pratense (Nutman et al. 1971), T. repens (Mytton & Jones 1971) and Desmodium intortum (Imrie 1975). Recent work by Barnes and colleagues (e.g. Viands & Barnes 1978) confirms that significant variation exists among M. sativa genotypes for factors associated with nitrogen fixation.

However, in most studies, advantages expressed under laboratory conditions are often not expressed in the field, or when plants are nodulated by a strain other than that used for selection. So far, no significant advances in nitrogen fixation in the field have been achieved.

Hoglund & Brock (1978) suggested another approach to attaining increases in total nitrogen fixation in T. repens. They pointed out the necessity of developing genotypes which have lowered sensitivity to mineral nitrogen in the soil so that fixation is not inhibited, thus maintaining growth at high and low levels of soil nitrogen. Alternatively, genotypes could be bred which respond rapidly to a fall in mineral nitrogen level, for example, as a result of active grass growth, and re-establish fixation.

iv) Compatibility with insect pollinators. Many of the widely used forage legumes are cross-pollinated, and thus insects are important as pollinators. However seed production in general is also an important breeding objective. Gladstones (1975a) reviewed the selection criteria of the west Australian T. subterraneum program: disease resistance and low oestrogen levels had been achieved and emphasis was largely on characters associated with seed production and germination, and specific adaptation. This has subsequently enabled this species to be used in improving pasture quality in new zones, through the introduction of hardseededness (Francis & Gladstones 1981).

To Williams' four compatibilities I would thus add "compatibility with the environment". As the breeder's knowledge becomes more refined, he can select populations that have particular edaphic and/or climatic adaptations.

The fore-going has been a brief examination of the general process of developing improved legumes, and shows that different characters are important at different stages. In the beginning, selection for legume quality is not important, but selection for pasture quality is. In some species, selection for quality is never important; frequently, it is only in the "fine tuning" stage that it receives attention.

NEGATIVE AND POSITIVE QUALITY FACTORS

Negative factors

As a family, legumes, because of their high protein content, are an attractive feed source for the grazing animal (and insects). Thus many legumes have evolved either physical (thorns) or chemical (toxins, unpalatability) defenses against grazing (see Hegarty 1982). In addition they may contain undesirable compounds which affect animal growth rate and/or reproduction. Problems with some of these biochemical factors may not become apparent until a species has been in use for some time. In the remainder of this section I will deal with some examples of reduction of these "negative" quality factors.

Oestrogenic isoflavones in *Trifolium subterraneum*

In 1963, following the recognition of differences in oestrogenicity among *T subterraneum* varieties in Western Australia (Davies & Bennett 1962) formononetin was identified as the major oestrogenic factor (Millington *et al*. 1964). Most of the parents used in the breeding program at that time were high in formononetin.

Artificial mutation techniques were applied to agronomically proven parent lines such as Geraldton and Yarloop. The Geraldton mutant L58 (Francis & Millington 1965b) was released in 1967 as cv. Uniwager, but proved to be too low yielding in terms of dry matter production to be commercially viable.

At the same time a rapid screening technique for isoflavones (Francis & Millington 1965a) made possible a survey of the then available ecotypes, many of which proved to have naturally low formononetin contents (Francis & Millington 1965a, Gladstones 1967). Two ecotypes which had already proved themselves in Western Australia in long-term agronomic trials conducted by the CSIRO (Rossiter 1966) were Daliak and Seaton Park, and these were released commercially in 1967. Screening for low isoflavone content is now routine in subclover breeding (Francis & Gladstones 1981).

Coumarin in *Melilotus alba*

All varieties of *M. alba* (sweet clover) grown commercially in the USA until the early 1950's had a high content of coumarin which gave the forage a bitter taste and caused low palatability. The cultivar Denta (Smith 1964) was bred to have low coumarin (typically 0.03 percent of dry matter compared to 3-4 percent in other varieties. Denta was bred from a cross of *M. alba* with coumarin free *M. dentata*. The F_1 hybrids had only a trace of chlorophyll and could be raised to maturity only through grafting on normal sweetclover plants. Backcrossing these F_1's to *M. alba* produced a few seeds, which grew into weak plants. However, following self-pollination, one progeny had a few normal green, weak, low coumarin seedlings. Vigour and productivity were combined with the low coumarin character by repeating four times the program of backcrossing the low coumarin segregates to vigorous high coumarin plants and selecting within the progeny of the backcross plants for low coumarin content, productivity, disease resistance, and forage quality. Coumarin content is governed by a single gene, with high coumarin being partly dominant (Goplen *et al*. 1957).

Subsequently, low coumarin varieties have also been developed in Canada. Goplen (1969) has found that the high coumarin varieties are usually superior in terms of seed yield, forage yield, and seedling vigour.

Bloat

Bloat in cattle grazing certain temperate legume-based pastures has long been a problem. Bloat is caused by the formation of a stable foam in the rumen which prevents eructation of gases produced by microbial fermentation of ingested food. Soluble leaf proteins are the principal foaming agents in legumes which cause bloat (Mangan 1959, Kendall 1966, Jones *et al*. 1970).

Studies have shown that legumes which do not cause bloat, such as sainfoin (*Onobrychis viciifolia*) and birdsfoot trefoil (*Lotus corniculatus*) contain protein precipitants, principally condensed tannins, in their leaves (Jones & Lyttleton 1971, Jones *et al*. 1973, Ross & Jones 1974). Tannins do not occur in the leaves of bloat inducing legumes. The tannins apparently act to reduce the level of soluble proteins in ingested legume leaves below that required to induce bloat (Miltimore *et al*. 1970).

In the hope of finding tannin-containing genotypes large screening programs have been undertaken in some species. In New Zealand no tannin-containing genotypes were discovered in a massive screening of *T. repens* (Pandey 1971, Williams 1981). Marshall *et al*. (1979) screened 530 ecotypes of *T. subterraneum*, but found no evidence of tannin production. In fact only five of 113 *Trifolium* species tested contained tannins in their leaves. This suggests that breeding of useful tannin-containing *Trifolium* would be difficult.

Screening large populations of *Medicago* for tannins also produced negative results (Goplen *et al*. 1980, Marshall *et al*. 1981). Marshall has suggested that mutagensis offers the greatest hope in producing both *Trifolium* and *Medicago* genotypes that contain tannin. Inter-specific hybridization is likely to be a more difficult approach.

Two protein fractions capable of producing stable foams have been implicated in the bloat syndrome in *M. sativa* (Howarth *et al*. 1973). A threshold level of 1.8 percent 18S protein has been proposed as a guide, above which bloating may occur (Miltimore *et al*. 1970). However recent results show no relationship between the content of fractions I and II proteins in lucerne and incidence and severity of bloat in grazing ruminants (Goplen & Howarth 1977). Rumbaugh (1969) used *in vitro* foam tests to study the inheritance of bloat-inducing qualities of *M. sativa*, in a 12 clone diallel cross. The heritability estimate of 0.73 suggested that selection should be successful. Howarth *et al*. (1973) investigated the soluble protein concentration of 289 plants of diverse genetic origin. Their conclusion was that it should be possible to select for low soluble protein concentration as a means of developing a low-bloat cultivar. Recently, resistance to mesophyll cell rupture has been proposed as a criterion for low bloat varieties (Howarth & Goplen 1978). The proportion of ruptured cells can be estimated indirectly by determining the soluble protein content.

The saponin complex in *M. sativa* has also been implicated in ruminant bloat. There is ample evidence that it should be possible to modify saponin content (Hanson *et al*. 1963). Selection for both high and low saponin has been possible (Jones 1969, cited in Elliott *et al*. 1972, Pedersen & Wang 1971), but no variety specifically selected for low saponin has been released.

Thus, in spite of extensive investigation, and the apparent existence of suitable genetic variability, to my knowledge no low-bloat cultivar of *M. sativa* has been produced.

In contrast with temperate legumes, bloat is not a problem in tropical legumes. Only *Lablab purpureus* has been reported to induce bloat (Hamilton & Ruth 1968), but the absence of bloat cannot be explained on the basis of tannin content (Hutton & Coote 1960).

Mimosine in *Leucaena leucocephala*

Leucaena leucocephala is a leguminous tree which is rapidly expanding in use as a forage. In some areas of the world, high proportions of leucaena in an animal's diet can lead to problems of lack of appetite, loss of weight, and thyroid enlargement (Jones 1979). These clinical signs are due to the presence in the plant of mimosine, a non-essential amino acid. This is metabolized in the rumen to 3-hydroxy-4(1H)-pyridone, which is the goitrogenic agent (Hegarty *et al*. 1976). Within *L. leucocephala*, although there appears to be some variation for mimosine (Brewbaker & Hylin 1965) much of this variation is associated with differences in growth rate and development (Bray, unpublished data). Consequently it has been necessary to seek genotypes with low mimosine concentration in related species. E.M. Hutton hybridized *L. leucocephala* and *L. pulverulenta* (a low mimosine species) and backcrossed to *L. leucocephala* to try to improve the self-fertility of the hybrids. After three backcross generations with selection for low mimosine, reasonable fertility was restored. Field trials of this material showed that mimosine content was about 60 percent of control *L. leucocephala* cultivars, but dry matter yield was not high (Bray, unpublished data). However, under conditions which give rise to extreme mimosine toxicity in *L. leucocephala* animals perform better on the low mimosine lines mainly because of their lower productivity and hence lower percentage in the diet (R.J. Jones, personal communication).

Current interest centres on the direct utilization of the F_1 interspecific hybrids which are extremely vigorous, and have intermediate levels of mimosine (Bray, unpublished).

The examples given above are by no means exhaustive. However they do serve to illustrate various approaches to the problem of antiquality factors.

Positive factors

Characters which will be considered in this context are levels of desirable nutrients. Although such characters receive considerable attention in the literature, most data have been obtained only from comparisons between varieties. This tells us little, except that differences in various characters, such as nitrogen content, do exist. This is not surprising, since the breeder's "eye-ball" selection often includes characters such as leafiness during early evaluation. Many of the differences between varieties are related to other characters, such as flowering date, and provide little useful information to the breeder.

Most detailed information in the literature is on *M. sativa*, and I will deal mainly with this species.

Digestibility

There are conflicting reports in the literature concerning the possibility of selecting for digestibility in lucerne. Although Thomas *et al.* (1968), Shenk & Elliott (1970) and Davies (1979) all suggest that progress should be possible, Kellogg *et al.* (1976) and Hill & Barnes (1977) found that little or no response to selection could be expected. These different views may have been due to differences in the germplasm pools used in the different experiments: heritability was relatively high when the sample of parents differed for only a few characters, but was low when many factors were involved. In one case where variation was significant (Gil *et al.* 1967) all parents had percentage digestibility exceeding 74 percent. Could any real improvement be expected at this level?

There are no strong indications of correlations between yield and digestibility, but data generally indicate that the best way to increase total yield of digestible dry matter is to increase forage yield rather than digestibility.

In other species there are reports of differences in digestibility (e.g. *Centrosema* - Clements 1977) but little information on inheritance is available.

In *Lespedeza cuneata*, high levels of tannin inhibit digestion (Donnelly & Anthony 1969, 1970, Cope & Burns 1971). Although low tannin concentration is a simply inherited character, there has not been a successful low tannin, high digestible cultivar registered in this species. However 'Serala' sericea was selected as having fine pliable stems "more suitable for animal consumption" (Donnelly 1965).

Nitrogen content

This also responds to selection in *M. sativa* (Gil *et al.* 1967, Heinrichs *et al.* 1969). In general, selection should be based both on "leafiness" and on chemical analysis. Hill & Barnes (1977) have suggested that selection for greater protein content or reduced ADF, NDF or lignin would be better than selection for greater digestibility.

Minerals

Levels of minerals are generally satisfactory in temperate forage legumes, except perhaps under conditions of very intensive management. In such situations some problems may occur; for example phosphorus levels may not be high enough for milking cows. Only with management practices that maximize yield are mineral imbalances likely to occur. Heinrichs *et al.* (1969) found variation for chemical constituents and minerals to be as great as for morphological characters. Hill & Jung (1975) found that additive genetic variance (and therefore possibility for selection) existed for most elements, and thought that phosphorus deficiency could be corrected by breeding. However, selection for any one element would influence the concentration of others, and may result in no improvement in mineral balance for cattle diets. Yield is generally negatively correlated with mineral content (Hill & Barnes 1977).

Relationships between quality factors

Since nutrient quality is the result of the interaction of many factors, selection for one aspect of higher nutritive value will undoubtedly have implications for others. I can do no better than to quote *verbatim* the comments of Elliott *et al*. (1972): "There is, in general, a negative relationship between yield and quality. High yields are often associated with mature and tall forage, which is usually high in fibre and lignin and low in protein, but not necessarily so. Number, length, and structure of stems contribute highly to yields of dry matter, while leaves contribute primarily to protein, ash, and nitrogen free extract. Leafiness and protein content are positively correlated, and percent protein usually decreases with maturity. Stem lignification proceeds with age, and there is more fibre in stems than in leaves. Consequently, dry matter yields increase and digestibility decreases with maturity".

Since the stem fraction of *M. sativa* becomes less digestible with age, it has been suggested that selection might be made for more digestible stems. Davies (1979) has shown that there is little advantage in analyzing stems instead of the whole plant.

In some plant species, such as *T. repens*, only leaves and petioles are eaten, and consequently stem factors became less important. Nutrient yield is then directly related to leaf yield.

The relationship between leaf diseases and quality

Although leaf diseases are generally of concern because of the reduction they cause in leaf yield, they may also affect fodder quality, including palatability, digestibility, and protein content (Schoth & Hyslop 1929, Hanson 1965, Raymond 1969). Thus *T. repens* plants infested with rust (*Uromyces*) contained 21 percent less carotene than non-rusted plants (Sullivan & Chilton 1941, quoted in Smith 1952). Brigham (1959) reported a 40 percent reduction in the protein content of *M. sativa* leaves infected with *Cercospora*. This type of effect may become more marked as plants age. Morgan & Parbery (1980) found that infection of *M. sativa* by *Pseudopeziza medicaginis* reduced the digestibility by 14 percent and crude protein by 16 percent. Recently R.J. Jones (personal communication) showed that *Uromyces* depressed digestibility in *Macroptilium atropurpureum* by about four percentage units (ie. 5.5 percent). There is also evidence that infection by fungi causes, or increases, the oestrogenic activity of *M. sativa* (Hanson *et al*. 1965).

The above results only serve to emphasize the importance of disease resistance breeding. The progress to be made by eliminating or reducing the effects of disease is likely to be far greater than could be made by selection for quality itself.

Are varieties that have been selected for higher quality likely to be more disease susceptible? There is little evidence available, but Ingham (1978) found that low coumarin *Melilotus alba* would not be expected to be more diseased, because levels of phytoalexin production remained high. However, low mimosine lines of *L. leucocephala* seem to harbour more glasshouse pests (such as mealy bugs) than high mimosine lines (Bray, unpublished data) and low saponin in *M. sativa* may result in increased susceptibility to pea aphids (Hanson *et al*. 1973).

FACTORS AFFECTING BREEDING FOR QUALITY

Now that breeding for quality has been put into reasonable perspective in legume breeding, we can examine some of the factors affecting the progress of breeding for quality.

What are the requirements for successful selection for quality? On the basis of past successes, I can suggest four pre-requisites.

i) The problem must have been accurately identified. An example of this is the early breeding work on _Indigofera spicata_ by E.M. Hutton. The aim of the breeding program was to breed lines free from toxin, but the work had to be abandoned when the chemical concerned could not be identified (Hutton 1965). An accurate analytical technique thus could not be developed. Subsequent work has identified both the amino acid indospicine, which is hepatotoxic to mammals but not to chicks, and 3-nitropropanoic acid which is toxic to chicks but not hepatotoxic to rats and mice (Britten _et al_. 1963, Hegarty & Pound 1970).

ii) Variation exists for the particular factor. Although variation for quantitative factors such as digestibility and nitrogen content may well exist, it may be more difficult to find variation for particular chemical substances. Thus in breeding _Melilotus_ and _Leucaena_ for lower levels of toxins it has been necessary to go to related species to find the necessary variation. In _T. repens_ none of the commercial species provided useful levels of tannins (Marshall _et al_. 1979). The use of mutation breeding has often been suggested in such a situation, and while this may be useful in a self-pollinated diploid such as _T. subterraneum_, it is of much less relevance to many of the other pasture legumes which are highly polyploid (e.g. leucaena is an octoploid with $2n = 104$).

iii) The factor is controlled by a simple genetic system. The examples of _T. subterraneum_ and _Melilotus alba_ given above are good examples of this, as is the significant progress in lupin breeding which has made extensive progress through the use of single gene control of sweetness and pod shattering characteristics (Gladstones 1975b).

It is of interest that it would be possible to select against cyanogenesis in _Trifolium repens_ since there is considerable variability, under simple genetic control (Corkill 1942). However, grazing animals appear not to be affected by the presence of cyanogenic glycosides, and consequently selection against cyanogensis has no priority as a breeding objective (Williams 1981).

iv) Successful programs have dealt with clear cut chemical products (e.g. coumarin) rather than with a biochemical that is involved in a complex series of reactions and interactions before producing an observable effect. An obvious example of a difficult problem is bloat, which is a complex interaction of plant, animal, and microbial factors. Selection against any "bloat causing factor" may well be effective, but any effect on actual incidence and severity may be less certain.

There is also a number of factors that may make breeding for quality difficult. Not the least of these is that most legumes already of high quality, although some legumes are not used because of the presence of compounds causing toxicity or unpalatability. Others are so good that they are used in spite of whatever problems they may cause - these are probably the ones that may be difficult to change.

Where complex characters such as positive quality factors are involved, as many as ten generations (or more) of selection appear necessary to ensure reasonable fixation and stability across a series of environments (Elliott et al. 1972). Even so, interactions with management practices (or even changes in accepted management practices) may largely negate any breeding progress.

In some cases, correlations between characters may limit progress. This may well be the case in Leucaena, where there is a positive genetic correlation between yield and mimosine content (Bray, unpublished data). This suggests it may be difficult to breed a high yielding, low mimosine line. Since mimosine is derived from lysine, an essential amino acid, selection for low mimosine may only result in selection for slow growing plants.

WHERE CAN PROGRESS BE MADE IN BREEDING FOR QUALITY?

While there are numerous alternative strategies for increasing quality, it seems that lowering levels of negative components deserves high priority. The presence of such compounds interferes with the utilization of the "positive" components. In addition, many of the negative components seem to be under simple genetic control, and therefore relatively easy to select against. In general, the more complex the character, the lower the heritability.

Most interest at present centres on tannins - either on the possibility of introducing tannins where bloat is important, or of greater utilization of those legumes with low levels of tannin that provide very high quality feed.

In Table 2 I have set out what I consider to be the possibilities for improving quality in major pasture legumes. I have endeavoured to put some index on the importance of each problem, relative to other breeding objectives in the species. I have also tried to estimate the likelihood of success in each case. These estimations are necessarily very subjective.

From the table it is obvious that I consider that, in temperate legumes, there are very few high priority objectives in breeding for quality, relative to the other selection criteria in each individual species. However, in species such as T. subterraneum, continued routine screening for oestrogens is essential.

The tropical legumes are, of course, in a less refined stage of development than the temperates, and there are more obvious problems. In the perennial Stylosanthes, quality of these pasture declines when the legumes shed their leaves in the dry (cool) season. Following any precipitation, the shed leaves rapidly mould and become unattractive to the grazing animal (Gardener 1980). Selection of genotypes which retain their leaves, flowers and pods longer into the dry season would be valuable. The presence of mimosine in leucaena is influencing its use, and is a problem of some importance.

In addition to species listed in Table 2, there are some that are not widely used (perhaps not at all) because of quality problems. For example, agronomists have reservations about the palatability of the otherwise promising S. viscosa. In Colombia, Desmodium ovalifolium is showing considerable promise, but intake is low when grown in sward with a low proportion of grass. Selection for increased palatability is thus a major objective (Anon. 1980). Again, the use of species such as Cassia, Crotolaria, Indigofera and Tephrosia is restricted by their toxicity. These species offer considerable promise, and major screening programs are underway (R.W. Strickland, personal communication).

TABLE 2

Prospects of selecting improved quality legumes for grazing.

Species	Character	Means of achieving objective	Priority*	Chance of success
a) Temperate				
M. sativa	bloat	reduce foaming proteins	medium	low
		increase tannins	medium	low
	digestibility	more digestible stems	medium	medium
T. pratense	bloat	tannins	medium	low
	digestibility	more digestible stems	medium	medium
	oestrogens		medium	high
T. repens	bloat	tannins	medium	low
	oestrogens		low	high
	cyanogenesis		nil	high
T. subterraneum	bloat	tannins	low	low
	oestrogens		high	certain
b) Tropical				
Stylosanthes spp. (perennial)	increased leaf retention		high	high
Leucaena leucocephala	mimosine		high	medium
Macroptilium atropurpureum	maintenance of high quality	rust resistance	high	high

* relative to other breeding objectives.

In this paper I have attempted to set selection for pasture quality and selection for legume quality in the proper perspective. The examples I have given are not intended to be a complete catalogue of quality breeding, but rather to illustrate the type of problems the breeder may be faced with, and to help in the realistic assessment of the probability of success of an improvement program.

REFERENCES

ANON (1980) CIAT Report, p. 83.

Bolton, J.L.; Goplen, B.P.; Baenziger, H. (1972) World Distribution and Historical Developments. In: Alfalfa Science and Technology. Editor C.H. Hanson. Madison, American Society of Agronomy pp. 1-34.

Bray, R.A.; Irwin, J.A.G. (1978) Selection for resistance to Phytophthora megasperma var. sojae in Hunter River lucerne. Australian Journal of Experimental Agriculture and Animal Husbandry 18, 708-713.

Brewbaker, J.L.; Hylin, J.W. (1965) Variations in mimosine content among Leucaena species and related Mimosaceae. Crop Science 5, 348-349.

Brigham, R.D. (1959) Effect of Cercospora disease on forage quality of alfalfa. Agronomy Journal 51, 365.

Britten, E.J.; Palafox, A.L.; Frodyma, M.M.; Lynd, F.T. (1963) Level of 3-nitropropanoic acid in relation to toxicity of Indigofera spicata in chicks. Crop Science 3, 415-416.

Brockwell, J.; Hely, F.W. (1966) Symbiotic characteristics of Rhizobium meliloti. An appraisal of the systematic treatment of nodulation and nitrogen fixation interactions between hosts and rhizobia of diverse origins. Australian Journal of Agricultural Research 17, 885-899.

Clements, R.J. (1977) Agronomic variation in Centrosema virginianum in relation to its use as a subtropical pasture plant. Australian Journal of Experimental Agriculture and Animal Husbandry 17, 435-444.

Donnelly, E.D. (1965) Serala sericea. Crop Science 5, 605.

Donnelly, E.D.; Anthony, W.B. (1969) Relationship of tannin, dry matter digestibility and crude protein in Sericea lespedeza. Crop Science 9, 361-362.

Donnelly, E.D.; Anthony, W.B. (1970) Effect of genotype and tannin on dry matter digestibility in Sericea lespedeza. Crop Science 10, 200-202.

Cope, W.A.; Burns, J.C. (1971) Relationship between tannin levels and nutritive value of sericea. Crop Science 11, 231-233.

Corkill, L. (1942) The inheritance of cyanogenesis. New Zealand Journal of Science and Technology 23, 178B-193B.

Davies, W.E. (1979) Investigations of in vitro digestibility of lucerne. Biuletyn Instytutu Hodowli i Aklimatyzacji Roślin, Nr. 135, pp. 138-149.

Davies, W. E.; Thomas, T.A.; Young, N.R. (1967) The assessment of herbage legume varieties. III. Annual variation in chemical composition of eight varieties. Journal of Agricultural Science, Cambridge 71, 233-241.

Davies, H. L.; Bennett, D. (1962) Studies on the oestrogenic potency of subterranean clover (Trifolium subterraneum L.) in south-western Australia. Australian Journal of Agricultural Research 13, 1030-1040.

Edye, L.A.; Grof, B. (1981) Selecting cultivars from naturally occurring genotypes : evaluating Stylosanthes species. In: Genetic resources of forage plants. Editors J.G. McIvor and R.A. Bray. Melbourne, CSIRO (in press).

Elliott, F.C.; Johnson, I.J.; Schonhorst, M.H. (1972) Breeding for forage yield and quality. In: Alfalfa science and technology. Editor C.H. Hanson. Madison, American Society of Agronomy. pp.319-333.

Francis, C.M.; Gladstones, J.S. (1981) Exploitation of the genetic resource through breeding Trifolium subterraneum. In: Genetic resources of forage plants. Editors J.G. McIvor and R.A. Bray. Melbourne, CSIRO (in press).

Francis, C.M.; Millington, A.J. (1965a) Varietal variation in the isoflavone content of subterranean clover: its estimation by a microtechnique. Australian Journal of Agricultural Research 16, 557-564.

Francis, C.M.; Millington, A.J. (1965b) Isoflavone mutations in subterranean clover. I. Their production, characteristics and inheritance. Australian Journal of Agricultural Research 16, 565-573.

Gardener, C.J. (1980) Diet selection and liveweight performance of steers on Stylosanthes hamata - native grass pastures. Australian Journal of Agricultural Research 31, 379-392.

Gil, H. Chaverra; Davies, R.L.; Barnes, R.F. (1967) Inheritance of in vitro digestibility and associated characteristics in Medicago sativa L. Crop Science 7, 19-21.

Gladstones, J.S. (1967) Naturalized subterranean clover strains in Western Australia: A preliminary agronomic evaluation. Australian Journal of Agricultural Research 18, 713-731.

Gladstones, J.S. (1975a) Legumes and Australian agriculture. Journal of the Australian Institute of Agricultural Science 41, 227-240.

Gladstones, J.S. (1975b) Lupin breeding in Western Australia : the narrow-leafed lupin (Lupinus angustifolius). Journal of Agriculture Western Australia 16, 44-49.

Goplen, B.P. (1969) Forage yield and other agronomic traits of high- and low-coumarin isosynthetics of sweetclover. Crop Science 9, 477-480.

Goplen, B.P.; Howarth, R.E. (1977) Breeding a bloat-safe alfalfa (Medicago sativa L.) cultivar. Proceedings of the 13th International Grassland Congress, Leipzig, 355-358.

Goplen, B.P.; Greenshields, J.E.R.; Baenziger, H. (1957). The inheritance of coumarin in sweet clover. Canadian Journal of Botany 35, 583-593.

Goplen, B.P.; Howarth, R.E.; Sarkar, S.K.; Lesins, K. (1980) A search for condensed tannins in annual and perennial species of Medicago, Trigonella, and Onobrychis. Crop Science 20, 801-804.

Hamilton, R.I.; Ruth, G. (1968) Bloat on Dolichos lablab. Tropical Grasslands 2, 135-136.

Hanson, C.H. (1965) Foliar diseases and forage quality. Proceedings of the 9th International Grassland Congress, Sâo Paulo, 1209-1213.

Hanson, C.H.; Kohler, G.O.; Dudley, J.W.; Sorensen, E.L.; Van Atta, G.R.; Taylor, K.W.; Pedersen, M.W.; Carnahan, H.L.; Wilsie, C.P.; Kehr, W.R.; Lowe, C.C.; Stanford, E.H.; Yungen, J.A. (1963) Saponin content of alfalfa as related to location, cutting, variety, and other variables. US Department of Agriculture, ARS 34-44, 38 pp.

Hanson, C.H.; Pedersen, M.W.; Berrang, B.; Wall, M.E.; Davis, K.H. (1973) The saponins in alfalfa cultivars. In: Antiquality components of forages. Editor A.G. Matches. Madison, Wisconsin. Crop Science Society of America. pp. 33-52.

Hegarty, M.P. (1982) Deleterious factors in forages affecting animal production. In: Nutritional limits to animal production from pastures. Editor J.B. Hacker. Farnham Royal, U.K., Commonwealth Agricultural Bureaux, pp. 133-150.

Hegarty, M.P.; Pound, A.W. (1970) Indospicine, a hepatotoxic amino acid from Indigofera spicata: isolation, structure and biological studies. Australian Journal of Biological Sciences 23, 831-842.

Hegarty, M.P.; Court, R.D.; Christie, M.D.; Lee, C.P. (1976) Mimosine in Leucaena leucocephala is metabolised to a goitrogen in ruminants. Australian Veterinary Journal 52, 490.

Heinrichs, D.H.; Troelsen, J.E.; Warder, F.G. (1969) Variation of chemical constituents and morphological characters within and between alfalfa populations. Canadian Journal of Plant Science 49, 293-305.

Hill, R.R.; Barnes, R.F. (1977) Genetic variability for chemical composition of alfalfa. II. Yield and traits associated with digestibility. Crop Science 17, 948-952.

Hill, R.R.; Jung, G.A. (1975) Genetic variability for chemical composition in alfalfa. I. Mineral elements. Crop Science 15, 652-657.

Hoglund, J.H.; Brock, J.L. (1978) Regulation of nitrogen fixation in a grazed pasture. New Zealand Journal of Agricultural Research 21, 73-82.

Howarth, R.E.; McArthur, J.M.; Hikichi, M.; Sarkar, S.K. (1973) Bloat investigations: denaturation of alfalfa fraction. II. Proteins by foaming. Canadian Journal of Animal Science 53, 439-443.

Howarth, R.E.; Goplen, B.P. (1978) Recent developments in breeding a bloat-safe alfalfa cultivar. Report of the 26th Alfalfa Improvement Conference, South Dakota State University, South Dakota. Editor D.K. Barnes. p.21.

Howarth, R.E.; McArthur, J.M.; Goplen, B.P. (1973) Bloat investigations: Determination of soluble protein concentration in alfalfa. Crop Science 13, 677-680.

Hutton, E.M. (1962) Siratro - a tropical pasture legume bred from Phaseolus atropurpureus. Australian Journal of Experimental Agriculture and Animal Husbandry 2, 117-125.

Hutton, E.M. (1965) A review of the breeding of legumes for tropical pastures. Journal of the Australian Institute of Agricultural Science 31, 102-109.

Hutton, E.M.; Coote, J.C. (1966) Tannin content of some tropical legumes. Journal of the Australian Institute of Agricultural Science 32, 139-140.

Imrie, B.C. (1975) The use of agar tube culture for early selection for nodulation of Desmodium intortum. Euphytica 24, 625-631.

Ingham, J.L. (1978) Phytoalexin production by high- and low-coumarin cultivars of Melilotus alba and Melilotus officinalis. Canadian Journal of Botany 56, 2230-2233.

Irwin, J.A.G.; Lloyd, D.L.; Bray, R.A.; Langdon, P.W. (1980) Selection for resistance to Colletotrichum trifolii in the lucerne cultivars Hunter River and Siro Peruvian. Australian Journal of Experimental Agriculture and Animal Husbandry 20, 447-451.

Jones, Merlyn (1969) Evaluation of alfalfa plants for saponin. Ph.D. thesis. Michigan State University, E. Lancing, Michigan.

Jones, R.J. (1979) The value of Leucaena leucocephala as a feed for ruminants in the tropics. World Animal Reveiw No. 31, 1-11.

Jones, W.T.; Lyttleton, J.W. (1971) Bloat in cattle. XXXIV. A survey of legume forages that do and do not produce bloat. New Zealand Journal of Agricultural Research 14, 101-107.

Jones, W.T.; Lyttleton, J.W.; Clarke, R.T.J. (1970) Bloat in cattle XXIII. The soluble proteins of legume forages in New Zealand, and their relationships to bloat. New Zealand Journal of Agricultural Research 13, 149-156.

Jones, W.T.; Anderson, L.B.; Ross, M.D. (1973) Bloat in cattle. XXXXIX. Detection of protein precipitants (flavolans) in legumes. New Zealand Journal of Agricultural Research 16, 441-446.

Kellogg, D.W.; Melton, B.A.; Watson, C.E.; Miller, D.D. (1976) Genetic and environmental effects on nutritive content of alfalfa. New Mexico Agricultural Experimental Station Research Report No. 308, 9pp.

Kendall, W.A. (1966) Factors affecting foams with forage legumes. Crop Science 6, 487-489.

Leach, G.J. (1968) The effectiveness of nodulation of a wide range of lucerne cultivars. Australian Journal of Experimental Agriculture and Animal Husbandry 8, 323-326.

Mangan, J.L. (1959) Bloat in cattle. XI. The foaming properties of proteins, saponins, and rumen liquor. New Zealand Journal of Agricultural Research 2, 47-61.

Marshall, D.R.; Broué, P.; and Munday, J. (1979) Tannins in pasture legumes. Australian Journal of Experimental Agriculture and Animal Husbandry 19, 192-197.

Marshall, D.R.; Broué, P.; Grace, J.; Munday, J. (1981) Tannins in pasture legumes. 2. The annual and perennial Medicago species. Australian Journal of Experimental Agriculture and Animal Husbandry 21, 55-58.

Millington, A.J.; Francis, C.M.; McKeown, N.R. (1964) Wether bioassay of annual pasture legumes. II. The oestrogenic activity of nine strains of Trifolium subterraneum L. Australian Journal of Agricultural Research 15, 527-536.

Miltimore, J.E.; McArthur, J.M.; Mason, J.L.; Ashby, D.L. (1970) Bloat investigations. The threshold fraction I (18S) protein concentration for bloat and relationships between bloat and lipid, tannin, Ca, Mg, Ni and Zn concentrations in alfalfa. Canadian Journal of Animal Science 50, 61-68.

Morgan, Wendy C.; Parbery, D.G. (1980) Depressed fodder quality and increased oestrogenic activity of lucerne infected with Pseudopeziza medicaginis. Australian Journal of Agricultural Research 31, 1103-1110.

Mytton, L.R.; Jones, D.G. (1971) The response to selection for increased nodule tissue in white clover (Trifolium repens L.). Plant and Soil. Special volume, Biological nitrogen fixation in natural and agricultural habitats, pp.17-25.

Nutman, P.S.; Mareckova, H.; Raicheva, L. (1971) Selection for increased nitrogen fixation in red clover. Plant and Soil. Special volume, Biological nitrogen fixation in natural and agricultural habitats, pp.27-31.

Pandey, K.K. (1971) Prospects of breeding non-bloating clovers for ruminants. Proceedings of the 1st Agronomy Society of New Zealand Conference, Christchurch, pp. 111-120.

Pedersen, M.W.; Wang, Li-Chun (1971) Modification of saponin content of alfalfa through selection. Crop Science 11, 833-835.

Raymond, W.F. (1969) The nutritive value of forage crops. Advances in Agronomy 21, 1-108.

Rogers, V.E.; Irwin, J.A.G.; Stovold, G. (1978) The development of lucerne with resistance to root rot in poorly aerated soils. Australian Journal of Experimental Agriculture and Animal Husbandry 18, 434-441.

Ross, M.D.; Jones, W.T. (1974) Bloat in cattle. XL. Variation in flavanol content in Lotus. New Zealand Journal of Agricultural Research 17, 191-195.

Rossiter, R.C. (1966) The success or failure of strains of Trifolium subterraneum L. in a mediterranean environment. Australian Journal of Agricultural Research 17, 425-446.

Rumbaugh, M.D. (1969) Inheritance of foaming properties of plant extracts of alfalfa. Crop Science 9, 438-440.

Schoth, H.A.; Hyslop, G.R. (1929) Alfalfa in western Oregon. Oregon Experimental Station Bulletin No. 246.

Shaw, N.H.; Whiteman, P.C. (1977) Siratro - a success story in breeding a tropical pasture legume. Tropical Grasslands 11, 7-14.

Shenk, John S.; Elliott, F.C. (1970) Two cycles of directional selection for improved nutritive value of alfalfa. Crop Science 10, 710-712..

Smith, D.C. (1952) Breeding for quality. Proceedings of the 6th International Grassland Congress, State College, Pennsylvania 1597-1606.

Smith, W.K. (1964) Denta sweet clover. Crop Science 4, 666-667.

Sullivan, J.T.; Chilton, J.P. (1941). The effect of leaf rust on the carotene content of white clover. Phytopathology 31, 554-557.

Thomas, J.W.; Campbell, J.L.; Tesar, M.B.; Elliott, F.C. (1968) Improved nutritive value of alfalfa by crossing and selection based on an in vitro method. Journal of Animal Science 27, 1783-1784.

Viands, D.R.; Barnes, D.K. (1978) Response from selection in alfalfa for factors associated with nitrogen fixation. Report of the 26th Alfalfa Improvement Conference, South Dakota State University, Brookings, South Dakota, 27.

Williams, W.M. (1981) Exploitation of the genetic resource through breeding: Trifolium repens. In: Genetic resources of forage plants. Editors J.G. McIvor and R.A. Bray. Melbourne, CSIRO. (in press).

SELECTING AND BREEDING BETTER QUALITY GRASSES

J.B. HACKER.

CSIRO, Division of Tropical Crops and Pastures, Cunningham Laboratory, St. Lucia, Queensland 4067, Australia.

ABSTRACT

Genetic variation in digestibility, mineral and toxin concentration, water soluble carbohydrates, fibre content and characteristics associated with intake is reviewed. Heritability estimates and phenotypic and genotypic correlations between quality characters are presented, and breeding programs designed to improve grass quality are discussed.

It is concluded that considerable variation exists within species for characteristics associated with quality and that this variation may be exploited by the plant breeder to overcome existing limits to animal production.

INTRODUCTION

Grasses always have provided and probably always will provide the major part of the diet of domesticated ruminants. It has long been known that grasses differ in their capacity to provide productive pastures. In recent years many of the characteristics of grasses which are associated with 'good quality' or 'poor quality' have been identified, but there is still no better definition of a good pasture than that of the simple shepherd Corin in "As you like it" in his statement "Good pasture makes fat sheep" (Shakespeare 1623).

The plant breeder cannot, of course, breed for animal production directly, but is obliged to breed for characters which are known to be associated with higher animal production. Before reviewing variation in factors contributing to pasture quality it is pertinent to enquire what evidence there is for variation within grass species in animal production potential, which is not simply due to differences in yield. Table 1 shows animal production data for high and low quality varieties of a number of species. Clearly there are substantial differences between varieties, in the order of 10-30 percent.

The characteristics that make a grass "better" vary from species to species. In the absence of toxic factors, and provided no essential element is in short supply, yield of milk or meat is the product of intake of digestible energy and efficiency of utilization. Selecting plants for higher digestibility will not lead to improved animal performance if the plant is seriously deficient in an essential element, or is susceptible to a toxin-producing organism such as the bacterium responsible for ryegrass toxicity in South Australia (Culvenor et al. 1978, Michelmore et al. 1980). Clearly the factor or factors in grass species which limit animal production must be defined before the plant breeder can be expected to produce an improved cultivar.

TABLE 1

Animal production (kg *per diem*) from varieties of the same species grazed under conditions in which availability did not limit intake.

Species	Stock	Low quality	High quality
Cynodon dactylon[1]	Cattle	0.67	0.74
	Cattle	0.50	0.67
Dactylis glomerata[2]	Cattle	0.60	0.66
Eragrostis curvula[3]	Cattle	0.44	0.49
Lolium perenne[2]	Cattle	0.66	0.74
Lolium spp.[4]	Sheep	0.09	0.14

1 Chapman *et al*. 1971
2 Evans *et al*. 1979
3 Voigt *et al*. 1970
4 Rae *et al*. 1964

It is also important for the breeder to have some estimate of the extent of genetic variation within the species he is working with, and the nature of any genetic correlations with desirable or undesirable plant characteristics. If a high proportion of the variation in a character is non-genetic, selection will not result in a marked improvement.

In this paper I will first consider genetic variation in characters associated with digestibility and intake. Efficiency of utilization will then be discussed, followed by sections on toxic and anti-quality characteristics and on essential elements. Where information is available the inheritance and genetic correlations between characters will be reviewed in an attempt to determine any underlying pattern or principle.

DIGESTIBILITY AND INTAKE

In general digestibility and intake are the most serious quality limitations to animal production from pastures. When a wide range of forages is compared there is an overall positive correlation between the two characteristics - the more digestible the forage, the more it is eaten. This relationship breaks down when the capacity of the feed to supply energy exceeds the energy requirements of the animal.

Digestibility depends on the proportion of cell contents, which are completely digested, and of cell wall, in which digestibility depends on the extent of lignification. As pasture grasses mature, digestibility decreases due to the decreasing proportion of the younger more digestible leaf, increasing proportion of older, less digestible leaves and increasing lignification of the stem with flowering and maturation. Thus improvement in digestibility by breeding could potentially operate through any one of a number of pathways, for example, later flowering, greater leafiness, reduced lignification or higher proportion of cell contents.

Digestibility and intake cannot be measured on small samples. Before 1960 digestibility of small samples was estimated from crude fibre or crude protein concentration and it was widely believed that differences in digestibility were due only to differences in maturity and leafiness. The in vitro technique was developed in the early 1960's (Clark 1958, Tilley and Terry 1963) and gave a much more accurate estimate of digestibility. At the same time Minson et al. (1960) showed that Lolium perenne and Dactylis glomerata differ in digestibility and that this is independent of maturity and leafiness. This suggested that variation might also occur within a species, a suggestion which has since been verified in a wide range of pasture species and genera.

The existence of a relationship between digestibility and intake indicates that selecting for improved digestibility should result in an improvement in intake (Cooper et al. 1962); however more recently it has been shown that forages of related cultivars may show very different intake levels at the same digestibility (Osbourne et al. 1966, Walters 1971, Minson 1971). This had led to renewed interest in factors, other than digestibility, which are associated with intake, such as the fibrousness index proposed by Chenost (1966) and rate of breakdown within the rumen (Jones 1980).

Variation in digestibility

Selection for digestibility may be based on material either of a particular calendar age, or physiological age related to flowering, depending on usage. Grasses intended for hay are frequently harvested at a physiological age, for example early anthesis. Grasses for grazing are more often cut at regular intervals, or "plucked" samples are harvested to simulate grazing.

Often the differences between genotypes are relatively small in young forage, increasing as the forage matures (Schank et al. 1973, Klock et al. 1975, Sleper & Mott 1976) but in some genera the range is similar in young and mature regrowth (Hacker & Minson 1972).

Tables 2 and 3 show ranges of heritabilities for digestibility in a number of pasture grass species. In Table 2 plants within each study were harvested on the same date, whereas in Table 3 they were harvested at a particular physiological age. Heritabilities are indicated as broad sense, that is the proportion of the variance which is genetic (including additive and non-additive), or narrow sense, the proportion of the total variance which is additive genetic.

The estimates obtained are based on differing units - for example single plant, plots, family means over one or several sites - and therefore should not be directly compared. However the values do give an overall picture of the extent of genetically useful variation.

In most studies there is a range of at least ten digestibility units, and about half of this variation is evidently genetic. Those few studies in which combining ability has been estimated indicate both general and specific effects may be important.

Trials in which genotype x environment interaction for digestibility has been studied give inconsistent results. Genotype x environment effects were relatively unimportant in studies by Hacker & Minson (1972), Hovin et al. (1974), Mason & Shenk (1976) and Bray & Pritchard (1976). In contrast significant genotype x management and genotype x year effects are noted by Carlson et al. (1969), Walters & Evans (1974), Sleper & Mott (1976) and Stratton et al. (1979). In a study on Lolium perenne a heritability of 0.69 in the year of planting dropped to 0.14 in the following year (Rogers & Thomson 1970).

TABLE 2

Variation in digestibility and its heritability for a range of grasses; harvests within species on the same date.

Species	Range	Correlation with leafiness	Correlation with yield	Combining ability	Heritability
Agropyron intermedium[1]	46-49			GCA,SCA	
Agropyron spp.[2]	(6 units)	0.89	+0.26 to -0.30		^{n}0.36-0.76
Andropogon gerardii[3]	45-52				^{n}0.72
Cenchrus ciliaris+[4]	53-64		-0.08 to -0.50		^{b}0.25-0.77
Cenchrus ciliaris[5]	55-64	0.33			
Chloris gayana[6]	51-65				^{b}0.35-0.48
Cynodon dactylon[7]	37-61				^{b}0.27-0.69
Cynodon dactylon[8]					^{n}0.53
Dactylis glomerata[9]			+0.14 to -0.84		^{n}0.49-0.91
Dactylis glomerata[10]			negative		
Digitaria spp.[11]	51-64				
Digitaria spp.[12]	62-68				
Hemarthria altissima[13]	38-66				
Lolium perenne[14]				GCA,SCA	
Phalaris aquatica[15]	(12 units)				0.60
Phalaris aquatica[16]	36-45		-0.47, +0.11		
Phalaris arundinacea[17]	64-73		^{g}-0.13		^{n}0.02-0.63
Phalaris arundinacea[18]			-0.45	GCA	^{b}0.58
Phalaris arundinacea[19]				GCA	^{n}0.71, 0.77
Setaria sphacelata[20]	55-65				
Setaria sphacelata[21]	56-60		^{g}0.08		
Mean	12 units				

+ plucked sample ^{g} genetic correlation ^{b} broad sense ^{n} narrow sense.
GCA - general combining ability. SCA - specific combining ability.

1 Thaden et al. 1975
2 Coulman & Knowle 1974
3 Ross et al. 1975
4 Bray & Pritchard 1976
5 Lovelace et al. 1972
6 Quesenberry et al. 1978
7 Burton & Monson 1972
8 McGehec 1976
9 Stratton et al. 1979
10 Mason & Schenk 1976
11 Klock et al. 1975
12 Schank et al. 1977
13 Schank et al. 1973
14 Rogers & Thomson 1970
15 Oram et al. 1974
16 Clements et al. 1970
17 Marum et al. 1979
18 Hovin et al. 1976
19 Hovin et al. 1974
20 Hacker 1974a
21 Bray & Hacker 1981

TABLE 3

Variation in digestibility and its heritability for a range of grasses; harvests within species at the same physiological age.

Species	Stage of growth	Range	Correlation with yield	Combining ability	Herit-ability
Agropyron sp.[1]	full bloom		-0.60		
Bromus inermis[2]	50% emergence	62-74			
[3]	emergence	55-62		GCA	^{n}1.06
[4]	full bloom		-0.78		
Dactylis glomerata[5]	50% emergence	49-68			
Panicum maximum[6]	early full head	41-72			
Phalaris aquatica[7]	heading	59-68			0.54
	mature	27-39			0.77
Phleum pratense[8]	early anthesis	56-59			
Sorghum bicolor[9]	flowering			GCA,SCA	
Mean	12 units				

1 Junk & Austenson 1971
2 Christie & Mowat 1968
3 Ross et al. 1970
4 Junk & Austenson 1971
5 Christie & Mowat 1968
6 Milot & Burton 1973
7 Clements 1973
8 Koch 1976
9 Dangi et al. 1979

Correlations between digestibility and other plant characteristics

Correlations between digestibility and morphological and chemical characteristics are highly variable, depending on species, sampling system and age of regrowth. Thus when compared at the same calendar date, late flowering varieties have the higher digestibility (Brown et al. 1968, Burton et al. 1968, Quesenberry et al. 1978). In contrast, when compared at flowering early varieties are usually the more digestible (Dent & Aldrich 1963, Clements 1973, Boonman 1978a).

Most studies indicate a negative correlation between digestibility and yield (Tables 2 and 3). However, the correlation is usually weak, and most authors consider that there is sufficient variation in both characters for digestibility to be improved without a sacrifice in yield (eg. Hovin et al. 1976).

The extent to which digestibility of a forage is related to leafiness will depend on the relative digestibility of leaf and stem. This will vary with stage of growth and with species. In Agropyron spp. there is a positive correlation between leafiness and whole plant digestibility (Coulman & Knowles 1974) whereas in Cenchrus ciliaris and Andropogon gerardii the correlation is poor (Lovelace et al. 1972, Williams 1972, Ross et al. 1975). In general there is less variation in leaf digestibility than there is in stem digestibility, and heritability estimates are lower (Table 4). However, at least some of the factors controlling digestibility of leaf and stem must be the same, as leaf and stem digestibility are usually positively correlated.

TABLE 4

Variation in leaf, stem and whole plant digestibility, heritabilities (narrow sense, ^{n}h) and correlations.

Species	Stem	^{n}h	Leaf	^{n}h	Plant	^{n}h	r leaf:stem
Andropogon gerardii[1]	42-50	0.64	55-60	0.12	45-52	0.72	0.30
Bromus inermis*[2]	58-73		66-76		62-74		0.43 to 0.46
Cenchrus ciliaris+[3]	51-61		62-69		55-64		
Chloris gayana+[4]	54-58	[i]0.14	60-63	[i]0.05	54-59	[i]0.15	
Dactylis glomerata*[5]	33-49		63-66		41-51		
Phalaris aquatica*[6]							0.43, 0.65
Setaria sphacelata+[7]	60-74		58-72		60-71		-0.04 to 0.21

* at same physiological age + at same calendar age [i] individual plant basis.

1 Ross et al. 1975
2 Christie & Mowat 1968
3 Lovelace et al. 1972
4 Boonman 1978b
5 Mowat et al. 1965
6 Clements et al. 1970
7 Hacker 1974b

Within a population harvested at the same age of regrowth, correlations between digestibility and cellular constituents (Table 5) tend to reflect the overall relationships shown in studies of forages harvested during their growth cycles (Johnson et al. 1973). Strong negative correlations occur between digestibility and lignin, and also cell wall content. Correlations with nitrogen concentration are frequently positive (Table 6) and with ADF and NDF and cellulose they are variable but tend to be negative. The extent of correlation between digestibility and cellular constituents may also be influenced by environmental effects, as, for example, in Dactylis glomerata, in which correlation between digestibility and both ADF and NDF are weak in spring but very strong in summer and autumn (Stratton et al. 1979). In contrast, in an earlier study on the same species by Dent & Aldrich (1963) there was a tendency for digestibility to be negatively correlated with crude fibre, but frequently correlations were not significant, and there was no seasonal trend.

TABLE 5

Correlations between digestibility and cellular components.

Species	ADF	NDF	Lignin	Si	Cellulose	CWC
Bromus inermis[g1] (stem)	nil		-0.93	0.23	nil	-0.78
Cenchrus ciliaris[2]		-0.81				
" (leaf)[2]		low				
" (stem)[2]		-0.66				
Dactylis glomerata[3]	-0.15 to-0.90	-0.03 to-0.89				
Phalaris arundinacea[g4]						-0.75, -0.90
Phalaris arundinacea[g5]	-0.82 to-1.03		-0.80 to -0.97	+0.51 to -0.74		-0.76 to-1.02

g - genotypic correlations

1 Bhat & Christie 1975
2 Lovelace et al. 1972
3 Stratton et al. 1979
4 Hovin et al. 1976
5 Marum et al. 1979

Water-soluble carbohydrates

Water-soluble carbohydrates may account for up to 50 percent of the dry matter of some forage species (Bugge 1978), and as they are totally digested, and related to palatability (Arnold 1964, Saiga & Kawabata 1975), plant breeders have shown interest in improving their concentration. They are markedly affected by environmental conditions (Jones 1961), being at the highest concentration in cool bright conditions.

Varietal differences in concentration of water soluble carbohydrates of almost 2-fold were recorded by Cooper (1962) both in L. perenne and D. glomerata, and by Bugge (1978) in Lolium multiflorum. Ecotypes from more temperate climates have the highest concentration of water soluble carbohydrates (Bugge 1978).

Water-soluble carbohydrates are generally negatively correlated with protein concentration (Table 6), although there is a sufficient degree of genotypic independence to allow simultaneous selection for both characteristics (Vose & Breese 1964).

TABLE 6

Correlations between nitrogen concentration and water soluble carbohydrates and digestibility in populations harvested at the same date.

Species	Water soluble carbohydrates	Digestibility
Agropyron spp.[1]		0.65
Bromus inermis[1]		0.79
Dactylis glomerata[2]	-0.57 to -0.35[P]	
D. glomerata[3]	-0.60 to 0.33[P]	0 to 0.63[P]
Festuca pratensis[3]	-0.19 to -0.14[P]	-0.10 to 0.70[P]
Lolium perenne[2]	-0.70 to -0.20[P]	
L. perenne[3]	-0.63 to -0.44[P]	-0.31 to 0.51[P]
L. multiflorum[4]	-0.50[P]	-0.10[P]
Phalaris aquatica[5]	-0.13	0.82
P. aquatica[6]		-0.19 to 0.20[g]
P. arundinacea[7]		0.16 to 0.42[g]
P. arundinacea[8]		0.77 to 0.95[g]
Phalaris spp.[9]		0.27[P]
Phleum pratense[3]	-0.42 to 0[P]	0.38 to 0.60[P]

[g] genotypic correlations [P] phenotypic correlations

1 Junk & Austenson 1971
2 Cooper 1962
3 Dent & Aldrich 1963
4 Bugge 1978
5 Clements 1969
6 Clements 1973
7 Carlson *et al*. 1969
8 Marum *et al*. 1979
9 Clements *et al*. 1970

Anatomy and digestibility

Anatomical differences between leaves of grass species have been known for a long time, and are partly responsible for the lower digestibility of C_4 as compared with C_3 grasses (Amos & Akin 1978, Wilson & Minson 1980). However, it is relatively recently that variation in anatomical characteristics within species has been documented.

Leaf mesophyll thickness responded to selection in *Lolium perenne*, and there was a correlated reduction in cell wall content, cellulose content and tensile strength (Wilson 1973). A relationship between digestibility and leaf thickness, number of vascular bundles and lacunae in the parenchyma has been found for *Poa pratensis* (Berg & Wilton 1978). A recent study in *Bromus inermis* has indicated that a number of anatomical characters, including number and frequency of vascular bundles, exhibit genetic variation, but the relationship between these characters and digestibility was not determined (Tan *et al*. 1976). In *Hemarthria altissima* low digestibility is associated with a greater proportion of vascular bundle in the stem (Schank *et al*. 1973). In contrast genotypes of *Cynodon dactylon* which differed markedly in digestibility were not anatomically distinguishable (Akin & Burdick 1973, Hanna *et al*. 1976) in either leaf or stem.

Major genes affecting characters associated with digestibility

In general digestibility of forage grasses is believed to be under the control of a large number of genes, but there are instances where single major genes control a major component of digestibility. All cases so far found are in annual forages, but there is no genetic reason why comparable genes should not occur in perennial pasture species.

In maize a series of "brown midrib" mutant genes is associated with a reduction in lignin percentage (Barnes *et al*. 1971, Gordon & Neudoerffer 1973) and silage made from lines which are homozygous recessive for the mutant gene, has a higher rate of digestion of dry matter, cellulose and hemicellulose (Muller *et al*. 1972). The low lignin content of the brown midrib lines results in higher digestibility of all parts of the plant, as shown in Table 7. Moreover the effect of the mutant genes is additive. *In vivo* studies with dairy cows have shown higher intake of the brown midrib genotype, and a reduction in weight loss for the same level of milk production (Rook *et al*. 1977).

TABLE 7
Digestibility of some plant fractions of normal maize and brown-rib mutants homozygons recessive for bm_1, bm_3 and both bm_1 and bm_3 (after Barnes *et al*. 1971).

	Normal	bm_1/bm_1	bm_3/bm_3	$bm_1/bm_1:bm_3/bm_3$
Whole plant	68.3	72.0	75.5	77.8
Stem	56.3	58.2	66.7	62.7
Leaf	60.9	63.0	68.9	68.8
Sheath	55.0	60.3	64.7	66.5

Other single gene effects are associated with the cuticle. Digestibility of leaf sections is markedly increased in 'bloomless' sorghum (Cummins & Dobson 1972, Hanna *et al*. 1974) a character usually inherited as a single gene (Ross 1972). A similar effect has been observed in "trichomeless" *Pennisetum americanum* (Powell & Burton 1971, Hanna *et al*. 1974), although the effect is presumably not caused by absence of the trichomes or leaf hairs but is more probably due to physical or chemical differences in cutinization (Hanna & Akin 1978).

The practical benefits of selecting for lower levels of cutinization will depend on the extent to which leaf cuticle inhibits digestion in the rumen, and the possible deleterious side-effects which might result. It is probable that leaf cuticle has little effect on digestibility *in vivo* due to the bruising and rupturing effects associated with mastication (Reid *et al*. 1962). Furthermore a reduction in effectiveness of the cuticle would be expected to increase transpiration (Hanna *et al*. 1974), reduce yield (Ross 1972) and increase disease susceptibility.

Pasture quality may also be improved by increasing the percentage of leaf. In *Pennisetum americanum* a recessive gene d_2 reduces stem internode length without affecting leaf yield. This results in an increase in protein content and digestibility of the whole plant. However, the gain in quality is at the expense of a substantial decrease in yield; cattle grazing the dwarf strain gain no more per hectare than cattle grazing a tall strain (Burton *et al*. 1969).

Disease resistance

Breeding for resistance to disease in pasture grasses has been primarily directed towards reducing loss of yield. However, there is increasing evidence that fungal infection may cause deterioration in quality. Different diseases have different effects on herbage quality, ranging from zero to a loss of more than 1 digestibility unit for a 10 percent increase in diseased area (Ross *et al*. 1973, Gros *et al*. 1975).

Decrease in quality due to disease may be due to various causes. Burton (1954) showed that leaves of Sudan grass infected with *Colletotrichum graminicola* had less protein and fat and more fibre than uninfected leaves. More recently fungal infection has been shown to cause decreases in water soluble carbohydrates (Hodges & Robinson 1977), amino acids and digestibility (Cagas 1976) and in some cases to cause toxicity (Porter *et al*. 1974, Bacon *et al*. 1977).

Little is known of the effects of disease on preference and intake by grazing animals. However, it would be reasonable to assume that any effects would be deleterious, if only because of the known general relationship between intake and digestibility.

Breeding for resistance to disease should be a primary objective in improving quality where disease is widespread and pastures are severely affected. In general resistance to fungal diseases is dominant and controlled by major genes (Person & Sidhu 1970). This is also the case in pasture grasses (Hayward 1977, Hacker & Bray 1981) indicating that breeding for resistance should be relatively simple. Improvement of quality through breeding for disease resistance should not result in any undesirable genetically correlated changes.

Breeding for digestibility

Despite ample evidence for genetic variation for digestibility, very few commercial cultivars have been bred for digestibility. The *Cynodon dactylon* cultivars Coastcross 1 and Tifton 44 were bred for high digestibility and this is reflected in increased animal weight gain (Burton 1972, Burton & Monson 1978). However, the physiologically unrelated rhizome character was lost in the breeding of Coastcross 1, and this is perhaps responsible for its lack of winter hardiness. In *Dactylis glomerata* selection for improved digestibility resulted in genetic gain, but the advance obtained in spaced plant populations was evident in aftermath growth in the sward, but not in primary growth (Walters & Evans 1974). The failure of spaced plant selections consistently to exhibit their superiority when grown as swards appears to be due to an association between digestibility and broad leaves and succulent tillers. These characters were not expressed in the sward situation (Cooper & Breese 1980). In another study selection for digestibility in *D. glomerata* did not result in improvement, despite evidence of genetic variation in the parents (Christie 1977). Similarly lines of *Bromus inermis* selected for high digestibility as spaced plants failed to show any improvement in digestibility or intake as swards (Kamstra *et al*. 1973). In an experiment on *Lolium perenne* general combining ability variance was significant in the year of planting, but disappeared the following year (Rogers & Thomson 1970). The evidence suggests that improvement in digestibility of a species should be possible, but that, if the species is not naturally sward-forming (like *Cynodon*), plants should be grown as swards and evaluated over two or more years before selections are made. The recent interest in anatomical differences associated with digestibility may be of value in accounting for differences in digestibility between existing cultivars, but is unlikely to provide a useful tool for the plant breeder, because of the practical problems associated with screening large populations.

INTAKE

Intake cannot be measured on small samples directly and the breeder must rely on correlated characters. A number of these have been suggested.

Leafiness

Leaf is almost invariably grazed in preference to stem (Arnold 1964, Stobbs 1975). It has generally been assumed that intake will be better from a leafy than a stemmy variety and in studies with separated leaf and stem of similar digestibility, the intake of leaf was 46 percent higher (Laredo & Minson 1973). Similarly at the same level of yield and digestibility differences in intake of *L. multiflorum* are associated with leafiness (Wilman & Omaliko 1978). However in a study of pairs of varieties of four temperate pasture species differing in digestibility - intake relationships, Walters (1971) considered leafiness to be an unimportant character.

Preference

Animal preference for a genotype or variety has been used in cafeteria trials for selecting more 'palatable' varieties, although marked variety x year interactions may occur (Mills & Boultwood 1978). *Phalaris arundinacea* varieties which are unpalatable when sheep are given a choice are eaten in small quantities in the absence of choice (O'Donovan *et al*. 1967, Marten *et al*. 1976). Unpalatability in this species is associated with high alkaloid content. In *Eragrostis curvula* selection for palatability gave rise to the cultivar Morpa, with improved animal production characteristics (Table 1). In *E. curvula* palatability is associated with low lignin and high cellulose content (Voigt *et al*. 1970). Palatability is also associated with lack of winter hardiness (cf. *C. dactylon*).

In contrast *Paspalum notatum* accessions which differed significantly in preference when fed in a cafeteria trial failed to exhibit differences in daily gain when grazed separately as pastures (Burton 1974).

Leaf flexibility

Palatability in *Festuca arundinacea* is associated with leaf flexibility which is highly heritable (Gillet & Jadas-Hecart 1966), and is more related to physical than chemical differences. Selection for leaf flexibility has resulted in the production of the cultivar Ludelle, with improved intake characteristics (Gillet & Huguet 1977).

Cellulose and leaf strength

In ryegrass higher weight gains are associated with low cellulose content (Bailey 1964) which is related to leaf tensile strength (Evans 1964). Leaf strength is highly heritable in ryegrass (Wilson 1965) but an experimental variety produced by selection for low cellulose showed only a small improvement in animal production (Lancashire & Ulyatt 1975). Interestingly, though, the selection was higher yielding, possibly due to resistance to weevils (Lancashire *et al.* 1977). Conversely in *E. curvula*, in which varietal differences in leaf strength also occur (Kneebone 1960), animals have a preference for the stronger leafed varieties (Kneebone 1961, cited by Evans 1964).

EFFICIENCY OF UTILIZATION

Apart from intake of digestible energy, efficiency of utilization of nutrients is the most important and least understood component of pasture quality. Ulyatt (1971) found that short-rotation ryegrass is more efficiently utilized than perennial ryegrass, possibly due to the higher rate of production of volatile fatty acids.

The importance of considering efficiency of utilization has been stressed by Cooper & Breese (1980) and is discussed by Armstrong (1982). There is very little information as to what are the critical *forage* characteristics which control efficiency of utilization, and its relative importance in comparison to intake and digestibility is uncertain. At this stage the breeder can only wait for the nutritionist to postulate a selection criterion.

TOXINS AND ANTI-QUALITY CHARACTERISTICS

Cyanogenetic glycosides and nitrates

The grasses as a family are relatively free of toxins, with the exception of cyanogenetic glycosides and nitrates which are accumulated by a wide range of species (see for example, Everist 1974 for Australian grasses). The occurence of nitrates is widespread and largely associated with management practices, especially nitrogen fertilization although there are some indications that particular species may accumulate large quantities (eg. Tanner grass *Brachiaria radicans*, Andrade *et al.* 1971) and that varieties within a species may differ in their nitrate concentration (cf. Griffith & Johnston 1960, Dotzenko & Henderson 1964).

Similarly cyanogenetic glycosides are produced in a wide range of grass genera. In *Sorghum*, considerable variation exists in cyanogenetic glycoside content (for example Gillingham *et al.* 1969) but the genetic control is probably not as simple as the two gene system which is believed to control cyanogenetic glycoside production in *Trifolium repens* (Atwood & Sullivan 1943).

If the production of cyanogenetic glycosides by grasses is controlled by a pair of genes, as is the case in white clover, elimination of the toxic effect by breeding should not be difficult. However the pronounced effect of environment on cyanogenetic glycoside accumulation (Kriedeman 1964) could be a complication. As cyanogenetic glycoside content may be controlled by grazing management breeding for low levels would not seem to be warranted, especially as there are indications that they may confer disease resistance (Vidhyasekaran *et al.* 1971).

Alkaloids

Toxic alkaloids have been identified in a number of temperate grasses including *Phalaris*, *Festuca* and *Lolium*. Eight different alkaloids have been identified in *P. arundinacea* (Williams *et al.* 1971, Simons & Marten 1971, Marten 1973) and they are associated with low palatability and intake of the forage although the effects differ for sheep and cattle (Woods & Clark 1974, Marten *et al*. 1976). Heritability values reported are high (Buckner *et al*. 1972, Barker & Hovin 1974, Coulman *et al*. 1977) and it has been suggested that control is through the action of two genes (Woods & Clark 1971, Marum *et al*. 1976). Alkaloids are also present in *P. aquatica*, *Festuca arundinacea* and to a lesser extent in *L. perenne*, and concentration is highly heritable, although the number of genes involved has not been determined (Butler 1962, Oram 1970, Webb 1973). The low alkaloid (perloline) concentration in *Lolium* leads to the possibility of reducing the concentration in *F. arundinacea* by intergeneric hybridization (Buckner *et al*. 1972).

Breeding for low alkaloid concentration would appear to be straight forward, and the risk of unforeseen correlated changes is low as alkaloid concentration is inherited independently of agronomic characters including digestibility (Hovin *et al*. 1974, Hovin & Buckner 1976). However, due to the changes which occur in concentration as a result of growth and development of the plant, selection should be made from populations at an equivalent stage of growth.

Tannins

Tannins are an ill-defined group of chemical compounds which have the effect of reducing *in vitro* digestibility by reducing cellulase activity (Griffiths & Jones 1977). High levels of tannin are more commonly a problem in legumes than in grasses, although in sorghum high levels of tannin in the grain are associated with bird resistance - and also with lower digestibility of the grain (Harris *et al*. 1970). In 67 varieties of sorghum Arora & Luthra (1974) found that the digestibility of the grain ranged from 50-96 percent and was negatively correlated with tannin percentage which ranged from 0.1-6.2 percent. Sorghum forage also contains tannin and species differ in tannin concentration. Again, there is a negative correlation with digestibility (Saini *et al*. 1977).

Although tannins have an adverse effect on digestiblity low levels may improve quality through protecting proteins from bacterial degradation and making them available for absorption in the duodenum (Reid *et al*. 1974). It is not completely clear whether the breeder should consider selecting for higher or lower concentrations of tannin, but available evidence suggests that concentration may be manipulated in those genera which contain tannins.

Oxalate

Oxalate accumulation is more common in the panicoid than the festucoid grasses and is particularly common in the genera *Pennisetum*, *Cenchrus*, *Setaria* and *Panicum* (Garcia-Rivera & Morris 1955, Jones & Ford 1972). High levels of oxalate in forage may lead to malformation of bone in non-ruminants, and to kidney damage in ruminants. Also, being a low energy carbohydrate, it contributes little to animal production when it is digested.

In *Pennisetum americanum* oxalate is associated with calcium and is insoluble. Broad sense heritabilities are low, 0 - 35.4 percent (Gupta & Sehgal 1971). In contrast in *Setaria sphacelata* oxalate is closely associated with sodium and potassium, and is soluble. Although in setaria there are wide differences in oxalate concentration between accessions, no accession has been found lacking in oxalate. In this species all accessions with low oxalate concentration are also low in sodium and breeding for low oxalate concentration would probably reduce sodium concentration to levels below those required by livestock. A further point to note about oxalate is the apparent relationship between oxalate and digestiblity (Hacker 1974a). The more highly digestible accessions have a higher concentration of the low energy carbohydrate oxalate and hence may be little better than the low digestibility accessions in terms of animal production.

MINERAL COMPOSITION

Low levels of essential elements frequently limit animal productivity (Little 1982) particularly in tropical grasses, in which nitrogen, phosphorus, copper, cobalt and sodium are often deficient (Jones 1963, Patil & Jones 1970, Fleming 1973, Gartner & Murphy 1974).

TABLE 8

Variation in mineral concentration in populations of grasses at the same age of regrowth.

	N	K	Ca	Mg	Na	P
Lolium perenne[1]		2.68-3.02	0.93-1.25		0.16-0.51	0.19-0.28
Lolium perenne[2]	1.96-3.78	2.65-3.85	0.34-0.56	0.14-0.21	0.07-0.15	0.32-0.47
Phalaris arundinacea[3]	2.01-2.61	2.60-3.10	0.33-0.53	0.31-0.46		0.37-0.46
Setaria sphacelata[4]		1.30-3.11	0.15-0.36	0.18-0.37	0.05-1.80	

1 Butler *et al*. 1962
2 Cooper 1973
3 Hovin *et al*. 1978
4 Hacker 1974a

Concentrations of major elements almost invariably show extensive genetic variation (Table 8), but frequently genetic correlations with yield are negative, as shown in Table 9. Some studies (eg. Butler 1962) have suggested that improvement in mineral content may be achieved without a sacrifice in yield. Genetic variation in mineral concentration is usually additive (Sleper *et al*. 1977, Hovin *et al*. 1978, Bray & Hacker 1981) and in consequence improvement by breeding should be readily achieved.

Heritability estimates for major elements are frequently high (Table 10) but values published will depend on the genetic characteristics of the usually small number of plants under study. Even within a species heritability may range from nil to a high value. Thus in *Setaria sphacelata* a heritability of 0.50 was obtained for sodium in a population in which families exhibited a two to six-fold range in sodium concentration (Bray & Hacker 1981). Had the study been carried out on the low sodium genotypes characteristic of populations from Kenya (Hacker 1974a) very different heritability values might be expected.

TABLE 9

Correlations between yield and concentration of major elements in pasture grasses.

	N	K	Ca	Mg	Na	P
Cynodon dactylon[1]	-0.10 to -0.22					
Lolium perenne+[2]		-0.64	-0.44		0.59	0.03
P. aquatica[3]	-0.88					
P. arundinacea[4]	-0.52, -0.68	-0.66, -0.09	-0.36, 0.01	0.32, -0.06		-0.79, -0.39
P. arundinacea[4]	-0.72 to -0.91					
Setaria sphacelata+[6]	-0.64	-0.16	-0.77	-0.30	-0.44	

+ genetic correlations

1 Burton *et al*. 1967
2 Butler *et al*. 1962
3 Clements 1969
4 Hovin *et al*. 1978
5 Asay *et al*. 1968
6 Bray & Hacker 1981

TABLE 10

Heritability estimates for mineral concentration in forage grasses. (b - broad sense; n - narrow sense)

	N	K	Ca	Mg	Na	P
Lolium perenne[1]	b	0	0.34		0.54	0.30
Lolium perenne[2]	n 0.63	0.80	0.78	0.86	0.55	0.68
Phalaris arundinacea[3]	b 0.28	0.54	0.64	0.73		0.32
	n 0.25	0.49	0.56	0.66		0.57
Pennisetum americanum[4]	b 0.55-0.66		0.62-0.84			0.45-0.80
Setaria sphacelata[5]	n 0.25	0.22	0.28	0.40	0.50	

1 Butler *et al*. 1962
2 Cooper 1973
3 Hovin *et al*. 1978
4 Gupta & Sehgal 1971
5 Bray & Hacker 1981

Evidence from phenotypic and genetic correlations between concentrations for different nutrients suggests that in any selection program changes in concentration of elements other than the one being selected for are to be expected. Some published data illustrating this are presented in Table 11. It is also clear from Table 11 that the direction and magnitude of correlations differ between species. Thus calcium and nitrogen are not correlated in *Phalaris arundinacea* but are strongly correlated in *Setaria sphacelata*. Similarly sodium and potassium are positively correlated in *Lolium perenne* and negatively correlated in *S. sphacelata*. Further studies on a wider range of species and populations are needed to confirm and extend these results.

Micro nutrients may also show heritable variation. Butler et al. (1962) obtained significant heritability estimates for manganese, aluminium, copper and zinc in L. perenne. In P. arundinacea Hovin et al. (1978) found copper to have a significant heritability in the broad sense, but not zinc and manganese. Both in P. arundinacea and in S. sphacelata (Bray & Hacker 1981) general combining ability effects for micronutrient concentration interacted with environments, suggesting that breeding for improved micronutrient concentration may not result in consistently superior cultivars.

TABLE 11
Correlation coefficients for concentration of various minerals.

		P	K	Ca	Mg	Na
N	a	0.42,0.50	0.23,0.36	-0.01,0.13	0.15,0.21	
	b					
	c		1.12	1.06	1.00	-0.76
P	a		0.67,0.61	0.12,0.14	0.38,0.46	
	b		0.28	0.22		0.20
	c					
K	a			0.13,0.41	0.38,0.59	
	b			0.67		0.37
	c			0.69	1.02	-0.88
Ca	a				0.71,0.60	
	b					-0.72
	c				0.79	-0.21
Mg	a					
	b					
	c					-0.73

a - phenotypic correlations Hovin et al. 1978, P. arundinacea
b - genetic correlations Butler et al. 1962, L. perenne
c - genetic correlations Bray & Hacker 1981, S. sphacelata

FUTURE PROSPECTS

The results reviewed in this paper indicate that there is considerable genetic diversity for most characters associated with pasture quality. Genetic variation for digestibility is evident in most species of pasture grass, but future work should pay more attention to selecting for more highly digestible leaf rather than whole plant, and to take account of the effects of plant spacing and management on plant morphological attributes which influence digestibility.

Selection for intake may be more difficult. Digestibility is still probably the most closely associated readily measured character. Other characters such as tensile strength and leaf flexibility may be suitable for particular species but are not likely to be generally applicable. Perhaps more attention should be directed towards selecting for leaf density within the sward and hence ease of prehension (Stobbs 1973, 1975). It should be pointed out, though, that the simulation of realistic pasture plant structure in nursery conditions is particularly difficult.

Extensive genetic variation is also evident for mineral concentration; however correlated changes in other minerals in a breeding program are likely to be species specific and perhaps even population specific. In those species in which low concentration of an element is the prime limitation to animal production, breeding for increased concentrations should be readily achieved.

It is concluded that there is a wealth of genetic variation available to the plant breeder for characteristics associated with quality. The best prospects for improvement are for those management systems in which the selector can sample material similar to that eaten by the animal - for example forages cut for hay or silage, or intensively and rotationally grazed. For more extensively managed forages adequate sampling is a considerable problem, and opportunity for genetic gain will depend on the extent of genetic correlation between the material sampled and the material eaten by the grazing animal.

REFERENCES

Akin, D.E.; Burdick, D. (1973) Micro-anatomical differences of warm season grasses revealed by light and electron microscopy. Agronomy Journal 65, 533-537.

Andrade, S.O.; Retz, L.; Marmo, O. (1971) Estudos sobre Brachiaria sp. (Tanner grass) III. Ocorrências de intoxicacoes de bovinos divante um ano (1970-71) e inveis de nitrato em amostras da gramminea. Arquivos do Instituto Biologico, Brazil 38, 239-252.

Armstrong, D.G. (1982) Digestion and utilization of energy. In: Nutritional limits to animal production from pastures. Editor J.B. Hacker. Farnham Royal, U.K., Commonwealth Agricultural Bureaux pp. 225-244.

Amos, H.E.; Akin, D.E. (1978) Rumen protozoal degradation of structurally intact forage tissues. Applied and Environmental Microbiology 36, 513-522.

Arnold, G.W. (1964) Factors within plant associations affecting the behaviour and performance of grazing animals. In: Grazing in terrestrial and marine environments. Editor D.J. Crisp. Oxford, Blackwell Scientific Publications pp.133-154.

Arora, S.K.; Luthra, Y.P. (1974) The in vitro digestibility of promising Indian varieties of sorghum and its relation with tannin content. Indian Journal of Nutrition and Dietetics 11, 233-236.

Asay, K.H.; Carlson, I.T.; Wilsie, C.P. (1968) Genetic variability in forage yield, crude protein percentage and palatability in reed canary grass, Phalaris arundinacea L. Crop Science 8, 568-571.

Atwood, S.S.; Sullivan, J.T. (1943) Inheritance of a cyanogenetic glucoside and its hydrolyzing enzyme in white clover. Genetics 28, 69.

Bacon, C.W.; Porter, J.K.; Robbins, J.D.; Luttrell, E.S. (1977) Epichloë typhina from toxic tall fescue grasses. Applied and Environmental Microbiology 34, 576-581.

Bailey, R.W. (1964) Pasture quality and ruminant nutrition. I. Carbohydrate composition of ryegrass varieties grown as sheep pastures. New Zealand Journal of Agricultural Research 7, 496-507.

Barker, R.E.; Hovin, A.W. (1974) Inheritance of indole alkaloids in reed canary grass (Phalaris arundinacea L.) Heritability estimates for alkaloid concentration. Crop Science 14, 50-53.

Barnes, R.F.; Muller, L.D.; Bauman, L.F.; Colenbrander, V.F. (1971) In vitro dry matter disappearance of brown midrib mutants of maize (Zea mays L.) Journal of Animal Science 33, 881-884.

Berg, C.C.; Wilton, A.C. (1978) Relationship of certain anatomical characteristics of Poa pratensis cultivars to their IVDMD. Agronomy Abstracts 121.

Bhat, A.N.; Christie, B.R. (1975) Plant composition and in vitro digestibility of brome grass genotypes. Crop Science 15, 676-679.

Boonman, J.G. (1978a) Rhodes grass breeding in Kenya. 1. Intra-variety variation and character relationships. Euphytica 27, 127-136.

Boonman, J.G. (1978b) Herbage quality in Rhodes grass (Chloris gayana Kunth.) 2. Intra variety variation in yield and digestibility of plants of similar heading date. Netherlands Journal of Agricultural Science 26, 337-343.

Bray, R.A.; Hacker, J.B. (1981) Genetic analysis in the pasture grass Setaria sphacelata. 2. Chemical composition, digestibility and correlations with yield. Australian Journal of Agricultural Research 32, 311-323.

Bray, R.A.; Pritchard, A.J. (1976) Some aspects of selection for herbage quality in buffel grass (Cenchrus ciliaris). Forage Research 2, 1-7.

Brown, R.H.; Blaser, R.E.; Fontenot, J.P. (1968) Effects of spring harvest date on nutritive value of orchard grass and timothy. Journal of Animal Science 27, 562-567.

Buckner, R.C.; Bush, L.P.; Burrus, P.B. (1972) Variability and heritability of perloline in Lolium spp., Festuca spp. and Lolium - Festuca hybrids. Agronomy Abstracts 24.

Bugge, G. (1978) Genetic variability in chemical composition of Italian rye grass ecotypes. Zeitschrift für Pflanzenzuchtung 81, 235-240.

Butler, G.W. (1962) Genetic differences in the perloline content of rye grass (Lolium) herbage. New Zealand Journal of Agricultural Research 5, 158-162.

Butler, G.W.; Barclay, P.C.; Glenday, A.C. (1962) Genetic and environmental differences in the mineral composition of rye grass herbage. Plant and Soil 16, 214-228.

Burton, G.W. (1954) Does disease resistance affect forage quality? Agronomy Journal 46, 99.

Burton, G.W. (1972) Registration of Coastcross 1 Bermuda grass. Crop Science 12, 125.

Burton, G.W. (1974) Improving forage quality by breeding. Proceedings of the 12th International Grasslands Congress, Moscow, 3, 705-714.

Burton, G.W.; Gunnells, J.B.; Lowrey, R.S. (1968) Yield and quality of early and late maturing near isogenic populations of pearl millet. Crop Science 8, 431-434.

Burton, G.W.; Hart, R.H.; Lowrey, R.S. (1967) Improving forage quality in bermuda grass by breeding. Crop Science 7, 329-332.

Burton, G.W.; Monson, W.G. (1972) Inheritance of dry matter digestibility in bermuda grass, Cynodon dactylon (L.) Pers. Crop Science 12, 375-378.

Burton, G.W.; Monson, W.G. (1978) Registration of Tifton 44 bermuda grass. Crop Science 18, 911.

Burton, G.W.; Monson, W.G.; Johnson, J.C.; Lowrey, R.S.; Chapman, H.D.; Marchant, W.H. (1969) Effect of the d_2 dwarf gene on the forage yield and quality of pearl millet. Agronomy Journal 61, 607-612.

Cagas, B. (1976) [The effect of Puccinia graminis subsp. graminicola on the contents of amino acids, oligosaccharides and digestible dry matter in timothy hay]. Ochrana Rostlin 12, 179-182.

Carlson, I.T.; Asay, K.H.; Wedin, W.F.; Vetter, R.L. (1969) Genetic variability in in vitro digestibility of full-saved reed canary grass, Phalaris arundinacea L. Crop Science 9, 162-164.

Chapman, H.D.; Marchant, W.H.; Burton, G.W.; Monson, W.G. (1971) Performance of steers grazing Pensacola bahia, coastal and Coastcross 1 bermuda grass. Abstract, Journal of Animal Science, 32, 374.

Chenost, M. (1966) Fibrousness of forages : its determination and its relation to feeding value. Proceedings of the 10th International Grassland Congress, Helsinki, 406-411.

Christie, B.R. (1977) Effectiveness of one cycle of phenotypic selection for in vitro digestibility in brome grass and orchard grass. Canadian Journal of Plant Science 57, 57-60.

Christie, B.R.; Mowat, D.N. (1968) Variability of in vitro digestibility among clones of brome grass and orchard grass. Canadian Journal of Plant Science 48, 67-73.

Clark, K.W. (1958) The adaptation of an artificial rumen technique to the estimation of the gross digestible energy of forages. Dissertation Abstracts 19(5), 926.

Clements, R.J. (1969) Selection for crude protein content in Phalaris tuberosa L. 1. Response to selection and preliminary studies on correlated responses. Australian Journal of Agricultural Research 20, 643-652.

Clements, R.J. (1973) Breeding for improved nutritive value of Phalaris tuberosa herbage : an evaluation of alternative sources of genetic variation. Australian Journal of Agricultural Research 24, 21-34.

Clements, R.J.; Oram, R.N.; Scowcroft, W.R. (1970) Variation among strains of Phalaris tuberosa L. in nutritive value during summer. Australian Journal of Agricultural Research 21, 661-675.

Cooper, J.P. (1962) Selection for nutritive value. Report of the Welsh Plant Breeding Station, 1961, pp.145-156.

Cooper, J.P. (1973) Genetic variation in herbage constituents. In: Chemistry and biochemistry of herbage. Editors G.W. Butler and R.W. Bailey. London, Academic Press Vol.2, pp.379-417.

Cooper, J.P.; Breese, E.L. (1980) Breeding for nutritive quality. Proceedings of the Nutrition Society 39, 281-286.

Cooper, J.P.; Tilley, J.M.A.; Raymond, W.F.; Terry, R.A. (1962) Selection for digestibility in herbage grasses. Nature, London, 195, 1276-1277.

Coulman, B.E.; Knowles, R.P. (1974) Variability for in vitro digestibility of crested wheat grass. Canadian Journal of Plant Science 54, 651-657.

Coulman, B.E.; Woods, D.L.; Clark, D.W. (1977) Distribution within the plant, variation with maturity, and heritability of gramine and hordenine in reed canary grass. Canadian Journal of Plant Science 57, 771-777.

Culvenor, C.C.J.; Frahn, J.L.; Jago, M.V.; Lanigan, G.W. (1978) The toxin of Lolium rigidum (annual rye grass) seedheads associated with nematode-bacterium infection. In: Effects of poisonous plants on livestock. Editors R.F. Keeler, K.R., van Kampen and L.F. James. New York, Academic Press, pp.349-352.

Cummins, D.G.; Dobson, J.W. (1972) Digestibility of bloom and bloomless sorghum leaves as determined by a modified in vitro technique. Agronomy Journal 64, 682-683.

Dangi, O.P.; Lodhi, G.P.; Luthra, Y.P. (1979) Inheritance of in vitro dry matter digestibility in forage sorghum. Forage Research 5, 75-77.

Dent, J.W.; Aldrich, D.T.A. (1963) The inter-relationships between heading date, yield, chemical composition and digestibility in varieties of perennial ryegrass, timothy, cocksfoot and meadow fescue. Journal of the National Institute of Agricultural Botany 9, 261-281.

Dotzenko, A.D.; Henderson, K.E. (1964) Performance of five orchard grass varieties under different nitrogen treatments. Agronomy Journal 56, 152-155.

Evans, P.S. (1964) A study of leaf strength in four rye grass varieties. New Zealand Journal of Agricultural Research 7, 508-513.

Evans, W.B.; Munro, J.M.M.; Scurlock, R.V. (1979) Comparative pasture and animal production from cocksfoot and perennial rye grass varieties under grazing. Grass and Forage Science 34, 64-65.

Everist, S.L. (1974) Poisonous plants of Australia. Sydney, Angus and Robertson Pty. Ltd. 684 pp.

Fleming, G.A. (1973) Mineral composition of herbage. In: Chemistry and biochemistry of herbage. Editors G.W. Butler and R.W. Bailey. London, Academic Press, Vol.1, pp.529-566.

Garcia-Rivera, J.; Morris, H.P. (1955) Oxalate content of tropical forage grasses. Science 122, 1089-1090.

Gartner, R.J.W.; Murphy, G.M. (1974) Evidence of low sodium status in beef cattle grazing Coloniao Guinea grass pasture. Proceedings of the Australian Society for Animal Production 10, 95-98.

Gillet, M.; Huguet, L. (1977) Bilan de la premiere variete de Fetuque elevee du Catalogue, selectionnee pour la valeur alimentaire. Annales de l'amelioration des Plantes 27, 331-339.

Gillet, M.; Jadas-Hecart, J. (1966) Leaf flexibility, a character for selection of tall fescue for palatability. Proceedings of the 9th International Grassland Congress, São Paulo, 155-157.

Gillingham, J.T.; Shirer, M.M.; Starnes, J.J.; Page, N.R.; McClain, E.F. (1969) Relative occurrence of toxic concentrations of cyanide and nitrite in varieties of sudan grass and Sorghum - sudan grass hybrids. Agronomy Journal 61, 727-730.

Gordon, A.J.; Neudoerffer, T.S. (1973) Chemical and in vivo evaluation of a brown midrib mutant of Zea mays. 1. Fibre, lignin and amino acid composition and digestibility for sheep. Journal of Science, Food and Agriculture 24, 565-577.

Griffiths, D.W.; Jones, D.I.H. (1977) Cellulase inhibition by tannins in the testa of field beans. Journal of the Science of Food and Agriculture 28, 983-989.

Griffith, G.ap; Johnston, T.D. (1960) The nitrate nitrogen content of herbage. 1. Observations on some herbage species. Journal of the Science of Food and Agriculture 11, 623-626.

Gros, D.F.; Mankin, C.J.; Ross, J.G. (1975) Effect of diseases on in vitro digestibility of smooth bromegrass. Crop Science 15, 273-275.

Gupta, V.P.; Sehgal, K.L. (1971) Genetic variation for chemical composition of bajra fodder in pearl millet. Indian Journal of Genetics and Plant Breeding 31, 416-419.

Hacker, J.B. (1974a) Variation in oxalate, major cations and dry matter digestibility of 47 introductions of the tropical grass setaria. Tropical Grasslands 8, 145-154.

Hacker, J.B. (1974b) The genetic relationship between the digestibility of leaf, stem and whole plant in Setaria. Australian Journal of Agricultural Research 25, 401-406.

Hacker, J.B.; Bray, R.A. (1981) Genetic analysis in the pasture grass Setaria sphacelata. 1. Dry matter yield and flowering. Australian Journal of Agricultural Research 32, 295-309.

Hacker, J.B.; Minson, D.J. (1972) Varietal differences in in vitro digestibility in Setaria and the effect of site, age and season. Australian Journal of Agricultural Research 23, 959-967.

Hanna, W.W.; Akin, D.W. (1978) Microscopic observations on cuticle from trichomeless, tr, and normal, Tr, pearl millet. Crop Science 18, 904-905.

Hanna, W.W.; Monson, W.G.; Burton, G.W. (1974) Leaf surface effects on in vitro digestion and transpiration in isogenic lines of sorghum and pearl millet. Crop Science 14, 837-838.

Hanna, W.W.; Monson, W.G.; Burton, G.W. (1976) Histological and in vitro digestion study of 1- and 4- week stems and leaves from high and low quality bermuda grass genotypes. Agronomy Journal 68, 219-222.

Harris, H.B.; Cummins, D.G.; Burns, R.E. (1970) Tannin content and digestibility of sorghum grain as influenced by bagging. Agronomy Journal 63, 500-502.

Haywood, M.D. (1977) Genetic control of resistance to crown rust (Puccinia coronata Corda) in Lolium perenne and its implications in breeding. Theoretical and Applied Genetics 51, 49-53.

Hodges, C.F.; Robinson, P.W. (1977) Sugar and amino acid content of Poa pratensis infected with Ustilago striiformis and Urocystis agropyri. Physiologia Plantarum 41, 25-28.

Hovin, A.W.; Buckner, R.C. (1976) Breeding to reduce anti-quality components in temperate grasses. Agronomy Abstracts, 108.

Hovin, A.W.; Stucker, R.E.; Marten, G.C. (1974) Inheritance of in vitro digestible dry matter in Phalaris arundinacea L. Proceedings of the 12th International Grassland Congress, Moscow, 3, 793-798.

Hovin, A.W.; Marten, G.C.; Stucker, R.E. (1976) Cell wall constituents of reed canary grass; genetic variability and relationship to digestibility and yield. Crop Science 16, 575-578.

Hovin, A.W.; Tew, T.L.; Stucker, R.E. (1978) Genetic variability for mineral elements in reed canary grass. Crop Science 18, 423-427.

Johnson, W.L.; Guerrero, J.; Pezo, D. (1973) Cell-wall constituents and in vitro digestibility of napier grass (Pennisetum purpureum). Journal of Animal Science 37, 1255-1261.

Jones, D.I.H. (1961) Studies of the soluble carbohydrate content of grasses. Report of the Welsh Plant Breeding Station, 1961, pp.157-164.

Jones, D.I.H. (1963) Mineral content of six grasses from a Hyparrhenia dominant grassland in Northern Rhodesia. Rhodesian Journal of Agricultural Research 1, 35-38.

Jones, D.I.H. (1980) Factors affecting the digestibility of forages. Report of the Welsh Plant Breeding Station, 1979, pp.114-117.

Jones, R.J.; Ford, C.W. (1972) The soluble oxalate content of some tropical pasture grasses in south east Queensland. Tropical Grasslands 6, 201-204.

Junk, R.J.G.; Austenson, H.M. (1971) Variability of grass quality as related to cultivar and location in Western Canada. Canadian Journal of Plant Science 51, 309-315.

Kamstra, L.D.; Ross, J.G.; Ronning, D.C. (1973) In vivo and in vitro relationships in evaluating digestibility of selected smooth brome grass synthetics. Crop Science 13, 575-576.

Klock, M.A.; Schank, S.C.; Moore, J.E. (1975) Laboratory evaluation of quality in subtropical grasses. 3. Genetic variation among Digitaria species in in vitro digestibility and its relationship to plant morphology. Agronomy Journal 67, 672-675.

Kneebone, (1960) Tensile strength variations in leaves of weeping lovegrass (Eragrostis curvula (Schrad.) Nees.) and certain other grasses. Agronomy Journal 52, 539-542.

Koch, D.W. (1976) In vitro dry matter digestibility in timothy (Phleum pratense L.) cultivars of different maturity. Crop Science 16, 625-626.

Kriedeman, P.E. (1964) Cyanide formation in Sorghum almum in relation to nitrogen and phosphorus nutrition. Australian Journal of Experimental Agriculture and Animal Husbandry 4, 15-17.

Lancashire, J.A.; Wilson, D.; Bailey, R.W.; Ulyatt, M.J. (1977) Improved summer performance of a 'low cellulose' selection from 'Grasslands Ariki' hybrid perennial rye grass. New Zealand Journal of Agricultural Research 20, 63-67.

Lancashire, J.A.; Ulyatt, M.J. (1975) Liveweight gains of sheep grazing rye grass pastures with different cellulose contents. New Zealand Journal of Agricultural Research 18, 97-100.

Laredo, M.A.; Minson, D.J. (1973) The voluntary intake, digestibility and retention time by sheep of leaf and stem fractions of five grasses. Australian Journal of Agricultural Research 24, 875-888.

Little, D.A. (1982) Utilization of minerals. In: Nutritional limits to animal production from pastures. Editor J.B. Hacker. Farnham Royal, U.K., Commonwealth Agricultural Bureaux, pp. 259-283.

Lovelace, D.A.; Holt, E.C.; Ellis, W.C.; Bashaw, E.C. (1972) Nutritive value estimates in apomictic lines of buffel grass (Cenchrus ciliaris L.). Agronomy Journal 64, 453-456.

Marten, G.C. (1973) Alkaloids in reed canary grass. In: Anti-quality components of forages. Editor A.G. Matches. Madison, Wisconsin, Crop Science Society of America Special Publication No. 4, 15-31.

Marten, G.C.; Jordan, R.M.; Hovin, A.W. (1976) Biological significance of reed canary grass alkaloids and associated palatability variation to grazing sheep and cattle. Agronomy Journal 68, 909-914.

Marus, P.; Hovin, A.W.; Marten, G.C. (1976) Inheritance of three groups of indole alkaloids that affect forage quality of reed canary grass. Agronomy Abstracts, 110.

Marum, P.; Hovin, A.W.; Marten, G.C.; Shenk, J.S. (1979) Genetic variability for cell wall constituents and associated quality traits in reed canary grass. Crop Science 19, 355-360.

Mason, N.W.; Schenk, J.S. (1976) The inheritance of forage quality traits in orchard grass (Dactylis glomerata L.). Agronomy Abstracts, 110.

McGehee, B.R. (1976) Heritability of in vitro dry matter digestibility in bermuda grass. Dissertation Abstracts International B 36, 5384.

Michelmore, A.; McKay, A.; Mackie, D. (1980) Annual rye grass toxicity. Fact Sheet 91/77, Department of Agriculture, South Australia.

Mills, P.F.L.; Boultwood, J.N. (1978) A comparison of Paspalum notatum accessions for yield and palatability. Rhodesian Agricultural Journal 75, 71-74.

Milot, J.C.; Burton, G.W. (1973) Variation in the morphology, IVDMD and reproduction of 158 ecotypes of Panicum maximum Jacq. Proceedings of the Association of Southern Agricultural Workers Inc. Atlanta, U.S.A. 70.

Minson, D.J.; Raymond, W.F.; Harris, C.E. (1960) The digestibility of grass species and varieties. Proceedings of the 8th International Grassland Congress, Reading, 470-474.

Minson, D.J. (1971) The digestibility and voluntary intake of six Panicum varieties. Australian Journal of Experimental Agriculture and Animal Husbandry 11, 18-25.

Mowat, D.N.; Christie, B.R.; Winch, J. (1965) The in vitro digestibility of plant parts of orchard grass clones with advancing stages of maturity. Canadian Journal of Plant Science 45, 503-507.

Muller, L.D.; Lechtenberg, V.L.; Bauman, L.F.; Barnes, R.F.; Rhykerd, C.L. (1972) In vivo evaluation of a brown midrib mutant of Zea mays L. Journal of Animal Science 35, 883-889.

O'Donovan, P.B.; Barnes, R.F.; Plumlee, M.P.; Mott, G.O.; Packett, L.V. (1967) Ad libitum intake and digestibility of selected reed canary grass (Phalaris arundinacea L.) clones as measured by fecal index method. Journal of Animal Science 26, 1144-1152.

Oram, R.N. (1970) Genetic and environmental control of the amount and composition of toxins in Phalaris tuberosa L. Proceedings of the 11th International Grassland Congress, Surfers Paradise, Australia, 785-788.

Oram, R.N.; Clements, R.J.; McWilliam, J.R. (1974) Inheritance of nutritive quality of summer herbage in Phalaris tuberosa. Australian Journal of Agricultural Research 25, 265-274.
Osbourn, D.F.; Thomson, D.J.; Terry, R.A. (1966) The relationship between voluntary intake and digestibility of forage crops using sheep. Proceedings of the 10th International Grassland Congress, Helsinki, 363-367.
Patil, B.D.; Jones, D.I.H. (1970) The mineral status of some temperate herbage varieties in relation to animal performance. Proceedings of the 11th International Grassland Congress, Surfers Paradise, Australia, 726-730.
Person, C.; Sidhu, G. (1970) Genetics of host parasite relationships. Proceedings of a panel on mutation breeding for disease resistance. Vienna, Austria. 12-16 Oct. 1970.
Porter, J.K.; Bacon, C.W.; Robbins, J.D. (1974) Major alkaloids of a Claviceps isolated from toxic bermuda grass. Journal of Agricultural and Food Chemistry 22, 838-841.
Powell, J.B.; Burton, G.W. (1971) Genetic suppression of shoot trichomes in pearl millet, Pennisetum typhoides. Crop Science 11, 763-765.
Quesenberry, K.H.; Sleper, D.A.; Cornell, J.A. (1978) Heritability and correlations of IVOMD, maturity and plant height in Rhodes grass. Crop Science 18, 847-850.
Rae, H.L.; Brougham, R.W.; Barton, R.W. (1964) A note on liveweight gains of sheep grazing different rye grass pastures. New Zealand Journal of Agricultural Research 7, 491-495.
Reid, C.S.W.; Lyttleton, J.W.; Mangan, J.L. (1962) Bloat in cattle XXIV. A method of measuring the effectiveness of chewing on the release of plant cell contents from ingested feed. New Zealand Journal of Agricultural Research 5, 237-248.
Reid, C.S.W.; Ulyatt, M.J.; Wilson, J.H. (1974) Plant tannins, bloat and nutritive value. Proceedings of the New Zealand Society of Animal Production 34, 82-92.
Rogers, H.H.; Thomson, A.J. (1970) Aspects of the agronomy and genetics of quality components in a diallel set of progenies of Lolium perenne L. Journal of Agricultural Science, Cambridge 75, 145-158.
Rook, J.A.; Muller, L.D.; Shank, D.B. (1977) Intake and digestibility of brown midrib corn silage by lactating dairy cows. Journal of Dairy Science 60, 1894-1904.
Ross, J.G.; Bullis, S.S.; Lin, K.C. (1970) Inheritance of in vitro digestibility in smooth brome grass. Crop Science 10, 672-673.
Ross, J.G.; Mankin, C.J.; Gross, D.F. (1973) The effects of foliar disease on in vitro digestibility of smooth brome grass. Proceedings of the South Dakota Academy of Science 52, 268.
Ross, J.G.; Thaden, R.T.; Tucker, W.L. (1975) Selection criteria for yield and quality in big blue stem grass. Crop Science 15, 303-306.
Ross, W.M. (1972) Effect of bloomless (bl bl) on yield in Combine Kafir-60. Sorghum Newsletter 15, 121.
Saiga, S.; Kawabata, S. (1975) [Variation of chemical components of orchard grass varieties and relationship between chemical components and palatability.] Journal of the Japanese Society of Grassland Science 21, 238-244.
Saini, M.L.; Paroda, R.S.; Goyal, K.C. (1977) Path analysis of quality characters in forage sorghum. Forage Research 3, 131-136.
Schank, S.C.; Kloch, M.A.; Moore, J.E. (1973) Laboratory evaluation of quality in sub-tropical grasses. II. Genetic variation in Hemarthrias in in vitro digestion and stem morphology. Agronomy Journal 65, 256-258.
Schank, S.C.; Day, J.M.; de Lucas, E.D. (1977) Nitrogenase activity, nitrogen content, in vitro digestibility and yield of 30 tropical forage grasses in Brazil. Tropical Agriculture 54, 119-125.
Shakespeare, W. (1623) As you like it. In: The complete works of William Shakespeare. 1905 Edition, Oxford, Oxford University Press, 217-242.
Sleper, D.A.; Garner, G.B.; Assay, K.H.; Boland, R.; Pickett, E.E. (1977) Breeding for Mg, Ca, K and P content in tall fescue. Crop Science 17, 433-438.
Sleper, D.A.; Mott, G.O. (1976) Digestibility of four Digitgrass cultivars under different harvest frequencies. Agronomy Journal 68, 993-995.
Simons, A.B.; Marten, G.C. (1971) Relationship of indole alkaloids to relationships of Phalaris arudinacea. Agronomy Journal 63, 915-919.
Stobbs, T.H. (1973) The effect of plant structure on the intake of tropical pastures. 1. Variation in the bite size of grazing cattle. Australian Journal of Agricultural Research 24, 809-819.
Stobbs, T.H. (1975) The effect of plant structure on the intake of tropical pasture. III. Influence of fertilizer nitrogen on the size of bite harvested by Jersey cows grazing Setaria anceps cv. Kazungula swards. Australian Journal of Agricultural Research 26, 997-1007.
Stratton, S.D.; Sleper, D.A.; Matches, A.G. (1979) Genetic variation and inter-relationships of in vitro dry matter disappearance and fiber content in orchard grass herbage. Crop Science 19, 329-334.
Tan, G.; Tan, W.; Walton, P.D. (1976) Genetics of vein and stomatal characters in Bromus inermis Leyss. Crop Science 16, 722-724.
Thaden, R.T.; Ross, J.G.; Akyurek, A. (1975) Variability of in vitro dry matter digestion in diallel and polycross progenies of intermediate wheat grass. Crop Science 15, 375-378.
Tilley, J.M.A.; Terry, R.A. (1963) A two stage technique for the in vitro digestion of forage crops. Journal of the British Grassland Society 18, 104-111.

Ulyatt, M.J. (1971) Studies on the causes of the differences in pasture quality between perennial ryegrass, short-rotation rye grass, and white clover. New Zealand Journal of Agricultural Research 14, 352-367.
Vidhyasekaran, P.; Chinnadurai, G.; Govindaswamy, C.V. (1971) HCN content of sorghum leaves in relation to rust resistance. Indian Phytopathology 24, 332-338.
Voigt, P.W.; Kneebone, W.R.; Mellvaine, E.H.; Shoop, M.S.; Webster, J.E. (1970) Palatability, chemical composition, and animal gains from selections of weeping lovegrass, Eragrostis curvula (Schrad.) Nees. Agronomy Journal 62, 673-676.
Vose, P.B.; Breese, E.L. (1964) Genetic variation in the utilization of nitrogen by ryegrass species Lolium perenne and L. multiflorum. Annals of Botany 28, 251-270.
Walters, R.J.K. (1971) Variation in the relationship between in vitro digestibility and voluntary dry matter intake of different grass varieties. Journal of Agricultural Science, Cambridge 76, 243-252.
Walters, R.J.K.; Evans, E.M. (1974) Digestibility and voluntary intake of Be6393 Cocksfoot. Report of the Welsh Plant Breeding Station, 1973, pp.43-44.
Webb, P.J. (1973) Alkaloid studies. Plant Breeding Institute Annual Report, 1972, pp.107-108.
Williams, J.M. (1972) An evaluation of plant materials and procedures for developing nutritionally superior lines of buffel grass (Cenchrus ciliaris L.). Dissertation Abstracts International B 33, 985.
Williams, M.; Barnes, R.F.; Cassady, J.M. (1971) Characterisation of alkaloides in palatable and unpalatable clones of Phalaris arundinacea L. Crop Science 11, 213-217.
Wilman, D.; Omaliko, C.P.E. (1978) Some varietal differences in factors affecting nutritive value and in recovery after cutting in Lolium multiflorum. Journal of Agricultural Science, Cambridge, 90, 401-416.
Wilson, D. (1965) Nutritive value and the genetic relationships of cellulose content and leaf tensile strength in Lolium. Journal of Agricultural Science, Cambridge 65, 285-292.
Wilson, D. (1973) Leaf anatomy, productivity and nutritive quality. Report of the Welsh Plant Breeding Station, 1973, p.15.
Wilson, J.R.; Minson, D.J. (1980) Prospects for improving the digestibility and intake of tropical grasses. Tropical Grasslands 14, 253-259.
Woods, D.L.; Clark, D.W. (1971) Genetic control and seasonal variation of some alkaloids in reed canary grass. Canadian Journal of Plant Science 51, 323-329.
Woods, D.L.; Clark, K.W. (1974) Palatability of reed canary grass pasture. Canadian Journal of Plant Science 54, 89-91.

ANIMAL BREEDING FOR IMPROVED PRODUCTIVITY

J.E. VERCOE, J.E. FRISCH

CSIRO, Tropical Cattle Research Centre, Division of Animal Production, Rockhampton, Queensland 4701, Australia.

ABSTRACT

Realized productivity is a consequence of two genetically determined factors, potential productivity (that which is measured in the absence of environmental stress) and resistance to environmental stresses. These two factors are negatively correlated between breeds and probably within breeds and selection for productivity will therefore be for different combinations of these attributes in different environments or in different seasons and years. This concept is developed in detail in relation to one component of productivity, growth rate, but can be applied to other components such as fertility and survival, and milk or wool production. The implications for breeding programs to improve productivity, particularly where environmental stresses are high and fluctuating, are discussed.

The relevance and possible importance of genetic improvement in such attributes as digestive efficiency, selectivity, drought tolerance and water usage, are mentioned in the light of the potential and adaptive factors which affect productivity.

The application of the principles outlined in the paper to the improvement of meat, milk and wool production are briefly presented.

INTRODUCTION

This paper is concerned with genetic and environmental factors which affect the productivity of domestic animals in grazing situations particularly in the tropics and subtropics. It analyzes the realized productivity into two factors, one of which is a result of the inherent potential of the animal and the other which is a consequence of the animal's inherent resistance to environmental stresses. Present knowledge suggests that across breeds and possibly within breeds these two factors are negatively correlated and it is this which presents difficulties firstly in breeding a highly productive breed for all environments and secondly in making rapid and continuing genetic gains in productivity in any particular environment.

The paper is not concerned with the quantitative genetics of livestock productivity improvement but with the physiological principles which must be considered if selection of breeds and animals for improved productivity is to be efficient.

PRODUCTIVITY IN RELATION TO ENVIRONMENT

Productivity of domestic animals is principally a function of fertility, mortality ahd growth rate. In the case of meat, milk or wool production, the quality and quantity of the product must also be considered as an integral part of the function. Productivity can be expressed on a per head or a per hectare basis. For products such as milk and wool the composition of the product is relatively unaffected by stocking rate and productivity can be meaningfully expressed on a per hectare basis. However, carcass composition ('finish') is markedly affected by stocking rate and both expressions of productivity are then relevant.

Productivity per head is influenced by the mature size of the animal so scaling factors, such as kg meat produced per 100 kg of breeding cow, are necessary in order to take account of this effect. Animals of similar mature size may differ in maintenance requirement (Frisch 1973, Frisch & Vercoe 1977) and this may alter the outcome of comparisons of productivity expressed either on a per animal or per hectare basis.

The level of each component of productivity which is realized in a particular environment is the result of genetic and environmental factors and the interaction between them. Environmental factors are related to climate, nutritional fluctuations and deficiencies, parasitic and other diseases, and managerial variables. Tropical and subtropical environments are characterized by high ambient temperatures and humidities, high solar radiation, a variety of external and internal parasites including tick-borne protozoal diseases, other diseases such as bovine infectious keratoconjunctivitis (BIK), and large seasonal and year to year fluctuations in the quantity and quality of the available nutrition. Each of these environmental stresses acts as a constraint on productivity, though not all genotypes are affected to the same extent at the same level of exposure to the environmental stress. The relative differences between breeds of cattle will be discussed in more detail later. Improvements to productivity can be achieved therefore by controlling or modifying the environment, changing to less affected, that is better adapted genotypes, or some combination of these two alternatives.

POTENTIAL AND REALIZED PRODUCTIVITY

Because the productivity of a genotype changes with the environment (e.g. Hamilton & Langlands 1969, Frisch & Vercoe 1980, 1981), it is most useful when comparing different genotypes to distinguish between potential productivity, that is productivity in the absence of stress, and the realized productivity, that is productivity in the presence of known environmental stresses. The extent to which the potential productivity is realized in a given environment is a measure of the adaptation of the genotype to that environment. Whilst it is difficult to define potential productivity in absolute terms, relative estimates can be made by comparing the productivity of two or more genotypes in a low stress environment. Likewise estimates of relative susceptibility to an environmental stress (that is, adaptation) can be made by measuring the decrease in productivity of the different genotypes when the stress is imposed. Physiological and pathological indices, for example tick numbers, rectal temperatures, BIK scores, which relate the severity of an environmental stress to its effect on an animal's productivity, can be used to analyse realized productivity into components associated with potential and adaptation. Such a procedure has more general application.

This type of analytical approach has been extensively developed at the CSIRO Tropical Cattle Research Centre, Rockhampton. The results from this type of approach are presented in more detail for one component of productivity, namely growth.

FACTORS AFFECTING GROWTH POTENTIAL

When different breeds of cattle are kept parasite and disease free and offered high quality feed _ad libitum_ in shaded pens, the breed differences in growth rate are closely correlated with differences in voluntary food intake (VFI) (Frisch & Vercoe 1977).

Between breeds, voluntary food intake is closely correlated with fasting metabolic rate (FM) and the ratio between them (the relative feeding level) appears to be similar for all breeds. Consequently, in a non-stressful environment breeds with the highest absolute voluntary food intake will have the fastest gains and any differences between breeds in the efficiency of gain will not be due to differences in relative feeding levels. This is illustrated in Table 1. Breeds with the lowest voluntary food intake and fasting metabolic rate have the lowest maintenance requirements (Frisch & Vercoe 1977, 1978) and consequently, on restricted feeding and in the absence of other stresses, breeds will rank for growth rate in the reverse order to that when fed _ad libitum_.

TABLE 1

Growth rates, voluntary food intakes and fasting metabolic rates of different breeds in pens.

Experiment	Breed*	Wt. Gain kg/day	Food Intake g/kg/day	FM KJ/kg/d	Ratio
1	BX	0.81	26.6	93.9	0.28
	AX	0.85	27.5	98.8	0.29
	HS	0.84	28.5	97.9	0.29
2	B	0.68	31.8	88.2	0.36
	BX	0.75	33.4	93.1	0.36
	HS	0.81	37.4	99.8	0.37

* B = Brahman; A = Africander; HS = Hereford x Shorthorn; X = HS.

The situation that exists between breeds also appears to exist between genotypes within a breed. Approximately 75 percent of the variation in growth rate between animals within a breed in the absence of stress is accounted for by differences in voluntary food intake (See Table 2) and similar relationships exist between voluntary food intake and fasting metabolism within a breed as between breeds (Frisch & Vercoe 1977, 1981).

Growth potential for breeds of similar mature size is therefore largely a reflection of voluntary food intake measured in the absence of stress and is closely related to maintenance requirement and fasting metabolic rate.

TABLE 2
Correlations between food intake and gain for penned animals of the same breed fed lucerne hay.

Breed	No. of Animals	Correlation Coefficient
Hereford Crosses	18	0.88
Hereford x Shorthorn (HS)	24	0.85
Brahman	9	0.94
Africander	9	0.82
Hereford Shorthorn (HS)	9	0.42

FACTORS AFFECTING REALIZED GROWTH

When different breeds of cattle are grazed together under conditions where environmental factors are largely uncontrolled (for example no parasite control and no supplementary feeding) ranking of breeds will be different from the ranking when parasites are controlled and feed is offered ad libitum. The change in rank can be accounted for in terms of different susceptibilities of the breeds to the environmental stresses which operate largely but not solely, by depressing food intake.

The growth rate of different breeds in different grazing environments is shown in Table 3. Experiments 1 and 2 are comparable; the animals were grazed together but one half of each breed was treated by dipping and drenching at three-weekly intervals to control ticks and gastrointestinal helminths (worms). The results presented for Experiment 3 are from a separate experiment comparing F_1 Brahman x Hereford (BH) and Simmental x Hereford (SimH) with Hereford x Hereford (HH) steers. The data demonstrate that low growth rate is associated with high values for tick numbers, worm burdens (estimated from faecal egg counts epg) BIK scores (a score from 2 to 12 with increasing severity of infection, Frisch 1975) and rectal temperatures.

TABLE 3
Growth rate of breeds in different grazing environments.

Expt.	Breed*	Growth rate kg/day	Tick no./ side	Worms epg	BIK score per animal	Rectal Temperature °C
	B	0.61	0	0	2.0	39.8
1 (T)†	BX	0.69	0	0	2.1	39.8
	HS	0.61	0	0	2.5	40.5
	B	0.61	1	197	2.0	39.6
2 (C)	BX	0.59	9	356	2.2	39.6
	HS	0.38	22	204	4.0	40.4
	BH	0.40	69	104	2.0	39.6
3	SimH	0.00	194	201	3.4	40.5
	HH	0.00	242	191	3.3	40.6

* See Table 1; SimH - Simmental x Hereford; †T = treated; C = control

The ranking of breeds for growth rate depends on the level of stress to which they are subjected. Thus, if ticks and worms are controlled in B, BX and HS cattle, but the animals are still subjected to high temperatures and BIK infection, the BX breed grows fastest. These rankings can be compared with Experiment 2 in Table 1 where, in the absence of stress, the HS breed grew fastest. Experiment 3 illustrates that under grazing conditions which enabled BH to gain 0.4 kg/day, SimH and HH steers only maintained weight; the difference is associated with higher levels of each stress in the unadapted breeds.

While at low levels of stress the growth rate of each breed is related to its growth potential, as the level of stress increases adaptive characters become increasingly important in determining growth rate. With increasing levels of environmental stress, breeds which are better adapted but which have lower growth potentials grow fastest (Frisch & Vercoe 1981).

ANTAGONISM BETWEEN PRODUCTIVE AND ADAPTIVE CHARACTERS

The previous section has illustrated that one component of productivity, namely growth, is influenced by two separate groups of factors. The first group is manifested in the absence of stress in such parameters as high voluntary food intake and fasting metabolic rate. The second group, the adaptive attributes, is manifested in the presence of stresses in such parameters as heat tolerance, resistance to or tolerance of parasites and disease, and low metabolic rate and maintenance requirements.

The combination of high levels of both these groups of factors is not found in any breed. Across breeds, the correlation of growth potential and degree of adaptation is strongly negative (Frisch & Vercoe 1978, 1981). To what extent this negative correlation occurs between animals within a breed is not known but is under investigation. However, it has been shown that selection for growth rate within an unadapted breed in the presence of stress has reduced growth potential and increased adaptation (Frisch 1981). This suggests that combining high growth potential with high levels of adaptation will be difficult to achieve by selecting for growth rate alone rather than for its components.

At present it appears that to achieve maximum realized growth in any given tropical or subtropical environment growth potential must be traded off against adaptation. To what extent a deficiency in adaptation can be overcome by incorporating higher levels of growth potential is not known.

To enable efficient methods of genetic improvement of animal productivity in tropical areas to be made, it is necessary to first of all ascertain the magnitude of the negative correlation between productive and adaptive traits and then to devise selection criteria which assess independently, productive and adaptive characters so that the two can be combined in the required amounts.

FERTILITY

Whilst the foregoing detailed account has centred on growth rate a similar analysis can be made for fertility although at present its components are less well defined. Realized fertility can then be explained in terms of potential fertility, which is the level attained in the absence of environmental stresses and reductions caused by environmental stresses. This is illustrated in Figure 1 for three different breeds.

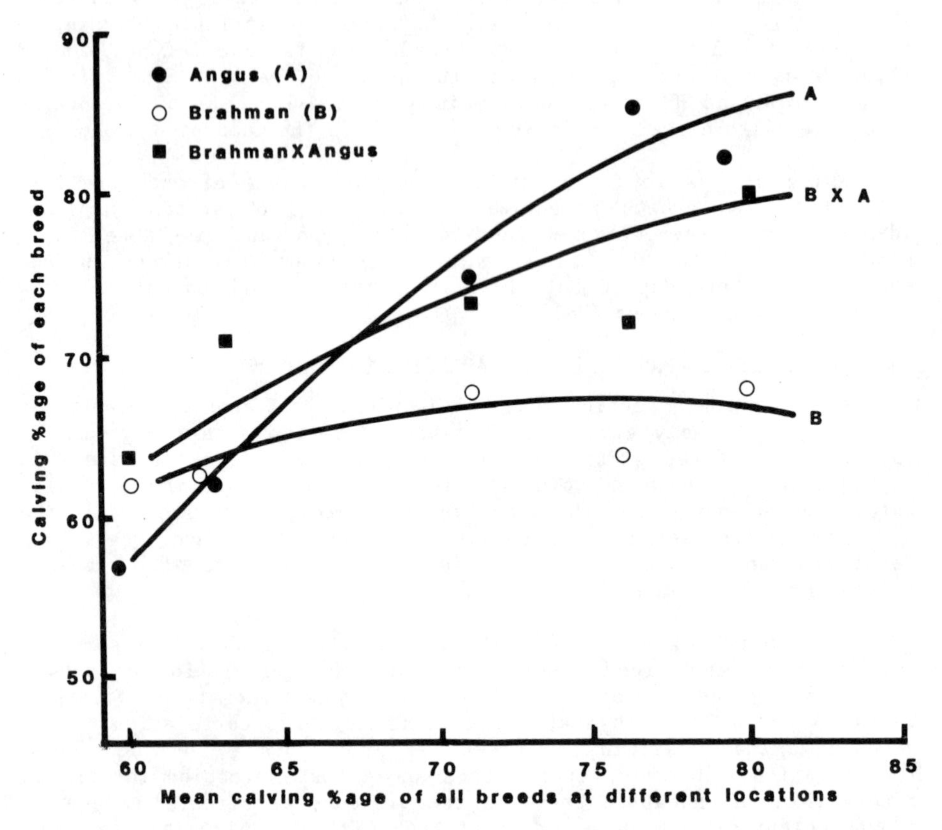

Fig. 1. The relative fertility of different breeds in environments which produce different mean calving percentages for all breeds (Data derived from Preston & Willis 1974).

In years and environments which resulted in a high mean calving percentage the Angus and Brahman x Angus have higher fertility than the Brahman but in years of low mean calving percentage the Brahman will be higher than the Angus and the crossbred. The variation between years and locations is lowest in the best adapted breed (Brahman) and highest in the breed that is potentially the most productive (Angus).

A similar concept to that explaining growth can be applied to explain the interaction between breeds in their levels of realized fertility; the differences in adaptation in the different breeds enable different proportions of different potential fertilities to be expressed in the different environments. Potentially highly fertile breeds only express this potential in non-stressful environments; breeds of lower potential fertility but with higher adaptation realize a relatively higher fertility than other breeds as environmental stresses increase. The principle is demonstrated by Turner (1979 and unpublished) in his studies of the effect of one environmental stress, namely heat, on fertility of different cattle breeds.

GENERAL IMPLICATIONS TO BREEDING PROGRAMS

In environments where breeds of high productive potential are unproductive due to lack of adaptation, crossbreeding with adapted breeds will improve realized productivity despite a lower production potential in the crossbred. Depending on the severity of the environmental stresses the crossbred may also have a higher realized productivity than the better adapted parent breed. In environments of moderate stress the major improvement to realized productivity will be associated with the first cross but, where environmental stresses are severe, further crossing with an adapted breed to produce a 3/4 bred may produce a significant improvement over the half-bred. The improvement in realized productivity will result entirely from the combining of adaptive traits with productive traits so that a greater amount of a lower production potential will be realized. The combination of these traits resulting from crossbreeding is only a "first approximation" and the development of the most desirable combination is then dependent on the use of selection.

The benefits of selection within the crossbred for productive performance will depend on whether environmental stresses are relatively constant or whether there are marked seasonal or annual variations. In an environment where stresses are relatively uniform and performance is a result of constant pressure from both productive and adaptive characters, selection for performance will result in optimum balance of these characters for that environment. However, transfer of those animals to an environment where the stresses individually or collectively are not the same will result in reduced performance, in some environments because the animals lack production potential and in others because they are not sufficiently well-adapted.

In an environment where the stresses are extremely variable, a situation which exists over much of the tropics and subtropics, the gains due to selection within a crossbred may be small or non-existent or at best, extremely slow. In some years performance will be a reflection of potential production associated with high metabolic rate and feed intake in the absence of stress. In others it will be a reflection of adaptive characters such as heat tolerance, tick resistance or BIK resistance. Selection will therefore be for different mixtures of two confounding influences in different seasons and years. Any long term genetic improvement in growth rate will therefore be small if, as appears likely, production potential and adaptation are negatively correlated. Provided this negative correlation is not absolute, simultaneous selection for production potential and adaptive attributes will result in genetic improvement of productivity. Methods for the direct selection for environmental adaptation to ticks, worms, heat tolerance, eye diseases and drought tolerance have been outlined elsewhere (Frisch & Vercoe 1980).

It has been shown that selection for growth rate in an unadapted breed of cattle (Hereford x Shorthorn) in a tropical environment has increased adaptation and reduced production potential characters relative to an unselected population (Frisch 1981). It is likely that the converse is also true, namely, selection for performance in an adapted breed (e.g. Brahman) will improve production potential and reduce adaptive characters to a level appropriate to the environment in which the selection is made. In the crossbred (e.g. Brahman x Hereford) selection for performance will favour Hereford-type characters when stresses are low and Brahman-type characters when stresses are high.

Selection for growth rate or voluntary food intake in the absence of stress will increase mature size and the rate at which maturity is approached, which in some environments and economic circumstances may be undesirable. Under these conditions methods must be devised to separate the rate and scale components of growth and voluntary food intake (H.G. Turner, personal communication).

The concept of two antagonistic groups of factors influencing productivity is crucial to the understanding of genetic improvement of productivity in the tropics and subtropics and research into their interrelationship is a major challenge to both animal physiologists and breeders.

THE RELEVANCE AND IMPORTANCE OF OTHER FACTORS

Whilst most of the difference in weight gains between breeds in non-stressful environments is accounted for by differences in intake, maintenance requirement, and perhaps efficiency of feed utilization (Frisch & Vercoe 1969, 1977, Vercoe 1970, Vercoe & Frisch 1974), within breeds, about 75 percent of the variation between animals is accounted for by differences in intake depending on breed and diet (Table 2, Frisch & Vercoe, 1969, 1977).

In adapted breeds, the correlation between animals for growth rate in pens on *ad libitum* feeding and the growth rate under grazing conditions is of the order of 0.9 (Frisch & Vercoe, unpublished). This similarity in ranking between animals for intake and growth rate in pens, and between growth rate in pens and when grazing, indicates that other differences between animals in nutritional parameters such as digestive efficiency, selectivity and 'foraging ability' are unlikely to contribute significantly to production differences within a breed. Langlands (1967) however, reported significant differences between sheep in diet selected at pasture for some trials. These differences were not consistent for pastures, nor repeatable between days, which suggests that attempts to select for this criterion would be fruitless and wasteful.

Under penned conditions, when environmental stresses are controlled, differences between breeds of cattle in digestive efficiency are small or non-existent (Vercoe 1967, Vercoe & Frisch 1970, Vercoe *et al.* 1972, Warwick & Cobb 1976), though when differences occur they usually favour the *Bos indicus* breeds. It seems likely that where differences exist they stem from the breed differences in resistance to those environmental stresses which affect digestion. No information is available about breed differences in selectivity or foraging ability and both these attributes are probably confounded with differences in voluntary food intake. In the absence of other environmental stresses there is no breed x diet interaction for BX, AX and HS cattle fed high or low quality roughages (Frisch & Vercoe 1977). It certainly would be difficult to select for either selectivity or foraging ability.

Drought tolerance, that is the ability to maintain body condition and survive when availability of feed is low, is correlated with fasting metabolic rate and maintenance requirement (Frisch 1973, Frisch & Vercoe 1977). Breeds with low fasting metabolic rate and maintenance requirement for example Brahman crossbred, had lower weight losses and mortality rates than Hereford x Shorthorns, which have higher metabolic rates and maintenance requirements. This is illustrated in Table 4.

TABLE 4

Differences between breeds in mortality rates in drought years, fasting metabolic rates and relative maintenance requirements.

Breed	Mortality rates in drought years %	Fasting metabolic rate - KJ/kg/day	Relative Maintenance Requirement
Brahman	0.5[a]	77	83
BX	1.5	83	92
AX	2.0	85	97
HS	5.6	89	100

[a] Small number of animals only

Survival is also a function of disease, parasite resistance and heat tolerance which, across breeds, are negatively correlated with fasting metabolic rate. However, in the absence of other stresses, survival is predominantly a function of fasting metabolic rate and maintenance requirement.

In view of the high positive correlation between fasting metabolic rate and voluntary food intake (Vercoe & Frisch 1977, Frisch & Vercoe 1977, 1978), drought tolerant animals are likely to have low potential productivity. There is evidence, however, that the variation in fasting metabolic rate associated with changes in the level of nutrition is greater in some breeds than others (Frisch & Vercoe 1977). For example the rate of reduction in fasting metabolic rate associated with a decline in food intake and diet quality is greater in AX cattle than in BX and HS cattle (Frisch & Vercoe 1977, 1978). Such a mechanism provides a way of uncoupling drought tolerance from the level of growth potential and developing an animal with a potentially high growth rate but with good drought tolerance.

Water metabolism is intimately related to food intake, heat production and temperature regulation. In a hot environment animals must evaporate water to maintain thermal equilibrium. Increases in body temperature resulting from high endogenous (fasting heat production and heat increment of feeding) or exogenous (solar and other radiation and ambient temperatures) heat loads are associated with increases in evaporative water losses (panting and sweating) and a reduction in food intake (Ragsdale _et al_. 1950, 1951, Brown & Hutchinson 1973, Robertshaw & Finch 1976). Water restriction aggravates these effects and the animal conserves water for evaporative needs by further reducing feed intake, decreasing urine volume and increasing the dry matter content of faeces (Thornton & Yates 1968, Shkolnik _et al_. 1972). Breeds of cattle differ in the way in which evaporative water is increased under heat stress; British breeds increase water intake and urine volume whereas AX cattle increase only water intake (Vercoe _et al_. 1972). There are marked breed differences in heat tolerance which relate primarily to differences in ability to dissipate heat by evaporative losses and fasting metabolic rate (Schleger & Turner 1965, Vercoe et al. 1972, Frisch & Vercoe 1978, Finch unpublished). From a physiological point of view, high productivity, which is associated with high food intake, high metabolic rate and heat tolerance, must be accompanied by high evaporative water loss and high water throughput. An increase in evaporative loss without concommitant increases in urine volume and faecal water is desirable for efficient water use, but under most Australian conditions selection of breeds or individuals for efficient water use could be considered "fine-tuning" and relatively unimportant. If such marginal grazing areas need to be stocked, the use of desert adapted species of animal would be a more realistic proposition. However the situation may be quite different in other countries where stock are taken to water every second or third day. Renal and intestinal conservation mechanisms to ameliorate effects of water deprivation on food intake and milk production will be important (Shkolnik et al. 1980) and direct selection for milk production or growth in that environment should induce the desired changes in water conservation and utilization.

APPLICATION OF PRINCIPLES TO IMPROVEMENT OF PRODUCTION

Beef production in the tropics and sub-tropics

Some of the principles of selection for growth rate have been outlined in detail earlier. In addition to the need to take into account separately the potential and adaptive components of growth rate, other correlated changes must also be considered, for example selection for rapid growth from birth to weaning may lead to undesirable consequences on fertility. This is indicated by the data of Seifert (1975), which show that cows which wean the heaviest calves lose most weight in early lactation, also by the data of T.B. Post (personal communication) which shows that they also have a longer post-partum anoestrus period. In addition Frisch (1982) has shown that cows which have "bottle teats" wean heavier calves than cows without "bottle teats". Thus selection for high weaning weight alone is likely to increase the frequency of "bottle teats" with its attendant problems of low fertility and high mortality rate of calves (Frisch 1982). The use of weaning weight as a selection criterion therefore needs to be considered in the light of other factors.

Provided the repeatability of calving for individual cows is high it is possible to maintain a phenotypically fertile herd of cows of an adapted breed which genetically may have a relatively low potential fertility (Rudder et al. 1976, Seifert et al. 1980). In this situation, culling cows which fail to calve will increase realized fertility of the herd without significantly increasing the genetically determined potential fertility.

Selection procedures must be devised to identify high producing animals which combine both high production potential and high resistance to environmental stresses and which do not improve one component of productivity at the expense of another.

Milk production

Similar principles apply to improvement in milk yield as for meat production. Unless major modifications are made to the natural environment, milk will continue to be produced in most tropical countries from adapted breeds and their crosses with European dairy breeds such as Friesian and Jersey.

Voluntary food intake is the major constraint to high milk production from cows with a potentially high yield (Hibbs & Conrad 1975). If potentially high yielding cows, for example European breeds and their crosses with native breeds, are unable to maintain food intake because of environmental stresses, the realized level of milk produced may be considerably less than the potential yield. Again, a high level of adaptation is needed so that the voluntary food intake potential for any given diet (and therefore milk yield) can be realized.

The most important stress likely to affect voluntary food intake for milk producing animals maintained under a "cut and carry" system is heat stress. Where cattle spend substantial portions of the day grazing, resistance to parasites and microbial diseases assumes a greater relative importance. Sustained genetic improvement of milk yield again depends on improving both potential milk yield and resistance to environmental stresses.

Improvement in heat tolerance poses a dilemma because it must be achieved without sacrificing the high metabolic rate and high voluntary food intake potential required for high milk production. On theoretical grounds, selection for high milk production under heat-stressful conditions is likely to reduce fasting metabolic rate and voluntary food intake potential unless simultaneous improvements are made to heat dissipating mechanisms. This is likely to occur because the realized milk production and food intake in heat stressful conditions will depend on heat tolerance and the heat tolerant animal is unlikely to be the one which has the highest voluntary food intake in the absence of stress. Some voluntary food intake potential will then be lost to achieve heat tolerance. It is therefore probable that the most efficient method of selecting for high milk production in hot environments will include selection for high sweating rate and other heat dissipating mechanisms to enable greater expressions of voluntary food intake and milk production potential.

Wool Production

Wool production (g/day) is markedly affected by food intake (Ferguson 1959, Schinkel 1960, Williams & Winston 1965). Environmental stresses which reduce food intake will therefore be important in determining wool growth under grazing conditions. Diet also differentially affects the components of fleece quantity and quality but this will not be considered here (Piper & Dolling 1969).

Selection for increased fleece weight under grazing conditions at the Wool Research Laboratory, NSW Department of Agriculture, Trangie, and at CSIRO National Field Station, Cunnamulla and Armidale, has resulted in increased fleece weight, a small increase in body weight and a higher proportion of dry ewes in the selected flock (Turner & Young 1969, McGuirk & Atkins 1976).

Measurements of food intake of sheep selected for wool growth under grazing conditions indicate that the differences between selected lines and unselected controls are small and not significant in pens. In contrast under grazing conditions sheep selected for high fleece weight (F+) have higher intake than sheep selected for low fleece weight (F-). This is shown in Table 5. Wool growth of F+

selected sheep is higher than F- or unselected sheep in both environments but is relatively higher under grazing conditions (Table 6). The higher wool production of F+ sheep is partly due to differences in efficiency of wool growth but there is a component attributed to higher food intake in a grazing situation, if it is assumed that the efficiency of wool growth is independent of the environment.

TABLE 5

Voluntary food intakes of selected and unselected sheep under grazing and penned conditions.

Genotype†	Weight kg	Food intake g/d/kg*	Reference
Pens			
F+	65	22.4	Williams
R	65	22.3	&
F-	66	21.1	Miller 1965
F+	65	16.7	Rams - Trangie
R	65	17.4	
F-	66	16.7	
F+	49	21.8	Dolling
R	50	21.1	Piper 1968
			Ewes - Cunnumulla
F+	39	38.4	Saville & Robards
F-	40	36.7	1972 - Ewes
			Trangie
Grazing			
F+	42	20.0	Hamilton &
F-	44	17.5	Langlands 1969
			Rams -Trangie
F+	47	34.2	(At Armidale)
F-	47	31.1	

† F+ = Selected for high fleece weight;
F- = selected for low fleece weight;
R = Unselected.

* Pens - dry matter basis;
Grazing - digestible organic matter basis.

The relative differences in food intake between the different lines under penned and grazing conditions suggest that selection for wool growth has in part been for increased adaptation to the grazing environment which has enabled a higher proportion of the potential intake to be realized. The observations in pens suggest that there has been no increase in potential food intake in the F+ line despite both higher wool growth and the relation between intake and wool growth.

TABLE 6
Wool growth of selected and unselected sheep under penned and grazing conditions.

Genotype†	Weight kg	Wool growth g/d	Wool growth relative	Reference
Pens				
F+	72	15.3	201	Williams 1966
R	70	11.1	146	Rams - Trangie
F-	71	7.6	100	
F+	49	13.8	113	Dolling & Piper
R	50	12.2	100	1968 - Ewes - Cunnummulla
F+	39	13.3	117	Saville & Robards
R	40	11.4	100	1972 - Ewes - Trangie
Grazing				
F+	47	13.5	270	Hamilton &
F-	47	5.0	100	Langlands 1969 Rams - Trangie (At Armidale)

† See Table 5.

There is evidence that improvement to wool growth through selection has reached a plateau (Pattie & Barlow 1974, Ferguson 1976, McGuirk & Atkins 1976, McGuirk 1980) and that the relation between food intake and efficiency which in unselected sheep is positive (Turner & Young 1969), has become negative in F+ sheep (Saville & Robards 1972). Whereas past estimates of the relative contribution of food intake and efficiency to the increase in wool growth by selection are of the order of 30:70, future estimates will probably show an increasing proportion of increased wool growth is associated with food intake.

Where there are seasonal and annual variations in environmental stresses, selection for the component of wool growth related to food intake will sometimes be for intake potential and sometimes for resistance to stress (always for realized food intake) and a plateau in wool growth will be reached because it will reflect different sources of variation in food intake in different years. A way out of this dilemma may be to select simultaneously for food intake potential (in the absence of environmental stresses) and resistance to environmental stresses which affect wool growth. Progress will depend on whether physiological processes associated with high food intake (per unit liveweight) are compatible with the physiological processes required to withstand environmental stresses.

Although wool growth is linearly related to food intake only 1-2 percent of ingested nutrients are converted to wool (Allden 1979) and this, coupled with the fact that selection for wool growth has resulted in an increased efficiency for wool growth, suggests the possibility of selecting a wool producing animal which has a relatively low maintenance requirement but has a high wool growth. Graham (1968) found that F+ rams had higher fasting metabolic rates than F- rams per unit of weight and if it is assumed that there were no differences in voluntary food intakes between the lines it is possible that the higher metabolic rate of the F+ rams was associated with more metabolically active skin and wool follicles. If the difference was due to more metabolically active general body tissues a higher voluntary food intake would be anticipated, similar to that in cattle (Frisch & Vercoe 1978, 1980). If the metabolic rates of skin and other body tissues can be independently modified, techniques and practical procedures by which this "switch" can be operated should be investigated.

CONCLUSIONS

The improvement of productivity by genetic means requires an understanding of factors which cause the differences in productivity between different genotypes in different environments. Whilst selection for productivity in a particular environment will improve productivity in that environment, the rate of improvement will be slower and will plateau at a lower level than those achieved by concurrent selection for potential production and adaptive traits, especially when the levels of stress vary markedly with seasons and years. In addition, the gains made in that environment will not be transferable to another environment where different levels of stress operate.

At present it is not known to what extent the negative correlation between production potential and adaptation observed between breeds, operates within breeds. Assuming that it is not -1.0, the opportunity exists to select those individuals that combine the highest levels of each. For growth rate this will necessitate within and between animal comparisons in two distinctly different environments, one in which potential food intake is the main contributor to differences in growth rate and the other in which adaptive characters are the major contributors.

At a time when technology could make rapid changes to the nutritional and parasitic components of the environment it is important to attempt to develop breeds of animals which can capitalize on the changes if they occur but be as productive as the existing environment allows if they do not. In the foreseeable future resistance to parasites and heat stress are likely to remain necessary for efficient animal production and breeding programs must still incorporate selection for these attributes. Attempts to combine a high level of production potential with a high level of resistance to environmental stress is a valid goal for future research.

ACKNOWLEDGEMENTS

We wish to thank our colleagues at Rockhampton who, over the years, have contributed to the thoughts expressed in this paper.

REFERENCES

Allden, W.G. (1979) Feed intake, diet composition and wool growth. In: Physiological and environmental limitations to wool growth. Editors J.L. Black and P.J. Reis. Armidale, Australia, University of New England Publishing Unit, pp. 61-78.

Brown, G.D.; Hutchinson, J.C.D. (1973) Climate and animal production. In: The pastoral industries of Australia. Practice and technology of sheep and cattle production. Editors G. Alexander and O.B. Williams. Sydney, Sydney University Press, pp. 336-370.

Dolling, C.H.S.; Piper, L.R. (1968) Efficiency of conversion of food to wool. III. Wool production of ewes selected for high clean wool weight and of random control ewes on restricted and unrestricted food intake in pens. Australian Journal of Agricultural Research 19, 1009-1028.

Ferguson, K.A. (1959) Influence of dietary protein percentage on growth of wool. Nature, London 184, 907.

Ferguson, K.A. (1976) Australian sheep breeding programs - Aims, achievements and the future. In: International Sheep Breeding Congress, Muresk and Perth, Western Australia. Editors G.J. Tomes, D.E. Robertson and R.J. Lightfoot, pp. 13-25.

Frisch, J.E. (1973) Comparative drought resistance of Bos indicus and Bos taurus crossbred herds in central Queensland. 2. Relative mortality rates, calf birth weights and weights and weight changes of breeding cows. Australian Journal of Experimental Agriculture and Animal Husbandry 13, 117-126.

Frisch, J.E. (1975) The relative incidence and effect of bovine infectious keratoconjunctivitis in Bos taurus and Bos indicus cattle. Animal Production 21, 265-274.

Frisch, J.E. (1981) Changes occurring in cattle as a consequence of selection for growth rate in a stressful environment. Journal of Agricultural Science, Cambridge 96, 23-38.

Frisch, J.E. (1982) The use of teat size measurements or calf weaning weight as an aid to selection against teat defects in cattle. Animal Production (in press).

Frisch, J.E.; Vercoe, J.E. (1969) Liveweight gain, food intake and eating rate in Brahman, Africander and Shorthorn x Hereford cattle. Australian Journal of Agricultural Research 20, 1189-1195.

Frisch, J.E.; Vercoe, J.E. (1977) Food intake, eating rate, weight gains, metabolic rate and efficiency of feed utilization in Bos taurus and Bos indicus crossbred cattle. Animal Production 24, 343-358.

Frisch, J.E.; Vercoe, J.E. (1978) Utilizing breed differences in growth of cattle in the tropics. World Animal Review 25, 8-12.

Frisch, J.E.; Vercoe, J.E. (1980) Genotype-environment interactions in growth of cattle - their occurrence, explanation and use in the genetic improvement of growth. Proceedings of the 4th World Conference on Animal Production, Buenos Aires, 615-622.

Frisch, J.E.; Vercoe, J.E. (1981) The physiological genetics of environmental adaptation. Proceedings of the World Congress on Sheep and Beef Cattle Breeding - New Zealand (in press).

Graham, N. McC. (1968) The metabolic rate of Merino rams bred for high or low wool production. Australian Journal of Agricultural Research 19, 821-824.

Hamilton, B.A.; Langlands, J.P. (1969) Efficiency of wool production of grazing sheep. I. Differences between Merino sheep selected for high and low fleece weight. Australian Journal of Experimental Agriculture and Animal Husbandry 9, 249-253.

Hibbs, J.W.; Conrad, H.R. (1975) Minimum concentrate feeding for efficient milk production. World Animal Review 15, 33-43.

Langlands, J.P. (1967) Studies on the nutritive value of the diet selected by grazing sheep. II. Some sources of error when sampling oesophageally fistulated sheep at pasture. Animal Production 9, 167-175.

McGuirk, B.J. (1980) The effects of selection for increased fleece weight. Proceedings of the Australian Society of Animal Production 13, 171-174.

McGuirk, B.J.; Atkins, K.D. (1976) Response to selection for increased fleece weight in Merino sheep. In: International Sheep Breeding Congress, Muresk and Perth, Western Australia. Editors G.J. Tomes, D.E. Robertson, R.J. Lightfoot. pp. 100-104.

Pattie, W.A.; Barlow, R. (1974) Selection for clean fleece weight in Merino Sheep. I. Direct response to selection. Australian Journal of Agricultural Research 25, 643-655.

Piper, L.R.; Dolling, C.H.S. (1969) Efficiency of conversion of food to wool. IV. Comparison of sheep selected for high clean wool weight with sheep from a random control group at three levels of dietary protein. Australian Journal of Agricultural Research 20, 561-578.

Preston, T.R.; Willis, M.B. (1974) Intensive Beef Production, 2nd Edition. New York, Pergamon Press, pp. 210-256.

Ragsdale, A.C.; Thompson, H.J.; Worstell, D.M.; Brody, S. (1950) Milk production and feed and water consumption responses of Brahman, Jersey and Holstein cows to changes in temperature, 50°F to 105°F and 50°F to 8°F. Research Bulletin, Missouri Agricultural Experimental Station No. 460.

Ragsdale, A.C.; Thompson, H.J.; Worstell, D.M.; Brody, S. (1951) Influence of increasing temperature, 40°F to 105°F on milk production in Brown Swiss cows, and on feed and water consumption and body weight in Brown Swiss and Brahman cows and heifers. Research Bulletin, Missouri Agricultural Experimental Station No. 471.

Robertshaw, D.; Finch, V.A. (1976) Effect of climate on the productivity of beef cattle. In: Beef cattle production in developing countries. Editor A.J. Smith. U.K., C.T.V.M. University of Edinburgh, pp. 281-293.

Rudder, T.H.; Seifert, G.W.; Maynard, P.J. (1976) Factors affecting reproduction rates in a commercial Brahman crossbred herd. Australian Journal of Experimental Agriculture and Animal Husbandry 16, 623-629.

Saville, D.G.; Robards, G.E. (1972) Efficiency of conversion of food to wool in selected and unselected Merino types. Australian Journal of Agricultural Research 23, 117-130.

Schinkel, P.G. (1960) Variation in feed intake as a cause of variation in wool production of grazing sheep. Australian Journal of Agricultural Research 11, 585-594.

Schleger, A.V.; Turner, H.G. (1965) Sweating rates of cattle in the field and their reaction to diurnal and seasonal changes. Australian Journal of Agricultural Research 16, 92-106.

Seifert, G.W. (1975) Effectiveness of selection for growth rate in Zebu x British crossbred cattle. I. Pre-weaning growth. Australian Journal of Agricultural Research 26, 393-406.

Seifert, G.W.; Bean, K.G.; Christensen, H.R. (1980) Calving performance of reciprocally mated Africander and Brahman crossbred cattle at Belmont. Proceedings of the Australian Society of Animal Production 13, 62-64.

Shkolnik, A.; Borut, A.; Choshniak, I. (1972) Water economy of Bedouin goats. Symposia of the Zoological Society of London 31, 229-235.

Shkolnik, A; Maltz, E.; Gordin, S. (1980) Desert conditions and goat milk production. Journal of Dairy Science 63, 1749-1754.

Thornton, R.F.; Yates, N.G. (1968) Effects of water restriction on apparent digestibility and water excretion of cattle. Australian Journal of Agricultural Research 19, 665-672.

Turner, H.G. (1979) Genetic correlation between heat tolerance and fertility in beef cows. Proceedings of the 1st Conference of the Australian Association of Animal Breeding & Genetics. 56-58.

Turner, H.N.; Young, S.S.Y. (1969) Quantitative genetics in sheep breeding. Melbourne, McMillan Co. of Australia, pp.195-211.

Vercoe, J.E. (1967) Breed and nutritional effects on the composition of faeces, urine, and plasma from Hereford and Brahman x Hereford steers fed on high and low quality diets. Australian Journal of Agricultural Research 18, 1003-1013.

Vercoe, J.E. (1970) Fasting metabolism and heat increment of feeding in Brahman x British and British cross cattle. In: 5th Symposium energy metabolism of farm animals. Editors A. Schürch and C. Wenk. Zurick, Juris Druck & Verlag, pp. 85-88.

Vercoe, J.E.; Frisch, J.E. (1970) Digestibility and nitrogen metabolism in Brahman, Africander and Shorthorn x Hereford cattle fed lucerne hay. Proceedings of the Australian Society of Animal Production 8, 131-135.

Vercoe, J.E.; Frisch, J.E. (1974) Fasting metabolism liveweight and voluntary feed intake of different breeds of cattle. In: 6th Symposium energy metabolism of farm animals. Editors K.H. Menke, J. Lantzch and J.R. Reichl. Hohenheim, W. Germany, Universität Hohenheim Dokumentationsstelle, pp. 131-134.

Vercoe, J.E.; Frisch, J.E. (1977) The importance of voluntary food intake and metabolic rate to production of different genotypes of cattle in different environments. 3rd International Congress, Society for the Advancement of Breeding Research in Asia and Oceania, Canberra 1(C).-42.

Vercoe, J.E.; Frisch, J.E.; Moran, J.B. (1972) Apparent digestibility, nitrogen utilization, water metabolism and heat tolerance of Brahman cross, Africander cross and Shorthorn x Hereford steers. Journal of Agricultural Science, Cambridge 79, 71-74.

Warwick, E.J.; Cobb, E.H. (1976) Genetic variation in nutrition of cattle for meat production. World Review of Animal Production 12, 75-81.

Williams, A.J. (1966) The efficiency of conversion of feed to wool during limited and unlimited feeding of flocks selected on clean fleece weight. Australian Journal of Experimental Agriculture and Animal Husbandry 6, 90-95.

Williams, A.J.; Miller, H.P. (1965) The voluntary feed intake of sheep genetically different in wool production. Australian Journal of Experimental Agriculture and Animal Husbandry 5, 385-389.

Williams, A.J.; Winston, R.J. (1965) Relative efficiencies of conversion of feed to wool at three levels of nutrition in flocks genetically different in wool production. Australian Journal of Experimental Agriculture and Animal Husbandry 5, 390-395.

OVERCOMING NUTRITIONAL LIMITATIONS THROUGH PASTURE MANAGEMENT

T.R. EVANS

CSIRO Division of Tropical Crops and Pastures, Cunningham Laboratory, St. Lucia, Queensland 4067, Australia.

ABSTRACT

Management of grazed pastures involves manipulation of plant resources to meet animal requirements to sustain a stable system of production. The options available for improving fodder resources are selection of adapted plant species and varieties that improve pasture yield, its seasonal distribution and quality; use of irrigation for special purpose pastures; manipulation of fertilizer input; control of grazing through stocking rate and method of grazing; changing the composition of the grazing population.

In extensive systems of animal production use of adapted legumes is the major basis for improvement in forage supply and its quality. In some arid and semi-arid areas improvement is restricted to sowing improved grasses on more fertile soils and management is mainly directed towards maintenance of a stable ecosystem.

Under more intensive land use better sown grasses and legumes are used to extend the grazing season and to partly compensate for seasonal fluctuation in food supply and quality. Improved utilization is attained by matching the nutritional requirements of different classes of stock to pastures of different quality.

Management for different intensities of animal production in a range of environments from arid to humid temperate and tropical is discussed. It is concluded that management strategies are available for improving the nutritional value of pastures in many developing regions. In highly intensive systems in temperate regions management criteria have yet to be defined to fully exploit the genetic potential of highly selected bred pasture cultivars, and the role of legumes deserves greater emphasis in pasture systems.

INTRODUCTION

Problems of animal production from pastures in temperate, mediterranean or tropical environments have been described earlier in this symposium by Reid & Jung (1982), Allden (1982) and Mannetje (1982). In each environment nutritional limitations occur through inadequacy in quantity of feed available or its nutritive value. These limitations may occur for longer or shorter periods depending on length of the growing season. Most ruminant production is based wholly or partly on herbage or fodder plants and dependance on grazing means that ruminants must derive nutrients from the diet available. Pasture growth is influenced by soil characteristics, fertility status and is controlled by climate, and thus the quantity of forage produced and its quality will vary within and between grazing seasons. For example, Snaydon (1981) cites year to year variation in pasture yields in New Zealand of from two- to six-fold; in semi-arid grasslands, yields may vary by more than three-fold (Williams 1960, Box 1977).

In most environments there are opportunities to change, manipulate and integrate feed resources to meet animal requirements at different stages of growth or for different levels of production and to improve the efficiency of utilization of these resources. The options available include the introduction of superior plant species into an existing ecosystem or its replacement by fully sown pastures; adjustment of soil nutrient resources by fertilizer input; control of pasture utilization through manipulation of stocking rate, method of grazing and composition of the ruminant population, and the integration of irrigated, conserved or special purpose pastures into a production system. In this paper pastures are defined in the broadest sense and encompass any grazed ecosystem where the vegetation is predominantly herbaceous. A detailed discussion of management for complex systems of animal production within each environment is not possible in this short review. This discussion is limited to a consideration of the management strategies adopted in different environments, so that the basic principles involved in improving pasture and animal production may be defined.

CHOICE OF SPECIES

The rationale for selection of species for improving pasture production is (i) that the species will increase forage yield or change the pattern of production to reduce deficiencies during critical periods of animal requirement, (ii) that the nutritive value of forage is improved and nutritional stress to the animal is removed or reduced, (iii) that animal production or the efficiency of production is increased, (iv) that any changes in management required to exploit the improved feed supply are practicable and within the resources of management and (v) that benefits are realizable in economic or other terms.

Grasses

The nutritive value of pasture species, both temperate and tropical, differs between species and cultivars (Milford 1960, Minson *et al*. 1960, Cooper *et al*. 1962, Minson & Milford 1966, Minson & McLeod 1970, Minson 1971, Burton 1972, Hacker & Minson 1972, McCawley & Dahl 1980, Ulyatt 1981). Changes in feeding value are also associated with stage of growth, and may occur seasonally or between cycles of regrowth. In a temperate environment Greenhalgh & Reid (1969) showed that *Lolium perenne* cv. S24 gave higher milk yields in summer than *Dactylis glomerata* cv. S37 although there was little difference between these species in spring. Wilson & McDowall (1966) reported higher milk yields from *L. perenne* cv. Ariki than from cv. Ruanui in spring, but the reverse in autumn. These differences in milk production are partly explained by differences in leaf/stem proportions and rates of change in digestibility between species and cultivars. Ulyatt (1981) also found large differences between and within species in sheep weight gains, but these were not so apparent in wool production. In some humid temperate environments the problem of increasing feed supply in early spring may be overcome by choice of genotypes or species adapted for growth under lower temperatures and radiation levels (Ollerenshaw *et al*. 1976). Where there are defined warm and cold seasons and no moisture limitations, fluctuations in seasonal pasture growth may be reduced and animal production improved by the integrated use of species adapted to the different seasonal conditions (Hann & Lazenby 1974, Lambert *et al*. 1979). In New Zealand use of the sub-tropical grasses *Pennisetum clandestinum* and *Paspalum dilatatum* in *Lolium perenne* pastures substantially reduced seasonal variation in pasture and animal production (Baars 1976). A comparison of L. perenne - Trifolium repens pastures with or without P. dilatatum or P. clandestinum for sheep production showed that in spring and early summer L. perenne supplied 55 percent of the metabolizable energy intake and P. clandestinum 42-48 percent in the summer period (Rumball & Boyd 1980). In the subtropics choice of grasses adapted for growth under lower winter temperatures and with cold or frost tolerance is used to extend the grazing season and improve pasture quality in this period (Oakes et al. 1980).

The prospects for improving pasture production in arid and semi-arid environments through reseeding with grasses are largely dependant on whether the existing grassland comprises annual or perennial grass species. Stable perennial pastures such as the Astrebla grasslands in Australia, which occur on inherently fertile soils, offer less opportunity for improvement by the introduction of other grasses into the ecosystem than communities of annual species (Burrows 1980). Because drought is a recurring phenomenon and rainfall variability is high, plant communities based on annual grasses are unstable. Oversowing with perennial grasses has the advantage of reducing the fluctuation in pasture production and a more rapid response to out of season rainfall (Williams 1960). Grasses within the genera Anthephora, Cenchrus, Ehrharta and Schmidtia have been used in overseeding arid grasslands and have increased herbage yields by up to 70 percent as compared with those from native pastures (Ebersohn 1970, O'Donnell et al. 1973). Reduction in fluctuation in feed supply and increased animal production has also been obtained by introducing non-gramineous plants such as Atriplex spp. into arid zone pastures (Williams 1960, Nemati 1978).

The contribution of improved species in some semi-arid environments is not only through improvement in pasture production, but also through the provision a more drought tolerant forage. Coaldrake et al. (1969) found that on fertile soils in the brigalow region of Queensland (mean annual rainfall 600 mm) over a three year drought period (annual rainfall of 457, 330 and 200 mm) sown pastures of Cenchrus ciliaris were able to carry twice as many beef as cattle, native pastures that comprised Chloris, Paspalidium, Dichanthium and Eragrostis species, and the sown pasture recovered more rapidly after drought. Similar responses were reported by Silvey et al. (1978) from Panicum maximum var. trichoglume pastures.

Differences in nutritive value between species, reviewed earlier in this symposium by Norton (1982) are an important basis for choice for improving animal production. A clear example of this is shown by McCawley & Dahl (1980) in a comparison of Cynodon dactylon (Coast cross-1), Chloris gayana and Panicum coloratum on cleared mesquite rangeland in southern Texas. Although the estimated intake by beef cattle of all species was similar, daily weight gains were highest from Cynodon dactylon and lowest from Chloris gayana and a similar ranking occurred for in vitro dry matter digestibility, digestible crude protein and digestible energy. Changes in phosphorus content over the growing season differed between species, decreasing from 0.16 percent to 0.06 percent in Rhodes grass, but only from 0.24 percent to 0.15 percent in the other two species.

Legumes

The role of legumes in improving pasture and animal production is well recognized. Differences between species in nutritive value occur in temperate legumes (Ulyatt 1981), tropical legumes (Milford 1967) and in those used in mediterranean environments (Purser 1981). Animal production per head is nearly always greater from legume based pastures than from pure grass pastures. For example Ulyatt et al. (1976) obtained 86 percent greater liveweight gain in lambs from T. repens than from L. perenne, a nine percent increase over that obtained from lucerne and 23 percent more than that from sainfoin (Onobrychis viciifolia). There is ample evidence of improvement in animal production from temperate legume based pastures over pure grass swards as shown by Rae et al. (1964), Grimes et al. (1967) and in the review by Thomson (1977). This improvement is partly due to a higher intake of legumes compared with grasses (Gibb & Treacher 1976) but also through a greater efficiency in utilization of metabolizable energy. A further important attribute in legumes is the slower rate of decline in digestibility with maturity compared with grasses. Differences occur between species and the rate of decrease is greater for the taller growing temperate legumes such as T. pratense and M. sativa (Ulyatt et al. 1976), partly because of the increase in stem component with maturity. However, these authors also showed that digestibility of leaf and stem may vary with species, for example the leaf of M. sativa had a higher digestibility than leaf of O. viciifolia but stem of M. sativa was lower in digestibility. Differences in nutritive value between some tropical legumes were reported by Milford (1967), who also found marked differences between species in changes in nutritive value following frosting.

The use of adapted legumes in improving the quantity and quality of tropical pastures has been demonstrated in many environments. In some the introduction of legumes into native pasture has substantially increased animal production without any fertilizer input. Clatworthy & Holland (1979) obtained a 53 percent increase in liveweight gain of cattle on Hyparrhenia filipendula - S. guianensis pastures in Zimbabwe; Stobbs (1966b, 1969) reported an increase of 11-49 percent from H. rufa - S. guianensis pastures in Uganda. Modest inputs of fertilizer (superphosphate) have further increased liveweight gain (Shaw 1961, 1978b; Shaw & Mannetje 1970, Woods 1970) and reproductive rate (Edye et al. 1971, Ritson et al. 1971, Holroyd et al. 1977). The improved levels of animal production are not only expressed in higher liveweight gains per head but also in improved carrying capacity (two to six fold), a reduction in time to turnoff of fat cattle (one to two years earlier than off native pastures) and improved preweaning growth rates and increased turn off weights of calves per cow mated (Mannetje & Coates 1976). Beef production of between 260 kg and 500 kg liveweight gain per hectare have been obtained from fully sown tropical grass-legume pastures (Evans 1970, Grof & Harding 1970, Aronivich et al. 1970, Evans & Bryan 1973, Vicente-Chandler et al. 1974, Jones & Jones 1982).

Milk production from sub-tropical pastures and the quality of milk produced may be influenced by legume species. Stobbs (1971, 1976) showed that the solids-not-fat content of milk was lower from siratro (Macroptilium atropurpureum) than from desmodium (Desmodium intortum) pastures and highest from T. semipilosum pastures, indicating a higher nutrient intake from pastures based on this species.

The relationship between legume content of tropical pasture and liveweight gain from beef cattle is not well understood on a quantitative basis, although close positive relationships have been shown between legume content of pastures and liveweight change (Norman 1970, Evans 1970, Evans & Bryan 1973). Gillard *et al.* (1980) found that increased weight gains occurred only when legume yields were above 600 kg dry matter per ha for some *Stylosanthes* species. Seasonal variation in climate may influence this relationship. Walker *et al.* (1982) found a poor correlation of weight gain and yield of the legume *M. atropurpureum* in years when moisture availability reduced the length of the growing season, but a high positive correlation with legume yield or percentage content in the sward in normal years when preferential grazing of the legume was evident, particularly in autumn. In a shorter growing season this did not occur; yields of pasture were reduced and nitrogen content of the grass was higher.

Some other legume characteristics of nutritional value are the generally higher mineral concentration than that of the grasses (except for sodium which varies with species of legume) and the potential contribution of seed retained on the plant for improving feed quality. For example, Playne (1969) and Robinson & Jones (1972) found that seed of *Stylosanthes humilis* contained 40-45 percent crude protein, 0.4-0.8 percent phosphorus, 0.3-0.5 percent sulphur and had a digestibility of 55-60 percent. At plant maturity seed comprised about 30 percent of the total dry matter and contained 62 percent of the total nitrogen, 72 percent of the total phosphorus and 53 percent of the total sulphur. This attribute may be even more important with perennial legumes. Retention of green leaf into the dry season has extended the period of weight gain and reduced the stress period on animals in a range of environments (Stobbs & Joblin 1966, Langlands & Bowles 1974, Mannetje & Ebersohn 1980, Gardener 1980) and should be considered in choice of species for improving animal production in regions with a pronounced dry season. In the subtropics where pasture production in winter from subtropical species is controlled mainly by low temperature, improvement in feed quantity and quality has been obtained by oversowing temperate legumes into these pastures (Knight 1970, Hoveland *et al.* 1978, Kalmbacher *et al.* 1980).

The problem of bloat on legeum-based pastures, could be overcome by use of legume species which do not induce bloat, such as *Lotus*, *Onobrychis viciifolia* or *Coronilla varia* in the pasture. The condensed tannins in these species form insoluble complexes with plant protein in the rumen and prevent the formation of a stable protein foam. Additional nutritional advantages are a reduction in deamination in the rumen and decreased loss of ammonia and more efficient utilization of plant protein after the breakdown of the protein-tannin complex in the small intestines.

Oestrogenic isoflavones in some *Trifolium subterraneum* cultivars are a problem that is best overcome by selection of low potency cultivars. A major factor leading to a high concentration is a deficiency in soil phosphate (McDonald 1981) which suggests a measure of control through use of phosphate fertilizers.

USE OF FERTILIZERS

Fertilizers will increase herbage production where the applied element is deficient in the soil. When these deficiencies have been corrected, the role of fertilizers is to further increase the quantity and quality of herbage, especially the legume contribution to feed supply, and to overcome periods of feed deficit by manipulation of quantity and frequency of fertilizer input. The fertilizers most widely used for these purposes are superphosphate and nitrogen.

Superphosphate

The evidence for improvement in animal production from superphosphate application to pastures has been referred to earlier in this paper and is clearly demonstrated in subtropical pastures by data from Evans & Bryan (1973), Bryan & Evans (1973) and Shaw (1978a,b). Increases in liveweight gains from beef cattle were attributed to an improvement in legume content of pastures in response to increased superphosphate input, associated with a higher level of phosphorus concentration in grasses and legumes and a higher calcium concentration in grasses and nitrogen in legumes. Digestibility of pasture was also increased by level of superphosphate application (Thornton & Minson 1973). The influence of the calcium, phosphorus and sulphur components of superphosphate on animal production have not been quantified, but the effects of calcium and sulphur on improving intake and digestibility have been demonstrated by Rees & Minson (1976, 1978), and may have been influential in improving animal production from these pastures. Although legumes adapted to low soil phosphorus availability may be quite productive in agronomic terms, their contribution to the diet and effect on animal production may not be realised in the absence of superphosphate application to pastures, as shown by McLean *et al*. (1981) for *Stylosanthes* based pastures in tropical Australia. It is notable that in the humid tropics where deficiencies of both phosphorus and sulphur occur, the use of 'biosuper' (rock phosphate and sulphur inoculated with *Thiobacillus thiooxidans*) has been more effective in increasing pasture production than superphosphate (Partridge 1980a), and that legume responses may differ between species; in this case *Stylosanthes* was more responsive than *M. atropurpureum*.

The effect of superphosphate in increasing legume yield and phosphorus concentration in the pasture has been shown for a number of species in different environments in the tropics and subtropics, but an associated increase in nitrogen concentration may not always occur, as shown by Gilbert & Shaw (1980) for some *Stylosanthes* spp. Superphosphate has been shown to influence diet selection in grazed grass-legume pastures. McLean *et al*. (1981) in northern Australia found that superphosphate increased the legume content of the diet five fold compared with a two fold increase in the proportion of legume in the sward. The phosphorus and nitrogen concentration in leaves of unfertilized or fertilized grass or legume were similar and preference was thus not related to phosphorus or nitrogen concentrations. The mechanisms influencing diet selection have yet to be defined, but the results indicate a possible use of superphosphate (or its constituent minerals) in manipulating diet selected. In a mediterranean environment Ozanne & Howes (1971) found that sheep preferentially grazed the dry residues from high phosphate treatments in summer but in this case phosphorus concentration was higher (range 0.06 to 0.17 percent); no preferences were apparent within the range of 0.10 to 0.15 percent phosphorus. In pen feeding experiments Ozanne *et al*. (1976) showed that increasing the phosphorus content of a grass legume mixture, increased dry matter intake, percentage of dry matter digested and increased bodyweight gains.

Nitrogen

In humid temperate environments high levels of nitrogen fertilizer have been used to maximise herbage production in intensive management systems, particularly for milk production. Such systems are vulnerable to problems in animal health from possible metabolic dysfunction through changes in mineral concentration in the plant, or from increased levels of endoparasite infection in the animal. The change in mineral supply may arise from botanical changes in the pasture; where nitrogen is applied to grass-legume pastures the legume content is reduced and mineral levels in the diet may decline, particularly calcium. Also, with the increased growth rate of nitrogen fertilized pasture a rapid reduction in mineral concentration in the plants will occur unless soil fertility is adequate to provide for plant requirements to maintain a high concentration in the tissues (Hartmans 1975). Magnesium availability to cattle is decreased at a young growth stage in pasture and by a high potassium content in herbage (Kemp et al. 1961); decreased copper availability may also occur where there is a high digestibility of the feed and high protein content relative to energy (Hartmans 1975). The form of nitrogen fertilizer used may influence mineral concentrations; nitrate nitrogen increasing calcium, magnesium, sodium and potassium, while ammonium nitrogen depresses the concentration of these cations (Hartmans 1975).

High levels of beef production (ca. 1000 kg liveweight gain per ha per annum) have been obtained from nitrogen fertilized tropical pastures (Evans 1969, Vicente-Chandler et al. 1974, Harding & Grof 1978), but gains per head are generally not as high as from grass-legume pastures.

An alternative use of nitrogen fertilizer is as a strategic application to improve pasture production and quality in periods of shortage. Autumn application of nitrogen to subtropical grasses has increased nitrogen content, digestible protein and weight gains of cattle (Blue et al. 1961, Kretschmer 1965) and intake and digestibility by sheep (Minson 1967). Increased milk production was reported by Cowan & Stobbs (1976) from *P. maximum* var. *trichoglume* - *Neônotonia wightii* pastures fertilized with 50 kg N per hectare in autumn and winter. Year round nitrogen fertilization of pure grass pastures produce similar daily milk yields per head as from grass-legume pastures (Stobbs 1976).

PASTURE UTILIZATION

Grazing methods

Grazing may be continuous set stocking or some form of rotational grazing, within which grazing pressure is manipulated to adjust the quality of forage available to meet stock requirements. The advantages of rotational grazing over continous set stocking in increasing animal production are rarely evident, except at high grazing pressures. Under intensive grazing the maximum feeding value of pastures in the young leafy stage of growth may be best utilized by rotational or strip grazing. For example Walshe et al. (1971a,b) obtained higher growth rates in rotationally grazed calves and also reported increased milk production from cows of from 13 to 20 percent (Walshe 1975); Brougham et al. (1975) reported annual production of more than 1000 kg per ha carcass weight from rotationally grazed grass-clover pastures.

However, most of the evidence from temperature pastures show no advantage from rotational grazing in increasing production per head. For example Hood (1974) found no differences in milk production between set stocked and rotationally grazed pastures at stocking rates of 5 cows per ha on pastures receiving 400 kg N per ha per annum. In Queensland Chopping et al. (1978) obtained higher milk production from continuously grazed irrigated D. decumbens pastures, and Davison et al. (1981) reported similar results from nitrogen fertilized Panicum maximum and Brachiaria decumbens pastures. Mowing of these pastures after rotational grazing increased leaf percentage, crude protein and digestibility but there was no increase in milk production. Rotational grazing resulted in a diet which was mainly stem at the end of a two week grazing period; in contrast continuously grazed animals consistently selected a high proportion of leaf. In southern Australia Willoughby (1970), concluded that under year round grazing there was no advantage in subdivision for rotational, strip grazing or creep grazing compared with continuous grazing, except in management of lucerne pastures. A review of grazing systems in southern Africa (Gammon 1978) and comparisons in latin America (Paladines et al. 1974, Paladines & Leal 1979) support the general evidence for higher animal production from continuously grazed pastures. The lack of any substantial difference between grazing systems is probably explained by different plant responses to defoliation under the continuous or rotational grazing system. In a continuously grazed pasture intervals between defoliation of individual plants may be similar to that imposed under rotational grazing. In the latter, herbage allowance is high at the commencement of grazing a subdivision and decreases during the grazing period. This effectively limits selection by the animal and the nutritional level of the pasture decreases from high to a lower level which reduces animal performance.

There are particular circumstances where rotational grazing is an essential component in pasture management. M. sativa productivity and persistence are markedly increased by rotational grazing (Leach 1978), and Leucaena leucocephala may not be productive under continuous grazing in some environments. Jones & Jones (1982) grazed leucaena under a rotation of four weeks grazing and four weeks rest over a 12 year period at 2.5 yearling cattle per hectare; the legume persisted and produced a mean of 310 kg weight gain per ha per year. Manipulating stocking rate to match the seasonal change in legume growth in this subtropical environment (1.3 beasts per ha in winter and 4.0 per ha in summer) increased production to over 400 kg per ha a year (R.M. Jones personal communication).

The benefits of rotational grazing systems are in general more related to a need for conservation of pasture as hay or silage for winter feed in temperate environments, or as a technique for controlling botanical composition in other environments, rather than as a means of increasing nutritional value per se and animal performance. In arid and semi-arid grasslands drought is a recurring phenomenon and rainfall variability is high, especially in the lower rainfall regions. In many of these grasslands perennial grasses are the main source of feed supply and if overgrazed are replaced by annual species of lower quality. Under these conditions a flexible grazing system is often used to maintain this resource. Grazing properties are extensive, often >1000 km^2, and cover many different land systems. Rotational grazing to apply equitable grazing pressure to all areas aids in maintaining a stable resource (Harrington 1981). Strategic heavy grazing is used in some environments for control of shrub regrowth (Laycock 1970, Burrows 1974). This may be the only practicable method of control in fragile environments where the use of fire may reduce stability of the pasture system (Burrows 1972, Booysen 1978). In some areas fire plays an essential role in maintenance of grassland communities (Tainton 1978) and requires rotational grazing management to obtain sufficient fuel to carry a fire. This may involve deferment of grazing for one or more growing seasons.

Fire is traditionally used in many environments to remove dead herbage and allow improved light conditions for regrowth of seedlings or new tillers and to improve accessibility of this regrowth to grazing animals.

A form of rotational grazing in which a portion of the farm area is reserved for grazing late in the growing season may have advantages in extending the grazing season (Gutman & Seligman 1979) and offset lower weight gains from a rotational grazing system. However, Wheeler (1965) found that deferment in autumn to provide feed in late winter for ewes lambing in spring, produced no overall benefit in production. Ewes lost weight on the area reduced for forage accumulation and regained it when fed the saved material. Lambing percentage, birth weights and wool weights were similar to those of a group grazed on a similar area where no deferment was used. These results suggest the possibility of utilizing the animal's energy reserves in periods of feed deficit without widely affecting subsequent performance.

Intensity of grazing

The long term stability of animal production from grass-legume pastures is dependant on persistence of the legume component. This is influenced by the legume response to the frequency and intensity of defoliation, the opportunities for seed production (and hence resources for plant replenishment in the population) and the competitive relationships between legumes and the associated grass components of the pasture. Stocking rate is a major determinant influencing the botanical composition of pastures and consequently the nutritive value of forage available, but there are marked differences between legume species in response to change in stocking rate. Those with a prostrate growth form such as *Trifolium repens*, *T. subterraneum*, *Lotononis bainesii*, *Stylosanthes humilis*, *Desmodium heterophyllum* are more tolerant of heavy defoliation than the erect species, such as *T. pratense*, *Medicago sativa* or the trailing or climbing tropical legumes such as *Macroptilium atropurpureum*, *Centrosema pubescens*, *Pueraria phaseoloides* or *D. intortum*. Stobbs (1970) showed that with *Hyparrhenia rufa* - *S. guianensis* pastures in Uganda a low grazing pressure caused a decline in legume and lower weight gains but the reverse occurred at higher grazing pressure. Similar results were reported by Eng *et al*. (1978a,b) in Malaysia and for *S. humilis* pastures in northern Australia (Woods 1970, Shaw 1978a,b). In Fiji *M. atropurpureum* declined with increase in stocking rate whereas D. *heterophyllum* increased (Partridge 1979, 1980b).

Persistence of tropical legume species under high stocking rates may be enhanced by deferred grazing over the period of seed production to allow regeneration from seed (Jones & Jones 1978). The length of the period of deferment, however, may have a marked influence on the nutritional value of the forage available for grazing, especially when this is towards the end of the growing season and quality is declining.

In mediterannean pastures regeneration of annual legumes is dependant on seed set and grazing pressure may have to be manipulated to achieve this (Rossiter 1978). The potential for manipulating grazing management to improve legume persistence depends on understanding how management affects aspects of persistence such as plant survival, seed set, seedling regeneration and seedling survival (Jones & Mott 1980).

Renovation techniques may help improve persistence in subtropical pastures, as shown by Bishop *et al.* (1982) where light cultivation of a *M. atropurpureum* - based pasture increased seedling regeneration about 300 percent, and subsequent yield of legume was increased from 260 to 1290 kg per ha. It is not possible to predict similar responses for other tropical species. The problem of persistence of some tropical legumes in the humid subtropics could well be overcome by using a system in which compatable pastures are integrated into a production system. For example, *Trifolium repens* will withstand heavy grazing and with a compatable grass form a complimentary pasture to one based on tropical legumes that requires reduced grazing pressure at certain periods.

Integration of pasture resources to meet animal requirements

In most extensive grazing systems in environments as contrasting as the high rainfall tropical savannahs of South America, arid or semi-arid regions in Africa and Australia or hill lands in temperate climates a major constraint to animal production is the low level of dry matter available, often less than 2000 kg per ha, and its poor quality.

Any improvement in the nutritional quality of forage supply by use of better species, is usually limited to a portion of the area. This is utilized for satisfying nutritional requirements of the animal, during those periods of nutritional stress which most influence production. For example, a major constraint in hill sheep production in temperate environments is nutritional stress during lactation, early lamb growth and in the mating period (Eadie 1976). Use of improved pastures in these periods and unimproved pasture during summer, after weaning, and in winter increases flock productivity. However as the area of improved pasture is increased stocking rates need to be increased to fully utilize them, which puts greater grazing pressure on the unimproved pasture at other times. The influence on the stability of this resource under increased pressure needs to be evaluated; similar problems may arise in other environments. Mannetje & Coates (1976) used sown pastures for winter-spring forage and native pastures for summer in a cow-calf system and obtained improved calf performance. Higher weaning weights of steers showed a potential for earlier turnoff, and of heifers the possibility of earlier mating at about 15 months of age.

In Fiji (Partridge 1976) avoided post weaning weight loss in calves on *Pennisetum polystachyon* pastures by using *L. leucocephala* as a feed resource in the dry season; Falvey & Mikled (1978) used *D. intortum* in north-east Thailand to supplement *Imperata cylindrica* pastures and obtained greater dry season weight gains than from a supplement of urea, molasses and minerals.

Improvement in the efficiency of pasture utilization may be achieved by matching the feeding value with the nutritional requirements of different classes of stock. For example a leader-follower system provides higher quality feed for the leader group (for example high producing dairy cows) and a lower quality feed for followers whose nutritional requirements are met by this feed supply. In this way management can manipulate changing feeding value to meet animal requirements (Bryant et al. 1961, Archibald et al. 1975, Blaser et al. 1976), and may be utilized for stock of different ages or levels of production.

Alternatively, groups of mixed animal species can be used to improve pasture utilization and overall level of animal production. Almost all studies with mixed animal species have been restricted to sheep and beef cattle. Ratios of sheep to cattle and stocking rates have been compared. On semi-arid pastures, Ebersohn (1966) found no advantage in liveweight production by grazing sheep and cattle together. However, on sown pastures of Lolium rigidum - Trifolium subterraneum in Victoria, Hamilton & Bath (1970) obtained greater production from mixed grazing. On similar pastures with a range of rates of from 20 to 80 percent sheep to cattle and a stocking rate of 7.4 ewes : 2.1 steers per ha, Hamilton (1976) reported that in years of normal rainfall (mean of 570 mm) performance of steers was unaffected by mixed stocking and sheep benefited to a greater extent as the ratio of sheep to cattle decreased. In a drought year (250 mm) sheep performance remained unchanged but cattle lost weight and there was little or no benefit from mixed stocking. Production from cool temperate pastures of Phalaris aquatica -Trifolium subterraneum showed seasonal differences in response, cattle gaining less in autumn and winter but at a faster rate in spring and this compenstory gain made up for the weight loss. Sheep produced more wool and lambs and weaning weights were increased. When grazed separately, steers had slightly better weight gains than in the mixed group and sheep lower production (Bennet et al. 1970). Integrating production systems (beef fattening and sheep breeding) in this manner could provide substantial advantages in increasing production per unit area. Similar advantages were reported by Heinemann (1970) in a study of irrigated Dactylis glomerata -Medicago sativa pastures rotationally grazed by steers followed by sheep.

Mixed grazing may also improve botanical composition and pasture production. Suckling (1976) used beef cattle to control unpalatable herbage in sheep pastures in New Zealand. In the absence of cattle a dense mat of unpalatable herbage formed and white clover content declined. Judicious use of cattle improved pasture utilization, and botanical composition.

Irrigated pastures

Management of irrigated pastures is outside the scope of this paper and their use is only considered as an aid in improving animal production primarily based on rain fed pastures. The advantages of incorporating irrigated pastures into a pasture production system are to reduce fluctuation in feed supply, ameliorate the effects of drought or for utilization as high quality pastures for milk or fat lamb production. In general the costs of installing an irrigation system are high and requires a high return from stock production to justify its use in a dryland farm system. In the subtropics of Australia irrigated pastures are utilized mainly for milk production and based on fully sown pastures of either ryegrass and use of nitrogen fertilizer or ryegrass with annual *Medicago* species. Attempts to use mixtures of summer and winter adapted grasses and legumes have not been very encouraging (Jones *et al*. 1968). Although total annual dry matter was increased the botanical composition of such pastures was unstable. In north Queensland irrigated pasture mixtures of tropical species of *Brachiaria mutica*, *Chloris gayana* grown with *Centrosema pubescens* or *Stylosanthes guianensis* have been successfully used in fattening beef cattle (Kleinschmidt & Skerman 1977). Irrigated nitrogen fertilized *Digitaria decumbens* pastures have produced about 2000 kg annual liveweight gains a hectare (Evans *et al*. 1978). This clearly indicates a potential use with dry land pastures for increase in animal production by improving feeding quality of the diet for the dry season.

The possibility of using tropical pasture species under irrigation in Mediterranean environments during the dry summer period also needs further evaluation. Roberts & Carbon (1969) in Western Australia obtained yields of 43,000 kg dry matter a hectare from *Pennisetum purpureum* and more than 20,000 kg a hectare from a range of subtropical grasses; yields were recorded over a 24 week period. There is obviously a potential for tropical species in this environment as part of a year round feed program.

Potential for improving animal production in developing countries

The genetic diversity within tropical pasture species has enabled selection of adapted plants, particularly legumes, with the potential for improving pasture production in many environments. Schultz-Kraft & Giacometti (1979) report that a number of species in the genera *Desmodium*, *Centrosema*, *Galactia*, *Macroptilium*, *Stylosanthes*, *Vigna*, and *Zornia* show promise in the acid infertile soils of the humid tropics of Latin America; Oyenuga & Olubajo (1966) and Keya *et al*. (1971) obtained increased pasture production by use of legumes sown into native pastures in west and east Africa respectively; Magadan *et al*. (1974) demonstrated improved animal production in the high rainfall areas of the Philippines. There is also a potential for improving animal production through grazing of introduced adapted legumes with native grasses under plantation crops. Steel & Humphreys (1974), Boonlinkajoorn & Duriyaprapan (1977), Thomas (1978) have demonstrated some of this potential under coconuts. In rice growing areas, encouraging results have been reported by Thomas & Humphreys (1970) and Shelton & Humphreys (1972, 1975) from use of *Stylosanthes* species sown into rice crops, but interactions of seeding rate, time of sowing and effect of rice variety have still to be elucidated.

Concluding remarks

The level of animal production from grazed pastures is an expression of the quantity and quality of the feed ingested and the efficiency of utilization by the animal. Improvements in production depends on some understanding of the interactions between the components of the soil-plant-animal system and the manner in which they may be manipulated. At a time of expanding world population the efficiency of utilizing the pasture resource assumes great importance. In intensive production systems such as those in humid temperate environments, two options for improving efficiency in animal production are (i) a better utilization of proven species, especially of nitrogen fertilized grasses and (ii) a greater use of legumes in pastures systems. Snaydon (1979) claimed that herbage production between grass species varied from 0-15 percent, between cultivars from 0-10 percent and of associated animal production from 0-50 percent. Management and environmental variables accounted for 95-100 percent of the variation in herbage production. These fluctuations may largely be explained by the lack of development of appropriate management systems that are able to exploit the species potential. This is an important area of research in all environments. It may well be true that even with a "management package" the potential production would never be attainable in broad commercial terms, because of environmental variations and the problems of practical day to day management of stock and pastures. It would therefore seem appropriate for some flexibility to be available within the pasture species, but that the limits be better defined.

In the tropics and subtropics there is clearly a valuable role for use of adapted legumes in improving pasture and animal production from native and fully sown pastures. More research is required on the nutrient requirements of improved tropical pasture species and on their response to grazing. The feeding value of some of these species has been studied but this needs to be expanded to the grazed situation; very little is known of the feeding value of many tropical legume species. As yet there is little knowledge of the quantitative aspects of legume content in pasture and animal response. Pasture management criteria applicable to the wide differences shown by plant growth form of tropical species needs to be defined. The problems of improving the nutrition of ruminants in either a pastoral system or that of a smallholder system of agriculture in developing countries are substantial. However, the increasing evidence for adaptation of different legume species to a wide range of tropical soils and climatic conditions indicate a clear possibility for improvement in nutritional status of forage for ruminants. The contribution of shrub legumes especially in drier environments deserves more detailed study in terms of selection of species and their management.

Some potential has been shown for improvement in animal production by management of rangelands, but the concept of range condition and its assessment needs to be more critically investigated. Most research is directed towards recovery of grasslands from previous degradation and if successful this would lead to considerable increases in animal production. It has yet to be proved however, that the climax grassland vegetation is the most stable or productive ecosystem.

REFERENCES

Allden, W.G. (1982) Problems of animal production from Mediterranean pastures. In: Nutritional limits to animal production from pastures. Editor J.B. Hacker. Farnham Royal, U.K., Commonwealth Agricultural Bureaux, pp. 45-65.

Archibald, K.A.E.; Campling, R.C.; Holmes, W. (1975) Milk production and herbage intake of dairy cows kept on a leader and follower system of grazing. Animal Production 21, 147-156.

Aronovich, S.; Serpa, A.; Ribeiro, H. (1970) Effect of nitrogen fertilizer and legume upon beef production of pangola grass pasture. Proceedings of the 11th International Grassland Congress, Surfers Paradise, Australia, 796-800.

Baars, J.A. (1976) Seasonal distribution of pasture production in New Zealand. New Zealand Journal of Experimental Agriculture 4, 151-156.

Bennett, D.; Morley, F.H.W.; Clark, K.W.; Dudzinski, M.L. (1970) The effect of grazing cattle and sheep together. Australian Journal of Experimental Agriculture and Animal Husbandry 10, 694-709.

Bishop, H.G.; Walker, B.; Rutherford, M.T. (1982) Renovation of tropical grass-legume pastures in northern Australia. Proceedings of the 14th International Grassland Congress, Lexington, Kentucky (in press).

Blaser, R.E.; Hammes, R.C.; Fontenot, J.P.; Polan, C.E.; Bryant, H.T.; Wolf, D.D. (1976) Forage animal production systems on hill land in the eastern United States. In: Hill Lands. Proceedings of an international symposium. Editors J. Luchok, J.D. Cawthon and M.J. Breslin. Morgantown, West Virginia University Books, pp.678-684.

Blue, J.; Gammon, L.A.; Lundy, G. (1961) Late summer fertilization for winter forage in north Florida. Proceedings of the Soil and Crop Science Society of Florida 21, 56-65.
Boonlinkajoorn, P.; Duriyaprapan, S. (1977) Herbage yields of selected grasses grown under coconuts in southern Thailand. Thai Journal of Agricultural Science 10, 35-40.
Booysen, P. de V. (1978) Range improvement opportunities. Proceedings of the 1st International Rangeland Congress, Denver, 14-16.
Box, T.W. (1977) Potential of arid and semi-arid rangelands for ruminant animal production. In: Potential of the world's forages for ruminant animal production. Arkansas, Winrock Report pp. 79-90.
Brougham, R.W.; Causley, D.C.; Madgwick, L.E. (1975) Pasture management systems and animal production. Proceedings of the Ruakura Farmers' Conference pp. 65-99.
Bryan, W.W.; Evans, T.R. (1973) Effects of soils, fertilizers and stocking rates on pastures and beef production on the Wallum of south-eastern Queensland. 1. Botanical composition and chemical effects in plants and soils. Australian Journal of Experimental Agriculture and Animal Husbandry 13, 516-529.
Bryant, H.T.; Blaser, R.E.; Hammes, R.C.; Hardison, W.A. (1961) Method for increased milk production with rotational grazing. Journal of Dairy Science 44, 1733-1739.
Burrows, W.H. (1972) Productivity of an arid zone shrub (Eremophila gilesii) community in south western Queensland. Australian Journal of Botany 20, 317-329.
Burrows, W.H. (1974) Vegetation management decision in Queensland's sheeplands. Proceedings of the 4th Workshop of the United States/Australia Rangeland panel pp. 202-218.
Burrows, W.H. (1980) Range management in the dry tropics with special reference to Queensland. Tropical Grasslands 14, 281-287.
Burton, G.W. (1972) Registration of Coastcross-1 Bermuda grass. Crop Science 12, 125.
Chopping, G.D.; Moss, R.J.; Goodchild, I.K.; O'Rourke, P.K. (1978) The effect of grazing systems and nitrogen fertilizer regimes on milk production from irrigated pangola-couch pastures. Proceedings of the Australian Society of Animal Production 12, 229.
Clatworthy, J.N.; Holland, D.G.E. (1979) Effects of legume reinforcement of veldt in the performance of beef steers. Proceedings of the Grassland Society of Southern Africa 14, 111-114.
Coaldrake, J.E.; Smith, C.A.; Yates, J.J.; Edye, L.A. (1969) Animal production on sown and native pastures on brigalow land in southern Queensland during drought. Australian Journal of Experimental Agriculture and Animal Husbandry 9, 47-56.
Cooper, J.P.; Tilley, J.M.A.; Raymond, W.F.; Terry, R.A. (1962) Selection for digestibility in herbage grasses. Nature, London 195, 1276-1277.
Cowan, R.T.; Stobbs, T.H. (1976) Effects of nitrogen fertilizer applied in autumn and winter on milk production from a tropical grass-legume pasture grazed at four stocking rates. Australian Journal of Experimental Agriculture and Animal Husbandry 16, 829-837.
Davison, T.M.; Cowan, R.J.; O'Rourke, P.K. (1981) Management practices for tropical grasses and their effects on pasture and milk production. Australian Journal of Experimental Agriculture and Animal Husbandry 21, 196-202.
Eadie, J. (1976) Animal production systems from hill country in the United Kingdom. In: Hill Lands. Proceedings of an international symposium. Editors J. Luchok, J.D. Cawthon and M.J. Breslin. Morgantown, West Virginia University Books, pp. 686-691.
Ebersohn, J.P. (1966) Effects of stocking rate, grazing method, and ratio of sheep to cattle on animal liveweight gains in a semi-arid environment. Proceedings of the 10th International Grassland Congress, Helsinki, 495-499.
Ebersohn, J.P. (1970) Herbage production from native grasses and sown pastures in south-west Queensland. Tropical Grasslands 4, 37-41.
Edye, L.A.; Ritson, J.B.; Haydock, K.P.; Davies, J.G. (1971) Fertility and seasonal changes in liveweight of Droughtmaster cows grazing a Townsville stylo - spear grass pasture. Australian Journal of Agricultural Research 22, 963-977.
Eng, P.K.; Kerridge, P.C.; Mannetje, L.'t (1978a) Effects of phosphorus and stocking rate on pasture and animal production from a guinea grass-legume pasture in Johore, Malaysia. 1. Dry matter yields, botanical and chemical composition. Tropical Grasslands 12, 188-197.
Eng, P.K.; Kerridge, P.C.; Mannetje, L.'t (1978a) Effects of phosphorus and stocking rate on pasture and animal production from a guinea grass-legume pasture in Johore, Malaysia. 2. Animal weight change. Tropical Grasslands 12, 198-207.
Evans, J.; Pulsford, J.S.; Ebersohn, J.P. (1978) Irrigated pasture studies at Parada, north Queensland. Steer gains on pangola grass in relation to stocking rates and nitrogen levels 1969-1972. Queensland Department of Primary Industries, Agricultural Branch Project Report No. P.19-27.
Evans, T.R. (1969) Beef production from nitrogen fertilized pangola grass. (Digitaria decumbens) on the coastal lowlands of southern Queensland. Australian Journal of Experimental Agriculture and Animal Husbandry 9, 282-286.
Evans, T.R. (1970) Some factors affecting beef production in the coastal lowlands of southeast Queensland. Proceedings of the 11th International Grassland Congress, Surfers Paradise, Australia, 803-807.
Evans, T.R.; Bryan, W.W. (1973) Effects of soils, fertilizers and stocking rates on pastures and beef production on the Wallum of south-eastern Queensland. 2. Liveweight change and beef production. Australian Journal of Experimental Agriculture and Animal Husbandry 13, 530-536.

Falvey, L.; Mikled, C. (1978) Dry season supplementation of cattle in northern Thailand. Proceedings of the Australian Society of Animal Production 12, 175.

Gammon, D.M. (1978) A review of experiments comparing systems of grazing management in natural pasture. Proceedings of the Grassland Society of South Africa 13, 75-82.

Gardener, C.J. (1980) Diet selection and liveweight performance of steers on Stylosanthes hamata - native grass pastures. Australian Journal of Agricultural Research 31, 379-392.

Gibb, M.A.; Treacher, T.T. (1976) The effect of herbage allowance on herbage intake and performance of lambs grazing perennial ryegrass and red clover swards. Journal of Agricultural Science, Cambridge 86, 355-365.

Gilbert, M.A.; Shaw, K.A. (1980) The effect of superphosphate application in establishment and persistence of three Stylosanthes spp. in native pasture on an infertile duplex soil near Mareeba, north Queensland. Tropical Grasslands 14, 23-27.

Gillard, P.; Edye, L.A.; Hall, R.L. (1980) Comparison of Stylosanthes humilis, with S. hamata and S. subsericea in the Queensland dry tropics. Effects on pasture composition and liveweight gain. Australian Journal of Agricultural Research 31, 205-220.

Greenhalgh, J.F.D.; Reid, G.W. (1969) The herbage consumption and milk production from cows grazing S24 ryegrass and S37 cocksfoot. Journal of the British Grassland Society 24, 98-103.

Grimes, R.C.; Watkin, B.R.; Gallagher, J.R. (1967) The growth of lambs grazing on perennial ryegrass, tall fescue and cocksfoot, with and without white clover as related to the botanical and chemical composition. Journal of Agricultural Science, Cambridge 68, 11-21.

Grof, B.; Harding, W.A.T. (1970) Dry matter yields and animal production of guinea grass (Panicum maximum) on the humid tropical coast of north Queensland. Tropical Grasslands 4, 85-95.

Gutman, M.; Seligman, N.G. (1979) Grazing management of mediterranean foothill range in the upper Jordan valley. Journal of Range Management 32, 86-92.

Hacker, J.B.; Minson, D.J. (1972) Varietal differences in in vitro dry matter digestibility in Setaria, and the effects of site, age and season. Australian Journal of Agricultural Research 23, 959-967.

Hamilton, D. (1976) Performance of sheep and cattle grazed together in different ratios. Australian Journal of Experimental Agriculture and Animal Husbandry 16, 5-12.

Hamilton, D.; Bath, J.G. (1970) Performance of sheep and cattle grazed separately and together. Australian Journal of Experimental Agriculture and Animal Husbandry 10, 19-26.

Hann, W.; Lazenby, A. (1974) Competitive interactions of grasses with contrasting temperature responses and water stress tolerances. Australian Journal of Agricultural Research 25, 227-246.

Harding, W.A.T.; Grof, B. (1978) Effect of fertilizer nitrogen on yield, nitrogen content and animal productivity of Brachiaria decumbens cv. Basilisk on the wet tropical coast of north Queensland. Queensland Journal of Agricultural and Animal Sciences 35, 11-21.

Harrington, G.N. (1981) Grazing arid and semi-arid pastures. In: Grazing animals. Editor F.H.W. Morley. Amsterdam, Elsevier Publishing Company, pp. 181-202.

Hartmans, J. (1975) Animal health in relation to intensive pasture use, especially in the Netherlands. In: Proceedings of the 3rd World Conference on animal production. Editor R.L. Reid. pp. 233-237.

Heinemann, W.W. (1970) Dual grazing of irrigated pastures by cattle and sheep. Proceedings of the 11th International Grassland Congress, Surfers Paradise, Australia, 810-814.

Holroyd, R.G.; Allan, P.J.; O'Rourke, P.K. (1977) Effect of pasture type and supplementary feeding on the reproductive performance of cattle in the dry tropics in north Queensland. Australian Journal of Experimental Agriculture and Animal Husbandry 17, 197-206.

Hood, A.E.M. (1974) Intensive set stocking of dairy cows. Journal of the British Grassland Society 29, 63-67.

Hoveland, C.S.; Anthony, W.B.; McGuire, J.A.; Starling, J.G. (1978) Beef cow-calf performance on coastal bermuda grass overseeded with annual clovers and grasses. Agronomy Journal 70, 418-420.

Jones, R.J.; Jones, R.M. (1978) The ecology of Siratro-based pastures. In: Plant relations in pastures. Editor J.R. Wilson. Melbourne, CSIRO, pp. 353-367.

Jones, R.J.; Jones, R.M. (1982) Observations on the persistence and potential for beef production of pastures based on Trifolium semipilosum and Leucaena leucocephala in subtropical coastal Queensland. Tropical Grasslands 16, (in press).

Jones, R.M.; Mott, J.J. (1980) Population dynamics in grazed pastures. Tropical Grasslands 14, 218-224.

Jones, R.J.; Davies, J.G.; Waite, R.B.; Fergus, I.F. (1968) The production and persistence of grazed irrigated pasture mixtures in south-eastern Queensland. Australian Journal of Experimental Agriculture and Animal Husbandry 8, 177-189.

Kalmbacher, R.S.; Mislevy, P.; Martin, F.G. (1980) Sod-seeding Bahia grass in winter with three temperate legumes. Agronomy Journal 72, 114-118.

Kemp, A.; Deijs, W.B.; Hemkes, O.J.; van Es, A.J.H. (1961) Hypomagnesaemia in milking cows: intake and utilization of magnesium from herbage by lactating cows. Netherlands Journal of Agricultural Science 9, 134-158.

Keya, N.C.O.; Olsen, F.J.; Holliday, R. (1971) Oversowing improved pasture legumes in natural grasslands of the medium altitudes of western Kenya. East African Agriculture and Forestry Journal 37, 148-155.

Kleinschmidt, F.H.; Skerman, P.J. (1977) Irrigation of tropical pasture legumes. In: Tropical forage legumes. Editor P.J. Skerman. Rome, F.A.O. Plant Production and Protection Series No. 2, pp.150-158.

Knight, W.E. (1970) Productivity of crimson clover and arrowleaf clover grown in a coastal bermuda grass forage. Agronomy Journal 62, 773-775.

Kretschmer, A.E. (1965) The effect of nitrogen fertilization of mature pangola grass just prior to utilization in the winter on yields, dry matter, and crude protein contents. Agronomy Journal 57, 529-534.

Lambert, J.P.; Rumball, P.J.; Boyd, A.F. (1979) Comparison of ryegrass-white clover pastures with and without paspalum and kikuyu grass. 1. Pasture production. New Zealand Journal of Experimental Agriculture 7, 295-302.

Langlands, J.P.; Bowles, J.E. (1974) Herbage intake and production of merino sheep grazing native and improved pastures at different stocking rates. Australian Journal of Experimental Agriculture and Animal Husbandry 14, 307-315.

Laycock, W.A. (1970) The effects of spring and fall grazing on sagebrush-grass ranges in eastern Idaho. Proceedings of the 11th International Grassland Congress, Surfers Paradise, Australia, 52-54.

Leach, G.J. (1978) The ecology of lucerne pastures. In: Plant relations in pastures. Editor J.R. Wilson. Melbourne, CSIRO, pp.290-308.

Magadan, P.B.; Javier, E.Q.; Madamba, J.C. (1974) Beef production on native (Imperata cylindrica (L.) Beauv.) and para grass (Brachiaria mutica (Forsk.) Stapf.) pastures in the Philippines. Proceedings of the 12th International Grassland Congress, Mowcow 3, 293-298.

McCawley, P.F.; Dahl, B.E. (1980) Nutritional characteristics of high yielding exotic grasses for seeding cleared south Texas brushland. Journal of Range Management 33, 442-445.

McLean, R.W.; Winter, W.H.; Mott, J.J.; Little, D.A. (1981) The influence of superphosphate on the legume content of the diet selected by cattle grazing Stylosanthes - native grass pastures. Journal of Agricultural Science, Cambridge 96, 247-249.

McDonald, I.W. (1981) Detrimental substances in plants consumed by grazing ruminants. In: Grazing animals. Editor F.H.W. Morley. Amsterdam, Elsevier Publishing Company, pp.349-360.

Mannetje, L.'t (1982) Problems of animal production from tropical pastures. In: Nutritional limits to animal production from pastures. Editor J.B. Hacker. Farnham Royal, U.K., Commonwealth Agricultural Bureaux, pp. 67-85.

Mannetje, L.'t; Coates, D.B. (1976) Effects of pasture improvement on reproductive and pre-weaning growth of Hereford cattle in central sub-coastal Queensland. Proceedings of the Australian Society of Animal Production 11, 257-260.

Mannetje, L.'t; Ebersohn, J.P. (1980) Relations between sward characteristics and animal production. Tropical Grasslands 14, 273-280.

Milford, R. (1960) Nutritive values for 17 subtropical grasses. Australian Journal of Agricultural Research 11, 138-148.

Milford, R. (1967) Nutritive values and chemical composition of seven tropical legumes and lucerne grown in sub-tropical south-eastern Queensland. Australian Journal of Experimental Agriculture and Animal Husbandry 7, 540-545.

Minderhoud, J.W.; van Burg, P.F.J.; Deinum, B.; Dirvan, J.G.P.; Hart, M.L.'t (1974) Effect of high levels of nitrogen fertilization and adequate utilization in grassland productivity and cattle performance, with special reference to permanent pastures in temperate regions. Proceedings of the 12th International Grassland Congress, Moscow 1, 99-121.

Minson, D.J. (1967) The voluntary intake and digestibility, in sheep, of chopped and pelleted Digitaria decumbens (pangola grass) following a late application of fertilizer nitrogen. British Journal of Nutrition 21, 587-597.

Minson, D.J. (1971) The digestibility and voluntary intake of six varieties of Panicum. Australian Journal of Experimental Agriculture and Animal Husbandry 11, 18-25.

Minson, D.J.; McLeod, M.N. (1970) The digestibility of temperate and tropical grasses. Proceedings of the 11th International Grassland Congress, Surfers Paradise, Australia, 719-722.

Minson, D.J.; Milford, R. (1966) The energy value and nutritive value indices of Digitaria decumbens, Sorghum almum and Phaseolus atropurpureus. Australian Journal of Agricultural Research 17, 411-423.

Minson, D.J.; Raymond, W.F.; Harris, C.E. (1960) Studies on the digestibility of herbage. VIII. The digestibility of S.37 cocksfoot, S.23 ryegrass and S.24 ryegrass. Journal of the British Grassland Society 15, 174-180.

Nemati, N. (1978) Range improvement practices in Iran. Proceedings of the 1st International Rangelands Congress, Denver, 631-632.

Norman, M.J.T. (1970) Relationships between liveweight gain of grazing beef steers and availability of Townsville lucerne. Proceedings of the 11th International Grassland Congress, Surfers Paradise, Australia, 829-832.

Norton, B.W. (1982) Differences between species in forage quality. In: Nutritional limits to animal production from pastures. Editor J.B. Hacker. Farnham Royal, U.K., Commonwealth Agricultural Bureaux, pp. 89-110.

Oakes, A.J.; Langford, W.R.; Schank, S.C.; Roush, R.D.; Hodges, E.M. (1980) Winter hardiness in new Digitaria germplasm. Agronomy Journal 72, 457-459.

O'Donnell, J.F.; O'Farrell, R.; Hyde, K.W. (1973) Plant introduction and reseeding in the mulga zone. Tropical Grasslands 7, 105-110.

Ollerenshaw, J.H.; Stewart, W.S.; Gallimore, J.F.; Baker, R.H. (1976) Extending the seasonality of growth of hill land pastures. In: Hill Lands. Proceedings of an international symposium. Editors J. Luchok, J.D. Cawthon, M.J. Breslin. Morgantown, West Virginia University Books, pp.583-586.

Oyenuga, V.A.; Olubajo, F.O. (1966) Productivity and nutritive value of tropical pastures at Ibadan. Proceedings of the 10th International Grassland Congress, Helsinki, 962-969.

Ozanne, P.G.; Howes, K.M.W. (1971) Preference of grazing sheep for pasture of high phosphate content. Australian Journal of Agricultural Research 22, 941-950.

Ozanne, P.G.; Purser, D.B.; Howes, K.M.W.; Southey, I. (1976) Influence of phosphorus content on feed intake and weight gains in sheep. Australian Journal of Experimental Agriculture and Animal Husbandry 16, 353-360.

Paladines, O.; Alarcon, E.; Hilton, J.; Spain, J.M.; Grof, B.; Perez, R. (1974) Development of a pasture program in the tropical savannah of Colombia. Proceedings of the 12th International Grassland Congress, Moscow, 3, 389-401.

Paladines, O.; Leal, J.A. (1979) Pasture management and productivity in the Llanos orientales of Colombia. In: Pasture production in acid soils of the tropics. Editors P.A. Sánchez and L.E. Tergos. Cali, Colombia, CIAT. pp. 311-325.

Partridge, I.R. (1976) Technically feasible, economically marginal and social a exercise: beef production from hill land in Fiji. In: Hill Lands. Proceedings of an International symposium. Editors J. Luchok, J.D. Cawthon, M.J. Breslin. Morgantown, West Virginia University Books, pp. 692-696.

Partridge, I.J. (1979) The improvement of Nadi blue grass (Dichanthium caricosum) with superphosphate and siratro on hill land in Fiji. Effects of stocking rate on beef production and botanical composition. Tropical Grasslands 13, 157-164.

Partridge, I.J. (1980a) The effect of grazing and superphosphate on a naturalised legume, Desmodium heterophyllum, on hill land in Fiji. Tropical Grasslands 14, 63-68.

Partridge, I.J. (1980b) The efficacy of Biosupers made from different forms of phosphate on forage legumes in hill land in Fiji. Tropical Grasslands 14, 87-94.

Playne, M.J. (1969) The nutritional value of intact seed pods of Townsville lucerne (Stylosanthes humilis). Australian Journal of Experimental Agriculture and Animal Husbandry 9, 502-507.

Purser, D.B. (1981) Nutritional value of Mediterranean pastures. In: Grazing animals. Editor F.H.W. Morley. Amsterdam, Elsevier Publishing Company, pp. 159-180.

Rae, A.L.; Brougham, R.W.; Barton, R.A. (1964) A note on liveweight gains of sheep grazing different ryegrass pastures. New Zealand Journal of Agricultural Research 7, 491-495.

Rees, M.C.; Minson, D.J. (1976) Fertilizer calcium as a factor affecting voluntary intake, digestibility and retention time of pangola grass (Digitaria decumbens) by sheep. British Journal of Nutrition 36, 179-187.

Rees, M.C.; Minson, D.J. (1978) Fertilizer sulphur as a factor affecting voluntary intake, digestibility and retention time of pangola grass (Digitaria decumbens) by sheep. British Journal of Nutrition 39, 5-11.

Reid, R.L.; Jung, G.A. (1982) Problems of animal production from temperate pastures. In: Nutritional limits to animal production from pastures. Editor J.B. Hacker. Farnham Royal, U.K., Commonwealth Agricultural Bureaux, pp. 21-43.

Ritson, J.B.; Edye, L.A.; Robinson, P.J. (1971) Botanical and chemical composition of a Townsville stylo-spear grass pasture in north-eastern Queensland in relation to conception rate of cows. Australian Journal of Agricultural Research 22, 993-1007.

Roberts, F.J.; Carbon, B.A. (1969) Growth of tropical and temperate grasses and legumes under irrigation in south-west Australia. Tropical Grasslands 3, 109-111.

Robinson, P.J.; Jones, R.K. (1972) The effects of phosphorus and sulphur fertilisation on the growth and distribution of dry matter, nitrogen, phosphorus and sulphur in Townsville stylo (Stylosanthes humilis). Australian Journal of Agricultural Research 23, 633-640.

Rossiter, R.C. (1966) Ecology of the Mediterranean annual type pasture. Advances in Agronomy 18, 1-57.

Rossiter, R.C. (1978) The ecology of subterranean clover based pastures. In: Plant relations in pastures. Editor J.R. Wilson. Melbourne, CSIRO, pp.325-339.

Rumball, P.J.; Boyd, A.F. (1980) Comparison of ryegrass-white clover pasture with and without paspalum and kikuyu grass. New Zealand Journal of Experimental Agriculture 8, 21-26.

Schultze-Kraft, R.; Giacometti, D.C. (1979) Genetic resources of forage legumes for the arid, infertile savannas of tropical America. In: Proceedings of a Seminar on Pasture Production in Acid Soils of the tropics. Editors P.A. Sanchez and L.E. Tergas. Cali, Colombia, CIAT, pp. 55-64.

Shaw, N.H. (1961) Increased beef production from Townsville lucerne (Stylosanthes humilis) in the spear grass pastures of central coastal Queensland. Australian Journal of Experimental Agriculture and Animal Husbandry 1, 73-80.

Shaw, N.H. (1978a) Superphosphate and stocking rate effects on a native pasture oversown with Stylosanthes humilis in a central coastal Queensland. 2. Animal Production. Australian Journal of Experimental Agriculture and Animal Husbandry 18, 800-807.

Shaw, N.H. (1978b) Superphosphate and stocking rate effects on a native pasture oversown with Stylosanthes humilis in a central coastal Queensland. 1. Pasture Production. Australian Journal of Experimental Agriculture and Animal Husbandry 18, 788-799.

Shaw, N.H.; Mannetje, L.'t (1970) Studies on a spear grass pasture in central coastal Queensland - the effect of fertilizer, stocking rate, and oversowing with Stylosanthes humilis on beef production and botanical composition. Tropical Grasslands 4, 43-56.

Shelton, H.M.; Humphreys, L.R. (1972) Pasture establishment in upland rice crops at Na Pheng, Central Laos. Tropical Grasslands 6, 223-228.
Shelton, H.M.; Humphreys, L.R. (1975) Undersowing rice (Oryza sativa) with Stylosanthes guyanensis. Experimental Agriculture 11, 89-111.
Silvey, M.W.; Coaldrake, J.E.; Haydock, K.P.; Ratcliff, D.; Smith, C.A. (1978) Beef cow performance from tropical pastures on semi-arid brigalow lands under intermittent drought. Australian Journal of Experimental Agriculture and Animal Husbandry 18, 618-628.
Snaydon, R.W. (1979) Selecting the most suitable species and cultivars. Proceedings, British Grassland Society Occasional Symposium No.10, York, 179-198.
Snaydon, R.W. (1981) The ecology of grazed pastures. In: Grazing animals. Editor F.H.W. Morley. Amsterdam, Elsevier Publishing Company, pp. 13-31.
Steel, R.J.H.; Humphreys, L.R. (1974) Growth and phosphorus response of some pasture legumes sown under coconuts in Bali. Tropical Grasslands 8, 171-178.
Stobbs, T.H. (1966a) Beef production from Uganda pastures containing Stylosanthes gracilis and Centrosema pubescens. Proceedings of the 9th International Grassland Congress, Sao Paulo, 939-942.
Stobbs, T.H. (1966b) Beef production from pasture leys in Uganda. Journal of British Grassland Society 24, 81-86.
Stobbs, T.H. (1969) Animal production from Hyparrhenia grassland oversown with Stylosanthes guyanensis. East African Agriculture and Forestry Journal 35, 128-134.
Stobbs, T.H. (1970) The use of liveweight gain trials for pasture evaluation in the tropics. Journal of the British Grassland Society 25, 73-77.
Stobbs, T.H. (1971) Production and composition of milk from cows grazing siratro (Phaseolus atropurpureus) and greenleaf desmodium (Desmodium intortum). Australian Journal of Experimental Agriculture and Animal Husbandry 11, 268-273.
Stobbs, T.H. (1976) Kenya white clover (Trifolium semipilosium) a promising legume for dairy production in subtropical environments. Proceedings of the Australian Society of Animal Production 11, 477-480.
Stobbs, T.H.; Joblin, A.D.H. (1966) The use of liveweight gain trials for pasture evaluation in the tropics. 1. An "animal" latin square design. Journal of the British Grassland Society 21, 49-55.
Suckling, F.E.T. (1976) A 20 year study of pasture development through phosphate and legume oversowing on north island hill country of New Zealand. In: Hill Lands. Proceedings of an International symposium. Editors J. Luchok, J.D. Cawthon, M.J. Breslin. Morgantown, West Virginia University Books, pp. 367-380.
Tainton, N.M. (1978) Fire in the management of humid grasslands in South Africa. Proceedings of the 1st International Rangeland Congress, Denver, 684-686.
Thomas, D. (1978) Pastures and livestock under tree crops in the humid tropics. Tropical Agriculture, Trinidad 58, 39-44.
Thomas, R.; Humphreys, L.R. (1970) Pasture improvement at Na Pheng, central Laos. Tropical Grasslands 4, 229-236.
Thomson, D.J. (1977) The role of legumes in improving the quality of forage diets. Proceedings of an International Meeting of Animal Production from Temperate Grasslands, Dublin pp. 131-135.
Thornton, R.F.; Minson, D.J. (1973) Effects of soils, fertilizers and stocking rates on pastures and beef production on the Wallum of south-eastern Queensland. 3. Relation of liveweight changes to chemical composition of blood and pasture. Australian Journal of Experimental Agriculture and Animal Husbandry 13, 537-543.
Ulyatt, M.J. (1981) The feeding value of temperate pastures. In: Grazing animals. Editor F.H.W. Morley. Amsterdam, Elsevier Scientific Publishing Company, pp.125-141.
Ulyatt, M.J.; Lancashire, J.A.; Jones, W.T. (1976) The nutritive value of legumes. Proceedings of the New Zealand Grassland Association 38, 107-118.
Vicente-Chandler, J.; Abruna, F.; Caro-Costas, R.; Figarella, J.; Silva, S.; Pearson, R.W. (1974) Intensive grassland management in the humid tropics of Puerto Rico. University of Puerto Rico, Bulletin 233.
Walker, B.; Rutherford, M.T.; Whiteman, P.C. (1982) Diet selection by cattle on tropical pastures in northern Australia. Proceedings of the 14th International Grassland Congress, Lexington, Kentucky, (in press).
Walshe, M.J. (1975) Grazing management and the productivity of grazing systems. Proceedings of the 3rd World Conference on Animal Production pp. 165-173.
Walshe, M.J.; Downey, N.E.; Connolly, J. (1971a) Calf rearing. 2. Comparison of three systems of grazing management including observations on the occurrence of parasitic bronchitis. Irish Journal of Agricultural Research 10, 161-172.
Walshe, M.J.; Kelleher, D.; Connolly, J. (1971b) Calf rearing. 1. Effect of age at putting out to pasture on performance of spring born calves. Irish Journal of Agricultural Research 10, 81-94.
Wheeler, J.L. (1965) The improvement of winter feed in year long grazing programs. Proceedings of the 9th International Grassland Congress, São Paulo, 975-980.
Willoughby, W.M. (1970) Grassland management. In: Australian grasslands. Editor R.M. Moore. Australian National University Press, pp. 392-397.
Williams, O.B. (1960) The selection and establishment of pasture species in a semi-arid environment - an ecological assessment of the problem. Journal of the Australian Institute of Agricultural Science pp. 258-265.

Wilson, G.F.; McDowall, F.H. (1966) Ryegrass varieties in relation to dairy cattle performance. I. The influence of ryegrass varieties on milk yield and composition. New Zealand Journal of Agricultural Research 9, 1042-1952.

Woods, L.E. (1970) Beef production from pastures and forage crops in a tropical monsoon climate. Proceedings of the 11th International Grassland Congress, Surfers Paradise, Australia, 845-849.

FORAGE CONSERVATION

R. JARRIGE, C. DEMARQUILLY, J.P. DULPHY

Institut National de la Recherche Agronomique, Centre de Recherches Zootechniques et Vétérinaires de Theix, 63110, Beaumont, France

ABSTRACT

Livestock production in the temperate zones of the Northern Hemisphere has been progressively freed from the seasonality of herbage production by using increasing proportions of conserved herbage and forage crops. The nutritive value and intake of these conserved forages are determined primarily by those of the fresh forage conserved and by the modifications caused by the processes of harvesting and conservation.

Silage-making has developed rapidly following the introduction of forage harvesters, strong polythene sheets for sealing, efficient additives, and machinery for cutting and feeding the silage. Chopping the crop, restricting fermentation and inhibiting clostridia result in forage silages with the same digestibility and nearly the same voluntary dry matter intake by cattle as the fresh herbage. Sheep eat larger quantities of fresh herbage than silage. Depressions in intake are shown to be related to the loss of protein and soluble carbohydrates in the silo and changes in the physical form of the feed. Silage making from tropical forages that are already of low nutritive value and intake involves more risk of bad fermentation, decreased nutritive value and enhanced losses.

Hay-making causes reductions in digestibility and intake. The extent of these reductions is shown to be related to the time cut herbage is exposed to leaching and other losses in the field. These losses can be reduced by barn-drying and high-temperature drying of the herbage. In the tropics standing dried mature grass and the various roughages derived from field crops, and sometime tree crops, will remain the cheapest fodders for the dry season in most situations.

EVOLUTION AND PRESENT SITUATION OF FODDER CONSERVATION

The growth of pasture plants ceases during cold or dry seasons, which can range from approximately two months per year in the more favourable areas of the World to ten months in the subdesertic areas. Farmers have been and are always faced with the problem of feeding their animals during these periods which are a major limitation to animal production. The solutions adopted differ widely, not only according to the climatic factors but also to the farming systems, which are the result of ecological as well as historical, economical and socio-cultural factors (Jarrige 1980).

The great transformation of European agriculture started in the middle of the 19th century as a consequence of industrialization. Since that time and mainly during the last three decades, meat and milk production has increased at an accelerated rate per animal, per hectare and per worker simultaneously with maize and grass silage assuming a major role in improving winter feeding of dairy cows and fattening animals (Table 1).

Silage

Silage was made in Egypt some 3,000 years ago and has since been recorded at many different periods, particularly in Italy (see Schukking 1976). The possibilities of using silage to take advantage of the heavy yield of forage maize began to be explored in the 1860's in Continental Europe. Goffart in France demonstrated the necessity of chopping the plant into small (1 cm) pieces and of excluding air by sealing his large bunker silos with a cover platform loaded at 400-500 kg/m^2. In 1877 he published his famous book "Manuel de la culture et l'ensilage du mais et des autres fourrages verts" which, when translated into English did much to stimulate the establishment of maize silage in the U.S.A. where around 100,000 silos, mainly tower silos, were built before 1900. At the beginning of the century silage-making was attempted in many countries and many different methods were proposed. In Italy, for instance, Samarini developed the ensiling of moist hay (60-70 percent dry matter) in special air-tight concrete silos (Cremasco silos).

Most of the basic principles of silage-making were established before the 2nd World War (see *inter alia* Watson & Nash 1960, Breirem & Ulvesli 1960, Schukking 1976). The necessity to exclude air, the beneficial effects of lactic acid fermentation (addition of carbohydrates), of high dry matter content and of laceration of forage, at least for consolidation of the ensiled mass, and the use of acids to prevent undesirable fermentation were all well recognized. Virtanen's A.I.V., and formic acid were first tested between 1925 and 1930. The main advantages of silage-making over hay-making were 1) it reduced weather risks and allowed herbage to be harvested at the chosen date independently of the meteorological conditions, 2) herbage could be cut at an earlier stage of growth when it has a higher nutritive value, 3) the level of nitrogen fertilization and the stocking rate could be increased and 4) grazing and mowing could be integrated in the utilization of grasslands. Despite these advantages, the development of ensiled herbage was restrained for decades by the high labour requirement for harvesting, conserving and feeding this voluminous aqueous feed and the discouraging frequency of bad fermentation, high wastage, low animal intake and objectionable odours encountered with direct cut silages.

These difficulties have been solved in the last two decades by the development of machinery and implements that enable the farmer to apply the previously known rules required for reliable preservation and complete mechanization of feeding to stock (see Raymond *et al.* 1972). The major contributions have been 1) the flail harvester, which originated in the U.S.A. around 1955, and later the double-chop and the precision-chop harvester; 2) the automatic distribution of additives through applicators fixed to the harvesters (around 1964 in Norway); 3) covering with strong polythene sheets to eliminate surface wastage thus allowing cheap unwalled silos (clamps) to be used; 4) the mechanical cutting and feeding of silage.

Hay

Despite the recent increase in production of silage, hay-making remains the dominant method of forage conservation in many areas (Table 1), particularly in mountainous regions. The main developments have focused on machinery that 1) accelerates the loss of water from the herbage in the early stages of drying through mechanical conditioning (Klinner & Shepperson 1975, Klinner & Hale 1980) and 2) mechanizes the system for bale handling. This last development has seen the introduction of the large bale (up to 500 kg) that can be stored outside.

TABLE 1

Production of fodders for the winter period : estimates for some European countries and the U.S.A. (1978-1980)

	Total production 10^6 t. DM			Conserved forages		Silage-making	
				%	%		
	Forages	Maize silage	Roots (+tops)	Hay	Silage	Method	Additives
Finland	2.6			65	35	DC	x x x x (near 100%)
Sweden	5.6			85	15		
Norway	1.8		-	30	70	DC	x x x(65%)
Denmark	1.5		1.5	25	75	DC or SW	ε
Ireland	6.0		0.1	60	40	-	-
U.K.	12.4	0.2	0.5	50	50	SW(25%DM)	x x
Netherlands	5.4	1.4	0.1	25	75	W(>35%DM)	0
Belgium-Lux.	4.0	1	1.1	35	65	DC or W	ε
Germany (F.D.)	19.6	7	6	55	45	W(35-45%DM)	ε
France	40	11	2	70	30	DC or SW	ε
Italy	15.2	3.5	0.3	60	40	-	-
U.S.A.	121	36		97	3	SW, W	-

Sources: Data on the amount conserved of forages in Europe compiled by Wilkinson (1980), Eurostat, USDA Agricultural Statistics, Personal communications.
Initials: DC=direct cut; SW=slightly wilted; W=highly wilted

Barn-drying is a process that allows moist hay (60-65 percent dry matter) to be put under cover at a earlier stage of drying thus reducing the risks of high field losses. The crop may also be cut nearer the optimum stage of growth and thus is of higher nutritive value (Table 2). Research and development work has resulted in efficient installations and techniques (Klinner & Shepperson 1975). However barn-drying remains of minor importance due to problems in handling the hay, low throughput, and the high cost of fuel to heat the air. Barn drying has been adopted on the dairy farms of mountain areas where silage-making is prohibited because the milk is made into cheese of the Gruyere type and on modernized farms that have new buildings specifically designed for the handling, drying and feeding of hay.

TABLE 2

Differences in the chemical composition and nutritive value to sheep of conserved forages compared with fresh herbage (Demarquilly, Dulphy *et al.* various publications).
FA: formic acid; F: formaldehyde.

		Change in content (%DM)			Change in nutritive value as % of the fresh		
		Number of samples	Crude fibre	Crude protein	O.M. digestibility	Net(1) energy content[1]	Voluntary D.M. intake
Dried forages	Legumes	4	- 1.8	+ 0.5	- 8.3	0	- 12
	Grasses	17	- 0.7	0	- 1.6	0	- 3
Grasses and meadow hays	Barn-dried	36	+ 2.7	- 0.6	- 6.1	- 7.7	- 12
	Field-dried, good weather	32	+ 1.6	-0.6	- 6.1	- 7.9	- 20
	Field-dried, < 10 rain	14	+ 3.9	- 1.0	- 8.4	-11.7	- 24
	Field-dried > 10 rain	10	+ 5.3	- 1.5	-13.1	-16.7	- 30
Lucerne hays	Barn-dried	16	+ 3.3	- 1.6	- 5.6	-10.7	- 15
	Field dried, good weather	10	+ 9.1	- 4.2	-12.2	-18.7	- 26
Grasses and meadow silages	Direct-cut	64	+ 3.2	+ 0.4	- 1.1	- 4.4	- 21
	Direct-cut + FA	96	+ 2.4	+ 0.2	+ 0.4	- 1.1	- 17
	Direct-cut + FA + F	24	+ 2.5	+ 0.3	- 2.2	- 4.3	- 23
	Wilted (30-35 % DM)	10	+ 0.9	- 0.3	- 3.1	- 5.6	- 30
Lucerne silages	Direct-cut	11	+ 3.1	- 0.6	- 1.4	- 3.6	- 13
	Direct-cut + FA	19	+ 1.9	- 0.7	- 1.5	- 3.5	- 14

1. Calculated in Feed units for milk production (INRA 1978), assuming that conservation does not modify the gross energy content; 2. Low temperature drier (125-150°);

High temperature drying

High temperature drying began to spread in the 1930's in the U.S.A., U.K., Germany, Netherlands. It increased sharply in the 1960's in France and Denmark which became the two main European producers of dried lucerne (900,000 and 200,000 tons respectively). The highest producer remained the U.S.A. (1,100,000 tons). Large scale enterprises have developed in arable areas for the production of lucerne pellets which go mainly to the compound feed industry. By about 1965, large forage drying cooperatives were established, particularly in France and Denmark. These developments ceased when fuel prices increased sharply.

THE NUTRITIVE VALUE AND POTENTIAL FOR ANIMAL PRODUCTION OF CONSERVED FORAGES

The energy value, protein value and intake of conserved forages are determined primarily by those of the fresh forage conserved. Conservation usually decrease these values. These losses are caused by respiration, leaf-shatter, leaching by rain during drying in the field and from oxidation, fermentation and effluents in the silo. All these factors adversely influence the most digestible components of the forage, that is the cell-contents, especially the water-soluble carbohydrates and protein. These losses lead to an increase in the proportion of cell walls, a reduction in organic matter and nitrogen digestibility and generally a reduced dry matter intake as a consequence of a slower rate of forage breakdown in the reticulo-rumen. Changes in the physical structure, moisture content and composition may also affect digestion and intake of conserved forages.

Changes in composition, energy value and protein value

Digestibility and energy utilization

The changes in organic matter digestibility which occur during harvesting and conservation are extremely variable (Table 2). The magnitude of the reduction in the organic matter digestibility of hays over fresh herbage is closely related to the dry matter losses during drying (Shepperson 1960, Demarquilly & Jarrige 1969) (Table 2). It is at a minimum (6 percent) for barn-dried hays and grass hays rapidly dried in the field. It increases sharply with the length of the drying period, and the amount of rain. It is higher for legumes, particularly for clovers, than for grasses and it increases with the digestibility of the fresh forage.

The organic matter digestibility of forages ensiled correctly in well-sealed silos is generally very similar to that of the fresh forage (Harris & Raymond 1963, Demarquilly 1973, Andrieu & Demarquilly 1974) provided the dry matter content of silages is correctly measured. There is a small increase in cell wall content but this is compensated for by a small increase in digestibility (Demarquilly 1973) resulting from modifications caused by acidic conditions and microbial activity (Morrison 1979). However the organic matter digestibility is slightly reduced by wilting prior to ensiling, by treatment with formaldehyde and as a consequence of effluent losses in high moisture silages (Table 2).

Artificial dehydration keeps these losses to a minimum. It reduces to a small extent the digestibility of legumes but has no effect on grasses dried at high or low temperatures (Henke & Laube 1968, Demarquilly 1970, Prym & Weissbach 1977).

The gross energy content is not changed by drying but it is increased in silages (Alderman et al. 1971). This increase is higher in direct-cut lactate silages (up to 9 percent) than in wilted silages (3-4 percent) (Van der Honing et al. 1973, McDonald & Edwards 1976). The conservation method appears to have no significant effect on the energy losses as methane, the conversion of digestible energy into metabolizable energy nor, at least provisionally, on the efficiency of utilization of metabolizable energy (McDonald & Edwards 1976, Greenhalgh & Wainmann 1980). Modifications occur in the composition of the digestible organic matter and in proportions of rumen volatile fatty acids but these changes do not appear to affect the efficiency of utilization of silage.

Protein utilization

Nitrogen redistribution in conserved forages markedly influences the extent of nitrogen degradation in the rumen, microbial protein synthesis and the quantity of amino-acid absorbed from the small intestine (see reviews by Beever 1980, Thomson & Beever 1980).

Natural drying reduces the crude protein content, the apparent digestibility of the protein and protein breakdown in the rumen assessed by the ammonia concentration in the rumen liquor. The proportion of total nitrogen retained in growing animals fed well-conserved hays is similar to that found with fresh herbage (Grenet & Demarquilly 1981). Artificial drying causes a marked depression in protein solubility, increases the quantity of dietary protein escaping rumen degradation and leads to an increased quantity of amino-acids absorbed from the small intestine (Beever et al. 1976) despite a reduction in the apparent digestibility of the nitrogen.

When forage is ensiled the plant proteases hydrolyse the forage protein, particularly the chloroplaste proteins (Bousset et al. 1972). Hydrolysis of protein ceases when the pH has fallen to 4, so that silages contain a large but variable proportion of non-protein nitrogen and soluble nitrogen, including ammonia. In comparison with fresh herbage, this nitrogen redistribution in direct-cut silages without additives results in 1) a breakdown of crude protein in the rumen of up to 80-85 percent, 2) a higher content of ammonia in the rumen liquor and urea in the plasma, 3) a reduced flow of amino acids into, and absorption from, the small intestine, and 4) increased nitrogen losses in urine and a lower nitrogen retention (Conrad et al. 1961). Additives such as formic acid, formaldehyde or both together, restrict the degradation of amino-acids by proteolytic clostridia in the silo, reduce rumen degradation of silage crude protein (55-70 percent and 40-50 percent respectively) and increase the quantity of amino-acid absorbed in the small intestine (Siddons et al. 1979). The changes lead to higher nitrogen retention (Waldo & Tyrrell 1980) and daily gain in growing animals particularly when there is a simultaneous increase in intake of the silage (see below). Wilting was shown to have the same beneficial effects in the experiment of Durand et al. (1968) on lucerne but not in that of Thomson & Beever (1980) on grass.

The efficiency of microbial protein synthesis in the rumen appears to be lower with silages than with dried forages (see Beever 1980, Thomas et al. 1980). This could be related, at least partly, to a lower content of rapidly available energy in silage resulting from the fermentation of water soluble carbohydrates. For 87 well-preserved grass silages studied at Theix by Elizabeth Grenet, the nitrogen retention of growing wethers was positively correlated with the soluble carbohydrate content (r = 0.50) and the organic matter digestibility (0.32) and negatively correlated with the contents of the silage fermentation products : ammonia nitrogen as a percentage of total nitrogen (-0.43), lactic acid (-0.31) and propionic acid (-0.33). It was positively correlated with the intake of buffer-insoluble nitrogen (0.72) total nitrogen (0.65) and digestible organic matter (0.32). The same trends were observed for 21 lucerne silages. Similar results were reported by Barry *et al.* (1978b) who found the nitrogen retention of young sheep decreased by 0.3 g/day for each 1 percent increase in ammonia nitrogen (percent of total nitrogen) in lucerne silages.

Intake and potential for animal production

The potential of forages for animal production is fundamentally determined by their digestible organic matter intake. Both voluntary dry matter intake and digestibility are depressed by the increase in the proportion of cell-wall with increasing plant maturity (see Jarrige *et al.* 1974) or as a consequence of loss of cell contents during field drying (Table 2). The high potential of early cut hays for milk production was clearly established at Cornell University in the 1950's (Trimberger *et al.* 1955) and has been confirmed in numerous experiments (Bertilsson *et al.* 1980). However the voluntary dry matter intake of direct cut silages appeared variable, unpredictable, frequently low and less than that of hay made from the same crop (Moore *et al.* 1960, Merril & Slack 1965). Most of these difficulties have now been explained and solved (see reviews by Wilkins 1975, Waldo 1977, 1978, Demarquilly & Dulphy 1977, Dulphy 1980, Ekern & Vik-Mo 1979).

Effect of chop length

The introduction and development of forage harvesters has led to the improvement of fermentation quality through the chopping and laceration of herbage (Woodward & Sheperd 1938, Martin & Buysse 1953, Murdoch *et al.* 1955). This has resulted in higher intake of dry matter compared with long material (see review by Marsh 1978).

With flail harvesters the average chop length is greater than 12-15 cm but with precision-chop harvesters it can be adjusted between 0.3 and 7 cm. In 20 comparisons (Dulphy & Demarquilly 1972, 1973) sheep ate much more finely chopped silages than flail-harvested silages made from the same crop : 42 percent for 3 grass silages made without additives; 64 percent for 12 grasses made with additives and 31 percent for 5 legume silages. Increases of 48 percent have been reported for 5 comparisons with lucerne (Barry *et al.* 1978a). This improvement in intake results from an improvement in fermentation quality and from reduction in particle size *per se*. Intake of flail-harvested silage has been increased by rechopping just before feeding (Dulphy & Demarquilly 1973, Dulphy *et al.* 1975). This depressing effect of long particles on intake is associated with a reduction in the eating rate, efficiency of rumination (Dulphy 1972) and longer retention time although the rumen contents are low (Dulphy *et al.* 1975). Deswysen (1980) has confirmed all these findings and suggested that the disturbance of rumination resulted from a delay in the backflow of the small particles from the ventral sac into the cranial sac, due to the presence of long interwoven particles (Deswysen & Ehrlein 1979).

Cattle are also sensitive to the fineness of chopping but to a smaller extent than sheep. In 8 comparisons (Dulphy & Michalet 1975 and unpublished data) one year old heifers ate 16 percent more finely chopped silage than flail harvested silage; sheep ate 55 percent more of the finely chopped silage. Deswysen (1980) has confirmed this lower sensitivity of cattle to long-chop silages and attributed it to anatomical and functional differences making both the backflow of particles from the cranial sac and their passage through the reticulo-omasal orifice easier.

Fine chopping increases production of animals fed silage with concentrates. In six experiments carried out in different countries (Dulphy & Demarquilly 1975b, De Brabander _et al_. 1976, Castle _et al_. 1979, Weiss _et al_. 1979), the mean increases by dairy cows were, voluntary intake (10 percent), milk (5.5 percent) and daily gain (110 g) in dairy cows. The production response appeared to be directly related to the increase in silage intake. The efficiency of utilization of the silage for milk production was not changed by fine chopping but with growing cattle there was a small increase (Dulphy & Demarquilly 1975a).

Intake by cattle of intermediate chop lengths (3-10 cm) produced by the double-chop harvester is intermediate between finely chopped and flail-harvested. This result agrees with the data obtained on an out-dated harvester producing medium chop (Dulphy _et al_. 1975) or by varying the chop length of a precision chop harvester (Castle _et al_. 1979).

Effects of additives

Many different silage additives, stimulants and inhibitors have been tested (reviewed by Thomas 1978). At the present time formic acid is by far the most widely used additive. Treatment of herbage with formic acid at ensiling increases the voluntary intake of silage by cattle to a larger extent than by sheep.

The effect of formic acid on the intake and growth rate of cattle has been reviewed by Waldo (1977) and the data is presented in Figs. 1 and 2. When formic acid was added during the making of direct cut grass silage, the voluntary intake and growth rate by cattle receiving only small quantities of concentrates was increased in 19 comparisons by 8.5 percent and 38 percent respectively. When concentrates were fed the formic acid increased intake by 7.5 percent but the increase in growth rate was only 17 percent.

The addition of formic acid during the making of legume silage had a larger effect on both intake and production than was found with grass silage; 13.5 percent and 65 percent respectively in the absence of concentrates and 11 percent and 24 percent respectively when concentrates were fed.

In dairy cows fed concentrates, the mean responses to addition of formic acid to direct cut silage were 12 percent higher intake of dry matter, 5 percent more milk and 115 g/d gain in weight. Smaller responses were recorded when formic acid was added when making wilted silage with 8 percent higher intake, 2 percent more milk and losses of 54 g/d in body weight. The magnitude of the increase in intake appears to be independant of the intake of non-treated silage. The observed increase in production response to the increased intake

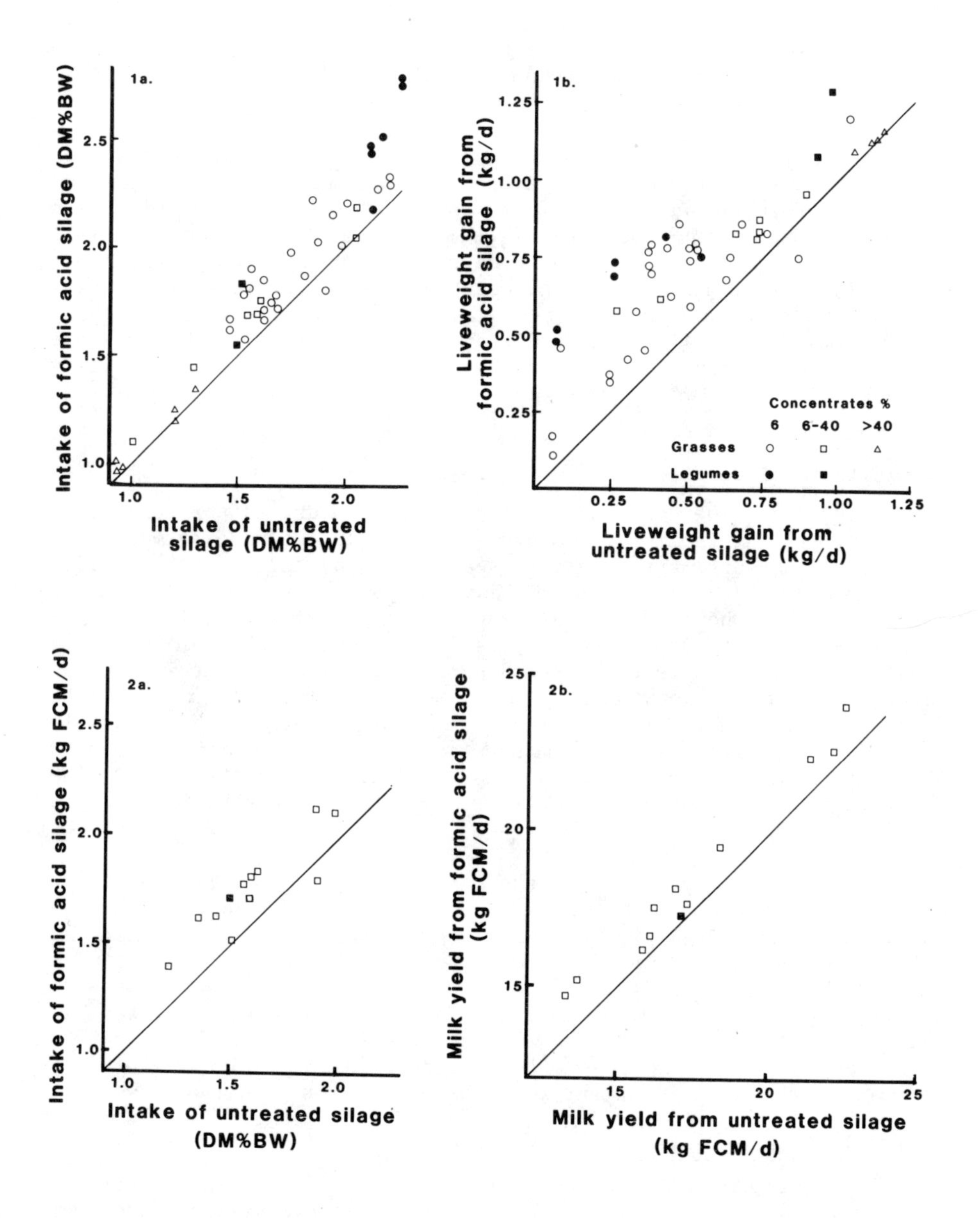

Figs. 1 and 2. Effect of adding formic acid at ensiling on the voluntary intake of silage by growing-fattening cattle (> 200 kg) (Fig. 1) and dairy cows (Fig. 2) and their performance.

implies that the nutritive value of the ensiled herbage is slightly improved by the formic acid treatment.

The use of formaldehyde with formic acid as a silage additive leads to increases in voluntary intake and daily gain in growing cattle : 9.5 and 50 percent respectively in 3 comparisons without concentrates and 7.5 and 9 percent respectively in 3 comparisons with concentrates. A mixture of formaldehyde-sulphuric acid is slightly less effective (Barry 1975). At Jealott's Hill it enhanced the daily gain of beef animals by 42 percent in 12 comparisons and the silage intake by approximately 10 percent (Owen & Blair 1977).

Efficient treatment with additives appears to increase silage intake by reducing the production of fermentation acids and the breakdown of amino-acids by proteolytic clostridia; as a result there is an increase in the quantity of amino-acid absorbed from the small intestine. Negative correlations have been found between silage dry matter intake and the acetic acid content, total volatile fatty acids content and ammonia nitrogen (percent total nitrogen) (Wilkins et al. 1971, Demarquilly 1973). In addition to these products the presence of naturally occuring chemical inhibitors of intake and rumen motility has been demonstrated in the juice of lucerne silage (Clancy et al. 1977).

Effects of wilting

Many studies have shown that increasing the dry matter content of silage by wilting the forage before ensiling improved the fermentation quality and the dry matter intake (see reviews by Merrill & Slack 1965 and Marsh 1979). From results shown in Figs. 3 and 4 it can be calculated that the increase in voluntary intake of silage in response to a 1 percent increase in the dry matter content of the forage before ensiling up to 30-35 percent is 2.8 percent and 1.6 percent in growing cattle without or with concentrates respectively. The response to wilting is at a maximum when fresh forages and long-cut or coarsely chopped forages are ensiled without additives. It continues beyond 30-35 percent dry matter but at a lower rate. Moreover, the substitution rate for concentrates is greater with wilted than with direct-cut silages.

The production response to the increased intake was generally low particularly with direct-cut silages treated with additives. The daily increase in gain of growing cattle was 0.11 kg and 0.02 kg without concentrates or with concentrates respectively and 0.1 kg milk and 0.21 kg daily gain in dairy cows in experiments summarized by R.J. Wilkins (personal communication). The corresponding increase in silage dry matter intake was 20, 22 and 12 percent. This reduced efficiency of utilization of wilted silages is not entirely explained by a lower gross energy content and reduced digestibility (see below).

Protein supplementation of silages

The extensive degradation of amino acids by clostridial activity is a major cause of the low intake of high moisture silages made from herbage that is flail-harvested and ensiled directly without additives. As for low protein dry roughages (Egan & Moir 1965) the dry matter intake of these silages of poor fermentation quality can be increased by abomasal infusion of casein or methionine (Rogers et al. 1979) or by intraperitoneal injections of methionine (Barry 1976, Lancaster 1976), whereas such a response is not observed for

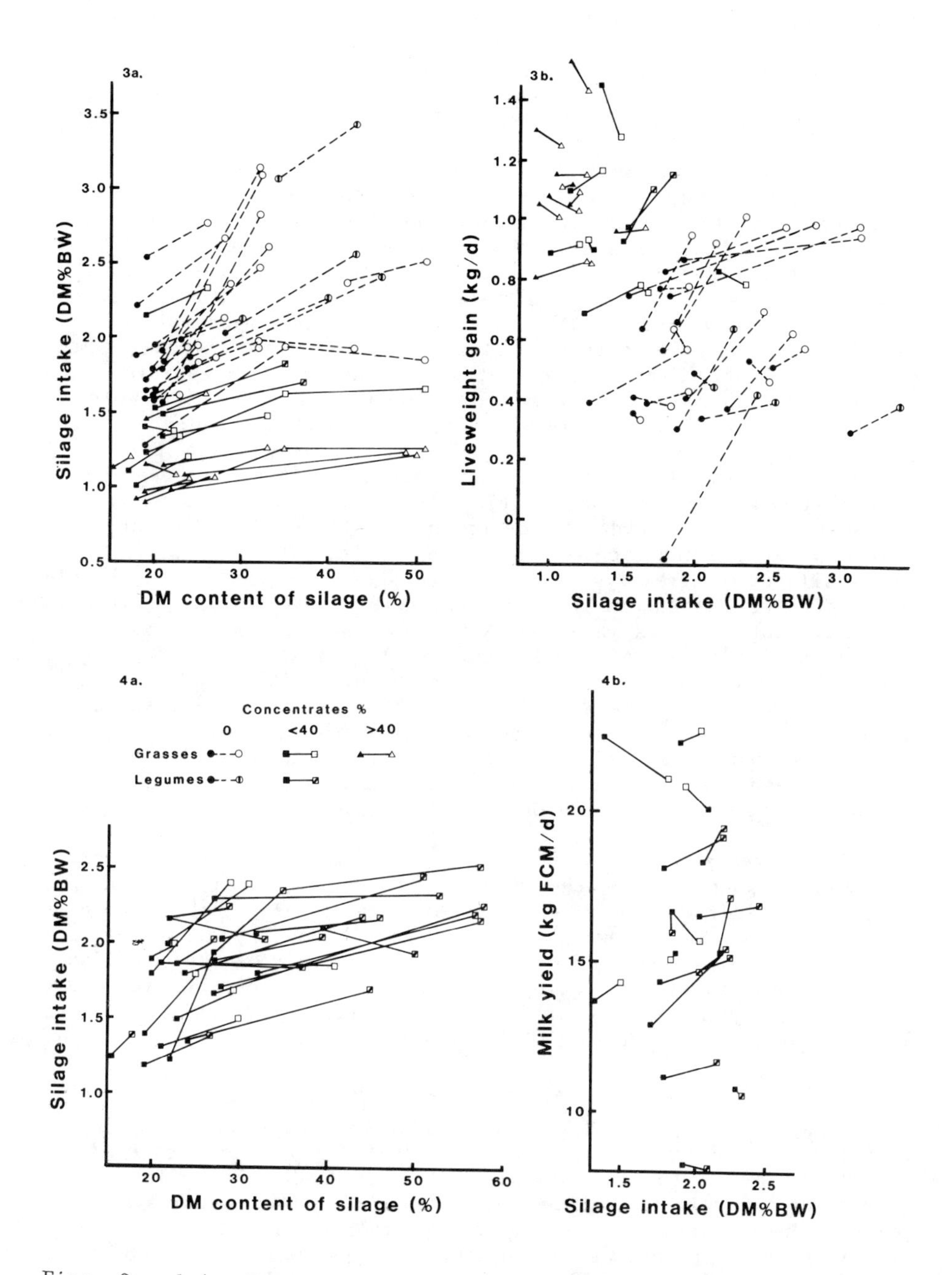

Figs. 3 and 4. Effect of wilting the herbage on the voluntary intake of silage by growing-fattening cattle (> 200 kg) (Fig. 3) and dairy cows (Fig. 4) and their performance.

precision-chopped silages (Hutchinson et al. 1971, Barry et al. 1978a). An improvement in the amino-acid supply in the duodenum and therefore in the protein status and appetite of the animals explains, at least in part, the 5-10 percent increase in silage intake which had been obtained by 1) preventing the clostridial breakdown of amino-acids by treating with additives or by wilting; 2) increasing the protein content of concentrates through substituting groundnut or soybean meals for a part of barley in the concentrate fed to dairy cows (Murdoch 1962, Castle & Watson 1969, Gordon 1979).

These different ways of protein supplementation essentially increase the amount of rumen-undegraded dietary protein which is absorbed from the small intestine, but they could also stimulate microbial protein synthesis in the rumen to a certain extent, at least with untreated meals. Animal production (milk, liveweight gain or wool) responds to these increased amounts of amino-acids and digestible energy. Milk production on high quality grass silages can be further enhanced without increased intake, by increasing the protein content of the concentrates and supplying digestible crude protein far above the recommended allowance as shown in work in Scotland (Castle & Watson 1969, 1974, 1976) and in Ireland (Gordon 1979, 1980a,b, Gordon & Murray 1979). Similarly Waldo & Tyrrell (1980) have obtained increased body gain and nitrogen retention in steers without increased digestible energy intake by the substitution of treated (formaldehyde + formic acid) silage for the untreated silage or by the addition of protected casein. The protein value of grass silages is greatly over-estimated when expressed in digestible crude protein. It should be expressed in terms of protein absorbed in the small intestine as in recently proposed systems (Roy et al. 1977, INRA 1978, Vérité et al. 1979, ARC 1980). In the INRA experiments increasing the supply of true protein digestible in the small intestine and the recommended allowance (50 g per kg milk) resulted in a slight but consistent increase in milk production on grass silage rations (Dulphy et al. 1980). This difference was probably caused by the lower efficiency of microbial protein synthesis on grass silages and/or the higher rumen degradability of nitrogen.

An increased absorption of forage amino-acids in the small intestine appears to be the major cause of the high responses to wilting or adding formic acid to lucerne silage that have been reported for cattle liveweight gain (Figs. 1b and 3b) as well as for nitrogen balance in sheep (Barry et al. 1978a, E. Grenet unpublished data). These treatments allow the high protein content, voluntary intake (Demarquilly & Jarrige 1974) and efficiency of utilization of legumes (Ulyatt 1971, Thomson 1977) to be exploited.

Comparisons of silages and hay cut simultaneously from the same crop

Silage and hay made from the same herbage have been compared in numerous feeding trials on dairy cows. In the experiments carried out in the U.S.A. up to the early 60's (review by Merril & Slack 1965) direct-cut or slightly wilted but otherwise untreated silages (<20 percent dry matter) were consumed in smaller quantities than high quality barn dried hays or low-moisture silages, while the differences in milk yield were generally small and rather conflicting. Therefore these high moisture silages appeared to have a greater nutritive value/kg dry matter for milk production. Similar differences were reported at the same period in Europe, expecially by Presegghe (1959) in Norway who obtained in a series of comparisons larger milk yields from cows receiving formic acid silages than from cows receiving the same calculated amount of net energy as hay. In more recent experiments in Scandinavia (Saue & Breirem 1969, Bertilsson 1980), cut or slightly wilted silages treated with formic acid, and hence well consumed, gave consistently higher milk yields than barn dried hays cut at the same time and had a 6-9 percent higher apparent nutritive value for milk production per kg dry matter. Higher percentages of milk fat have generally been reported for silage fed cows. In the less numerous comparisons on growing heifers or steers the animals consumed markedly less dry matter as silage than as hay but their daily gain was only slightly inferior (*inter alia* Thomas *et al*. 1969, Waldo *et al*. 1969).

Direct-cut or slightly wilted silages have a greater efficiency per kg dry matter than good hay from made the same herbage, both for milk production and liveweight gain in growing cattle; low moisture silages show an intermediate value. This results partly from 1) a slightly greater digestibility; 2) a higher gross energy content per kg dry matter (see above) and also 3) an underestimation of the dry matter content of silages in the older studies. Nevertheless high moisture silage still appears to have a greater efficiency for animal performance per kg digestible dry (or organic) matter in the limited number of trials where digestibility was measured (Gordon *et al*. 1961, Roffler *et al*. 1967, Byers 1965, Gordon *et al*. 1963, Thomas *et al*. 1969, Waldo *et al*. 1969 for growing cattle). Dairy cows utilize preferentially the digestible energy of high moisture silages for milk secretion at the expense of body gain, as shown by their lower daily gain compared to those of cows fed the corresponding hay or low moisture silage. Growing cattle could deposit more protein and less fat from these high moisture silages. These differences in the partition of energy may be associated with differences in the digestion end-products. However a lower proportion of acetic acid in the rumen volatile fatty acids and a greater proportion of butyric acid and higher acids have generally been observed on high moisture silages than on hays from the same herbage (Roffler *et al*. 1967 and Demarquilly *inter alia*). Differences in the supply of total amino-acids and methionine absorbed from the small intestine should be considered.

Table 3 summarizes the voluntary intake potential of some typical conserved forages when they are fed *ad libitum* with a minimum amount of concentrates in well balanced rations for cattle. They can be predicted satisfactorily using the INRA bulk unit system (Jarrige *et al*. 1979).

Silage-related diseases

Silages have commonly been considered to be less wholesome feeds than dried forages and accused of causing various health disorders, at least in the case of badly preserved silages (see review by Vetter & Von Glan 1978).

The end products of clostridial fermentations in grass silages may be responsible for health disorders, especially when the ration is not correctly balanced. High butyric acid content may result in an increase in blood ketone concentration and be a predisposing factor in the occurrence of bovine ketosis (Roffler *et al*. 1967) especially in cows in early lactation, all the more so as the propionic acid production in the rumen is reduced. High concentrations of ammonia and rapidly degradable nitrogenous compounds facilitate an excessive

TABLE 3

Approximate values for the voluntary intake, production potential and substitution rate of some well-conserved 1st cut forages in production rations for cattle, and for the corresponding production/ha.

		GRASSES (*Lolium*)			LEGUMES (Lucerne)			MAIZE SILAGE	
		Ear emergence stage		Early flowering	Bud stage		Early flowering	Milk-dough stage	Glaze stage
		Wilted silage	F.A. silage	Well-cured hay	Wilted silages	F.A. silage	Barn-dried hay	25% D.M.	35% D.M.
D.M. cut tons/ha		6.8	6.8	8.2	5.4	5.4	6.2	11	13
O.M. digestibility		70	72	62	63	65	60	70	71
DAIRY COWS	DMI (1)	2.25	2.15	1.85	2.35	2.20	2.15	1.95	2.45
25 kg milk	Milk (2)	13.0	14.0	6.5	11.5	12.0	8.5	11.0	18.0
	S (4)	0.65	0.60	0.45	0.65	0.60	0.60	0.40	0.80
	d/ha (5)	425	440	620	320	340	405	845	795
	milk/ha (5)	5,500	6,150	4,050	3,700	4,100	3,450	9,300	14,300
FATTENING	DMI (1)	1.45	1.35	0.80	1.60	1.45	0.95	1.30	1.55
YOUNG BULLS	F E (3)	65	65	30	60	60	35	65	80
1.25 kg/d	S (4)	0.65	0.65	0.70	0.70	0.65	0.85	0.50	0.70
GROWING	DMI (1)	2.25	2.15	1.85	2.45	2.30	2.35	-	-
HEIFERS	F E (3)	100	100	80	100	100	100	-	-
0.75 kg/d	d/ha (5)	840	830	1,250	615	660	745	-	-
	LWG/ha (5)	630	705	315	460	560	485	-	-

(1) VDMI : voluntary dry matter intake kg/100 kg liveweight; (2) Milk production/cow/day from forage net energy above maintenance; (3) Proportion of total net energy supplied by forages; (4) Substitution rate; (5) Maximum values per ha (16-17% DM losses - Fig. 6) for number of animal days, kg milk and kg liveweight gain. Values for early cuts should be increased by the production (around 900 kg milk or 100 kg liveweight gain) from the 3 weeks regrowth up to the date of flowering. F.A. - formic acid.

absorption of ammonia from the rumen and may lead to disturbances in the acid-base status and subclinical disorders. The likelihood of these diseases is increased when the liver is damaged by parasites or degenerative processes. A few reports, mainly from the Netherlands, describe outbreaks of cattle botulism. Amines, arising from the decarboxylation of amino-acids, are produced by Clostridium botulinum which multiplies in wilted grass silages (Notermans et al. 1979).

In addition, highly digestible silages made from young herbage give rise to soft faeces which have sometimes been claimed to favour the development of mastitis, endometritis and lameness in dairy animals. However, in a multi-lactation experiment in France (Hoden et al. 1981), no difference was found in the health and reproduction of dairy cows fed grass silage as the sole forage during the winter period in comparison with those fed hay or grass silage + hay, a result in agreement with a New Hampshire (Holter et al. 1976).

Other health disorders are associated with fungal and bacterial contamination of silage (see review by Vetter & Von Glan 1978). Moulds develop in the silage pockets and in the silage front during the feed out period. More than 50 species have been isolated, about half of which are capable of producing metabolites toxic to animals and man (Escoula et al. 1972). Aspergillus fumigatus is one of the predominant fungi in moulded maize silage (Cole et al. 1977) and it has been shown to induce abortion, diarrhoea and pneumonia in cattle. A bacterial infection commonly associated with silage feeding is listeriosis, especially in sheep, causing meningo-encephalitis and abortion in late pregnancy. However Listeria monocytogenes can neither develop nor survive in good quality silages (Gouet et al. 1977). The ingestion of toxic plants present in the original crop is more likely to occur in silages than in hays.

Finally, most of the disorders associated with silage feeding can be prevented if the well-known rules of good conservation, supplementation and hygiene are applied and animals are fed according to their requirements.

SOME ECONOMICAL ASPECTS OF CONSERVATION METHOD CHOICE

The choice of a forage conservation method, and of the place attributed to conserved forages in winter feeding, involves many agro-climatic, biological and economical factors. The relative importance of each of them varies from farm to farm. Depending on the case, the decisive factor can be the farm size, the type of herd, the available labour, the machinery or the buildings, etc. Therefore, the relative proportions of hay and grass silages and the method used, especially for silage making, differ widely between countries (Table 1).

Influence of herd type and of forage growth stage

Beef cows and ewes have relatively low nutrient requirements. They can live on low quality forages and crop residues. The basic method for feeding this class of stock thus remains that of waiting for the end of the first pasture growth cycle and good weather to harvest a maximum amount of hay (or a maximum number of daily maintenance rations), at the lowest possible cost because these animals have a low profitability. However, ensiling part of the pastures at an earlier stage allows more grass to be available at the beginning of the summer drought and thus a higher stocking rate. Furthermore, it provides a good quality forage for females in early lactation and for young animals.

Dairy cows and fattening animals require forages of high nutritive value, harvested at early heading for first cycle grasses, at the leafy stage for regrowths, and at budding for legumes. In humid temperate climates, in spring, ensiling or else barn drying is necessary. Hay-making allows high nutritive value forages to be harvested only in summer (aftermaths), especially in irrigated zones of Mediterranean climate. However it remains the basic method for the small dairy farms that cannot resort to ensiling owing to difficulties in harvesting silage on sloping mountain pastures and/or in feeding the silage to stall-tied animals.

Despite the smaller quantity of harvested dry matter, silage made from young herbage provides higher milk or meat production per hectare than forages harvested at flowering, 3 weeks later, as shown by calculations in Table 3 (see also Blaxter & Wilson 1963).

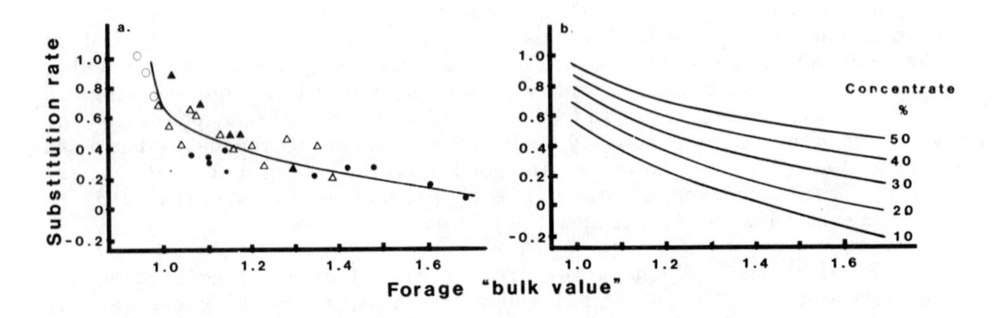

Fig. 5. Relationship between the substitution rate and the bulk value of forages in dairy cows (5a) and growing-fattening cattle (5b) (see Jarrige et al. 1979). The bulk value of a forage is obtained by dividing 123.6 by the VDM intake of the forage expressed in g/kg $W^{0.75}$. 123.6 g kg $W^{0.75}$ is the average value of VDM intake of young pasture grass by standard dairy cows.

The difference is even greater if the production given by animals at pasture from the 3-week regrowths after early cutting of silage is taken into account. However, to measure the profitability of early cutting many other factors must be considered, notably harvest and storage costs, the ratio between conserved forages and concentrates, the substitution rate of concentrates for forages, etc. The latter is directly related to the voluntary intake of forages, and therefore is higher for young than for mature forages (Table 3, Fig. 5).

Cost of harvest and storage

The cost of harvest and conservation of grass by ensiling always appears high, especially compared to grain. In France the field cost is about 600-700 francs/hectare (100 to 110 US dollars) and the total cost from cutting to trough is from 0.40 to 0.45 francs (0.065 to 0.075 US dollar) per kg DM consumed (Lienard & Dulphy 1981). Silage-making requires high machinery investments whose cost per kg of forage varies inversely with the area harvested per year.

Barn-drying is also very expensive, because of the great amount of electricity used. Nevertheless, use of solar collectors to heat air could be a way of reducing costs. Dehydration is even more costly, in spite of the improved efficiency of drying plants which recycle exhaust gas. White (1980) calculated the support energy input for forage conservation using the methodology of energy analysis.

The proportion of crop nutrients which is finally consumed by the animal is of great economic importance. It results from the decrease in the nutrient contents previously examined and from the dry matter losses. Fig. 6 summarizes the total dry matter and nutrient losses that could be expected in different situations when proper techniques are utilized at each stage from the field to the animal. The losses have been calculated from the partial or total dry matter losses reported by Zimmer (1951), Honig (1957), Watson & Nash (1960), Papendick (1974), Dulphy & Andrieu (1976) and from the decrease in nutritive value shown in Table 2. The losses encountered on the farm are frequently higher because of bad weather conditions and careless techniques.

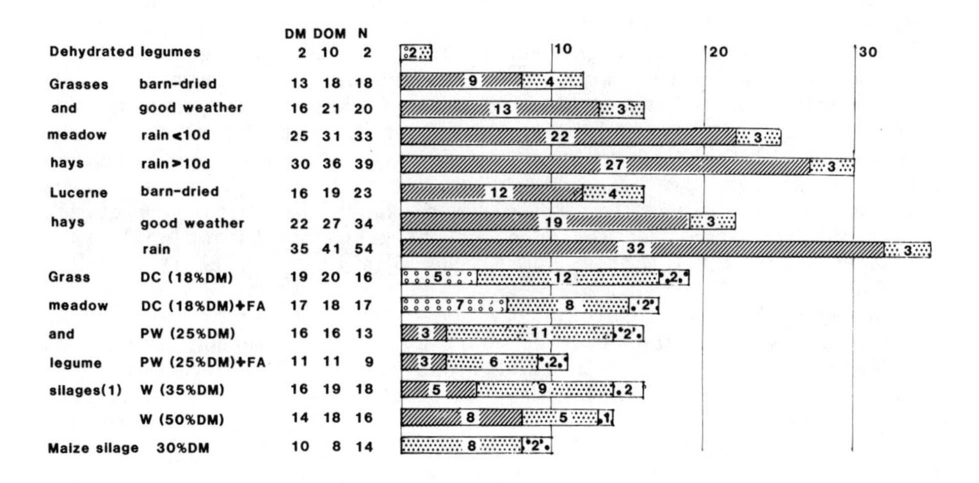

Fig. 6. Diagram of comparative dry matter and nutrient losses from the field to the trough under very good conditions in harvesting, conservation and unloading. (1) Forage ensiled in conventional silos except for W (50% DM) in airtight silos, DC : direct cut, PW : prewilted, W : wilted, FA : formic acid.

field losses, effluent losses, gas losses, '2' spoiled forages

SOME FEATURES AND PROBLEMS OF FORAGE CONSERVATION IN THE TROPICS

Improved nutrition during the dry period is a prerequisite to enhanced milk production and meat production in the tropics, more particularly in the small holdings of the populated areas. In addition to a more efficient utilization of standing hays as discussed by Siebert & Hunter (1982) and the various roughages derived from field crops (rice, maize, sugar-cane, ground-nut, pineapple) and sometimes tree crops (cacao), it is tempting to ensile highly productive grasses (*Digitaria decumbens*, *Pennisetum purpureum*, *Cynodon dactylon*) cultivated with legumes or grown alone with nitrogen fertilizers. Unfortunately, these tropical grasses generally have a low nutritive value and intake (Minson & McLeod 1970, Moore & Mott 1973, Chenost 1975) caused by early lignification.

Dried mature grass, often referred to as "standing hay" is the cheapest form of conserved forage and may form the bulk of the ration of grazing animals during the dry season. The grasses at this stage are stemmy, highly lignified and of low nutritive value, as they are deficient in digestible energy, protein and minerals, especially phosphorus (Todd 1956, Butterworth 1967). All these factors, especially protein deficiency, result in low dry matter and nutrient intake and this is often insufficient to provide for the maintenance requirements of animals.

In the humid tropics hay-making is out of the question during the rainy season. Suitable conditions do not occur before the nutritive value of the grasses had reached a low level. Furthermore, in many environments hay is almost certain to mould during storage. Satisfactory hays can be made at certain periods in Mediterranean and semi-arid zones, especially from forage crops such as *Medicago sativa*, *Trifolium alexandrinum*, *Vigna unguiculata* and cereal-legume mixtures such as oats and vetch. When these legumes are cured on the ground, considerable loss of leaves results because they dry very rapidly and crumble. If sufficient labour is available, stacking on tripods reduces these losses and results in more leafy hays that are higher in digestible energy, protein, phosphorus and carotene than ground-cured hays, as shown in Egypt for *Trifolium alexandrinum* (Danasoury *et al*. 1971a, 1971b). Some crop residues such as groundnut haulms could be a valuable source of hay (Miller *et al*. 1964, Diallo *et al*. 1976).

Silage-making is theoretically the only method of harvesting the surplus forage in the tropics during the wet season and at a high-nutritive growth stage. Sufficient experimental data have been published, especially in Florida, the West Indies, Brazil, Nigeria, Ghana, Rhodesia, India, and Australia to show some features and problems of silages made from tropical herbage plants. The main conclusions have been reviewed by Catchpoole & Henzell (1971).

Tropical herbage species are generally unsuitable for ensiling: 1) Their coarseness and structural rigidity impede the compaction of the silage and the exclusion of air (Hamilton *et al*. 1978) particularly in the drier tropics, and chopping is necessary. 2) The water soluble carbohydrate contents of most tropical grass species are frequently low. The small accumulation of starch does not offset the negligible content of fructosans (Hunter *et al*. 1970, Noble & Lowe, 1974). Addition of molasses improves preservation but the quantities required (up to 8 percent) are much higher than in temperate countries. 3) At maximal growth and nutritive value, the dry matter content is low (less than 18 percent) and the risks of soil contamination and bad fermentation are high. Wilting improves silage preservation (Davies 1965) but it is difficult to achieve or to control and wilted silages often suffer extensive mould growth. The high ambient temperature may result in clostridial fermentation (McDonald *et al*. 1966) and instability at least in silage of low dry matter content, especially as the soluble carbohydrate content is low (Ohyama *et al*. 1974). However, reasonably well-preserved silages have been obtained from several tropical grasses without additives. They are of the acetic-acid type rather than of the lactic-acid type (Catchpoole & Williams 1969, Aguilera 1975).

Well-preserved silages from tropical grasses cut at a leafy stage (5-6 weeks) rarely have an organic matter digestibility and a crude protein content exceeding 60 percent and 10 percent respectively (Miller et al. 1963, Semple et al. 1966, Butterworth 1967, Melotti et al. 1968,1969). According to the limited data available they would provide little more than the maintenance requirements of ruminants and should be supplemented with protein. Marked and variable decreases in digestibility in comparison with fresh or frozen herbage have been reported in Australia (Levitt et al. 1964, Morris & Levitt 1968, Hamilton et al. 1978), and the French West Indies (Xandé 1978). The addition of ground cassava scrapings (Ferreira et al. 1974) or of ground maize (Tuah & Okyere 1974) at ensiling has increased dry-matter intake.

Silages of higher energy value, protein content and intake could be prepared from legumes in the presence of efficient additives and or wilting prior to ensiling, as shown, for instance, from T. alexandrinum in Tunisia, Egypt (Abdel-Malik et al. 1973) and India (Verma & Mukherjee 1972). Whole cereal crops (maize, sorghum) are more suitable for ensilage than grasses. However they cannot always be grown satisfactorily and often compete with the production of human food. For instance highest yield and satisfactory fermentation quality of sorghum silages have been obtained by harvesting at the milk or early dough stage (Catchpoole 1962, Playne & Skerman 1964). However at these stages the organic matter digestibility and the crude protein content of the crop do not exceed about 55 and 5-6 percent respectively (Melotti et al. 1969). These values can be enhanced by intercropping the cereals with legumes.

Compared to the situation in cool temperate zones, silage-making from tropical herbage appears to involve more risks of bad fermentation and of low nutritive value, enhanced losses and spoilage through seepage, soaking by rain and aerobic deterioration and moulding on exposure to hot air during the feed-out period. Nevertheless the implements necessary for efficient ensiling are the same as in cool temperate zones.

Given the obstacles that hinder the development of reliable silage-making and feeding methods, it is probably more efficient to rely on irrigation and on crops that can be used directly without conservation during the dry season, such as sugar cane and cassava. The major contraints limiting the usefulness of sugar cane for ruminants have been identified and solved (Preston & Leng 1978,1980). When chopped and properly supplemented, especially with by-pass nutrients, cane has brought about satisfactory animal production in feeding experiments.

CONCLUSIONS

Over the last century, livestock production in the temperate zones of the Northern Hemisphere has been progressively freed from the seasonality of herbage production by using increasing proportion of conserved herbage and forage crops (maize). This improvement in winter feeding has allowed animals of increased milk production potential and muscular growth to be used. Silage is the most efficient method of forage conservation, but is costly and depend on fossil energy. It requires capital, adequate implements, training and organization of farmers and high market prices for animal products.

All these conditions are rarely met in tropical zones, and practically never in developing countries. Nevertheless, animal feeding during the dry period could be improved first by making the most of all available fodder resources without conservation, and, as a supplement, by adapting to local conditions some of the conservation techniques used in temperate countries. In most tropical countries, both genetic potential of the animals and their management will continue to be fitted to the low level of available resources during the dry season.

ACKNOWLEDGEMENTS

The authors are indebted to Dr D.J. Minson and Dr J.B. Hacker for revising and contracting the paper.

REFERENCES

ARC (1980) The nutrient requirements of ruminants livestock. Technical review by an Agricultural Research Council working party. Farnham Royal, U.K., Commonwealth Agricultural Bureaux, 351 pp.

Abdel-Makik, W.H.; Abou-Raya, A.K.; Makky, A.M.; Hathout, M.K. (1973) Effect of chopping and molasses addition on the nutritional qualities of trench silage prepared from the first cut of clover-ryegrass mixture. 4. Studies on silage consumption and milk production with Friesan cows. Agricultural Research Review 51, 81-89.

Aguilera, G.R. (1975) Dynamics of the fermentation of tropical grass silage. 1. Elephant grass (P. purpureum) without additives. Cuban Journal of Agricultural Science 9, 227-235.

Alderman, G.; Collins, F.C.; Dougall, H.W. (1971) Laboratory methods of predicting feeding value of silage. Journal of the British Grassland Society 26, 109-111.

Andrieu, J.; Demarquilly, C. (1974) Valeur alimentaire du mais fourrage. III. Influence de la composition et des caractéristiques fermentaires sur la digestibilité et l'ingestibilité des ensilages de mais. Annales de Zootechnie 23, 27-43.

Barry, T.N. (1975) Effect of treatment with formaldehyde, formic acid, and formaldehyde-acid mixtures on the chemical composition and nutritive value of silage. 1. Silage made from immature pasture compared with hay. New Zealand Journal of Agricultural Research 18, 285-294.

Barry, T.N. (1976) Effects of intraperitoneal injections of DL-methionine on the voluntary intake and wool growth of sheep fed sole diets of hay, silage and pasture differing in digestibility. Journal of Agricultural Science, Cambridge 86, 141-149.

Barry, T.N.; Cook, J.E.; Wilkins, R.J. (1978a) The influence of formic acid and formaldehyde additives and type of harvesting machine on the utilization of nitrogen in lucerne silages. 1. The voluntary intake and nitrogen retention of young sheep consuming the silages with and without intraperitoneal supplements of DC-methionine. Journal of Agricultural Science, Cambridge 91, 701-715.

Barry, T.N.; Mundell, D.C.; Wilkins, R.J.; Beever, D.E. (1978b) The influence of formic acid and formaldehyde additives and type of harvesting machines on the utilization of nitrogen in lucerne silages. 2. Changes in amino-acid composition during ensiling and their influence on nutritive value. Journal of Agricultural Science, Cambridge 91, 717-725.

Beever, D.E. (1980) The utilization of protein in conserved forage. In: Forage conservation in the 80's. Editor C. Thomas. Occasional Symposium No. 11, British Grassland Society, Brighton, pp. 131-143.

Beever, D.E.; Thomson, D.J.; Cammell, S.B. (1976) The digestion of frozen and dried grass by sheep. Journal of Agricultural Science, Cambridge 86, 443-452.

Bertilsson, J. (1980) Hay and silage fed to dairy cows on restricted feeding regimes. In: Forage conservation in the 80's. Editor C. Thomas. Occasional Symposium No. 11, British Grassland Society, Brighton, pp. 375-378.

Bertilsson, J.; Burstedt, E.; Knuttson, P.G. (1980) The voluntary intake of hay and silage fed to dairy cows with different levels of concentrate. In: Forage conservation in the 80's. Editor C. Thomas. Occasional Symposium No. 11, British Grassland Society, Brighton, pp. 372-374.

Blaxter, R.L.; Wilson, R.S. (1963) The assessment of a crop husbandry technique in terms of animal production. Animal Production 5, 27-42.

Bousset, J.; Fatianoff, N.; Gouet, Ph.; Contrepois, M. (1972) Ensilages gnotoxéniques de fourrages. I. Catabolisme des gluicides et métabolisme fermentaire dans les ensilages gnotoxéniques de luzerne, fétuque et ray-grass. Annales de Biologie Animale, Biochimie, Biophysique 12, 453-477.

Breirem, K.; Ulvesli, O. (1960) Ensiling methods. Herbage Abstracts 30, 1-18.

Byers, J.H. (1965) Comparison of feeding value of alfalfa hay, silage and low-moisture silage. Journal of Dairy Science 48, 206-208.

Butterworth, M.H. (1967) The digestibility of tropical grasses. Nutrition Abstracts and Reviews 37, 349-368.

Castle, M.E.; Watson, J.N. (1969) The effect of level of protein in silage on the intake and production of dairy cows. Journal of the British Grassland Society 24, 187-192.

Castle, M.E.; Watson, J.N. (1974) Red clover silage for milk production. Journal of the British Grassland Society 29, 101-108.

Castle, M.E.; Watson, J.N. (1976) Silage and milk production. A comparison between barley and dried grass as supplements to silage of high digestibility. Journal of the British Grassland Society 31, 191-195.

Castle, M.E.; Retter, W.C.; Watson, J.N. (1979) Silage and milk production : comparisons between grass silage of three different chop lengths. Grass and Forage Science 34, 293-301.

Catchpoole, V.R. (1962) The ensilage of sorghum at a range of crop maturities. Australian Journal of Experimental Agriculture and Animal Husbandry 2, 101-105.

Catchpoole, V.R.; Henzell, E.F. (1971) Silage and silage-making from tropical herbage species. Herbage Abstracts 41, 213-221.

Catchpoole, V.R.; Williams, W.T. (1969) The general pattern in silage fermentation in two subtropical grasses. Journal of the British Grassland Society 24, 317-324.

Chenost, M. (1975) La valeur alimentaire du Pangola (Digitaria decumbens) et ses facteurs de variation en zone tropicale humide. Annales de Zootechnie 24, 317-349.

Clancy, M.; Wangsness, P.J.; Baumgardt, B.R. (1977) Effect of silage extract on voluntary intake, rumen fluid constituents and rumen mobility. Journal of Dairy Science 60, 580-590.

Cole, R.J.; Kirksey, J.W.; Wilson, D.M.; Johnson, J.C.Jr; Johnson, A.N.; Bedell, D.M.; Springer, J.P.; Chexal, K.K.; Clardy, J.C.; Cox, R.H. (1977) Mycotoxins produced by Aspergillus fumigatus species isolated from molded silage. Journal of Agricultural Food Chemistry 25, 826-830.

Conrad, H.R.; Hibbs, J.W.; Pratt, A.D.; Davis, R.R. (1961) Nitrogen metabolism in dairy cattle. I. The influence of grain and meadow crops harvested as hay, silage or soilage on the efficiency of nitrogen utilization. Journal of Dairy Science 44, 85-95.

Danasoury, M.S.; El-Nouby, H.M.; Makky, A. (1971a) Yield and losses of dry matter and nutrients in berseem hay (Egyptian clover) cured by ground or tripod methods. Agricultural Research Review 49, 131-139.

Darmarquilly, C. (1970) Influence de la déshydration à basse température sur las valeur alimentaire des fourrages. Annales de Zootechnie 19, 45-51.

Demarquilly, C. (1973) Composition chimique, caractéristiques fermentaires, digestibilité et quantité ingérée des ensilages de fourrages verts : modifications par rapport au fourrage vert initial. Annales de Zootechnie 22, 1-35.

Demarquilly, C.; Dulphy, J.P. (1977) Effect of ensiling on feed intake and animal performance. Proceedings of the International Meeting on Animal Production from Temperate Grasslands, Dublin, 53-61.

Demarquilly, C.; Jarrige, R. (1969) The effect of method of forage conservation on digestibility and voluntary intake. Proceedings of the 11th International Grassland Congress, Surfers Paradise, Australia, 733-737.

Demarquilly, C.; Jarrige, R. (1974) The comparative nutritive value of grasses and legumes. In: Quality of herbage, Växtodling 28, 33-41.

Deswysen, A.; Ehrlein, H.J. (1979) Radiography of intake and (pseudo) rumination behaviour. Annales de Recherches Vétérinaires 10, 208-210.

Deswysen, A. (1980) Influence de la longueur des brins et de la concentration en acides organiques des silages sur l'ingestion volontaire chez les ovins et les bovins. Thése Doctorat Université catholique de Louvain. 254 pp.

Diallo, S.; Pugliese, P.L.; Calvet, H. (1976) Nutrition des bovins tropicaux dans le cadre des élevages extensifs sahéliens. Mesures de consommation et appréciation de la digestibilité et de la valeur alimentaire des fourrages. Revue d'Elevage et de Médecine Vétérinaire des Pays Tropicaux 29, 233-246.

Dulphy, J.P. (1972) Etude de quelques relations entre le mode de conservation du fourrage ingéré et le comportement alimentaire et mérycique des moutons. Annales de Zootechnie 21, 429-442.

Dulphy, J.P.; Demarquilly, C. (1975a) Influence de la machine de récolte sur la valeur des ensilages de graminées pour les génisses de race laitiée. Annales de Zootechnie 24, 351-362.

Dulphy, J.P.; Demarquilly, C. (1975b) Influence de la machine de récolte sur les quantités d'ensilage ingérées et les performances des vaches laitières. Annales de Zootechnie 24, 363-371.

Dulphy, J.P. (1980) The intake of conserved forages. In: Forage conservation in the 80's. Editor C. Thomas. Occasional Symposium No. 11. British Grassland Society, Brighton, pp. 107-121.

Dulphy, J.P.; Andrieu, J.P. (1976) Bilan de conservation des ensilages d'herbe. Bulletin Technique C.R.Z.V. de Theix, Institut National de la Recherche Agronomique 40, 27-33.

Dulphy, J.P.; Demarquilly, C. (1972) Influence de la finesse de hachage des ensilages de graminées sur le comportement alimentaire des moutons. Annales de Zootechnie 21, 443-451.

Dulphy, J.P.; Demarquilly, C. (1973) Influence de la machine de récolte et de la finesse de hachage sur la valeur alimentaire des ensilages. Annales de Zootechnie 22, 199-217.

Dulphy, J.P.; Demarquilly, C. (1975) Influence de la machine de récolte sur la valeur des ensilages de graminées pour les génisses de race laitière. Annales de Zootechnie 24, 351-362.

Dulphy, J.P.; Michalet, Brigitte (1975) Influence comparée de la machine de récolte sur les quantités d'ensilage ingérées par des génisses et des moutons. Annales de Zootechnie 24, 757-763.

Dulphy, J.P.; Bechet, G.; Thomson, E. (1975) Influence de la structure physique et la qualité de conservation des ensilages de graminées sur leur ingestibilité. Annales de Zootechnie 24, 81-94.

Dulphy, J.P.; Andrieu, J.P.; Demarquilly, C. (1980) Etude de la valeur azotée d'herbe additionné ou non d'acide formique pour les vaches laitières. Bulletin Technique C.R.Z.V. de Theix, Institut National de la Recherche Agronomique 40, 27-33.

Durand, Michelle; Zelter, S.Z.; Tisserand, J.L. (1968) Influence de quelques techniques de conservation sur l'efficacité de l'azote de la luzerne chez le mouton. Annales de Biologie Animale, Biochimie, Biophysique 8, 45-67.

Egan, A.R.; Moir, R.J. (1965) Nutritional status and intake regulation in sheep. 1. Effects of duodenally infused single doses of casein, urea, and propionic upon voluntary intake of a low-protein roughage by sheep. Australian Journal of Agricultural Research 16, 437-449.

Ekern, A.; Vik-Mo, L. (1979) Conserved forage as feeds for dairy cows. In: Feeding strategy for the high yielding dairy cow. Editors W.H. Broster and H. Swan. London, Granada Publications, pp. 322-373.

Escoula, L.; Le Bars, J.; Larrieu, G. (1972) Etudes sur la mycoflore des ensilages mycoflore des fronts de coupe d'ensilage de graminées fourragéres. Annales de Recherches Vétérinaires 3, 469-481.

Ferreira, J.J.; Silva, J.F.C. Da; Gomide, J.A. (1974) Effect of growth stage, wilting and the addition of cassava scrapings on the nutritive value of elephant grass silage (*Pennisetum purpureum* Schum). From Herbage Abstracts 44, 3807.

Gordon, F.J. (1979) The effect of protein content of the supplement for dairy cows with access *ad libitum* to high digestibility, wilted grass silage. Animal Production 28, 183-189.

Gordon, F.J. (1980a) The response of spring-calving cows to a high level of protein in the supplement given with grass silage during early lactation. Animal Production 30, 23-28.

Gordon, F.J. (1980b) The effect of silage type on the performance of lactating cows and the response to high levels of protein in the supplement. Animal Production 30, 29-37.

Gordon, C.H.; Derbyshire, J.C.; Wiseman, H.G.; Kane, E.A.; Melvin, C.G. (1961) Preservation and feeding value of alfalfa stored as hay, haylage, and direct-cut silages. Journal of Dairy Science 44, 1299-1311.

Gordon, C.H.; Derbyshire, J.C.; Jacobson, W.C.; Wiseman H.G. (1963) Feeding value of low-moisture alfalfa silage from conventional silos. Journal of Dairy Science 46, 411-415.

Gordon, F.J.; McMurray, C.H. (1979) The optimum level of protein in the supplement for dairy cows with access to grass silage. Animal Production 29, 283-291.

Gouet, Ph.; Girardeau, J.P.; Riou, Y. (1977) Inhibition of *Listeria monocytogenes* by defined lactic microflora in gnotobiotic silages of lucerne, fescue, rye-grass and maize-influence of dry matter and temperature. Animal Feed Science and Technology 2, 297-305.

Greenhalgh, J.F.D.; Wainmann, F.W. (1980) The utilization of energy in conserved forages. In: Forage conservation in the 80's. Editor C. Thomas. Occasional Symposium No. 11, British Grassland Society, Brighton, pp. 122-130.

Grenet, Elizabeth; Demarquilly, C. (1981) Nitrogen value of forages. Proceedings of the 14th International Grassland Congress, Lexington, Kentucky (in press).

Hamilton, R.I.; Catchpoole, V.R.; Lambourne, L.J.; Kerr, J.D. (1978) The preservation of a Nandi Setaria silage and its feeding value for dairy cows. Australian Journal of Experimental Agriculture and Animal Husbandry 18, 16-24.

Harris, C.E.; Raymond, W.F. (1963) The effect of ensilage on crop digestibility. Journal of the British Grassland Society 18, 204-212.

Henke, G.; Lambe, W. (1968) Untersuchungen zur Heiblufttrochung von Grüfutter. 2. Mitteilung Verluste an verdaulichen Nährstoffen. Archiv. für Tierernährung 18, 437-446.

Hoden, A.; Garel, J.P.; Journet, M.; Lienard, G.; Wegat-Litré, Erna (1981) Foin, ensilage d'herbe ou ration mixte pour les vaches laitiéres en zone de demi-montagne et niveau optimum de complémentation. Bulletin Technique C.R.Z.V. de Theix. Institut National de la Recherche Agronomique 43, 15-20.

Holter, J.B.; Urban, W.E.; Davis, H.A. (1976) Hay crop silage versus hay in a mixed ration for lactating cows. Journal of Dairy Science 59, 1087-1099.

Honig, H. (1967) Gärvorgang, verluste und qualität bei der konservierung von mähweidegras mit unterschiedlichem trocken-substanzgehalt. Das Wirtschaftseigene Futter 13, 287-298.

Hunter, R.A.; McIntyre, B.L.; McIlroy, R.J. (1970) Water-soluble carbohydrates of tropical pasture grasses and legumes. Journal of the Science of Food and Agriculture 21, 402-405.

Hutchinson, K.J.; Wilkins, R.J.; Osbourn, D.F. (1971) The voluntary intake of silage by sheep. III. The effects of post-ruminal infusions of casein on the intake and nitrogen retention of sheep given silage *ad libitum*. Journal of Agricultural Science, Cambridge 77, 545-547.

INRA (1978) Alimentation des Ruminants. Versailles, I.N.R.A. Publications, pp. 89-128.

Jarrige, R. (1980) Place of herbivores in the agricultural ecosystems. In: Digestive physiology and metabolism in ruminants. Editors Y. Ruckebusch and P. Thivend. Lancaster, U.K., M.T.P. Press Limited, pp. 763-823.

Jarrige, R.; Demarquilly, C.; Dulphy, J.P. (1974) The voluntary intake of forages. In: Quality of herbage, Växtodling 28, 98-106.

Jarrige, R.; Demarquilly, C.; Hoden, A.; Journet, M.; Béranger, C.; Geay, Y.; Malterre, C.; Micol, D.; Petit, M.; Robelin, J. (1979) Le système des Unité d'encombrement. Bulletin Technique C.R.Z.V. de Theix-Institut National de la Recherche Agronomique 38, 57-79.

Klinner, W.E.; Hale, O.D. (1980) Engineering developments in the field treatment of green crops. In: Forage conservation in the 80's. Editor: C. Thomas; Occasional Symposium No. 11, British Grassland Society, Brighton, pp. 224-228.

Klinner, W.E.; Shepperson, G. (1975) The state of haymaking technology. A review. Journal of the British Grassland Society 30, 259-266.

Lancaster, R.J. (1976) Effect of formaldehyde-treated casein and methionine on the intake of silage by sheep. Proceedings of the New Zealand Society of Animal Production 36, 127.

Levitt, M.S.; Hegarty, A.; Radel, M.J. (1964) Studies on grass silage from predominantly Paspalum dilatatum pastures in south-eastern Queensland. 2. Influence on length of cut on silages with and without molasses. Queensland Journal of Agricultural Science 21, 181-192.

McCullough, M.E. (1978) Fermentation of silage. A review. West Des Moines, Iowa, National Feed Ingredients Association, 332 pp.

McDonald, P.; Edwards, R.A. (1976) The influence of conservation methods on digestion and utilization of forages by ruminants. Proceedings of the Nutrition Society 35, 201-211.

McDonald, P.; Henderson, A.R.; Whittenbury, R. (1966) The effect of temperature on silage. Journal of the Science of Food and Agriculture 17, 476-480.

Marsh, R. (1978) A review of the effects of mechanical treatments of forages on fermentation in the silo and on feeding value of the silages. New Zealand Journal of Experimental Agriculture 6, 271-278.

Marsh, R. (1979) The effects of wilting on fermentation in the silo and on the nutritive value of silage. Grass and Forage Science 34, 1-9.

Martin, J.; Buysse, F. (1953) Proefnemingen over de invloed van hakselen in voordrogen op het on bewaringsproces van groenvoeders. Mededeling van de Landbouwhogeschool Gent 18, 565-591.

Melotti, L.; Boin, C.; Schneider, B.H. (1968) Trials on the apparent digestibility of silages of sorghum, maize and Napier grass. Boletin Industria Animal 25, 187-195.

Melotti, L.; Boin, C.; Labao, A.O. (1969) Digestion trial to determine the nutritive value of sorghum silage (Sorghum vulgare, Pers) Santa Eliza - variety. Boletin Industria Animal 26, 321-334.

Merrill, W.G.; Slack, S.T. (1965) Feeding value of perennial forages for dairy cows. A review. Animal Science mimeo. Series No. 3. Cornell University.

Miller, T.B.; Blair Rains, A.; Thorpe, R.J. (1963) The nutritive value and agronomic aspects of some fodders in northern Nigeria. 2. Silages. Journal of the British Grassland Society 18, 223-229.

Miller, T.B.; Blair Rains, A.; Thorpe, R.J. (1964) The nutritive value and agronomic aspects of some fodders in northern Nigeria. III. Hays and dried crop residues. Journal of the British Grassland Society 19, 77-80.

Minson, D.J.; McLeod, M.N. (1970) The digestibility of temperate and tropical grasses. Proceedings of the 11th International Grassland Congress, Surfers Paradise, Australia, 719-722.

Moore, J.E.; Mott, G.O. (1973) Structural inhibitors of quality in tropical grasses. In: Antiquality components of forages. Editor A.G. Matches. Madison, Wisconsin, Crop Science Society of America, pp. 53-98.

Moore, L.A.; Thomas, J.W.; Sykes, J.F. (1960) The acceptability of grass/legume silage by dairy cattle. Proceedings of the 8th International Grassland Congress, Reading, 701-704.

Morrison, I.M. (1979) Changes in the cell-wall components of laboratory silages and the effect of various additives on these changes. Journal of Agricultural Science, Cambridge 93, 581-586.

Murdoch, J.C.; Balch, D.A.; Holdsworth, M.C.; Wood, M. (1955) The effect of chopping, lacerating and wilting of herbage on the chemical composition of silage. Journal of the British Grassland Society 10, 181-188.

Noble, A.; Lowe, K.F. (1974) Alcohol-soluble carbohydrates in various tropical and temperate pasture species. Tropical Grasslands 18, 179-187.

Notermans, S.; Kosaki, S.; Van Schothorst, M. (1979) Toxin production by Clostridium botulinum in grass. Applied and Environmental Microbiology 38, 767-771.

Ohyama, Y.; Masaki, S. (1974) Effects of ensilage temperature on silage fermentation with special reference to the stage of the temperate treatment and soluble carbohydrate content of the material. Japanese Journal of Zootechnical Science 45, 419-423.

Owen, T.R.; Blair, T. (1977) Forage conservation research. Journal of the Science of Food and Agriculture 28, 317.

Papendick, K. (1974) Losses in the handling and conservation of forages. In: Quality of herbage. Växtodling 28, 81-89.

Playne, M.J.; Skerman, P.J. (1964) Influence of height of cutting on yield, nitrogen content and ensilage of sweet sorghum. Empire Journal of Experimental Agriculture 32, 325-330.

Presteghe, K. (1959) Forsøk med grasprodukter til storfe. Föringsforsøkene. Beretning No. 193.
Preston, T.R.; Leng, R.A. (1978) Sugar cane as a cattle feed : nutritional constrains and perspectives. World Animal Review 27, 2-12.
Preston, T.R.; Leng, R.A. (1980) Utilization of tropical feeds by ruminants. In: Digestive physiology and metabolism in ruminants. Editors Y. Tuckebusch and P. Thivend. Lancaster, England, M.T.P. Press Limited, pp. 621-640.
Prym, R.; Weissbach, F. (1977) Changes in feeding value of preserved grass and legumes by the action of high temperatures. Proceedings of the 13th International Grassland Congress, Leipzig, 1387-1389.

Raymond, F.; Shepperson, G.; Waltham, R. (1972) Forage conservation and feeding. Ipswich, U.K., Farming Press Limited, 175 pp.
Roffler, R.E.; Niedermeier, R.P.; Baumgardt, B.R. (1967) Evaluation of alfalfa-brome forage stored as wilted silage, low moisture silage, and hay. Journal of Dairy Science 50, 1805-1813.
Rogers, G.L.; Bryant, A.M.; McLeay, L.M. (1979) Silage and dairy cow production. 3. Abomasal infusions of casein, methionine, and glucose and milk yield and consumption. New Zealand Journal of Agricultural Research 22, 533-541.
Roy, J.H.B.; Balch, C.C.; Miller, E.R.; Orskov, E.R.; Smith R.H. (1977) Calculation of the N-requirements for ruminants from nitrogen metabolism studies. Proceedings of the 2nd International Symposium on Protein Metabolism and Nutrition. EAAP Publication No. 22, 126-129.
Saue, O.; Breirem, K. (1969) Comparison of formic acid silage with other silages and dried grassland products in feeding experiments. In: Crop conservation and grassland. Proceedings of the 3rd General Meeting of the European Grassland Federation pp. 282-284.
Schukking, S. (1976) The history of silage making. Stikstof 19, 2-11.
Semple, J.A.; Grieve, C.M.; Osbourn, D.F. (1966) The preparation and feeding value of Pangola grass silage. Tropical Agriculture, Trinidad 43, 251-255.
Shepperson, G. (1960) Effect of time of cutting and method of making on the feed value of hay. Proceedings of the 8th International Grassland Congress, Reading, 704-708.
Siddons, R.C.; Evans, R.T.; Beever, D.E. (1979) The effect of formaldehyde treatment before ensiling on the digestion of wilted grass silage by sheep. British Journal of Nutrition 42, 535-545.
Siebert, B.D.; Hunter, R.A. (1982) Supplementary feeding of grazing animals. In: Nutritional limits to animal production from pastures. Editor J.B. Hacker. Farnham Royal, U.K., Commonwealth Agricultural Bureaux, pp. 409-426.
Thomas, J.W.; Brown, L.D.; Emergy, R.S.; Benne, E.J.; Huber, J.T. (1969) Comparisons between alfalfa silage and hay. Journal of Dairy Science 52, 195-204.
Thomas, J.W. (1978) Preservatives for conserved forage crops. Journal of Animal Science 47, 721-735.
Thomas, P.C.; Kelly, N.C.; Chamberlain, D.G. (1980) Silage. Proceedings of the Nutrition Society 39, 257-264.
Thomson, D.J. (1977) The role of legumes in improving the quality of forage diets. In: Proceedings International Meeting on Animal Production from Temperate Grassland, Dublin. Editor B. Gilsenan. 131-135.
Thomson, D.J.; Beever, D.E. (1980) The effect of conservation and processing on the digestion of forages by ruminants. In: Digestive physiology and metabolism in ruminants. Editors Y. Ruckebusch and P. Thivend. Lancaster, England, M.T.P. Press Limited, pp. 291-308.
Todd, J.R. (1956) Investigations into the chemical composition and nutritive value of certain forage plants at medium altitudes in the tropics. Journal of Agricultural Science, Cambridge 47, 225-231.
Trimberger, G.W.; Kennedy, W.K.; Turk, K.L.; Loosli, J.K.; Reid, J.F.; Slack, S.T. (1955) Effect of curing methods and stage of maturity upon feeding value of roughages. 1. Some levels of grain. Cornell University Agricultural Experiment Station. Bulletin 910.
Tuah, A.; Okyere, O. (1974) Preliminary studies on the ensilage of some species of tropical grasses in the Ashanti forest belt of Ghana. Ghana Journal of Agriculture Science 7, 81-87.
Ulyatt, M.J. (1971) Studies on the causes of differences in pasture quality between perennial rye-grass, short-rotation rye-grass and white clover. New Zealand Journal of Agricultural Research 14, 352-367.
Van der Honing, Y.; Van Es, A.J.H.; Nijkamp, H.J.; Terluin, R. (1973) Net energy content of Dutch and Norwegian hay and silages in dairy cattle rations. Zeitschrift für Tierphysiologic, Tierernährung und Futtermittelkunde 31, 149-158.
Vérité, R.; Journet, M.; Rémond, B. (1979) Niveaux de complémentation énergétique des rations d'ensilage de mais pour les vaches laitières. Euromais 79, Cambridge (in press).
Verma, N.C.; Mukherjee, R. (1972) Ensilage of berseem (Trifolium alexandrinum). Indian Journal of Dairy Science 25, 295-296.
Vetter, R.L.; Von Glan, K.N. (1978) Abnormal silages and silage related disease problems. In: Fermentation of silage. A review. Editor M.E. McCullough. West Des Moines, Iowa, National Feed Ingredients Association, pp. 281-332.
Waldo, D.R.; Smith, L.W.; Miller, R.W.; Moore, L.A. (1969) Growth, intake and digestibility from formic acid, silage versus hay. Journal of Dairy Science 52, 1609-1616.

Waldo, D.R. (1977) Potential of chemical preservation and improvement of forages. Journal of Dairy Science 60, 306-326.

Waldo, D.R. (1978) The use of direct acidification in silage production. In: Fermentation of silage. A review. Editor M.E. McCullough. West Des Moines, Iowa, National Feed Ingredients Association, pp. 117-182.

Waldo, D.R.; Tyrrell, H.F. (1980) The relation of insoluble nitrogen intake to gain, energy retention and nitrogen retention in Holstein steers. Proceedings of the 3rd International Symposium on Protein Metabolism and Nutrition, Braunscheweig. Editors H.J. Oslage and K. Rohr. EAAP Publication No. 27, 572-577.

Watson, S.J.; Nash, M.J. (1960) The conservation of grass and forage crops. Edinburgh, Oliver and Boyd, 758 pp.

White, D.J. (1980) Support energy use in forage conservation. In: Forage Conservation in the 80's. Editor C. Thomas. Occasional Symposium No. 11. British Grassland Society, Brighton, pp.33-45.

Wilkins, R.J.; Hutchinson, K.J.; Wilson, R.F.; Harris, C.E. (1971) The voluntary intake of silages by sheep. 1. Interrelationship between silage composition and intake. Journal of Agricultural Science, Cambridge 77, 531-537.

Wilkins, R.J.(1975) The nutritive value of silages. In: Nutrition Conference for Feed Manufacturers No.8. Editors M.H. Butterworth, H. Swan and D. Lewis. pp. 167-189.

Wilkinson, J.M. (1980) Production of hay and silage in Europe : estimates for the countries of the EEC and Scandinavia 1978/79. In: Forage conservation in the 80's. Editor C. Thomas. Occasional Symposium No. 11. British Grassland Society, Brighton, Appendix 457.

Woodward, T.E.; Sheperd, J.E. (1938) Methods of making silage from grasses and legumes. Technical Bulletin USDA 611.

Xandé, A. (1978) L'ensilage d'herbe, une technique de conservation de l'herbe permettant de pallier le déficit alimentaire des ruminants durant la période du carème. I. Aspects théorique et pratique. Particularité des fourrages tropicaux. Nouvelles Agronomiques Antilles-Guyane 4, 63-80.

Zimmer, E. (1967) Nährstoffverluste bei der vergärung von futterpflanzen. 1. Mitteilung : der einfluss der siloform auf die höhe der verluste. Das Wirtschaftseigene Futter 13, 271-286.

IMPROVING FORAGE QUALITY BY PROCESSING

R.J. WILKINS

Grassland Research Institute, Hurley, Maidenhead, Berks. U.K. SL6 5LR.

ABSTRACT

Physical, chemical and biological treatments to improve the feeding value of forage grasses and legumes are discussed. With unfractionated forage, energy value may be increased by grinding and pelleting, by treatment with alkali, and probably also by fungal treatment. Protein value may be increased by heat treatment. Formaldehyde addition can reduce forage nitrogen degradability in the rumen, but may have adverse effects on nitrogen digestibility and on fibre digestion in the rumen. Calculations on processing mature grass indicated that the efficiencies of use of land and support energy were higher for treatment with ammonia than for either sodium hydroxide treatment or grinding and pelleting. When compared with field-cured hay + barley grain, ammonia-treated grass gave higher liveweight gain/ha and similar liveweight gain/GJ support energy input. There is the possibility that urea may be added to forage to provide an indirect source of alkali in a procedure with minimum hazard to the operator. Juice can be extracted from forages and used in pig feeding or for the production of leaf-protein concentrate. However, in terms of the use of land or support energy, there appears to be little incentive to change from systems involving feeding ruminants during the growing season principally on grazed grass and feeding pigs with grain and conventional protein supplements. Where cattle are stall-fed through the year and animal production is organized on a large scale, fractionation becomes relatively more attractive. Fractionated forage can be used as a source of xanthophyll, ethanol and other fermentation products. Developments in demand for such products, rather than for feed energy and protein, are likely to be important for any possible future development of forage fractionation.

INTRODUCTION

Forage diets fed without supplements will not generally give levels of production approaching the animal's genetic and physiological limits. The short-fall from potential production increases as forages mature and have reduced concentrations of metabolizable energy (ME) and nitrogen (N). Thus, for example, Lonsdale & Tayler (1969) reported empty-weight gains by young cattle of 0.79 kg/day when fed dried ryegrass (*Lolium perenne*) with 11.5 MJ metabolizable energy/kg dry matter dry matter but only 0.17 kg/day when the ryegrass was cut at a more mature stage of growth with metabolizable energy of only 9.3 MJ/kg dry matter.

This paper discusses the contribution that forage processing can make to improving animal production. The term processing is taken to cover physical, chemical and biological treatments imposed at or subsequent to harvesting with the objective of improving feeding value. Thus, silage and hay additives applied principally to improve preservation are not discussed here. Consideration is given initially to the treatment of unfractionated forage and then to forage fractionation. The effects of processing on the efficiency of use of support energy and land are discussed in order to identify situations in which processing is most likely to be adopted in practice. Whenever possible information is taken from experiments with perennial grasses and legumes, but reference is made to work with straw and other low-quality feeds in some instances.

UNFRACTIONATED FORAGE

Mechanical, chemical and biological treatments to improve energy value are discussed initially, followed by consideration of treatments to improve protein value.

Energy value

Mechanical treatment

It has long been known that mechanical treatment of forages to reduce particle size results in decreases in digestibility but increases in voluntary intake and the efficiency with which digested nutrients are utilized. These effects are associated with small particle size facilitating increased rate of passage of ingesta from the rumen. The magnitude of the responses to grinding and pelleting varies between experiments; Greenhalgh & Wainman (1972) noted that the response in the intake of forage with dry matter digestibility (DMD) of 55 percent ranged from 20 to 110 percent. For a particular forage, the extent of the changes in digestibility and intake are proportional to the extent of reduction in particle size, at least down to a mean particle size of 0.5 mm (Wilkins *et al*. 1972, Jarrige *et al*. 1973, Tetlow & Wilkins 1974). The response in intake is greater with sheep than with cattle and generally greater with forages of low digestibility (Greenhalgh & Wainman 1972) although the response may be small with forages of very low nitrogen content, unless supplementary nitrogen is provided (Minson 1967). Herbage composition also influences the change in digestibility, with depressions being particularly large with forages of high water-soluble carbohydrate concentration and low buffering capacity (Osbourn *et al*. 1976a).

As with intake, the effect of grinding and pelleting on net energy is greater with forages of low than high digestibility. Improved efficiency of energy utilization in pelleted as compared to chopped forage may arise from the higher proportion of energy absorbed as amino acids and lower proportion as volatile fatty acids (Osbourn *et al*. 1976b), also to the lower energy expenditure in feed ingestion and rumination.

It is unlikely that the inputs required for grinding and pelleting can be justified with high-quality forages, but the effects of grinding and pelleting mature grass crops are discussed further in a later section.

Chemical treatment

Chemical treatments have been imposed with the objective of improving the accessibility of structural carbohydrate to microbial enzymes in the rumen. Most attention has been given to treatment with alkalis but digestibility may also be increased by treatment with oxidizing agents such as chlorine gas, ozone, hydrogen peroxide and sodium peroxide (Jackson 1977). Ben Ghedalia & Miron (1981) have recently reported that treatment of straw with gaseous sulphur dioxide at 70°C increased the *in vitro* organic matter digestibility (OMD) of wheat straw to 80 percent compared with 67 percent for treatment with sodium hydroxide. Further work with oxidizing agents and sulphur dioxide appears warranted, but such techniques are a long way from commercial application and the remainder of this section concentrates on treatment with alkali.

Briefly, work on the treatment of straw with alkalis has shown:

(i) The *in vitro* digestibility of straw increases with increase in sodium hydroxide application rate following a quadratic relationship (Wilson & Brigstocke 1977). The response in digestibility with ammonia treatment is somewhat less than with sodium hydroxide (Jackson 1978).

(ii) The increase in digestibility is greater with straws of low than with high initial digestibility (Owen 1978).

(iii) Increases in *in vivo* digestibility are less than increases *in vitro*, particularly at levels of application of sodium hydroxide above 60 g sodium hydroxide/kg DM (Thomsen *et al*. 1973). The mean responses in organic matter digestibility or dry matter digestibility in ten experiments with sodium hydroxide applied at 40-50 g/kg DM were 12.5 percentage units for *in vivo* digestibility compared with 19.9 percentage units for *in vitro* digestibility (Klopfenstein *et al*. 1972, Thomsen *et al*. 1973, Rexen & Thomsen 1976, Owen 1978, Sundstøl *et al*. 1978).

(iv) The response in *in vivo* digestibility is reduced at high feeding levels and when treated straw is fed with concentrates (Dulphy, quoted by Jackson 1978).

(v) Treatment with sodium hydroxide results in increased voluntary intake. The mean response in organic matter intake in 11 experiments in which straw comprised more than 70 percent of the total diet was 36 percent (calculated from Donefer *et al*. 1969, Hasimoglu *et al*. 1969, Koers *et al*. 1970, Saxena *et al*. 1971, Carmona & Greenhalgh 1972, Jayasuriya & Owen 1975, Dulphy, quoted by Demarquilly 1977, Xande, quoted by Demarquilly 1977, Coombe *et al*. 1979, Garrett *et al*. 1979).

(vi) Adverse effects on animal health are avoided when sodium hydroxide addition is limited to 40 g/kg total diet dry matter (Jackson 1978).

Homb *et al*. (1977) described systems which are in practical use. For straw, the Beckmann process, in which excess alkali was removed by washing, has been succeeded by "dry" processes that do not involve washing and use sodium hydroxide in solid form or as a concentrated solution. Additional detail on systems for ammonia treatment is given by Sundstøl *et al*. (1978).

Although commercial treatment of straw is now well established, there has been little commercial treatment of grass or legume forage. The *in vitro* digestibility of *Lolium perenne* harvested at different stages of growth, field dried and treated with 60 g of NaOH per kg grass organic matter was studied by Mwakatundu & Owen (1974). As with straw, the response in *in vitro* organic matter digestibility (Y) was inversely proportional to the percentage digestibility of the untreated grass (X).

$$Y = 53.8 - 0.62\,x \pm 0.04$$

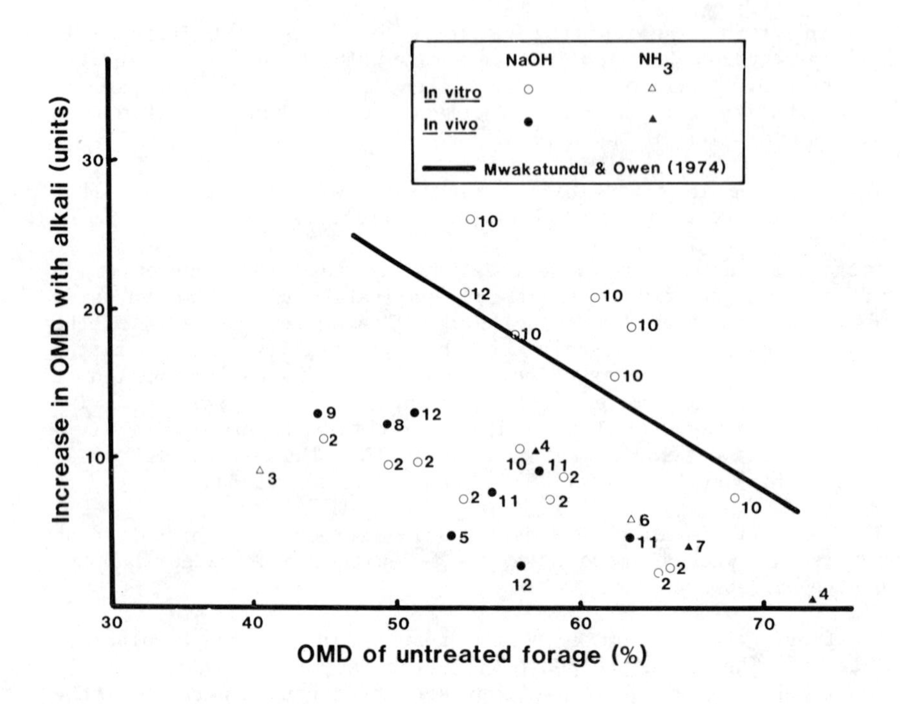

Fig. 1. The effect of alkali addition on the digestibility of grasses and whole-crop cereals.

1 Mwakatundu & Owen (1974), *Lolium perenne*, NaOH *in vitro*
2 Thomas (1978), *Setaria sphacelata*, *Pennisetum purpureum*, *Chloris gayana*, NaOH *in vitro*
3 Coxworth *et al.* (1978), grass
4 Richter *et al.* (1980), permanent grass
5 Siebert (1974), *Heteropogon contortus*
6 Kuntzel *et al.* (1980), grass
7 Mølle & Winther (1979), grass
8 Kategile (1981), *Hyparrhenia*
9 Meissner *et al.* (1973), *Cymbopogon*, *Themeda*
10 Bolsen & Tetlow (private communication), wheat, barley, maize
11 Tetlow & Deschard (private communication), wheat, barley
12 Tetlow (private communication), *Lolium perenne*, *Festuca arundinacea*

Figure 1 illustrates this relationship, together with data from other experiments with grasses and whole-crop cereals. The responses obtained by Mwakatundu & Owen (1974) are higher than those obtained by most other authors. It is encouraging that some *in vivo* responses in digestibility of around 10 percentage units have been reported with treatment by either sodium hydroxide or ammonia in grasses which had organic matter digestibility before treatment of 40 to 60 percent.

There is some indication that the responses with whole-crop cereals are particularly large and those for tropical grasses less than with temperate grasses, but further comparisons are required. Experiments with legumes were not included in Figure 1, because McManus (1978) and Coxworth *et al*. (1978) found only very small responses with legume forages and crop residues, presumably because of differences in cell-wall structure between grasses and legumes.

Recent research at the Grassland Research Institute has shown in five experiments mean increases of 52 and 74 percent in intakes of dry matter and digestible dry matter respectively by sheep fed mature grasses or whole-crop cereals treated with sodium hydroxide at 50-70 g/kg DM in comparison with the untreated forages (Tetlow & Deschard, personal communication). Meissner *et al*. (1973), with hay made from *Themeda* and *Cymbopogon*, reported increased digestible energy intake of 67 and 89 percent with sodium hydroxide additions of 33 and 67 g/kg DM respectively. Thus, it appears that the response in intake for mature grasses treated with sodium hydroxide is at least as great as with straws, despite somewhat smaller increases in digestibility.

Systems designed for the treatment of stored straw may also be used with grasses, but treatment at harvest would have advantage through avoiding double handling and permitting the alkali to contribute to improved preservation. Wilkinson (1977) described a simple system in which solid sodium hydroxide is applied in the field to grass with 50-70 percent dry matter; the treated grass is stored in a silo. Another simple in-field application system using sodium hydroxide solution to which non-protein nitrogen and minerals were also added was described by Kellaway *et al*. (1978). Although biologically efficient, these systems present some hazard in practical application.

Anhydrous ammonia has been injected into polythene-covered stacks of bales of moist hay and this treatment has given increased stability in aerobic and anaerobic conditions as well as enhanced digestibility (Mølle & Winther 1979, Kuntzel *et al*. 1980).

Urea may be added as an indirect source of ammonia. Thomsen (1980) reported work by Rashiq in which the addition of urea at 35 g/kg to straw with 35 percent dry matter led to the release of ammonia and increase *in vitro* digestibility to 65 percent. Saadullah *et al*. (1981) reported an increase in the *in vivo* organic matter digestibility of rice straw of 9 percentage units through treatment with urea at 30 g/kg DM; a similar increase resulted from the addition of urine. Forages have been treated with urea by Ghate & Bilanski (1979) in Canada and Tetlow (personal communication) in Britain. In both cases the conversion of urea to ammonia was rapid, even without the addition of urease, and in Tetlow's work the addition of urea at 60 g/kg DM resulted in successful storage of grass with 60 percent dry matter in aerobic conditions without heating or mould growth. More information on the effects of urea treatment on feeding value and on the consistency with which rapid conversion of urea to ammonia can be achieved is required, but there is promise that the addition of urea could provide a simple and hazard-free means of upgrading grasses.

Biological treatment

Possible biological treatments to increase the digestibility of lignocellulosic waste materials were reviewed by Linko (1977). Most attention has been given to treatment with white-rot fungi which preferentially metabolize lignin rather than structural carbohydrates. Responses in digestibility with straw have, however, been modest and there have been large differences between materials treated with different fungi. Latham (1979) examined 134 isolates of white-rot fungi and after incubation for 28 days at room temperature only five of these improved the *in vitro* dry matter digestibility of straw by more than two percentage units, with the largest increase being six units. Larger increases in *in vitro* digestibility have resulted from treatment with organisms other than the white-rot fungi. Thomson & Poole (1979) found that a culture comprising a basidiomycete and a yeast increased *in vitro* digestibility by 14 percentage units and Zadrazil (1979), examining the effect of various Basidiomycetes on wheat straw, found increases in *in vitro* digestibility of up to 34 units. Fungal dry matter could have made substantial contributions to digestible dry matter in both these studies.

A successful biological treatment would have obvious attractions, because the system would not involve the use of caustic chemicals and inputs of support energy would probably be low. The large variability in response between different fungal cultures encourages further research on the identification of cultures with greater effects on digestibility. The organisms that have been investigated are, however, aerobic. They could be used with stored dried hay; with silages, there may, in the long term, be the possibility of the addition at ensiling of enzymes prepared from aerobic delignifying organisms.

Protein value

For low-quality forages, deficiencies in rumen degradable nitrogen can be rectified by the addition of non-protein nitrogen. This section concentrates on the use of forage in the diet of animals with high protein requirements for which there is a need for feed nitrogen which is digested in the intestines rather than in the rumen.

Recent research has been concerned with identifying treatments to reduce forage nitrogen degradability in the rumen. It must be remembered, though, that reduction in rumen degradability will not improve protein value in all dietary situations. Treatments which reduce nitrogen degradability in the rumen may also reduce degradability in the intestines. Also, reductions in rumen degradability may lead to a shortage of soluble nitrogen in the rumen with adverse effects on digestion and intake.

Heat treatment

The effect of high-temperature drying on nitrogen digestion and utilization has been reviewed by Goering & Waldo (1979) and discussed by Beever (1980). Heat treatment results in a reduction in rumen ammonia concentration, an increase in nitrogen passing to the intestines and an increase in faecal nitrogen loss. The net effect is normally increased amino-acid absorption as typified by increases given by Proud (1972) and Beever *et al*. (1976) of 32 and 36 percent respectively.

Goering & Waldo (1979) drew attention to the large reductions in nitrogen digestibility which can occur when forages are dried at high temperatures at low moisture content. They recommended that *Medicago sativa* should not be dried to a moisture content below 50 g/kg and noted that high values for acid-detergent insoluble nitrogen indicated heat damage. In forages dried at temperatures of 60, 130, 160 and 180°C, nitrogen retention was highest at the two intermediate temperatures and the lowered nitrogen retention at the highest temperature was associated with an increase in acid-detergent insoluble nitrogen from 70 to 170 g/kg nitrogen.

In all of these experiments, heat treatment has been associated with forage drying. The high energy costs in high-temperature drying arise largely from the latent heat of evaporation of water. There is the need for work on the possibilities of controlled heating to reduce nitrogen degradability in the rumen, without the energy costs needed for water evaporation.

Mechanical treatment

Grinding and pelleting high-temperature dried forage has been found to further reduce nitrogen degradation in the rumen, in line with reduced rumen retention time, but Beever (1980) noted that the availability of amino acids in the small intestine may be reduced in ground and pelleted forage with the result that there may be little net improvement in amino acid uptake. He attributed the effect to the extra heat produced during grinding and pelleting further reducing nitrogen digestibility.

The effect of grinding and pelleting on nitrogen digestion in field-dried forages has not been extensively investigated. Reduced rumen nitrogen degradability would be expected, because of reduced rumen retention time and the effects of heating during the pelleting operation, but Hogan & Weston (1967), working with sun-cured *M. sativa* and wheat hays, found that grinding and pelleting did not alter the proportion of feed nitrogen which was apparently digested in the stomach.

Chemical treatment

There has been a large research effort on the use of formaldehyde to improve forage nitrogen use. Hemsley *et al*. (1970) found that the treatment of a clover-grass hay with formaldehyde reduced rumen ammonia concentration and increased the quantity of nitrogen leaving the stomach, the quantity of nitrogen digested in the intestines and clean wool production. Subsequent work with formaldehyde addition to forages at ensiling has consistently shown reduced protein breakdown in the silo, reduced rumen ammonia concentration and reduced nitrogen digestibility *in vivo*. Experiments by Beever *et al*. (1977) and Siddons *et al*. (1979) with sheep showed increases, compared with untreated silages, in amino-acid absorption of 13 and 80 percent when formaldehyde was applied at 60 and 35 g/kg crude protein respectively. Rumen degradability of amino acids in the formaldehyde-treated silages was 23 and 43 percent in the two studies compared with 85 percent for the untreated silages.

Formaldehyde-treated silages have given higher rates of animal growth than untreated silages in many experiments (Fig. 2), but when formaldehyde treatment has been compared with formic acid treatment of silages, mean values for liveweight gain and milk production have been similar. Where the analysis was restricted to include comparisons in which formaldehyde was applied at less than 25 g/kg crude protein, liveweight gain was six percent higher than with formic acid (Kaiser (1979). In many of these experiments, though, the animals would have been insensitive to increased amino-acid supply.

In addition to the responses in animal production being modest, there are problems in exploiting formaldehyde as a silage additive. Low (less than 50 g HCHO/kg crude protein) rates of formalin applied alone may induce clostridial fermentation in the silo (Wilkins et al. 1974), although this may be avoided if formaldehyde is used in mixture with other preservatives such as acids (Wilson & Wilkins 1980). Also, fibre digestion rates in the rumen of animals fed formaldehyde-treated silage may be reduced (Wilkins et al. 1975, Kaiser et al. 1982). In the experiments of Kaiser et al. (1982) the time for the disappearance of 50 percent of the silage cellulose in the rumen was as follows:

formic acid only	26.5 h
formic acid + 31 g formaldehyde per kg crude protein	33.0 h
formic acid + 123 g formaldehyde per kg crude protein	38.3 h

Voluntary intake was depressed at the high rate of formaldehyde. The addition of urea to the diet to increase rumen ammonia levels had only a small effect on cellulose digestion rate in the rumen and did not increase intake. This contrasts with experiments of Lonsdale et al. (1977) and Tayler et al. (1979), in which urea additions significantly increased intake of formaldehyde-treated silages. Kaiser et al. (1982) concluded that digestion rate was reduced due to some other adverse effects of formaldehyde on the rumen environment. If formaldehyde application is reduced below the level of 31 g HCHO/kg crude protein used by Kaiser et al. (1982), it is doubtful whether there would be any substantial reduction in protein breakdown during ensiling.

There is need for alternative treatments to reduce rumen protein degradability in silage. It is pertinent that in the experiment of Hemsley et al. (1970) the treated material was dried by ventilation prior to storage and feeding. In these circumstances excess formaldehyde could be lost from the feed, in contrast to the situation with ensiled forages. Nevertheless, when lucerne sprayed with formaldehyde prior to harvest was given to rumen-fistulated cows, Mangan et al. (1980) found there to be toxic effects on rumen micro-organisms, particularly protozoa. In contrast, glutaraldehyde-treated lucerne did not have adverse effects on the rumen microflora. The reaction of glutaraldehyde with forage proteins is more rapid than that of formaldehyde and the work of Mangan et al. (1980) indicated the possibility of using glutaraldehyde as a pasture spray to reduce nitrogen degradability. The use of glutaraldehyde as a silage additive has been investigated by Wilson & Jordan (1982) and this compound appears to reduce protein breakdown in the rumen, whilst having no effect on acid production by fermentation. Further in vivo studies with glutaraldehyde are in progress, but, in view of the markedly higher cost of glutaraldehyde than formaldehyde, it is

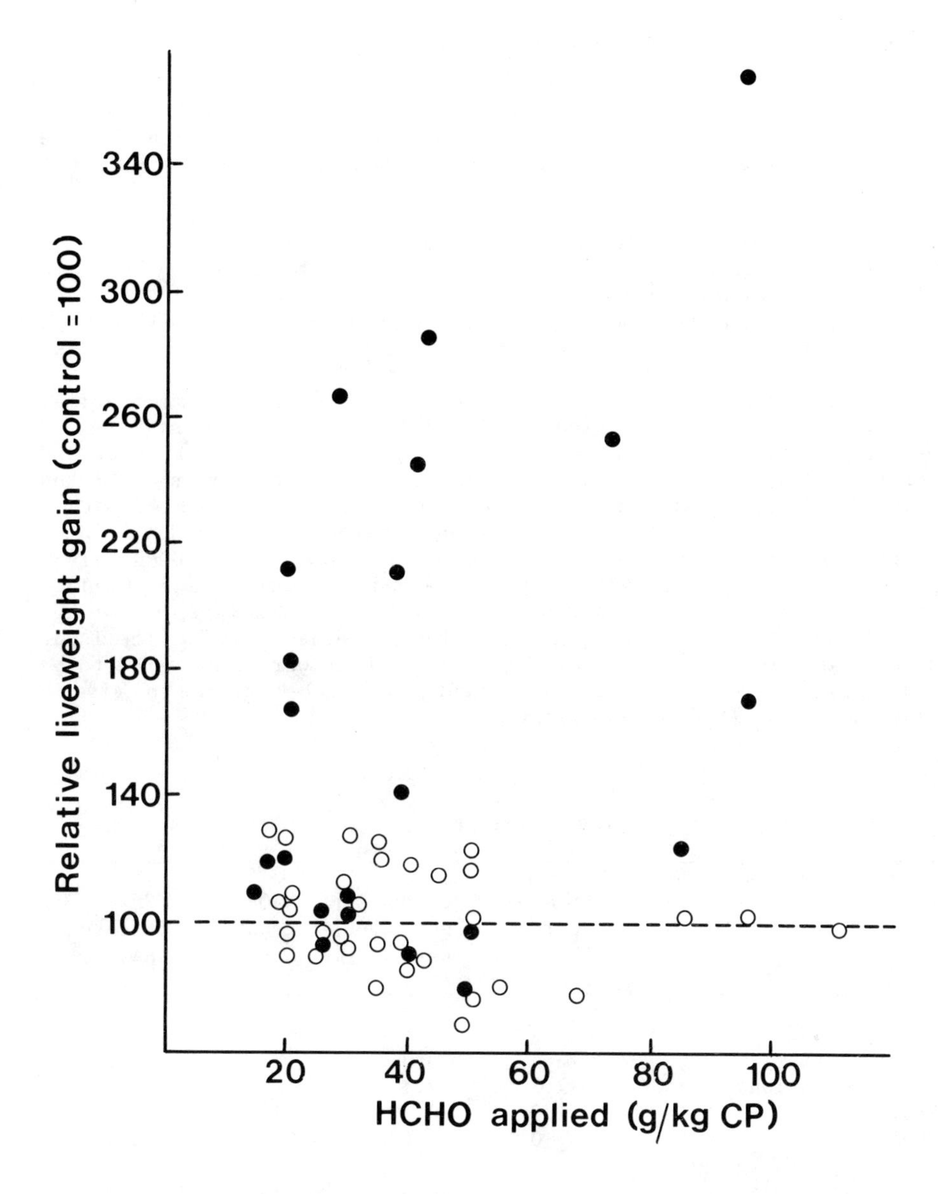

Fig. 2. Relative liveweight gain with formaldehyde-treated silages (from Kaiser 1979)
• Untreated control o Formic acid control

doubtful whether treatment with glutaraldehyde will be commercially attractive.

Support energy and land use with processed forages

The methods which are at present most attractive for unfractionated forage appear to be the treatment of mature grasses by grinding and pelleting or with sodium hydroxide or ammonia. The following systems are now discussed in more detail:

(i) Field haymaking
(ii) Grinding and pelleting field-cured hay
(iii) Sodium hydroxide (60 g/kg DM) of field-cured hay using a stationary machine
(iv) Ammonia treatment (30 g NH_3/kg DM) of herbage wilted to a moisture concentration of 40 percent.

Inputs and outputs of energy are given in Table 1 and animal production from the feeds given in Table 2, together with estimates of metabolizable energy production/GJ of support energy (SE) input and per ha of land. The calculations were made in relation to a cut taken from the first growth of Lolium perenne in England harvested with a yield of 11.5 t DM/ha, and an organic matter digestibility of 63 percent and digestible organic matter in the dry matter (DOMD) of 57 percent (Green et al. 1971). The conserved products were assumed to be given ad libitum to wether sheep with a liveweight of 40 kg. Calculations were made with this animal type because most of the input data were from sheep feeding experiments; sheep were considered to be fed forages alone or with supplements of barley grain to give a liveweight gain (LWG) of 110 g/day.

TABLE 1.

Energy balances in processing mature grass.

	Field-cured hay	Ground and pelleted hay	NaOH-treated hay	NH_3-treated hay
SE input (GJ/ha)				
Fertilizers[1]	13.5	13.5	13.5	13.5
Field operations[2]	2.1	2.1	2.1	2.1
Storage[3]	1.0	0.3	1.0	1.2
Chemical[4]	0	0	24.9	15.1
Processing energy and equipment[5]	0	14.4	2.3	0
	16.6	30.3	43.8	31.9
Output				
Dry matter (t/ha)[6]	8.0	8.0	8.5	10.1
In vivo DOMD[7]	51	41	60	60
ME (MJ/kg DM)[8]	8.06	7.08	9.60	9.60
ME (GJ/ha)	65.3	56.6	81.6	97.0
ME output/SE input (GJ/GJ)	3.93	1.87	1.86	3.04

Footnotes

1 N, P_2O_5 and K_2O at 150, 60 and 90 kg/ha with energy inputs/kg of 80, 14 and 9 MJ respectively.

2 Including fuel used and allowance for tractor and machinery depreciation (from White 1980).

3 126 MJ/t DM stored for long forage (from White 1980) and 30% of that figure for pelleted material.

4 NaOH at 60 kg/t DM and anhydrous NH_3 at 30 kg/t DM with energy inputs/kg of 51.8 and 51.5 MJ respectively.

5 Fuel and power used and allowance for machinery depreciation. NaOH treatment using JF6275. Electricity for grinding and pelleting taken as 87 kwh/t DM (based on Butler & Hellwig 1973).

6 Grass DM losses taken at 30% for field hay making and 15% with grass harvested at 40% moisture content. DM added in chemicals included.

7 Loss in DOMD for field curing and harvesting at 40% moisture content based on Shepperson (1960). Increase in DOMD through NaOH and NH_3 treatment based on Mwakatundu and Owen (1974) but with the response *in vivo* taken to be 2/3 that *in vitro* and the response to NH_3, 2/3 that of NaOH at any initial digestibility.

8 ME taken as DOMD x 16 but ME of ground and pelleted hay increased by 8% to allow for improved efficiency of energy utilization (Ministry of Agriculture, Fisheries and Food 1975).

TABLE 2.
Animal output, support energy and land use with processed mature grass.

	Field-cured hay	Ground & pelleted hay	NaOH-treated hay	NH_3-treated grass
Fed without supplement				
DM intake (kg/head)[1]	0.70	1.22	1.05	1.05
ME intake (MJ/head)	5.71	8.64	10.1	10.1
LWG (g/head)[2]	< 0	55	110	110
LWG (kg/ha)	-	360	890	1060
LWG (kg/GJ SE)	-	11.9	20.3	33.2
Fed to give LWG of 110 g/day				
DM intake (kg/head)[3]				
Grass	0.48	1.09	1.05	1.05
Barley	0.45	0.25	0	0
Area barley required (ha/ha grass)[4]	1.88	0.46	0	0
LWG (kg/ha grass)	1833	807	890	1060
LWG (kg/ha grass + barley)[5]	637	552	890	1060
LWG (kg/GJ SE)[6]	31.7	20.0	20.3	33.2

Footnotes

1 Field-cured hay from Agricultural Research Council (1980) increased by 75% for grinding and pelleting (from Agricultural Research Council, 1980 and Tetlow, personal communication) and by 50% for treatment with NaOH. The increase in intake for NH_3-treated grass was taken as 33%, but in this case the untreated material was of higher ME and intake potential.

2 Calculated from Agricultural Research Council (1980).

[3] Assuming a substitution rate (fall in grass intake per unit intake of barley grain) of 0.5.

[4] Barley yield of 4.0 t DM/ha.

[5] Ignoring any production from grass regrowths and from straw.

[6] SE input in barley production taken as 21.9 GJ/ha (16.9 GJ for production and 5.0 GJ for processing (based on Wilkins & Bather 1981).

Processing involved large increases in support energy input, particularly with sodium hydroxide treatment. The ratio of metabolizable energy output to support energy input was highest for field-cured hay. However, this treatment could not support the maintenance requirement of the sheep, whereas the alkali-treated forages provided energy sufficient for a liveweight gain of 110 g/day. It was calculated that the field-cured hay required supplementation with 450 g barley DM/sheep/day to achieve a liveweight gain equal to that of the unsupplemented alkali-treated forages. When allowance was made for the area of land to produce barley as well as grass, the LWG/ha was still highest for the alkali-treated feeds, but there was little difference in LWG/GJ support energy between field-cured hay plus barley and ammonia-treated grass; LWG/GJ support energy was lower with sodium hydroxide treatment. Processing by grinding and pelleting was inferior to alkali-treatment, for both LWG/ha and LWG/GJ support energy.

Another option would be to harvest grass at an earlier stage of growth to achieve liveweight gain without supplementation or processing. Using data from Shepperson (1960), it was calculated that to produce field-cured hay with similar DOMD to the alkali-treated mature grass, it would be necessary to cut at DOMD of 68 percent. At Hurley, this would involve cutting some five weeks earlier at a yield of only 6.8 t DM/ha compared with 11.5 t DM/ha at the later cutting date (Green _et al_. 1971); the yield for five-weeks regrowth from the earlier cut would be around 2.5 t DM/ha (Corrall 1974).

The situation examined here was one involving relatively modest rates of animal production. If higher rates of production were required, necessitating the use of supplementary concentrates, the beneficial effects of alkali treatment on digestibility and animal production are likely to be reduced. In summary, it appears that the case for alkali treatment of mature forage is strongest in situations in which:

(i) a modest rate of animal production per head is required

(ii) the relative price (or availability) of alkali relative to alternative feeds is low

(iii) the relative yield of alkali-treated conserved forage is much higher than that which could be achieved from unprocessed forage harvested at a less advanced stage of growth

(iv) alkali can reduce conservation losses, through facilitating the harvest of moist forage, as with the ammonia treatment examined here

More work on the treatment of tropical grasses with ammonia or urea as a source of ammonia seems warranted, particularly as such grasses for much of the year are low both in digestibility and protein content.

FRACTIONATED FORAGE

The high yield of protein/ha that can be otained from forage crops and the apparent excess of nitrogen for ruminants in young herbage has sustained research into systems involving the fractionation of forage into juice, which can be fed to monogastric animals either directly or after separation to give a leaf-protein concentrate (LPC), and a fibrous residue (pressed forage) for ruminant animals.

In view of the high degradability of forage protein in the rumen, there would be advantages in removing some of that protein for feeding monogastric animals irrespective of whether there is a quantitative excess of protein in the forage; the pressed forage could be supplemented with rumen degradable nitrogen as urea or ammonia. This whole subject has recently been discussed (Wilkins 1977, Howarth 1979, Jones 1981).

The following main conclusions may be drawn from work in which currently available equipment (screw press or pulper and belt press) has been used to extract some 15-35 percent of the dry matter from forages.

(i) the pressed forage may be used fresh or after ensiling or dehydration. Digestibility is slightly lower than that of unfractionated grass, but levels of intake and animal production are affected little, particularly when the extraction ratio (Wilkins 1977) is relatively low. It appears that the reduction in particle size resulting from processing compensates for the increase in concentration of cell-wall components and facilitates maintenance of intake levels similar to those for unfractionated forage (Jones 1981).

(ii) Grass and lucerne juice can satisfactorily replace the protein supplement in the diets of pigs of above three months of age; with younger pigs partial replacement is possible (Braude *et al*. 1977). Fresh juice is unstable and this generally necessitates preservative treatment to prevent a rapid fall in true protein content.

(iii) Leaf-protein concentrate can, with proper care in preparation, give animal production results consistent with its amino-acid composition (Morris 1977).

(iv) A soluble leaf-protein concentrate with good functional properties can be prepared for use as an ingredient in human food (Fremery & Kohler 1979).

The commercial adoption of forage fractionation is, however, at present negligible. Economic studies by Enochian *et al*. (1977) and Heath *et al*. (1979) indicated that the value attributed to the xanthophyll in leaf-protein concentrate is of crucial importance to the economics of large-scale systems involving the production of dried leaf-protein concentrate. Also farm development studies of systems producing fresh juice and fresh pressed forage have highlighted the problems of organizing regular supply of crop and utilization of these unstable products (Jones 1977).

An attempt has been made to appraise various systems of forage fractionation in terms of animal production and use of support energy and land. It was assumed that the crop fractionated was *Lolium perenne* grown with high nitrogen inputs and cut at 4-weekly intervals to give annual dry matter yields of 10.8 t/ha (Wilkins *et al*. 1981) with a nitrogen concentration of 2.88 percent, DOMD of 72 percent and metabolizable energy concentration of 11.2 MJ/kg DM. The grass was fractionated to give a dry matter extraction ratio of 20 percent. The whole and pressed grass were considered to be given to 300 kg Friesian steers and the grass juice used as a supplement to barley in feeding 80 kg pigs.

Support energy inputs were lowest with zero grazing, increased with fractionation and massively increased when the whole grass or pressed grass was dehydrated (Table 3). Metabolizable energy yield is not increased by fractionation, so that ME/GJ SE was also highest with zero grazing. In order to compare the systems in terms of production/ha, it was assumed that barley grain was grown to produce 4 t DM/ha with a support energy input for production and processing of 21.9 GJ/ha and soya was grown to produce 2 t DM/ha with a support energy input of 14.6 GJ/ha (derived from Pimental 1980). The carcass gain/ha total land area given in Table 4 of 920 kg with zero grazing + juice was somewhat lower than 1020 kg for zero grazing alone, but higher than 810 kg calculated for barley + soya. This contrasts with calculations of Jones (1981) of higher carcass gains with zero grazing + juice than zero grazing alone, but the liveweight gain/head predicted for beef production here was higher than the values used by Jones (1981) and the dry matter yield for grass relative to barley may also have been higher. Systems involving ensiling either grass or pressed grass led to lower carcass outputs/ha, because of losses in ensiling and reduced predicted rates of LWG.

Carcass gains/GJ SE were particularly low for ground and pelleted dehydrated grass and, of the systems in Table 4, highest at 24.8 kg for zero grazing + juice. However, more efficient use of support energy at 39.1 kg/GJ resulted from using barley + soya and the value for unfractionated grazed grass (assuming that 80 percent of the grass grown was utilized by grazing), at 24.1 kg/GJ, was similar to that for zero grazing + juice. Thus, in terms of use either of land or of support energy, there appears to be little incentive to change from the existing normal system in the UK of feeding ruminants during the growing season principally on grazed grass and feeding pigs with barley and conventional protein supplements.

Clearly there are opportunities for improving the efficiency of existing fractionation systems. Straub *et al*. (1979) noted possibilities for markedly reducing energy input in juice extraction and the energy required could be further reduced if the juice was used fresh rather than after preservation. Also systems for heat recycling can reduce energy inputs in dehydration, as discussed by Heath *et al*. (1979). A large reduction in energy requirement would be necessary to produce a major incentive for the adoption of fractionation.

Fractionation is, perhaps, more attractive in situations where cattle are already stall fed during the forage growing season. The adoption of zero grazing + juice may be particularly appropriate in situations, such as in some eastern European countries, where, in addition to cattle being stall fed, there are large farming enterprises operating on sufficient scale to justify the high

TABLE 3
Energy balances in fractionated grass.

	Unfractionated grass			Fractionated grass			
	Zero graze	Silage	Dry and pellet	Zero graze + juice	Silage + juice	Dry and pellet + juice	Dry and pellet + LPC[1]
SE input (GJ/ha)							
Grass production[2]	34.1	34.1	34.1	34.1	34.1	34.1	34.1
Harvesting[3]	18.2	18.2	18.2	18.2	18.2	18.2	18.2
Preservation and processing[4]	0	8.0	213.0	32.9	37.6	149.2	179.3
Total	52.3	60.3	265.3	85.2	89.9	201.5	231.6
Output							
Dry matter (t/ha)[5]							
Juice or LPC	0	0	0	2.2	2.2	2.2	0.9
Other products	10.8	8.6	10.3	8.6	6.9	8.2	9.4
ME (MJ/kg DM)							
Juice or LPC[6]	-	-	-	12.3	12.4	12.4	12.4
Other products[7]	11.2	10.9	11.2	10.9	10.6	10.9	11.2
ME (GJ/ha)	121	94	115	121	100	117	116
ME output/SE input (GJ/GJ)	2.31	1.55	0.43	1.42	1.12	0.58	0.50

Footnotes

1 DPJ returned to pressed grass

2 Calculated from Wilkins & Bather (1981) with N, P_2O_5 and K_2O inputs of 353, 120 and 210 kg/ha

3 From White (1980) adapted to six cuts/year

4 Silage includes for silo and application of formic acid at 2 l/t of grass with 180 g DM/kg (from White 1980) Drying and pelleting unprocessed grass from White (1980) and for other systems calculated in relation to drying and pelleting on the basis of energy costs, capital requirements and preserving chemicals for the systems given by Wilkins *et al*. (1977)

5 Assuming loss of DM during ensiling of 200 g/kg and during grinding and pelleting of 5 g/kg

6 Calculated on a residual basis so that fractionation *per se* does not alter total feed ME taken as DOMD x 16 (Ministry of Agriculture, Fisheries and Food 1975)

7 Calculated on the basis that the OM in the juice had an OMD of 88 percent and that the ash extraction ratio is 30%. ME concentration reduced by 3% by ensiling, but ME assumed to be unaltered by grinding and pelleting, to allow for more efficient utilization of ME

TABLE 4
Animal output, support energy and land use with fractionated grass.

	Unfractionated grass			Fractionated grass		
	Zero graze	Silage	Dry and pellet	Zero graze + juice	Silage + juice	Dry and pellet + juice
Grass or pressed grass fed to beef cattle						
DM intake (kg/head)[1]	6.5	5.8	7.1	6.5	5.8	7.1
ME intake (MJ/head)	72.8	63.2	79.5	70.8	61.5	77.4
Liveweight gain (kg/head)[2]	1.12	0.92	1.24	1.08	0.90	1.20
Liveweight gain (kg/ha grass)	1860	1360	1800	1430	1070	1390
Carcass gain[3] (kg/ha grass)	1020	750	990	790	590	760
Juice fed to pigs[4]						
No. pigs days/ha grass	0	0	0	4230	4230	4230
Quantity of barley required (t/ha grass)	0	0	0	12.6	12.6	12.6
Area barley required (ha/ha grass)[5]	0	0	0	3.15	3.15	3.15
Carcass gain (kg/ha grass + barley)[6]	0	0	0	730	730	730
Total carcass output (kg/ha grass + barley)	1020	750	990	920	870	910
Carcass output (kg/GJ SE)[7]	19.5	12.4	3.7	24.8	22.7	14.7

Footnotes

1 Based on Agricultural Research Council (1980) with intake of silage reduced by 10% and that of dried and pelleted feeds increased by 10%

2 From Agricultural Research Council (1980)

3 Killing-out proportion of 0.55

4 Based on Tables 7 and 8 in Braude *et al.* (1977). An 80 kg pig is fed 2.98 kg barley DM and 0.52 kg juice DM to give a LWG of 1 kg/day

5 Barley yield of 4t DM/ha

6 Killing-out proportion of 0.72

7 SE input in barley produced as in Table 2

management skill requirement for successful integration of daily forage supply and the production and utilization of unstable feeds for both cattle and pigs. The opportunities for production of alternative protein feeds for pigs and the relative demands for products from monogastric and ruminant animals will also influence the case for forage fractionation.

The future importance of fractionation may well depend on the requirements to supply factors other than feed energy and protein. The effect of xanthophyll value on the economics of factory-scale forage production has already been mentioned. Forages give high yields of gross energy/ha and there are possibilities for using the pressed forage as a source of fuel either directly or after anaerobic digestion. Worgan & Wilkins (1977) noted that deproteinized forage juice could be fermented to produce methane, ethanol or biomass food products. Thus although the utilization of grass by fractionation currently requires higher support energy inputs than utilization by grazing or zero grazing, there are opportunities to adapt fractionation systems to contribute to supplies of support energy as well as to provide feeds for ruminant and monogastric animals.

CONCLUSIONS

Although the feeding value of forages for ruminants may be increased by grinding and pelleting and by treatment with heat or chemicals, and protein utilization may be improved by feeding extracted forage juice to pigs, once the possibility of using a supplementary non-forage feed is considered, these treatments do not provide outstanding advantages in terms of the use either of land or support energy. The range of situations in which forage processing is at present justified is quite limited. The most attractive processing method is probably the treatment of mature grass with ammonia, or materials such as urea used as a source of ammonia, and further work on this approach is justified. Biological techniques for delignification hold long-term promise, but such techniques are more likely to be developed initially with by-products such as straw or wood-waste rather than with forage grasses or legumes.

The demand for products that can be derived from forages, such as xanthophyll and ethanol, rather than the requirements for feed energy and protein, are likely to strongly influence future developments in forage fractionation, although some opportunities already exist for the adoption of fractionation for animal feed purposes in large-scale intensive agriculture.

REFERENCES

Agricultural Research Council (1980) The nutrient requirements of ruminant livestock. Technical review by an Agricultural Research Council working party. Farnham Royal, U.K., Commonwealth Agricultural Bureaux, 351 pp.

Beever, D.E. (1980) The utilisation of protein in conserved forage. In: Forage conservation in the 80's. Editor C. Thomas. Occasional Symposium No. 11. British Grassland Society, Brighton, pp.131-143.

Beever, D.E.; Thomson, D.J.; Cammell, S.B. (1976) The digestion of frozen and dried grass by sheep. Journal of Agricultural Science, Cambridge 86, 443-452.

Beever, D.E.; Thomson, D.J.; Cammell, S.B.; Harrison, D.G. (1977) The digestion by sheep of silage made with and without the addition of formaldehyde. Journal of Agricultural Science, Cambridge 88, 61-70.

Ben-Ghedalia, D.; Miron, J. (1981) Effect of sodium hydroxide, ozone and sulphur dioxide on the composition and in vivo digestibility of wheat straw. Journal of the Science of Food and Agriculture 32, 224-228.

Braude, R.; Jones, A.S.; Houseman, R.A. (1977) The utilization of juice extracted from green crops. In: Green crop fractionation. Editor R.J. Wilkins. Occasional Symposium No. 9, British Grassland Society, Harrogate, pp.47-55.

Butler, J.L.; Hellwig, R.E. (1973) Dehydrating, grinding and pelleting forage grasses. Proceedings of the 1st International Grass Crop Drying Congress, Oxford, 266-276.

Carmona, J.F.; Greenhalgh, J.F.D. (1972) Nutritive value of NaOH treated straw. Journal of Agricultural Science, Cambridge 78, 477-485.

Coombe, J.B.; Dinius, D.A.; Wheeler, W.E. (1979) Effect of alkali treatment on intake and digestion of barley straw by beef steers. Journal of Animal Science 49, 169-176.

Corrall, A.J. (1974) The effect of interruption of flower development on the yield and quality of perennial ryegrass. Växtodling 29, 39-43.

Coxworth, E.P.; Kuelman, J.K.; Darrach, W. (1978) Effect of ammonia treatment on speciality crop residues and forages. Canadian Journal of Animal Science 58, 817-818.

Demarquilly, C. (1977) Utilisation des pailles et autres sous-produits vegetaux cellulosiques de grandes cultures dans les systems de production animale intensifs: comparaison avec les systems classiques. FAO Animal Production and Health Paper 4, pp. 61-86.

Donefer, E; Adeleye, I.O.A.; Jones, T.A.O. (1969) Effect of urea supplementation on the nutritive value of NaOH treated oat straw. In: Cellulases and their applications. Editor R.F. Gould. Advances in Chemistry Series 95, pp. 328-339.

Enochian, R.V.; Edwards, R.H.; Zuzmicky, D.J.; Kohler, G.O. (1977) Protein concentrate (Pro-Xan) for alfalfa: an updated economic evaluation. Annual Society of Agricultural Engineers Paper 77-6538. Winter Meeting.

Fremery, D. de; Kohler, G.O. (1979) Progress towards production of soluble food-grade LPC. Proceedings of the 2nd International Green Crop Drying Congress, Saskatoon, 68-80.

Garrett, W.N.; Walker, H.G.; Kohler, G.O.; Hart, R.J. (1979) Response of ruminants to diets containing NaOH or NH_4OH treated rice straw. Journal of Animal Science 48, 92-103.

Ghate, S.R.; Bilanski, W.K. (1979) Treating high-moisture alfalfa with urea. Transactions of the American Society of Agricultural Engineers 22, 504-506.

Green, J.O.; Corrall, A.J.; Terry, R.A. (1971) Grass species and varieties. Relationships between stage of growth, yield and forage quality. Technical Report 8, Grassland Research Institute.

Greenhalgh, J.F.D.; Wainman, F.W. (1972) The nutritive value of processed roughages for fattening cattle and sheep. Proceedings of the British Society of Animal Production 1972, 61-72.

Goering, H.K.; Waldo, D.R. (1979) The effects of dehydration on protein utilization in ruminants. Proceedings of the 2nd International Green Crop Drying Congress, Saskatoon, 277-291.

Hasimoglu, S.; Klopfenstein, T.J.; Doane, T.H. (1969) Nitrogen source with sodium hydroxide treated wheat straw. Journal of Animal Science 29, 160.

Heath, S.B.; Wilkins, R.J.; Windram, A.; Foxell, P.R. (1979) Green crop fractionation - an economic analysis. Proceedings of the 2nd International Green Crop Drying Congress, Saskatoon, 98-111.

Hemsley, J.A.; Hogan, J.P.; Weston, R.H. (1970) Protection of forage protein from ruminant degradation. Proceedings of the 11th International Grassland Congress, Surfers Paradise, Australia, 703-706.

Hogan, J.P.; Weston, R.H. (1967) The digestion of chopped and ground roughages by sheep. II. The digestion of nitrogen and some carbohydrate fractions in the stomach and intestines. Australian Journal of Agricultural Research 18, 803-819.

Homb, T.; Sundostøl, F.; Arnason, J. (1977) Chemical treatment of straw at commercial and farm levels. FAO Animal Production and Health Paper 4, pp. 25-37.

Howarth, R.E. (Ed.) (1979) Proceedings of the 2nd International Green Crop Drying Congress, Saskatoon.

Jackson, M.G. (1977) Review article: the alkali treatment of straws. Animal Feed Science and Technology 2, 105-130.

Jackson, M.G. (1978) Treating straw for animal feeding. FAO Animal Production and Health Paper 10.

Jayasuriya, M.C.N.; Owen, E. (1975) Sodium hydroxide treatment of barley straw; effect of volume and concentration of solution on digestibility and intake by sheep. Animal Production 21, 313-322.

Jarrige, R.; Demarquilly, C.; Journet, M.; Beranger, C. (1973) The nutritive value of processed dehydrated forages with special reference to the influence of physical form and particle size. Proceedings of the 1st International Green Crop Drying Congress, Oxford, 99-118.

Jones, A.S. (1977) A report of the joint Scottish Agricultural Development Council, Rowett Research Institute and North of Scotland College of Agriculture development programme on grass fractionation. British Grassland Society Occasional Symposium, 9, 143-148.

Jones, A.S. (1981) Potential changes in animal output from grassland: production from fractionated forage. In: Grassland in the British Economy, CAS Paper 10. Editor J.L. Jollans. Reading, U.K., Centre for Agricultural Strategy, pp. 496-510.

Kaiser, A.G. (1979) The effects of formaldehyde application at ensiling on the utilization of silage by young growing cattle. Ph.D. thesis, University of Reading.

Kaiser, A.G.; Osbourn, D.F.; England, P. (1982) Intake and digestion of formaldehyde-treated red clover silages offered to calves either alone or with a urea supplement. Journal of Agricultural Science, Cambridge, (in press).

Kategile, J.A. (1981) Simultaneous cutting and alkali treatment in a modified harvester for ensilaging. Workshop on utilization of low quality roughage in Africa, Arusha, Tanganyika Jan. 1981.

Kellaway, R.C.; Crofts, F.C.; Thiago, L.R.L.; Redman, R.G.; Leibholz, J.M.L.; Graham, C.A. (1978) A new technique for upgrading the nutritive value of roughages under field conditions. Animal Feed Science and Technology 3, 201-210.

Klopfenstein, T.J.; Krause, V.E.; Jones, M.J.; Woods, W. (1972) Chemical treatment of low quality roughages. Journal of Animal Science, 55, 418-422.

Koers, W.; Woods, W.; Klopfenstein, T.J. (1970) Sodium hydroxide treatment of corn stover or cobs. Journal of Animal Science 31, 1030.

Kuntzel, J.; Lesham, Y.; Pahlow, G. (1980) Anhydrous ammonia as a moist hay preservative. In: Forage conservation in the 80's. Editor C. Thomas. Occasional Symposium No. 11, British Grassland Society, Brighton, pp.252-256.

Latham, M.J. (1979) Pretreatment of barley straw with white-rot fungi to improve digestion in the rumen. In: Straw decay and its effect on diposal and utilization. Editor E. Grossbard. Chichester, U.K. John Willey and Sons Ltd, pp. 134-137.

Linko, M. (1977) Biological treatment of lignocellulose materials. FAO Animal Production and Health Paper 4, 39-50.

Lonsdale, C.R.; Tayler, J.C. (1969) The effect of stage of maturity of artificially dried ryegrass and method of processing on the growth of young cattle. Animal Production 11, 273.

Lonsdale, C.R.; Thomas, C.; Haines, M.J. (1977) The effect of urea on the voluntary intake by calves of silages preserved with formaldehyde and formic acid. Journal of the British Grassland Society 32, 171-176.

Mangan, J.L.; Jordan, D.J.; West, J.; Webb, P.J. (1980) Protection of leaf protein of lucerne (Medicago sativa L.) against degradation in the rumen by treatment with formaldehyde and glutaraldehyde. Journal of Agricultural Science, Cambridge 95, 603-617.

McManus, W.R. (1978) Alkali effects on agricultural wastes and their cell wall fractions. Australian Journal of Experimental Agriculture and Animal Husbandry 18, 231-242.

Meissner, H.H.; Franck, F.; Hofmeyr, H.G. (1973) 'N kort mededeling oor die invloed van natriumhidroksied op die vertebaarheid en inname van winterveldgras van swak kwaliteit. South African Journal of Animal Science 3, 51-52.

Ministry of Agriculture, Fisheries and Food (1975) Energy allowances and feeding systems for ruminants. Technical Bulletin 33. London; Her Majesty's Stationery Office.

Minson, D.J. (1967) The voluntary intake and digestibility in sheep of chopped and pelleted Digitaria decumbens (pangola grass) following a late application of fertilizer N. British Journal of Nutrition 21, 587-597.

Mølle, K.R.; Winther, P. (1979) Effect of ammonia and ammonium compounds as a hay preservative. Proceedings of the 2nd Hay Research Discussion Group, Burchetts Green, January 1979.

Morris, T.R. (1977) Leaf-protein concentrate for non-ruminant farm animals. In: Green crop fractionation. Editor R.J. Wilkins. Occasional Symposium No. 9, British Grassland society, Horrogate, pp.67-82.

Mwakatundu, A.G.K.; Owen, E. (1974) In vitro digestibility of sodium hydroxide-treated grass harvested at different stages of growth. East African Agricultural and Forestry Journal 40, 1-10.

Osbourn, D.F.; Terry, R.A.; Outen, G.E.; Cammell, S.B. (1976a) The significance of a determination of cell walls as the rational basis for the nutritive evaluation of forages. Proceedings of the 11th International Grassland Congress, Moscow. Vol.3, 374-380.

Osbourn, D.F.; Beever, D.E.; Thomson, D.J. (1976b) The influence of physical processing on the intake, digestion and utilization of dried herbage. Proceedings of the Nutrition Society 35, 191-200.

Owen, E. (1978) Processing of roughages. Proceedings of the 12th Nutrition Conference Feed Manufacturers, Nottingham, 127-148.

Pimental, D. (1980) Handbook of energy utilization in agriculture. Boca Raton, Florida; CRC Press, Inc., 475 pp.

Proud, C.J. (1972) A study of the digestion of nitrogen in the adult sheep. Ph.D. thesis, University of Newcastle-upon-Tyne.

Rexen, F.; Thomsen, K.V. (1976) The effect on digestibility of a new technique for alkali treatment of straw. Animal Feed Science and Technology 1, 73-83.

Richter, W.I.F.; Gross, F.; Beck, T.H. (1980) Behandlung von Feuchtheu mit Ammoniak. Proceedings of the European Association of Animal Production, Munich, Paper N 6.10.

Saadullah, M.; Haque, A.M.; Dolberg, F. (1981) Practical methods for chemical treatment of rice straw for ruminant feeding in Bangladesh. Workshop on utilization of low quality roughages in Africa, Arusha, Tanzania. Jan. 1981.

Saxena, S.K.; Otterby, D.E.; Donker, J.P.; Good, A.L. (1971) Effect of feeding alkali treated oat straw supplemented with soy bean meal or non-protein nitrogen on growth of lambs and on certain blood and rumen liquor parameters. Journal of Animal Science 33, 485-490.

Shepperson, G. (1960) Effect of time of cutting and method of making on the feed value of hay. Proceedings of the 8th International Grassland Congress, Reading, 704-708.

Siddons, R.C.; Evans, R.J.; Beever, D.E. (1979) The effect of formaldehyde treatment prior to ensiling on the digestion of wilted grass silage by sheep. British Journal of Nutrition 42, 535-545.

Siebert, D.B. (1974) The treatment of a tropical roughage with alkali, nitrogen and sulphur in relation to the nutritional limitations of pastures in northern Australia. Proceedings of the Australian Society of Animal Production 10, 86-90.

Straub, R.J.; Nelson, F.W.; Bruhn, H.D.; Koegel, R.G. (1979) Wet fractionation research at the University of Wisconsin - Madison. Proceedings of the 2nd International Green Crop Drying Congress, Saskatoon, 24-34.

Sundstøl, F.; Coxworth, E.; Mowat, D.N. (1978) Improving the nutritive value of straw and other low quality roughages by treatment with ammonia. World Animal Review 26, 13-21.

Tayler, J.C.; Aston, K.; Daley, S.R. (1979) Milk production from diets of silage and dried forage. 3. Effect of formalin-treated ryegrass silage of high digestibility given ad libitum with and without urea. Animal Production 28, 171-181.

Tetlow, R.M.; Wilkins, R.J. (1974) The effects of method of processing on the intake and digestibility by lambs of dried perennial ryegrass and tall fescue. Animal Production 19, 193-200.

Thomas, C. (1978) The effect of sodium hydroxide treatment on the organic matter digestibility of hay from three tropical grasses. Tropical Agriculture, Trinidad 55, 325-327.

Thomsen, K.V. (1980) The nutritional improvement of low quality forages. In: Forage conservation in the 80's. Editor C. Thomas. Occasional Symposium No. 11, British Grassland Society, pp.164-174.

Thomsen, K.V.; Rexen, F.; Kristensen, F.V. (1973) Forsøg med natrium hydroxid-behandling af halm. Ugeskrift for Agronomer og Hortonomer 118, 436-437.

Thomson, I.J.M.; Poole, N.J. (1979) Improving the nutritional value of straw. In: Straw decay and its effect on disposal and utilization. Editor E. Grossbard. Chichester, U.K., John Willey and Sons Ltd., p. 261.

White, D.J. (1980) Support energy use in forage conservation. British Grassland Society Occasional Symposium 11, 33-45.

Wilkins, R.J. (Editor) (1977) Green crop fractionation. British Grassland Society Occasional Symposium 9.

Wilkins, R.J.; Bather, M. (1981) Potential for changes in support energy use in animal output from grassland. In: Grassland in the British economy. CAS Paper 10. Editor J.L. Jollans. Reading, Centre for Agricultural Strategy, pp. 511-520.

Wilkins, R.J.; Heath, S.B.; Roberts, W.P.; Foxell, P.R.; Windram, A. (1977) Green crop fractionation. An economic analysis. Technical Report 19, Grassland Research Institute, Hurley, U.K.

Wilkins, R.J.; Lonsdale, C.R.; Tetlow, R.M.; Forrest, T.J. (1972) The voluntary intake and digestibility by cattle and sheep of dried grass wafers containing particles of different sizes. Animal Production 14, 177-188.

Wilkins, R.J.; Morrison, J.; Chapman, P.F. (1981) Potential production from grasses and legumes. In: Grassland in the British Economy. CAS Paper 10. Editor J.L. Jollans. Reading; Centre for Agricultural Strategy, pp. 390-413.

Wilkins, R.J.; Wilson, R.F.; Cook, J.E. (1975) Restriction of fermentation during ensilage: the nutritive value of silages made with the addition of formaldehyde. Proceedings of the 12th International Grassland Congress, Moscow. Vol 3, 674-690.

Wilkins, R.J.; Wilson, R.F.; Woolford, M.K. (1974) The effects of formaldehyde in silage fermentation. Växtodling 29, 197-201.

Wilkinson, J.M. (1977) Ensiling alkali-treated straws. In: Report on straw utilisation conference, Oxford, Feb. 1977. London, Ministry of Agriculture, Fisheries and Food, pp. 32-35.

Wilson, P.N.; Brigstocke, T. (1977) The commercial straw process. Process Biochemistry 12, 17-20.

Wilson, R.F.; Jordan, D.J. (1982) Glutaraldehyde as a silage additive: effects on fermentation and protein degradation in the silo. Proceedings of the British Society of Animal Production and British Grassland Society Symposium, Leeds, Sept. 1981.

Wilson, R.F.; Wilkins, R.J. (1980) The effects of mixtures of formalin and acid on silage fermentation. Proceedings of the 13th International Grassland Congress Leipzig, 12891292.

Worgan, J.T.; Wilkins, R.J. (1977) The utilisation of deproteinised forage juice. In: Green crop fractionation. Editor R.J. Wilkins. Occasional Symposium No. 9, British Grassland Society, Harrogate, pp.119-129.

Zadrazil, F. (1979) Screening of Basidiomycetes for optimal utilization of straw. In: Straw decay and its effect on disposal and utilization. Editor E. Grossbard. Chichester, John Willey and Sons Ltd., pp. 139-146.

SUPPLEMENTARY FEEDING OF GRAZING ANIMALS

B.D. SIEBERT, R.A. HUNTER

CSIRO, Division of Animal Production, Tropical Cattle Research Centre, P.O. Box 545, North Rockhampton, Queensland 4701, Australia.

ABSTRACT

The supplementation of grazing animals is carried out during periods of nutrient inadequacy to enable animals to survive, to produce or to reproduce. These aims can be achieved by providing whole feeds, or by providing specific nutrients which enable the animal to consume more of the available forage, digest or metabolize the same quantity of forage more efficiently, or which overcome a nutrient deficiency *per se*. In situations where intake is limited by a low availability of feed a supplement must take the place of forage. In other circumstances however, where intake, digestion, absorption or metabolism are adversely affected by nutrient deficiency despite a surplus of forage, then a supplement acts as an additive to the diet.

Whole feeds can be supplied as hays or grains, or treated forms of both, such as silage. In these situations the consumption of paddock forage declines with increasing level of feeding. Protein supplementation can be as true protein from animal or plant sources, as concentrates, or as legumes sown into the pasture, or as non-protein nitrogen. Both protein or non-protein sources of nitrogen are most effective when the nitrogen content of the pasture is less than that required for satisfactory microbial activity in the rumen and yet a potential source of digestible energy is available. Above such levels, their value is dubious and their effectiveness, if any, is often confounded with the energy value of the supplement.

Non-protein nitrogen and macro-minerals such as sulphur, phosphorus or sodium are usually supplied as licks in Australia, although in some other countries they may be fed together with a general feed supplement. Micro-elements such as cobalt and selenium are usually administered individually to animals. The frequency of distribution of feeds supplements to grazing animals can determine their effectiveness, continual input being the most desirable goal, but the most costly in terms of labour input. Cost efficiency is greatest in the area of trace element supplementation. The number of elements which may be administered by devices, such as pellets or capsules which lodge in the intestinal tract and deliver the supplement over a long period of time, is increasing.

The success of supplementation is judged in the final analysis on the return in terms of increased animal survival or increased productivity per unit of supplement input.

INTRODUCTION

Supplementation of grazing animals is carried out during periods of nutrient inadequacy to enable animals to survive, to produce or to reproduce. The nature of the deficiency may vary as may its degree. The term "supplementation" is sometimes used inappropriately, since the feed supplied may comprise the entire diet. For the purposes of this paper, a supplement is an addition to the diet which supplies deficient nutrients. At times it substitutes for part of the diet, while at others it increases forage consumption. The gross deficiency of energy will be discussed firstly, followed by those of protein, the macro-minerals then the micro-minerals.

ENERGY DEFICIENCY

Feeding for survival

During severe drought in pastoral areas or in semi-arid rangeland when the forage available is in very short supply, supplementary feeding is necessary for animals to survive. The principles involved in such situations approach those used in the stall-feeding of animals. Forages conserved as hay or silage can obviously be used; but if feeds must be purchased, cereal grains, particularly oats, wheat or sorghum are generally more economical. To maintain constant bodyweight, 6.6 MJ of metabolizable energy from approximately 0.5 kg grain are required for a 35 kg sheep and 40 MJ of metabolizable energy from approximately 3.0 kg of grain are required for a 350 kg steer. These quantities can probably be reduced by approximately 40 percent and still be expected to keep animals alive for six months. Investigations have been carried out feeding low-cost waste products such as sawdust with concentrates in attempts to reduce the grain input and thus lower the cost (Hunter 1978). In terms of the rate of loss of liveweight, little advantage has been found, although sawdust diets were found to increase survival rate.

Feeding for production

During dry periods in many pastoral areas of the world there is usually sufficient forage for animals to survive, but they lose liveweight due to low intakes of energy. For this reason in Australia in particular, some form of supplementation is desirable for animals grazing summer Mediterranean pastures, winter tropical pastures, cereal grain stubbles or semi-arid rangeland. Grain supplementation can be used to maintain liveweight and even support growth if desired, but lower cost supplements which aim primarily at maximizing the intake of the available forage are those most commonly sought. Many investigations have attempted to define the types of supplements required, as some supplements have been used with varying success. With present-day knowledge of the factors which limit feed consumption and utilization, some of the inconsistencies can now be explained. Since grazing animals suffer from energy deficits at such time, energy supplements are the most obvious, but it is not always these that necessarily increase the intake of forage.

Substitution

It has been appreciated for some time that when concentrates are fed to grazing stock, the supplement substitutes for part of the diet formerly grazed. Gulbransen (1974) showed that the degree of substitution is greater with poor quality forage than with high quality forage, although total intake of energy is greater with the latter. Allden & Jennings (1962) showed that this effect was quite marked for grazing sheep and Allden & Tudor (1976) demonstrated that cattle grazing Mediterranean summer pasture probably substituted supplement for herbage.

As a result of 'substitution', the objective of utilizing the pasture to its maximum and supplementing the animal's requirement with additional feed cannot be achieved. Indeed, the greater the supplementation the less the intake of forage. The reason is that a forage difficult to consume, digest and passage is being supplemented by a diet which is palatable, easy to consume and highly digestible. The animal chooses the supplement which is generally more digestible than the fibrous basal diet. Henning *et al*. (1980) have found that during such circumstances there is a change in the rumen microflora resulting in a decline in the number of cellulolytic bacteria.

Physico-chemical limitations to rumen function

Intake of fibrous forages is limited by the long retention time of indigestible residues in the rumen (see Minson 1982). The low concentration of readily digestible cellular components means that digestible energy per unit of feed eaten is low. Further, low concentrations of protein, or nitrogen and sulphur may limit the rate of digestion and hence the physical and chemical characteristics of the forage are compounded to keep digestion, flow of digesta and feed consumption low. In plants, low protein, low soluble carbohydrate and high cell wall content tend to be highly correlated (Fig. 1, Deinum & Dirven 1975). Digestibility tends to decline with increasing levels of cell wall, but the relationship is modified by longer retention in the rumen and thus more time is available for microbial fermentation of mature than immature forages. Forage availability, however, tends to be correlated with those properties associated with rapid growth and high yield. Tropical forages have higher cell wall concentrations than those found in temperate species which grow less rapidly under lower environmental temperatures and less radiative inputs (Deinum & Dirven 1975). For similar reasons there can be substantial differences in forage quality between years in both temperate and tropical regions. Differences are indicated in the types of supplementation required by grazing animals forages grown under varying climatic conditions.

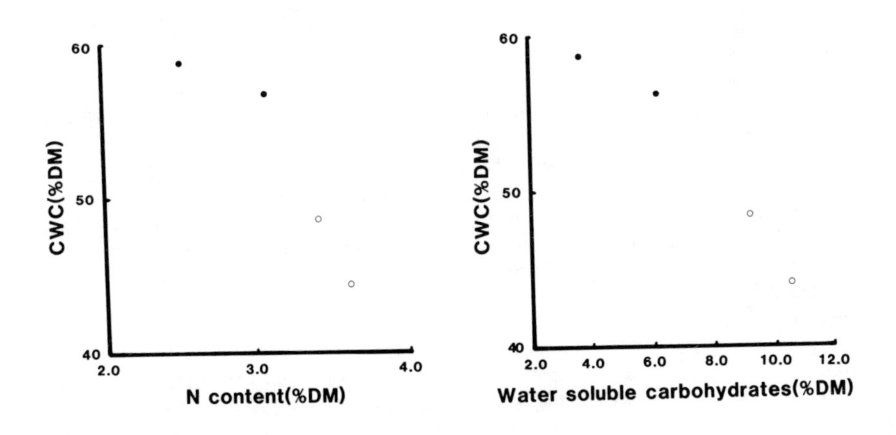

Fig. 1. The relationships between the physical components of plants (cell wall constituents (CWC)) and more soluble components (total N or carbohydrate). Each plot is the mean of 16 values from two temperate (o) and two tropical (•) plant species.

Physical limitations to the intake of the forage itself cannot be overcome by supplementation, although improvement of poor quality forage diets with protein alters the indigestible residue component, lessens the retention time and enables feed consumption to increase (Siebert & Kennedy 1972). The only alternative approach is to increase the intake of the grazing animal by various physico-chemical treatments (Pigden & Bender 1972). The most popular of these is alkali treatment and although each situation needs to be assessed in its own economic environment, the process developed by Kellaway (1980) combines the benefits of increases in digestibility with low cost supplementation with non-protein nitrogen and non-protein sulphur.

Apart from such treatment, however, nutrient supplementation will have most effect when chemical rather than physical factors limit animal productivity. If forage availability is adequate, supplements of protein, non-protein nitrogen and non-protein sulphur, soluble carbohydrate and minerals will be most useful where such nutrients are limiting, and plants are not highly fibrous.

PROTEIN DEFICIENCY

Protein can be supplemented in two forms - soluble compounds such as those found in fresh herbage and casein, and less soluble forms such as seed meals, meat meal and fish meal, particularly if these products that have been heated. Synthetically 'protected' proteins such as formaldehyde-treated casein fall into the second category (Ferguson et al. 1967). Soluble protein supplementation is successful when the forage is very low in protein concentration (or nitrogen and sulphur), and where the forage is sufficiently digestible to act as an adequate source of energy for potential microbial protein production. In this situation protein supplementation assists forage intake by first increasing the rate of digestion (Weston & Hogan 1973) and second by increasing the rate of passage of digesta. Even with the poorest of forages, energy intake is increased by 15 percent when as little as 10 percent of the diet consists of a supplement of hay of high protein content (Siebert & Kennedy 1972).

Less soluble protein, sometimes called 'by-pass' protein, is also able to increase the feed intake of cattle and sheep. A response of this nature was first noticed when Egan & Moir (1965) infused casein into the duodenum of sheep fed poor quality roughage. The mechanism by which feed intake is increased is not clear, but certainly in some circumstances protected proteins have caused positive liveweight responses in cattle and sheep (Kempton et al. 1977). In other experiments, however, such as those of Redman et al. (1980), who used poor quality oaten chaff supplemented with urea, casein or protected casein, it was concluded that increases in amino acid availability in the intestines did not increase feed intake or liveweight gain, although there were increases in oaten chaff consumption in all supplemented diets.

Protein thus acts as a source of nitrogen and sulphur for the rumen microflora, it alters the flow rate of digesta and may provide an additional source of amino acids at the tissues. Its net effect is one of increasing forage intake.

Protein supplements have been used in both tropical and temperate areas to offset liveweight loss or to increase weight gain of animals. Williams (1979) has summarized the response of some of these in terms of the efficiency of use of the quantity of the supplement fed. The supplements used include various crop meals such as peanut, cottonseed, linseed and others of animal origin such as meat meal. The best responses have occurred where there has been high availability of very poor pasture (1 kg gain/kg supplement) (Norman 1963). Less response has been recorded in temperate grassland (0.32 kg gain/kg), although this value was raised when the supplement was treated with formaldehyde (0.67 kg gain/kg) (Langlands & Bowles 1976). As little as 400 g of cottonseed meal fed to a 300 kg steer increased the intake of poor quality *Heteropogon contortus* by 20 percent and the total intake of energy by 35 percent (Hunter & Siebert 1980). In terms of nitrogen flow, such an amount of cottonseed will increase the amount of non-ammonia nitrogen leaving the rumen from 35 to 72 g and turn a negative nitrogen balance of -16 g/d into a positive balance of 9 g/d, equivalent to approximately 200 g of lean meat. Under dry seasonal conditions of tropical Australia where pasture is plentiful, such a form of supplementation is attractive if the cost of the supplement is not high. In southern Australia, the proposition is less attractive if, due to lack of pasture, the effect cannot be expressed.

A separate approach to supplementation is to provide the supplement as a pasture or crop species, either within the forage or available near-by. In temperate climates introduced pasture legumes grow simultaneously with grass species during winter/spring, and cannot act as a supplement to nitrogen-deficient mature forage or crop residues during summer. Allden & Geytenbeek (1980) have investigated a number of grain legumes as sources of nitrogen (and energy) for sheep and cattle grazing crop residues in summer. Such legumes include peas, beans, lupins and vetches. The field bean (*Vicia faba*) and lupin grain (*Lupinus angustifolius*) are two of the most promising. Hawthorne (1980) supplemented steers fed, or grazing, perennial pasture. The addition of 3.5 to 4.0 kg of lupin per day increased the liveweight gain of steers at pasture from 127 to 695 g/d. Such an approach to supplementation attempts to increase the use of forage while supplying a large proportion of the animal's requirement by additional (supplemental) feed. However, in these systems the supplement can become the main dietary source of energy, substituting for the forage.

In large extensive properties of northern Australia, such an approach is usually uneconomic and most supplements have been low-cost. The research approach of pasture improvement with tropical legumes such as *Stylosanthes* spp. is unlike the temperate approach as it attempts to maintain green legume plant species amongst the mature native grass species during the dry season. A number of varieties are available and have been released (Eyles 1979), their potential differing in relation to the time of availability and preference by the grazing animal. It is common for the leaves of these legumes to provide forage with more than 2 percent nitrogen content and to be 60 percent digestible (McIvor 1979). Such a source of nutrient sown amongst grasses which fall to less than 0.7 percent nitrogen content and to approximately 45 percent dry matter digestibility, is a useful supplement, increasing the quantity of poor quality forage that can be consumed. Romero & Siebert (1980) found that steers grazing low quality native grass pastures could not consume sufficient energy for their maintenance requirements, but when two different *Stylosanthes* spp. were sown into the pasture their nitrogen intake doubled and they exceeded their maintenance energy requirement. The higher digestibility of the grass/legume forage consumed was a major factor in increasing energy intake, so that apart from being a protein supplement, the legumes acted as an energy supplement. C.J. Gardiner (personal communication) has found that the digestibility of the leaf and stem fractions of nine accessions of various *Stylosanthes* spp. of varying ecological significance was usually in excess of 60 percent, except for older stem portions which fell below 40 percent.

Among browse plants the tree legume *Leucaena leucocephala* is one of the most prodigious fixers of nitrogen known and is used as a feed for ruminants in the tropics (Jones 1979). Its leaf protein content is similar to that of *Medicago sativa* and has a dry matter digestibility in excess of 75 percent. It has been used as cut forage to supplement pen-fed animals and can support growth rates of 0.6 kg/d with highly digestible roughages and lesser rates when fed with poor quality rice straw. Its greatest value could be as a grazing supplement for beef cattle on grass pastures in the tropics. Strategies under investigation include planting leucaena in rows with grass forming the 'understorey'. In Fiji liveweight gains of 215 g/d for steers on grass pasture were raised to 500 g/d on pastures where 20 percent of the area was planted with fertilized leucaena (Partridge & Ranacou 1974).

Microbial fixation of ammonia and sulphide into bacterial cells in the rumen is the basis for the use of non-protein nitrogen and non-protein sulphur supplements with ruminants. Non-protein nitrogen (mostly urea) has been used with variable success in pen and field experiments. Not all of the conflicting results are explainable, but at least some are associated with the absence or presence of sufficient digestible energy and sulphur for microbial protein production. Forages with a low digestibility due to a high fibre content can only produce 134 g microbial organic matter (65 percent protein) from a kg of feed (Weston & Hogan 1973). This requires 14 g N/kg feed from dietary and internal recycled sources. If the feed source of nitrogen is too little, nitrogen from non-protein nitrogen can make up the difference, provided sufficient energy and sulphur are available. If not, the energy must be provided from other sources, for example readily digestible carbohydrate such as molasses, and the sulphur from a source such as sulphate, sulphide or even elemental sulphur. Molasses usually contains sufficient sulphur for this purpose. When grazing beef cattle are fed with urea significant reductions in liveweight loss have been recorded when pastures reached very low protein concentrations (Winks & Laing 1972). However, with forages in which the fibre is highly resistant to fermentation, supplements of non-protein nitrogen, non-protein sulphur and energy, while increasing rate of digestion may not increase the rate of passage of digesta sufficiently to increase the intake of forage. The effect of different supplements on the voluntary feed consumption of a forage is shown in Fig. 2.

With steers on low quality forage, but receiving adequate energy and nitrogen, supplementation with sulphur lowers blood urea and increases feed intake. In some plants the protein concentrations are sufficient for adequate rumen function, but there is an imbalance of nitrogen and sulphur. For instance, legumes grown on low sulphur soil may have nitrogen:sulphur ratios of 15:1 (Hunter *et al.* 1979). Sulphur supplementation in these situations has increased feed intake by 42 percent and the liveweight gain of steers from -9 kg to +26 kg over 5 months.

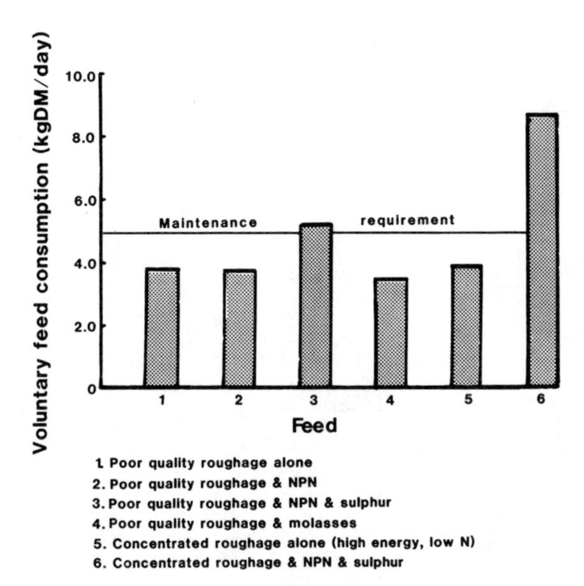

Fig. 2. The voluntary feed consumption of roughages by cattle in relation to their maintenance requirement when fed alone or with sources of non-protein nitrogen and non-protein sulphur.

DECIDING ENERGY AND PROTEIN SUPPLEMENTATION

Feed availability, fibre content, protein content and the ratio of nitrogen to sulphur thus provide criteria on which to judge the probable success or failure of supplementation of energy, protein, non-protein nitrogen and non-protein sulphur. The expected responses of cattle and sheep fed such supplements, when the characteristics of the forage are at different levels, are shown in Table 1.

During the absence or low availability of feed, supplements of energy will obviously result in a marked response in animal growth, particularly so if the feed available is high in fibre. Alternatively if there is plentiful feed, there will only be a response if the feed is highly fibrous.

In a situation where there is low availability (but not absence) of relatively non-fibrous, low protein forage such as that in a good quality but mature Mediterranean ryegrass pasture, it could be expected that a response to a protein supplement might be achieved. It is less likely if the forage is highly fibrous, although there may be some response due to the energy value of the protein supplement. Where there is a high availability of low protein forage such as in a tropical grassland, some response to protein might be noted, especially with relatively low fibre forage. Similar responses would be noted with non-protein nitrogen and non-protein sulphur

TABLE 1

The expected response in digestible energy intake and liveweight of cattle and sheep fed supplements of energy, protein or non-protein nitrogen (NPN) or sulphur (NPS) while grazing forage at different levels of availability, fibre content, protein content, or nitrogen and sulphur.

Forage characteristic	Level (Low (L) or High (H))															
Availability	L								H							
Fibre content	L				H				L				H			
Protein content	L		H		L		H		L		H		L		H	
Ratio N : S	L	H	L	H	L	H	L	H	L	H	L	H	L	H	L	H
Supplement	**Response (Nil (0), Small (+), Medium (++) or Large (+++))**															
Energy	+	+	+	+	++	++	++	++	0	0	0	0	+	+	+	+
Protein	+	+	0	0	+	+	+	+	+++	+++	0	0	++	++	+	+
NPN & NPS	+	+	0	0	0	0	0	0	++	++	0	0	+	+	0	0
NPS	0	+	0	0	0	0	0	0	0	++	0	0	0	+	0	0

supplements, and with sulphur alone when the ratio of nitrogen to sulphur in the forage is high. These supplements of course have no energy value.

MACRO-MINERAL DEFICIENCIES

Depressed growth or ill-health can occur in grazing animals due to deficiencies of some of the macro elements such as sulphur, calcium, phosphorus, sodium and magnesium. Sulphur supplementation has been discussed above, under protein deficiency, in relation to depressed rates of digestion and limited feed intake. Sulphur may be supplemented directly to animals or it may reach the animal indirectly by way of pasture fertilizer. The same means may be used to supply calcium and phosphorus. Superphosphate fertilizer incorporates all three elements. However, there are two modes of action when elements are provided by this means, the primary effect of the element on the animal, and the secondary effect of the element on pasture growth, sward composition and nutritional value (Rees & Minson 1976). In the case of sulphur, 65 percent of a difference in voluntary intake of *Digitaria decumbens* by sheep fed a control diet or a sulphur fertilized diet was attributable to the primary effect (Rees *et al*. 1974). The remaining secondary effect was most probably due to the high proportion of leaf present in the fertilized feed - it is known that there is a positive relationship between leafiness and intake (Laredo & Minson 1973).

In contrast, the action of calcium on forage intake is purely by its action on plant growth. Direct supplementation of sheep with calcium had no effect on feed intake (Rees & Minson 1976), but calcium fertilized *Digitaria decumbens* had an 11.3 percent higher voluntary intake than unfertilized grass. This was apparently due to an increase in digestibility brought about by decrease in the acid detergent fibre fraction of the pasture, despite a three-fold increase in calcium concentration. The effect was therefore secondary, not primary. Calcium does have an important role in animal metabolism however, chiefly in relation to phosphorus metabolism and bone formation.

Deficiencies of phosphorus in cattle have been recognized for many years, particularly in South Africa and northern and eastern Australia. Osteomalacia occurs in cattle, particularly lactating cows, following long dry seasons. Bone chewing is not uncommon and rib-bones are fragile (Rose 1954). Deficiency has also been associated with low reproduction rates (Edye *et al*. 1971), but the physiological connection is not clear. Deficiencies are not as common in sheep as in cattle, possibly due to the fact that cattle more frequently inhabit phosphorus deficient areas. Further, sheep are able to consume proportionally more phosphorus and nitrogen than are cattle when grazing the same pasture, particularly when the proportion of green herbage in the pasture is low (Langlands & Sanson 1976). There are examples of both primary and secondary effects of phosphorus on animal productivity. Primary effects are not common but have been seen in legume-based pastures in both temperate and tropical regions. Ozanne *et al*. (1976), using sheep fed *Trifolium subterraneum* low in phosphorus, and Little (1968), using cattle fed *Stylosanthes humilis* low in phosphorus, found that phosphorus supplementation increased organic matter intake. Secondary effects are more common. Where there is sufficient nitrogen relative to digestible organic matter in a forage the effect of phosphorus fertilization has been to increase the quantity of forage available, to increase the intake of energy and to increase the liveweight gain of livestock (Edye et al. 1972, Winks et al. 1974, Wadsworth & Cohen 1976). In some of these situations the phosphorus status of animals is improved along with increased growth and, in females, oestrus activity. Thus the association of aphosphorosis and infertility can be confounded with energy intake, and in situations when protein deficiency limits energy intake, aphosphorosis can also be confounded with protein deficiency. The effects of high rainfall and high radiation input in some tropical regions produce pasture on low phosphorus soils that is low in protein and phosphorus and is not very digestible.

Sodium and magnesium deficiencies in grazing livestock are not common, although salt is often made available freely, in the belief that animals require sodium, without specific proof of its need. Frequently it is used as a regulator of intake of supplements. Morris (1980), in reviewing the sodium requirements of beef cattle, pointed out that narrow salivary ratios of sodium to potassium are the best measures of sodium deficiency and that production responses have been recorded in some situations (e.g. Murphy & Plasto 1973, Leche 1977). Although there have been cases where responses have not occurred despite broad ratios of sodium to potassium. When the concentration of sodium in forage is less than 0.08 percent supplementation should be considered, although the sodium concentration in drinking water should be taken into account. Grass tetany or hypomagnesaemia, a nervous disorder, is sometimes found in grazing ruminants, particularly lactating cattle. It is a complex disease brought about not only by a deficiency of magnesium, but also possibly involving many other factors, including sodium, potassium, calcium, phosphorus, protein and the ratio of energy to protein (Martens & Rayssiguier 1980). Prophylaxis is difficult as the animal must be supplied with sufficient sodium to stablize the ratio of sodium to potassium in the rumen, sufficient energy must be supplied to keep the rumen ammonia level low and magnesium losses must be decreased by practices such as provision of shelters against cold.

TRACE-MINERAL DEFICIENCIES

There are 15 trace minerals that are currently considered to be essential for animals. Until recently only four - cobalt, copper, iodine and selenium - had been shown to be of economic importance to grazing ruminants. Deficiencies of these elements were recognized because of the catastrophic nature of symptoms in affected animals, for example goitre in iodine deficient lambs, 'falling disease' in copper deficient cattle. Although the past ten years has seen a virtual disappearance of clinical trace element deficiencies in developed countries, there has been a growing awareness of the importance of subclinical deficiencies that arise due to suboptimal supply of a particular element. In this context production responses have been achieved by supplementation with the four elements referred to above even though there have been no overt signs of deficiency. The list of trace elements of known importance to grazing animals has also been increased with the recent discoveries that supplemental manganese (Egan 1972), molybdenum (D.B. Purser, personal communication) and zinc (Egan 1972, Spais & Papasteriadis 1974) have improved animal performance.

With the exception of cobalt the major effect of trace mineral supplementation is not mediated through more efficient rumen function, and hence a stimulation of intake, as is the case of some nutrients aforementioned. For example, iodine functions through its role as a constituent of the hormone thyroxine and selenium as a constituent of the enzyme glutathione peroxidase (Rotruck et al. 1973). When such minerals are in inadequate supply, supplementation aims to supply the missing nutrient to affected tissues so that their metabolism can function normally. Hence, it is sometimes possible to bypass the gut and inject the nutrient in soluble form subcutaneously or directly into muscle or blood. Although selenium is incorporated into rumen micro-organisms (Whanger et al. 1978), and there is evidence that in animals fed a selenium deficient diet there is a shift in the pattern of volatile fatty acid production towards a butyrate fermentation (Hidiroglou & Lessard 1976), it has been found that feed intake by mildly selenium deficient animals is not significantly less than selenium adequate animals. In contrast, cobalt is an integral constituent of vitamin B_{12} which is synthesized in the rumen by several micro-organisms, and supplementary cobalt only relieves clinical symptoms of deficiency after conversion to vitamin B_{12}. An inadequacy of cobalt has a deleterious effect on rumen function by changing the quantitative relationship between the various micro-organisms.

Mild deficiencies of trace minerals in grazing animals are difficult to detect because often the only effect on the animal is reduced growth or reduced fertility. Mineral analysis of the diet is often an unreliable diagnostic procedure. Many interactions between trace minerals, where one mineral affects the absorption or the biological availability of the other, have been documented (see Underwood 1977). Thus an animal may suffer an inadequacy of a certain mineral even though the concentration of that mineral in the diet may appear adequate. The most reliable method of diagnosing subclinical deficiencies is by biochemical analysis of appropriate body tissues and fluids. Values for each mineral of the functional unit of the mineral (thyroxine for iodine, B_{12} for cobalt) below which a deficiency state may exist are given in authorative texts (e.g. Underwood 1977). However, care should be taken in interpretation of analytical results as the evidence suggests that breeds and strains of animals differ in their nutrient requirements for trace elements for the prevention of disorder and possibly for optimum performance (Wiener 1979).

Much trace mineral supplementation has been on an individual animal basis, though application of the minerals in fertilizer is widespread. The method of supplementation of a particular mineral depends to a large extent on whether or not it is stored in the body. Cobalt, copper, selenium, manganese and zinc are stored in a number of tissues, but principally in the liver. Supplementation with those minerals thus need only be at intervals, any deficit between dietary supply and demand being met by body reserves. On the other hand, molybdenum and iodine are not stored to any appreciable extent in the body so supplementation should, at best, be daily. This can be achieved by fertilization of pastures, by licks or by slow release devices located within the body of the animal. Supplementation of minerals which are stored in the body can also be achieved by the same methods or by pulse dosing.

Pulse dosing

Pulse doses are given orally, or by subcutaneous or intramuscular injection. Selenium as sodium selenate or selenite can be administered by either method. Copper is usually given orally as copper sulphate or by injection as copper glycinate. Cobalt may be administered by drench but it is more usual to inject vitamin B_{12}. Zinc and manganese are given orally as inorganic salts.

Pulse dosing has the advantage that such animal receives the desired amount of supplement. It has the disadvantage that it is time consuming and labour intensive. For this reason other forms of supplementation are of greater practical importance.

Salt licks

Salt licks have been used successfully in many countries as a vehicle for mineral supplementation of livestock (Paulson *et al*. 1968, Jenkins *et al*. 1974). The intake of minerals is regulated because animals apparently regulate their intake of salt. The licks should be protected from the weather as toxicities may occur if rain accumulates in hollows of the lick and dissolves toxic quantities of mineral. Leaching of nutrients from the lick may also occur in other situations.

One disadvantage of licks, especially with sheep, is the variable intake between animals. In one study 35 percent of sheep in a flock failed to consume supplement (Wheeler *et al*. 1980). Incorporation of 4 or 8 percent molasses in salt licks has been shown to improve acceptability (Rocks *et al*. 1980), but in other experiments non-consumers have been detected in flocks and herds fed supplements containing molasses (Nolan *et al*. 1974, Lobato & Pearce 1978).

Trace element fertilization

Trace elements may conveniently be mixed with fertilizer when top-dressing of pastures is practised. It is an inefficient method of supplementation, but may be justified economically when pastures are improved and able to support a high stocking rate. Under extensive grazing conditions other methods of supplementation are more suitable.

The frequency of application depends on the mineral and the soil type. The Department of Agriculture in Western Australia recommends that on most soil types copper and zinc should be applied no more than once (Gartrell & Glencross 1969). though there is not total agreement on this recommendation, it is generally accepted that these materials, especially copper, should not be applied more frequently than at five yearly intervals. On the other hand, cobalt and selenium require more frequent application. Skerman *et al*. (1959) found that cattle responded to supplementation by cobalt pellet, even though the pasture was topdressed with cobalt 16 months previously. In the USA, Allaway *et al*. (1966), and in Western Australia (D.W. Peter, personal communication), found that addition of selenium to soil resulted in pasture that protected sheep from selenium responsive disorders for at least two years. However, in New Zealand it appears that it is necessary to reapply selenium fertilizer every 12 months (Hupkens van der Elst & Watkinson 1977).

Absorption of selenium onto clay prills and application to only 5 percent of the paddock area is an innovation which has been shown by Hupkens van der Elst & Watkinson (1977) in New Zealand to result in elevated blood selenium concentrations in sheep for approximately one year. When this procedure was tested in Australia (D.W. Peter, personal communication), no significant difference in blood selenium concentrations was found between sheep grazing pasture when selenized prills applied to 5 percent of the area and sheep grazing pasture fertilized with selenium applied to the whole area. Both treatments had the same amount of selenium applied per hectare and both treatments resulted in elevated blood selenium concentrations for at least two years.

Heavy intraruminal pellets

These pellets are introduced into the reticulo-rumen via the mouth. There they lodge and liberate the desired trace mineral into the rumen solution from which it is absorbed across the gut wall. Cobalt and selenium pellets for sheep and cattle are available commercially, while a zinc pellet (Masters & Moir 1980) has only been used experimentally. The selenium pellet for sheep consists of 5 percent by weight elemental selenium, the sheep cobalt pellet 5 percent cobalt oxide and the zinc pellet 50 percent zinc shot. The other constituent of all pellets is iron, either finely divided or as iron filings, held together by compression. The release of cobalt and selenium from pellets is greatest in the first weeks after administration. The rate of release as determined by selenium concentration in plasma (Hunter et al. 1981) or cobalt excretion in faeces (Dewey et al. 1969) thereafter tapers off and reaches a crude equilibrium. Masters & Moir (1980) state that the zinc pellet functions as a shorted voltaic cell and should release zinc at a constant rate, however there is some preliminary evidence which suggests that the release rate slows due to coating of the pellet with zinc phosphate. There have also been reports of cobalt pellets being coated with calcium phosphate. Therefore in practice a grinder - usually a short length of threaded steel - is administered with the pellets. This abrades the pellet in the rumen, removing the accumulated chemical coating.

The mechanism by which selenium and cobalt enter the rumen was not understood until very recently. Iron was included in the original cobalt pellet prototypes (Dewey et al. 1958), mainly to ensure that the pellet was sufficiently heavy for it not to be regurgitated. Following the success of the cobalt pellet, the selenium pellet was designed similarly (Kuchel & Buckley 1969) with little effort directed at understanding the mechanism of release of selenium. With recent concern about sub-optimal production, workers in Western Australia have been studying the selenium pellet as a model for future intraruminal pellet technology. They have found that grain size of the elemental selenium particles is important in the functional longevity of pellets (Hudson et al. 1981a). Pellets with grains approximately five μm in diameter do not maintain an elevated selenium status in treated sheep as long as pellets with grain size of approximately 50 μm (Fig. 3). The mechanism of release of selenium appears to be electrochemical (Peter et al. 1981) with some selenium being released into the rumen and some being incorporated within the pellet as an iron-selenium compound which is probably iron selenide (Hudson et al. 1981b).

The commercial manufacturers of the cobalt and selenium pellets state that they are effective for one and three years respectively. The zinc pellet in its present initial stage of development, is only effective for about six weeks (Masters & Moir 1980) in part because the animal requires quantitatively more zinc than cobalt or selenium.

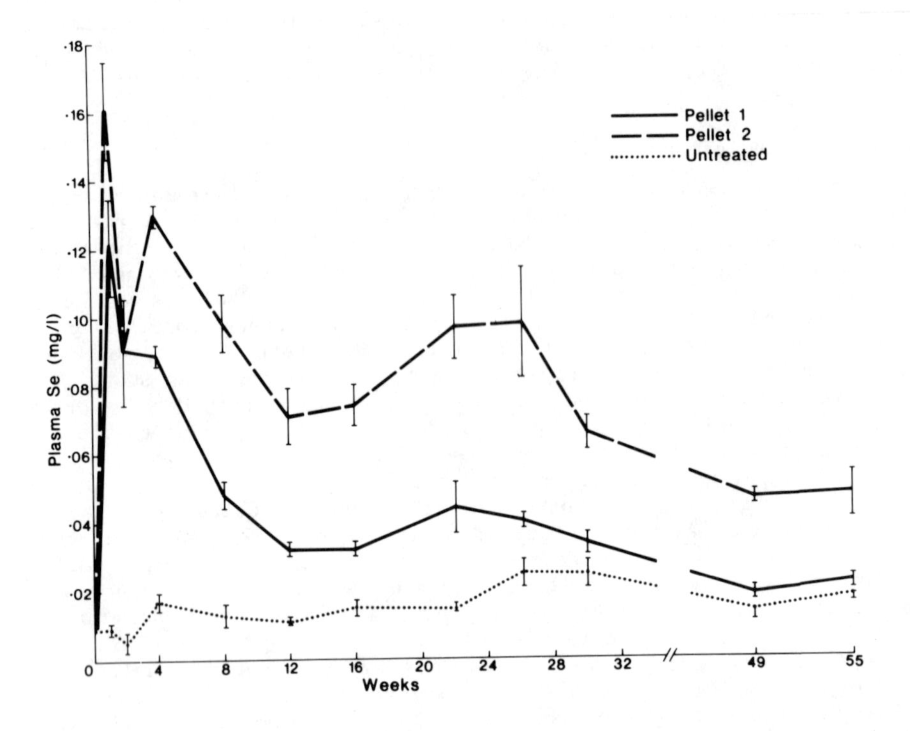

Fig. 3. The plasma selenium concentration (± standard error of mean) of untreated sheep and sheep treated with fine-grained (pellet 1) and coarse-grained (pellet 2) pellets (reprinted from Hudson et al. 1981a, with permission of Australian Academy of Sciences).

Copper oxide needles (Dewey 1977), which are administered by mouth and which lodge in folds of the abomasum, are gaining widespread attention as a method for copper administration. In the acid medium of the abomasum trace amounts of copper are solubilized in ionic form, absorbed from the gut to meet current copper requirements, or stored in the liver. The remainder of the needles are eventually excreted in faeces. The success of this form of supplementation compared with pulse dosing depends on the length of time particles are retained in the abomasum. In sheep the retention time is limited to about 30 days (Dewey 1977), in calves to about 40 days (Suttle 1979), and in mature cattle to at least 100 days (Costigan & Ellis 1980). Thus, although the release rate of copper is not controlled, it is sustained for a period of time and is ideally suited for use with animals which may require supplementation for only a few months each year.

Controlled release devices

Biotechnology has now reached the stage where, in the near future, it will be possible to administer many trace minerals by controlled release intraruminal devices. There are two main types that have been tested experimentally: the iodine capsule (Figure 4a) and the general purpose capsule (Fig. 4b) which can be used for single or multi element supplementation. In the iodine capsule (Ellis 1980) solid iodine is surrounded by a plastic envelope. The rate of release of iodine into the rumen depends on the vapour pressure of iodine (constant at constant temperature) and the solubility of iodine in the plastic envelope. The 'wings' of the capsule open in the rumen and prevent regurgitation. The concept of the general purpose capsule is equally simple. The capsule consists of a hollow plastic cylinder containing a core of matrix. The supplemental mineral is homogeneously dispersed through the matrix. As the carrier matrix dissolves at the open end the core is moved to the orifice by the spring to maintain a constant area of dissolution. Hence an approximately constant release rate is achieved.

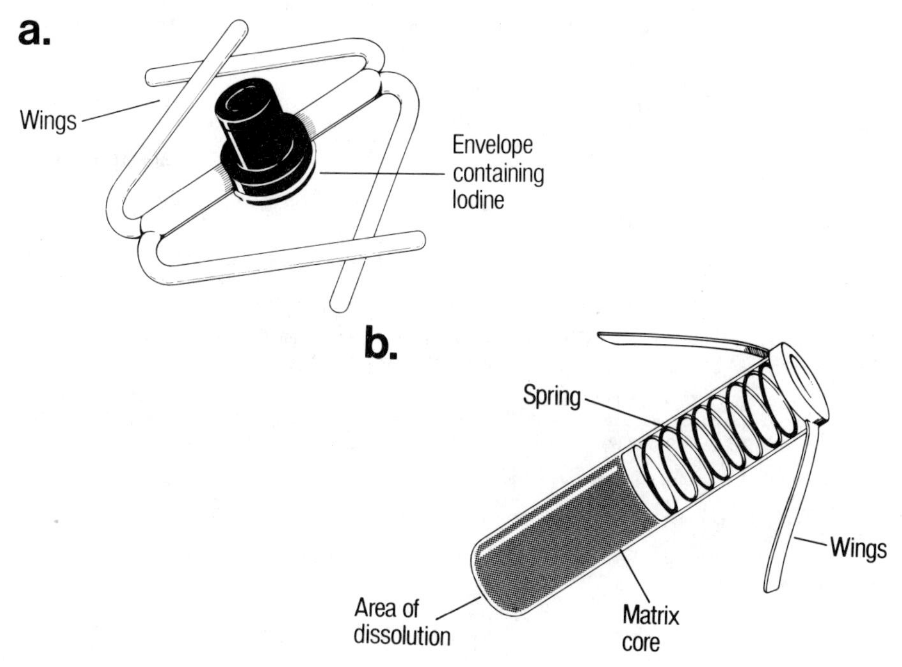

Fig. 4. (a) Illustration of iodine capsule with wing release after administration.
(b) Illustration of general purpose capsule showing wings in expanded state.

CONCLUSION

Deficiencies ranging from trace element to digestible energy limit the growth of grazing livestock. These deficiencies can be overcome with supplements which may provide an alternative energy source, or, ideally, increase the consumption of a forage. In some cases a specific supplement may overcome the limitation but in others more than one nutrient is involved and appropriate remedies must be used. For example, low mineral states can be recognized in animals but supplementation might well be ineffective if, for instance protein is limiting as well. Such circumstances are not unusual as pastures which have senesced are usually low in many nutrients simultaneously. Further, if feed consumption is raised to higher production levels, other nutrient deficiencies then may become apparent because of the increased requirement for that higher production. That is, maximizing production by supplementation is often a step-wise process where, when the nutrient most critically lacking in the diet is supplied, another may then become limiting. The commercial livestock producer must decide at what stage of this step-wise process, supplementation is no longer justified in terms of increased animal production per unit of supplement input.

REFERENCES

Allaway, W.H.; Moore, D.P.; Oldfield, J.E.; Muth, O.H. (1966) Movement of physiological levels of selenium from soils through plants to animals. Journal of Nutrition 88, 411-418.

Allden, W.G.; Geytenbeek, P.E. (1980) Evaluation of nine species of grain legumes for grazing sheep. Proceedings of the Australian Society of Animal Production 13, 249-252.

Allden, W.G.; Jennings, A.C. (1962) Dietary supplements to sheep grazing mature herbage in relation to herbage intake. Proceedings of the Australian Society of Animal Production 4, 145-153.

Allden, W.G.; Tudor, G.C. (1976) Energy and nitrogen supplementation of grazing beef cattle in a Mediterranean environment. Proceedings of the Australian Society of Animal Production 11, 345-348.

Costigan, P.; Ellis, K.J. (1980) Retention of copper oxide needles in cattle. Proceedings of the Australian Society of Animal Production 13, 451.

Deinum, B.; Dirven, J.G.P. (1975) Climate, nitrogen and grass. 6. Comparison of yield and chemical composition of some temperate and tropical grass species grown at different temperatures. Netherlands Journal of Agricultural Science 23, 69-82.

Dewey, D.W. (1977) An effective method for the administration of trace amounts of copper to ruminants. Search 8, 326-327.

Dewey, D.W.; Lee, H.J.; Marston, H.R. (1958) Provision of cobalt to ruminants by means of heavy pellets. Nature, London 181, 1367-1371.

Dewey, D.W.; Lee, H.J.; Marston, H.R. (1969) Efficacy of cobalt pellets for providing cobalt for penned sheep. Australian Journal of Agricultural Research 20, 1109-1116.

Edye, L.A.; Ritson, J.B.; Haydock, K.P.; Griffiths Davies, J. (1971) Fertility and seasonal changes in liveweight of Droughtmaster cows grazing a Townsville stylo-spear grass pastures. Australian Journal of Agricultural Research 22, 963-977.

Egan, A.R. (1972) Reproductive responses to supplemental zinc and manganese in grazing Dorset Horn ewes. Australian Journal of Experimental Agriculture and Animal Husbandary 12, 131-135.

Egan, A.R.; Moir, R.J. (1965) Nutritional status and intake regulation in sheep. I. Effects of duodenally infused single doses of casein, urea, and propionate upon voluntary intake of a low-protein roughage by sheep. Australian Journal of Agricultural Research 16, 437-449.

Ellis, K.J. (1980) Controlled release in animal production. Proceedings of the Australian Society of Animal Production 13, 5-17.

Eyles, A.G. (1979) Forage cultivars released for use in Queensland. Tropical Grasslands 13, 176-177.

Ferguson, K.A.; Hemsley, J.A.; Reis, P.J. (1967) Nutrition and wool growth. The effect of protecting dietary protein from microbial degradation in the rumen. Australian Journal of Science 30, 215-217.

Gartrell, J.W.; Glencross, R.N. (1969) Copper, zinc and molybdenum fertilizers for new land crops and pastures. Western Australian Department of Agriculture, Bulletin No. 3614.

Gulbransen, B. (1974) Utilization of grain supplements by roughage-fed cattle. Proceedings of the Australian Society of Animal Production 10, 74-77.

Hawthorne, W.A. (1980) Lupin grain as a supplement for grazing or penned steers. Proceedings of the Australian Society of Animal Production 13, 289-292.

Henning, P.A.; Linden, Y. van der; Mattheyse, M.E.; Nauhaus, W.K.; Schwartz, H.M.; Gilchrist, F.M.C. (1980) Factors affecting the intake and digestion of roughage by sheep fed maize straw supplemented with maize grain. Journal of Agricultural Science, Cambridge 94, 565-573.

Hidiroglou, M.; Lessard, J.R. (1976) The effect of selenium or vitamin E supplementation on volatile fatty acid content of rumen liquor in sheep fed a purified diet. International Journal for Vitamin and Nutrition Research 46, 458-463.

Hudson, D.R.; Hunter, R.A.; Peter, D.W. (1981a). Effect of Se grain-size on efficacy of selenium pellets. In: Trace element metabolism in animals and man, 4. Editors J.M. Howell, J. Gawthorne, C.L. White. Berlin, Heidelberg, New York, Springer-Verlag. (in press).

Hudson, D.R.; Hunter, R.A.; Peter, D.W. (1981b) Studies with the intraruminal selenium pellet. 2. The effect of grain size of selenium on the functional life of the pellets in sheep. Australian Journal of Agricultural Research (in press).

Hunter, R.A. (1978) The performance of pregnant and lactating Merino ewes and their lambs fed survival rations of wheat grain and sawdust. Australian Journal of Experimental Agriculture and Animal Husbandry 18, 34-40.

Hunter, R.A.; Siebert, B.D. (1980) The utilization of spear grass (*Heteropogon contortus*). IV. The nature and flow of digesta in cattle fed on spear grass alone and with protein or nitrogen and sulphur. Australian Journal of Agricultural Research 31, 1037-1047.

Hunter, R.A.; Siebert, B.D.; Webb, C.D. (1979) The positive response of cattle to sulphur and sodium supplementation while grazing *Stylosanthes guianensis* pastures in north Queensland. Australian Journal of Experimental Agriculture and Animal Husbandry 19, 517-521.

Hunter, R.A.; Peter, D.W.; Hudson, D.R.; Chandler, B.S. (1981) Studies with the intraruminal selenium pellet. 1. Some factors influencing the effectiveness of the pellet for selenium supplementation of sheep. Australian Journal of Agricultural Research (in press).

Hupkens van der Elst, F.C.C.; Watkinson, J.H. (1977) Effect of topdressing pasture with selenium prills on selenium concentration in blood of stock. New Zealand Journal of Experimental Agriculture 5, 79-83.

Jenkins, K.J.; Hidiroglou, M.; Wauthy, J.M.; Proulx, J.E. (1974) Prevention of nutritional muscular dystrophy in calves and lambs by selenium and vitamin E additions to the maternal mineral supplement. Canadian Journal of Animal Science 54, 49-60.

Jones, R.J. (1979) The value of Leucaena leucocephala as a feed for ruminants in the tropics. World Animal Review 31, 13-23.

Kellaway, R.C. (1980) Improving the nutritive value of low quality forages. In: Advances in animal nutrition. Editor D.J. Sarrell. Armidale, Australia, University of New England Publication Unit, pp.10-19.

Kempton, T.J.; Nolan, J.V.; Leng, R.A. (1977) Principles for the use of non-protein nitrogen and bypass protein in diets of ruminants. World Animal Review 22, 2-10.

Kuchel, R.E.; Buckley, R.A. (1969) The provision of selenium to sheep by means of heavy pellets. Australian Journal of Agricultural Research 20, 1099-1107.

Langlands, J.P.; Bowles, J.E. (1976) Nitrogen supplementation of ruminants grazing native pastures in New England, New South Wales. Australian Journal of Experimental Agriculture and Animal Husbandry 16, 630-635.

Langlands, J.P.; Sanson, J. (1976) Factors affecting the nutritive value of the diet and the composition of rumen fluid of grazing sheep and cattle. Australian Journal of Agricultural Research 27, 691-707.

Laredo, M.A.; Minson, D.J. (1973) The voluntary intake, digestibility, and retention time by sheep of leaf and stem fractions of five grasses. Australian Journal of Agricultural Research 24, 875-888.

Leche, T.F. (1977) Effects of sodium supplement on lactating cows and their calves on tropical native pastures. Papua New Guinea Agricultural Journal 28, 11-18.

Little, D.A. (1968) Effect of dietary phosphate on the voluntary consumption of Townsville lucerne (Stylosanthes humilis) by cattle. Proceedings of the Australian Society of Animal Production 7, 376-380.

Lobato, J.F.; Pearce, G.R. (1978) Variability in intake of supplements by grazing sheep. Proceedings of the Australian Society of Animal Production 12, 164.

Martens, H.; Rayssiguier, Y. (1980) Magnesium metabolism and hypomagnesaemia. In: Digestive physiology and metabolism in ruminants. Editors Y. Ruckebusch and P. Thivend. Lancaster, U.K., M.T.P. Press Ltd., pp. 447-468.

Masters, D.G.; Moir, R.J. (1980) Provision of zinc to sheep by means of an intraruminal pellet. Australian Journal of Experimental Agriculture and Animal Husbandry 20, 547-551.

McIvor, J.G. (1979) Seasonal changes in nitrogen and phosphorus concentrations and in vitro digestibility of Stylosanthes species and Centrosema pubescens. Tropical Grasslands 13, 92-97.

Minson, D.J. (1982) Effects of chemical and physical composition of herbage eaten upon intake. In: Nutritional limits to animal production from pastures. Editor J.B. Hacker. Farnham Royal, U.K., Commonwealth Agricultural Bureaux pp. 167-182.

Morris, J.G. (1980) Assessment of sodium requirements of grazing beef cattle : a review. Journal of Animal Science 50, 145-152.

Murphy, G.M.; Plasto, A.W. (1973) Liveweight responses following sodium chloride supplementation of beef cows and their calves grazing native pasture. Australian Journal of Experimental Agriculture and Animal Husbandry 13, 369-374.

Nolan, J.V.; Ball, F.M.; Murray, R.M.; Norton, B.W.; Leng, R.A. (1974) Evaluation of a urea-molasses supplement for cattle. Proceedings of the Australian Society of Animal Production 10, 91-94.

Norman, M.J.T. (1963) Dry season protein and energy supplements for beef cattle on native pastures at Katherine, N.T. Australian Journal of Experimental Agriculture and Animal Husbandry 3, 280-283.

Ozanne, P.G.; Purser, D.B.; Howes, K.M.W.; Southey, I. (1976) Influence of phosphorus content on feed intake and weight gain of sheep. Australian Journal of Experimental Agriculture and Animal Husbandry 16, 353-360.

Partridge, I.J.; Ranacou, E. (1974) The effects of supplemental Leucaena leucocephala browse on steers grazing Dichanthium caricosum in Fiji. Tropical Grasslands 8, 107-112.

Paulson, G.D.; Broderick, G.A.; Baumann, C.A.; Pope, A.L. (1968) Effect of feeding sheep selenium fortified trace mineralized salt: effect of tocopherol. Journal of Animal Science 27, 195-202.

Peter, D.W.; Mann, A.W.; Hunter, R.A. (1981) Effect of Zn-containing pellets on selenium pellet function. In: Trace element metabolism in animals and man, 4. Editors J.M. Howell, J. Gawthorne and C.L. White. Berlin, Heidelberg, New York, Springer-Verlag. (in press).

Pigden, W.J.; Bender, F. (1972) Utilization of lignocellulose by ruminants. World Animal Review 4, 7-10.

Redman, R.G.; Kellaway, R.C.; Leibholz, Jane (1980) Utilization of low quality roughages : effects of urea and protein supplements of differing solubility on digesta flows, intake and growth rate of cattle eating oaten chaff. British Journal of Nutrition 38, 1-12.

Rees, M.C.; Minson, D.J. (1976) Fertilizer calcium as a factor affecting the voluntary intake, digestibility and retention time of pangola grass (Digitaria decumbens) by sheep. British Journal of Nutrition 36, 179-187.

Rees, M.C.; Minson, D.J.; Smith, F.W. (1974) The effect of supplementary and fertilizer sulphur on voluntary intake, digestibility, retention time in the rumen, and site of digestion of pangola grass in sheep. Journal of Agricultural Science, Cambridge 82, 419-422.

Rocks, R.L.; Wheeler, J.L.; Hedges, D.A. (1980) Effect of molasses, aniseed and other additives on the acceptability of salt licks to sheep. Proceedings of the Australian Society of Animal Production 13, 293-296.

Romero, A.; Siebert, B.D. (1980) Seasonal variations of nitrogen and digestible energy intake of cattle on tropical pasture. Australian Journal of Agricultural Research 31, 393-400.

Rose, A.L. (1954) Osteomalacia in the Northern Territory. Australian Veterinary Journal 30, 172-177.

Rotruck, J.J.; Pope, A.L.; Somther, H.E.; Swanson, A.B.; Hafeman, D.G.; Hoekstra, W.G. (1973). Selenium : biochemical role as a component of glutathione perioxidase. Science 179, 588-590.

Siebert, B.D.; Kennedy, P.M. (1972) The utilization of spear grass (*Heteropogon contortus*) I. Factors limiting intake and utilization by cattle and sheep. Australian Journal of Agricultural Research 23, 35-44.

Skerman, K.D.; Sutherland, A.K.; O'Halloran, M.W.; Bourke, J.M.; Munday, B.L. (1959) The correction of cobalt or vitamin B_{12} deficiency in cattle by cobalt pellet therapy. American Journal of Veterinary Research 79, 977-984.

Spais, A.G.; Papasteriadis, A.A. (1974) Zinc deficiency in cattle under Greek conditions. In: Trace element metabolism in animals, 2. Editors W.G. Hoekstra, J.W. Suthi, H.E. Ganther and W. Mertz. Baltimore, U.S.A. University Park Press, pp. 628-631.

Underwood, E.J. (1977) Trace elements in human and animal nutrition. New York, Academic Press Inc., 543 pp.

Wadsworth, J.C.; Cohen, R.D.H. (1976) Phosphorus utilization by ruminants. In: Reviews in Rural Science III. Armidale, Australia, University of New England Publication Unit, pp. 143-154.

Weston, R.H.; Hogan, J.P. (1973) Nutrition of herbage-fed ruminants. In: The pastoral industries of Australia. Editors G. Alexander and O.B. Williams. Sydney, Australia, Sydney University Press, pp. 233-268.

Whanger, P.D.; Weswig, P.H.; Oldfield, J.E. (1978) Selenium, sulphur and nitrogen levels in ovine rumen micro-organisms. Journal of Animal Science 46, 515-519.

Wheeler, J.L.; Rocks, R.L.; Hedges, D.A. (1980) Intake of mineral supplements and productivity of sheep grazing sorghum. Proceedings of the Australian Society of Animal Production 13, 297-300.

Weiner, G. (1979) Review of genetic aspects of mineral metabolism with particular reference to copper in sheep. Livestock Production Science 6, 223-232.

Williams, R.D. (1976) Supplementation of beef cattle on native pasture. M.Sc. thesis. University of New England, Armidale, Australia.

Winks, L.; Laing, A.R. (1972) Urea, phosphorus and molasses supplements for grazing beef weaners. Proceedings of the Australian Society of Animal Production 9, 253-257.

Winks, L.; Lamberth, F.C.; Moir, K.W.; Pepper, P.M. (1974) Effects of stocking rate and fertilizer on the performance of steers grazing Townsville stylo-based pasture in north Queensland. Australian Journal of Experimental Agriculture and Animal Husbandry 14, 146-154.

MODIFICATION OF RUMEN FERMENTATION

R.A. LENG

Department of Biochemistry and Nutrition, University of New England, Armidale, NSW 2351, Australia.

ABSTRACT

Modification of rumen fermentation is discussed in relation to the theoretical aspects of rumen function. The objectives of manipulation are discussed in relation to fermentative heat losses, methane production and changing proportions of volatile fatty acids. Emphasis is given to the potential for manipulation of microbial protein availability to ruminants. The maximum theoretical efficiency of microbial cell yield is discussed in relation to the maintenance ATP requirement and microbial cell turnover in the rumen. The role of protozoa in the rumen is discussed in relation to their possible effect on microbial cell yields. Emphasis is given to this aspect since in animals grazing fresh pastures protozoal biomass may be large. Distinctions are made between the large and small ciliate protozoa and their role in rumen function. The data is related to the grazing system and the scope for manipulation is discussed.

INTRODUCTION

Research into the modification of rumen metabolism has often attempted to manipulate a single factor in isolation. For instance a decrease in methane production is attempted without consideration of the train of changes that must result in a disturbed rumen ecosystem before it comes to another equilibrium. A modification of the rumen by administration of a chemical in the feed is followed by a period of instability which may obscure or remove any subsequent beneficial effects.

In general, researchers have failed to recognize that the first factor to limit production will govern the outcome of any attempt to change rumen function. For example, increasing the metabolizable energy available from a ration will be of no avail if the primary limitation to production is the supply of essential amino acids. Similarly, increasing the amino acid availability to ruminants will not result in increased production, if dietary energy or any other nutrient is the primary limitation to production.

The aims of research on rumen manipulation have been mainly to increase productivity and efficiency of utilization of feed, whereas it may often be more economic to reduce the level of a costly supplement such as protein, for example by increasing microbial protein synthesis from non-protein nitrogen (NPN) and maintaining production at the same level.

Other points to be taken into consideration are that feed intake is often a dominant variable and reduced feed intake may overshadow a manipulation that is potentially beneficial. Procedures that stimulate feed intake are likely to have large effects on production as also is technology that increases the proportion of dietary nutrients that avoid rumen fermentation (termed here dietary bypass nutrients).

In the context of this symposium manipulation of rumen fermentation must be related to the grazing ruminant. However, there is a dearth of information on manipulation of the rumen in grazing animals and discussion of rumen fermentation in animals under pen conditions and on typical concentrate diets is necessary in order to understand the objectives that should be applied to grazing ruminants.

Since the efficiency of feed utilization in grazing animals is lower than that in animals on grain based diets, there appears to be considerably more scope for responses to manipulation under field conditions.

MANIPULATION: OBJECTIVES, THEORETICAL ASPECTS AND CONSEQUENCES:

Rumen manipulative procedures may be aimed at achieving the following:

(1) increased feed intake

(2) decreased losses of feed energy and protein in the rumen by causing dietary nutrients to bypass the rumen, or decreasing losses of fermentative heat and methane

(3) increased efficiency of rumen microbial protein synthesis.

Depending on the diet objectives (2) and (3) may result in increased feed intake which will have by far the greatest effect.

Feed intake

Appetite control in ruminants is complex (see Baile & Mayer 1970, Baile 1975) and depends on diet, climate and physiological state of the animal. Until recently the intake of low digestibility forages has generally been assumed to be primarily controlled by bulk-distension of the rumen. However, in a number of situations with relatively indigestible feeds, increases in feed intake have resulted from feeding bypass proteins (see Leng *et al*. 1977, Hennessy *et al*. 1981), by subjecting animals to cold stress (Kennedy & Milligan, 1978) or by changing physiological states such as from pregnancy to lactation (Weston 1979). Factors affecting intake have been discussed earlier in this symposium by Hodgson (1982), Minson (1982) and Weston (1982).

Although the mechanisms which control feed intake are complex, three factors seem to predominate in practical situations. These are:

(1) distension of the rumen, as a result of slow fermentation of roughage;

(2) the effect of high concentrations (or absorption rates) of volatile fatty acids in the rumen;

and (3) the amounts of amino acids absorbed from the small intestine.

(1) and (3) are probably more important in grazing ruminants and (2) may be dominant in animals on high energy/starch based diets.

Under some circumstances factor (3) modifies the effects of (1) and (2) as food intake of cattle with very high concentrations of volatile fatty acids in the rumen is stimulated on high energy/low protein diets by supplements of bypass protein (see Ørskov et al. 1973, Preston & Willis 1974, Meyreles et al. 1979, Clay & Satter 1979, Preston & Leng 1980). In addition supplementation with dietary bypass protein has increased the intake of low quality forage (46 percent digestible) by cows where rumen distension was believed to be the primary limitation to feed intake. Point (3) appears to be a major factor in the control of feed intake of ruminants, particularly on low protein diets and therefore the major effector of the efficiency of utilization of feed intake. Feed intake on low protein diets may therefore be influenced markedly by the efficiency of microbial protein production in the rumen and this appears to have scope for manipulation. This is particularly so since the mean level of microbial protein production in the rumen of 30 g N/kg of organic matter apparently fermented (FOM app)* (see Roy et al. 1977) is 30-40 percent of the theoretical maximal microbial protein yield (see Stouthamer 1979). Similarly any manipulation that decreases proteolysis in the rumen and/or amino acid deamination will increase feed intake under specified conditions.

Bypass protein in ruminant diets may also affect the hormonal balance of ruminants, leading to higher growth hormone levels in blood (see for discussion Oldham 1980), this can lead to increased efficiency of utilization of amino acids (Bines et al. 1980). An increased availability of microbial amino acids on any diet could be expected to have similar effects.

Manipulation of rumen function to give rise to greater quantities of amino acids available for absorption is of major practical significance as discussed later.

Increasing efficiency of utilization of energy and protein

There are a number of means of increasing metabolizable energy from a feed. These include

(1) avoiding fermentation of intestinally digestible carbohydrate and protein largely through physical preparation of a supplement

(2) decreasing amino acid or protein degradation by inhibiting specific deaminase or proteolytic activity in the rumen

(3) decreasing losses of energy by increasing 'hydrogen' utilization in end product production in the rumen (seen as decreased methane production).

Bypass energy and protein

Although the major advantages of ruminant digestion are the ability to digest, through microbial fermentation, structural components of plant materials and to use ammonia for microbial protein synthesis, it has the disadvantage of fermenting intestinally-digestible carbohydrates and protein. Carbohydrates yield 11-30 percent more energy to the animal when digested postruminally than when fermented in the rumen. In addition fermentation of proteins in the rumen reduces amino acid availability.

*FOM is used to indicate the organic matter fermented to volatile fatty acids and microbes, FOM app is the apparent fermented organic matter in the rumen and is not corrected for the microbial cells leaving the rumen.

An increase in the proportion of digestible carbohydrates which bypass rumen fermentation will benefit the animal provided no other nutrients, in particular essential amino acids, are deficient. If these carbohydrates are absorbed as, or are readily converted to, glucose the energy is used with further efficiency since the need for gluconeogenesis from propionate (which requires ATP) is reduced.

There are few known methods for increasing starch flow to the intestines. Fine grinding of grain supplements fed to sheep on roughage diets appears to allow starch to enter the duodenum (Morgan 1977); maize appears to bypass rumen fermentation to a greater extent than barley or other cereals (Waldo 1973); on high sugar based diets given to cattle, supplements of rice-starch apparently pass totally to the duodenum (Elliott et al. 1978, Ferreiro et al. 1979). It seems likely that antibiotic chemicals may decrease fermentation rate and allow some carbohydrate to enter the intestines of ruminants on concentrate diets.

Manipulation of protein fermentation

There are a number of chemicals which inhibit either proteolysis or deaminase activity in the rumen. These chemicals have been used in attempts to increase the availability of dietary amino acids to the animal. The concept is simple. If amino acids are fermented in the rumen they are converted to volatile fatty acids and ammonia. The production of ATP from fermentation of amino acids is low, probably half that provided by fermentation of carbohydrate (see Demeyer & Van Nevel 1979). From 100 g amino acid fermented about 1.5 moles of ATP may be available and at a Y-ATP* of 12, about 18 g of microbial cell (about 9 g protein) may be synthesized.

Rumen organisms are capable of efficient utilization of ammonia for amino acid synthesis. There is, however, some suggestion that a slow release of amino acids into the rumen may increase the efficiency of microbial cell synthesis (Maeng et al. 1976) and even increase digestibility of high concentrate diets (Oldham 1980). Thus there is potential for increasing the availability of dietary amino acids to both the microbes (thereby increasing microbial protein synthesis) and the animal by processing feeds to decrease protein fermentation; this should result in more highly efficient utilization of dietary protein.

A further approach is to decrease amino acid degradation in the rumen using chemicals (e.g. diaryliodenuum compounds, see Chalupa 1980) that inhibit proteolysis and/or deaminase activity. The addition of these chemicals may increase the quantity of amino acids that enter the small intestine and through increased concentrations of amino acids in the rumen increase the efficiency of microbial protein synthesis as postulated by Oldham (1980). Both effects potentially increase efficiency of utilization of the end products of digestion in the animal (Oldham 1980, Gordon 1980).

Energy losses in the rumen

The energy transactions occurring in the rumen are described in Fig. 1.

* Y-ATP is defined as the g of dry cells produced per mole of ATP available.

TABLE 1

The composition of bacterial cells and the ATP requirements for microbial cell synthesis (adapted from Stouthamer 1979).

Macromolecule synthesis	Amount g/100 g cells	ATP Requirements (10^4 mole/g cells formed) Glucose	Pyruvate	Acetate
Polysaccharide	16.6	20.6	71.8	92
Protein	52.4	204.9	339.4	427
Lipid	9.4	1.4	27	50
RNA	15.7	43.7	71.2	101
DNA	3.2	24.4	29.8	19
Transport of nutrients into cells (carbon ammonium ions, K and PO_4^{-3}		52.0	200	306*
Total ATP requirements		347	740	995
Y-ATP (g drycells/mole ATP)		28.8	13.5	10.0

*A large proportion of the transport costs of nutrients is the transport of ammonia, acetate and pyruvate into cells.

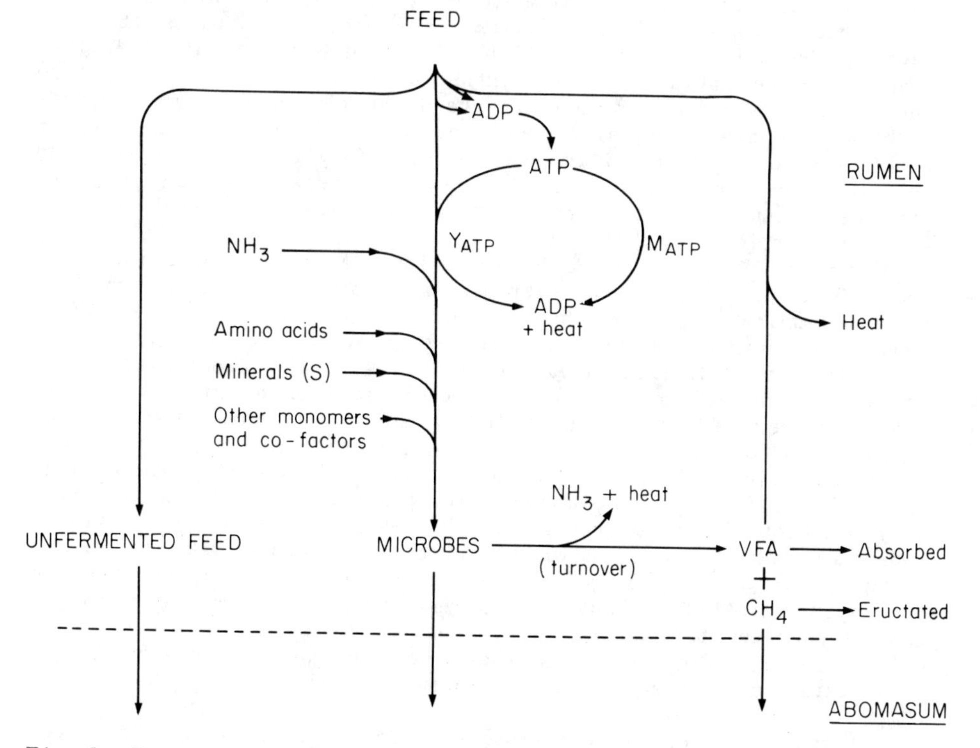

Fig. 1. Energetics of rumen fermentation

In bypassing rumen fermentation with intestinally digestible carbohydrate the energy losses associated with the following are not apparent: heat of fermentation, the formation of methane and the synthesis of microbes.

Stoichiometry of rumen fermentation and microbial cell synthesis

Rumen microorganisms use ATP for essentially two purposes, for synthesis of cells and to provide the energy for maintenance of the microorganisms.

The anaerobic conditions of the rumen limit the availability of ATP and the majority of the energy of the substrate is retained in the end products of the fermentation (which include volatile fatty acids, methane and microbes). The ATP available for microbial growth is dependent on the extent of the ATP dissipation to meet the maintenance requirements of the organisms. The efficiency of utilization of the ATP for growth may also depend on the monomers available to be used as 'building blocks' for cellular material.

Substrates for microbial amino acid synthesis

The synthesis of cellular material of microorganisms in the rumen has not been studied extensively. As protein is 40-60 percent of the dry matter of microbial cells, the synthesis of amino acids and their polymerisation are the major reactions requiring ATP. The mechanisms of synthesis of amino acids in rumen microbes are not clearly understood. Ferredoxin-dependent carboxylation reactions involved in the synthesis of 2-keto acids (major precursors of amino acids in rumen bacteria) have been shown to exist in some strains of rumen bacteria by Allison & Peel (1971). This led Sauer _et al_. (1975) to examine the potential for the synthesis of amino acids from acetate and propionate. These researchers used the continuous rumen culture technique of Slyter _et al_. (1964) and added ^{14}C labelled acetate or propionate to a culture of rumen organisms. They isolated bacteria and measured incorporation of radioactivity into protein and amino acids. From these studies they suggested that "the rumen flora synthesized a large proportion of their amino acids _de novo_ from volatile fatty acids (Sauer _et al_., 1975). Studies with ^{15}N ammonia have indicated that 60-70 percent of the nitrogen of rumen microbes arises from ammonia, which supports the concept of _de novo_ synthesis of a substantial proportion of bacterial amino acids (see Pilgrim _et al_. 1970, Mathison & Milligan 1971, Nolan & Leng 1972). Stouthamer (1979) calculated the ATP requirements for the formation of microbial cells from a number of substrates including acetate, glucose and pyruvate and showed that synthesis commencing from acetate is approximately one third as efficient as that commencing from glucose (Stouthamer 1979) (see Table 1). Thus the substrate for synthesis of cellular material of rumen bacteria is important. If the results of Sauer _et al_. (1975) indicate acetate as a major precursor this would make it highly unlikely that microbial cell synthesis could be much higher than 7 g dry cells/ mole ATP. Thus two extreme models of rumen fermentation can be considered. One in which acetate is the major precursor of cell material (Fig. 2b) and the other in which intermediates in fermentation of glucose are used for polymer synthesis (Fig. 2a).

A re-examination of the results of Sauer _et al_. (1975) indicates that only a small proportion of the total amino acids in rumen bacteria is actually synthesized from acetate (i.e. 1 to 6 percent).

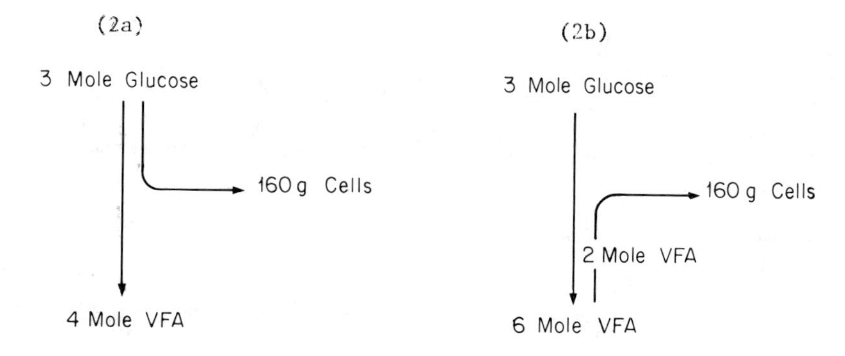

Fig. 2. Alternative models for rumen fermentation.

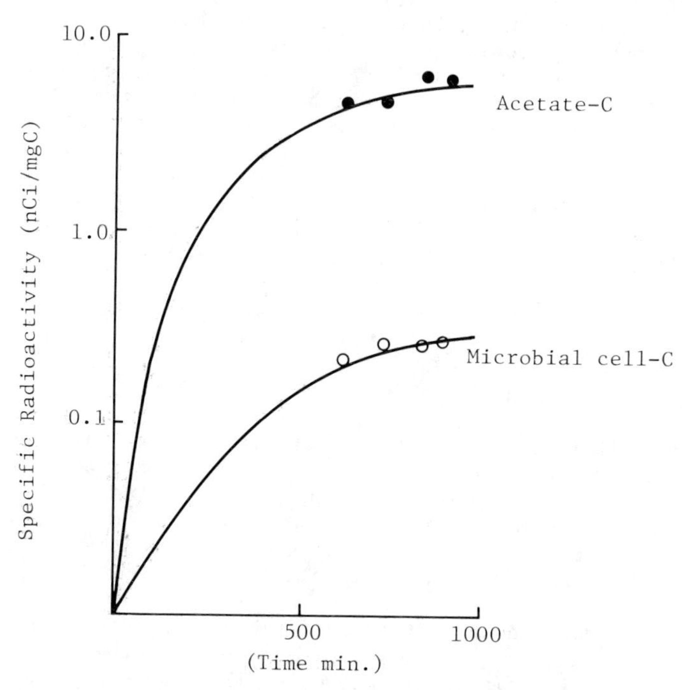

Fig. 3. The incorporation of ^{14}C into bacteria during a constant infusion of ^{14}C acetate into the rumen of a sheep (from Cottle 1981).

In studies in this laboratory (Cottle 1981) continuous infusions of ^{14}C acetate (or ^{14}C-propionate) were made into the rumen of sheep on a lucerne chaff diet and the specific radioactivities of rumen acetate-C (or propionate-C) and microbial-C were estimated (see Fig. 3). The synthesis of microbial carbon from acetate and propionate carbon was about 4 percent and 2 percent respectively.

There appears to be always a small number of secondary fermentation bacteria (or sludge fermentors) in the rumen (M.P. Bryant, personal communication, 1980) which primarily utilize acetate in cell synthesis. However the majority of cell components appear to be synthesized from intermediates produced in the fermentative pathways shown in Fig. 2a.

The composition of rumen microorganisms reported by different authors has been quite variable depending on diet (see Czerkawski 1976). However, for the purposes of this discussion I have accepted that rumen microorganisms are similar to *Escherichia coli* for which there are detailed data. The composition of *E. coli* is shown in Table 1.

Assuming that the monomers needed for the synthesis of the major components of microbes arise from intermediates of fermentation, some ATP will be made available in the reactions leading to their synthesis. To calculate this ATP a series of assumptions is necessary:

(1) that no ATP is available when glucose is used for polysaccharide synthesis (2) that the majority of the amino acids are synthesized from pyruvate, oxaloacetate, α-oxoglutarate, 3-phosphoglycerate or phosphoenol pyruvate (see Gottschalk 1979), (3) that only half the reducing equivalents that arise when pyruvate is produced are oxidized in reactions leading to synthesis of cellular material the other half are converted to methane; and (4) that RNA, DNA and lipids are synthesized from simple carbon and nitrogen compounds.

Synthesis of amino acids may be represented as follows:

$$\text{CHO} + 2\text{NAD} + 2\text{ADP} \longrightarrow 2\text{ pyruvate} + 2\text{ NAD(H}_2) + 2\text{ ATP}$$

$$2\text{ pyruvate} + 2\text{ NH}_4^+ + 2\text{NAD(H}_2) \longrightarrow 2\text{ amino acids} + 2\text{ NAD}^+$$

- -

$$\text{Overall, CHO} + 2\text{ ADP} \longrightarrow 2\text{ amino acids} + 2\text{ ATP} \quad \ldots \quad \text{Equation 1}$$

It is considered that in the conversion of carbohydrate to amino acids no free hydrogen is generated, whereas in the synthesis of RNA, DNA and other cellular components hydrogen is liberated to be converted to methane. It appears from the composition of microbial cells that about 0.25 mole of methane is generated per mole of carbohydrate incorporated into microbial organic matter.

Although there are many uncertainties associated with the overall availability of ATP it appears that for each mole of carbohydrate incorporated into microbial cells about 2.25 mole ATP is generated and 0.25 mole of methane is produced.

A model of fermentation in the rumen

For the purposes of the present discussion a model for a 200 kg steer will be used to illustrate the various arguments. The steer consumes 4 kg (25 mole anhydroglucose) of organic matter which is fermented in the rumen.

It is assumed (1) that a mole of carbohydrate gives rise to 2 mole acetate, or 2 mole of propionate or 1 mole of butyrate and (2) that the production of individual volatile fatty acids is proportional to its concentration (Leng & Brett 1966) (3) that one third of the organic matter fermented is converted to microbial cells and (4) that the moles ATP generated per mole end product produced are acetate 2, butyrate 3, propionate 3 and methane 1 (Isaacson et al. 1975).

The equation for fermentation is as follows:

$$16.7 \text{ CHO} \rightarrow 21 \text{ HAc} + 6\text{H Prop} + 3\text{H But} + 7.5 \text{ CH}_4 + 78 \text{ ATP}$$

$$8.3 \text{ CHO} \rightarrow 1.4 \text{ polysaccharide} + 13.8 \text{ pyruvate} + 2.0 \text{ CH}_4 + 17 \text{ ATP}$$

Overall

$$25 \text{ CHO} \rightarrow 21 \text{ HAc} + 6\text{H Prop} + 3\text{H But} + 9.5 \text{ CH}_4 + 1300 \text{ g dry cells} \quad \text{Equation 2}$$

In the example, one third of the carbohydrate is incorporated into cells and about 1300 g dry cells (assuming that 1 g fermentable carbohydrate is converted to 1 g bacterial cells) are produced for a Y-ATP of about 14. At Y-ATP higher than this volatile fatty acid production is decreased and cell production is increased (see later). The model is represented by the following diagram (Fig. 4).

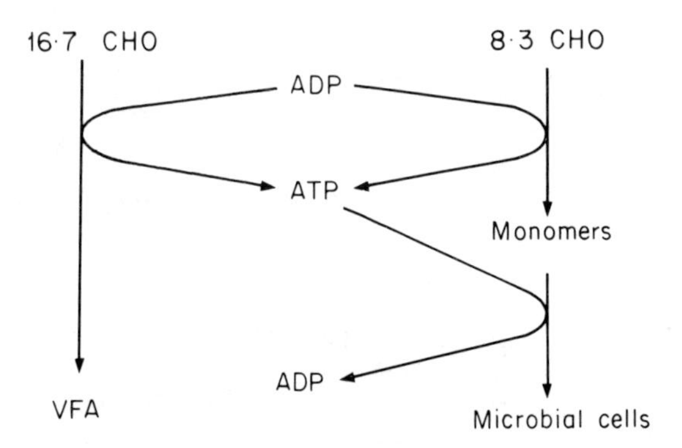

Fig. 4. Flow of carbohydrate into VFA and microbial cells in rumen fermentation

Using this model but allowing Y-ATP to vary, the percentage of the carbohydrate metabolized in the rumen that is converted to microbes or volatile fatty acids and methane is shown in Fig. 5. At Y-ATP 0 the microbial cell production should be zero but in the figure some discrepancy from this is apparent indicating that microbial cell production is overestimated by the model in Fig. 4. The point to be stressed here is that, depending on the efficiency of utilization of ATP, the carbohydrate converted to microbial cells can approach the amount fermented to volatile fatty acids. This graph was calculated from the above stoichiometry assuming that 1 g cells is formed from 1 g of anhydro-carbohydrate fermented but it fits exactly the value found by Baldwin (1970) using a detailed biochemical approach.

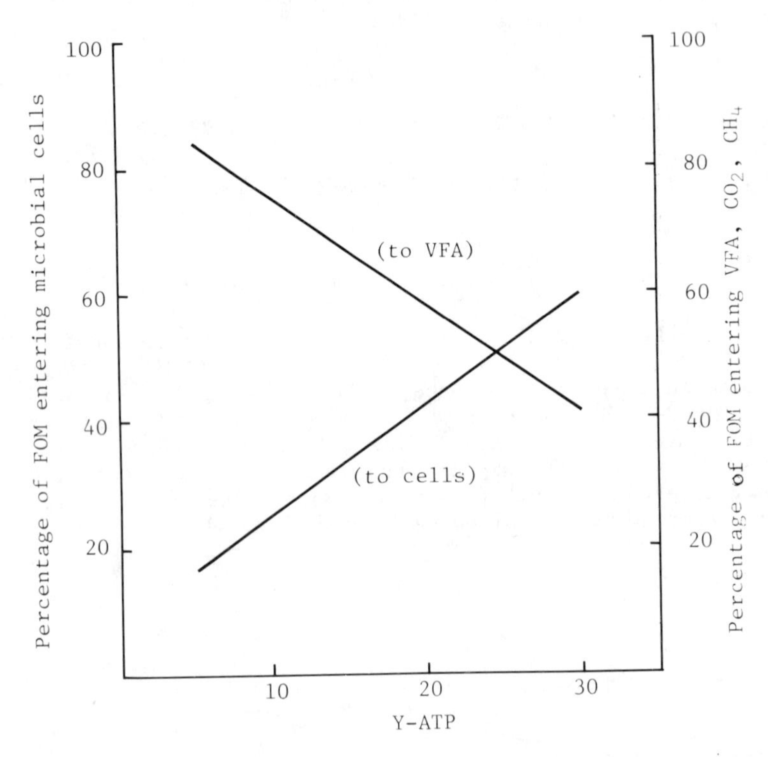

Fig. 5. Relationship between Y-ATP and the percentage of FOM entering cells or VFA, CO_2 and CH_4.

Heat of fermentation

Energy losses as heat in the rumen result from the conversion of carbohydrate to volatile fatty acids and microbial cells. The extent of energy losses as heat in the rumen is always small but depends to some extent on the Y-ATP. In order to demonstrate this, the division of digested carbohydrate between volatile fatty acids production and cell synthesis was estimated from Fig. 5. The production of volatile fatty acids, CH_4, and cells was then calculated from the stoichiometry given in equations 1 and 2 and the energy balance in the rumen calculated (see Table 2). These calculations show that as the efficiency of cell synthesis increases there is a decrease in methane production and heat of fermentation. Thus any manipulation which increased microbial cell yield may lead to an increased availability of metabolizable energy. However microbes are between 79 and 95

TABLE 2
The influence of the efficiency of microbial cell synthesis on the heat of fermentation in the rumen and the metabolisable energy availability to an average steer consuming 4 kg FOM

	Y-ATP 8		Y-ATP 14		Y-ATP 19		Y-ATP 25	
	Moles	MJ	Moles	MJ	Moles	MJ	Moles	MJ
Carbohydrate fermented - to VFA	19.75	55.5	16.66	46.8	14.5	40.8	12.4	34.9
- to cells	5.25	14.8	8.33	23.4	11.5	29.5	12.1	35.4
VFA Produced								
H Ac	25.3	22.2	21.6	18.7	18.5	16.3	15.9	14.0
H Prop	7.1	10.9	6.0	9.2	5.2	8.0	4.5	6.9
H But	3.6	7.8	3.0	6.5	2.6	5.7	2.2	4.8
Methane	10.1	9.4	9.4	8.5	8.9	8.0	8.5	7.6
Total methane + VFA		50.3		42.9		38.0		33.3
Approximate energy in microbes*		13.5		22.2		28.1		33.7
Microbial cells produced (kg)		0.83		1.33		1.68		2.02
Heat of fermentation (MJ)		6.4		5.1		4.3		3.1
Indigestible energy of microbes (assuming 80% digestible) (MJ)		2.7		4.4		5.6		6.7
Total energy loss (MJ)		9.1		9.6		9.9		9.8
Metabolisable Energy (M.E.)(MJ)		61.2		60.7		60.4		60.5

*Assuming 20.9 kj/g of organic matter

percent digestible in the intestines (Roy *et al*. 1977, Hagemeister *et al*. 1980). The extra energy losses in the faeces removes any energetic advantage of such a manipulation, but protein to energy ratios in absorbed nutrients would be increased markedly. If, however, rumen microorganisms were 100 percent digestible, as suggested by recent ^{15}N studies (J.V. Nolan, personal communication), then an increase in Y-ATP would result in a significant increase in energy to the animal. Hagemeister *et al*. (1980) reported convincing evidence for microbial protein being at least 95 percent digestible.

Changes in proportions of volatile fatty acids

Changing proportion of volatile fatty acids (in particular increasing propionic acid proportions) is likely to have only a small effect on the yield of microbial cells. The higher ATP yield per mole of propionate relative to acetate and/or butyrate produced in fermentation is offset to a large extent by the increased requirement for monomers for cell synthesis. The small increase in synthesis of cells reduces volatile fatty acid production and decreases methane production.

The only point to make here is that changes in proportions of volatile fatty acids are likely to have only a minor effect on microbial cells synthesis and this will be small relative to changes in maintenance ATP requirement (M-ATP) (see later) and/or microbial cell turnover in the rumen (see later).

ATP yield to the animal from an increased availability of propionate in total volatile fatty acids

The beneficial effects of a 10 percent increase in propionate production in the rumen, at the expense of acetate can be calculated in terms of ATP available to the tissues of the animal. In the example given (equation 2) acetate and propionate production rates are 21 and 6 mole/ day respectively. If this is changed to 17 and 10 mole/day respectively the availability of ATP to the animals tissues can be calculated as follows: Assuming that oxidation of a mole of acetate and propionate gives rise to 10 and 18 mole ATP respectively then the oxidation in the animal of acetate and propionate produced gives rise to 321 mole ATP prior to manipulation and 350 mole ATP with the altered volatile fatty acid pattern whereas the energy available from butyrate and the oxidation of the products of digestion of microbial cells remains about the same.

The total ATP available from volatile fatty acid oxidation would be of the order of 391 against about 421 if there is a shift towards propionate of 10 percent, increasing the availability to the animal of ATP by about 7.5 percent.

This calculation should be compared with a manipulation which allows say 8 percent (the same amount of glucose that is directed to propionate rather than acetate in the above example) of the carbohydrate to escape fermentation and be digested in the small intestine to glucose. Under these circumstances the production of microbial cells and volatile fatty acid will be 8 percent less and the ATP available from volatile fatty acid oxidation will be 360 moles, but an extra 76 mole ATP will be available from glucose oxidation (assuming 38 moles ATP/mole glucose). In addition the synthesis of glucose will be decreased by about the same amount as is absorbed (see Judson & Leng 1972) and therefore a further 2 ATP/mole glucose will be available. Thus in the animal effectively 440 moles of ATP will be available from oxidation of volatile fatty acids and glucose, increasing the ATP yield by 10 percent but decreasing the cell yield by 8 percent. Thus avoiding rumen fermentation would have a marked effect on the amino acid/energy availability (i.e. P:E ratio) and, provided protein is not deficient, would have a major effect on productivity.

Increasing propionate production in the rumen could have the added advantage of providing glucose. At times when glucose demand is high, as in peak lactation (see Leng 1970 for review) the animal may have difficulty in obtaining sufficient glucogenic precursors. On the majority of diets, however, glucose precursors do not appear to be deficient (see Leng 1970, Lindsay 1970). The response to a manipulation which increases ATP at the tissue level will only become translated to a production response if the availability of amino acids to the animal's tissues is not deficient. It appears that even on diets with only moderate levels of dietary bypass protein, extra energy available due to a manipulation will be used efficiently. In studies in these laboratories it has been shown that growth rates of lambs on a suboptimal level of bypass protein (i.e. levels at which intake would be stimulated by extra bypass protein) could be stimulated by infusions of acetate, propionate or glucose into the duodenum (see Table 3). This must mean that in the control animals which were without infusion of substrate, more essential amino acids were available than required. It therefore appears that a manipulation to a higher propionate and lower methane production or to an increase in the bypass of dietary-intestinally-digestible carbohydrate is always likely to increase production at any dietary protein content.

TABLE 3

The effects of continuous infusion of substrates into the abomasum of lambs given oaten chaff/sugar based diets (3 percent crude protein) with 0 and 6 percent fish meal to supply dietary bypass protein. Lambs were approximately 25 kg. Infusions were made over a period of 4-6 weeks (unpublished observations).

Supplement	Infused substrate	Feed intake (g/d)	Growth rate (g/d)
Experiment 1			
0	0.9% NaCl	890	97
	80 g/d glucose	770	129
6% Fishmeal	0.9% NaCl	1139	202
	80 g/d glucose	1074	257
Experiment 2			
6% Fishmeal	0.9% NaCl	657	123
	60 g glucose	773	188
	60 g acetate*	740	175
	44 g propionate*	684	160

*Isocaloric amounts relative to glucose

Gaseous energy losses

The major emphasis in research on manipulation of rumen function has been on reducing the loss of energy as methane from the rumen. A lower methane production must be accompanied by either an increased synthesis of more reduced end-product of rumen fermentation such as propionate or butyrate or the evolution of hydrogen. The reduction of sulphate to sulphide and or unsaturated long chain fatty acids are also "sinks" for hydrogen produced in metabolism. The energy loss in methane may represent some 8 percent of the gross energy of a feed. The main effect of methane inhibition must be to increase the energy in other end products, but it may at times result in a decrease in microbial protein synthesis through suppression of the growth of methanogens.

Applied studies with propionate enhancers/methane inhibitors

This area has been reviewed by Chalupa (1980). Monensin (Rumensin) is the only commercial feed additive in use as a rumen manipulator. Originally it was identified as a propionate enhancer but it appears to have other properties. Chalupa (1980) claims a consistent effect of monensin is to increase propionate, whilst decreasing acetate and butyrate production. In addition, methane production is partially inhibited but without accumulation of hydrogen. Microbial nitrogen production in the rumen is either decreased or not changed, which supports the suggestion that the change of volatile fatty acids proportions has little or no effect on microbial cell production. This in turn is supported by Raun *et al*. (1976) who found in rumen fluid that monensin decreased total volatile fatty acid production whilst propionate proportion increased with no change in cell synthesis.

Manipulation of microbial protein (cells) availability from the rumen

Although progress has been great in attempting to quantify the availability of microbial protein from the rumen of cattle and sheep, there are as yet no clear cut generalizations that can be made. Some confusion arises because of the lack of uniformity in the way that microbial protein production is expressed (see for discussion Czerkawski 1976). It is often difficult to ascertain whether published results are expressed as an apparent or true digestibility of organic matter across the rumen. The relationship between the efficiency of microbial synthesis (Y-ATP) and the g cells produced/kg apparent (FOM app) or true fermented organic matter (FOM) is shown in Figure 6. These graphs emphasize the extent of the confusion that can arise if the efficiency of microbial growth is variable between diets.

In reality the estimation of microbial protein production in the rumen is relatively inaccurate since it depends on a number of rather inaccurate measurements and in the discussion to come of microbial protein production, values have been rounded off because of a lack of confidence in the accuracy of the estimates.

The methods of detecting microbial protein in duodenal contents have relied upon a number of internal markers (DAPA, RNA and amino acid proportions) or markers incorporated into microbes during a continuous infusion of labelled compounds (i.e. $^{35}SO_4^{=}$ ^{15}N ammonia, $^{14}CO_2$ or ^{14}C acetate). The different markers can give quite different results (Siddons *et al*. 1979). Despite this, the mean of a very large number of estimates of microbial protein leaving the rumen is about 30 gN/kg of FOM app and this value is used as a standard in determining protein requirements of ruminants (Roy et al. 1977).

In a notable comparison of maize and barley, as the grain in a concentrate/hay mixture given to dairy cows, the calculated microbial nitrogen synthesis in the rumen for the cattle on the two diets were markedly different, indicating the variability that might exist under practical feeding conditions (see Oldham 1980). In continuous culture systems increasing turnover of fluid has given very large increases in microbial cell yield (Isaacson et al. 1975). These together are probably the strongest evidence that there is scope for manipulating the availability of microbial protein to ruminants.

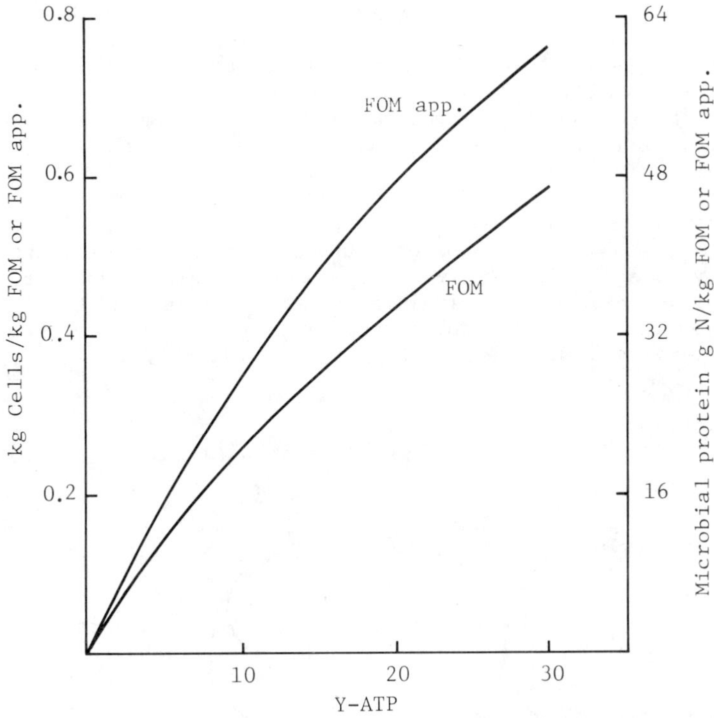

Fig. 6. The relationship between Y-ATP and the quantity of bacterial cells or protein leaving the rumen per kg of FOM or FOM app.

Factors affecting the yield of rumen microorganisms to the animal

The major factors that affect microbial protein synthesis are:

(1) the availability and/or concentration in rumen fluid of precursors e.g. glucose, nucleic acids, amino acids, ammonia and minerals (including S, K, P etc.)

(2) the maintenance ATP requirements (M-ATP)* of the microbes

*M-ATP is defined here as the ATP required for maintenance of the pool of microorganisms in the rumen and it is not synonymous with maintenance energy requirements which usually includes the ATP expended in the synthesis of cells which are degraded in the rumen.

(3) the turnover of microbial cells

(4) the destruction of bacteria by predatory protozoa in the rumen.

Availability of substances for microbial cell synthesis

There seems little doubt that the continuous availability of all necessary cell precursors has a marked effect on the efficiency of microbial cell yield. The continuous supply of fermentable carbohydrates to maintain fermentation and precursors is paramount in efficient utilization of ATP. The rate of fermentation must be synchronized to the rate of uptake of ammonia and/or peptides and amino acids. Maximum microbial synthesis rate apparently occurs at ammonia concentrations between 5 and 8 mg N/100 ml (Satter & Slyter 1974) although various optimum levels have been found by different researchers (see Stern & Hoover 1979) suggesting that possibly the diets of the experimental animals influence the optimum ammonia level. The high ammonia concentration for maximum cell growth suggests that the rumen microorganisms probably have similar mechanisms for incorporation of ammonia as those in microbes grown in chemostat culture. In these cultures and at high ammonia concentrations, N assimilation occurs via glutamate dehydrogenase, however the same cultures grown under low ammonia concentrations fix ammonia in a two step process involving glutamine synthetase and glutamate synthase. These reactions involve amidation of glutamate to glutamine and then a reductive transfer of the amide-N of glutamine to 2-oxoglutarate that requires ATP (Meers *et al*. 1970, Tempest *et al*. 1970, Brown *et al*. 1974).

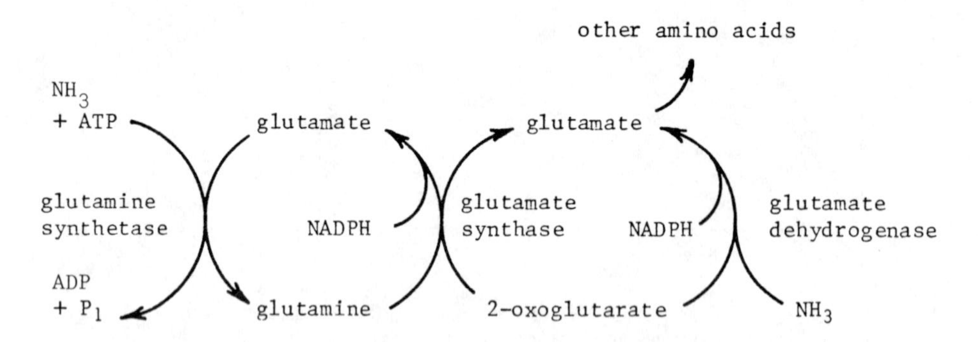

Hume (1970) found increased microbial protein synthesis in the rumen of sheep on diets with casein as the nitrogen source compared to urea based diets and Maeng *et al*. (1976) recorded marked effects on the efficiency of microbial protein synthesis in rumen microorganisms held in flasks, depending on whether urea or amino acids supplied the nitrogen for microbial growth. Oldham (1980) used the information of Maeng *et al*. (1976) to argue that 'bypass' protein or so called 'rumen non-degraded protein' increased microbial protein production in the rumen through the slow release of amino acids stimulating microbial growth. However according to Stouthamer (1979) and Hespell (1979) this is not a result of the relative supply of amino acids or ammonia, since the theoretical growth yields are influenced to a relatively small extent by nitrogen source. The findings of Maeng *et al*. (1976), where the replacement of urea with amino acids increased microbial growth by up to 80 percent, may be explained if under their *in vitro*

conditions the uptake of ammonia was inhibited, for instance, due to the low concentration of NH_3 as against NH_4^+ at low pH (J.V. Nolan, personal communication, 1981) or the enzymes fixing ammonia are ineffective at low pH in rumen fluid. If either of these mechanisms was operative this would also rationalize the variations in optimum ammonia concentrations found by a number of workers and the low microbial growth rates in the rumen of cattle on finely ground maize (rapidly fermented and therefore low pH) vs. cracked barley (more slowly fermented and therefore higher pH) (see Oldham et al. 1979, Sutton et al. 1980). The low microbial growth in the rumen of dairy cows on high concentrate diets is attributed to low pH and lactic acid production (Hagemeister et al. 1980).

The support of high milk yields in dairy cows on starch/sugar/cellulose/urea based diets (Virtanen 1966) indicates a high efficiency of microbial protein production on such diets. This efficiency can be explained on the basis of the various carbohydrate sources, with differing rates of fermentation, allowing a constant supply of precursors for microbial synthesis. Under these circumstances ammonia is apparently a nitrogen source that can be used very efficiently. The Finnish research emphasizes the need for a long period of adaptation to these diets before high feed intake and milk production can be achieved. This period is possibly needed to avoid low pH, through too rapid fermentation rate, and also to allow salivary flow rates to increase and thus the establishment of an efficient rumen microbial ecosystem.

Maintenance ATP requirements of microorganisms (M-ATP) in the rumen

It is difficult to define accurately the M-ATP requirements of microorganisms during growth. The definition of M-ATP used here is the ATP available that is directed from utilization in growth to utilization in other processes not involved in growth, for instance osmo-regulation. The associated effects of an increase in M-ATP of bacteria in the rumen is a small increase in heat production but a larger increase in volatile fatty acid production and a decrease in cell yield (or more succinctly, a decrease in Y-ATP).

The M-ATP has a number of components including (a) ATP needed for motility (b) requirements for ATP and nutrients to replace cellular component turnover (i.e the dynamic state) (c) ATP for production of both extracellular proteins (mainly enzymes) and polysaccharides (d) ATP for active transport.

In this presentation the inefficiency of utilization of ATP in resynthesis of cells following lysis in the rumen is considered as being separate from M-ATP. Hespell & Bryant (1979) considered that motility, turnover, extracellular polymer formation and transport would probably account for no more than a 15 percent lowering of theoretical Y-ATP or of the total ATP needed for cell formation. Hespell & Bryant (1979) also summarized estimates for M-ATP of a wide range of cultures and showed that the M-ATP varied from 50-0.25 m moles ATP/g cells/hr. Mixed rumen microorganisms have been estimated to have an M-ATP of about 1.6 m moles ATP/g cells/hr (Isaacson et al. 1975, Kennedy & Milligan 1978). For organisms grown in complex media with glucose limiting growth, 102 m moles ATP/g cells/hr was needed for maintenance with Y-ATP of from 24-28 (see Harrison & McAllan 1980). In an average steer's rumen with a bacterial pool size of say 1 kg of cells, at a maximum the M-ATP may be 20-25 percent of the available ATP (or the ATP available from the fermentation of 1.1 kg hexose) (Owens & Isaacson 1977) and therefore the rumen should have the potential for a Y-ATP as high as 25. M-ATP can vary considerably up to 25 fold in some aerobic organisms (Hempfling & Mainzer 1975) and therefore large variations may occur in mixed rumen organisms.

No one seems to have calculated the M-ATP of protozoa even though this may be excessively high because of their almost continuous motion and their apparent retention within the rumen (Weller & Pilgrim 1974) (see Leng et al. 1981) and their sometimes very large biomass in the rumen.

The effects of dilution rate on M-ATP

Studies using continuous cultures of rumen organisms appear to show that, depending on the specific growth rate of microorganisms, the Y-ATP can change. The concept here is that a large pool of microorganisms growing at a slow rate will be less efficient than a smaller pool of microbes turning over rapidly. For instance Isaacson et al. (1975) showed that the Y-ATP for a continuous culture of ruminal bacterial with a dilution rate of the contents of the culture of .02-0.5, .06-1.5 or .12-3.0 the Y-ATP for the bacteria was 7.5, 11.6 and 16.7 respectively. This has been used to suggest that in the rumen such increases in dilution rate might increase microbial cell yields (see Harrison & McAllan 1980), and studies specifically aimed at increasing rumen fluid turnover by feeding salts or using infusions of salts into the rumen tend to support these contentions (Helmsley 1975, Harrison et al. 1976). However, where flow rate of digesta through the rumen has been varied by feeding level (Hagemeister et al. 1980) or changed at ad libitum intakes by using supplements (Kempton et al. 1979) little or no apparent change in microbial protein availability has resulted. In recent studies with cattle on derinded sugar cane diets, the microbial protein leaving the rumen was 30 g N/kg FOM (Y-ATP about 13) and was unaffected by supplementation of the diet which induced a 2.5 fold increase in feed intake and a similar increase in rumen turnover (T.J. Kempton, J.V. Nolan, J.B. Rowe, S.J. Stachiw, M. gill, M. Bobadilla, T.R. Preston and R.A. Leng, unpublished observations). In 75 trials with dairy cows Hagemeister et al. (1980) showed an apparent relationship between organic matter apparently fermented in the rumen and microbial protein availability (see Fig. 8) with 221 g protein available/kg FOM app. (Y-ATP about 11). Where microbial protein production in the rumen has been measured in sheep on green pastures either cut and fed in pens (Walker et al. 1975) or measured under grazing conditions (Corbett 1981) microbial protein synthesis has been about 30 g N/kg FOM (or a Y-ATP about 12) despite high feed intake and therefore high flow rates from the rumen. Thus it appears that, up to now and under practical conditions, no measurable increase in microbial cells flowing to the duodenum occurs with increasing dilution rate of rumen contents in cattle and sheep. The most likely cause of this is an increase in the turnover of the more efficient bacteria in the rumen through a greater effect of lytic factors.

Microbial cell turnover in the rumen

A factor that appears to have a major effect on the availability of microbial protein is the degradation of bacteria in the rumen. The results of the studies of Nolan & Leng (1972) indicated that some 30 percent of microbial nitrogen is recycled within the rumen of sheep on a lucerne diet. Microbial cell degradation in the rumen can result from death of bacteria due to lack of substrate (starvation). For instance Hespell (1979) reported that about 60 percent of rumen bacteria die and about 30 percent are lysed within 2 hr in the absence of substrate. The presence of lytic organisms such as bacteriophages (Adams *et al*. 1966, Hoogenraad *et al*. 1967) or mycoplasma (see Leng 1974) or the ingestion of bacteria by protozoa (Coleman 1975) also cause a major loss of bacterial protein to the animal. Invasive agents of bacteria are likely to account for a fairly constant loss of organisms in the rumen. In the studies of Nolan & Stachiw (1979) the protozoa biomass in the sheep was relatively small but the apparent recycling of nitrogen was high, suggesting that the value of 30 percent reported previously is a minimal value.

The theoretical Y-ATP of rumen microorganisms appears to be much higher than the actual Y-ATP in the rumen of cattle and sheep even on high quality diet. The Y-ATP for the rumen is almost always less than that in continuous cultures of rumen organisms. There is therefore scope for manipulation to increase the availability of protein from rumen fermentation but before this can be achieved a great deal of research is needed to understand the factors that are involved in the M-ATP and the turnover of organisms in the rumen. However, since an increase in microbial cells (protein) availability is accompanied by a concomitant decrease in volatile fatty acid production, an increase in microbial protein availability to the host animal will only be of benefit if the animal requires the extra amino acids made available. The response to such manipulations will be high if the increase in microbial protein results in an increase in feed intake.

The effects of protozoa in the rumen on microbial protein availability to ruminants

A number of authors have given constructive arguments in support of the concept of a large protozoal biomass reducing bacterial protein leaving the rumen (see Leng 1976, Bergen & Yokoyama 1977, Owens & Isaacson 1977). The major evidence for this is that protozoa may be largely responsible for the turnover of bacterial protein in the rumen since they ingest bacteria in large numbers (Coleman 1975) and their removal from the rumen has been shown to increase the microbial protein leaving the rumen of sheep by 20 percent (Lindsay & Hogan 1972). The variable responses of ruminants to defaunation in terms of production are therefore confusing. The only studies showing marked responses to defaunation have been with cattle (Bird & Leng 1978) and sheep (Bird *et al*. 1979, W. Burggraaf & R.A. Leng unpublished data) fed on sugar/ roughage diets which were suboptimal in protein. However defaunation cannot be regarded as a single influence and in all manipulation practices, interactive changes occur which may or may not lead to increased availability of protein to the animal. In some situations an increase in rumen volume (and therefore pool size of bacteria) occurs on defaunation and this possibly results in an increased M-ATP of rumen organisms due to slower turnover of the bacterial pool.

Studies of the turnover of protozoa have recently been made using protozoa labelled with ^{14}C-choline (see Fig. 7) (Coleman *et al*. 1980, Leng *et al*. 1981). The turnover rate of large ciliate protozoa in cattle fed sugar based diets was extremely slow indicating a preferential retention of large ciliate protozoa in the rumen (see Table 4). Studies using small ciliate protozoa (mainly *Entodinium* spp.) labelled by incubation with ^{14}C-choline and injected back into the rumen of sheep on roughage based diets or cattle on molasses diets have indicated that protozoa are lost from the rumen at only a slightly slower rate than liquid (see Table 4).

The isotope studies, together with growth studies with defaunated sheep reported from this laboratory, indicate a decreased availability of metabolizable nutrients in ruminants which have a large protozoal biomass in the rumen.

An interesting observation reported by Hagemeister et al. (1980) was the relationship between microbial protein production and FOM app intake from 75 trials with dairy cows (see Fig. 8). They interpreted their data as indicating a constant efficiency of microbial cell synthesis independent of feed intake, and therefore flow rate from the rumen. The data were fitted to a linear relationship in which most of the data obtained at low intakes fell below the line of best fit and most of the data at high intakes fell about the line. It appears that perhaps the data should be re-examined with the possibility in mind that there may be at least two relationships as tentatively shown in Fig. 8. To justify the two relationships shown it is necessary to consider two distinct rumen microbial ecosystems. The diets were based on concentrate/roughage mixtures and were varied from ad lib. intake to a restricted intake. The apparent change to a more efficient rumen system occurs at the feed intake (4-5 kg FOM app) which is restricted. Under restricted concentrate intakes the populations of protozoa in the rumen are likely to increase to substantial biomasses (see Eadie et al. 1970, Whitelaw et al. 1972).

TABLE 4

Fluid turnover and dynamics of protozoa in the rumen of cattle and sheep. The results were obtained using single injections of polyethylene glycol or CrEDTA as fluid markers and protozoa were labelled in vitro with ^{14}Ccholine before being returned to the rumen.

Species	No. of animals	Diet	Rumen volume (ℓ)	t½ rumen fluid (day)	Protozoa No. (10^{-5}/ml)	Protozoa Pool size (gN)	Protozoa t½ (day)	Protozoa Appt. I.L.* (gN/d)
Cattle	3	Sugar cane/ urea	54	0.3	0.3	32	8.4	2.8
Cattle	4	Molasses/ urea	41	0.4	7.0	28	0.6	34
Sheep	8	Lucerne/ molasses	4.8	0.4	7.0	1.2	0.5	1.8

*Appt. I.L. is the apparent irreversible loss of protozoal N. Results are taken from Leng et al. (1981) and unpublished studies of Leng, R.A.; Ffoulkes, D. & Leng, R.A.; Cumming, G., Graham, C.A. & Leng, R.A. Only large ciliates (Isotricha and Dasytricha spp.) are considered in the cattle on sugar cane, whilst the cattle on molasses had mainly small ciliates, (Entodinium spp.) in the rumen with some Polyplastron spp. In the sheep Entodinium and Epidinium spp. were the largest populations of protozoa.

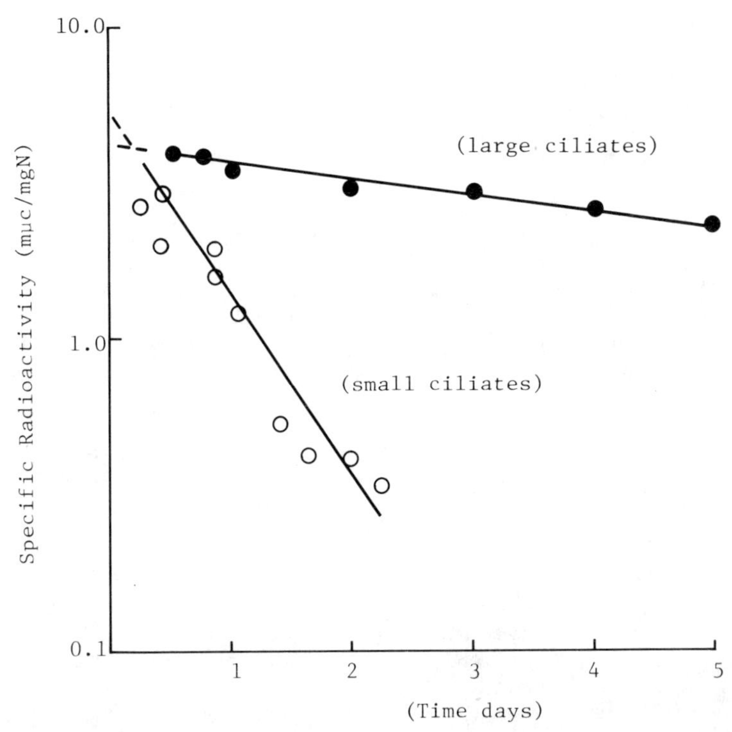

Fig. 7. The relationship between specific radioactivity of large ciliate (from Leng et al. 1981) or small ciliates (Ffoulkes & Leng 1981) and time after injection of ^{14}C labelled protozoa into the rumen of a 350 kg bull given sugar cane or a 200 kg steer given molasses based diets respectively. (Values are adjusted to 100 μc injected).

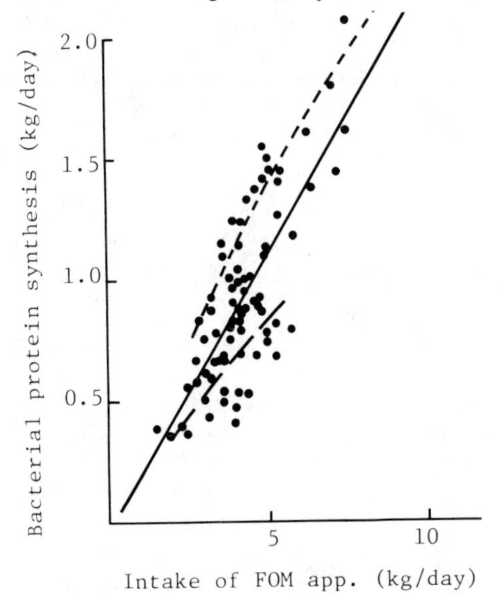

Fig. 8. The relationship between fermentable organic matter intake (FOM app.) and a microbial protein synthesis in dairy cows (from Hagemeister et al. 1980). The solid line is the best fit to the data as given by Hagemeister et al. (1980); the dotted lines tentatively indicate the possibility of two distinct microbial systems.

The apparent change occurs at the feed intake at which protozoal numbers in the rumen are often at a maximum and at which the rumen is likely to become defaunated with higher feed intakes (see Christiansen *et al*. 1964) (possibly due to lower or transient depressions of pH in the rumen, see Purser & Moir 1959). If the explanations here are correct then the presence of a large protozoal biomass in the studies of Kaufmann and his colleagues may have a marked effect on protein availability to the animal. At low feed intakes of dairy cows W. Kaufmann & H. Hagemeister (1981 personal communication) have shown a negative relationship between the bacterial protein entering the duodenum and the percentage of protozoal protein (measured using 2-amino ethyl phosphonic acid) in the total microbial protein in the duodenum, supporting the suggestions that a large protozoal biomass in the rumen will substantially decrease microbial protein availability to the animal.

The major effect of defaunation should be an increase in bacterial protein availability, since a large protozoal biomass may result in a high M-ATP and a low Y-ATP. In addition, due to predation by protozoa, a high bacterial cell turnover in the rumen may occur. This will only result in increased production if the animal was unable to meet its amino acid requirements. However, if defaunation was accompanied by an increase in propionate in the total volatile fatty acids this may result in further increases in production in addition to any response to extra protein which may explain the results of Bird & Leng (1978). An increased breakdown of bacteria resulting in fermentation of microbial protein, or an increased rumen volume which may increase M-ATP, could eliminate the advantages of defaunation.

CONCLUSIONS

The above theoretical discussion highlights the interactive nature of the rumen ecosystem and emphasizes that, irrespective of the apparent rationality of a manipulation procedure, the changes in the rumen and the effect on production of the animal are not readily predictable. There appears to be scope to decrease methane production and increase propionate production which appears to increase the availability of energy with a probable increase in efficiency of utilization of metabolizable energy.

Although the theoretical microbial protein production is considerably higher than the average value recorded in cattle and sheep on production diets, in practice increased flow rate from the rumen has not apparently increased microbial protein output. Defaunation has apparently increased microbial protein availability (Lindsay & Hogan 1972) and increased growth and wool growth in sheep on diets suboptimal in protein (Bird *et al*. 1979). There is a strong possibility that increased microbial protein production will be accompanied by increased metabolizable energy in the feed through decreased heat of fermentation and decreased methane loss. The response to increased microbial protein availability depends largely on the protein content of the basal diet.

Manipulation of rumen fermentation under grazing conditions

The extensive nature of farming in Australia means that the scope for manipulation of the rumen system of sheep and cattle is much less than under lot-feeding conditions. There is a basic and fundamental difference between the feed-lot and animal production under grazing. In the intensive industry with high-capital investment and small margins (or often economic loss) the small increases in efficiency of conversion of feed to liveweight gain, that have so far resulted from the use of chemical manipulators, may be economic since little or no extra labour is involved as a chemical can be incorporated into the feed mix. However, with grazing systems there is a large labour cost as the chemical must be administered through a molasses or mineral block (see De Muth et al. 1977); included in a slow release capsule (Laby 1973) or given to animals in a supplementary feed (Wilkinson et al. 1980).

A major problem associated with this area is that in general there is a reluctance to publish negative results and therefore a literature survey is likely to be biased. However, in field trials in U.S.A. and Europe responses to low level of monensin supplementation (50-200 mg/d) have resulted in from 0.08-0.12 kg/d increases in growth rates in cattle grazing low to average quality forages (see for examples Wilkinson et al. 1980, Horn et al. 1981). In Australia responses to monensin given in slow release capsules to cattle on short green pastures were 180 g/day in one trial (Watson & Laby 1978) but there was no response in a second trial (M.J. Watson personal communication). In a major trial in N.S.W. with cattle, the overall response to monensin in Hereford steers (290 kg) receiving 57 mg/day through a slow release capsule was 86 g/day or a weight advantage of 11 kg after 4 months on the supplement (File et al. 1980). Similarly with crossbred lambs the increase in growth rate on a lucerne:oaten hay (4:1) feed mix was 15 g/day (File et al. 1980). Such small increases in growth rate and efficiency of feed utilization are not likely to be economic. Fundamentally the question is whether a manipulation to increase the efficiency of feed utilization under field conditions is rational, when under the majority of production systems green pasture biomass is underutilized. However, there are many situations where production from grazing sheep and cattle is well below that expected from experience (see Archer 1980) and where there is considerable scope for increasing production.

The changing nutrients in pastures in response to changing conditions for plant growth also make manipulative measures much more uncertain. The one major difference between the feed lot and grazing system on green pastures is that the rumen pH is usually continuously high in grazing animals. This results in conditions almost always conducive to high protozoal populations which may build up to a very large biomass at times, particularly on highly productive temperate pastures (ryegrass/ clover)(see Clarke 1965). Whilst high populations of small ciliate protozoa in the rumen occur under restricted grain feeding, only on sugar cane based diets or pastures do both the large and small ciliates occur consistently in large numbers (see Clark 1965, Valdez et al. 1977). These extensive populations should reduce microbial protein production (see earlier discussion) and therefore their control may have major effects on production. However, the field studies await the development of a protozoal toxin which is effective at low concentration and can be administered in a 'Laby-slow release device'.

The effects of feeding "bypass protein" meals to cattle and sheep on dry native pastures in Australia indicates that the supplement acts as a catalyst in increasing pasture intake (see earlier discussion). No increase in digestibility of the poor quality roughage has been reported, contrary to Oldham's studies (Oldham 1980). However, these effects have not been specifically examined. If there was an effect of bypass protein on digestibility this would explain the often reported lack of effect of formaldyhyde treated casein (protected protein) as compared to heat treated cottonseed meal (which is only 75 percent protected) as nitrogen supplements for cattle and sheep on poor quality feeds.

REFERENCES

Adams, J.C.; Gazaway, J.A.; Brailsford, M.D.; Hartman, P.A.; Jacobson, N.L. (1966) Isolation of bacteriophages from the bovine rumen. Experientia 22, 717-718.

Allison, H.J.; Peel, J.L. (1971) The biosynthesis of valine from isobutyrate by Peptostreptococcus elsdenii and Bacteroides ruminicola. Biochemical Journal 121, 431-437.

Archer, K.A. (1980) Low productivity of lambs on improved pasture. In: Recent advances in animal nutrition - 1980. Editor D.J. Farrell. Armidale, Australia, University of New England Publishing Unit, pp.20-27.

Baile, C.A. (1975) Control of feed intake in ruminants. In: Digestion and metabolism in the ruminant. Editors I.W. McDonald and A.C. Warner. Armidale, Australia, University of New England Publishing Unit, pp.333-350.

Baile, C.A.; Mayer, J. (1970) Hypothalmic centres: feedbacks and receptor sites in the short-term control of feed intake. In: Physiology of digestion and metabolism in the ruminant. Editor A.T. Phillipson. Newcastle-Upon-Tyne, England, Oriel Press, pp.254263.

Baldwin, R.L. (1970) Energy metabolism in anaerobes. American Journal of Clinical Nutrition 23, 1508-1513.

Bergen, W.G.; Yokoyama, M.T. (1977) Productive limits to rumen fermentation. Journal of Animal Science 45, 573-584.

Bines, J.A.; Hart, I.C.; Morant, S.V. (1980) Endocrine control of energy metabolism: the effect on milk yield and levels of some blood constituents of injecting growth hormone and growth hormone fragments. British Journal of Nutrition 43, 179-188.

Bird, S.H.; Leng, R.A. (1978) The effects of defaunation of the rumen on the growth of cattle on low protein high energy diets. British Journal of Nutrition 40, 163-167.

Bird, S.H.; Hill, M.K.; Leng, R.A. (1979) The effects of defaunation of the rumen on the growth of lambs on low protein high energy diets. British Journal of Nutrition 42, 81-87.

Brown, C.M.; MacDonald-Brown, D.S.; Meers, J.L. (1974) Physiological aspects of microbial inorganic nitrogen metabolism. Advances in Microbial Physiology 11, 1-52.

Chalupa, W. (1980) Chemical control of rumen microbial metabolism. In: Digestive physiology and metabolism in the ruminant. Editors Y. Ruckebush and P. Thivend. Lancaster, U.K., MTP Press Ltd., pp.325-347.

Christiansen, W.C.; Woods, W.; Burroughs, W. (1964) Ration characteristics influencing rumen protozoal populations. Journal of Animal Science 23, 984-988.

Clarke, R.T.J. (1965) Diurnal variation in the numbers of rumen ciliate protozoa in cattle. New Zealand Journal of Agricultural Research 8, 1-6.

Clay, A.B.; Satter, L.D. (1979) Milk production response to dietary protein and methionine hydroxy-analog. Journal of Dairy Science 66 (Supplement 1) 75-76.

Coleman, G.S. (1975) Interrelationship between rumen ciliate protozoa and bacteria. In: Digestion and metabolism in the ruminant. Editors I.W. McDonald and A.C. Warner. Armidale, Australia, University of New England Publishing Unit, pp. 140-164.

Coleman, G.S.; Dawson, R.M.C.; Grime, D.W. (1980) The rate of passage of ciliate protozoa from the ovine rumen. Proceedings of the Nutrition Society 39, 6A.

Corbett, J.L. (1981) Determination of the utilization of energy and nutrients by grazing animals. In: Forage evaluation: concepts and techniques. Editors J.L. Wheeler and R.D. Machrie. Melbourne, Australia, CSIRO and American Forage and Grassland Council, pp. 383-398.

Cottle, D. (1981) The synthesis, turnover and outflow of ruminal microorganisms. Ph.D. Thesis. University of New England, Armidale, Australia.

Czerkawski, J.W. (1976) Chemical composition of microbial matter in the rumen. Journal of the Science of Food and Agriculture 27, 621-632.

Demeyer, D.; Van Nevel, C. (1979) Protein fermentation and growth by rumen microbes. Annales de Recherches Veterinaires 10, 277-279.

De Muth, L.; Essig, H.W.; Smithson, L.J.; Withers, F.T.; Mormson, E.G.; Chapman, H. (1977) Block consumption on performance of cattle provided Monensin (Rumensin ®) in molasses-mineral blocks. Mississippi Agricultural and Forestry Experimental Station Research Report. 3(8), 1-4.

Eadie, J.M.; Hyldgaard-Jensen, J.; Mann, S.O.; Reid, R.S.; Whitelaw, F.G. (1970) Observations on the microbiology and biochemistry of the rumen in cattle given different quantities of a pelleted barley ration. British Journal of Nutrition 24, 157-177.

Elliott, R.; Ferreiro, H.M.; Priego, A.; Preston, T.R. (1978) Rice polishings as a supplement in sugar cane diets: the quantities of starch (α-linked glucose polymers) entering the proximal duodenum. Tropical Animal Production 3, 30-35.

Ferreiro, H.M.; Priego, A.; Lopez, J.; Preston, T.R.; Leng, R.A. (1979) Glucose metabolism in cattle given sugar cane based diets supplemented with varying quantities of rice polishings. British Journal of Nutrition 42, 341-347.

File, G.C.; Weston, R.H.; Margan, D.E.; Laby, R.H. (1980) Performance and digestion responses to monensin sodium by herbage-fed cattle and sheep. Proceedings of the Australian Society of Animal Production 13, 486.

Gordon, F.J. (1980) Feed input-milk output relationships in the spring-calving dairy cow. In: Recent advances in animal nutrition - 1980. Editor W. Haresign. London, Butterworths, pp. 15-31.

Gottschalk, G. (1979) Bacterial metabolism. New York, Springer-Verlag.

Hagemeister, H.; Lupping, W.; Kaufmann, W. (1980) Microbial protein synthesis and digestion in the high-yielding dairy cow. In: Recent advances in animal nutrition - 1980. Editor W. Haresign. London, Butterworths, pp. 67-84.

Harrison, D.G.; McAllan, A.B. (1980) Factors affecting microbial growth yields in the reticulo-rumen. In: Digestive physiology and metabolism in ruminants. Editors Y. Ruckebusch and P. Thivend. Lancaster, U.K., MTP Press Limited, pp. 205-226.

Harrison, D.G.; Beever, D.E.; Thomson, D.J.; Osbourn, D.F. (1976) Manipulation of fermentation in the rumen. Journal of the Science of Food and Agriculture 27, 617-620.

Hemsley, J.A. (1975) Effect of high intakes of sodium chloride on the utilization of a protein concentrate by sheep. I. Woolgrowth. Australian Journal of Agricultural Research 26, 709-714.

Hempfling, W.P.; Mainzer, S.E. (1975) Effects of varying the carbon source limiting growth on yield and maintenance characteristics of Escherichia coli in continuous culture. Journal of Bacteriology 123, 1076-1087.

Hennessy, D.W.; Williamson, P.J.; Lowe, R.F.; Baigent, D.R. (1981) The role of protein supplements in nutrition of young grazing cattle and their subsequent productivity. Journal of Agricultural Science, Cambridge 96, 205-212.

Hespell, R.B. (1979) Efficiency of growth by ruminal bacteria. Federation Proceedings 38, 2707-2712.

Hespell, R.B.; Bryant, M.P. (1979) Efficiency of rumen microbial growth; influence of some theoretical and experimental factors on Y-ATP. Journal of Animal Science 49, 1640-1649.

Hodgson, J. (1982) Influence of sward characteristics on diet selection and herbage intake by the grazing animal. In: Nutrition limits to animal production from pastures. Editor J.B. Hacker. Farnham Royal, U.K., Commonwealth Agricultural Bureaux, pp. 153-166.

Hoogenraad, N.J.; Hird, F.J.R.; Holmes, I.; Millis, N.F. (1967) Bacteriophages in rumen contents of sheep. Journal of General Virology 1, 575-576.

Horn, G.W.; Mader, T.L.; Armbruster, S.L.; Fratim, R.R. (1981) Effect of monensin on ruminal fermentation, forage intake and weight gains of wheat pasture stocker cattle. Journal of Animal Science 52, 447-454.

Hume, I.D. (1970) Synthesis of microbial protein in the rumen. III. The effect of dietary protein. Australian Journal of Agricultural Research 21, 305-314.

Isaacson, H.R.; Hinds, F.C.; Bryant, M.P.; Owens, F.N. (1975) Efficiency of energy utilization by mixed rumen bacteria in continuous culture. Journal of Dairy Science 58, 1645-1659.

Judson, G.J.; Leng, R.A. (1972) Estimation of the total entry rate and re-synthesis of glucose in sheep using glucoses uniformly labelled with ^{14}C or variously labelled with ^{3}H. Australian Journal of Biological Sciences 25, 1313-1332.

Kempton, T.J.; Nolan, J.V.; Leng, R.A. (1979) Protein nutrition of growing lambs. 2. Effect on nitrogen digestion of supplementing a low protein cellulosic diet with either urea, casein or formaldehyde treated casein. British Journal of Nutrition 42, 303-315.

Kennedy, P.M.; Milligan, L.P. (1978) Effects of cold exposure on digestion, microbial synthesis and nitrogen transformations in sheep. British Journal of Nutrition 39, 105-117.

Laby, R.H. (1973) The anti-bloat capsule and detergents for bloat control. In: Reviews in rural science - I: Bloat. Editors R.A. Leng and J.R. McWilliam. Armidale, Australia, University of New England Publishing Unit, pp. 81-83.

Leng, R.A. (1970) Glucose synthesis in ruminants. Advances in Veterinary Science and Comparative Medicine 14, 209-260.

Leng, R.A. (1974) Salient features of the digestion of pastures by ruminants and other herbivores. In: Chemistry and biochemistry of herbage. Editors G.W. Butler and R.W. Bailey. London and New York, Academic Press, Vol.3, pp.81-129.

Leng, R.A. (1976) Factors influencing net protein production by the rumen microbiota. In: Reviews in rural science - II: From plant to animal protein. Editors T.M. Sutherland, J.R. McWilliam and R.A. Leng. Armidale, Australia, University of New England Publishing Unit, pp.85-91.

Leng, R.A.; Brett, D.B. (1966) Simultaneous measurements of the rates of production of acetic, propionic and butyric acids in the rumen of sheep on different diets and the correlation between production rates and concentrations of these acids in the rumen. British Journal of Nutrition 20, 541-548.

Leng, R.A.; Kempton, T.J.; Nolan, J.V. (1977) Non-protein nitrogen and by-pass proteins in ruminant diets. Australian Meat Research Committee Reviews No. 33, 1-21.

Leng,, R.A.; Gill, M.; Kempton, T.J.; Rowe, J.B.; Nolan, J.V.; Stachiw, S.J.; Preston, T.R. (1981) Kinetics of large ciliate protozoa in the rumen of cattle given sugar diets. British Journal of Nutrition, (in press).

Lindsay, D.B. (1970) Carbohydrate metabolism in ruminants. In: Physiology of digestion and metabolism in the ruminant. Editor A.T. Phillipson. Newcastle-upon-Tyne, England, Oriel Press, pp.438-451.

Lindsay, J.R.; Hogan, J.P. (1972) Digestion of two legumes and rumen bacterial growth in defaunated sheep. Australian Journal of Agricultural Research 23, 321-330.

Maeng, W.J.; Van Nevel, C.J.; Baldwin, R.L.; Morris, J.G. (1976) Rumen microbial growth rates and yields; effect of amino acids and protein. Journal of Dairy Science 59, 68-79.

Mathison, G.W.; Milligan, L.P. (1971) Nitrogen metabolism in sheep. British Journal of Nutrition 25, 351-366.

Meers, J.L.; Tempest, D.W.; Brown, C.M. (1970) 'Glutamine (amide): 2-oxoglutarate amino transferase oxido-reductase (NADP)', an enzyme involved in the synthesis of glutamate by some bacteria. Journal of General Microbiology 64, 187-194.

Meyreles, L.; Rowe, J.B.; Preston, T.R. (1979) The effect on the performance of fattening bulls of supplementing a basal diet of derinded sugar cane stalk with urea, sweet potato forage and cottonseed meal. Tropical Animal Production 4, 255-262.

Minson, D.J. (1982) Effects of chemical and physical composition of herbage eaten upon intake. In: Nutritional limits to animal production from pastures. Editor J.B. Hacker. Farnham Royal, U.K., Commonwealth Agricultural Bureaux, pp.167-182.

Morgan, P.J.K. (1977) The flowpaths taken by ground supplements in the stomachs of sheep. South African Journal of Animal Science 7, 91-95.

Nolan, J.V.; Leng, R.A. (1972) Dynamic aspects of ammonia and urea metabolism in sheep. British Journal of Nutrition 27, 177-194.

Nolan, J.V.; Stachiw, S. (1979) Fermentation and nitrogen dynamics in Merino sheep given a low quality roughage diet. British Journal of Nutrition 42, 63-80.

Oldham, J.D. (1980) Amino acid requirements for lactation in high yielding dairy cows. In: Recent advances in animal nutrition -1980. Editor W. Haresign. London, Butterworths, pp.33-65.

Oldham, J.D.; Broster, W.H.; Napper, D.J.; Sivites, J. (1979) The effect of low protein ration on milk yield and plasma metabolites in Friesian heifers during early lactation. British Journal of Nutrition 42, 149-162.

Ørskov, E.R.; Fraser, C.; Pirie, R. (1973) The effects of bypassing the rumen with supplements of protein and energy on intake of concentrates by sheep. British Journal of Nutrition 30, 361-367.

Owens, F.N.; Isaacson, H.R. (1977) Ruminal microbial yields: factors influencing synthesis and bypass. Federation Proceedings 36, 198-202.

Pilgrim, A.F.; Gray, F.V.; Weller, R.A.; Belling, G.B. (1970) Synthesis of microbial protein from ammonia in the sheep's rumen and the proportion of dietary nitrogen converted into microbial N. British Journal of Nutrition 24, 589-598.

Preston, T.R.; Leng, R.A. (1980) Utilization of tropical feeds by ruminants. In: Digestive physiology and metabolism in ruminants. Editors Y. Ruckebusch and P. Thivend. Lancaster, U.K., MTP Press Limited, pp.621-640.

Preston, T.R.; Willis, M.B. (1974) Intensive beef production. Oxford, Pergamon Press, 544 pp.

Purser, D.B.; Moir, R.J. (1959) Rumen flora studies in the sheep, 9: the effect of pH on the ciliate population of the rumen in vivo. Australian Journal of Agricultural Research 10, 555-564.

Raun, A.P.; Cooley, C.O.; Potter, E.P.; Rathmacher, R.P.; Richardson, L.F. (1976) Effect of monensin on feed efficiency of feedlot cattle. Journal of Animal Science 43, 670-677.

Roy, J.H.B.; Balch, C.C.; Miller, E.L.; Ørskov, E.R.; Smith, R.H. (1977) Calculation of the N-requirement for ruminants from nitrogen metabolism studies. In: Protein metabolism and nutrition. EAAP, Wageningen, Pudoc. pp.126-129.

Satter, L.D.; Slyter, L.L. (1974) Effect of ammonia concentration on rumen microbial protein production in vitro. British Journal of Nutrition 32, 199-208.

Sauer, F.D.; Erfle, J.D.; Mahadevan, S. (1975) Amino acid biosynthesis in mixed rumen cultures. Biochemical Journal 150, 357-372.

Siddons, R.C.; Beever, D.E.; Nolan, J.V.; McAllan, A.B.; MacRae, J.C. (1979) Estimation of microbial protein in duodenal digesta. Annales de Recherches Veterinaires 10, 286-287.

Slyter, L.L.; Nelson, W.O.; Wolin, M.J. (1964) Modifications of a device for maintenance of the rumen microbial population in continuous culture. Applied Microbiology 12, 374-377.

Stern, M.D.; Hoover, W.H. (1979) Methods for determining, and factors affecting rumen microbial protein synthesis: a review. Journal of Animal Science 49, 1590-1603.

Stouthamer, A.H. (1979) The search for correlation between theoretical and experimental growth yields. International Review of Biochemistry 21, 1-46.

Sutton, J.D.; Oldham, J.D.; Hart, I.C. (1980) Products of digestion, hormones and energy utilization in milking cows given concentrates containing varying proportions of barley or maize. In: Energy metabolism. Editor L.E. Mount. London, Butterworths, pp. 303-306.

Tempest, D.W.; Meers, J.L.; Brown, C.M. (1970) Synthesis of glutamate in aerobacter aerogenes by a hitherto unknown route. Biochemical Journal 117, 405-407.

Valdez, R.E.; Alvarez, F.J.; Ferreiro, H.M.; Guerra, F.; Lopez, J.; Priego, A.; Blackburn, T.H.; Leng, R.A.; Preston, T.R. (1977) Rumen function in cattle given sugar cane. Tropical Animal Production 2, 260-272.

Virtanen, A.I. (1966) Milk production of cows on protein free feed. Science 153, 1603-1614.

Waldo, D.R. (1973) Extent and partition of cereal grain starch digestion in ruminants. Journal of Animal Science 37, 1062-1074.

Walker, D.J.; Egan, A.R.; Nader, C.J.; Ulyatt, J.M.; Storer, G.B. (1975) Rumen microbial protein synthesis and proportions of microbial and non-microbial nitrogen flowing to the intestines of sheep. Australian Journal of Agricultural Research 26, 699-708.

Watson, M.J.; Laby, R.H. (1978) The response of grazing cattle to monensin administered from a controlled release capsule. Proceedings of the Nutrition Society of Australia 3, 86.

Weller, R.A.; Pilgrim, A.F. (1974) Passage of protozoa and volatile fatty acids from the rumen of the sheep and from a continuous in vitro fermentation system. British Journal of Nutrition 32, 341-351.

Weston, R.H. (1979) Digestion during pregnancy and lactation in sheep. Annales de Recherches Veterinaires 10, 442-444.

Weston, R.H. (1982) Animal factors affecting feed intake. In: Nutritional limits to animal production from pastures. Editor J.B. Hacker. Farnham Royal, U.K., Commonwealth Agricultural Bureaux, pp. 183-198.

Whitelaw, F.G.; Hyldgaard-Jensen, J.; Reid, R.S.; Kay, M.G. (1972) Some effects of rumen ciliate protozoa in cattle given restricted amounts of a barley diet. British Journal of Nutrition 27, 425-437.

Wilkinson, J.I.D.; Appleby, W.G.C.; Shaw, C.J.; LeBas, G.; Pflug, R. (1980) The use of monensin in European pasture cattle. Animal Production 31, 159-162.

INTEGRATION OF GRAZING WITH OTHER FEED RESOURCES[1/]

J.C. BURNS

USDA, ARS and Department of Crop Science and Animal Science, North Carolina State University, Raleigh, North Carolina, USA 27650

ABSTRACT

In most environments, where year-round grazing is practised, there is a need for feed supplements during periods when forage is inadequate. Stress may vary in duration and range from simple droughts to complete winter close down. Skilled producers can obtain desired production by controlling the quality of supplemental feed. The quality of supplemental feed must relate to the desired animal response. Preserved feeds including hay, silage and crop residues are considered. Special purpose pastures and management practices to obtain specific responses from grazing above that from base pasture are also discussed.

INTRODUCTION

The potential diet of grazing ruminants depends on the digestibility of the component parts of plants and on the quantity offered. Their diet results from animals expressing their innate drive to consume high quality plant parts and interaction of animal and plant characteristics which affect ease of prehension.

Ruminants often control their own production responses by ranging and selecting their ration. Selective grazing is greater in more extensive enterprises with less management and economic inputs than in more intensive ones where such inputs intermittently alter type, quality and quantity of available forage. Selective grazing may not always favour the economics of ruminant production, but it should be exploited when appropriate. Utilizing such behaviour to attain desired responses can be accomplished by selecting special-purpose forages, or by the use of supplemental feed sources.

Except for most feed-lot and certain dairy enterprises most ruminants still derive the major portion of their yearly energy intake from grazing. Yet, variation in available forage throughout the year causes major fluctuations in animal responses. This results from changes in plant growth due to weather patterns in association with soil characteristics. Theoretically, producers can provide nutrients for anticipated occurrences (summer slump or perennial snow cover), but not for unexpected droughts, fires, early freezes, etc. Periods when base pasturage is limited or nonexistent require a range from partial supplementation to provision of all feed inputs if animal production is to be maintained. This paper examines such supplementation of base pastures by fibrous feeds and by special purpose pastures. Emphasis is directed to periods of primary dependence on supplemental feeds. Because the subject of this paper

---

1/ Paper No. 6882 of the Journal series of the North Carolina Agricultural Research Service, Raleigh, N.C. 27650, in cooperation with USDA,ARS.

focuses on situations of stress feeding or on intensive grazing, efficiency of utilization will generally be high and subsequently, will not be considered further.

AN APPROACH TO INTEGRATING SUPPLEMENTAL FEEDS INTO BASE GRAZING SYSTEMS

Frequently, in forage-based animal enterprises, animal production coincides with the growth curve of a producer's base forage supply. Certainly, ease of operation supports such a system. However, the ease of altering economic advantages by applying current technology makes that approach overly restrictive and perhaps even archaic.

The key question for the producer concerns the periodic production per animal that he desires for the enterprise. That decision determines the quality of forage needed at various times and the number of animals to be held determines the quantity. The quality or quantity needs may not correspond to the natural forage supply. Further, because an animal's *ad libitum* daily dry matter intake is rather constant (assuming no adverse factors) the quality of the consumed forage essentially determines the resulting animal response. A simplified expression of the above relationship is as follows:

$$\underset{\text{(Mcal/day)}}{\text{Energy required}} = \underset{\text{(kg/day)}}{\text{Dry matter intake}} \times \underset{\text{(Mcal/kg)}}{\text{digestible energy}}$$

This expression estimates forage quality in terms of a particular maintenance, growth, reproduction or production function when limits are set on dry matter intake. Altering to determine forage digestibility based on the relationship between dry matter digestibility and digestible energy gives:

$$\underset{\text{(percent)}}{\text{Dry matter digestibility}} = \frac{\text{Energy required (units/day)}}{\text{Dry matter intake (units/day)}} \times 100$$

Managing for a desired animal response instead of accepting the animal's best effort in fixed circumstances allows several alternatives. Incorporating a supplemental feed into a base system is similar regardless of the stress factor (drought, fire, frosts, seasonal close down or simply over stocking). The supplemental feed must be readily available and digestibility and dry matter intake must be known, or approximated by thumb rule (i.e. intake of 1.8 to 3 percent of body weight of growing animals and up to 4 percent for lactating animals). Once the desired animal response is determined the daily energy needs can be obtained from nutritional requirement tables (National Research Council, 1976). With this information the above expression is useful for deciding whether supplemental feed should be included. Although estimates of digestibility and intake need only be approximations, refinements may allow more economy.

ALTERING THE QUALITY AND QUANTITY OF POTENTIAL SUPPLEMENTAL FEEDS

Feed to supplement base grazing systems should be provided in advance of the need. For stored feed, producers can control the combination of quality and quantity harvested from an area and further, control the mean daily dry matter intake per animal. Hay is the major supplemental feed used during stress periods in much of the world. The effects of maturity on its quality and quantity are well known. However, the manipulation of these effects in animal feeding systems in order to achieve desired responses has not been widespread. The relationship between dry matter digestibility, plant growth, dry matter intake and maturity for the tropical, annual sorghum hybrid, Sweet Sioux [Sorghum bicolor x S. sudanense] fed to goats are given in Fig. 1.

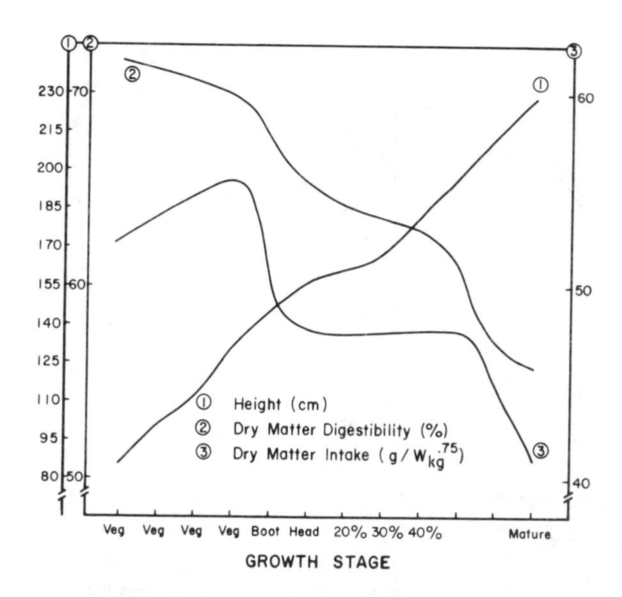

Fig. 1. Maturity changes in plant height, dry matter digestibility and intake of the tropical, annual Sorghum bicolor x S. sudanense hybrid, Sweet Sioux. Taken from Ademosum et al. (1968).

Such characterization permits estimation of the digestibility desired for a specific animal response at various maturities with reasonable estimates of both yield and dry matter intake. This information allows one to manage for the animal response desired. The rather drastic changes in both digestibility and dry matter intake that occur at specific maturity stages (Fig. 1) are striking. The harvest point selected to meet animal needs will affect dry matter yield appreciably.

For legume, Medicago sativa, the relationship between digestibility, dry matter yield and maturity is similar to the tropical annual, except that change in digestibility is much smoother over the maturities examined (Fig. 2).

The quality changes associated with maturity for a temperate and a tropical perennial (Dactylis glomerata and Cynodon dactylon, respectively) and for an ensiled tropical annual (Zea mays) are shown in Fig. 3. Generally, yield maturity relationships for species in Fig. 3 will be similar to those for the Sweet Sioux and M. sativa.

Such information permits a producer to store forage of a particular digestibility to achieve the animal performance he desires for his enterprise. When forage digestibility exceeds the level needed for the particular animal response desired, then dry matter intake must be limited. Selection of the desired animal performance should be given careful attention because maximum economic returns may not occur at maximum animal responses.

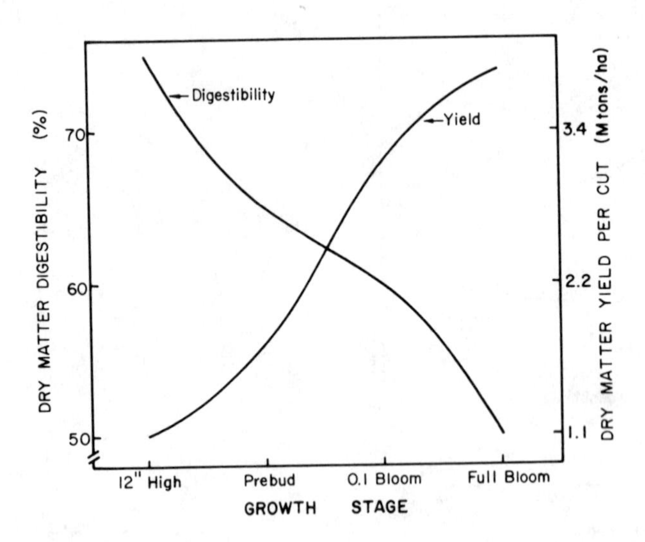

Fig. 2. The effect of maturity on dry matter digestibility and yield of Medicago sativa. Taken from Blaser et al. (1969).

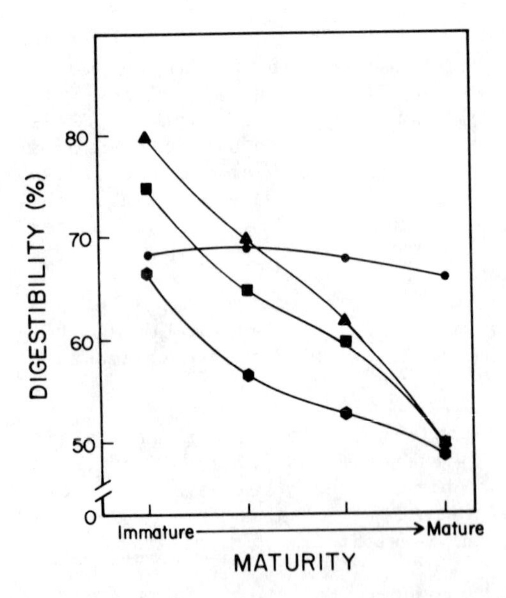

Fig. 3. Maturity and digestibility relationships of species representative of a temperate perennial [Dactylis glomerata (▲), taken from Jung & Baker (1973)]; a tropical perennial (Cynodon dactylon cv. Coastal (◆), J.C. Burns, unpublished data); a legume [Medicago sativa (■) taken from Blaser et al. (1969)]; and a tropical annual silage [corn (●), taken from Colovos et al. (1970)].

FITTING SUPPLEMENTAL FEEDS TO ANIMAL RESPONSES

The challenge lies in putting the above relationships into practice. Digestibility of fibrous supplemental feeds is of primary importance in achieving animal responses and will be considered for a range of beef cattle gains and lactation and maintenance (plus reproduction) responses. Although a model that allows changes in intake with changes in digestibility is required for best estimates, approximations can be made by simply assuming a constant daily dry matter intake of 2 percent of body weight for beef cows and 2.5 percent for yearlings and calves. Obviously intakes could be much lower when digestibility falls below 45 percent. Further, *in vivo* dry matter digestibility and *in vitro* dry matter disappearance (IVDMD) are assumed similar. With these assumptions one can determine the stage of maturity at which hay should be harvested to obtain the digestibility needed for a particular animal response. The opportunity to obtain immature stages of legumes is generally limited because of stand weakening.

Growing cattle

Several examples of various sizes and rates of gain of calves and yearlings will be used to determine which forage at what maturities can satisfy the specific animal energy requirements.

A 272-kg stocker (equivalent to a store in Australia) overwintering to gain 0.5 kg/day needs forage of at least 48 percent digestibility (Fig. 4). This can be provided by either the mature perennial tropical (*C. dactylon*) or the temperate (*D. glomerata*) hays. The tropical annual (*S. bicolor* x *S. sudanense*) hay (Fig. 5) and ensiled corn (Fig. 6) are more digestible than needed, whereas mature *M. sativa* hay (Fig. 6) essentially meets the need. Feeding corn or sorghum would have to be restricted for the daily gains desired (0.5 kg).

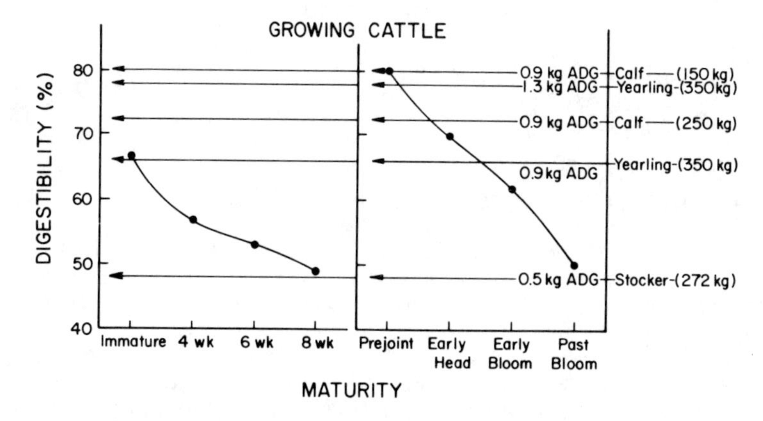

Fig. 4. Potential of a tropical (*Cynodon dactylon* cv. Coastal, left) and a temperate (*Dactylis glomerata*, right) perennial hay to give various average daily gains (ADG).

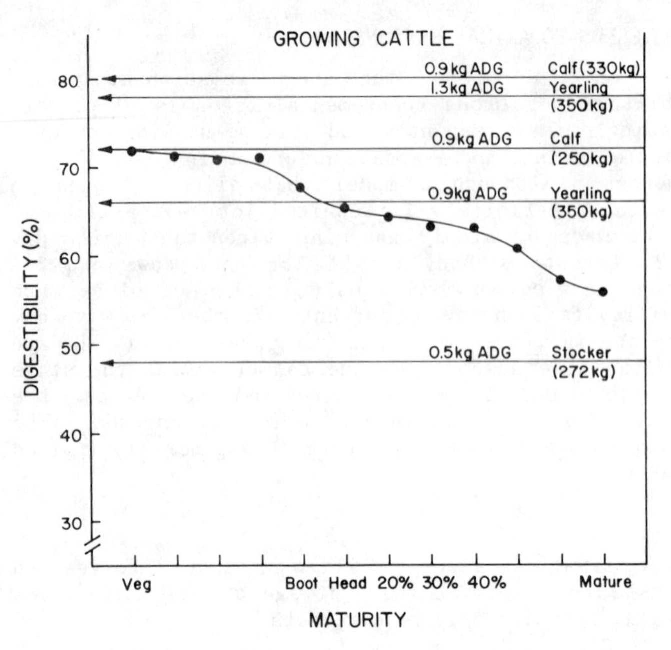

Fig. 5. Potential of a tropical annual (S. bicolor x S. sudanense) to give various average daily gains (ADG).

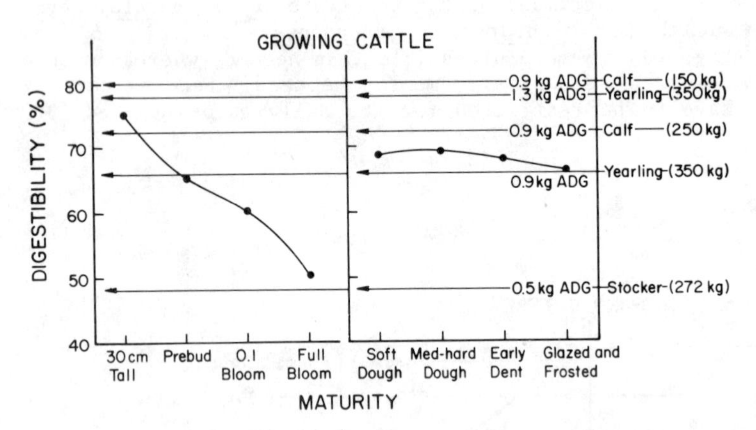

Fig. 6. Potential of a legume (Medicago sativa, left) or a silage (corn, right) to give various average daily gains (ADG).

For higher performance levels with larger animals, forage digestibilities of over 66 percent are required. Only the most immature tropical perennial hay provided the digestibility level needed for 350 kg yearlings to gain 0.9 kg/day (Fig. 4). The temperate perennial hay would allow the desired response if harvested somewhat after early head, as would the tropical annual (Figs. 4 and 5) and corn silage at any maturity (Fig. 6). Also, prebud is shown to be of adequate digestibility (Fig. 6). However legumes are generally consumed in larger quantities per unit of body weight than are grasses (Thornton & Minson 1973), although some tropical legumes may be exceptions. Therefore, prebud M. sativa if fed ad libitum might provide excess energy intake.

Greater levels of performance (350 kg yearling to gain 1.3 kg/day) require a digestibility of 78 percent. Only the temperate perennial appears to satisfy this requirement if harvested up to the prejoint stage (Fig. 4). Obviously supplemental grain (corn, barley, wheat, etc) would be needed to meet the energy needs for this response.

For lighter weight calves to gain 0.9 kg/day requires a forage digestibility in excess of 70 percent. Calves averaging 150 kg could approach this when fed the temperate perennial harvested at prejoint (Fig. 4). Heavier calves (250 kg) could not gain 0.9 kg/day from even the most immature tropical perennial (Fig. 4), but could from a temperate perennial (Fig. 4) or tropical annual (Fig. 5) if harvested by the boot or early head stage. The legume (Fig. 6) would probably be of adequate quality up to prebud while corn silage appears adequate through early dent (Fig. 6).

Cow lactation and maintenance (plus reproduction)

A different animal type and physiological state requires a different forage digestibility. A lactating beef cow of superior ability (10 kg milk/day) requires forages of 67 percent digestibility. All the immature harvests are adequate for all feeds (Figs. 7, 8 and 9) as are the mid-maturity stages of all hays except the tropical perennial (Fig. 7). The only mature crop with suitable digestibility is corn silage (Fig. 9); however considerable supplementary proteins would be required.

Forage digestibility requirements for the average lactating cow (5 kg milk/day) is only 53 percent. Consequently, all hay digestibilities (except the tropical perennial, Fig. 7) are adequate up to and including the mature stage. Digestibility of the tropical perennial was adequate through the six-weeks stage.

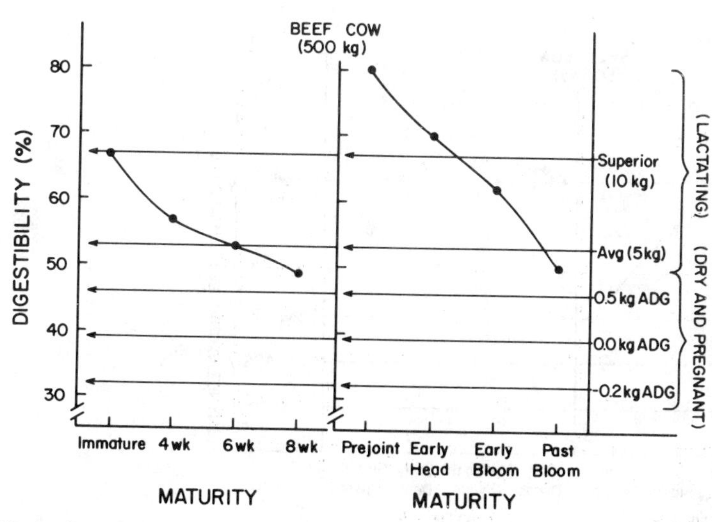

Fig. 7. Potental of a tropical (Cynodon dactylon cv. Coastal, left) and a temperate (Dactylis glomerata) perennial hay to give various breeding cow responses.

Generally, all the representative forages have suitable digestibilities at full maturity to allow breeding cow gains of 0.5 kg/day. Corn silage is more digestible than needed and intake would need to be limited as would that of other hays when only maintenance or some weight loss is desired.

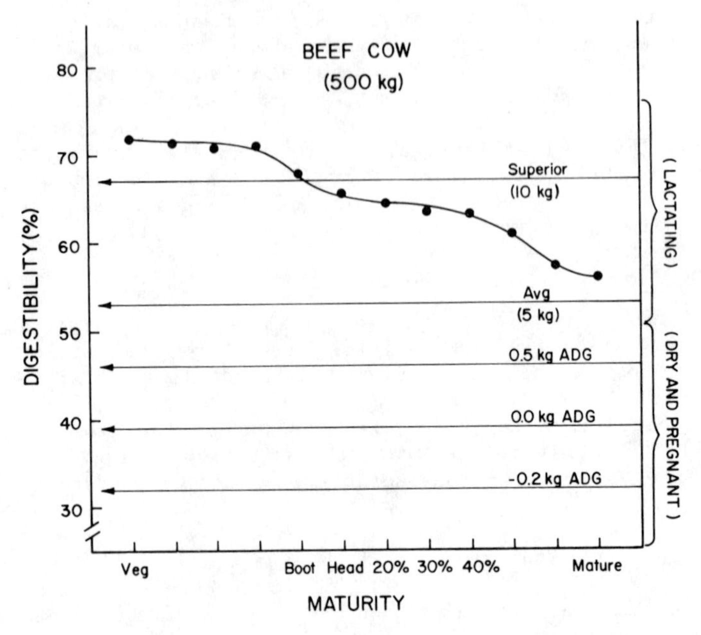

Fig. 8. Potential of a tropical annual (sorghum hybrid) to give various breeding cow responses.

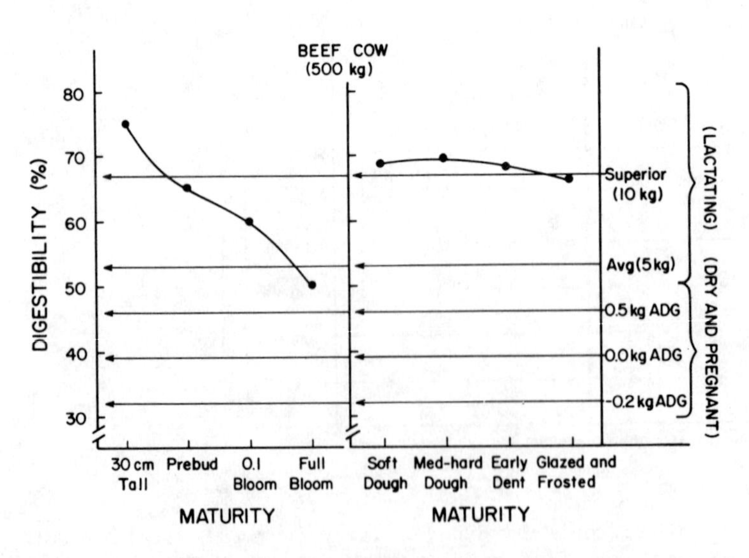

Fig. 9. Potential of a legume (*Medicago sativa*, left) or a silage (corn, right), to give various breeding cow responses.

These relationships, although developed from forages harvested for hay, apply to other fibrous supplemental feeds (such as crop residues or special purpose pastures) if dry matter intake are similar.

INNOVATIVE USE OF THE BASE GRAZING SYSTEM OR OF SUPPLEMENTAL PASTURES

Simple management that advantageously utilizes pastures in the base system or the sowing of special purpose species or mixtures offers valuable alternatives for changing animal responses at little additional cost. Exploiting these agronomic opportunities in conjunction with anticipated selective grazing behaviour allows one to take advantage of several practices. A discussion of these possibilities follows.

Forward creep grazing

Forward creep grazing is easily achieved when base pastures are rotationally grazed. The offspring is given access to the next pasture in the rotation before the breeding herd has access. The breeding animals are retained on forage of lower digestibility which is adequate for their requirements. This method was successfully used in the mideastern USA and consistently gave high calf daily gains (0.96 kg) while dam live weights changed appreciably (437 kg to 524 kg) (Blaser _et al_. 1976). Stocking rates of 1.2 and 1.5 cow-calf units/ha did not appreciably alter calf daily gains (0.96 and 0.94 kg/day), or weaning weights (244 and 235 kg). This practice allowed efficient utilization of forage by cows and offset reduced forage intake by calves at heavy grazing pressures (Jamieson & Hodgsen 1979).

First and last grazers

This management tool is similar to the forward creep grazing principal, but is applied to a mixture of animal classes. Animals with the highest production requirement (first grazers) are allowed first access to a pasture and are then followed by clean-up animals (last grazers) having lesser energy requirements. The technique is well suited to intensive enterprises, but it can also be used in extensive operations. First grazer steers gain more per day than last grazers and first grazer lactating cows produce more milk per day than last grazers (Table 1).

TABLE 1

Comparison of responses from first and last grazers using steers and lactating Holstein cows. Taken from Blaser _et al_. (1969).

Item	Steer liveweight gain			Lactating cows
	Daily	Seasonal		
		Group	Total	Milk/day
	(kg)	(kg/ha)		(kg)
Combined group	0.49	416	416	--
First grazers	0.61	267	428	16.6
Last grazers	0.37	161		11.5

The complexity of the method can be varied depending on the management skills of the producer. In a dairy enterprise, the lactating cow might be given first access, the replacement heifers second, followed by dry cows. In the case of beef or sheep enterprises, the growing animal might be given first access, followed by replacement females with clean-up achieved by the dry cow or ewe or by cows or ewes with nursing offspring having access to creep grazing.

Creep grazing

Straight creep grazing can be used in either extensive or intensive enterprises where rotational grazing is not practical. Creep areas can be simply fenced off from the base pasture and fitted with slip-through openings for the offspring. An alternative is to sow adjacent special creep areas using forage of high digestibility. In both cases the offspring can select a diet of high digestibility.

Midsummer drought in the mountains of North Carolina reduced daily gains of winter calves to 0.54 kg when confined to the base pasture of Kentucky bluegrass (*Poa pratensis*) and white clover (*Trifolium repens*). A supplemental creep area of millet [*Pennisetum americanum*] resulted in daily gains of 0.82 kg (R.W. Harvey & J.C. Burns, unpublished). However, in years without moisture stress calf daily gains were similar (1.1 kg). Increasing the stocking rate (cow and calves) by 40 percent for the creep grazing treatment changed calf daily gains little (1.10 *vs* 1.05 kg) but reduced cow daily gains to -0.41 kg from the 0.4 kg at the original rate.

Results from the use of supplemental *Lolium perenne* - *Trifolium repens* pasture or grain supplementation showed benefit only when heavy grazing pressure occurs (Vadiveloo & Holmes 1979). Such pressure can be continuously forced through management decisions or it can occur naturally under various stress conditions.

Early weaning

Weaning at five to six months can be practised routinely, or used in an emergency in both extensive or intensive enterprises. The practice is most useful for winter born offspring because milk secretion is appreciably reduced after four to six months and subsequent energy needs of the mother animal are appreciably reduced. The offspring's rumen has become functional and it is dependent on forage for at least 60 percent of its energy needs at a time when forage is frequently limited.

Early weaning of the offspring and providing it with supplemental grain while grazing a high quality forage separate from the dam allows the maximum production possible (neither feed quality nor quantity limited) while the dry cows or ewes can be heavily stocked to merely maintain body weight or allow some weight loss. In an experiment of Harvey *et al*. (1975) calf daily gains in mid-summer in North Carolina averaged 0.87 kg from early weaning compared with 0.57 kg from nursing calves. Dry cows maintained body weight when stocked at 8.6 hd/ha.

Stockpiled forage

The carryover of the latter portion of a season's growth for utilization during a stress period should be considered as an important potential supplement to the base system. Pasture growth made during favourable periods may be the major source of feed during subsequent stress periods. Preservation of the stockpiled forage depends on the particular species used and on the weather conditions following accumulation.

Two specific examples of stockpiled forage serving a major role in animal enterprises follow. First, in the mild, temperate eastern USA, tall fescue (Festuca arundinacea) is readily stockpiled after mid-August, with dry matter yields of 1,900 (Rayborn et al. 1979) to 3,000 kg/ha accumulating by late November (Taylor & Templeton 1976). Digestibility (IVDMD) is adequate (65 percent) from October to December (J.C. Burns & D.S. Chamblee, unpublished data), but declines to 54.7 percent by February. This decline is associated with the increase in brown (dead) leaves (34.5 percent digestibility) from about 18 percent at frost to 84 percent by March (Taylor & Templeton 1976). Generally, stockpiled fescue is satisfactory for overwintering most cattle and sheep through January with some supplemental hay required for the late winter period (Van Keuren & Stuedemann 1979).

In the more temperate zones, freezing temperatures and frequent snow clover in winter make it necessary to store feed to meet all the animals winter needs. Such stores may include crop residues, in addition to conventional hays and silages.

The situation in the dry tropics is very different from that in the temperate zone, but there is a similar lack of forage growth and declining digestibility during stress periods (Stobbs 1975). In such areas a four-month wet season produces abundant, lush forage and high liveweight gains (Gillard 1979). Excess forage (mainly grasses) accumulates, flowers and matures. A transition period occurs (about two months) when gains vary with digestibility of forage and the ability of the animal to select. Then follows an extended (six-month) dry period in which the unused forage cures on the stalk. The inherently lower quality of tropical grasses compared with temperate grasses (Minson & McLeod 1970), and the increasing maturity of the stockpiled grass, results in liveweight losses (Norman 1968, Gillard 1979). Rather extensive grazing enterprises have developed in these regions that minimize weight losses through selective grazing and the provision of supplemental feed is not considered economically feasible (Willoughby 1970).

Avoiding liveweight losses in the dry season requires supplemental feed such as hay, silage or grain, as in the temperate zones, or the upgrading of the quality of the holdover stockpile. The introduction of Townsville stylo (Stylosanthes humilis) into native range at Katherine, N.T. Australia, is an excellent example of the latter. It increased animal gains during the wet season and provided a high quality component in the stockpiled range for consumption in the dry season. Yet, further south at Townsville, the same forage was not effective in the dry season because frequent light rains resulted in its deterioration and spoilage (Gillard & Fisher 1978). Recent evaluation of a more widely adapted perennial stylo (Stylosanthes hamata cv. Verano) gave higher dry matter yields and liveweight gains during the wet season and gain in liveweight continued four to six weeks into the dry season (Gillard et al. 1980). Verano also decomposed by the late dry season when a decline in liveweight occurred.

CROP RESIDUES AS FEED SUPPLEMENTS TO GRAZING

Crop residues represent a large, but heterogeneous and diverse supply of supplemental energy for ruminants (Streeter & Horn 1980). Their digestibilities and crude protein content differ appreciably (Table 2). As anticipated, residue yields and crude protein concentrations are positively influenced by nitrogen fertilization (Table 3). Delaying utilization after grain harvest generally results in reduced yields and in some cases digestibility (Table 4), while crude protein changes little.

TABLE 2

Dry matter digestibility and crude protein content of crop residues (percent). Taken from Streeter & Horn (1980).

Crop residue	Digestibility	Crude protein
Grain straw		
Wheat	33 to 44	4.3
Barley	38	4.1
Oats	48	5.2
Rice	46	4.5
Corn		
Stover	48	6.4
Stalklage	51	--
Husklage	57	--
Cobs	47	2.1
Sorghum stover	54	7.8
Sorghum straw	40	9.2
Bagasse	23	1.7

TABLE 3

Effect of nitrogen (N) fertilization on yield and quality of residues. Taken from Perry & Smith (1975).

N rate (kg/ha)	Residue yield		IVDMD		Protein	
	Corn	Sorghum	Corn	Sorghum	Corn	Sorghum
	kg/ha		%			
0	4213	4446	55	48	3.3	5.1
90	5416	4959	54	46	3.7	5.7
180	5985	5003	52	47	4.0	6.5

Low digestibility (33 to 48 percent) of most untreated straw residues resticts their use, for livestock, to breeding animal maintenance with some gain possible during early access. The better quality stovers and husklage (54 to 57 percent digestible) may permit overwintering of yearling animals with daily gains of 0.3 to 0.5 kg (Figs. 4 through 9). Protein and mineral supplementation (especially phosphorus) are required when residues are fed alone (Streeter & Horn 1980).

TABLE 4

Effect of harvest date on yield and quality of residues. Taken from Perry & Smith (1975).

N rate (kg/ha)	Residue yield		IVDMD		Protein	
	Corn	Sorghum	Corn	Sorghum	Corn	Sorghum
	kg/ha		%			
Oct. 22	6877	4726	46	47	5.7	6.7
Dec. 6	6138	4010	41	43	5.6	6.4
Feb. 5	5264	3450	39	46	5.8	6.9

Grazing is the most economical method of residue utilization but use is generally restricted to 60 to 90 days following grain harvest. Collecting residues either at grain harvest or after mowing to a short stubble is far more efficient than grazing and can be done with available haying equipment (Manor & Butchelder 1980). Harvesting reduces weather effects, avoids tillage delay, trampling and high waste, allows the use of residues in critical periods and provides opportunity for treatment to improve quality prior to feeding. Either physical or chemical treatments, or both, have improved residue digestibility (see also Wilkins 1982). Off-farm pelleting and on-the-farm chemical treatment or combinations have been used. If residues are to be moved to the animal, pelleting is a major advantage because of increased density and ease of transportation (Klopfenstein et al. 1978). Local use of residues allows on-farm mixing with a chemical and adding water to raise the moisture up to 65 percent and ensiling directly or holding 24-48 hr after treatment before feeding (Klopfenstein et al. 1978). Broiler litter incorporation into residues prior to ensiling is a potential source of nitrogen (deficient in residues, Table 2) with proper fermentation (Goode & Harvey 1980). Relative costs of treatment and effects on digestibility and intake are summarized in Table 5.

TABLE 5
Treatment influences on crop residue quality and relative costs. Taken from Streeter and Horn (1980).

Treatment	Changes in digestion (% units)	Changes in intake (%)	Relative cost per kg gain	Special problems
Grinding/pelleting	-3	25-100	1.0	Equipment cost
Steam-pressure	+10	50	--	Equipment cost
Sodium-hydroxide (4%)	+15	10-50	0.5	Soil contamination
Ammonia (3%)	+15	10-50	0.2	Application losses

Applying the 15 percentage units increase in digestibility from chemical treatment (Table 5) to the digestibility values of untreated residues in Table 2 gives values of 62 and 72 percent for cobs and husklage respectively; feed of this digestibility would be expected to support steer and calf daily gains of 0.9 kg (Fig. 4). These projections agree well with research showing gains for steer of about 1 kg/day from treated cobs, and for calves of 0.75 kg/day from treated husklage (Klopfenstein _et al_. 1978). Feeding sodium hydroxide-treated corn stalks or wheat straw at 70-75 percent of the total ration resulted in lamb gains of 90 and 121 g/day compared with gains of 18 and -5 g/day, respectively, from untreated controls (Klopfenstein _et al_. 1978). Grinding and pelleting increases intake with some reduction of digestibility, however the cost per kg of liveweight is appreciably higher than with chemical treatments (Table 5). However, sodium hydroxide application results in high sodium excretion and subsequent soil contamination.

When crop residues are available they should be considered as a supplement to base pastures. Their chemical treatment is promising for both short-and long-term needs and will become more attractive as grain and other feed prices rise. However, the effect of residue harvesting on soil losses and water preservation must be considered (Tindall & Crabtree 1980).

REALISTIC APPRAISAL OF THE USE OF SUPPLEMENTAL FEEDS IN GRAZING SYSTEMS

In reality, the use of supplemental feeds in grazing systems may depend more on philosophy than on rational thinking. Generally, enterprises are either extensive or intensive, with few being a mixture. Producers tend toward one philosophy or the other and often react to situations without questioning other alternatives.

In extensive enterprises, supplemental feeding will be invoked for survival and the economic advantages for production purposes are of secondary importance. The producer maintains a reasonable buffer between forage available and animal numbers. This response is due to climatic conditions which can be quite extreme in regions where extensive enterprises develop. Survival feeding periods may be short- or long-term and the feed allocated may not be the most economic source of energy. Availability or ease of introducing the supplemental feed may be the major criterion for its selection. Frequently, severe feed shortages occur during stress periods in wool production enterprises and no supplemental feed is supplied because some wool growth continues even when animals are in a weight loss situation (Sharkey _et al_. 1962). Further, moderate weight losses can easily be replaced in periods of pasture excess and total wool production is not altered greatly. In such situations, producers accept death losses which are unacceptable in more intensive enterprises.

In intensive enterprises, supplemental feeding is more prevalent. A producer develops a philosophy of making management changes that return additional income to the enterprise. In these situations, the use of supplemental feed for other than survival seem likely. The products from intensive enterprises are frequently gain or fluid milk. Inadequate energy for even short periods may markedly reduce or essentially eliminate the product. On the other hand, slight energy increases above maintenance greatly increase productive responses. This places a higher value on a supplemental feed.

TABLE 6

Likely use and advantage of potential supplemental feeds to base grazing systems[1/]

Supplemental feed	Likely use	Animal type	Main advantage	Main disadvantage	Duration Short	Duration Long
I Stored						
Grain	Emergency; high daily response	All	Easy	Cost	X	X
Hay	Emergency; W and S close down	All	Reliable	Curing, handl.	X	X
Silage and haylage	Emergency; W and S close down	All	Reliable	Equipment	X	X
Crop residues (untreated)						
Cereals	Emergency; W and S close down	1,2,9	Available	Handl., Qual.	X	--
Corn and sorghum	Emergency; W and S close down	1,2,3	Available	Handl., Qual.	X	--
Crop residues (treated)	Emergency; W and S close down, and portion of finishing rations	4,5,6,7	Available	Handl., Tmt.	X	--
II Grazed						
Creep grazing (CG)	During growing season	8	High gain	Mgt., Labour	X	X
Forward CG	Year-round	7,8	High gain	Mgt., Labour	X	X
Early weaning	Stress periods	8	High gain	Grain, Labour	X	X
First and last grazers	Year-round	All	High gain	Mgt., Labour	X	X
Stockpiling	Wintering and droughts	All	Available	Losses; Qual.	X	--
Crop residues (first 45 days)						
Cereals	Emergency; S close down	9	Available	Qual.	X	--
Corn and sorghum	Emergency; W close down	2,3,7,9	Available	Losses	X	--
Crop residues (after 45 days)	Emergency; W close down	9	Available	Qual., Losses	X	--

1/ Abbreviations used are: W = winter; S = summer; 1 = dry female; 2 = stocker; 3 = replacement female; 4 = lactating beef cow; 5 = lamb; 6 = calf; 7 = yearling; 8 = offspring; 9 = mature animal; Handl. = handling; Qual. = quality; Mgt. = management, and Tmt. = treatment.

Stored supplemental feeds and grazing management practices that are believed to provide opportunities to improve the productivity of the enterprise during stress of base pastures are summarized in Table 6. In enterprises where the product is liveweight, compensatory gains can be expected following periods of stress (Preston & Willis 1974). The extent of compensation will depend on the level and duration of supplemental feeding. Consequently feeding of supplements over a short period of pasture stress may not alter seasonal productivity, whereas feeding over a long period of stress may markedly improve seasonal production. Compensating recovery in fluid milk enterprises is limited. The animal response desired during periods of partial or complete lack of grazing from base pastures, and the anticipated compensatory response should determine the digestibility and amount of the supplemental feed. These factors must be considered for a local economic setting for each available supplemental feed opportunity (Table 6) in arriving at the ultimate profit potential for a particular enterprise.

REFERENCES

Ademosum, A.A.; Baumgardt, B.R.; Scholl, J.M. (1968) Evaluation of a sorghum-sudangrass hybrid at varying stages of maturity on the basis of intake, digestibility and chemical composition. Journal of Animal Science 27, 818-823.

Blaser, R.E.; Bryant, H.T.; Hammes, R.C.; Boman, R.L.; Fontenot, J.P.; Polan, C.E. (1969) Managing forages for animal production. Virginia Polytechnic Institute Research Division Bulletin 45. 88 pp.

Blaser, R.E.; Hammes, Jr. R.C.; Fontenot, J.P.; Polon, C.E.; Bryant, H.T.; Wolf, D.D. (1976) Forage-animal production systems on hilland in the Eastern United States. In: Hill lands. Editors J. Luchak, J.D. Cowthon and M.J. Breslin. Morgantown, West Virginia, West Virginia University Books, U.S.A. pp. 674-685.

Colovos, N.F.; Holter, J.B.; Koes, R.M.; Urban, W.E.; Davis, H.A. (1970) Digestibility, nutritive value and intake of ensiled corn plant (Zea mays) in cattle and sheep. Journal of Animal Science 30, 819-824.

Gillard, P. (1979) Improvement of native pasture with Townsville stylo in the dry tropics of sub-coastal northern Queensland. Australian Journal of Experimental Agriculture and Animal Husbandry 19, 325-336.

Gillard, P.; Edye, L.A.; Hall, R.L. (1980) Comparison of Stylosanthes humilis and S. hamata and S. subsericea in the Queensland dry tropics: Effects on pasture composition and cattle liveweight gain. Australian Journal of Agricultural Research 31, 205-220.

Gillard, P.; Fisher, M.J. (1978) The ecology of Townsville stylo based pastures in northern Australia. In: Plant relations in pastures. Editor J.R. Wilson. Melbourne, CSIRO, pp. 340-352.

Goode, Lemuel; Harvey, R.W. (1980) The use of corn stover ensiled with either soybean meal or broiler litter for wintering stocker steers. Annual Report, Department of Animal Science, North Carolina State University, Raleigh, 24-25.

Harvey, R.W.; Burns, J.C.; Blumer, T.N.; Linnerud, A.C. (1975) Influence of early weaning on calf and pasture productivity. Journal of Animal Science 41, 740-756.

Jamieson, W.S.; Hodgsen, J. (1979) The effect of daily herbage allowance and sward characteristics upon the ingestive behaviour and herbage intake of calves under strip grazing management. Grass and Forage Science 34, 261-271.

Jung, G.A.; Baker, S. (1973) Orchardgrass. In: Forages. Editors M.E. Heath, D.S. Metcalf and R.F. Barnes. Ames, Iowa, Iowa State University Press, pp. 285-296.

Klopfenstein, T.; Berger, L.; Paterson, J. (1978) Performance of animals fed crop residues. In: Extending use of crop residues by ruminants (19th Annual Ruminant Nutrition Conference presented at the 62nd annual meeting of the Federation of American Societies for Experimental Biology, Atlantic City, New Jersey), pp.1939-1943.

Manor, G.; Butchelder, D. (1980) Crop residue management in livestock production and conservation systems. III. Methods of harvesting, transporting, storing and feeding crop residues. Oklahoma Experiment Station Special Report P-797. 38 pp.

Minson, D.J.; McLeod, M.N. (1970) The digestibility of temperate and tropical grasses. Proceedings of the 11th International Grassland Congress, Surfers Paradise, Australia, 719-722.

National Research Council (1976) Nutrient requirements of beef cattle. In: Nutrient Requirements of Domestic Animals. Washington D.C., National Academy of Science, No. 4, 56 pp.

Norman, M.J.T. (1968) The performance of beef cattle on different sequences of Townsville lucerne and native pastures at Katherine, N.T. Australian Journal of Experimental Agriculture and Animal Husbandry 8, 21-25.

Perry, L.J.; Smith, D.H. (1975) Composition of corn and grain sorghum residues. In: Proceedings of the 8th Research-Industry Conference. Editor G. Johnson. Lincoln, University of Nebraska Press, pp. 46-63.

Preston, T.R.; Willis, M.B. (1974) Growth and efficiency: Nutrition. In: Intensive beef production. Editors T.R. Preston and M.B. Willis. New York, Pergamon Press Inc., pp. 306-310.

Rayborn, E.B.; Blaser, R.E.; Wolf, D.D. (1979) Winter tall fescue yield and quality with different accumulation periods and N rates. Agronomy Journal 71, 959-963.

Sharkey, M.J.; Davis, I.F.; Kenny, P.A. (1962) The effect of previous and current nutrition on wool production in south Victoria. Australian Journal of Experimental Agriculture and Animal Husbandry 2, 160-169.

Stobbs, T.H. (1975) Factors limiting the nutritional value of grazed tropical pastures for beef and milk production. Tropical Grasslands 9, 141-150.

Streeter, L.; Horn, G.W. (1980) Crop residue management in livestock production and conservation systems. I. The use of crop residues as feedstuff for ruminant animals. Oklahoma Experiment Station Special Report P-795. 64 pp.

Taylor, T.H.; Templeton, W.C. (1976) Stockpiling Kentucky bluegrass and tall fescue forage for winter pasturage. Agronomy Journal 68, 235-239.

Thornton, R.F.; Minson, D.J. (1973) The relationship between apparent retention time in the rumen, voluntary intake and apparent digestibility of legume and grass diets in sheep. Australian Journal of Agricultural Research 24, 889-898.

Tindall, T.A.; Crabtree, R. (1980) Crop residue management in livestock production and conservation systems. II. Agronomic considerations of crop residue removal. Oklahoma Experimental Station Special Report P-796. 49 pp.

Vadiveloo, J.; Holmes, W. (1979) Supplementary feeding of grazing beef cattle. Grass and Forage Science 34, 173-179.

Van Keuren, R.W.; Stuedemann, A. (1979) Tall fescue in forage-animal production systems for breeding and lactating animals. In: Tall fescue. Editors R.C. Buckner and L.P. Bush. Madison, Wisconsin, American Society of Agronomy, pp. 201-232.

Wilkins, J.R. (1982) Improving forage quality by processing. In: Nutritional limits to animal production from pastures. Editor J.B. Hacker. Farnham Royal, U.K., Commonwealth Agricultural Bureaux, pp. 389-408.

Willoughby, W.M. (1970) Grassland management. In: Australian grasslands. Editor R.M. Moore. Canberra, Australian National University Press, pp. 392-397.

ROLE OF COMPUTER SIMULATION IN OVERCOMING LIMITATIONS TO ANIMAL PRODUCTION FROM PASTURES

J.L. BLACK, G.J. FAICHNEY

CSIRO, Division of Animal Production, P.O. Box 239, Blacktown, NSW 2148, Australia.

R.E. SINCLAIR

CSIRO, Division of Computing Research, P.O. Box 333, Wentworth Building, Sydney University, NSW 2006, Australia.

ABSTRACT

The nutritional value of pasture plants varies widely between species and also within species depending upon their stage of maturity and the conditions in which they are grown. In addition, the nutritional value of a particular pasture plant is not constant, but varies with the amount eaten, animal factors including species, breed and physiological state, and the environment. Although a great deal is now known about the determinants of the nutritional value of pastures, the complexity of the interacting factors often makes it difficult both to diagnose with confidence the reasons for limitations to animal production and to formulate appropriate quantitiatve solutions. By integrating current data and concepts, computer simulation can be used not only to indicate the relative importance of the limiting factors in specific situations but also to assess the value of alternative strategies for alleviating deficiencies.

The philosophy behind the computer simulation technique and the procedures involved in modelling physiological systems are briefly outlined. When establishing the nutritional limits to animal production from pastures, computer programs based on conventional nutritional principles using correlation and regression techniques are adequate for some assessments. However, to understand the reasons for differences between pastures and to determine the likely benefits from the addition of supplements, such as inorganic sulphur or non-protein nitrogen, a more complete biochemical representation of rumen function is required.

A computer program, which simulates protein and energy utilization in sheep, is used to determine the significance of various factors which affect (i) the rate of removal of organic matter from the rumen and, hence, voluntary feed intake, and (ii) the flow of true protein from the stomach. These comparisons are used to postulate reasons for differences observed in the nutritional value of grasses and clovers, and to suggest ways by which nutritional value of pasture plants might be improved.

INTRODUCTION

Pasture species vary widely in their capacity to support production in ruminants. The value for liveweight gain in sheep when grazing some temperate species in the vegetative stage of growth is twice that from others (Ulyatt 1981). Annual pasture plants which produce liveweight gains in sheep of 200-300 g/day in their vegetative stage are often unable to maintain liveweight when mature (Purser 1981). Fertilizer application to the soil (Ozanne et al. 1976, Rees & Minson 1976, 1978) and the physical or chemical treatment of conserved forages (Thomson & Beever 1980) can also markedly affect the nutritional value of pasture plants. Many reasons for these differences have been proposed, including variations in the physical structure and chemical composition of forages which lead to differences in voluntary feed consumption, in rates of digestion and times of retention of material in the digestive tract, in nutrient supply to the animal, and in palatability (Minson 1981, Ulyatt 1981).

Moreover, the nutritional value of a feed, defined as its ability to satisfy the requirements of an animal, can vary substantially with the circumstances in which it is eaten. Differences in the nutrients required to achieve potential rates of production between animals of varying species, breed, sex, age and physiological state, will result in differences in the nutritional value of a pasture plant even when the nutrients made available from its digestion are similar in each situation. Conversely, changes in digestion and nutrient supply without alteration in animal requirements also will affect the nutritional value of a feed. Both animal requirements and the supply of nutrients to the animal vary with almost every situation. For example, the ratio of absorbed amino acid nitrogen to metabolizable energy required for nitrogen retention in lambs weighing 15 kg is twice that required by lambs weighing 35 kg when energy intake is three times that required for maintenance, but body weight has little effect when intake is at maintenance (Black & Griffiths 1975). The flow of protein from the stomach of sheep and the relative availability of protein and energy are affected also by level of feeding in relation to body weight (Ørskov & Fraser 1973, Weston & Margan 1979). The way in which these changes in nutrient requirements and nutrient supply interact to influence the nutritional value of a feed in a particular situation is difficult to ascertain. Similarly, the states of pregnancy and lactation, and the effects of cold exposure, are known to alter both the nutrient requirements of the animal and the digestion of its feed. Differences also occur between sheep and cattle in nutrient requirements and in the digestion of feedstuffs.

Clearly, many plant and animal factors interact to determine the nutritional value of pasture plants. Although a great deal is now known about these factors, the complexity of their interactions makes it difficult to diagnose with confidence the reasons for limitations to animal performance in specific situations and to determine a quantitative solution. A major role of computer simulation is to integrate quantitatively current data and concepts so that they may be used to indicate the relative importance of factors limiting production and to assess the value of alternative management strategies for alleviating deficiencies. In this paper, the philosophy behind the computer simulation technique and the procedure involved in modelling physiological systems are briefly outlined. Computer programs which simulate protein and energy utilization in sheep are then used to show how the technique can help to indicate the relative importance of factors contributing to the differences observed between pasture plants in nutritional value.

COMPUTER SIMULATION PROCEDURE

The philosophy underlying the use of computers to simulate physiological systems in animals, and details of the procedures adopted, have been clearly described by Baldwin & Koong (1980). The technique is merely the formalization of the process, used by all research scientists, by which ideas and concepts about the operation and control of the system under study are developed from data obtained during experimentation. This process alone often leads to the identification of critical experiments and the further advancement of knowledge. However, as physiological systems are more fully understood, it becomes clear that many factors normally interact to affect their operation. Often the consequences of manipulations to individual components of the system can be resolved only qualitatively and the temporal effects of such manipulations are generally impossible to comprehend.

However, if the concepts and data are transformed into mathematical equations which can be solved readily using computers, a quantitative and dynamic appraisal of the proposed operation of the system can be obtained. The validity of the perceived mechanisms can then be tested by comparing predictions with experimental results. When these agree, some confidence in the understanding of the system is obtained. However, a major disparity leads either to a re-evaluation of the factors thought to control the system or highlights the need for additional or more reliable experimental data. When predictions fail to follow the general patterns observed in a system, the assumptions and concepts upon which they are based are usually faulty. Alternative hypotheses can then be developed and evaluated within the overall framework of the perceived system before critical experiments are designed to advance knowledge. On other occasions, the general behaviour of the predictions may be satisfactory, but numerical accuracy is lacking. Concepts are then most likely suitable, but the numerical inputs to their mathematical representation require refinement.

Baldwin & Koong (1980) emphasize that the computer simulation technique is most valuable when it is closely associated with experimental research. Computer programs constructed in isolation from experimentation are often judged by subject specialists as naive or inaccurate representations of current concepts. Similarly, the effectiveness of research programs can often be improved when hypotheses are first evaluated quantitatively within the framework of available knowledge. Data are obtained then because their close definition has been shown to be required for the satisfactory description of the system under study.

COMPUTER PROGRAMS FOR EVALUATING NUTRITIONAL LIMITS FROM PASTURES

As with any research program a computer simulation project must have a clearly defined objective. It is particularly important to set limits to a computing exercise and to define the scope of the information required because the amount of detail which must be incorporated into a computer program depends upon the reason for its development. Generally, a minimum number of well-proven concepts are initially used in a program and additional complexity is included only when the model resolution does not fulfil the original objective. To the authors' knowledge, no programs have been designed specifically to evaluate the nutritional limits to animal production from pastures. However, existing programs which simulate protein and energy utilization in sheep contain a number of features which allow them to be used to indicate the importance of some factors affecting the nutritional value of pastures.

Empirical equations describing conventional nutritional principles

One program, described by Graham *et al*. (1976b), adopts the conventional nutritional approach and uses empirical equations established from experimental results by correlation and regression techniques to predict nitrogen and energy utilization, body composition, body weight change and wool growth in sheep of any age or physiological state and accounts for the effects of energy expenditure associated with cold stress, feeding activities and locomotion. From initial information about the animal and the amount and composition of the feed eaten, the program calculates metabolizable energy and amino acid nitrogen availability and then partitions these between various body functions depending upon the respective requirements for the particular conditions simulated. The program was developed to test both the adequacy of existing knowledge of protein and energy utilization for describing the performance of sheep, and the conventional means of its representation. Predictions from the program and experimental observations agree closely in many situations, but a number of limitations to the program have been described (Graham *et al*. 1976a).

TABLE 1

Observed and predicted performance of lambs fed white clover and ryegrass *ad libitum*

	White clover		Ryegrass	
	Observed	Predicted	Observed	Predicted
Crude protein (%)	25.9	-	20.5	-
Dry matter intake (g/day)	774	-	711	-
Empty body weight gain (g/day)	123	117	86	79
Wool growth (g/day)	13	12	10	8
Energy retention (MJ/day)	1.63	1.58	1.10	1.20
Nitrogen retention (g/day)	3.5	3.5	2.2	2.3[a]

[a] Amino acid nitrogen absorption predicted to be insufficient to meet the needs of lambs with the particular energy availability. Table reproduced from Black *et al*. (1976).

This program was used (Black *et al*. 1976) to aid the interpretation of an experiment in which lambs given white clover (*Trifolium repens*) pasture grew faster than those fed perennial ryegrass (*Lolium perenne*) (Rattray & Joyce 1969). A number of factors had been suggested as being likely to contribute to the superior performance of sheep grazing clover pastures. These included more feed eaten and a higher efficiency of energy utilization resulting from less time spent in feeding activities (Lancashire & Keogh 1966) and more amino acids absorbed from the intestine (MacRae & Ulyatt

1974). However, the relative importance of these factors was unclear. When the empirical equations of MacRae & Ulyatt (1974) describing amino acid nitrogen absorption for sheep eating ryegrass or white clover were used in the computer program, the predictions of animal performance corresponded with those observed (Table 1). The simulation indicated that there was insufficient absorption of amino acids to meet the needs of the lambs eating ryegrass. Information was also provided about the magnitude of the deficiency and how it changed as the protein requirements of the animal declined relative to its energy needs with increasing body weight. At the beginning of the experiment when the lambs weighed 20 kg, an extra 10 g of a protein similar in amino acid composition to casein was predicted to be required each day to prevent protein availability from limiting growth. However, 94 days later, at the end of the experiment, only 3.5 g of extra protein was needed. Similar predictions that amino acid absorption limited the growth of early weaned lambs grazing grass pastures have been verified by experimentation (Black et al. 1979).

The program was then used to illustrate the consequence of differences in feed intake, amino acid absorption and feeding activities on animal performance (Table 2). When the intake of digestible energy of the lambs given ryegrass was assumed to be similar to that of the lambs eating clover, their predicted body weight gain increased from 79 to 91 g/day. Similarly, when feed intake was assumed to remain at its original level, but amino acid nitrogen absorption was made non-limiting, the predicted growth rate of the lambs given ryegrass increased from 79 to 94 g/day. The combined effect of an increase in both feed intake and amino acid nitrogen absorption was predicted to lift empty body weight gain to 111 g/day. The difference between this value and the 117 g/day predicted for white clover reflected the differences in energy expenditure associated with feeding activities.

Although the computer program of Graham et al. (1976b) provides useful insights into the consequences for animal performance of observed differences in feed intake and amino acid absorption, it cannot provide an explanation for these observations. The empirical equations upon which it is based provide only descriptions of the relationships observed between two or more variables and imply nothing about the underlying mechanisms. In addition, the above analysis of the effect on animal performance of differences in the nutritional value of clover and ryegrass depended upon the availability of a suitable equation relating the absorption of amino acid nitrogen to the intake of each forage. Only few such equations exist and, although they are satisfactory when used for predictions under conditions similar to those in which they were established, they are unreliable in other situations (Egan 1974). It is seldom possible to obtain an empirical relationship which will describe the behaviour of a complex, interacting system over a wide range of situations. This applies particularly to equations which are used to represent rumen function and predict the flow of amino acids from the stomach of sheep. Furthermore, existing empirical equations do not describe well the benefits which may result from certain manipulations of rumen function and how these may affect voluntary feed consumption and the nutritional value of pasture plants. We believe that the possible benefits resulting from the addition of non-protein nitrogen or inorganic sulphur to the diet could be assessed better when information on the determinants of microbial growth and the flow of material from the rumen is taken into account. Likewise, accurate prediction of the benefits arising from genetic, agronomic, physical or chemical manipulation of the composition and structure of pasture plants depends upon a detailed consideration of interactions occurring within the rumen.

TABLE 2

Predicted effects of altering energy expenditure, digestible energy intake and amino acid nitrogen absorption on the empty body weight gain of lambs given ryegrass *ad libitum*

Condition simulated	Predicted empty body weight gain (g/day)
White clover *ad libitum*	117
Ryegrass *ad libitum*	79
Ryegrass - feeding and ruminating activities reduced to equal those for white clover	82
Ryegrass - digestible energy intake increased to equal that from white clover	91
Ryegrass - nitrogen absorption not limiting	94
Ryegrass - nitrogen absorption not limiting, digestible energy intake increased	111
Ryegrass - nitrogen absorption not limiting, digestible energy intake increased, activity reduced	117

Table reproduced from Black *et al*. (1976).

Simulation of rumen function and voluntary feed consumption

The limitations outlined above have led to the development of computer programs which simulate rumen function (Baldwin *et al*. 1977, Black *et al*. 1980-81). These programs include few empirical equations and are largely based on theoretical equations which represent current thoughts on the factors controlling events within the rumen. They predict the degradation of dietary components, the products of fermentation, microbial growth yields, accumulation of material within the rumen and the flow of protein and other materials from the stomach. From information about the intake of individual dietary components, their potential degradability within the rumen and calculations of their relative rates of degradation and outflow, the amounts of each substance fermented in the rumen can be assessed. Knowledge of the stoichiometry of fermentation and of synthetic reactions is then used to calculate the supply of adenosine triphosphate (ATP) and to determine microbial growth in relation to the availability of ATP or other potentially limiting factors such as amino acids, non-protein nitrogen or inorganic sulphur.

One representation of rumen function (Black *et al*. 1980-81) is now being used to predict the flow of amino acids from the stomach of sheep in a computer program containing much of the information on protein and energy utilization described by Graham *et al*. (1976b). Another major difference between the new program and that described by Graham *et al*. (1976b) is the inclusion, as an option, of a section representing the control of voluntary feed consumption. The concepts incorporated into the new program are based on the proposition (Weston 1981) that the amount of feed eaten by a ruminant is primarily determined by an interaction between the energy demand of the animal and the amount of digesta which can accumulate in the rumen. Energy demand associated with fasting losses, activity, cold exposure, pregnancy and lactation, is calculated in relation to the body weight and physiological state of the animal and climatic conditions on the basis of the equations described by Graham et al. (1976b). The energy demand for wool growth is predicted for different genotypes from possible potential rates of wool growth suggested by Hogan et al. (1979). In addition, the potential rate of energy deposition into body tissue is estimated in relation to empty body weight for different genotypes and sexes of sheep from information derived from experiments where sheep were given free access to high quality diets, in some cases until they ceased to increase in weight (Hodge 1974, Blaxter 1976, R.M. Butterfield personal communication).

Estimates of the efficiency of utilization of metabolizable energy for each body function together with estimates of the metabolizability of the diet and its energy content are then used to determine the potential dry matter intake required to satisfy the energy demand. However, when consumption of this amount of feed is calculated to result in an excessive accumulation of organic matter in the rumen, intake is reduced until the predetermined rumen digesta load is achieved. Thus, the possible effect on voluntary feed consumption and subsequent animal performance of factors which influence the energy demand of the animal, the rate of removal of organic matter from the rumen or the limit to rumen digesta load can readily be investigated. However, it is not yet possible to use this program to predict voluntary feed consumption of sheep in all situations because the ways in which dietary and animal factors interact to influence the upper limit to rumen digesta load have not been elucidated. Nevertheless, the program can be used to assist in the determination of the significance of factors which may be responsible for the difference observed between pasture plants in voluntary feed consumption and nutritional value.

EVALUATION OF THE RELATIVE SIGNIFICANCE OF FACTORS AFFECTING THE NUTRITIONAL VALUE OF PASTURES

Factors affecting nutritional value

The nutritional value of a forage given to an animal with constant requirements is a function of the amount of forage eaten and the nutrients made available during its digestion. Dry matter intake and the supply of energy and amino acids to ruminants are closely correlated with the digestibility of dietary organic matter (Freer 1981). However, there are significant departures from this general relationship. When forage materials of similar digestibility are compared, clovers are eaten in greater quantity than grasses (Ulyatt 1971, Thornton & Minson 1973) and plant leaves in greater quantity than plant stem fractions (Laredo & Minson 1973, 1975, Poppi et al. 1981a). Provided the energy demand of the animal is not satisfied, differences in intake of plant material may result from either differences in the rate of removal of organic matter from the rumen and/or differences in the amount of digesta which can accumulate in the rumen. Thornton & Minson (1973) have demonstrated that a highly significant relationship exists between the intake of digestible organic matter from a variety of temperate and tropical pasture species and the rate of organic matter removal from the rumen of sheep. Similarly, differences in the intake of leaf and stem fractions of plants have been correlated with differences in the apparent mean retention time of organic matter in the rumen of both sheep and cattle (Poppi et al. 1981a). The effect on the intake of pasture plants of differences in the upper limit to rumen digesta load is less clear. Thornton & Minson (1973) found 14 percent more organic matter in the rumen of sheep fed tropical legumes than in sheep fed tropical grasses of similar digestibility, whereas Ulyatt (1971) measured significantly more organic matter in the rumen of sheep grazing perennial ryegrass than in animals grazing white clover in one experiment but not another. No differences between grass stem or leaf fractions have been observed in rumen dry matter content when given to sheep or cattle ad libitum (Laredo & Minson 1973, Poppi et al. 1981a).

Organic matter can be removed from the rumen by digestion and by passage to the lower gut. The rate of digestion of plant material is influenced by its chemical composition. Soluble carbohydrates can be degraded by rumen microbes about 30 times faster than storage carbohydrates which, in turn, are degraded about 5 times faster than structural carbohydrates (Maeng & Baldwin 1976). Thus, large differences in the proportions of these broad groups of plant constituents can markedly affect the rate of organic matter removal from the rumen. Because structural carbohydrates are degraded at the slowest rate, they contribute most to dietary organic matter accumulation within the rumen. Thus, variations in either the potential degradability or degradation rate of structural carbohydrate could affect substantially the rate of organic matter disappearance from the rumen. Passage of organic matter to the lower gut is likely to be affected both by the outflow from the rumen of water which carries soluble materials and small particles (Faichney et al. 1980-81) and by the rate of breakdown of large particles to sizes small enough to pass readily from the rumen (Poppi et al. 1980). Furthermore, the growth of microbes in the rumen and the amount of dietary protein which escapes degradation in the rumen are functions of the rates of digestion of organic matter and its outflow from the rumen. Thus, the availability of amino acids for absorption, which may limit animal performance, is affected by the factors which influence these rates.

Computer simulation of nutritional value

The role of computer simulation in assessing the relative importance of some determinants of the nutritive value of pasture plants is illustrated by using the computer program described above to compare the predicted voluntary feed consumption, nutrient availability and performance of adult sheep given either white clover or perennial ryegrass. The chemical composition and digestion characteristics of the pasture plants and information about the animal used in the simulations are given in Table 3. Organic matter digestibility of the two pasture species was assumed to be identical. The upper limit to organic matter accumulation in the rumen was initially set at 500 g on the basis of the experiment of Thornton & Minson (1973) where values for sheep of similar weight to that simulated were 489 and 558 g for legumes and grasses, respectively.

TABLE 3

Data used to simulate the effects of pasture type on the voluntary feed consumption and performance of sheep

	Pasture	
	White clover	Perennial ryegrass
Diet composition (% dry matter)		
Structural carbohydrates[a]		
Hemicellulose	9.7	12.6
Cellulose	20.9	25.4
Lignin	8.2	6.8
	38.8	44.8
Storage carbohydrate[b]		
Pectin	7.1	1.4
Water soluble polysaccharides	0.7	1.4
	7.8	2.8
Soluble carbohydrates		
Water soluble sugars[a]	7.9	12.6
True protein[ac]	22.0	17.4
Non-protein nitrogen[ac]	0.622	0.492
Lipid[d]	9.3	6.6
Ash[a]	10.3	12.7
Inorganic sulphur[d]	0.26	0.26
Digestion characteristics (fraction)		
Potential degradability of		
Structural carbohydrates[e]	0.75	0.75
Protein[f]	0.80	0.80
Organic matter digestibility[a]	0.74	0.74
Animal characteristics		
Empty body weight (kg)	40	40
Fleece weight (kg)	1	1
Age (years)	3.0	3.0
Time feeding (h/kg dm intake)[g]	5.3	10.2
Time ruminating (h/kg dm intake)[g]	4.2	5.5

a Rattray & Joyce (1969).
b Ulyatt (1971).
c Assume crude protein to contain 15% non-protein nitrogen (Lyttleton 1973).
d Leche & Groenendyk (1978).
e Based on Wilkins (1969).
f Estimated from Beever *et al*. (1980-81).
g From Black *et al*. (1976)

Predictions of voluntary feed consumption and of several aspects of digestion and animal performance are presented in columns 1 and 2 of Table 4 for sheep given clover and ryegrass. Values given in column 3 are for a simulation when the intake of clover was set equal to the voluntary consumption of ryegrass. Although there are no experiments to compare directly with the simulations, a number of published observations suggest that the general magnitude of the predictions is satisfactory. Ulyatt (1971) observed that voluntary feed consumption of one year old sheep was, on average, 1.2 times (1.14 and 1.26 respectively, for two experiments) greater when grazing white clover than when grazing perennial ryegrass; the predicted voluntary feed consumption of clover was 1.21 times greater than that of ryegrass. When the upper limit to organic matter accumulation in the rumen was assumed to be 500 g, the predicted intakes were 150 and 200 g greater than those observed by Ulyatt. However, dry matter accumulation in the experimental sheep did not exceed 453g. When the upper limit to organic matter accumulation in the rumen was set at 450 g, the predicted voluntary feed consumption was reduced to 1,370 and 1,186 g/day, respectively for clover and ryegrass, compared with 1,374 and 1,144 g/day observed by Ulyatt (1971).

TABLE 4

Predicted voluntary feed consumption, digestion and production characteristics of sheep fed white clover and perennial ryegrass pastures

	Voluntary feed consumption		Adjusted feed intake
	White clover	Perennial ryegrass	White clover
Pasture dry matter intake (g/day)	1574	1298	1298
Organic matter in rumen (g)	500	500	432
Apparent mean retention time of organic matter[a] (h)	8.9	11.0	9.3
Fractional outflow rate (per day)			
Water	3.51	1.94	3.10
Particulate matter	0.95	0.79	0.91
Protein flowing from stomach (g/day)			
Microbial	140	110	114
Total	227	160	183
Metabolizable energy intake (MJ/day)	17.3	13.4	14.3
Empty body weight gain (g/day)	174	103	134
Clean wool growth (g/day)	15.6	11.3	14.0

[a] Apparent mean retention time of organic matter is total organic matter accumulated in the rumen divided by the daily intake of organic matter and expressed in hours.

The predicted relationship between the voluntary consumption of digestible organic matter and apparent mean retention time of organic matter in the rumen is also similar to that developed for temperate and tropical forages by Thornton & Minson (1973). In addition, the predicted flow of total protein from the stomach was 1.14 times greater for clover than for ryegrass when given at the same intake, a factor similar to that of 1.11 observed by MacRae & Ulyatt (1974) when similar forages were fed at a rate of 800 g/day. Although the predicted gains in body weight were about 100 g/day less than those observed by Ulyatt (1971), the relative difference between the pasture species (1.69) was of the same order as that observed (1.51). Since the body weights of the sheep used by Ulyatt were not specified, this discrepancy could be explained if the experimental sheep were lighter than the simulated sheep. These comparisons between predicted and observed results, together with those of Beever *et al.* (1980-81), suggest that the program can be used satisfactorily to assess the sensitivity of voluntary feed consumption and animal performance to the factors, outlined above, which influence the nutritional value of pasture plants.

Assessment of factors affecting nutritional value

Diet composition and potential degradability of structural carbohydrate

Predicted effects on voluntary feed consumption and on some aspects of animal performance of changes from 0.5 to 0.95 in the potential degradability of structural carbohydrate in the rumen for both clover and ryegrass are shown in Fig. 1. The digestibiltiy of organic matter was changed for each simulation in proportion to the alteration in potential degradability and to the amount of structural carbohydrate in the plant. The results of three simulation sets are presented; one each for clover and ryegrass in which the fractional outflow rates of water were fixed at the respective values derived from the first simulation (Table 4), and an additional one for ryegrass with the fractional outflow rate of water fixed at the value obtained for clover. The comparison between clover and ryegrass, with the same fractional outflow rate of water, illustrates the effect of differences in degradation rate of organic matter resulting from differences in chemical composition of the plants.

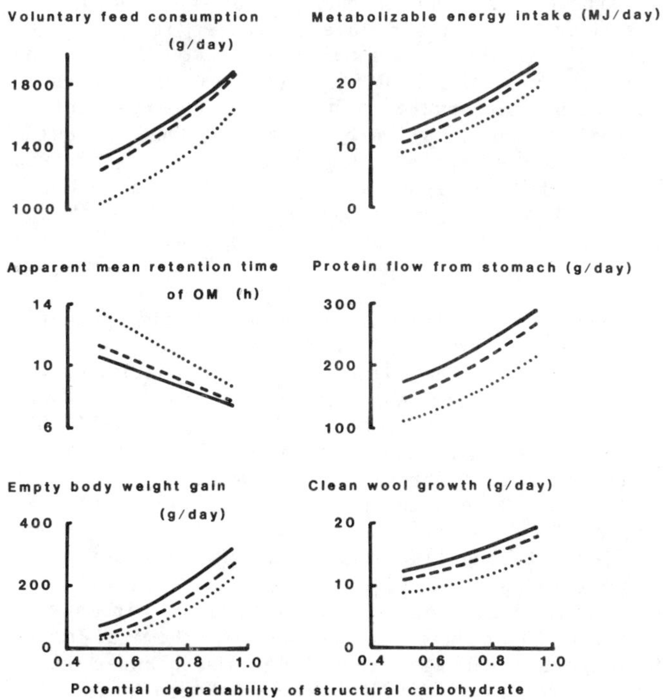

Fig. 1. Sensitivity of computer predictions to changes in the potential degradability of structural carbohydrate in white clover (——) and perennial ryegrass (-----) when the fractional outflow rate of water was held constant at 3.51 per day, and in ryegrass when the fractional outflow rate of water was 1.95 per day (.....).

Ulyatt (1981) suggested that the lower ratio of structural to readily fermentable carbohydrate in clover than in ryegrass is the major reason for the reduced apparent mean retention time of organic matter in the rumen and hence greater voluntary feed consumption of sheep eating clover. However, the predictions suggest that only small differences in apparent organic matter retention in the rumen and in voluntary feed consumption result from the differences in plant composition. However, the effect increases as the potential degradability of the structural carbohydrates declines. Wilkins (1969) found that the potential degradability of cellulose in ryegrass fell from about 0.85 to 0.7 as the plant matured. Over this range of potential degradability, the predicted differences between plants with the composition of clover and ryegrass in voluntary feed consumption and apparent mean retention time of organic matter in the rumen do not exceed, respectively, 50 g/day and 0.5 h.

Conversely, changes in the potential degradability of structural carbohydrate were predicted to have substantial effects on the apparent rate of organic matter removal from the rumen and on voluntary feed consumption. The potential degradability of structural carbohydrates in forage plants ranges at least from 0.92 for oats in its vegetative state to 0.39 for mature *Lotus corniculatus* and *Dactylis glomerata* (Smith *et al*. 1972). Even when the chemical composition of the plant is held constant, as in the example with ryegrass, a change of this magnitude was predicted to cause a difference in voluntary feed consumption of about 600 g/day. The decline in potential degradability of structural carbohydrate is normally associated with an increase in the ratio of structural to readily fermentable carbohydrate in a forage (Smith *et al*. 1972) and, therefore, the effects on voluntary feed consumption would be further enhanced. Nevertheless, substantial differences in the potential degradability of structural carbohydrates have been reported between forage materials of similar structural carbohydrate content. Poppi *et al*. (1981b) measured the potential degradaility of structural carbohydrate to be 0.66 and 0.60, respectively, in the leaf and stem fractions of two tropical grasses. A small difference like this is predicted to increase the intake of ryegrass in the simulated example by about 65 g/day.

Relatively few measurements have been made of the potential degradability of structural carbohydrate in forage plants, but those of Smith *et al*. (1972) suggest that the cell wall constituents of legumes have a lower degradability than those in grasses. Recent observations (A. Dunlop, R.C. Kellaway and J.C. Wadsworth 1981, personal communication) show that the potential degradability of structural carbohydrates in cereal straws can be increased from 0.6 to 0.73 by treatment with alkali. This change alone would be responsible for a significant increase in the nutritional value of cereal straw. The above simulations indicate the likely magnitude of improvements to the nutritional value of pasture plants that will result from increases in the potential degradability of structural carbohydrate brought about by genetic, agronomic, mechanical or chemical means. Alternatively, they suggest advantages resulting from changes in the ratio of structural to readily fermentable carbohydrate in plant material are likely to be less significant.

Degradation rate of structural carbohydrate

Predicted effects of changes in the degradation rate of structural carbohydrate on voluntary feed consumption and on apparent mean retention time of organic matter in the rumen are shown in Fig. 2a for sheep given ryegrass. The fractional outflow rate of water was held constant at 1.94 per day. The in vitro studies of Smith et al. (1972) indicate that the degradation rate of cell wall constituents can vary by almost eightfold across forage materials, but commonly the variation is three- to four-fold as plants mature. Clearly, differences of these magnitudes would have significant effects on voluntary feed consumption and animal performance. A three-fold variation in the rate of degradation of structural carbohydrate in the ryegrass example simulated would result in a predicted change in intake of 200 g/day. The results of Smith et al. (1972) also indicate that the rate of cell wall digestion is greater in legumes than in grasses. With both forage classes, the rate of digestion was positively correlated with the soluble dry matter content of the plant material which declines as plants mature. Despite the large differences observed by Smith et al. (1972), differences have not been shown in the rate of degradation of structural carbohydrate between leaf and stem fractions of tropical grasses or legumes (Poppi et al. 1981b, Hendricksen et al. 1981). However, the alkali treatment of cereal straws significantly increased the fractional rate of degradation of structural carbohydrate, estimated using nylon bags incubated in vivo, from 0.81 to 1.42 per day (A. Dunlop, R.C. Kellaway, J.C. Wadsworth, 1981, personal communication). The above predictions indicate that the nutritional value of pasture plants is affected when the rate of degradation of structural carbohydrate changes several fold, but the small differences observed between leaf and stem fractions, or clovers and grasses, are likely to have little effect on voluntary feed consumption.

Fractional outflow rate of structural carbohydrate

Predicted effects of changes in the fractional outflow rate from the rumen of structural carbohydrate on voluntary feed consumption and on apparent mean retention time of organic matter in the rumen are shown in Fig. 2b for sheep given ryegrass. The fractional outflow rate of water was held constant at 1.94 per day for these simulations. Measurements of the fractional rate of disappearance of lignin from the rumen reflect the fractional outflow rate of structural carbohydrate if lignin is not digested (Faichney 1980). Significant differences have been observed between leaf and stem fractions of both tropical grasses (Poppi et al. 1981b) and legumes (Hendricksen et al. 1981) in the fractional disappearance rate of lignin from the rumen of sheep and cattle. In the study with tropical grasses, there was little change in the fractional outflow rate of water between the plant fractions, but the lignin disappearance rate was 1.2 times faster in sheep eating the leaf component. With the legume *Lablab purpureus*, the fractional disappearance rate of lignin was 1.45 times greater when sheep were given the leaf rather than the stem, but the fractional outflow rate of water was also 1.2 times greater. Moseley (1981), when comparing white clover and perennial ryegrass, noted that the passage of particulate matter during the first three hours after feeding was faster when sheep were given clover. He also observed that clover was broken down to smaller sizes than ryegrass during eating and suggested that differences in plant morphology may be responsible for the more extensive comminution and faster rate of outflow of particulate matter from the rumen of sheep fed clover. Hendricksen et al. (1981), in their study with the tropical legume, *L. purpureus*, also observed a greater breakdown during eating of leaf than of stem particles to a size which readily passes from the rumen.

In addition, the substantial improvement in the nutritional value of forages resulting from grinding and pelleting could be due, in part at least, to an increase in the fractional outflow rate of particulate matter from the rumen. Laredo & Minson (1975) showed that the grinding and pelleting of tropical grasses increased the apparent disappearance rate of organic matter from the rumen of sheep by up to 1.35 times. Since the digestibility of organic matter declined, this observation suggests that an increase in the fractional outflow rate of feed particles did indeed occur. It can be seen in Fig. 2b that a 1.3 fold increase in the fractional outflow rate of structural carbohydrate would result in an increase in the intake of ryegrass of 200 g/day and a decrease of 1.5 h in the apparent mean retention time of organic matter.

Voluntary feed consumption (g/day)

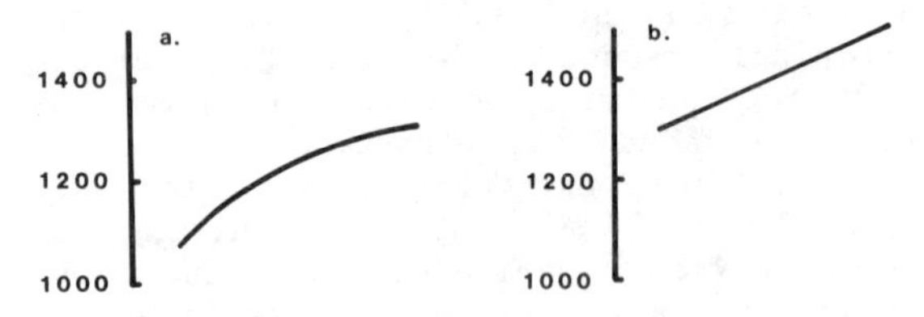

Apparent mean retention of organic matter in rumen (h)

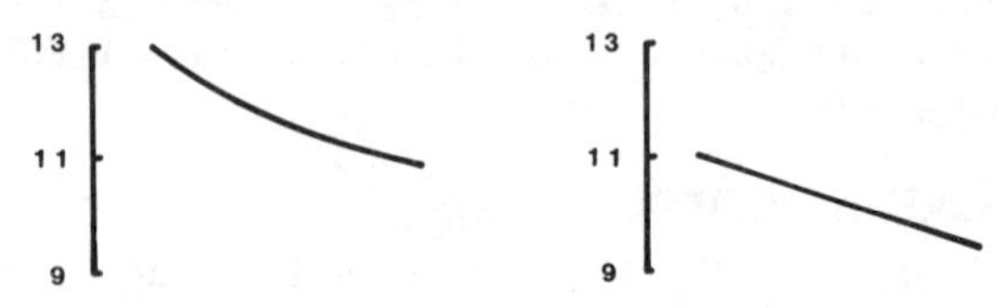

Protein flow from stomach (g/day)

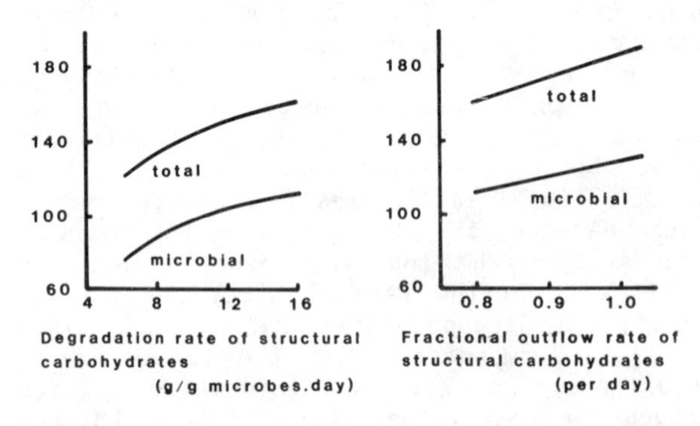

Fig. 2. Sensitivity of computer predictions to changes in a) the rate of degradation and b) the fractional outflow rate of structural carbohydrate in perennial ryegrass when the fractional outflow rate of water was held constant at 1.94 per day.

These simulations suggest that changes in the physical nature of plant material brought about by either genetic manipulation to plant morphology, which affects the rate of comminution during eating and ruminating, or by physical means, will have a substantial effect on nutritional value if they result in a significant reduction in the size of particles entering the rumen. In addition, differences between animals in their ability to break down feed particles during eating may be partially responsible for the observed variations in voluntary feed consumption and animal performance.

Fractional outflow rate of water

Predicted effects of a two-fold change in the fractional outflow rate of water on voluntary feed consumption and on apparent mean retention time of organic matter in the rumen are shown in Fig. 3a for a sheep given ryegrass. There are two sets of simulations: one in which the fractional outflow rate of structural carbohydrate was held constant at the value obtained during the initial simulation (Table 4), and the other in which it was allowed to change in association with changes in the fractional outflow rate of water.

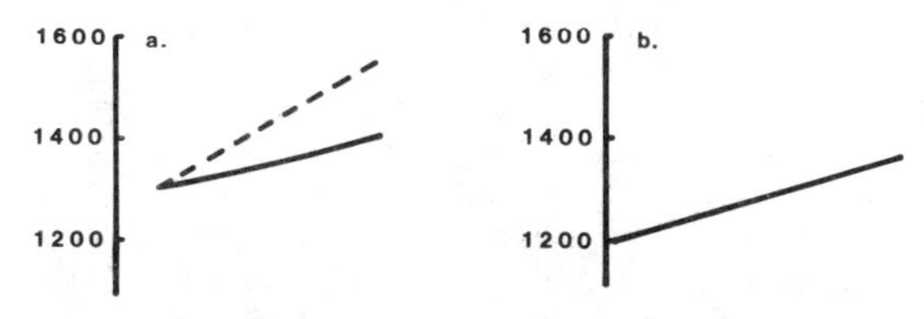

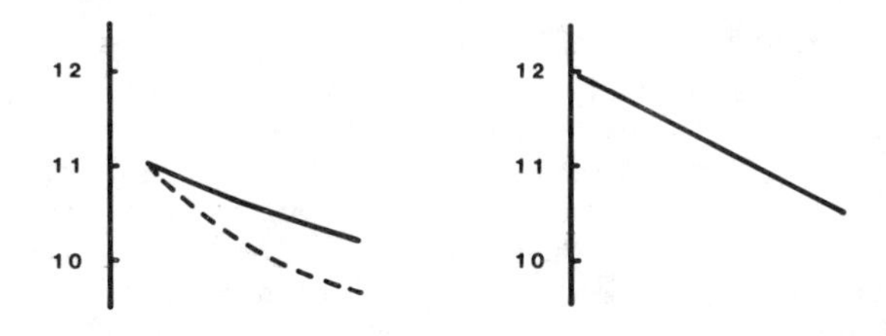

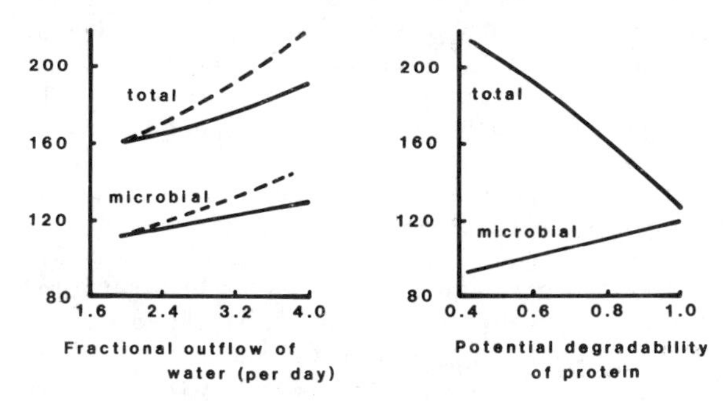

Fig. 3. Sensitivity of computer predictions to changes in a) the fractional outflow rate of water for perennial ryegrass when the fractional outflow rate of structural carbohydrate was either held constant at 0.794 per day (———) or allowed to change with changes in the outflow rate of water (-----), and b) the potential degradability of dietary protein in ryegrass when the fractional outflow rate of water was held constant at 1.94 per day.

Poppi et al. (1981b) and Hendricksen et al. (1981) observed small but significant differences in the fractional outflow rate of water when either leaf or stem fractions of tropical forages were given to cattle; the differences were not significant for sheep. A comparison between the experiments of Weston & Hogan (1968, 1971) for sheep given dried ryegrass or Trifolium subterraneum at different stages of maturity suggests little difference in the fractional outflow rates of water between these species when dried; values ranged from 2.05 to 2.67 per day for ryegrass and from 2.02 to 2.70 per day for T. subterraneum clover. However, M.J. Ulyatt (cited by Ulyatt & MacRae 1974) obtained values for the fractional outflow rate of water of 3.0 and 4.1 per day when sheep were given, respectively, fresh perennial ryegrass and white clover.

Results of the simulations presented in Fig. 3a suggest that factors which change the osmolarity of rumen contents, and thus the fractional outflow rate of water (Faichney et al. 1980-81), would have a relatively small effect on the apparent mean retention time of organic matter in the rumen and on voluntary feed consumption. A change in fractional outflow rate of the magnitude observed by Ulyatt would result in an increase in intake of the ryegrass of only 50 g/day. However, if there is an associated change in the outflow rate of structural carbohydrate, the effect on voluntary feed consumption is greatly increased. A doubling in the fractional outflow rate of water was associated with a predicted increase of 1.18 times in the fractional outflow rate of structural carbohydrate and these together resulted in an increase in intake of ryegrass of more than 250 g/day.

Protein flow from the stomach

The simulation technique has already been used to indicate that differences between pasture plants in amino acid absorption can influence their nutritional value for growing lambs (Tables 1 and 2). Most of the factors which influence voluntary feed consumption also improve the flow of protein from the stomach (Figs. 1 to 4). An increase in either the rate of organic matter degradation or the rate of passage of material from the rumen will increase protein flow independent of effects on either feed consumption or the potential degradability of dietary protein in the rumen (Table 4, Beever et al. 1980-81). Nevertheless, the potential degradability of plant proteins in the rumen is known to vary and Ulyatt (1981) suggests that this may further influence the nutritional value of pasture plants. The sensitivity of the predictions of voluntary feed consumption and apparent mean retention time of organic matter in the rumen to changes in the potential degradability of protein in ryegrass are shown in Fig. 3b. As the potential degradability of protein was increased, the apparent mean retention time of organic matter declined, voluntary feed consumption increased, and there was an increase in microbial growth and thus in the flow of microbial protein from the stomach. However, the increase in potential degradability caused a steep decline in the flow of dietary protein and, therefore, total protein from the stomach. Thus, in situations where amino acid availability is limiting animal performance, and where voluntary feed consumption is restricted by the amount of digesta which can accumulate in the rumen, a decrease in the potential degradability of dietary protein may be either advantageous or disadvantageous. The ultimate effect on animal performance and, hence, the nutritional value of a forage, depends upon the relative response of the animal to the increase in amino acid supply compared to the simultaneous decrease in voluntary feed consumption and energy availability. The animal response to a decrease in the potential degradability of dietary protein would be further complicated when the deficiency in amino acid absorption was such that it reduced the energy demand of the animal to such an extent that rumen digesta load was no longer limiting feed intake. In situations where amino acid availability is not limiting animal performance, a decrease in the potential degradability of dietary protein would decrease voluntary feed consumption and, therefore, the nutritional value of the pasture plant. In the particular situation simulated, wool growth was predicted to increase as the potential degradability of protein was increased to 0.65, but declined thereafter.

Upper limit to organic matter accumulation in the rumen

Predicted effects on voluntary feed consumption of varying the upper limit to organic matter accumulation in the rumen for sheep given either white clover or ryegrass are shown in Fig. 4. In these simulations, the fractional outflow rates of material from the stomach were held constant at the values obtained during the initial simulation (Table 4). Although feed intake affects fractional outflow rates, the program predicts the fractional outflow rate of water from an empirical equation (Faichney et al. 1980-81) and the data from which it was developed did not cover the range in intakes generated by these simulations.

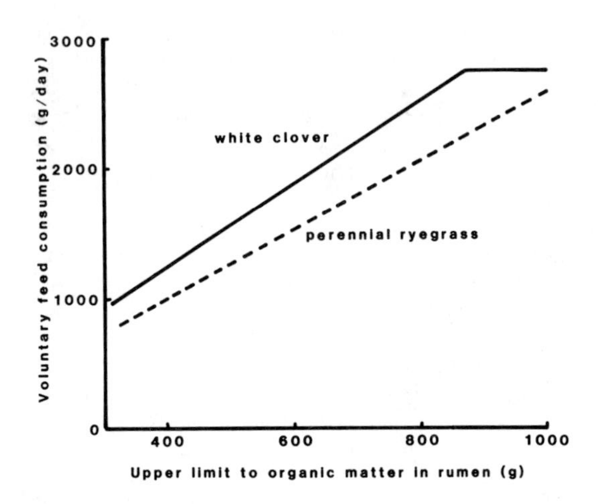

Fig. 4. Sensitivity of predicted voluntary feed consumption to changes in the upper limit to organic matter accumulation in the rumen of sheep given either white clover (———) or perennial ryegrass (-----).

The predictions indicate that a change to the upper limit of organic matter accumulation in the rumen could substantially affect voluntary feed consumption and animal performance. The simulations indicate that the energy demand of a 40 kg strongwool Merino ram eating white clover would be satisfied when dry matter intake reaches about 2.7 kg. At this intake, the predicted total organic matter accumulation in the rumen was 870 g and empty body weight gain was 340 g/day. Several reported values for the amount of organic matter in the rumen of sheep fed *ad libitum* range from about 300 to 900 g (Ulyatt 1971, Thornton & Minson 1972, 1973, Laredo & Minson 1973, Poppi *et al.* 1981a). However, small differences are generally observed within experiments between either grasses and clovers or leaf and stem fractions. Plant and animal factors probably both interact to affect the limit to digesta load. Thornton & Minson (1973) and Van Soest (1975) suggest that differences in packing density of the plant material in the rumen could cause some of the variation in digesta organic matter load. Plant materials which contain large quantities of cell wall constituents tend to accumulate less organic material in a given space. Weston (1981) suggests that pregnancy or the extent of energy deficit in an animal could affect the upper limit to rumen digesta load for animals of similar weight.

These simulations indicate that manipulations which result in an increase in the packing density, particularly of pasture material high in cell wall constituents, could have a substantial effect on its nutritional value. The grinding of tropical grasses, which was reported to increase voluntary feed consumption 1.6 to 1.9 times, was associated with a three-fold increase in bulk density (Laredo & Minson 1975). Furthermore, A. Dunlop, R.C. Kellaway and J.C. Wadsworth (personal communication) observed that a doubling of the intake of cereal straws by sheep resulting from treatment with alkali was associated with an increase of dry matter accumulation in the rumen from 480 to 660 g.

CONCLUSIONS

By using the computer simulation technique, it has been possible to assess, within the framework of current concepts, the relative importance of several factors thought to influence the nutritional value of pasture plants. Even within the narrow limits of the examples used, it is apparent that some factors are likely to have more influence than others on the nutritional value of plants. Differences between plants in chemical composition with respect to the ratio of readily degradable to structural carbohydrates, and differences in the rate of digestion of structural carbohydrates, are probably less important than differences in either the potential degradability or fractional outflow rate of structural carbohydrate. The simulations also suggest that the value of reducing the potential degradability of plant protein within the rumen depends on the situation in which the pasture plant is eaten; an improvement in nutritional value will occur only when the effect of the greater supply of amino acids outweighs that resulting from any reduction in voluntary feed consumption. Finally, factors which determine the upper limit to digesta load in the rumen have an important bearing on nutritional value of plants.

The computer simulation programs outlined should be of particular value when assessments are made of the effects on nutritional value of manipulations to the chemical composition or physical structure of plant material by genetic, agronomic, chemical or physical means. Such manipulations generally result in the alteration of a number of factors which influence nutritional value, and the simulation technique allows the importance of each to be isolated and the likely overall effect determined.

The major advantage of the simulation approach is that it allows the determinants of the nutritional value of a feed (that is, animal requirements and nutrient supply) to be considered simultaneously. By using this technique, many of the factors which may interact to affect the nutritional value of a specific plant material for a specific production situation can be integrated to isolate the most likely limitations in that system. This precludes undue emphasis being put on particular factors when they are considered in isolation. It also allows an assessment to be made of the likely value in terms of animal production of manipulations to a particular system. For example, if non-protein nitrogen was shown to be limiting microbial growth in the rumen, the simulation technique can be used to estimate the likely requirement for additional nitrogen and the effects of supplying varying quantities on voluntary feed consumption, nutrient supply and animal performance. The practical advantage resulting from such a manipulation can then be evaluated.

Lack of information about the chemical composition and digestion characteristics of plant materials currently hinders the extensive use of existing computer programs for identifying and overcoming the limitations to animal production from pastures. Few estimates have been made of the potential degradability or rate of degradation of structural carbohydrates in the plant materials selected by animals and there is little information on the potential degradability within the rumen of plant proteins. Of the 1,600 feedstuffs on file in the Australian Feeds Information Centre, only 20.4 percent have recordings for cell wall constituents, 9.5 percent for soluble carbohydrates, 0.7 percent for starch and in only 1 percent of the feedstuffs is nitrogen separated into protein and non-protein nitrogen (Leche & Groenendyk 1978).

REFERENCES

Baldwin, R.L.; Koong, L.J. (1980) Mathematical modelling in analyses of ruminant digestive function: philosophy, methodology and application. In: Digestive physiology and metabolism in ruminants. Editors Y. Ruckerbusch and P. Thivend. Lancaster, U.K., M.T.P. Press, pp. 251-268.

Baldwin, R.L.; Koong, L.J.; Ulyatt, M.J. (1977) A dynamic model of ruminant digestion for evaluation of factors affecting nutritive value. Agricultural Systems 2, 255-288.

Beever, D.E.; Black, J.L.; Faichney, G.J. (1980-81) Simulation of the effects of rumen function on the flow of nutrients from the stomach of sheep: Part 2 - Assessment of computer predictions. Agricultural Systems 6, 221-241.

Black, J.L.; Griffiths, D.A. (1975) Effects of liveweight and energy intake on nitrogen balance and total N requirements of lambs. British Journal of Nutrition 33, 399-413.

Black, J.L.; Graham, N.McC.; Faichney, G.J. (1976) Simulation of protein and energy utilization in sheep. In: From plant to animal protein. Reviews in Rural Science No. 2. Editors T.M. Sutherland, J.R. McWilliam, R.A. Leng. Armidale, Australia, University of New England, pp. 161-166.

Black, J.L.; Dawe, S.T.; Colebrook, W.F.; James, K.J. (1979) Protein deficiency in young lambs grazing irrigated summer pasture. Proceedings of the Nutrition Society of Australia 4, 126.

Black, J.L.; Beever, D.E.; Faichney, G.J.; Howarth, B.R.; Graham, N.McC. (1980-81) Simulation of the effects of rumen function on the flow of nutrients from the stomach of sheep: Part 1 - Description of a computer program. Agricultural Systems 6, 195-219.

Blaxter, K.L. (1976) Experimental obesity in farm animals. Publication of the European Association of Animal Production No. 19, 129-132.

Egan, A.R. (1974) Protein-energy relationships in the digestion products of sheep fed on herbage diets differing in digestibility and nitrogen concentration. Australian Journal of Agricultural Research 25, 613-630.

Faichney, G.J. (1980) Measurement in sheep of the quantity and composition of rumen digesta and of the fractional outflow rates of digesta constituents. Australian Journal of Agricultural Research 31, 1129-1137.

Faichney, G.J.; Beever, D.E.; Black, J.L. (1980-81) Prediction of the fractional rate of outflow of water from the rumen of sheep. Agricultural Systems 6, 261-268.

Freer, M. (1981) The control of food intake by grazing animals. In: Grazing animals. Editor F.H.W. Morley. Amsterdam, Elsevier, pp. 105-124.

Graham, N.McC.; Black, J.L.; Faichney, G.J. (1976a) Computer simulation of energy and nitrogen utilization in sheep. In: Feed composition, animal nutrient requirements and computerization of diets. Editors P.V. Fonnesbeck, L.E. Harris and L.C. Kearl. Utah State University, pp. 447-455.

Graham, N.McC.; Black, J.L.; Faichney, G.J.; Arnold, G.W. (1976b) Simulation of growth and production in sheep - model 1: A computer program to estimate energy and nitrogen utilization, body composition and empty liveweight change, day by day for sheep of any age. Agricultural Systems 1, 113-138.

Hendricksen, R.E.; Poppi, D.P.; Minson, D.J. (1981) The voluntary intake, digestibility and retention time by cattle and sheep of stem and leaf fractions of a tropical legume (Lablab purpureus). Australian Journal of Agricultural Research 32, 389-398.

Hodge, R.W. (1974) Efficiency of food conversion and body composition of the preruminant lamb and the young pig. British Journal of Nutrition 32, 113-126.

Hogan, J.P.; Elliott, N.M.; Hughes, D.A. (1979) Maximum wool growth rates expected from Australian Merino genotypes. In: Physiological and environmental limitations to wool growth. Editors J.L. Black and P.J. Reis. Armidale, University of New England, pp. 43-59.

Lancashire, J.A.; Keogh, R.G. (1966) Some aspects of the behaviour of grazing sheep. Proceedings of the New Zealand Society of Animal Production 26, 22-35.

Laredo, M.A.; Minson, D.J. (1973) The voluntary intake, digestibility, and retention time by sheep of leaf and stem fractions of five grasses. Australian Journal of Agricultural Research 24, 875-888.

Laredo, M.A.; Minson, D.J. (1975) The effect of pelleting on the voluntary intake and digestibility of leaf and stem fractions of three grasses. British Journal of Nutrition 33, 159-170.

Leche, T.F.; Groenendyk, G.H. (1978) Composition of animal feedstuffs in Australia (draft version), Australian Feeds Information Centre. Blacktown; C.S.I.R.O., Division of Animal Production, 254 pp.

Lyttleton, J.W. (1973) Proteins and nucleic acids. In: Chemistry and biochemistry of herbage. Editors G.W. Butler and R.W. Bailey. London, Academic Press, Vol. 1, pp. 63-103.

MacRae, J.C.; Ulyatt, M.J. (1974) Quantitative digestion of fresh herbage by sheep. II. The sites of digestion of some nitrogenous constituents. Journal of Agricultural Science, Cambridge 82, 309-319.

Maeng, W.J.; Baldwin, R.L. (1976) Dynamics of fermentation of a purified diet and microbial growth in the rumen. Journal of Dairy Science 59, 636-642.

Minson, D.J. (1981) Nutritional differences between tropical and temperate pastures. In: Grazing animals. Editor F.H.W. Morley. Amsterdam, Elsevier, pp. 143-157.

Moseley, G. (1981) Herbage quality and physical breakdown in the foregut of sheep. Proceedings of the New Zealand Society of Animal Production 41, 142-151.

Ørskov, E.R.; Fraser, C. (1973) The effect of level of feeding and protein concentration on disappearance of protein in different segments of the gut in sheep. Proceedings of the Nutrition Society 32, 68A-69A.

Ozanne, P.G.; Purser, D.B.; Howes, K.M.W.; Southey, I.N. (1976) Influence of phosphorus content on feed intake and weight gain in sheep. Australian Journal of Experimental Agriculture and Animal Husbandry 16, 353-360.

Poppi, D.P.; Norton, B.W.; Minson, D.J.; Hendricksen, R.E. (1980) The validity of the critical size theory for particles leaving the rumen. Journal of Agricultural Science, Cambridge 94, 275-280.

Poppi, D.P.; Minson, D.J.; Ternouth, J.H. (1981a) Studies of cattle and sheep eating leaf and stem fractions of grasses. I. The voluntary intake, digestibility and retention time in the reticulo-rumen. Australian Journal of Agricultural Research 32, 99-108.

Poppi, D.P.; Minson, D.J.; Ternouth, J.H. (1981b) Studies of cattle and sheep eating leaf and stem fractions of grasses. II. Factors controlling the retention of feed in the reticulo-rumen. Australian Journal of Agricultural Research 32, 109-121.

Purser, D.B. (1981) Nutritional value of mediterranean pasture. In: Grazing animals. Editor F.H.W. Morley. Amsterdam, Elsevier, pp. 159-180.

Rattray, P.V.; Joyce, J.P. (1969) The utilization of perennial ryegrass and white clover by young sheep. Proceedings of the New Zealand Society of Animal Production 29, 102-113.

Rees, M.C.; Minson, D.J. (1976) Fertilizer calcium as a factor affecting voluntary intake, digestibility and retention time of pangola grass (Digitaria decumbens) by sheep. British Journal of Nutrition 36, 179-187.

Rees, M.C.; Minson, D.J. (1978) Fertilizer sulphur as a factor affecting voluntary intake, digestibility and retention time of pangola grass (Digitaria decumbens) by sheep. British Journal of Nutrition 39, 5-11.

Smith, L.W.; Goering, H.K.; Gordon, C.H. (1972) Relationships of forage compositions with rates of cell wall digestion and indigestibility of cell walls. Journal of Dairy Science 55, 1140-1147.

Thomson, D.J.; Beever, D.E. (1980) The effects of conservation and processing on the digestion of forages by ruminants. In: Digestive physiology and metabolism in ruminants. Editors Y. Ruckebusch and P. Thivend. Lancaster, U.K., M.T.P. Press, pp. 291-308.

Thornton, R.F.; Minson, D.J. (1972) The relationship between voluntary intake and mean apparent retention time in the rumen. Australian Journal of Agricultural Research 23, 871-877.

Thornton, R.F.; Minson, D.J. (1973) The relationship between apparent retention time in the rumen, voluntary intake, and apparent digestibility of legume and grass diets in sheep. Australian Journal of Agricultural Research 24, 889-898.

Ulyatt, M.J. (1971) Studies on the causes of the differences in pasture quality between perennial ryegrass, short-rotation ryegrass, and white clover. New Zealand Journal of Agricultural Research 14, 352-367.

Ulyatt, M.J. (1981) The feeding value of temperate pastures. In: Grazing animals. Editor F.H.W. Morley. Amsterdam, Elsevier, pp. 125-141.

Ulyatt, M.J.; MacRae, J.C. (1974) Quantitative digestion of fresh herbage by sheep. I. The sites of digestion of organic matter, energy, readily fermentable carbohydrate, structural carbohydrate, and lipid. Journal of Agricultural Science, Cambridge 82, 295-307.

Van Soest, P.J. (1975) Physico-chemical aspects of fibre digestion. In: Digestion and metabolism in the ruminant. Editors I.W. McDonald and A.C.I. Warner. Armidale, University of New England, pp. 351-365.

Weston, R.H. (1981) Animal factors affecting feed intake. In: Nutritional limits to animal production from pastures. Editor J.B. Hacker. Farnham Royal, U.K., Commonwealth Agricultural Bureaux, pp. 183-198.

Weston, R.H.; Margan, D.E. (1979) Herbage digestion in the stomach and intestines of weaner lambs at different stages of maturity. Australian Journal of Agricultural Research 30, 543-549.

Weston, R.H.; Hogan, J.P. (1968) The digestion of pasture plants by sheep. II. The digestion of ryegrass at different stages of maturity. Australian Journal of Agricultural Research 19, 963-979.

Weston, R.H.; Hogan, J.P. (1971) The digestion of pasture plants by sheep. V. Studies with subterranean and berseem clovers. Australian Journal of Agricultural Research 22, 139-157.

Wilkins, R.H. (1969) The potential digestibility of cellulose in forage and faeces. Journal of Agricultural Science, Cambridge 73, 57-64.

PART 7

SUMMATIVE ADDRESS

NUTRITIONAL LIMITS TO ANIMAL PRODUCTION FROM PASTURES; SYMPOSIUM HIGHLIGHTS AND THEIR RELEVANCE TO FUTURE RESEARCH

G.W. BUTLER

Head Office, Department of Scientific and Industrial Research, Wellington, New Zealand.

ABSTRACT

Twenty-four invited papers presented to an international symposium on "Nutritional Limitations to Animal Production from Pastures" are discussed briefly in relation to their main conclusions and hence priorities for future research.

INTRODUCTION

The papers presented to this symposium have given us a timely, wide-ranging and valuable assessment of where the limitations are in the provision of good nutrition for grazing animals, both in our understanding of the processes involved and in our ability to manipulate those processes to overcome perceived limitations.

In this paper I have attempted to integrate the themes which were developed and to identify research areas which should have emphasis, because good opportunities for progress are seen and because substantial increases in ruminant production would result.

THE PRESENT SITUATION

In his opening paper, Mahadevan (1982) brought the question of animal production in developing regions of the world into sharp focus, with his quotation from Blair-Rains and Kassan (1979) - "Half a century of animal disease prevention and eradication measures, several decades of grassland and pastures investigations and of research into most aspects of animal nutrition and genetics have scarcely affected animal production." The principle shortcomings in major regions and the means seen by Mahadevan for overcoming them are summarized in Table 1.

Mahadevan concluded that "where livestock production is based on pastoralist systems, it is necessary to ensure that thresholds of resource tolerance are not exceeded by human and animal pressures, especially in drought years. On arable lands the most successful approaches are likely to be those based on integrated crop/livestock farming systems, including a more efficient utilization of crop residues and agro-industrial by-products for animal feeding. Where land is not a limiting factor to livestock development, the greatest need is for systems of grazing management that fit into traditional patterns of farming."

In considering temperate pastures, Reid & Jung (1982) pointed out that temperate grasslands constitute 30 percent of the world's total area in range and pasture and maintain 35 percent of the world's ruminant livestock units, but produce 65-70 percent of the world's beef, 50-55 percent of mutton and lamb and 75 percent of the world's whole milk. They emphasized the wide range of environments within temperate grasslands, varying from year-round grazing in areas in the lower south of the United States and in parts of New Zealand to periods of pasture growth of only 125 days in Finland. They concluded that the productivity of temperate grasslands had increased during the last 40 years with development of pastures with improved yield, adaptation and quality, and that much depended on the agrarian sociologist and political scientist to create the right socio-economic environments for increased production. Because of increased energy costs, the more favourable input-output energy relationships for grass-legume pastures rather than nitrogen fertilized grass pastures is becoming a key factor.

Two major limitations identified by Reid & Jung were:-

1. The need to select pasture species of reasonable quality adapted to acid and problem soils, since increasing populations will require pasture production to extend into areas now defined as marginal because of soils, climate and topography.

2. The need to minimize losses in animal production imposed by the apparently increased incidence of nutritional imbalances and metabolic disorders in animals grazing improved pastures.

TABLE 1

Principle limiting factors for pastoral production in the grasslands in developing countries and means of overcoming them (after Mahadevan 1982)

Region	Limiting Factor	Solution
AFRICA Arid zone	Over-stocking	Education in grazing management
Semi-arid zone	Forage shorage through cropping expansion	Improved pastures
Sub-humid and humid zones	Animal disease	Tsetse control, trypanotolerant breeds and trypanocidal drugs
Highlands	No major problems	Successful farming of temperate animal breeds on temperate pastures
ASIA AND FAR EAST	Shortage of available grazing lands	Skilled use of fibrous agricultural wastes
LATIN AMERICA		
Infertile savannah	Inadequate nutrition	Improved pasture management
Fertile soils	Management constraints	Improved pastures
NEAR EAST	Severe feed shortage	Better herbage production on range; introduction of cultivated pastures into cereal fallow rotation; better use of cereal strains

With regard to animal production from Mediterranean pastures, Allden (1982) summarized to use of pastures based on legume species of Mediterranean origin (subterranean clover and *Medicago* spp.) integrated with cereal operations. Irrigation is not usually feasible and systems must be geared to hot dry summers inimical to plant growth and to mild wet winters. The crucial requirement is for legume persistence through good seed carryover.

Mannetje (1982), in considering the limitations of tropical pastures, pointed out that the genetic potential for growth of large and small bovines is 1.5 and 0.5 kg/day respectively, but the maximum average daily gain on unsupplemented tropical pastures is only about half this. Furthermore, "if tropical pastures cannot supply the energy for the genetic potential of beef cattle, it follows that they are even less capable of fulfilling the needs of dairy cows." As far as fibre production goes, the situation is somewhat better because wool growth continues at the expense of body growth when feed intake is below requirements. Mannetje considers that there is great scope for improvement of tropical pastures, in particular by the development of new cultivars which possess "small rates of decline in DMD and crude protein, despite the rapid rate of growth and development under high temperatures" (Wilson & Minson 1980).

The general constraints to the nutrition of ruminants in the major grassland regions have thus been described, on a regional basis, in the four opening papers (Mahadevan 1982, Reid & Jung 1982, Allden 1982, Mannetje 1982). The components of ruminant nutrition which led to these major constraints were then identified in subsequent papers.

LIMITATIONS UPON RUMINANT NUTRITION IMPOSED BY VARIOUS FACTORS

The various plant and animal factors which inter-relate to determine the efficiency of ruminant nutrition from pastures were summarized in a series of ten papers. Amongst the salient points and highlights, I would select the following:-

Environmental effects on herbage quality

(a) The effect of high growth temperatures in causing large decreases in digestibility of grasses and smaller decreases in the digestibility of legumes is of great importance, especially for sub-tropical and tropical grasslands (Wilson 1982).

(b) Wilson discussed the effects on herbage quality, and hence animal production, of intermittent droughts of light to moderate severity. He concluded that the information available suggested that moderately water-stressed herbage of improved pasture varieties is likely to be of high quality and that provided low yield did not limit intake, daily liveweight gain could be better than average from this type of herbage. If this can be substantiated by further critical animal data, it would have significance for pasture management in many situations.

Deleterious quality factors

These have been identified in many pasture plants and much is known about their occurrence and metabolism by ruminants. Hegarty (1982) pointed out the considerable economic costs to the pastoral industry of deleterious herbage constituents. The first step in overcoming such losses is carefully to identify the causative agents, and while this can be relatively simple for some agents, such as cyanoglucosides or oestrogenic isoflavones, the problem may be more complex when several causative agents act in concert, as for example in bloat. The chemical modifications of deleterious herbage constituents which can occur through metabolism in the rumen also adds another dimension to the problem.

Besides removing them by plant breeding (Hacker 1982, Bray 1982), it is often possible to nullify or reduce their deleterious effects by management techniques, special supplementation etc. Whether one is aiming at the long-term objective of removal by plant breeding, or at the more immediate objective of control of deleterious quality factors through management, a knowledge of the chemistry and biochemistry of the factors is essential. Also, where a deleterious quality factor arises from a multiplicity of causal contributing factors (as, for example, in bloat), the chances for control by management are better than through breeding.

The principle of modifying rumen metabolism of deleterious herbage constituents is an important one, which is likely to be more widely practicable in future. The possibility has been illustrated by the finding that mimosine, present in *Leucaena* species, is only degraded as far as the goitrogen DHP (3-hydroxy-4 (1H)-pyridone) by rumen microflora in Queensland, but is degraded beyond DHP by rumen microflora in Hawaii and Indonesia, (Jones 1981, poster papers by R.J. Jones & R.G. Megarrity, and by J.B. Lowry *et al*. p.511). Thus countries having a toxicity problem with *Leucaena* may be able to achieve a biological solution by inoculation or modification of micro-organisms.

Herbage composition

Norton (1982) reviewed the differences in herbage composition between temperate and tropical legumes and grasses and several important features emerged. The fact that most tropical grasses are C_4 plants and exhibit the characteristic Kranz leaf anatomy has profound effects on the composition of tropical grass herbage. A high proportion of thick bundle sheath cell walls of high tensile strength and a close packing of mesophyll cells leads to greater difficulties in the harvesting of leaf material by grazing cattle and digestion by rumen micro-organisms. The basic difference in photosynthetic pathway also results in lower amounts of protein in tropical grasses and in a lower proportion of soluble protein.

Interesting trends in mineral composition between temperate and tropical species are also observed. Tropical grasses and legumes were generally lower in phosphorus, calcium and sodium. Conversely, the magnesium content of tropical grasses and legumes is generally higher.

Limitations to intake

(1) In considering sward characteristics in relation to diet selection and intake, Hodgson (1982) concluded that the vertical distribution of foliage in the sward canopy is the major factor in temperate swards, whereas in tropical swards variables associated with leaf density and leaf/stem ratio are of dominant importance. He emphasized that work is still at the survey stage and that more quantitative information on the botanical and morphological composition of vegetation and distribution of components through the canopy is required. The chances of identifying chemical constituents influencing preference - "palatability factors" -for plant breeding programs are not high (Arnold 1981).

(2) The effects of physical and chemical composition of herbage eaten upon intake were discussed by Minson (1982). He identified the main factors as the proportion of indigestible matter in the feed, the transit time of this residue through the rumen and the size of the rumen. Feeds vary in the time required to be broken down to particles of less than 1 mm, which are sufficiently small to leave the rumen. Thus:-

(a) intake of pelleted forages is greater than intake of chopped forages because they have been ground before pelleting.

(b) intake of stem is less than intake of leaf, because of a longer retention time in the rumen, probably because of the greater resistance of stem to physical breakdown.

(c) the intake of legumes is generally greater than that of grasses, again because of shorter retention time in the rumen and higher packing density.

Thus where mineral and protein contents are greater than levels likely to cause deficiencies, physical factors in the diet (usually related to stage of growth and fibre content) will control intake. Because our knowledge of the physical attributes is still at an early stage, and therefore laboratory screening techniques are generally not well-developed, measurements made using animals are the only satisfactory alternative.

(3) With regard to animal factors affecting intake, Weston (1982) considered that although there was genetic variation in capacity to consume feed, the diet available to the grazing animal would not permit this limitation to be expressed in general - only perhaps with the highest quality pasture and minimum environmental stress. In physiological states with a high nutrient demand, herbage intake is increased, for example in lactation, young animals, or following mild cold exposure. With pregnancy, intake in cattle and sheep may rise in early and mid-pregnancy, but often declines in late pregnancy. Environmental stresses of climate and disease (infectious, parasitic and metabolic) limit feed intake because of their effects on a range of physiological processes.

Limitations to digestion and utilization

(1) The contribution by Akin (1982) on microbial breakdown of feed in the digestive tract focussed particularly on the interactions between rumen micro-organisms and forage cell wall degradation, and discussed the microbial ecology involved in digestion of cell walls by colonies of attached bacteria, unattached bacteria and by protozoa. Of particular interest was the section in which differences in the anatomical characteristics and *in vitro* digestibilities of various herbage tissues were described. This is clearly an area of research of considerable significance, which could well point the direction in which herbage cultivar improvement should be directed. The techniques are quite straight forward for measuring the percentages of various herbage tissues, measuring their *in vitro* digestibilities, and assessing features of their cell wall composition. Emphasis on such work would seem appropriate, especially in genera where there is considerable variation in plant anatomical structure, such as *Panicum*.

(2) The processes of digestion and utilization of herbage organic matter were discussed by Armstrong (1982), Hogan (1982) and Leng (1982). As a result of the considerable progress made in understanding ruminant nutritional processes over the last 30 years, we now have quite detailed knowledge of what are the main digestive processes occurring along the digestive tract and the way the emphasis on different sites for digestion changes with nature of the ingested feed. Most of the work has been carried out under controlled conditions indoors, but techniques are now available for carrying out experiments on grazing animals and there should be emphasis on such experiments in future. Major factors limiting the nutritive value of ingested herbage per unit of intake can be summarized as:-

(a) the loss of approximately 25 percent digestible energy as methane and heat as a result of rumen and caecal fermentations,

(b) the loss of protein nitrogen as ammonia through protein fermentation in the rumen and caecum.

Hence any shift of digestion from the rumen to the small intestine should lead to better overall utilization of herbage, especially for higher quality feed.

In considering the possibilities for modification of rumen fermentation, Leng (1982) developed a theoretical treatment which emphasized that "irrespective of the apparent rationality of a manipulation procedure, the changes in the rumen and the effect on production of the animal are not readily predictable," because of the highly interactive nature of the rumen ecosystem. Furthermore he pointed out that the scope for manipulation of the rumen system of sheep and cattle under grazing conditions is much less than under lot-feeding conditions. Reported grazing trials with monensin have given only small increases in growth rate and efficiency of feed utilization, which are not likely to be economic. Thus, the possibilities at this stage for improving the nutrition of grazing animals by manipulating rumen fermentation through the use of chemical additives are small.

Leng also emphasized the significance of the large rumen protozoal biomass in animals grazing fresh pastures, in relation to reductions in microbial protein leaving the rumen. With defaunated sheep, evidence was obtained from both growth studies and isotope experiments of a decreased availability of metabolizable nutrients in ruminants which have a large protozoal biomass in the rumen.

(3) Our knowledge of the utilization of minerals from herbage was discussed by Little (1982). In general we have reasonably firm guidelines as to the desirable levels for ruminant nutrition of various minerals in herbage and can make firm recommendations to managers on the need for supplementation. In practice, the limitations to pastoral production arise through operational difficulties in diagnosing mineral deficiencies, especially when these are sub-clinical.

Little pointed out the marked influence on mineral utilization associated with the ingestion of soil and the incidence of gastro-intestinal parasitism. He considered both these warranted much more research work. A particularly striking illustration in relation to soil ingestion was the effect of increased dietary iron in reducing the apparent availability of copper.

OVERCOMING NUTRITIONAL LIMITATIONS BY PLANT BREEDING AND SELECTION

Plant breeding and selection represent the best long-term option for overcoming nutritional limitations of herbage, especially where animal yield per hectare is important. (Where yield per animal is more important, management systems which employ alternative feeds or special purpose pastures at critical periods will often be preferable).

Bray (1982) considered that legume selection progressed through three broad stages -

(a) a preliminary development stage, where selection is for general adaptation and yield, i.e. for productivity and persistence under grazing;

(b) a developmental stage, where the limitations of the plant are discovered and overcome e.g. specific pest or disease resistances and selection against any deleterious quality factors;

(c) a "fine-tuning" stage where desirable quality characteristics can be enhanced, regional varieties adapted to particular environments, and compatibilities improved with associated grass, insect pollinators etc.

This is a useful philosophic framework, although in practice the order in which selection characters are worked on tends to be specific to the species and the breeder.

Bray's wide-ranging review of pasture legume breeding emphasized that nutritional quality has not been a priority problem and the main breeding problem in legumes has been persistence, especially under intensive grazing. This does suggest that greater emphasis on legume breeding relative to grass breeding could be a fruitful approach to the whole quality question. If breeders had worked on legumes as much in the past as they have done on grasses, we would possibly have productive legume cultivars with greater persistence.

In many cases, deleterious quality factors in herbage plants can be either removed or their levels can be reduced. Both Bray & Hacker (1982) described a range of examples where the approach was successful on the one hand, or impracticable because of multiple causal factors, insufficient genetic variation or genetic linkage with other important factors.

The introduction of novel genetic techniques will reduce the screening and selection processes necessary in conventional plant breeding, and so make breeding to remove anti-quality factors much more practicable. For example the high mutation rates observed in repeated tissue culture (Larkin & Scowcroft 1981) and in the technique of using irradiated pollen (Pandey 1980, Jinks _et al_. 1981) are most encouraging.

The review by Hacker of breeding grasses for improved nutritional quality showed that although there were promising genetic possibilities in many characters, there is a low success rate in output of nutritionally superior cultivars. Even Coastcross 1 bermuda grass - surely one of the success stores - has apparently paid a price in the loss of rhizomes and of winter hardiness. A forage grass is required to be a very complex farming tool and the breeding of such plants is similarly complex.

The development of practicable procedures for screening plants for the anatomical and physical features which are deemed important in controlling the rate of digestion of herbage is of the highest importance. As emphasized by Minson (1982), "only animals are capable of responding to all the physical characters of the pasture that control intake" and the developing and refining of a range of simple laboratory procedures for screening plant selections, based on a thorough knowledge of the key physical and chemical attributes, is essential for a higher success rate in breeding nutritionally superior cultivars.

An important aspect is that breeding for disease resistance usually has the additional benefit of providing herbage of better quality. For example, Edwards et al. (1981), in a histological examination of healthy and rust-infested orchard grass (cocksfoot) leaves which had been subjected to digestion in rumen fluid found that the digestion of the diseased tissues was markedly impaired.

Enough is now known of the physical and chemical factors which are required for high nutritional quality in herbage for general prescriptions to be developed as targets for plant breeders and agronomists to aspire to. Such presciptions would serve as general guidelines in the management and design of plant improvement programs, always bearing in mind that they would represent only one of several inputs to plant improvement targets. They would be refined by an iterative process for particular programs, in the same way as Feed Requirement Tables.

In discussing positive attributes which should be incorporated into ideal herbage, Ulyatt (unpublished work presented in the Hector and Andrew stewart Memorial Lecture, University of Western Australia, Perth, 1980) suggested the following characteristics:- high protein content, particularly increased sulphur amino-acids; high levels of soluble carbohydrates; a feature such as thicker cell walls in the soft tissues, or presence of tannins, that will either slow the release of soluble protein or render it less soluble; an easily ruptured epidermis; vascular tissue that is sufficient to maintain agronomic merit but is fragile in terms of shear stress; concentrations of minerals sufficient to maintain animal health.

OVERCOMING NUTRITIONAL LIMITATIONS BY ANIMAL BREEDING

What can be done through animal breeding to overcome nutritional limitations from herbage? Vercoe & Frisch (1982) pointed out that realized productivity is a consequence of two genetically determined factors, potential productivity (measured in the absence of environmental stress) and resistance to environmental stress. These factors are found to be negatively correlated between breeds and probably within breeds. Resistance to environmental stresses such as tropical climatic conditions and parasites is the first priority, but it is a valid goal to attempt to combine it with a high production potential. Such an approach is consistent with Mannetje's views (1982), on the relatively greater gains to be made from improvement of the nutritional quality of pastures, in the context of tropical grasslands.

In practice, the principle of mating males of high performance test or potential productivity with females exposed to the various environmental stresses (usually using artificial insemination programs) represents an approach capable of achieving genetic gain.

In considering anti-quality factors in herbage the question of selecting for tolerance within ruminants can be a useful approach. For example, with the liver disease "facial eczema" caused by the mycotoxin sporidesmin, considerable selection for tolerance is being achieved with sheep (Mortimer 1981) and Cockrem & Mackintosh (1981) have a program of selection of cows for resistance to bloat, using salivary mucoprotein patterns as markers for screening cattle. In both cases, wide adoption of the approach depends on developing practicable screening techniques using biochemical "markers" which can be applied to large numbers of animals.

OVERCOMING LIMITATIONS BY AGRONOMIC MANAGEMENT PRACTICES

Agronomic approaches to improving the nutrition of grazing animals will usually be accomplished in a much shorter time scale than the improvement of pasture plants and grazing animals by breeding. As pointed out by Mahadevan (1982), Allden (1982) and Mannetje (1982), the technology is often known and has been demonstrated in the particular environment, but cultural, socio-economic and educational barriers all have to be surmounted.

Several papers (Evans 1982, Reid & Jung 1982; Allden 1982 and Mannetje 1982) described the principles of integrating different species and pastures to overcome seasonal deficits in meeting the feed requirements of different classes of stock, manipulating the annual pattern of pasture growth by judicious use of fertilizer, manipulating stocking rate or class of animal and so on.

The large increases in animal production which are achievable through incorporation of improved legume cultivars of good persistence offer the best prospect for low-cost grassland production systems in many temperate and tropical regions. Thus during the last decade, the selected varieties of *Stylosanthes guianensis*, *S. hamata* and *S. scabra* have provided the basis of viable production systems in many tropical countries. Similarly, *Lotus pedunculatus* selections are now significant in grass/legume associations on acid infertile soils in several temperate countries.

Jarrige *et al*. (1982), Siebert & Hunter (1982) and Burns (1982) contributed papers on aspects of supplementary feeding. Many of the principles of supplementation were summarized in the sentence: "Feed availability, fibre content, protein content and the ratio of nitrogen to sulphur thus provide a sequence of criteria which decides the success or failure of supplementation of energy, protein, non-protein nitrogen and non-protein sulphur," (Siebert & Hunter 1982).

Siebert & Hunter also emphasized the cost efficiencies which can result from trace element supplementation, and pointed to the developments in technology for the controlled slow release of an increasing number of such elements from pellets and capsules which can be lodged in the intestinal tract. Developments in the use of similar devices for controlling internal parasites by the slow release of anthelmintics will also improve animal production, provided that the strategy of using different anthelmintics in sequence prevents the development of resistant strains of parasites.

In considering forage conservation, Jarrige _et al_. (1982) reviewed the sophisticated procedures for silage-making, particularly in temperate grasslands of the northern hemisphere, which have progressively freed livestock production from the seasonality of herbage production by using increasing proportions of conserved herbage and forage crops, especially maize. They emphasized that, while silage is the most efficient method of forage conservation in many temperate grasslands, it is costly and dependent on fossil energy and requires adequate implements, good training and organization of farmers and high market prices for animal products.

In general, silage making in the tropics was seen by Jarrige _et al_. to be an unsatisfactory option, because of the unsuitability of most tropical species to production of good silage and the infra-structural obstacles to developing reliable silage-making and feeding methods. The preferable options were to rely on irrigation and on crops that can be used directly during the dry season, such as sugar cane and cassava.

All these approaches depend on a very sound agronomic appreciation by the pastoral farmer of the possibilities for manipulating pasture production and the provision of supplements in his own environment in order to yield the best rate of return in animal production per unit invested. The modelling of alternative production systems can be of great assistance in this regard; many farmers who operate intensive pastoral production farming systems now use modelling approaches to optimize animal production, as affected by variation from normal expectations in herbage production. Black's paper (1982) showed the directions in which simulation models should be increased in sophistication, in order to be of greater value for both research and for the management of pastoral farming systems.

THE ROLE OF HERBAGE PROCESSING

Wilkins (1982) pointed out that, although large quantities of poor quality feedstuffs are available world-wide and a number of processes are available for improving the feeds, economics govern their utilization. He concluded that the range of situations in which forage processing by mechanical, chemical or biological means is at present justified is quite limited. The most attractive processing method was seen to be the chemical treatment of mature grass with ammonia, or materials such as urea used as a source of ammonia. This has advantages over the use of sodium hydroxide because of improved nitrogen content and because of fears of sodium balance effects on soils. Emphasis on widespread study and adoption of this approach would seem very appropriate, particularly for tropical grasses.

The fractionation of high-quality forages to give a leaf protein concentrate for monogastric animals and a fibrous residue for ruminant animals has reached the stage where opportunities already exist for the adoption of such processes in large-scale intensive agriculture,

(Wilkins 1982). However, the extent of adoption of forage fractionation technology is at present small and future adoption will be strongly influenced by the demand for by-products such as xanthophyll and ethanol, rather than requirements for feed energy and protein. Forage fractionation technology will probably be adopted steadily rather than spectacularly in intensive pastoral farming systems in temperate grasslands. The future emphasis on biotechnology and the sustainable use of biological resources will favour the further development of forage fractionation systems.

CONCLUSION

The 24 papers presented have reviewed an impressive body of knowledge on nutritional limitations to pastoral production. The science of agriculture advances on many inter-linked fronts, but in particular environments and farming systems, certain problems will assume highest priority. Nevertheless, I suggest that the following research areas having wide applicability should be emphasized in our future endeavours.

(1) Devising pasture systems which give the grazing animal continuing access to herbage of high nutritive value or to herbage which has been appropriately supplemented.

(2) Developing the use of modelling procedures and other techniques of operational research as management and research tools, particularly for use in intensive pastoral farming systems and also for systems where socioeconomic problems are severe impediments to increasing production.

(3) Refining the prescriptions of the physical and chemical attributes of herbage required for high nutritional quality, for the use of both agronomists and plant breeders.

(4) Selection of improved pasture species of high nutritional quality for the acid and problem soils which will be used more for pastoral farming in future.

(5) Emphasis in all grassland ecosystems on breeding for legume persistence, with consequent improvement of pasture quality.

(6) Development of disease-resistant cultivars, which will generally have higher nutritional quality than disease-susceptible cultivars.

(7) Development of rapid screening procedures for attributes of high nutritional quality.

(8) Use of novel genetic techniques for plant improvement, particularly the removal of deleterious chemical factors.

(9) Selection of grazing animals which are tolerant or resistant to environmental stresses arising from climate, disease etc. and which also express their potential productivity.

(10) Improvement of low quality herbage by chemical treatment, particularly improvement of mature tropical grasses with ammonia or urea.

(11) Mineral utilization as affected by soil ingestion.

(12) Effect of gastrointestinal parasitism on nutrient and mineral absorption.

ACKNOWLEDGEMENTS

I am indebted to Drs J. Rumball, M.J. Ulyatt and D.E. Wright for valuable discussion.

REFERENCES

Akin, D.E. (1982) Microbial breakdown of feed in the digestive tract. In: Nutritional limits to animal production from pastures. Editor J.B. Hacker. Farnham Royal, U.K., Commonwealth Agricultural Bureaux, pp. 201-223.

Allden, W.G. (1982) Problems of animal production from Mediterranean pastures. In: Nutritional limits to animal production from pastures. Editor J.B. Hacker. Farnham Royal, U.K., Commonwealth Agricultural Bureaux, pp. 45-65.

Armstrong, D.G. (1982) Digestion and utilization of energy. In: Nutritional limits to animal production from pastures. Editor J.B. Hacker. Farnham Royal, U.K., Commonwealth Agricultural Bureaux, pp. 225-244.

Arnold, G.W. (1981) Grazing behaviour. In: Grazing animals. World animal science, B 1. Editor F.H.W. Morley. Amsterdam, Elsevier, pp.79-104.

Black, J.L.; Faichney, G.J.; Sinclair, R.E. (1982) Role of computer simulation in overcoming limitations to animal production from pastures. In: Nutritional limits to animal production from pastures. Editor J.B. Hacker. Farnham Royal, U.K., Commonwealth Agricultural Bureaux, pp. 473-493.

Blair-Rains, A.; Kassan, A.H. (1979) Land resources and animal production. Working Paper No. 8. FAO/UNFPA project INT/75/P13. Rome, AGLS, FAO. 28 pp.

Bray, R.A. (1982) Selecting and breeding better legumes. In: Nutritional limits to animal production from pastures. Editor J.B. Hacker. Farnham Royal, U.K., Commonwealth Agricultural Bureaux, pp. 287-303.

Burns, J.C. (1982) Integration of grazing with other feed resources. In: Nutritional limits to animal production from pastures. Editor J.B. Hacker. Farnham Royal, U.K., Commonwealth Agricultural Bureaux, pp. 455-471.

Campbell, A.G. (1979/80) Genetic resistance to facial eczema. Annual Report, Ministry of Agriculture and Fisheries, pp. 94-95.

Cockrem, F.R.M.; Mackintosh, J.T. (1981) Objectives and progress in breeding cows of low susceptibility to bloat. Proceedings Ruakura Farmers' Conference.

Edwards, M.T.; Sleper, D.A.; Leogering, W.Q. (1981) Histology of healthy and diseased orchard grass leaves subjected to digestion in rumen fluid. Crop Science 21, 341-343.

Evans, T.R. (1982) Overcoming nutritional limitations through pasture management. In: Nutritional limits to animal production from pastures. Editor J.B. Hacker. Farnham Royal, U.K., Commonwealth Agricultural Bureaux, pp. 343-361.

Hacker, J.B. (1982) Selecting and breeding better quality grasses. In: Nutritional limits to animal production from pastures. Editor J.B. Hacker. Farnham Royal, U.K., Commonwealth Agricultural Bureaux, pp. 305-326.

Hegarty, M.P. (1982) Deleterious factors in forages affecting animal production. In: Nutritional limits to animal production from pastures. Editor J.B. Hacker. Farnham Royal, U.K., Commonwealth Agricultural Bureaux, pp. 133-150.

Hodgson, J. (1982) Influence of sward characteristics on diet selection and herbage intake by the grazing animal. In: Nutritional limits to animal production from pastures. Editor J.B. Hacker. Farnham Royal, U.K., Commonwealth Agricultural Bureaux, pp. 153-166.

Hogan, J.P. (1982) Digestion and utilization of protein. In: Nutritional limits to animal production from pastures. Editor J.B. Hacker. Farnham Royal, U.K., Commonwealth Agricultural Bureaux, pp. 245-257.

Jarrige, R.; Demarquilly, C.; Dulphy, J.P. (1982) Conservation. In: Nutritional limits to animal production from pastures. Editor J.B. Hacker. Farnham Royal, U.K., Commonwealth Agricultural Bureaux, pp. 363-387.

Jinks, J.L.; Caligari, P.D.S.; Ingram, N.R. (1981) Gene transfer in Nicotiana rustica using irradiated pollen. Nature, London 291, 586-588.

Jones, R.J. (1981) Does ruminal metabolism of mimosine explain the absence of Leucaena toxicity in Hawaii? Australian Veterinary Journal 57, 55-56.

Larkin, P.J.; Scowcroft, W.R. (1981) Somoclonal variation - a novel source of variability from cell culture for plant improvement. Theoretical and Applied Genetics 58, (in press).

Leng, R.A. (1982) Modification of rumen fermentation. In: Nutritional limits to animal production from pastures. Editor J.B. Hacker. Farnham Royal, U.K., Commonwealth Agricultural Bureaux, pp. 427-453.

Little, D.A. (1982) Utilization of minerals. In: Nutritional limits to animal production from pastures. Editor J.B. Hacker. Farnham Royal, U.K., Commonwealth Agricultural Bureaux, pp. 259-283.

Mahadevan, P. (1982) Pastures and animal production. In: Nutritional limits to animal production from pastures. Editor J.B. Hacker. Farnham Royal, U.K., Commonwealth Agricultural Bureaux, pp. 1-17.

Mannetje, L. 't (1982) Problems of animal production from tropical pastures. In: Nutritional limits to animal production from pastures. Editor J.B. Hacker. Farnham Royal, U.K., Commonwealth Agricultural Bureaux, pp. 67-85.

Minson, D.J. (1982) Effects of chemical and physical composition of herbage eaten upon intake. In: Nutritional limits to animal production from pastures. Editor J.B. Hacker. Farnham Royal, U.K., Commonwealth Agricultural Bureaux, pp. 167-182.

Norton, B. (1982) Differences between species in forage quality. In: Nutritional limits to animal production from pastures. Editor J.B. Hacker. Farnham Royal, U.K., Commonwealth Agricultural Bureaux, pp. 89-110.

Pandey, K.K. (1980) Further evidence for egg transformation in Nicotiana. Heredity 45, 15-29.

Reid, R.L.; Jung, G.A. (1982) Problems of animal production from temperate pastures. In: Nutritional limits to animal production from pastures. Editor J.B. Hacker. Farnham Royal, U.K., Commonwealth Agricultural Bureaux, pp. 21-43.

Siebert, B.D.; Hunter, R.A. (1982) Supplementary feeding of grazing animals. In: Nutritional limits to animal production from pastures. Editor J.B. Hacker. Farnham Royal, U.K., Commonwealth Agricultural Bureaux, pp. 409-426.

Weston, R.J. (1982) Animal factors affecting feed intake. In: Nutritional limits to animal production from pastures. Editor J.B. Hacker. Farnham Royal, U.K., Commonwealth Agricultural Bureaux, pp. 183-198.

Wilkins, R.J. (1982) Improving forage quality by processing. In: Nutritional limits to animal production from pastures. Editor J.B. Hacker. Farnham Royal, U.K., Commonwealth Agricultural Bureaux, pp. 389-408.

Wilson, J.R. (1982) Environmental and nutritional factors affecting herbage quality. In: Nutritional limits to animal production from pastures. Editor J.B. Hacker. Farnham Royal, U.K., Commonwealth Agricultural Bureaux, pp. 111-131.

Wilson, J.R.; Minson, D.J. (1980) Prospects for improving the digestibility and intake of tropical grasses. Tropical Grasslands 14, 253-259.

Vercoe, J.E.; Frisch, J.E. (1982) Animal breeding for improved productivity. In: Nutritional limits to animal production from pastures. Editor J.B. Hacker. Farnham Royal, U.K., Commonwealth Agricultural Bureaux, pp. 327-342.

POSTERS

Archibald, K.A.E. (Trinidad) Status of pasture research in the Caribbean Commonwealth.

Ankrah, P.; Kellaway, R.C.; Leibholtz, Jane (Australia) Effects of physical processes on intake and utilization of alkali-treated oat straw by sheep.

Cheeke, P.R. (USA) Utilization of tropical forages by rabbits.

Corbett, J.L.; Pickering, F.S.; Furnival, E.P.; Inskip, M.W. (Australia) Digestion in grazing sheep.

Gammon, D.M. (Zimbabwe) Pattern of defoliation of rangeland by cattle during continuous grazing.

Gartner, R.J.W.; Blaney, B.J.; McKenzie, R.A. (Australia) Tropical grasses, oxalate and calcium deficiency in grazing horses.

Hall, D.G. (Australia) Seasonal variation in diet, quality, southern New South Wales.

Jones, R.J.; Megarrity, R.G. (Australia) Leucaena toxicity in ruminants.

Lemerle, C.; Barret, L.; Murray, R.M. (Australia) Animal and pasture changes in the seasonally dry tropics.

Lowry, J.B.; Yates, N.; Tang, B. (Indonesia) Mimosine and DHP both detoxified in Indonesia cattle fed leucaena.

Marais, J.P.; De Figueldo, M.C. (South Africa) Effect of nitrate on digestibility *in vitro* of pasture grasses.

McLennan, S.R.; Shepherd, R.K. (Australia) Relationship between urea concentration, intake of dry season supplements and liveweight change.

Murphy, M.R.; Jaster, E.W. (USA) Quantifying chewing patterns in dairy heifers.

Nicol, D.C.; Smith, L.D. (Australia) Cobalt and levamisole treatment for yearling cattle.

Petheram, J. (Indonesia) Village pasture systems and livestock nutrition.

Smeaton, D. (New Zealand) Feeding ewes for mating on hill country.

Till, A.R.; Ibrahim, T.; Rudolf, W.; Blair, G.T. (Indonesia) Cattle production on pastures in eastern Indonesia.

Tudor, G.D.; Minson D.J. (Australia) Utilization of the dietary energy of pangola and setaria by cattle.

Van Gylswyk, N.O.; Vander Linden, Yvonne (South Africa) The effect of feeding sheep different levels of concentrate together with maize straw on the numbers of types of fibrolytic bacteria in the rumen.

Watson, M.W.; Onwupkie, O.C. (Nigeria) The nutritional limits to voluntary feed intake in West African dwarf sheep given dry season Panicum maximum.

Williams, R.D.; Leng, R.A.; Nolan, J.V. (Australia) Response of soybean/meat meal supplement by grazing beef cattle.

Yates, J.J. (Australia) Introduced pasture species on the central plateau of Tasmania.

LIST OF PARTICIPANTS

AUSTRALIA

Mr. E. Ajileye, 12/59 Sandford Street, ST. LUCIA, QLD 4067.

Mr Rafat A.M. Aljassim, School of Wool & Pastoral Science, University of New South Wales, P.O. Box 1, Kensington NSW 2033

Dr. W. G. Allden, Waite Agricultural Research Institute, Private Bag, GLEN OSMOND, SA 5064.

Mr. K. Amaning-Kwarteng, M.C. Franklin Laboratory, University Farms, CAMDEN, NSW 2570.

Mr. P. Ankrah, University Farms, CAMDEN, NSW 2570.

Prof. E.F. Annison. Department of Animal Husbandry, Sydney University, SYDNEY, NSW 2006.

Dr. K.A. Archer, Agricultural Research Station, GLEN INNES, NSW 2370.

Mr. A. J. Ash, Queensland Department of Primary Industries, Brian Pastures Research Station, GAYNDAH, QLD 4625.

Mr. A. Ayied, 1/139 Bell Street, PRESTON, VIC 3072.

Mr. A. Bamualim, Tropical Veterinary Science Department, James Cook University, TOWNSVILLE, QLD 4811.

Dr. I.F. Beale, Box 282 Post Office, CHARLEVILLE, QLD 4470.

Mr. A. K. Bell, P.O. Box 547, TAMWORTH, NSW 2340.

Dr. A. W. Bell, School of Agricultural, La Trobe University, BUNDOORA, VIC 3083.

Dr. P. R. Bird, P.O. Box 180, HAMILTON, VIC 3300.

Mr. H. A. Birrell, P.O. Box 180, HAMILTON, VIC 3300.

Dr. J. L. Black, P.O. Box 239, BLACKTOWN, NSW 2148.

Dr. R. A. Bray, CSIRO Division of Tropical Crops and Pastures, Davies Laboratory, Aitkenvale, TOWNSVILLE, QLD 4814.

Mr. K. L. Butler, CSIRO Division of Mathematics & Statistics, Cunningham Laboratory, ST. LUCIA, QLD 4067.

Mr. R. R. Carter, c/- Bayer Aust. Ltd., P.O. Box 159, BOTANY, NSW 2019.

Mr. G. Chopping, Queensland Department of Primary Industries, Mutdapilly Research Station, M.S.825, IPSWICH, QLD 4305.

Mr. N. M. Clarkson, Queensland Department of Primary Industries, WARWICK, QLD 4370.

Mr. D.B. Coates, CSIRO, Kimberley Research Station, KUNUNURRA, WA 6743.

Dr. J.B. Coombe, CSIRO, Division of Plant Industry, CANBERRA CITY, ACT 2601.

Dr. J. L. Corbett*, CSIRO Division of Animal Production, ARMIDALE, NSW 2350.

Dr. T. Cowan, Queensland Department of Primary Industries, Mutdapilly Research Station, M.S. 825, IPSWICH, QLD 4305.

Mr. S. W. Cridland, Department of Biochemistry & Nutrition, University of New England, ARMIDALE, NSW 2351.

Ms P. N. Crisp, c/- School of Agriculture, La Trobe University, BUNDOORA, VIC 3083.

Mr. A. B. Davies, 121 Pinelands Street, SUNNYBANK HILLS, QLD 4109.

Mr. T. Davison, Queensland Department of Primary Industries, Kairi Research Station, via ATHERTON, QLD 4883.

Mr. O.L.P. De Oliveira, University of New South Wales Field Station, P.O. Box 217, WELLINGTON, NSW 2820.

Mr. G. Denney, Agricultural Research Centre, WOLLONGBAR, NSW 2480.

Dr. R.M. Dixon, Eungai Creek MARKVILLE NSW 2492

Mr. W. Dixon, Department of Biochemistry & Nutrition, University of New England, ARMIDALE, NSW 2351.

Dr. H. Dove, CSIRO, Division of Plant Industry, P.O. Box 1600, CANBERRA CITY, ACT 2601.

* Conference chairman

Dr. P. Doyle, School of Agriculture & Forestry, University of Melbourne, PARKVILLE, VIC 3052.

Mr. G. Dryden, Department of Animal Science, Queensland Agricultural College, LAWES, QLD 4343.

Mr. L. A. Edye, CSIRO Division of Tropical Crops and Pastures, Davies Laboratory, Aitkenvale, TOWNSVILLE, QLD 4810.

Dr. A. R. Egan, Department of Agronomy, Waite Agricultural Research Institute, GLEN OSMOND, SA 5064.

Mr. J. P. Egan, Turretfield Research Centre, ROSEDALE, SA 5350.

Mr. A. R. Eggington, Northern Territory Department of Primary Production, c/- Post Office, ADELAIDE RIVER, NT 5783.

Mr. T. R. Evans, CSIRO Division of Tropical Crops and Pastures, Cunningham Laboratory, ST. LUCIA, QLD 4067.

Mr. A.G. Eyles, CSIRO Division of Tropical Crops & Pastures, Cunningham Laboratory, ST. LUCIA QLD 4067.

Dr. G.J. Faichney, CSIRO Division of Animal Production, P.O. Box 239, BLACKTOWN, NSW 2148.

Mr. D. Ffoulkes, Department of Biochemistry & Nutrition, University of New England, ARMIDALE, NSW 2351.

Mr. P. C. Flinn, Pastoral Research Institute, P.O.Box 180, HAMILTON, VIC 3300.

Dr. J. Z. Foot, P.O. Box 180, HAMILTON, VIC 3300.

Dr. M. Freer, CSIRO Division of Plant Industry, P.O. Box 1600, CANBERRA CITY, ACT 2601.

Dr. C. J. Gardener, CSIRO Division of Tropical Crops and Pastures, Davies Laboratory, TOWNSVILLE, QLD 4810.

Mr. R.J.W. Gartner*, Queensland Department of Primary Industries, Animal Research Institute, YEERONGPILLY, QLD 4105.

Dr. G.L.R. Gordon, P.O.Box 239, BLACKTOWN, NSW 2148.

Mr. C. A. Graham, Merrindale Research Station, Dorset Road, CROYDON, VIC 3136.

Dr. N. Graham, P.O. Box 239, BLACKTOWN, NSW 2148.

Mr. P.J. Groves, c/- Elanco Products Co., Wharf Road, WEST RYDE, NSW 2114.

Dr. J. B. Hacker, CSIRO Division of Tropical Crops & Pastures, Cunningham Laboratory, ST. LUCIA, QLD 4067.

Mr. D. G. Hall, New South Wales Department of Primary Production, Agricultural Research Station, P.O. Box 242, COWRA, NSW 2794.

Mr. B. Hamilton, Agricultural Research Centre, WOLLONGBAR, NSW 2480.

Mr. R. E. Harrison, Mail Service 185, BEAUDESERT, QLD 4285.

Mr. K. P. Haydock, CSIRO Division of Mathematics & Statistics, Cunningham Laboratory, ST. LUCIA, QLD 4067.

Mr. I.G. Hazelton, CSIRO Division of Animal Production, P.O. Box 239, BLACKTOWN, NSW 2148.

Dr. M. P. Hegarty, CSIRO Division of Tropical Crops and Pastures, Cunningham Laboratory, ST. LUCIA, QLD 4067.

Dr. K.R. Helgar, Agricultural Research Centre, New South Wales Department of Agriculture, WOLLONGBAR, NSW 2480.

Mr. R.E. Hendricksen, Queensland Department of Primary Industries, Brian Pastures Research Station, GAYNDAH, QLD 4625.

Mr. D. W. Hennessy, Department of Agriculture, Agricultural Research Station, GRAFTON, NSW 2460.

Dr. E. F. Henzell*, CSIRO Division of Tropical Crops & Pastures, Cunningham Laboratory, ST. LUCIA, QLD 4067.

Mr. E.D. Higgs, Department of Agriculture, Box 1671 G.P.O. Adelaide, SA 5000.

* Conference chairman

Mr. M. Hillard, Department of Biochemistry & Nutrition, University of New England, ARMIDALE, NSW 2351.

Dr. R. Hodge, Victorian Department of Agriculture, Agricultural Research Institute, WERRIBEE, VIC 3030.

Dr. J.P. Hogan, CSIRO Division of Animal Production, P.O. Box 239, BLACKTOWN, NSW 2148.

Dr. J. Holmes, School of Agriculture and Forestry, University of Melbourne, PARKVILLE, VIC 3052.

Mr. A.D. Hughes, South Australia Department of Agriculture, Turretfield Research Centre, SA 5350.

Dr. L.R. Humphreys, Department of Agriculture, University of Queensland, ST. LUCIA, QLD 4067.

Dr. R. A. Hunter, CSIRO Division of Animal Production, P.O. Box 545, NORTH ROCKHAMPTON, QLD 4701.

Mr. J. I'ons, South African Embassy, Rhodes Place, YARRALUMLA, ACT 2600.

Mr. R.L. Ison, Department of Agriculture, University of Queensland, ST. LUCIA, QLD 4067.

Mr. A. Johnson, 44 Fitzroy Street, KIRRIBILLI, NSW 2061.

Dr. R.J. Jones, CSIRO Division of Tropical Crops and Pastures, Davies Laboratory, Aitkenvale, TOWNSVILLE QLD 4814.

Mr. R. M. Jones, CSIRO Division of Tropical Crops and Pastures, Cunningham Laboratory, ST. LUCIA, QLD 4067.

Mr. M.W. Jones, Australian Feeds Information Centre, CSIRO Division of Animal Production, P.O. Box 239, BLACKTOWN, NSW 2148.

Dr. R. Kellaway, University of Sydney, CAMDEN, NSW 2570.

Dr. T.J. Kempton, Department of Biochemistry & Nutrition, University of New England, ARMIDALE NSW 2350.

Mr. P. A. Kenney, Rutherglen Research Station, RUTHERGLEN, VIC 3685.

Dr. E.A. Kernohan, University of Sydney Farms, Weombi Road, CAMDEN, NSW 2570.

Dr. P.C. Kerridge, CSIRO Division of Tropical Crops and Pastures, Cunningham Laboratory, ST. LUCIA, QLD 4067.

Mr. F.H. Kleinschmidt, Queensland Agricultural College, LAWES, QLD 4345.

Mr. D. L. Lamela, 69 Durham Street, ST. LUCIA, QLD 4067.

Dr. J.P. Langlands, CSIRO Division of Animal Production, Private Mail Bag, ARMIDALE, NSW 2350.

Dr. T.F. Leche, CSIRO Division of Animal Production, P.O. Box 239 BLACKTOWN, NSW 2148.

Dr. J. Leibholz, University of Sydney, CAMDEN, NSW 2570.

Mr. B. G. Lemcke, Coastal Plains Research Station, P.O. Box 5160, DARWIN, NT 5794.

Ms. C. Lemerle, c/- Animal & Irrigated Pastures Research Institute, KYABRAM, VIC 3620.

Prof. R.A. Leng, University of New England, ARMIDALE, NSW 2350.

Mr. C.J. Levick, Department of Agriculture, P.O. Box 9, EAST MAITLAND, NSW 2223.

Mr. J. R. Lindsay, CSIRO Division of Animal Production, Private Bag P.O., WEMBLEY, WA 6014.

Dr. J. A. Lindsay, Swans Lagoon, Millaroo, AYR, QLD 4807.

Mr. D. A. Little, CSIRO Division of Tropical Crops and Pastures, Cunningham Laboratory, ST. LUCIA, QLD 4067.

Mr. D. Little, Department of Agriculture, PARNOANA, SA 5000.

Prof. H. Lloyd Davies, School of Wool & Pasture Sciences, University of New South Wales, P.O. Box 1, KENSINGTON, NSW 2033.

Mr. R.J. Lucas, c/- Margaret Mason, 34 Glen Road, TOOWONG, QLD 4066.

Dr. L. 't Mannetje, CSIRO Division of Tropical Crops and Pastures, Cunningham Laboratory, ST. LUCIA, QLD 4067.

Mr. D. E. Margan, CSIRO Division of Animal Production, 11 Jordan Road, WAHROONGA, NSW 2076.

Mr. P. R. Martin, Queensland Department of Primary Industries, 665 Fairfield Road, YEERONGPILLY, QLD 4105.

Mr. T.H. McCosker, Mt. Bundey Station, P.O. Box 3, ADELAIDE RIVER, NT 5783.

Dr. G. M. McKeon*, Agriculture Branch, Queensland Department of Primary Industries, P.O. Box 46, GPO, BRISBANE, QLD 4001.

Mr. R. McLean, CSIRO Division of Tropical Crops and Pastures, ST. LUCIA, QLD 4067.

Mr. S. R. McLennan, Queensland Department of Primary Industries, Swans Lagoon, MILLAROO, QLD 4807.

Mr. M. N. McLeod, CSIRO Division of Tropical Crops and Pastures, Cunningham Laboratory, ST. LUCIA, QLD 4067.

Dr. N. McMeniman, Queensland Department of Primary Industries, Box 46, G.P.O., BRISBANE, QLD 4001.

Mr. R. G. Megarrity, CSIRO Division of Tropical Crops and Pastures, Davies Laboratory, Aitkenvale, TOWNSVILLE, QLD 4814.

Mr. C.P. Miller, Queensland Department of Primary Industries, P.O. Box 149, MAREEBA, QLD 4880.

Dr. D.J. Minson, CSIRO Division of Tropical Crops and Pastures, Cunningham Laboratory, ST. LUCIA, QLD 4067.

Mr. A. P. Moesdradjad, School of Agriculture, University of Melbourne, PARKVILLE, VIC 3052.

Mr. R. Moss, Queensland Department of Primary Industries, Mutdapilly Research Station, M.S. 825, IPSWICH 4305.

Dr. R.M. Murray, Department of Tropical Veterinary Science, James Cook University, TOWNSVILLE, QLD 4811.

Dr. J. Murtagh, Agricultural Research Centre, WOLLONGBAR, NSW 2480.

Mr. D. C. Nicol, Queensland Department of Primary Industries, P.O. Box 1143, BUNDABERG, QLD 4670.

Mr. S. Noel, 1-25 Schonell Drive, ST. LUCIA, QLD 4067.

Dr. J. Nolan, Department of Biochemistry & Nutrition, University of New England, ARMIDALE, NSW 2351.

Prof. M.J.T. Norman*, Department of Agronomy and Horticultural Science, University of Sydney, SYDNEY, NSW 2006.

Dr. B.W. Norton, Department of Agriculture, University of Queensland, ST. LUCIA, QLD 4067.

Mr. P. O'Rourke, Queensland Department of Primary Industries, P.O. Box 1085, TOWNSVILLE, QLD 4810.

Mr. D.M. Orr, Queensland Department of Primary Industries, P.O. Box 282, CHARLEVILLE, QLD 4470.

Mr. R.K. Orton, Department of Biochemistry & Nutrition, University of New England, ARMIDALE, NSW 2351.

Mr. S. Pachirat, Department of Animal Production, School of Agriculture & Forestry, University of Melbourne, PARKVILLE, VIC 3052.

Dr. P. M. Pepper, Queensland Department Primary Industries, Mineral House, 41 George Street, BRISBANE, QLD 4000.

Dr. D. Peter, CSIRO Division of Animal Production, Post Office, WEMBLEY, WA 6014.

Mr. F. S. Pickering, CSIRO Division of Animal Production, Private Bag, P.O., ARMIDALE, NSW 2350.

Dr. M. J. Playne, CSIRO Division of Chemical Technology, P.O. Box 310, SOUTH MELBOURNE, VIC 3205.

Mr. S. Prasetyo, Department of Animal Production, School of Agriculture, University of Melbourne, PARKVILLE, VIC 3052.

Mr. B. Prawiradiputra, 207 Schonell Drive, ST. LUCIA, QLD 4067.

* Conference chairman

Dr. D. Ratcliff, CSIRO Division of Mathematics and Statistics, Cunningham Laboratory, ST. LUCIA, QLD 4067.
Mr. M. C. Rees, CSIRO Division of Tropical Crops and Pastures, Cunningham Laboratory, ST. LUCIA, QLD 4067.
Dr.K. G. Rickert, Queensland Department of Primary Industries, Brian Pastures Research Station, GAYNDAH, QLD 4625.
Dr. B. R. Roberts, School of Applied Science, Darling Downs Institute of Advanced Education, TOOWOOMBA, QLD 4350.
Dr. J. S. Russell, CSIRO Division of Tropical Crops and Pastures, Cunningham Laboratory, ST. LUCIA, QLD 4067.
Mr. D. M. Ryan, New South Wales Department of Agriculture, P.O. Box 477, WAGGA WAGGA, NSW 2650.
Mr. I.K. Saka, School of Agriculture & Forestry, University of Melbourne, PARKVILLE, VIC 3052.
Miss A. Santamaria, P.O. Box 180, HAMILTON, VIC 3300.
Mr. P. J. Saunders, Western Australia Department of Agriculture, KUNUNURRA, WA 6743.
Mr. A.C. Schlink, c/- Elanco Products Co., Wharf Road, WEST RYDE, NSW 2114.
Dr. T. W. Scott*, 41 Arnold Avenue, KELLYVILLE, NSW 2153.
Mr. T. W. Searle, CSIRO, Division of Animal Production, P.O. Box 239, BLACKTOWN, NSW 2148.
Dr. P. H. Selle, c/- Bayer . Ltd., P.O.Box 159, BOTANY, NSW 2019.
Mr. J.C.P. Severo, CSIRO Division of Tropical Crops and Pastures, Cunningham Laboratory, ST. LUCIA, QLD 4067.
Mr. D. Sevilla, Flat 9, 35 The Esplanade, ST. LUCIA, QLD 4067.
Dr. H. M. Shelton, Department of Agriculture, University of Queensland, ST. LUCIA, QLD 4067.
Dr. B. D. Siebert, CSIRO Division of Animal Production, P.O. Box 545, NORTH ROCKHAMPTON, QLD 4701.
Dr. M. W. Silvey, CSIRO Division of Tropical Crops & Pastures, Cunningham Laboratory, ST. LUCIA, QLD 4067.
Mr. M Simao Neto, CSIRO Division of Tropical Crops and Pastures, Cunningham Laboratory, ST. LUCIA, QLD 4067.
Mr. R. H. Skerman, Rippley Park, MS 501, DALBY, QLD 4405.
Mr. G. H. Smith, Rutherglen Research Institute, RUTHERGLEN, VIC 3685.
Mr. J. C. Spragg, M.C.Franklin Laboratory, University of Sydney, CAMDEN, NSW 2570.
Mr. I. B. Sudana, Department of Biochemistry & Nutrition, University of New England, ARMIDALE, NSW 2351.
Mr. J. L. Sullivan, 10 Heyington Place, TOORAK, VIC 3142.
Mr. E. Teleni, Department of Animal Husbandry, University of Sydney, CAMDEN, NSW 2570.
Dr. J.H. Ternouth*, Department of Animal Production, University of Queensland, ST. LUCIA, QLD 4067.
Mr. R. Thompson, P.O. Box 611, MUSWELLBROOK, NSW 2333.
Mr. J. Throckmorton, Department of Biochemistry & Nutrition, University of New England, ARMIDALE, NSW 2351.
Mr. P. N. Thurbon, Queensland Department of Primary Industries, NEWSTEAD, QLD 4006.
Dr. A. R. Till, CSIRO Division of Animal Production, ARMIDALE, NSW 2351.
Dr. P. Tow, Roseworthy College, ROSEWORTHY, SA 5371.
Mr. M. V. Tracey, CSIRO, P.O. Box 225, DICKSON, ACT 2602.
Mr. M. Trevino, 9/30 Sisley Street, BRISBANE, QLD 4000.
Dr. T. E. Trigg, Animal & Irrigation Pastures Institute, KYABRAM, VIC 3620.
Mr. G.D. Tudor, Queensland Department of Primary Industries, Animal Research Institute, YEERONGPILLY, QLD 4105.

* Conference chairman

Mr. S. C. Valentine, Box 1671, G.P.O., ADELAIDE, SA 5001.

Dr. J. E. Vercoe, CSIRO Division of Animal Production, P.O. Box 545, NORTH ROCKHAMPTON, QLD 4701.

Mr. J. M. Vieira, CSIRO Division of Tropical Crops and Pastures, Cunningham Laboratory, ST. LUCIA, QLD 4067.

Dr. J.C. Wadsworth, CSIRO Division of Animal Production, P.O. Box 239, BLACKTOWN, NSW 2148.

Mr. B. Walker, Queensland Department of Primary Industries, P.O. Box 689, ROCKHAMPTON, QLD 4700.

Mr. G. R. Want, New South Wales Department of Agriculture, P.O. Box 426, LISMORE, NSW 2480.

Dr. M.J. Watson, Pastoral Research Institute, Department of Agriculture, HAMILTON, VIC 3300.

Mr. S. L. Weise, 12/15 Jackes Street, ARMIDALE, NSW 2350.

Mr. M.C. Weller, 659 Boundary Road, WACOL, QLD 4076.

Mrs. N. Wenten, Agricultural Extension Section, Faculty of Agriculture & Forestry, University of Melbourne, PARKVILLE, VIC 3052.

Mr. R.H. Weston, CSIRO Division of Animal Production, Private Bag, P.O., ARMIDALE, NSW 2350.

Dr. J.L. Wheeler, CSIRO Division of Animal Production, ARMIDALE, NSW 2350.

Dr. C. White, CSIRO Division of Animal Production, Private Bag, P.O., WEMBLEY, WA 6014.

Dr. J. R. Wilson, CSIRO Division of Tropical Crops and Pastures, Cunningham Laboratory, ST. LUCIA, QLD 4067.

Mr. L. Winks, Queensland Department of Primary Industries, P.O. Box 46 G.P.O., BRISBANE, QLD 4001.

Mr. W. A. Wright, 173 Nicholson Parade, CRONULLA, NSW 2230.

Dr. J.J. Yates, Faculty of Agricultural Science, University of Tasmania, Box 252 C, G.P.O., HOBART, TAS 7001.

BOTSWANA

Mr. E.T. Dintho, Box 69, LABATSE, BOTSWANA

BRAZIL

Mr. C. L. Duarte, Estacao Exp. de Lages, Caixa Postal 181, 88.500 LAGES, SC. BRAZIL.

Dr. C.P. Moore, CPAC, Caixa Postal 70.0023, 73.300 Planaltina, D.F. BRAZIL.

Mr. W. Ritter, Estacao Exp. de Lages, Caixa Postal 181, 88.500 LAGES SC. BRAZIL.

Mr. M. Salviano, Pacifico da Luz, 900 Petrolina - PE. BRAZIL.

CANADA

Dr. P. M. Kennedy, Department of Animal Science, The University of Alberta, EDMONTON, CANADA.

COLOMBIA

Dr. Carlos E. Lascano, CIAT AA 6713, CALI, COLOMBIA.

ETHIOPIA

Dr. L. J. Lambourne, c/- ILCA, P.O. Box 5689, ADDIS ABABA, ETHIOPIA.

FIJI

Mr. I. J. Partridge, Sigatoka Research Station, SIGATOKA, FIJI.

* Conference Chaiman

FRANCE

Dr. R. Jarrige, I.N.R.A.-C.R.Z.V. de Theix, 63110 - BEAUMONT, FRANCE.

GHANA

Mr. K. Amankwaa-Duah, P.O. Box 29, MPRAESO, GHANA

Mr. K. S. Awuma, Department of Animal Husbandry, P.O. Box 3820, KAMASI, GHANA.

INDIA

Dr. P.J. Georgekunju, Senior Project Executive (FO&AH), National Dairy Development Board, ANAND 388001, INDIA.

Dr. A.K. Tripathi, FO&AH Division, National Dairy Development Board, ANAND 388001, INDIA.

INDONESIA

Dr. J. B. Lowry, Centre for Animal Research & Development, P.O. Box 123, BOGOR, INDONESIA.

Mr. J. Petheram, Centre for Animal Research & Development, P.O. Box 123, BOGOR, INDONESIA.

Mr. M. E. Siregar, Jl. Raya Pajajaran LPP, BOGOR, INDONESIA.

Mr. M. Tjandraatmadja, Centre for Animal Research & Development, P.O. Box 123, Bogor, INDONESIA.

Mr. M. Winugroho, Centre for Animal Research and Development, P.O. Box 123, BOGOR, INDONESIA.

ITALY

Dr. P. Mahadevan, Animal Production & Health Division, FAO, 00100, ROME, ITALY.

JAPAN

Prof. I. Goto*, Kyushu University, 46-06, FUKUOKA-SHI 812, JAPAN.

Dr. Y. Nada, c/- Kyushu Agricultural Experimental Station Nishigushi, KUMAMOTO, JAPAN.

MALAWI

Mr. M.W. Mfitilodze, Bunda College of Agriculture, P.O. Box 219, LILONGWE, MALAWI.

MALAYSIA

Mr. I. B. Ahmad, MARDI, SERDANG, Selangor, MALAYSIA.

Mr. F. Y. Chin, 35, Road SS 21/18, Damansara Utama, PETALING JAYA, Selangor, MALAYSIA.

Mr. H. S. Darum, Department of Agriculture, MIRI, Sarawak, MALAYSIA.

Mr. M.J.F. Luxton, 15 Jln Damansara Permai, KUALA LUMPUR, MALAYSIA.

MEXICO

Dr. R. Elliott, c/ Dr W. Thorpe Apartado 116-D, University of Merida, MERIDA, Yucatan, MEXICO.

Mr. J. F. Flores, Roberto Gayol 87, Col. Gpe. Insurgentes, MEXICO 14, DF, MEXICO.

MOZAMBIQUE

Mr. A. C. Dionisio, DINAP, Ministry of Agriculture, MAPUTO, MOZAMBIQUE.

Mr. C. A. Negrier, DINAP, Ministry of Agriculture, MAPUTO, MOZAMBIQUE.

Mr. H. M. Servoz, FAO, LINDP, P.O.Box 4595, MAPUTO, MOZAMBIQUE.

Mr. G. H. van Rootselaar, c/- UNDP, P.O. Box 4595, MAPUTO, MOZAMBIQUE.

* Conference chairman

NEPAL

Mr. L. P. Sharma, Livestock Development Project, Marihar Bhawan, KATAMANDU, NEPAL.

NEW ZEALAND

Dr. G. W. Butler, DSIR, Private Bag, WELLINGTON, NEW ZEALAND.

Dr. A.W.F. Davey, Massey University, PALMERSTON NORTH, NEW ZEALAND.

Mr. T. P. Hughes, P.O. Box 85, Lincoln College, CANTERBURY, NEW ZEALAND.

Dr. D. P. Poppi, Department of Animal Science, Lincoln College, CANTERBURY, NEW ZEALAND.

Dr. C.S.W. Reid, Applied Biochemistry Division, DSIR, Private Bag, PALMERSTON NORTH, NEW ZEALAND.

Dr. G. Rodgers, c/- Dairy Research Institute, ELLINBANK, 3820 NEW ZEALAND.

Mr. D. C. Smeaton, Whatawhata Research Station, Private Bag, HAMILTON, NEW ZEALAND.

Prof. A. R. Sykes*, Department of Animal Science, Lincoln College, CANTERBURY, NEW ZEALAND.

Dr. M. J. Ulyatt*, Division of Applied Biochemistry, Private Bag, PALMERSTON NORTH, NEW ZEALAND.

NIGERIA

Mr. R. A. Adeola, Federal Livestock Department, P.M.B. 2012 KADUNA, NIGERIA.

Mr. A. S. Muhamed, F.L.D., Sheep Project P.M.B. 2045, KATSINA, Kaduna State, NIGERIA.

PAPUA NEW GUINEA

Mr. A. K. Benjamin, Sheep Research Centre, P.O. Box 766, GOROKA, Eastern Highlands, PAPUA NEW GUINEA.

Mr. J. H. Schottler, c/- Department of Primary Industries, ERAP, P.O. Box 1434, LAE, PAPUA NEW GUINEA.

Mr. G. J. Tupper, Department of Primary Industries, ERAP, P.O. Box 1434, LAE, PAPUA NEW GUINEA.

PAKISTAN

Mr. M. L. Bhatti, Chakwal District, JHELUM, PAKISTAN.

Mr. C. W. Khan, Barani Agricultural Development Project, JHELUM, PAKISTAN.

PHILIPPINES

Mr. P. A. Aquino, San Miguel, Norala, Catabato, PHILIPPINES.

Mr Lorenzo J. Curayag, Central Mindanao University Musuan, Bukidnon PHILIPPINES.

Mr. J. L. Marbella, Pumpunga Breeding Station, Magalang, Pampanga, PHILIPPINES.

Mr. V. A. Ramos Jr, Isabela Breeding Station, UPI, Gamu, Isabela, PHILIPPINES.

Miss G. M. Reginalde, Mariano Marcos State Univesity, Batac, Ilocos Morte, PHILIPPINES.

Miss J. A. Salvador, San Agustin, Pili Camarines Sur, PHILIPPINES.

Mr. T. Villaver, Central Luzon State University, Munoz, Nueva Ecija, PHILIPPINES.

* Conference chairman

SOUTH AFRICA

Mr. T.D. Daines, Private Bag X15, STUTTERHEIM, SOUTH AFRICA.

Mr. C.S. Dannhauser, Private Bag X804 POTCHEFSTROOM, 2520, SOUTH AFRICA.

Dr. E. Engels, Department of Agriculture, P.O. GLEN, 9360, SOUTH AFRICA.

Dr. H. S. Hofmeyr, Animal and Dairy Science Research Institute, P/B X 2, IRENE, 1675, SOUTH AFRICA.

Mr. J.P. Marais, Department of Agriculture Cedara, Private Bag X9059, PIETERMARITZBURG, SOUTH AFRICA.

Dr. P. Theron*, P.O. Box 260, HOWICK, 3290, SOUTH AFRICA.

Dr. N. O. van Gylswyk, N.C.R.L., P.O. Box 395, PRETORIA 0001, SOUTH AFRICA.

SOLOMON ISLANDS

Mr. D.C. Macfarlane, Ministry of Agriculture & Lands, Box G11, HONIARA, SOLOMON ISLANDS.

Mr. M. A. Smith, Ministry of Agriculture & Lands, Box G11, HONIARA, SOLOMON ISLANDS.

SRI LANKA

Mr. E. A. Dias, Education & Training Division, Department of Agriculture, PERADENIYA, SRI LANKA.

Mr. M. N. Ibrahim, Coconut Research Institute, Bandirrippuna Estate, CUNUWILA, SRI LANKA.

Mr. G.S. Muttettuwegama, No 72 Ward Place Colombo-7, SRI LANKA.

Mr. H. B. Perdok, 42 Gallande Road, Anniewatte, KANDY, SRI LANKA.

Mr. N. Sriskandarajah, Department of Animal Science, University of Peradeniya, PERADENIYA, SRI LANKA.

THAILAND

Mr. P. Hengmichai, Tribal Research Centre, CHIANGMAI, THAILAND.

Mr. A. Isarasenee, Department of Agriculture, CHIANGMAI, THAILAND.

Mr. S. Pratumsuwan, Dairy Farming Promotion Organisation of Thailand, Mitrapab Road, Muakleg, SARABARI, THAILAND.

Prof. U. Termeulen, Department of Animal Husbandry, Faculty of Agriculture, Chiangmai University, CHIANGMAI, THAILAND.

Dr. P. Tinnimit*, Faculty of Natural Resources, Prince of Songkhla University, HAY YAI, THAILAND.

UNITED KINGDOM

Prof. D. G. Armstrong, Department of Agricultural Biochemistry & Nutrition, The University, NEWCASTLE-UPON-TYNE, UNITED KINGDOM.

Prof. R. C. Campling*, Wye College, University of London, ASHFORD, Kent, UNITED KINGDOM.

Dr. J. Hodgson, Hill Farming Research Organization Bush Estate, PENICUIK, Midlothian, EH26 OPY, UNITED KINGDOM.

Dr. L. J. Peel, Department of Agriculture and Horticulture, University of Reading, Earley Gate, READING, UNITED KINGDOM.

Dr. R.J. Wilkins, Grassland Research Institute, Hurley, MAIDENHEAD, Berks, SL6 5LR, UNITED KINGDOM.

* Conference chairman

USA

Dr. D. E. Akin, P.O. Box 5677, Russell Research Center, ATHENS, Georgia, 30613, USA.

Prof. R.E. Blaser*, 704 York Drive, BLACKSBURG, Va. 24060, USA.

Dr. J. C. Burns, North Carolina State University, 1316 Williams Hall, RALEIGH, N.C., 27650, USA.

Dr. P. R. Cheeke, Department of Animal Science, Oregon State University, CORVALLIS, Oregon, 97331, USA.

Dr. M. R. Murphy, Department of Dairy Science, 315 Animal Sciences Laboratory, 1207 West Gregory Drive, URBANA, IL 61801, USA.

Prof. R.L. Reid, Route 10, Box 9, MORGANTOWN, West Virginia, USA.

Dr. J. P. Telford, Department of Animal Science, Texas A & M University College Station, TEXAS 77840, USA.

VENEZUELA

Mr. M. Capriles, Facultad Agronomica Universiad Central, MARACAY, VENEZUELA.

Mr. E. Chacon, Facultad de Ciencias Veterinarias, Universidad Central, MARACAY, VENEZUELA.

Mr. A. Hernandez, P.O. Box 2835, CARACAS 1010, VENEZUELA.

Mr. G. Raga, P.O. Box 266, BARQUISIMETO, VENEZUELA.

Mr. C. H. Rodriguez, Apartado 62428, CARACAS 1060, VENEZUELA.

Mr. M. Ventura S., Urb. Doral Sur Calle 49C #12-91, MARACAIBO, VENEZUELA.

WEST INDIES

Dr. K.A.E. Archibald, Department of Livestock Science, University of the West Indies, ST. AUGUSTINE, TRINIDAD, WEST INDIES.

ZIMBABWE

Dr. D. M. Gammon*, P.O.Box 363, BULAWAYO, ZIMBABWE.

* Conference chairman

GLOSSARY OF COMMON NAMES

Alfalfa	*Medicago sativa*
Bahia grass	*Paspalum notatum*
Barley grass	*Hordeum leporinum*
Bean	*Vicia faba*
Bent	*Agrostis*
Bermuda grass	*Cynodon dactylon*
Berseem	*Trifolium alexandrinum*
Birdsfoot trefoil	*Lotus corniculatus*
Blady grass	*Imperata cylindrica*
Blue stem grass, big	*Andropogon gerardii*
Blue stem grass, Old World	*Bothriochloa caucasica*
Bromegrass, soft	*Bromus inermis*
Budda Pea	*Aeschynomene indica*
Buffel grass	*Cenchrus ciliaris*
Canary grass, reed	*Phalaris arundinacea*
Capeweed	*Arctotheca calendula*
Centro	*Centrosema pubescens*
Clover, Alsike	*Trifolium hybridum*
Clover, Bokhara	*Melilotus alba*
Clover, Egyptian	*Trifolium alexandrinum*
Clover, Kenya white	*Trifolium semipilosum*
Clover, red	*Trifolium pratense*
Clover, rose	*Trifolium hirtum*
Clover, subterranean	*Trifolium subterraneum*
Clover, white	*Trifolium repens*
Cocksfoot	*Dactylis glomerata*
Columbus grass	*Sorghum almum*
Coolah grass	*Panicum coloratum*
Cowpea	*Vigna unguiculata*
Crab grass	*Digitaria sanguinalis*
Creeping indigo	*Indigofera spicata*
Dallis grass	*Paspalum dilatatum*
Desmodium, green leaf	*Desmodium intortum*
Desmodium, hetero	*Desmodium heterophyllum*
Desmodium, silverleaf	*Desmodium uncinatum*
Elephant grass	*Pennisetum purpureum*
Fescue, meadow	*Festuca pratensis*
Fescue, tall	*Festuca arundinacea*
Finger grass, creeping	*Digitaria pentzii*
Glycine	*Neonotonia wightii*
Green panic	*Panicum maximum* var. *trichoglume*
Guinea grass	*Panicum maximum*
Halogeton	*Halogeton glomeratus*
Heliotrope, common	*Heliotropium europaeum*
Horse gram	*Macrotyloma uniflorum*
Kale	*Brassica oleracea*
Kentucky bluegrass	*Poa pratensis*
Kikuyu grass	*Pennisetum clandestinum*
Kleingrass	*Panicum coloratum*
Lablab bean	*Lablab purpureus*
Lespedeza, perennial	*Lespedeza cuneata*
Leucaena	*Leucaena leucocephala*
Limpo grass	*Hemarthria altissima*
Lotononis	*Lotononis bainesii*
Lovegrass, weeping	*Eragrostis curvula*
Lucerne	*Medicago sativa*
Lupin, narrowleaf	*Lupinus angustifolius*

Meadowgrass	*Poa pratensis*
Medic, barrel	*Medicago truncatula*
Medic, black	*Medicago lupulina*
Millet	*Pennisetum americanum*
Mission grass	*Pennisetum polystachyon*
Mitchell grass	*Astrebla* spp.
Orchard grass	*Dactylis glomerata*
Pangola grass	*Digitaria decumbens*
Panic, blue	*Panicum antidotale*
Panic, green	*Panicum maximum* var. *trichoglume*
Panic, millet	*Panicum miliaceum*
Para grass	*Brachiaria mutica*
Paspalum	*Paspalum dilatatum*
Pearl millet	*Pennisetum americanum*
Perennial lespedeza	*Lespedeza cuneata*, *Lespedeza sericea*
Phalaris	*Phalaris aquatica*
Phasey bean	*Macroptilium lathyroides*
Puero	*Pueraria phaseoloides*
Reed Canary grass	*Phalaris arundinacea*
Rhodes grass	*Chloris gayana*
Ryegrass, annual	*Lolium rigidum*
Ryegrass, annual	*Lolium multiflorum*
Ryegrass, Italian	*Lolium multiflorum*
Ryegrass, perennial	*Lolium perenne*
Ryegrass, Wimmera	*Lolium rigidum*
Sabi grass	*Urochloa mosambicensis*
Sainfoin	*Onobrychis viciifolia*
Saltbush, Oldman	*Atriplex numularia*
Setaria	*Setaria sphacelata*
Signal grass	*Brachiaria decumbens*
Siratro	*Macroptilium atropurpureum*
Sorghum	*Sorghum bicolor*
Soursop	*Oxalis pes-caprae*
Speargrass, black	*Heteropogon contortus*
St. John's Wort	*Hypericum perforatum*
Stylo	*Stylosanthes guianensis*
Stylo, African	*Stylostanthes fruticosa*
Stylo, Caribbean	*Stylosanthes hamata*
Stylo, shrubby	*Stylosanthes scabra*
Stylo, sticky	*Stylosanthes viscosa*
Stylo, Townsville	*Stylosanthes humilis*
Sudan grass	*Sorghum sudanense*
Switch grass	*Panicum virgatum*
Tambookie grass	*Hyparrhenia filipendula*
Tanner grass	*Brachiaria radicans*
Tef	*Eragrostis tef*
Thatchgrass	*Hyparrhenia rufa*
Timothy	*Phleum pratense*
Trefoil	*Lotus*
Trefoil, birdsfoot	*Lotus corniculatus*
Veldt grass	*Ehrharta* spp.
Vetch, crown	*Coronilla varia*
Wallaby grass	*Danthonia*
Wheatgrass, intermediate	*Agropyron intermedium*

INDEX